Innovative Bridge Design Handbook

Innovative Bridge Design Handbook

Construction, Rehabilitation and Maintenance

Edited by

Alessio Pipinato

ELSEVIER

AMSTERDAM • BOSTON • HEIDELBERG • LONDON
NEW YORK • OXFORD • PARIS • SAN DIEGO
SAN FRANCISCO • SINGAPORE • SYDNEY • TOKYO

Butterworth-Heinemann is an imprint of Elsevier

Butterworth Heinemann is an imprint of Elsevier
The Boulevard, Langford Lane, Kidlington, Oxford OX5 1GB, UK
225 Wyman Street, Waltham, MA 02451, USA

Notices
Knowledge and best practice in this field are constantly changing. As new research and
experience broaden our understanding, changes in research methods, professional practices,
or medical treatment may become necessary.

Practitioners and researchers must always rely on their own experience and knowledge
in evaluating and using any information, methods, compounds, or experiments described
herein. In using such information or methods they should be mindful of their own safety
and the safety of others, including parties for whom they have a professional responsibility.

To the fullest extent of the law, neither the Publisher nor the authors, contributors, or
editors, assume any liability for any injury and/or damage to persons or property as a
matter of products liability, negligence or otherwise, or from any use or operation of
any methods, products, instructions, or ideas contained in the material herein.

British Library Cataloguing in Publication Data
A catalogue record for this book is available from the British Library

Library of Congress Cataloging-in-Publication Data
A catalog record for this book is available from the Library of Congress

For information on all Butterworth Heinemann publications
visit our website at http://store.elsevier.com/

ISBN: 978-0-12-800058-8

Working together
to grow libraries in
developing countries

www.elsevier.com • www.bookaid.org

Dedication

To Laura, Francesca, Annamaria and Francesco

Contents

Contributor details

Adriaenssens S.
Princeton University, United States of America
Sigrid Adriaenssens is a structural engineer and assistant professor at the Department of Civil and Environmental Engineering at Princeton University in Princeton, New Jersey, where she directs the Form Finding Lab. She has a PhD in lightweight structures from the University of Bath (UK). She worked as a project engineer for Jane Wernick Associates, (London) and Ney+Partners (Brussels), where she was responsible for a series of award-winning bridge projects. Her current research interests include numerical form-finding techniques and lightweight structures. She has coauthored two books, *Shaping Forces: Laurent Ney* (2010) and *Shell Structures for Architecture: Form Finding and Optimization*" (2014), and has published over 30 journal papers. She is the recipient of the Alfred Rheinstein '11 Award 2015.

Agrawal A.K.
City College of New York University, United States of America
Anil Kumar Agrawal is Professor of Structural Engineering at the City College University of New York. He is currently involved in structural control systems research, such as passive dampers, tuned mass dampers, active/hybrid and semi-active control systems, have been widely accepted as effective means for protection of civil engineering structures against earthquakes and wind loads. He is also Editor of the Journal of Bridge Engineering, the most world-renowned international scientific journal on bridge engineering, edited by ASCE. He is ASCE and IABSE member. He is involved in international technical committee, as ASCE Committee on Engineering Mechanics Member, ASCE Committee on Structural Control Member, ASCE Sub-Committee on Performance of Structures.

Amjadian M.
City College of New York University, United States of America
Mohsen Amjadian is a Research Assistant with Department of Civil and Environmental Engineering at the City College of the City University of New York, United States. He received his B.S. in Civil Engineering from Razi University in 2006, M.S. in Earthquake Engineering from International Institute of Earthquake Engineering and Seismology (IIEES) in 2010, both in Iran. He worked as a professional engineer in construction and consulting companies in Iran from 2010 till 2013. His research interests are Nonlinear Analysis of Structural Systems, Structural Control and Health Monitoring, Structural Risk, Reliability and Safety, Performance-Based Seismic Design, Soil-Structure Interaction, and Seismic Pounding.

De Backer H.
Ghent University, Belgium
Hans De Backer, born in 1978, received his civil engineering degree from Ghent University in 2002, and obtained his doctorate, about the fatigue behavior of orthotropic steel decks, in 2006. He is currently an assistant professor in the Department of Civil Engineering of Ghent University and heads the Bridge Research Group. His research focuses on fatigue effects, orthotropic steel decks, tubular structures, and nondestructive in situ testing of bridge construction.

Balázs G.L.
Budapest University of Technology and Economics, Hungary
György L. Balázs is a professor at the Budapest University of Technology and Economics in Hungary. His main fields of activity are concrete, reinforced concrete, and prestressed concrete structures; fiber-reinforced concrete (FRC). fiber-reinforced polymers (FRPs) as internally bonded reinforcements; externally bonded reinforcements or near surface mounted reinforcements; durability; service life; fire behavior and design; bond and cracking; high-performance concrete (HPC); and sustainability. He serves as chairman of the Fédération de l'Industrie du Béton (FIB) Commission on Dissemination of Knowledge, including FIB courses and FIB Textbook on Advanced design of concrete structures. He has been a member of the FIB Presidium since 2002 and was elected president of the organization in 2011 and 2012. He served as the immediate past president of FIB for 2013 and 2014 and has continued thereafter as honorary president.

Bharil R.K.
URS Corporation, United States of America
Rajneesh K. "Raj" Bharil is a licensed professional civil and structural engineer and a practicing principal bridge engineer with URS Corporation, Santa Ana, California. He obtained his masters in structural engineering from the University of Michigan in Ann Arbor, and completed his bachelor's degree in civil engineering from Maulana Azad National Institute of Technology in Bhopal, India. His 30-year engineering career encompasses cofounding his own bridge engineering specialty consulting firm, CES, Inc. Engineering, and serving as the president, vice president, director, principal, project manager, and lead engineer in both the private and public sectors. During his practice, he also served as an adjunct professor, teaching university courses in bridge engineering. He is a member of the American Society of Civil Engineers (ASCE), recipient of the AASHTO Value Engineering Award, author of numerous technical papers on bridges, and the lead designer of numerous bridges in the western United States.

Bhattacharya B.
Indian Institute of Technology, India
Baidurya Bhattacharya obtained his BTech in civil engineering from the Indian Institute of Technology (IIT) Kharagpur in 1991 and his MS (1994) and PhD (1997) in civil engineering from Johns Hopkins University, Baltimore, Maryland. He was an assistant professor at the Univeristy of Delaware, Newark (2001–2006) before returning to

IIT Kharagpur in 2006, where he became a professor in 2011. He has been a member of visiting faculty at Stanford (2005) and Johns Hopkins (2012). He works in probabilistic mechanics and explores how random atomic scale structural defects and fluctuations affect material properties at the microscale and how that randomness, coupled with uncertainties in the environment, affect the performance and safety of structural components and systems. He works on probability-based design and reliability analyses of civil infrastructure systems in structures such as nuclear power plants, ships and offshore structures, buildings, and bridges. He was a speaker at the Indo-American Frontiers of Engineering Symposium of the National Academy of Engineering, Washington, DC, in 2012 and has been an associate editor of the ASCE *Journal of Bridge Engineering* since 2010.

Boegle A.

HafenCity Universität Hamburg, Germany

Annette Boegle is Full Professor for Design & Analysis of Structures at the HafenCity University of Hamburg, Germany. She studied structural and civil engineering at the University of Stuttgart, where she also received her PhD (Dr.-Ing.). She works and teach in the fields of: construction history, conceptual design, design methods in engineering, parametric design, biomimetic structures, analysis of lightweight structures. Actually she is initiator of an Erasmus+ Strategic Partnership around the Baltic Sea Region on "Intersections in Build Environment". She also has been curator of several exhibitions e.g. "Leicht Weit – Light Structures" at the DAM Frankfurt, Germany. As a member of several scientific boards she is active participating in the scientific community, e.g. she is member of the scientific board of the "Bautechnik" (Journal for Civil and Structural Engineering), Vice Chair of the IABSE working Commission WC5 "Design Methods and Processes" and Vice Chair of the "IngenieurBaukunst e.V." (Association of Structural Art).

Brühwiler E.

École Polytechnique Fédérale de Lausanne, Switzerland

Eugen Brühwiler's activities as a professor of structural engineering at the Swiss Federal Institute of Technology (EPFL) in Lausanne, Switzerland, are motivated by the following principle: "Methods for the examination of existing structures ("Examineering") must be developed with the ultimate goal to limit construction intervention to a strict minimum. If interventions are necessary, their objective is to improve the structure." His activities as researcher, teacher, and consultant include existing civil structures, particularly bridges of great cultural value, the fatigue, dynamic and structural behaviour of bridges, and the use of ultra-high-performance fiber-reinforced cement-based composites for the improvement of structures.

Caetano E.

University of Porto, Portugal

Born in 1965 in Porto, Portugal, Elsa Caetano received her civil engineering degree from the Faculty of Engineering of the University of Porto (FEUP) in 1988. In 1989, she joined FEUP herself, where she is presently an associate professor. She has been involved in the creation and development of the Laboratory of Vibrations and Monitoring of FEUP. In the context of the activities of this laboratory, she has conducted research and consultancy work in bridges and special structures. Some relevant studies include the dynamic testing of the Vasco da Gama, Millau (in collaboration with the Centre Scientifique et Technique du Bâtiment) and Humber (in collaboration with the University of Sheffield in the UK), the dynamic design studies for the new stadium of Braga's cable roof for the Euro 2004 Football Championship, the vibration assessment, design, and instrumentation of tuned mass dampers (TMDs) at the new Coimbra footbridge, and the measurement of cable forces on the London 2012 Olympic Stadium roof.

Chouw N.
University of Auckland, New Zealand
Dr. Nawawi Chouw is associate professor and director of the University of Auckland Centre for Earthquake Engineering Research. Prior to joining the University of Auckland, he worked at universities in Europe, Japan, and Australia. He earned his diploma in civil engineering from Ruhr University-Bochum, Germany. After working in a group of consulting engineers in Germany, he returned to the Research Centre for Structural Dynamics at the Ruhr University-Bochum, and in 1993, he was awarded his doctorate. He has been awarded the Gledden Fellowship of the University of Western Australia twice, the Fritz-Peter-Mueller Prize of the Technical University of Karlsruhe, Germany, the Best Research Award of Chugoku Denryoku Research Foundation, Japan, and twice received recognition for excellence in research supervision from the China Scholarship Council. He has been invited to teach at several universities and is an editorial board member of a number of international journals.

Cooling T.
AECOM, United States of America
Thomas Cooling is a licensed professional civil engineer and geotechnical engineer with over 40 years of experience. He is recognized as a Diplomate of Geotechnical Engineering by the ASCE Academy of Geo-Professionals and is vice president of geotechnical services at AECOM, in St. Louis, Missouri. He holds a BS in civil engineering from the University of Illinois and an MS in civil engineering from the University of California, Berkeley. His bridge engineering experience includes major river crossings of the Mississippi, Ohio, Potomac, and Hudson rivers in the United States, as well as numerous other smaller bridge projects.

Dicleli M.
Middle East Technical University at Ankara, Turkey
Murat Dicleli is currently a professor and department head at the Department of Engineering Sciences, Middle East Technical University (METU). Dr. Dicleli received his PhD in structural engineering from the University of Ottawa, Canada, in 1993, and his M.Sc. and B.Sc. degrees from the civil engineering department of METU in 1987 and 1989, respectively. Dr. Dicleli's academic experience include employment both in

Illinois, at Bradley University, and in Ankara, Turkey, at METU. His research interests include seismic behavior and retrofitting of buildings and bridges, passive control systems, behavior of integral bridges under thermal and gravity loading, and behavior of steel and reinforced concrete structures under monotonic and cyclic loads. He has considerable industrial experience. He has worked as a structural and head design engineer at MNG Inc. in Ankara, Turkey, as the director of the design and planning division at MITAŞ in Ankara, Turkey, as a structural design consultant at Morrison Hershfield Ltd., in Toronto, Canada, and as senior design engineer and project manager at the Ontario Ministry of Transportation, Toronto-St. Catharines, Canada. He has been involved in the design and rehabilitation of residential and commercial buildings, industrial structures, grain storage silos, power transmission lines, and communication structures, as well as highway and railway bridges. Dr. Dicleli is also the inventor and patent holder of a recently developed torsional hysteretic damper. He serves as an associate editor for the ASCE *Journal of Bridge Engineering* and is an editorial board member of *Earthquake and Structures, American Journal of Civil Engineering, Journal of Civil Engineering and Architecture, ISRN Civil Engineering, International Journal of Engineering and Applied Sciences*, and *The Open Construction & Building Technology Journal*. He is the author of more than 160 technical publications and is also the recipient of the 2006 outstanding paper award from the Earthquake Engineering Research Institute (EERI) and 2012 thesis of the year award from the M. Parlar Foundation of METU.

Farkas G.
Budapest University of Technology and Economics, Hungary
Graduated in 1971 from the Faculty of Civil Engineering at the Budapest University of Technology and Economics (BME), he earned his Dr. Tech. in 1976, his PhD in 1994, and his Dr. Habil. in 1999 at the BME. Since 1971, he has worked at the Faculty of Civil Engineering of BME. Now a full professor, he also was head of the Department of Structural Engineering from 1995–2010, and dean of the Faculty of Civil Engineering between 1997 and 2005. In addition, he is a member of the Hungarian group of Fédération de l'Industrie du Béton (FIB) and a member of the Hungarian Academy of Engineers. He is the author of more than 200 publications in the field of reinforced concrete structures.

Fidler P.R.A.
Cambridge University, United Kingdom
P.R.A. Fidler joined the Department of Engineering at the Cambridge Centre for Smart Infrastructure and Construction in 1995, where he has worked with Professor Campbell Middleton on software for yield-line analysis of concrete slab bridges. In 2007, he began working at the department on a project funded by the Engineering and Physical Science Research Council (EPSRC) called "Smart Infrastructure–Wireless Sensor Networks for Condition Monitoring and Appraisal. This project studied potential benefits and challenges of using wireless sensor networks (WSNs) to monitor key aspects of civil infrastructure, including bridges, tunnels, and water pipes. He was involved in developing much of the embedded software for these wireless

devices. He was part of a team awarded the Telford Gold Medal (2010) from the Institution of Civil Engineers for this work. His work on WSNs continued with a trial deployment on a bridge in Wuxi, China, in 2010, and then with deployments for the Cambridge Centre for Smart Infrastructure and Construction.

Gastineau A.J.
KPFF, United States of America
Andrew Gastineau is currently a design engineer at KPFF Consulting Engineers in Seattle, Washington, where he designs waterfront and bridge structures. He earned his BA in mathematics and physics in 2007 from St. Olaf College in Northfield, Minnesota, and subsequently his MS and PhD in civil engineering from the University of Minnesota in Minneapolis in 2013. He has been published in the *Journal of Bridge Engineering* and the *Journal of Engineering Mechanics* and has written a variety of conference publications and technical reports relating to the response modification and service life extension of existing bridge structures. He also has written about bridge health monitoring.

Ramos O.R.
University of Cantabria, Spain
Óscar Ramón Ramos Gutiérrez is a MSc. civil engineer with sixteenth years of experience as head of the Bridges Division at Louis Berger´s International Design Center (formerly APIA XXI). He has acted as the lead bridge engineer in most of the major projects developed by the company, including the hundreds of viaducts that the International Design Center has designed. Since 2006, he combines his professional duties with his work as a professor at the Department for Mechanical and Structural Engineering in the University of Cantabria (Spain). In 2011, he was given the FIB Achievement Award for Young Engineers.

Hegemier G.
University of California at San Diego, United States of America
After witnessing the devastation caused by the 1971 San Fernando Valley earthquake and the 1972 Nicaragua earthquake, Gilbert Hegemier, then an aerospace engineer, decided to focus his research on developing systems to retrofit bridges, roadways, and buildings. He helped assemble a team of experts at the University of California, San Diego (UCSD). He and his colleagues have succeeded in creating and testing full-scale models of bridge column retrofit systems, which have been applied by the California Department of Transportation. These systems stood the ultimate test in the 1994 earthquake that hit Los Angeles, when 114 retrofitted bridges received only minor damage from the quake while several bridges scheduled for retrofit failed. Today, he is working with industry partners to develop and use lightweight fiber-reinforced composites (FRCs) to prevent earthquake damage and restore components of the nation's aging infrastructure. He is also working on blast mitigation techniques using FRCs to protect critical structures such as embassies from terrorist attacks.

Humpf K.

Leonhardt, Andrä, und Partner, Germany

Karl Humpf graduated as Dipl.-Ing. Structural Engineering from the University of Aachen, Germany, in 1975. He started his career as a project engineer for Ibering S.A. in Spain. In 1976, he went to Leonhardt, Andrä, und Partner, and he was appointed as director of international projects in 1993. He has extensive experience in bridge engineering from numerous bridge projects, including some of the firm's largest cable stayed, concrete, and composite bridges worldwide, particularly in Spain and Latin America. He is a registered Professional Engineer in Germany and in the U.S. states of Arizona, Georgia, Kentucky, and Massachusetts; he is also a member of the International Associatio of Bridge and Structural Engineers (IABSE) and the American Society of Civil Engineers (ASCE). He is the author or coauthor of numerous publications on long-span bridge problems in various German and international technical journals.

Kimura K.

Tokyo University of Science, Japan

Kichiro Kimura is professor of structural engineering at the Department of Civil Engineering, Faculty of Science and Technology, Tokyo University of Science, Japan. He earned his PhD at Ottawa University, and his ME in Civil engineering at Tokyo University in 1987. He was a visiting researcher at the Boundary Layer Wind Tunnel Laboratory, Faculty of Engineering Science, University of Western Ontario, Canada, in 1991–1992. For his work on wind engineering, he has been given awards by the Japan Association for Wind Engineering in the outstanding publication category in 2012, the research paper category in 2008, and the research potential category in 1994. He is a member of the Japan Association for Wind Engineering.

Kovács T.

Budapest University of Technology and Economics, Hungary

Tamás Kovács earned his PhD in 2010 at the Budapest University of Technology and Economics, while being an assistant professor in the Department of Structural Engineering at the same university; in 2013, he became an associate professor. His research interests include dynamic-based damage assessment of concrete structures, life-cycle analysis of structures, reliability of structures, high-performance concrete (HPC) for bridges, modeling of prestressed structures, strengthening of bridges, and concrete pavements. He has been honored to receive the Scholarship of the Scientia et Conscientia Found, 1997; the Tierney Clark Award 2010 for the development of the FI-150 bridge girder family, 2011; and the Innovation Award 2010 of the Hungarian Intellectual Property Office for the development of the FI-150 type bridge girder family, 2011.

Malo K.A.

Norwegian University of Science and Technology, Norway

Kjell A. Malo got his PhD from the Norwegian Institute of Technology in Trondheim. His professional background is in steel-aluminium and timber structures. His current research topics and fields of interest are material models for wood, strength and stiffness of connections for timber structures, vibrations and comfort issues in multistory timber buildings, and design of timber bridges. Since 2002, he has taught university courses on timber engineering and basic mechanics and is supervisor for MSc and PhD students in timber engineering. He is the author of more than 40 professional publications in the field of timber engineering, and he is a national delegate to the European standardization committee on timber structures. In addition, he is the convenor for the committee responsible for the new Eurocode EN 1995-2 Timber Bridges, and he is the coordinator of the European ERA-NET Woodwisdom project DuraTB – Durable Timber Bridges.

Martin B.T.
Modjeski and Masters, United States of America
Barney T. Martin received his undergraduate degree in civil engineering in 1974 from Louisiana State University, Baton Rouge, Louisiana, and his master's and PhD degrees from Tulane University in New Orleans, Louisiana, in 1981 and 1992, respectively. Dr. Martin is active on the Transportation Research Board, having recently served as chairman of the Concrete Bridge Committee and the Steel Bridge Committee. He has had extensive highway bridge design experience, having been the managing engineer on bridge design projects ranging from simple girder spans to projects involving major suspension bridges. He has significant experience in the evaluation and design of long-span bridges, particularly the inspection and evaluation of parallel wire main cables of suspension bridges. In addition, he has significant experience in the design, structural evaluation, load rating, repair, and construction support of bridges of all types, both fixed and movable. President and CEO of Modjeski and Masters.

Middleton C.R.
Cambridge University, United Kingdom
Campbell Middleton is the Laing O'Rourke Professor of Construction Engineering and director of the Laing O'Rourke Centre for Construction Engineering and Technology at Cambridge University, Cambridge, UK. He joined the staff at Cambridge in 1989, having previously worked for nearly 10 years in bridge and highway construction and design in Australia and with Arup in London. He is chairman of the UK Bridge Owners Forum, established in 2000 by representatives of the major bridge-owning organizations in the UK to identify research needs and priorities for bridge infrastructure. He has been awarded the Diploma of the Henry Adams Award of the Institution of Structural Engineers twice (in 1999 and 2014) and the Telford Premium Award (1999) and Telford Gold Medal (2010) from the Institution of Civil Engineers. He was elected a fellow of the Transport Research Foundation in 2005, has been involved in the development of bridge codes of practice, and acts as a specialist bridge consultant to clients in the UK and abroad.

De Miranda M.
Studio de Miranda Associati, Technical Director, Italy
Mario de Miranda obtained his civil engineering degree from the Politecnico di Milano, Italy, in 1979. His work, experience, and research are mainly related to the design and

construction of cable stayed and suspension bridges, wind engineering, and the history of construction. He is a partner of Studio de Miranda Associati–Consulting Engineers and has experience in the design and construction of bridges and structures. He has been involved with many major projects, most of these as lead designer, including large cable stayed bridges in Italy, the Dominican Republic, Brazil, Algeria, and India, as well as with the construction engineering of the Storebaelt suspension bridge in Denmark. He has given lectures on bridge design and construction in many countries and is the author of 60 papers and chapters of books on the same subject. Since 2006, he has been an Invited Professor at the University IUAV of Venice, where he teaches structural design and steel construction.

Modeer V.
AECOM, United States of America
Victor is a Senior Geotechnical Program Manager for AECOM in St. Louis, Missouri. He has 38 years of geotechnical experience mainly in the US, but also in Europe and the Middle East. He is a PE and has been awarded certification by the ASCE as a Diplomate in Geotechnical Engineering (D.GE). He has a Master of Science Degree with emphasis in geotechnical engineering from Purdue University and a Bachelor of Science in Civil Engineering from Louisiana State University. He is a US Navy Civil Engineer Corps veteran. Victor served as a Committee Chairman for the Transportation Research Board Committee on Earthworks and served on the Bridge Foundation committee. He served as Co-Chairman of the Illinois Joint Research and Technology Center. He has managed geotechnical foundation investigation, design and construction projects for cable stayed, long span deep girder and truss bridges. He has also designed cofferdams for bridge foundation construction including evaluation of support for a floating cofferdam system in the Mississippi River. Victor has performed seismic analyses of new and existing bridge foundations for effects from liquefaction and lateral loads. He has published peer reviewed papers including "Foundation Selection and Construction Performance - Clark Bridge Replacement" that is cable stayed.

Nowak A.
Auburn University, United States of America
Andrzej Nowak is a professor of structural engineering and chair of the Samuel Ginn College of Engineering at Auburn University, in Auburn, Alabama. In addition, he is vice-chair of the Transportation Research Board-Task Group for LTBP Bridge Traffic and Truck Weight and a member of two American Concrete Institute (ACI) committees: the ACI 343 Committee on Concrete Bridges and ACI 348 Committee on Concrete Bridges. He is the author of more than 100 papers in renowned scientific journals. His research interests include the analysis and design of structures; code calibration procedures for load and resistance factor design (LRFD); the ultimate, serviceability, and fatigue limit states; load models for bridges, including extreme events and their combinations; resistance models for materials and structural components; evaluation of existing structure diagnostics, field testing, and proof loading for bridges; weigh-in-motion procedures for bridges; and mechanical properties and design criteria for lightweight concrete structures.

Patton R.

Norfolk Southern Railway Corporation, United States of America

Ronald D. Patton is a Division Engineer at Norfolk Southern Railway and has over 39 years of engineering experience involving the maintenance, construction and design of railway bridges and other structures. He is currently Chairman of American Railway Engineering and Maintenance of Way Association Committee 10 and a member of ASCE.

Pipinato A.

AP&P, Italy

Alessio Pipinato obtained a bachelor's degree in building and structural engineering from the University of Padua, and a bachelor's degree in architecture from the University of Venice-IUAV. He earned his PhD at the University of Trento in structural design. He served as an adjunct professor, teaching university courses in bridge engineering and structural design, and has been a research collaborator at the University of Padua for more than ten years in the structural engineering sector (ICAR09-08B3). His twelve years of engineering career encompasses founding his own engineering consulting firm, AP&P, serving as the CEO, scientific and technical director; and providing bridge, structural engineering, research and development (R&D) services. He is/has been a member of the American Society of Civil Engineers (ASCE), Structural Engineering Institute (SEI), International Association for Bridge and Structural Engineering (IABSE), Associazione Italiana Calcestruzzo Armato e Precompresso (AICAP), International Association of Railway Operations Research (IAROR), Collegio Tecnici dell'Acciaio (CTA), International Association for Life Cycle Civil Engineering (IALCCE), International Association for Bridge Maintenance and Safety (IABMAS), Collegio Ingegneri Ferroviari Italiani (CIFI), European Convention for Constructional Steelwork (ECCS), and American Institute of Architects (AIA). He is also the author of more than 200 scientific and technical papers on structures and bridges, the chair of international conference sessions (including IABMAS 2010, Philadelphia; and IABMAS 2012, Milan). In addition, he is peer revisor of many international structural engineering journals, including the ASCE *Journal of Bridge Engineering, Engineering Structures, Structure and Infrastructure Engineering, International Journal of Fatigue,* and *Journal of Structural Engineering.* He has participated in a number of international research projects. His research interests includes the design, analysis, and assessment of bridges; structural analysis and design; fatigue and fracture of steel bridges; reliability analysis; life cycle assessment; probabilistic analysis; design of innovative structure and application of new materials in structures; construction control design, and fast bridge construction. He has won many international and national awards during his professional and academic career, and he served as a volunteer in the evaluation of structures during seismic emergencies for the National Service of the Civil Protection (L'Aquila 2009, Emilia Romagna 2012).

Reiner S.

Leonhardt, Andrä, und Partner, Germany

Saul Reiner was born at Lünen, Westphalia, Germany in 1938 and graduated as Dipl.-Ing. in structural engineering from the Technical University of Hannover in 1963. He started his career with steel contractor Hein Lehmann AG Düsseldorf. In 1968, he went to Leonhardt, Andrä, und Partner, where he was appointed managing director in 1992. After his retirement in 2003, he became a consultant. From 1993 to 2006, he was licensed as a Legally Authorized Checking Engineer in Germany. In 1994, he was appointed a lecturer on steel bridges at the University of Stuttgart; in 2003, he received an honorary doctorate in structural engineering from the Technical University Carolo-Wilhelmina Braunschweig. In 2005, he became an honorary member of the Argentine Society for Structural Engineering (AIE). During his professional career, he has been involved in the design, site direction, or checking of about 40 cable stayed and suspension bridges and numerous other bridges, mainly with steel or steel composite girders. He is the author of numerous papers, mainly on steel, steel composite, and cable stayed bridges and related problems like cables and protection against ship impact.

Rosignoli M.
Dr. Ing., PE, United States of America
Marco Rosignoli has 32 years of experience in the design and construction engineering of complex bridges, the industrialization of large-scale bridge projects, and the design review and forensic engineering of bridge construction machines. Working with bridge contractors, designers, and owners in 21 countries on four continents, he has served as designer, reviewer, or technical leader for the construction of five cable stayed bridges, nine incrementally launched bridges, multiple balanced-cantilever bridges, and well over 50 km of light-rail and high-speed railway bridges. An international authority on mechanized bridge construction, he is the author of four books published worldwide, four book chapters, and over 80 publications and presentations, and he holds 32 patents on bridge construction methods.

Schanack F.
Austral University, Chile
Frank Schanack studied civil engineering at TU-Dresden, Germany, in 2003, and received his doctorate at Universidad Cantabria in Spain in 2008. Since then, he has been a professor on bridges and structures at Universidad Austral de Chile, where he is currently the director of the Institute of Civil Engineering. His research has an integrated focus on all aspects of bridge engineering, including conceptual design, analysis details and erection methods, and inspection and maintenance. He has worked as a consultant for the design, construction, and maintenance of over 100 bridge projects in Germany, Spain, Argentina, and Chile.

Schultz A.E.
University of Minnesota, United States of America
Arturo Ernest Schultz is a structural engineering researcher and educator. He holds a bachelor's degree in civil engineering from Southern Methodist University in Dallas, Texas, as well as master's and doctoral degrees in civil engineering from the University of Illinois at Urbana-Champaign. He is a fellow of The Masonry Society (TMS)

and member of the Precast/Prestressed Concrete Institute (PCI), the American Concrete Institute (ACI), and the American Society of Civil Engineers (ASCE). He is past recipient of the John B. Scalzi Award (TMS), the C.T. Grimm Award (Canada Masonry Design Centre), and the Charles C. Zollman and Martin P. Korn awards (PCI).

Stewart L.
Georgia Institute of Technology, United States of America
Dr. Lauren K. Stewart, a renowned expert in blast research, came to the School of Civil and Environmental Engineering (CEE), in Atlanta, Georgia, from the University of California, San Diego (UCSD). She earned her bachelor's and doctoral degrees in structural engineering from UCSD, where she was a postdoctoral scholar and lecturer. She is also a National Defense Science and Engineering Graduate Fellow and holds a P.E. license. She has been involved with many blast and earthquake experimental projects, including the blast testing of steel structural columns, steel stud wall systems, and high performance concrete (HPC) panels using the UCSD blast simulator. She has also conducted advanced finite element analysis for the World Trade Center 7, AFRL Munitions Directorate small munitions program, and programs supported by the Technical Support Working Group. She is considered by many to be among the top blast researchers in the US, and has served as a senior blast engineering consultant to a number of organizations since 2007.

Ferretti Torricelli L.
SPEA Ingegneria Europea, Italy
Lucio Torricelli received the master degree in civil structural engineering at the Politecnico di Milano in 1991. In 1992 he joined the design office of the construction company of Italstrade S.p.a. focusing on prestressed concrete bridges realized with various construction methods (frontal launching, movable scaffolding, balanced cantilever). Then expanded the field of expertise with the involvement as senior engineer and team leader, in the design of other transportation infrastructures, such as tunnels, and underground stations. In early 2000, joined the company SPEA Ingegneria Europea, engineering company partner of Autostrade per l'Italia-Atlantia Group, focusing the interest on highway infrastructures, as senior bridge engineer. He has been leading structural engineer charged of the design of some of the major bridges structures of the "Variante di Valico" project (A1 Highway); other noticeable works includes the preparation of the conceptual guidelines for "widening and seismic retrofitting of existing structures of A14, A9 and A1 highways"; participated in the implementation of the new structural Eurocodes, with special reference to packages 2, 3 and 4, developing a set of comparative studies in order to best fit the new design requirements; team member of the "Gronda di Genova" project, in charge of the structural design of the new Genova cable stayed viaduct; since july 2011 is the Head of Structural Engineering Department of SPEA Ingegneria Europea.

Vardanega P.J.
University of Bristol, United Kingdom
Paul J. Vardanega studied at the Queensland University of Technology in Brisbane, Australia, and earned a bachelor's degree in engineering with First Class Honours and a University Medal and a Master of Engineering Science in 2007 and 2008, respectively. He is a member of Engineers Australia and a member of the American Society of Civil Engineers (ASCE). He holds a PhD from Cambridge University in geotechnical engineering, completed under the supervision of Professor Malcolm Bolton, Fellow of the Royal Academy of Engineering (FREng). From April 2012 to September 2013, he worked as a research associate at the Laing O'Rourke Centre for Construction Engineering and Technology at Cambridge (under the direction of Professor Campbell Middleton, FICE) working on the project: "Best Practice Guide for Structural Monitoring over the Whole Life of Assets,". In September 2013, he took up the position of lecturer on civil engineering at the University of Bristol.

Webb G.T.
Parsons Brinckerhoff, United Kingdom
G.T. Webb recently completed his PhD on structural health monitoring (SHM) at Cambridge University, UK. His research focused on ways in which data can be interpreted to provide useful information, an area in which surprisingly little work has yet been published. He has developed a new classification system to aid users of SHM systems to clearly understand how data is used and what information can realistically be obtained. These new findings will help to better target investments in SHM so that results with a genuine impact can be delivered. Now working for Parsons Brinckerhoff in London, he is part of a team developing a long-term SHM strategy for the Hammersmith Flyover in London. Findings from his research are being used to ensure that a useful and beneficial system is delivered.

Foreword

I acknowledge all the special men and women I have met during my life. Special people believe in the young, believe in their dreams, and cultivate their good intentions and their small and big ambitions; special people are not selfish, believe in the next, helps and not leave others with indifference. Special people work in a transparent and fair way, believe in a better future, and do the best to change it during their lives. Special people truly believe in science, research, and the culture, and do their work seriously, not for personal gain. Special people want to live intensely this great opportunity that is life, and does it with the other, spreading positivity, courage, respect, selflessness, integrity, and honesty. Let's do that every day, and the world will be better!

Preface

Bridges represent the top level of the intellectual capacity of the construction sector and the structural engineering field: new materials, new construction innovations, and a wide variety of studies are focused in the sector that is very near the boundary of other innovative engineering and scientific field (aerospace, materials engineering etc.). Moreover, an increasing demand for new and retrofitted infrastructure is taking place worldwide, so the interest in the bridge engineering field is remarkable from both an economic and a political point of view.

This book is the culmination of much long and hard work, which began four years ago, when I realized that a comprehensive work on the state of the art of bridges, including theory, design, construction, research and development (R&D) innovation was not present in the worldwide panorama. I haven't found any existing manuals with useful content on the market, as these usually include a lot of content without precise answers on the most crucial questions arising from the everyday experience in the theory and practice of bridge engineering and design. Instead, I realized I wanted to create an innovative handbook, a reference book that could be updated regularly in the pursuit of innovation. First, I have tried to make a monograph on the matter on my own, spending some years to research books and articles during my doctoral and post-doctoral studies on bridge engineering. Second, I realized that a lot of colleagues among academics and prominent engineers from all over had the same thoughts and trusted in the proposal to write an innovative monograph on bridge engineering and design—not a manual, but a reference book in which students, academics, and engineers could find useful information on topics arising both from the studies, but also from the practice and from research works. The preparation work of this book has been very intensive, with thousands of communications passing between me and the other authors.

I hope that this final work has successfully expressed our thoughts and goals.

All the chapters in this book have been "built"—I love that term, which highlights the fatigue and the hard time spent by contributors preparing every chapter—and presented by leaders in the specific area of expertise in question, engineers or academics who have made a very deep and appropriate preparation in their arguments. So if you are searching for the best design and research tool in this area, here you can find everything you need to know about bridge design, engineering, construction, and R&D.

Why do I consider this not to be a conventional book? All the chapters have been realized with the specific mood of going over the present and the past knowledge including the best, most forward-looking information we have on. We have tried to look into the future as well, and for this reason, this book is quite different from the traditional literature on the matter. Most of the chapters includes R&D information on the specific issue, which describe research and innovations, or where research is going and what the market is asking for. Sometimes the two aspects coincide, but other times not at all.

I have personally chosen every contributor, trying at the same time to have in the same study the most prominent authority in the fields and representative authors from all over, in order to prepare a leading, innovative book.

I want to acknowledge all the authors and their collaborators, more than 100 persons from all over the world, who have worked to create what is now a real, innovative handbook.

Note

The views and opinions expressed in the following chapters are those of the authors and not necessarily reflect those of the organizations they belong to or of Elsevier. The reader is cautioned that independent professional judgment must be exercised when information set forth in this handbook is applied. Anyone using this information assumes all liability arising from such actions.

Section I

Fundamentals

The history, aesthetics, and design of bridges

1

Pipinato A.
AP&P, Technical Director, Italy

1 History of bridge structures

Bridge structures represent a challenge in the built environment: they are the crystallization of forces finalized to keep someone in an unreachable place. Bridges provide the most appropriate connection of what nature has divided: a river, a valley, or something that is impossible to be reached. The first bridge was a natural gift to humanity: probably a tree that fell across a small river or the observation of rock bridges. This suggested to the first prehistoric builders that it is possible to overpass obstacles. And from this simple structures, a relevant part of the entire structural engineering worldwide has been produced over the centuries. In this chapter, a synthesis of the history of bridge construction is presented, to be followed by deeper information in subsequent chapters.

1.1 Pre-roman era

The first bridge was a simply supported beam made of wood. This was probably developed in the Paleolithic age. In the Mesolithic period, an increasing amount of bridge structures were built. For example, consider the Sweet Track, 1800 m long, which was recently discovered at Somerset Levels in Great Britain and harked to the early stage of the Neolithic period (3806 B.C.), according to dendrocronological analysis (Figure 1.1). In Egypt, such small examples have been found as the stone bridge at Gizah (2620 B.C.) (Figure 1.2). Meanwhile, in Greece, the Kasarmi Bridge, at Argolide (1400 B.C.), was one of the first type of Miceneus bridges (Figure 1.3). It is a common historical belief that Etruschi taught the Romans how to build arch bridges, even if they left no relevant bridges behind to document this. In fact, the Romans learned about this from defense and hydraulic buildings such as the Volterra arch (fourth century B.C.), which certainly was a masterpiece of the Etruschans that was later altered by the Romans (Figure 1.4). Finally, some wooden structures from the Celtic period have been found: for instance, Figure 1.5 shows the Rodano Bridge in Geneve (58 B.C.). The presence of these bridges were documented in the first century B.C. by Cesare (50 B.C.) in the book *De Bello Gallico*, which listed a large number of wooden bridges in the Gallia territory.

Innovative Bridge Design Handbook. http://dx.doi.org/10.1016/B978-0-12-800058-8.00001-3

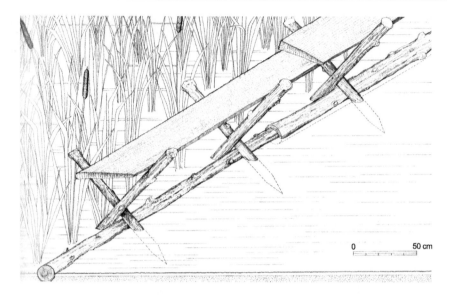

Figure 1.1 Graphic reconstruction of the 1800-meter-long Sweet Track (3806 B.C.).

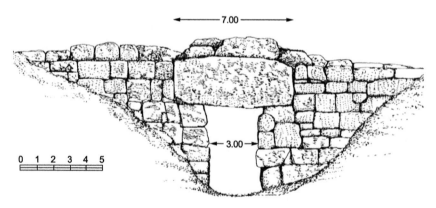

Figure 1.2 Stone bridge, Gizah (2620 B.C.).

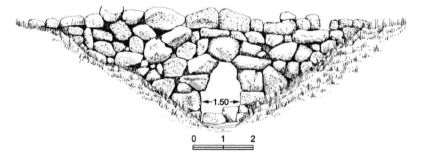

Figure 1.3 Kasarmi Bridge, Argolide (1400 B.C.).

Figure 1.4 Volterra Arch, Volterra (fourth century B.C.).

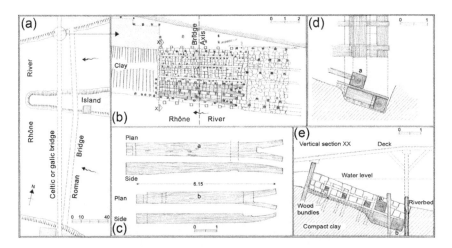

Figure 1.5 Rodano Bridge, Geneve (58 B.C.): (A) plan view, (B) plan of the first pile, (C) wooden platform for the first pile, (D) section of C, (E) built pile section.

1.2 Roman era

Although wooden bridges were common at first, stone bridges (especially arch bridges) increasingly dominated until the Middle Ages; as Palladio said: "Stone bridges were built for their longer life, and to glorify their builder" (Palladio, 1570). One of the most incredible period of bridge construction was started during the Roman Empire, in which stone arch bridge building techniques were developed. Two fundamental elements form the basis of this development: the first was geopolitical, as the military and political objective to grow faster and faster as an empire required a large amount of infrastructure; the second was technological, lying on the discovery and growing popularity of the *pozzolana*, as this fact made a strong turning in these construction types. Two notable structures pertaining to this period have been reported (Figures 1.6 and 1.7): the Sant'Angelo Bridge (in the year 136), and the Milvio Bridge (100), both in Rome. One construction improvement made by the Romans was the solution of the foundation in soft soils, by the innovative use of cofferdam, in which concrete could be poured. A relevant surviving monument of this period is the Pont du Gard aqueduct near Nîmes in southern France (first century B.C.), which measures 360 m at its longest point, was built as a three-level aqueduct standing more than 48 m high (Figure 1.8).

Figure 1.6 Sant'Angelo Bridge, Rome (136 B.C.).

Figure 1.7 Milvio Bridge, Rome (first century B.C.).

Figure 1.8 Pont Du Gard aqueduct, Nimes (first century B.C.).

1.3 Middle ages

The fall of the Roman Empire put a stop to the accelerated development of bridge construction for a long time. In the Middle ages, a particular type of bridge started to be built: the inhabited bridge. One of the most relevant and oldest of these was the Old London Bridge (Figure 1.9), finished in 1209 in the reign of King John and initially built under the direction of a priest named Peter of Colechurch; the bridge was replaced at the end of the 18th century, having stood for six hundred years with shops and houses on it. But the larger number of these bridge types are Italian inhabited bridges, such as the Ponte Vecchio in Florence.

Figure 1.9 Old London Bridge, London (1209).

1.4 The renaissance

A refined use of stone arch bridges came up during the Renaissance. The large variety and quantity of bridges that were constructed in this period make it impossible to keep a complete list of what was built. However, some masterpieces can be cited, which represent innovations of the time. The first of these was the inhabited Ponte Rialto in Venice (Figure 1.10), an ornate stone arch made of two segments with a span of 27 m and a rise of 6 m. The present bridge was designed by Antonio da Ponte, the winner of a design competition, who overcame the problem of soft and wet soil, by drilling thousands of timber piles straight down under each of the two abutments, upon which the masonry was placed in such a way that the bed joints of the stones were perpendicular to the line of thrust of the arch (Rondelet, 1841). Other notable structures of this period include the Pont de la Concorde in Paris, designed by J. R. Perronet at the end of the 18th century; London's Waterloo Bridge (Figure 1.11), by J. Rennie started in 1811; and finally, the New London Bridge (1831).

Figure 1.10 Ponte Rialto, Venice (1588).

Figure 1.11 Waterloo Bridge, London (1811).

1.5 The period of modernity from 1900 to present

The Industrial Revolution, which began in the late 18th century, completely changed the use of material not only in traditional buildings, but also in bridges. Wood and masonry constructions were replaced by iron. The famous bridge in Coaldbrookdale, an English mining village along the Severn River, was probably the first to be completely erected with iron (opened in 1779; Figure 1.12): it is a single-span bridge made of cast-iron pieces, a ribbed arch with a nearly semicircular 30-m span. The great reputation of this bridge, earned for its shape and robustness (for instance, it was the only one that successfully resisted against a disastrous flood in 1795), spurred the master engineer Thomas Telford to design a great number of arched metal bridges, including the surviving Craigellachie Bridge (1814) over the River Spey in Scotland, a 45-m flat arch made of two curved arches connected by X-bracing and featuring two masonry towers at each side (Figure 1.13). Another innovation fostered by the use of iron in construction was the opportunity to build lighter structures and such new structural components as cables. The first structural application in a bridge was probably the Menai Bridge (started in 1819, opened in 1826), another of Telford's constructions (Figure 1.14), spanning 305 m and with a central span of 177 m. This was the world's longest bridge at the time. In 1893, its timber deck was replaced with a steel one, and in 1940, steel chains replaced the corroded wrought-iron ones, in 1999 the road deck has been strengthen, and in 2005 the bridge was repainted fully for the first time since 1940; the bridge is still in service today.

Another innovation during the Industrial Revolution was the invention of the Portland cement, patented first in 1824, which, in conjunction with the recent iron industrialization, boosted the reinforced concrete (r.c.) era. François Hennébique saw Joseph Monier's (a French gardener) reinforced concrete tubs and tanks at the Paris Exposition of 1867 and began experimenting to apply this new material to building construction. Some years later, in 1892, Hennébique patented a complete building system using r.c. The first large-scale example of an r.c. bridge was the Châtellerault Bridge (1899), a three-arched structure with a 48-m central span. Subsequently, Emil Mörsch designed the Isar Bridge at Grünewald, Germany in 1904 (with a maximum span of 69 m); Eugène Freyssinet the Saint-Pierre-du-Vauvray Bridge over the Seine in northern France (built in 1922, with a maximum span of 131 m); the same Freyssinet also the Plougastel Bridge (Figure 1.15) over the Elorn Estuary near Brest, France (built in 1930 with a maximum span of 176 m); and finally, the Sandö Bridge in northern Sweden (built in 1943 with a maximum span of 260 m). Some of the first problems that arose with these medium-size structures with vehicle loadings included creep and fatigue.

A wide amount of innovations started in this period. For instance, in 1901, Robert Maillart, a Swiss engineer, started using concrete for bridges and other structures, adopting non-conventional shapes. Throughout his life, he built a wide variety of structures still known for their slenderness, and aesthetic expression. Some examples include the Tavanasa bridge over the Vorderrhein at Tavanasa, Switzerland (built in 1905), with a span of 51 m; and the Valtschielbach Bridge in 1926, a deck-stiffened arch with a 40-m span. However, undoubtedly the best-known structure is the Salginatobel Bridge, a 90-m, three-hinged hollow-box arched span in Graubünden, Switzerland. Maillart probably was the first engineer to merge engineering with the most functional form of architecture, reaching a very high quality in unconventional constructions.

Dealing with r.c. innovative solutions, industries who developed pre-stressing solutions lately, started paramount and experimental constructions: this is the case of the railway bridges near Kempten, Germany (1904), the longest span of which was 64.5 m. It was built by Dywidag Bau (now called Dyckerhoff & Widmann AG). It is also interesting to note that in 1927, the Alsleben Bridge in Saale was built with prestressed iron ties designed by Franz Dischinger, a predecessor of today's prestressing technique. And only one year later, in 1928, Freyssinet patented the first prestressing technology. Then, other bridges were completely realized in prestressed r.c.: e.g., the Luzancy Bridge (completed in 1946), with a span of 54 m (Figure 1.16). Other notable bridges were the bridge over the Rhine at Koblenz, Germany, completed in 1962 with thin piers and a central span of 202 m, designed by Ulrich Finsterwalder; and more recently, the Reichenau Bridge (1964) over the Rhine, a deck-stiffened arch with a span of 98 m designed by Christian Menn, a Swiss engineer who made great use of prestressing in bridge construction. More recently, Menn built the Ganter Bridge in 1980, a curved bridge crossing a deep valley in the canton of Valais, a cable-stayed structure with a prestressed girder, with the highest column rising 148 m and a central span of 171 m.

A wide variety of innovations arising in the late 20th century, together with the use of metal and reinforced concrete, consisted of pursuing increasing span length. This led to the first suspension bridges: the first such structure was the Brooklyn Bridge (Figure 1.17), by John Roebling and, in the final phase, by his son, Washington Roebling, which opened in 1883. This was the first suspension bridge with steel wires, with a total span of 1596 m, and a central span of 486 m. Subsequently, in New York, two other bridges were built to accommodate the increasing traffic, the Williamsburg and the Manhattan bridges. The first, spanning 2227 m, was the longest in the world in 1903, after its completion; the second, spanning 1762 m, was completed in 1910.

The Manhattan and Williamsburg bridges were the first two such structures in which the deflection theory were adopted while making the calculations, considering the relationship between deck and cable deflection and the required stiffness for increasing spans. Then, when Ralph Modjeski erected the Philadelphia-Camden Bridge in 1926 (today known as the Benjamin Franklin Bridge), reaching 2273 m that became the longest span in the world. And that was soon exceeded by the Ambassador Bridge (1929) in Detroit and the George Washington Bridge (1931) in New York.

It is an author's opinion that this latter bridge contained the most astonishing innovations, making it a masterpiece of engineering and architecture. Designed by Othmar Amman, the George Washington Bridge was long enough (1450 m) to shatter the previous record for bridge central span, reaching the 1067 m. At the same time, it was not built using the deflection theory; rather, it adopted the stabilization of the deck by its own weight. In addition, the girder depth ratio was innovative for that time, at nearly 1:350. Other similar structures followed, such as the Golden Gate (Figure 1.18), spanning 2737 (central span 1280 m) m and built in 1937; and the Bronx-Whitestone, spanning 1150 m (central span 701 m) and opened in 1939. The designers of these and other bridges learned a powerful lesson from the collapse of the Tacoma Narrows Bridge, which was destroyed by only a moderate wind in 1940, principally because its deck lacked torsional stiffness. As a result, most of the new bridges were reinforced to prevent another such disaster, adding new bracing systems or inclined suspenders to form a network of cables.

Figure 1.12 Coaldbrookdale Bridge, Coaldbrookdale (1779).

Figure 1.13 Craigellachie Bridge, Scotland (1814).

Figure 1.14 Menai Bridge, Wales (1816).

Figure 1.15 Plougastel Bridge, Brest (1930).

Figure 1.16 Luzancy Bridge, Luzancy (1946).

Figure 1.17 Brooklyn Bridge, New York (1883).

Figure 1.18 Golden Gate Bridge, San Francisco (1937).

1.6 Recent masterpieces

In contemporary times, a large number of bridges have been built. It is not easy to choose the most innovative recent structures around the world, however the presence of these elements helps in the choice: new materials (lighter, more resistant, easier to be reused); new construction methods, finalized to increase the productivity; new structural shapes (probably the most fascinating and most difficult task of a bridge engineer), and finally, elegance, which is a kind of synthesis of the aforementioned characteristics. For each of these categories, a project has been cited as an example:

- For the new materials category, the Ulsan Grand Harbor Bridge (Figure 1.19) for its innovative use of materials, such as the super high-strength steel cables (1960 MPa)
- For the construction methods category, the Providence River Bridge (Figure 1.20), built in a yard and then lifted on site
- For the innovative structural shape category, the Sunnibergbrücke (Figure 1.21), combining the cable stayed scheme with a curved plan, and featuring astonishing bifurcated columns
- For the elegant category, the Erasmus Bridge (Figure 1.22), a masterpiece of construction reflecting with a very simple shape the industrial character of Rotterdam.

Figure 1.19 Grand Harbor bridge, Ulsan (2015).

Figure 1.20 Providence River Bridge, Providence (2008).

Figure 1.21 Sunnibergbrücke, Klosters (1998).

Figure 1.22 Erasmus Bridge, Rotterdam (2003).

2 Bridge design and aesthetic

2.1 Bridge design

The bridge design phase is probably the most fascinating and most difficult task for an experienced senior engineer, if this is an original design and not an industrial/repetitive work. The definition of the bridge design process, the various steps required, and the bureaucratic procedures involved are unnecessary to explain in this context. Instead, it should be stated that the bridge is a complex structure that introduces into the surrounding landscape relevant variations, dealing with a number of specialist fields: for example, hydraulic, geotechnical, landscaping, structural, architectural, economic, and socio-political. For this reason, before starting the design of a bridge, a concept should be developed, with the realization of a scaled model, as a simulation of the three-dimensional (3D) overview of the construction and of all the considered alternatives. From this initial concept, some parametric considerations need to be performed to estimate the costs. This preliminary analysis is the basis for an open discussion with the client, the managing agencies, and any relevant local government agency on the most suitable solution. Only when the costs and the concept will be shared can the design stage start: the successive steps of the preliminary plan, finally culminating in a construction project that deals with the actual erection of the bridge. For large-scale projects, the preliminary stage includes economic and financial studies as well. It should be known that the large number of variables included in the design stage are mostly not fixed, as they depend on the precise place and time of the realization: e.g., there is not the best finite element method (FEM); rather, the FEM software most suitable for the specific bridge design must be chosen, and the same applies to codes and standards, the amount of human resources, and the hardware instrumentation required. The best project is a perfect mix of these various components. Surely, a good project must include an architectural consciousness, the structural engineering knowledge, the professional experience, and a strong informatic infrastructure.

2.2 Bridge aesthetics

There is no one rule to conceive the most perfect or most aesthetically pleasing bridge. However, awful bridges can be found anywhere. A good and well-known definition of the term *aesthetic* could be "pleasant architecture": consequently, it could be helpful to give the basic components of architecture. These, according to Vitruvio (27 B.C.), are:

- *Firmitas:* This is a key element for infrastructure, and is surely the most relevant for bridge structures; it is the ability of a bridge to preserve its physical integrity, surviving as an integral object, at least for its service life.
- *Utilitas:* The practical function of a structure is a common rule; however, it is often not applied; the simple requirement that set the spaces and the components of a bridge structure includes the usefulness for the specific purpose for which the bridge was intended for.
- *Venustas:* The sensibilities of those who see, or use the bridge structure may arise from one or more factors, including the symbolic meaning; the chosen shape and forms; the materials, textures, and colors; and the elegance to solve practical and programmatic problems. This is obviously a subjective factor that could cause delight in the observers, or not.

3 Research and innovation in bridge design

Research and development (R&D) are expected in the coming years in this particular and fascinating bridge engineering field. This activities are expected to be carried out by industries, universities, and specialized firms: the R&D field in this sector is expected to grow faster and faster, expanding into other fields of construction in the future.

The most prominent problems to be faced are the following:

- *Sustainable bridges*: As generally could be said about the construction sector, a reduction in the use of materials is expected in bridge construction, together with the possibility of conceiving new construction modes and new bridge types that can reduce the need for raw materials, and at the same time, the construction, operation, maintenance and decommissioning energy and cost consumption.
- *Intelligent bridges*: Bridges will be more like machines in the future, rather than a fixed and completely crystallized construction, and eventually, intelligent systems able to control in real time the bridge status (such as material decay, unexpected stress/strain levels, and external dangers) will be developed at a reasonable commercial cost, and integrated during the construction process in all new bridges at both large and small scales.
- *Intelligent bridge-net*: Managing authorities are nowadays concerned about managing and limiting the maintenance costs of old infrastructure, where bridges are reaching 100 years of age. There is no single answer, as different solutions can apply depending on the situation: however, in the future, ideally all bridges will be monitored as a net of constructions, where every maintenance cost could be planned and where a maintenance program could be easily performed and updated.
- *Life-long solutions:* A wide amount of research should be done in the specific sector of materials, as they can easily contribute to build life-longer and more sustainable bridges.

References

Cesare, G., 50 B.C. De Bello Gallico. Original Latin Version. Mursia Editorial Group, Milan, Italy.

Palladio, A., 1570. I quattro libri dell'architettura. Translation of the original, 1945, 4 vols. Ulrico Hoepli Editore, Milan, Italy.

Rondelet, A., 1841. Saggio storico sul ponte di Rialto in Venezia. Negretti Edition, 1841, Mantova, Italy.

Vitruvio, P.M., 27 B.C. De Architectura. Translation of the original, 1997, 2 vols. Einaudi, Milan, Italy.

Section II

Loads on bridges

Loads on bridges

2

Nowak A.[1], Pipinato A.[2]
[1]Department of Civil Engineering, Auburn University, Auburn, AL
[2]AP&P, Technical Director, Italy

1   Introduction

In this chapter, information regarding loads is presented: this includes models and load values associated with road traffic, pedestrian activities, rail traffic, dynamic and centrifugal effects, braking and acceleration actions. Imposed loads defined in codes and standards are intended to be used for the design of new bridges, including piers, abutments, upstand walls, wing walls, flank walls, and their foundations. An open question remains on existing bridges, where reduced traffic loads could be used during the structural assessment and imposed in new traffic limitations, to avoid for the bridge retrofit or reconstruction. Infact, in this case, only some nations provide detailed guidelines or codes on the assessment of existing bridges. For instance, the United States (AASHTO, 2013), the United Kingdom (Highways Agency, 2006; Network Rail, 2006), Denmark (Danish Road Directorate, 1996), Switzerland (Societè Suisse des Ingenieurset des Architects, 2011), and Canada (Canadian Standards Association, 2006).

2   Primary loads

2.1   Permanent loads: self-weight of structural elements

The self-weight or dead load consists of weight of structural components and non-structural elements that are permanently attached to the structure, including for example, noise and safety barriers, signals, ducts, cables and overhead line equipment (except the forces due to the tension of the contact wire etc.). The self-weight is generally estimated in the first design phase, and then it is updated analytically in the detailed design phase. The actual value can also be estimated using empirical formulae, or it can be assumed based on the designer's past experience. Special care is required in the analysis of self-weight during the construction period of the bridge, including consideration of the erection equipment (Figure 2.1).

2.2   Permanent loads: self-weight of non-structural elements

Road and railway equipment, sidewalks, parapets, barriers, channels or pipework, noise wall luminaires and sign supports are considered as non-structural elements. The magnitude of load are usually determined using mass/volume unit values specified in design codes and standards.

Innovative Bridge Design Handbook. http://dx.doi.org/10.1016/B978-0-12-800058-8.00002-5

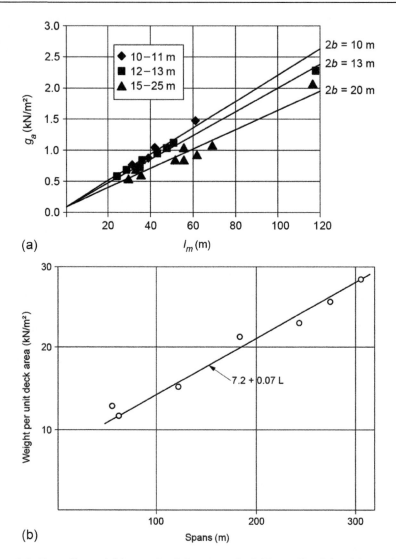

Figure 2.1 (a) small-span bridges, twin girder composite bridge, self weight of the steelwork (Leben and Hirt, 2012); (b) medium-span bridges—prestressed concrete bridge self weight (O'Connor, 1971).

2.2.1 Traffic loads: Eurocode

EN 1991-2 (2003) is intended to be used in conjunction with EN 1990 (especially A2). Section 1 gives definitions and symbols. Section 2 defines loading principles for road bridges, footbridges (or bicycle-track bridges), and railway bridges. Section 3 covers design and provides guidance on multiple presence of traffic load and on load combinations with non-traffic actions. Section 4 defines: loads (models and representative values) due to traffic action on road bridges and load combinations including combinations with pedestrian and bicycle traffic as well as other actions specific for the design of road bridges. Section 5 defines loads (models and representative values) on footways, bicycle tracks, and footbridges and other actions specific for the design of footbridges. Sections 4 and 5 also define loads transmitted to the structure by vehicle restraint systems, pedestrian parapets. Section 6 defines actions due to rail traffic on bridges and other actions specific for the design of railway bridges and structures adjacent to railway. Characteristic loads are intended for the determination of road traffic effects associated with the ultimate limit state and with particular serviceability limit states.

Characteristic values are determined from the analysis of data collected in several countries. The design values were calculated as corresponding to a probability of being exceeded annually and are adjusted using the coefficients α_{Qi} and α_{qi}. These coefficients for the traffic load model can be nationally adjusted (in the so-called National Annexes). The code EN 1991-2 (2003) specifies two principal load models for the normal highway bridge traffic. For instance, Load Model 1 (LM1) consists of a double axle, called *tandem system (TS)*, together with a uniformly distributed load, and is intended to cover "most of the effects of the traffic of lorries and cars". It is necessary, first, to define notional lanes. The normal basic lane width is 3 m, with the exception that roadway widths of 5.4–6 m is assumed to carry two lanes. Generally, a roadway is divided into an integral number of 3 m lanes, that may be positioned transversely so as to achieve the worst effect. Of these lanes, the one causing the most unfavorable effect is called Lane 1, the one causing the second most unfavorable effect is Lane 2, and so on. These lanes need not to correspond to the marked lanes on the bridge; indeed, a demountable central safety barrier is ignored in locating the traffic lanes. Space not occupied by the lanes is called *remaining area*. The total load models for vertical loads is represented by the following traffic effects:

- *Load Model 1 (LM1):* Concentrated and uniformly distributed loads that cover most of the effects of the traffic of trucks and cars. This model should be used for general and local verifications (Figure 2.2).
- *Load Model 2 (LM2):* A single-axle load applied on specific tire contact areas that covers the dynamic effects of the normal traffic on short structural members (Figure 2.3).
- *Load Model 3 (LM3):* A set of assemblies of axle loads representing special vehicles (e.g., for industrial transport), that can travel on routes permitted for abnormal loads. It is intended for general and local verifications.
- *Load Model 4 (LM4):* A crowd loading, intended only for general verification.

2.2.2   Traffic loads: AASHTO

Highway bridge design loads are established by the American Association of State Highway and Transportation Officials (AASHTO). For many decades, the primary bridge design code in the United States has been the AASHTO "Standard Specifications for Highway Bridges" (Specifications), as supplemented by agency criteria as applicable. During the 1990s, AASHTO developed and approved a new bridge design

Location	TS	UDL system
	Axle loads Q_{ik} (kN)	q_{ik} (or q_{rk}) (kN/m²)
Lane 1	300	9
Lane 2	200	2,5
Lane 3	100	2,5
Other lanes	0	2,5
Remaining area (q_{rk})	0	2,5

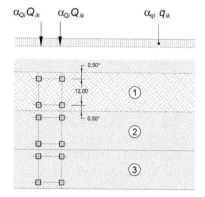

Key
(1) Lane Nr.1: Q_{1k} = 300 kN; q_{1k} = 9 kN/m²
(2) Lane Nr.2: Q_{2k} = 200 kN; q_{2k} = 2,5 kN/m²
(3) Lane Nr.3: Q_{3k} = 100 kN; q_{3k} = 2,5 kN/m²
* For w_l = 3,00 m

Figure 2.2 Load Model 1 (EN 1991-2, 2003).

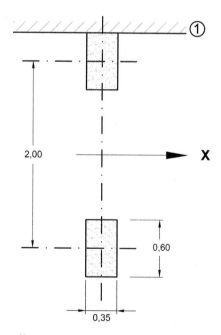

Key
X Bridge longitudinal axis direction
1 Curb

Figure 2.3 Load Model 2 (EN 1991-2, 2003).

code, entitled "AASHTO LRFD Bridge Design Specifications" (AASHTO, 2014). It is based upon the principles of load and resistance factor design (LRFD). Section 3 deals with Loads and Load Factors and includes information on permanent loads (dead load and earth loads), live loads (vehicular load and pedestrian load), and other loads (wind, temperature, earthquake, ice pressure and collision forces). The basic vehicular live loading for highway bridges is designated as HL-93 and it consists of a combination of the:

- Design truck or design tandem
- Design lane load

Each design lane under consideration is occupied by either the design truck or tandem, superimposed with the lane load. The live load is assumed to occupy 10.0 ft width within a design lane. The total live load effect resulting from multilane traffic can be reduced for sites with lower ADTT using the multilane reduction factors. A careful consideration is required in case of site-specific exceptional situations if any of the following conditions apply:

- The legal load of a given jurisdiction is significantly greater than the code specified load.
- The roadway is expected to carry exceptionally high percentages of truck traffic.
- Traffic flow control devices, such as a stop sign, traffic signal, or tollbooth, causes trucks to congregate on certain areas of a bridge.
- Exceptional industrial loads occur at the considered location of the bridge.

The live load model, consisting of either a truck or tandem coincident with a uniformly distributed load, was developed as a notional representation of a group of vehicles routinely permitted on highways in various states under "grandfather" exclusions to weight laws. The vehicles considered to be representative of these exclusions were based on a study conducted by the Transportation Research Board (Cohen, 1990). The load model is called "notional" because it is not intended to represent any particular truck. The weights and spacing of axles and wheels for the design truck is as specified in Figure 4. A dynamic load allowance is to be considered by increasing the static effects of the design truck or tandem, other than centrifugal and braking forces, by 33% of the truck load effect. That percentage is 75% for deck joints and 15% for fatigue and fracture limit state. The spacing between two 32.0-kip axles can vary between 4,3 m(14.0 ft) and 9 m (30.0 ft) to produce the extreme force effect. The design tandem consists of a pair of 100 kN (25.0-kip) axles spaced 1.2 m (4.0 ft) apart. The transverse spacing of wheels 1.8 m (6.0 ft). The design lane load consists of a load of 0.64 klf (9.3 kN/m) uniformly distributed in the longitudinal direction. Transversely, the design lane load is assumed to be uniformly distributed over a 3.05 m (10.0 ft) width. The force effects from the design lane load are not be subject to a dynamic load allowance.

2.2.3 Traffic loads: AREMA

The standard loading scheme incorporated by North American Railways and the American Railway Engineering and Maintenance-of-Way Association (AREMA) is the Cooper E-Series loading: AREMA (2013) recommends E-80

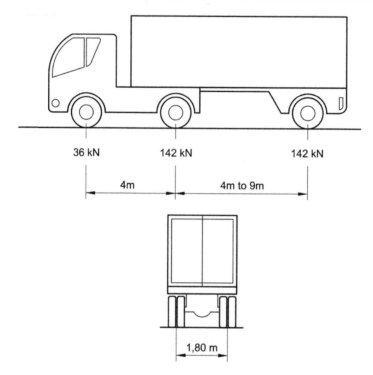

Figure 2.4 Characteristics of the design load (AASHTO 2014).

loadings (two locomotives coupled together in doubleheader fashion, with the maximum axle load of 335 kN) to be used for the design of steel, concrete, and most other structures. Yet the designer must verify the specific loading to be applied from the railway, as this may require a design loading other than the E-80 Cooper E-Series. More information is given in the specific chapter dedicated to railway bridges (Chapter 20).

2.2.4 Traffic loads: Australian standard

The Australian Standard (AS) (AS5100, 2004) normal design traffic load includes the following components, each considered separately:

- W80 wheel load, that comprises of an 80 kN load applied over a contact area of 400 mm wide x 250 mm long anywhere on the road surface (Figure 2.5).
- A160 axle load, comprising of two W80 wheels spaced 2 m apart between the center of the wheel contact areas (Figure 2.5).

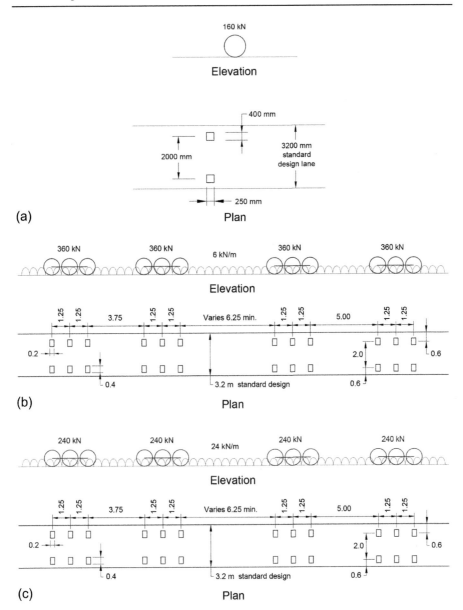

Figure 2.5 AS loading schemes (AS5100, 2004): (a) AS5100.2 W80 wheel load and A160 axle load configuration; (b) AS5100.2 M1600 moving traffic load configuration; (c) AS5100.2 M1600 stationary traffic load configuration.

- M1600 moving load, comprising a combination of axle group and uniformly distributed lane load (UDLs), as illustrated in Figure 2.5. The lane width is taken as 3.2 m. The lane UDL is either continuous or discontinuous to produce the most adverse effect, and the truck variable length is to be adjusted so as to produce a most adverse effect.
- S1600 stationary load, comprising the combination of axle group and lane UDL (Figure 2.5), applied in a similar fashion to the M1600 load.

In addition, where required by the authority, bridges are to be designed for heavy load platforms (HLPs). There are two forms of these alods: the HLP 320 load and the HLP 400 load (Figure 2.6). These loads are described as follows:

- 16 rows of axles spaced at 1.8-m center-to-center
- Total load per axle: 200 kN for the HLP 320 and 250 kN for the HLP 400
- Eight tires per axle row
- Overall width of axles: 3.6 m for the HLP 320 and 4.5 m for the HLP 400
- Tire contact area: 500 mm wide x 200 mm long for each set of dual wheels
- Tire contact areas are centered at 250 mm and 1150 mm from each end of each axle
- For continuous bridges, the load is considered as separated into two groups of eight axles, each with a central gap of between 6 m and 15 m, chosen to give the most adverse effect.

AS5100 (2004) defines the standard design lane width as 3.2 m, with the number of design lanes calculated as: $n = b/3.2$ (rounded down to the next integer), where n = the number of lanes and b = width between traffic barriers, in meters. These lanes are to be positioned laterally on the bridge to produce the most adverse effect.

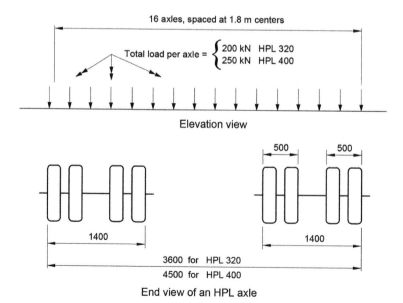

Figure 2.6 AS loading schemes for heavy load platform (AS5100, 2004).

3 Environmental effects

3.1 Wind

Wind forces must be considered in the design of bridges in two different conditions: during operations, when the bridge is completed, and during construction. Wind loads depend on geometrical form, size, and on constituent material of the structure. Design, Codes and standards provide numerical values and procedures to determine the wind loads to be applied to structures. An entire chapter of this book is dedicated to wind loads (Chapter 3).

3.1.1 Eurocode

EN-1991-1-4 (2005), "Part 1-4: General actions - Wind actions", provides guidance to determine the characteristic wind actions over the entire structure, some parts of the structure, or a single member of the structure. This code provides a platform to determine the wind action acting on any land-based structures. Eurocode 1 is used as a guidance for almost all member countries. Therefore, it is recommended to use the National Annex (NA). The National Annex provides specific data and methods based on the geological, topographical and meteorological characteristics of the considered country. The current version of the code can only be used for the structures with span lengths of not more than 200 m or heights of 200 m.

3.1.2 AASHTO

According to AASHTO standards, wind loads is assumed to be uniformly distributed over the area exposed to the wind. The exposed area is a sum of the areas of all components, including the floor system, railings, and sound barriers, as seen in elevation taken perpendicular to the assumed wind direction. This direction is to be selected to determine the extreme force effect in the structure or in its components. Areas that do not contribute to the extreme force effect under consideration can be neglected in the analysis. Base design wind velocity varies significantly due to local conditions. For small and low structures, wind usually does not govern. For large, tall bridges and sound barriers, however, the local conditions should be considered. The pressures on the windward and leeward sides are to be taken simultaneously in the assumed direction of the wind. Typically, a bridge structure should be examined separately under wind pressures from two or more different directions in order to determine if windward, leeward, or side pressure produces the most critical load on the structure.

3.2 Temperature

Two forms of temperature effect can be considered in bridges:

- Overall temperature changes are to be considered in the design of moving bearings and in the selection of their location.

- Differential temperature effects may occur such that at a particular time, the temperature at one point in a structure is not the same as at another, and these temperature differences may cause locked-in stresses and possible failure.

Codes and standards specify temperature changes and variations to be considered during the design stage. If a bridge deck is free to expand, the variation of temperature ΔT implies deformations in the longitudinal direction that has to be accommodated by the expansion joints. Such deformation effects can be calculated using the following equation:

$$\Delta l = \alpha_T * l * \Delta T$$

where α_T is the thermal expansion coefficient, l is the length of the considered element and ΔT the uniform temperature variation.

Concerning overall temperature changes, in a single-span bridge, it is conventional to permit longitudinal, horizontal movement in the bearings at one end of a span so that the bridge can expand or contract freely under the action of temperature changes or other related effects, such as concrete shrinkage or creep, and elastic strains in the structure under load. They also allow for foundation movement. In the case of a multi-span bridge, if it is continuous over its full length or over a number of spans, these longitudinal movements can add up at one location. Alternatively, if the bridge consists of a number of simply supported spans, there can be a moving bearing at one end of each span (O'Connor and Shaw, 2000). Concerning differential temperatures, these can cause damage to bridges (e.g. the major Newmarket Viaduct in Auckland, New Zealand, in the period following its completion in 1966); see Buckle and Lanigan (1971), Leonhardt et al. (1965), Priestley (1972), and White (1979). Temperatures may vary within a cross section: if the variation is linear and the structure is statically determinate, then it can be adopted a deflected shape without the development of stresses due to these temperature differences. If either of these conditions is not satisfied, then stresses can develop due to temperature. These stresses can take the form of longitudinal direct stresses. For example, in a bridge continuous over three spans, it can be expected that the temperatures in the upper flange in the morning are higher than in the lower flange. If the structure was freed from its central piers, these temperatures, when considered alone, would cause the girder to rise. However, it is, in fact, restrained from doing so and additional downward reaction components are applied to the structure at its intermediate supports. Differential temperatures of this kind, can therefore tend to cause restraint tensile stresses in the lower flange. This may not be a problem at the supports themselves, but they will add to other design stresses at midspan. Not only that, but the hold-down reactions developed at the intermediate piers will cause vertical end reactions that add to the end shears in the members. The combination of these effects—the effects of nonlinearity in the temperature distributions and the effects of restraint forces—may cause cracking in a concrete structure, and possibly greater distress as well (O'Connor and Shaw, 2000).

3.3 Snow

Design values of the snow load are provided in codes and standards. The designer can consider additional load combinations for greater safety if the region in which the bridge is built is subjected to heavy snowfall.

3.4   Earthquake

Earthquake events in the vicinity of an existing bridge structure can cause permanent failure. Not all regions are subjected to the seismic risk, however, many countries had to face this problem over time. Codes and standards providing procedures for the design and evaluation of bridges with regard to earthquakes are becoming a sort of *nightmare* for bridge designers, due to useless and long procedures to gain results. However, often a simple elastic design procedure can be adopted. The most damaging excitations are horizontal motions, in the longitudinal and transverse direction of the bridge: these can often lead to a partial collapse (e.g. a single span falling down), bearing damages, abutment and pier damage or a complete collapse. Seismic devices are used to maintain the superstructure and the structure as a whole in service also during and after strong earthquake, as discussed in the book chapter dedicated to this problem. In particular, structural details are very important in designing an earthquake resistant structure.

4   Dynamic amplification

The dynamic load effects of bridge structures include the following main aspects:

- Impact: the maximum vertical loads induced by a moving load will often exceed those produced by an equivalent static loads. This is commonly called Impact (where I=impact factor) defined as the ratio of the additional load (total dynamic minus static) divided by the equivalent static load; experimental studies in the past have revealed the common values of the impact factor (Figure 2.7).
- Braking/accelerating vehicles: Longitudinal loads can be applied by braking or accelerating vehicles.
- Transverse horizontal centrifugal forces: These forces are expressed by mV^2/R, where m is the mass of a body moving with tangential velocity V around a circle of radius R. Transverse horizontal centrifugal forces relate to a curved bridge, or when a vehicle changes its direction of movement;
- Earthquake effects

5   Bridge redundancy

Bridge redundancy can be defined as the capability of a bridge structural system to carry loads after damage to or the failure of one or more of its members (AASHTO 2013). There are three types of redundancy: load path redundancy, structural redundancy, and internal redundancy. A member is considered load path redundant if an alternative and sufficient load path is determined to exist. The alternative load paths must have sufficient capacity to carry the load redistributed to them from an adjacent failed member.

A member is considered *structurally redundant* if its boundary conditions or supports are such that failure of the member merely changes the boundary or support conditions but does not result in the collapse of the superstructure.

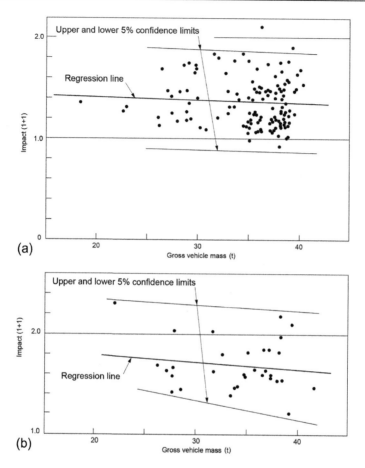

Figure 2.7 Values of impact, I+1: (a) First impact study of Six Mile Creek Bridge (Pritchard, 1982; O'Connor and Pritchard, 1985); b) second impact study of Six Mile Creek Bridge (Pritchard, 1982; O'Connor and Pritchard, 1985);

Internal redundancy is when a structural member has alternative and sufficient load paths existing within the member itself. For example, a riveted steel member connection is considered internally redundant if it has multiple plies.

6 Conclusions

Consideration of loads is very important during the design stage of a bridge, however, it is mostly considered as a routine step of the project. Apart from a detailed analysis of live loads and other types of loads, it is recommended to evaluate the actual site-specific loads. This is a crucial issue as magnitude of real traffic loads is often larger

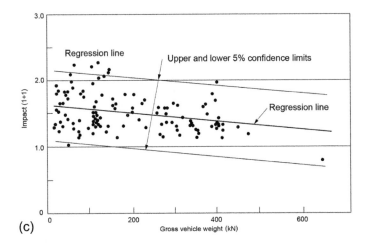

Figure 2.7 continued.   c) Six Mile Creek Bridge (Chan, 1988; Chan and O'Connor, (1990).

than the code specified traffic loads. Traffic loads should be revisited in design codes, and in particular this applies to highway loads, and to a less extent to railway loads, that are more closely checked by the managing authorities. Although each country have legal restrictions on the vehicle weight and geometries, in practice, the law enforcement is often not effective (O'Connor and Shaw 2000). Heywood (1992) reported a study of Australian road traffic loads, with measured values of average extreme daily axle loads for various axle groups, for two classes of sites. The corresponding legal limits for the complete axle groups are (i) single axle–steer 6.0 t (58.8 kN); (ii) tandem–steer 11.0 t (53.9 kN per axle); (iii) tandem–nonsteer 16.5 t (80.9 kN per axle); and (iv) tri-axle group 20.0 t (65.4 kN per axle). The ratios of measured values to legal limits (short, medium and long spans) were (1.29, 1.14), (1.13, 1.09), (1.26, 1.27), and (1.24, 1.22) for these four axle configurations. As can be seen, all the categories were exceeded, which agrees with what has been reported by the recent weight-in-motion (WIM) studies (FHWA, 2007). Therefore, for example, bridges are subjected to a greater damage than analytically predicted in the fatigue evaluation performed during the design stage, and of course to premature deterioration. For this reason, there are significant cost benefits in the use of increased design live loads. However, prior to introduction of more specific changes, it is necessary to perform a cost analysis, and then consider three alternatives:

a) existing load limits stay as they are , so as to safeguard existing bridges;
b) load limits are increased if it is perceived that older design procedures have resulted in bridges with a sufficient reserve of strength;
c) economic analysis of benefits of use of heavier vehicles justifies the construction of new bridges to a higher standard, accompanied by a program for the strengthening of existing bridges.

References

AASHTO, 2014. AASHTO LRFD Bridge Design Specifications. American Association of State Highway and Transportation Officials, Washington, D.C.

AASHTO, 2013. Manual for Bridge Evaluation, second ed with 2015 Interim Revisions American Association of State Highway and Transportation Officials, Washington, D.C.

AREMA, 2013. Practical Guide to Railway Engineering. AREMA–American Railway Engineering and Maintenance-of-Way Association, Lanham, Maryland (US).

AS5100, 2004. Australian Standard (AS) 5100, Bridge Design. Department of Main Roads, Road Planning and Design Manual, Sydney, Australia.

Buckle, I.G., Lanigan, A.G., 1971. Transient thermal response of box girder bridge decks. In: Third Australasian Conference on the Mechanics of Structures and Materials, Auckland, New Zealand.

Canadian Standards Association, 2006. Canadian Highway Bridge Design Code. Mississauga, Ontario, Canada.

Chan, T.H.T., 1988. Highway Bridge Impact. PhD thesis, University of Queensland, Brisbane, Australia.

Chan, T.H.T., O'Connor, C., 1990. Wheel loads from highway bridge strains: field studies. J. Struct. Eng. 116 (7), 1751–1771.

Cohen, H., 1990. Truck Weight Limits: Issues and Options, Special Report 225. Trans. Res. Board, National Research Council, Washington, D.C.

Danish Road Directorate, 1996. Calculation of Load-Carrying Capacity for Existing Bridges. Guideline Document, Report 291, Ministry of Transport, Denmark.

EN 1990, 2006. Basis of structural design. European Committee for Standardization (CEN), Brussels, Belgium.

EN 1991-2, 2003. Actions on structures—Traffic Loads on Bridges. European Committee for Standardization (CEN), Brussels, Belgium.

EN 1991-1-4, 2005. Eurocode 1: Actions on Structures—Part 1–4: General Actions—Wind Actions. European Committee for Standardization (CEN), Brussels, Belgium.

Federal Highway Administration (FHWA), 2007. Effective Use of Weigh-in-Motion Data: The Netherlands Case Study. U.S. Department of Transportation, Washington, D.C.

Heywood, R.J., 1992. Bridge Live Load Models from Weigh-in-Motion Data. PhD thesis, University of Queensland, Brisbane, Australia.

Highways Agency, 2006. Design Manual for Roads and Bridges—Highway Structures Inspection and Maintenance, vol. 3. Sec. 4, London.

Laman, J.A., Nowak, A.S., May 1997. Site-specific truck loads on bridges and roads. Proc. Inst. Civ. Eng., Transp. (123), 119–133.

Leben, T., Hirt, M., 2012. Steel Bridges: Design and Dimensioning of Steel and Steel-Concrete Composite Bridges. EPFL Press, Lausanne, Switzerland.

Leonhardt, F., Kolbe, G., Peter, J., 1965. Temperature differences endanger prestressed concrete bridges. Beton Stahlbetonbau 60 (7), 231–244.

Nassif, H., Nowak, A.S., 1995. Dynamic load spectra for girder bridges. Transport. Res. Rec. (1476), 69–83.

Network Rail, 2006. Structural Assessment for Underbridges. Guidance Note, London.

Nowak, A.S., 1993. Live load model for highway bridges. J. Struct. Safety 13 (1+2), 53–66.

Nowak, A.S., Lutomirska, M., Sheikh Ibrahim, F.I., 2010. The development of live load for long span bridges. Bridge Struct. J. 6, 73–79.

Nowak, A.S., Rakoczy, P., 2013. WIM-based live load for bridges. KSCE J. Eng. 17 (3), 568–574.

O' Connor (1971) Design of Bridge Superstructures, Wiley, New York.

O'Connor, C., Pritchard, R.W., 1985. Impact studies on a small composite girder bridge. J. Struct. Div. 111 (3), 641–653.

O'Connor, C., Shaw, P.A., 2000. Bridge Loads. Spon Press, Taylor and Francis Group, New York.

Priestley, M.J.N., 1972. Temperature gradients in bridges: some design considerations. New Zealand Eng. 27 (7), 228–233.

Pritchard, R.W., 1982. Service Traffic Loads on Six Mile Creek Bridge, Queensland. MEngSc thesis, University of Queensland, Brisbane, Australia.

Societè Suisse des Ingenieurset des Architects (SIA), 2011. SIA 269: Bases pur la Maintenance des Structures Porteuses. Zurich, Switzerland.

White, I.G., 1979. Non-linear Differential Temperature Distributions in Concrete Bridge Structures: A Review of Current Literature. Cement and Concrete Association, Wexham Springs, UK.

Wind loads

3

Kimura K.
Department of Civil Engineering, Tokyo University of Science, Noda-shi, Chiba, Japan

1 Introduction

Wind loading is one of primary horizontal loads acting on bridges, and its appropriate consideration is necessary to satisfy the design requirements. The dynamic wind effects are also important particularly for long-span bridges, which may induce significant vibrations not only in along wind direction but also in vertical and torsional directions, and they have to be avoided. In this chapter, wind effects on bridges are overviewed, and a typical procedure for wind resistant design of a long-span bridge is described. Design wind speeds and wind loads in codes, some examples of field measurements of full-scale bridges, and research results on stay cable vibrations are introduced.

2 Overview of wind effects on bridges

Wind effects have to be carefully considered in the design of long-span bridges. The effects are generally dynamic because the fluctuation of wind velocities due to turbulence and the vortices formed around the bridge generate a time-varying wind load. The most dominant component of the wind load is often in the along-wind direction, and its maximum value is mostly taken as the design wind load. For relatively flexible bridges with longer spans and cable-supported bridges, considerations of wind-induced dynamic responses are also important because they can be harmful to the safety and serviceability of the bridge.

The collapse of the old Tacoma Narrows Bridge in Washington in 1940 defined an epoch, because it clearly demonstrated the fatal impact of wind-induced vibration as recorded in a film. And since then, bridge engineers have been paying attention to the wind-induced dynamic response, which can be hazardous for long-span bridges. Also, buffeting (a random response caused by turbulence in natural wind) has been considered since the pioneering research by Davenport (1962) who originally presented a buffeting prediction procedure based on random vibration theory. In order to predict and analyze wind-induced responses more accurately, extensive research on bridge aerodynamics has been conducted (e.g., Simiu and Scanlan, 1996; Sockel, 1994; Simiu and Miyata, 2006; Holmes, 2007; Stathopoulos, 2007; Jurado et al., 2011; Fujino et al., 2012; Xu, 2013; Tamura and Kareem, 2013). As another example, there is much construction of long-span bridges in China, and active developments of bridge aerodynamics in these projects were introduced in a summary paper (Ge, 2008).

In smooth flow (i.e., wind with very small wind velocity fluctuations), mainly two types of wind-induced vibration of bridges occur: vortex-induced vibration and self-

Innovative Bridge Design Handbook. http://dx.doi.org/10.1016/B978-0-12-800058-8.00003-7

excited vibration. For both of these types of vibrations, the response is affected not only by the aerodynamic forces, but also by the motion of the bridge itself. This is because the flow around the bridge is influenced by the motion of the bridge, and thus, it significantly changes the aerodynamic forces acting on the bridge. Particularly for self-excited vibrations, the response is also called *aeroelastic* because the elastic motion, or vibration, of the structure plays a significant role in generating these forces. Both vortex-induced vibration and self-excited vibration of bridge decks are mainly caused by fluctuating aerodynamic forces due to vortices around the deck, as shown in Figure 3.1 (Kubo et al., 1992) for a shallow rectangular cylinder model where the flow from the left-hand side was visualized by smoke.

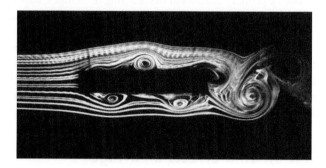

Figure 3.1 Vortices formed around a rectangular cylinder.

The first type of vibration is caused by vortices formed from separate flows around the bridge, and the wind speed range at which it occurs is limited and usually lower than the onset wind speed of self-excited vibration. The dominant motion of vortex-induced vibration is in the across-wind direction, which is vertical for a bridge deck. However, vortex-induced vibration in torsional motion may also occur. The maximum amplitude of the vibration is sensitive to the structural damping, and the amplitude often becomes much smaller if the damping can be increased. The response amplitude is also sensitive to the turbulence intensity. *Turbulence intensity* is defined as the ratio of standard deviation of fluctuating wind velocity to the mean wind speed, and it represents the intensity of wind velocity fluctuation. In many cases, the maximum amplitude of vortex-induced vibration decreases in a flow with larger turbulence intensity. However, there are exceptions, such as a flat hexagonal cross section where the response amplitude even increases slightly in a more turbulent flow (Fujino et al., 2012).

Self-excited vibration is caused by self-excited aerodynamic forces that are generated due to the vibrating motion of the structure itself. The self-excited vibration is further classified into two types depending on the direction of the vibration. Galloping occurs in an across-wind direction, and for a bridge deck, it is in the vertical direction against the horizontal wind. Flutter occurs mainly in the torsional direction, such as observed in the collapse of the old Tacoma Narrows Bridge. Usually, once the self-excited vibration starts to occur at an onset wind speed, the response amplitude grows more and more if the wind speed is increased. Therefore, it is very important to prevent self-excited vibration for the bridge safety. The allowable design wind speed against the self-excited vibrations are usually set higher than the normal design wind speed by considering their response characteristics that may directly result in the collapse of the bridge. Schematic relationships between the abovementioned responses and wind speed are shown in Figure 3.2.

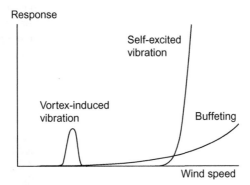

Figure 3.2 Wind-induced vibrations of bridges.

3 Procedure of wind-resistant design

A typical procedure of the wind-resistant design for a long-span bridge can be briefly summarized as follows. First, the design wind speed and necessary wind characteristics have to be determined. They may be provided in a regional code, but if that is not the case, they should be determined based on meteorological data or simulation of the strong wind speed. Wind speed changes with height and its profile, that is a distribution of mean wind speed as a function of height, depends on the surface roughness around the site. Therefore the height and surrounding roughness of the terrain have to be considered when determining the design wind speed. Then the wind loading due to buffeting is estimated based on buffeting analysis or a simplified formula based on the buffeting analysis. The structure must be confirmed to withstand the maximum response caused by buffeting. Then, when the structure also is considered to be sensitive to dynamic responses other than buffeting, it must be confirmed that the dynamic response occurs neither under the design wind speed nor with an amplitude larger than the allowable one. A simplified judgment as to whether the dynamic response should be considered may be made based on some design rules, such as are found in the United Kingdom (Highways Agency, UK, 2001) or in Japan (Fujino et al., 2012).

4 Design wind speeds provided in design codes

In this section and the next one, descriptions in several codes (EN 1991-1-4:2005, ISO 4354, and ASCE Standard ASCE/SEI 7-10) are briefly introduced. In addition, design rules with respect to the dynamic responses are summarized.

EN 1991-1-4 (European Committee for Standardization, 2010) gives guidance on the determination of natural wind actions for the structural design of civil engineering works. For bridges, it is applicable to those whose spans are not greater than 200 m. Neither bridge deck vibrations from transverse wind turbulence, wind action on cable-supported bridges, nor vibrations in which more than the fundamental mode is important, are considered in EN 1991-1-4.

The basic wind speed, v_b, is defined as the 10-min mean wind speed with an annual risk of exceedance of 0.02 (= 2%) at a height of 10 m above flat and open country terrain. The annual risk of exceedance corresponds to a mean return period of 50 years. Then, if necessary, it is modified to account for the directional and seasonal effects. Also, it is modified to account for the effect of terrain roughness and orography. Then, the mean wind speed at a height z above the terrain at the site, $v_m(z)$, which depends on the terrain roughness and orography, is expressed with the basic wind speed as follows:

$$v_m(z) = c_r(z) \times c_o(z) \times v_b, \tag{1}$$

where $c_r(z)$ is the roughness factor and $c_o(z)$ is the orography factor. The roughness factor is given based on a logarithmic profile at a height above the minimum height as follows:

$$c_r(z) = k_r \times \ln(z/z_0), \tag{2}$$

where k_r is the terrain factor depending on the roughness length, z_0; z_0 is tabulated with five different terrain categories, from 0 (above a sea or coastal area) to IV (an area in which at least 15% of the surface is covered with buildings with average height exceeding 15 m). In this case, $c_o(z)$ has to be used when the orography (e.g., hills, cliffs, etc.) increases wind speed by more than 5%. Also, the effects of any large and considerably higher neighboring structures must be considered.

The ISO 4354 standard, "Wind Actions on Structures" (ISO, 2009), describes the actions of wind on structures and specifies the methods of calculating wind loads. The peak design wind speed at the site, V_{site}, is given as follows:

$$V_{site} = V_{ref} \times C_{exp}, \tag{3}$$

where the exposure factor, C_{exp}, is determined based on the height above ground level of the structure, the roughness of the terrain, and the topography; and V_{ref} is the maximum wind speed averaged over 3 s referenced to a height of 10 m over flat and open country terrain. V_{ref} for any probability of exceedance in one year shall be determined from regionally derived reference wind speeds. The probability of exceedance is determined based on the importance of the structure, and an example of the classification of importance levels is provided. The storm type, such as synoptic storm, tropical cyclone storm, or thunderstorm, has to be appropriately accounted for in a way that is most applicable to both the ultimate limit state and the serviceability design.

In ASCE Standard ASCE/SEI 7-10, "Minimum Design Loads for Buildings and Other Structures" (ASCE, 2013), Chapters 26–31 describe wind loads in detail, mainly applicable to buildings. The basic wind speed (which is expressed in terms of 3 s gust speed at 10 m above the ground in open terrain) is provided in maps according to the risk category of the structures, ranging from I (low risk to human life in the event of failure) to IV (in which the failure could pose a substantial hazard to the community). On the maps, special wind regions are shown where unusual wind conditions have to be examined. The same applies for a location in mountainous terrain and gorges. In areas outside hurricane-prone regions, regional climatic data may be used instead of the maps to obtain the basic wind speed. In hurricane-prone regions, wind speeds derived from approved simulation techniques may be used instead of the maps, but using regional wind speed data is not permitted. This is because a Monte Carlo simulation model is more appropriate to estimate the hurricane wind speeds of which recurrence rates are much less than nonhurricane wind speeds. Exposure categories from B (urban and suburban areas) to D (flat, unobstructed areas and water surfaces) shall be determined within each 45° sector of wind direction based on ground surface roughness, and the effects of exposure and height are accounted for when the velocity pressure is determined. Wind speed-up effects at isolated hills, ridges, and escarpments constituting abrupt changes in the general topography shall be included in the determination of the wind loads when necessary.

5 Wind loads provided in design codes

In this section, descriptions related to wind loads on bridges from two of the codes described in the previous section are introduced. Also, a few design rules on the dynamic responses of bridges are briefly mentioned.

In EN 1991-1-4 (European Committee for Standardization, 2010), the wind action is represented by a simplified set of pressures or forces whose effects are equivalent to the extreme effects of the turbulent wind. For single-deck bridges with constant depth, guidance of wind actions is provided. When a dynamic response need not be considered, the wind force parallel to the deck width and perpendicular to the span direction (i.e., the x-direction) is expressed as

$$F_w = 0.5 \times \rho \times v_b^2 \times C \times A_{ref,x}, \tag{4}$$

where ρ is air density, $A_{ref,x}$ is reference area, and C is the wind load factor and $C = C_e \times C_{f,x}$. Here, C_e is the exposure factor that is the ratio of peak pressure at height z and mean pressure caused by v_b, and it is expressed as

$$C_e = [1 + 7I_v(z)] \times 0.5 \times \rho \times v_m^2(z) / (0.5 \times \rho \times v_b^2), \tag{5}$$

where $I_v(z)$ is the turbulence intensity at height z. A recommended expression of $I_v(z)$ is provided. $C_{f,x}$ represents the force coefficients of bridge decks and typical values are provided together with the definition of $A_{\text{ref},x}$. Also, recommended values of C are tabulated based on some assumptions for simplicity. Wind forces in the z-direction (i.e., vertical when the deck is horizontal) and y-direction (i.e., along the span) are also described. Wind effects on piers have to be considered as well.

Some information about wind-induced vibration is also provided in EN 1991-1-4. For vortex-induced vibration, empirical formulas of the critical wind speed and largest amplitude are provided, which may be used for bridge members. To determine galloping and flutter, expressions for onset wind speeds are provided, but one should get expert advice when dealing with bridges for which wind-induced dynamic effects are significant. Divergence (i.e., a static torsional instability resulting in huge torsional displacement caused by small torsional rigidity and large aerodynamic moment) is also discussed. For a very-long-span cable-supported bridge or a unique bridge structure with very low torsional rigidity, safety against divergence has to be confirmed. Simple formulas to estimate the dynamic characteristics, such as a fundamental natural frequency, are also provided.

In ISO 4354 (ISO, 2009), equivalent static wind loads that are obtained assuming linear elastic structural behavior are given in two forms:

$$F = q_{\text{site}} \times C_{\text{F}} \times C_{\text{dyn}} \times A_{\text{ref}} \qquad (6)$$

$$F = q_{\text{site, m}} \times C_{F_{\text{m}}} \times C_{\text{dyn, m}} \times A_{\text{ref}}. \qquad (7)$$

Eqs. (6) and (7) are formulated based on peak and mean wind speed, respectively. In Eq. (6), q_{site} is the site peak dynamic pressure (as discussed next), C_{F} is a force coefficient, C_{dyn} is a peak dynamic response factor, and A_{ref} is the reference area for force on the overall structure or a part of structure. The site peak dynamic pressure is determined from the regionally derived reference wind speed, V_{site}, as follows:

$$q_{\text{site}} = 0.5 \times \rho \times (V_{\text{site}})^2. \qquad (8)$$

In Eq. (7), $C_{F_{\text{m}}}$ is a mean force coefficient, and some examples are provided in the standard. $C_{\text{dyn,m}}$ is expressed as

$$C_{\text{dyn, m}} = C_{\text{dyn}} \times (1 + g_{\text{V}} I_{\text{V}})^2 \cong C_{\text{dyn}} \times (1 + 2g_{\text{V}} I_{\text{V}}), \qquad (9)$$

where g_{V} is a wind speed peak factor and I_{V} is turbulence intensity. g_{V} is defined as

$$V = V_{\text{m}} \times (1 + g_{\text{V}} I_{\text{V}}), \qquad (10)$$

where, V and V_{m} are the peak and mean wind speeds, respectively. The approximation on the right side of Eq. (9) is for low turbulence intensity; and $q_{\text{site,m}}$ is the site mean dynamic pressure that is obtained similarly to q_{site} in Eq. (8) based on $V_{\text{site,m}}$, the mean design wind speed. The dynamic response factors take into account the dynamic action of random wind gusts, fluctuating pressures induced by the wake of the structure, and fluctuating forces induced by the motion of the structure due to wind. Expressions for $C_{dyn,m}$ are given based on buffeting analysis, and those for C_{dyn} are derived based on them.

For certain wind-sensitive structures (such as long-span bridges), special supplementary studies are recommended. Simple empirical expressions for the critical wind speeds of flutter, galloping, and vortex-induced response are provided based on a reference. Wind tunnel tests are often conducted, and standard procedure is briefly given in an annex of ISO 4354.

For the dynamic responses of bridges, more detailed descriptions are given in such documents as in BD 49/01 (Highways Agency, UK, 2001) and a design manual for highway bridges in Japan (Fujino et al., 2012). BD 49/01, "Design Rules for Aerodynamic Effects on Bridges" (Highways Agency, UK, 2001), sets out the design requirements for bridges with respect to aerodynamic effects, including provision for wind tunnel testing. It first provides simple criteria to determine the susceptibility of a bridge to aerodynamic excitation based on the size, mass, natural frequency, and design wind speed. Then the empirical formulas for the critical wind speed and amplitude for vortex-induced vibration, a criterion for buffeting, as well as onset wind speed of galloping and flutter, are given for a number of bridge types. Also, a procedure is specified to estimate the fatigue damage due to vortex-induced vibration. Requirements for wind tunnel tests are given in an annex.

In "Wind-Resistant Design of Bridges in Japan—Developments and Practices," (Fujino et al., 2012), a design manual for highway bridges in Japan is summarized with other bridge related wind codes and their background. The design manual provides empirical formulas for wind-induced responses. To estimate the vortex-induced vibration amplitude, the effects of turbulence intensity of the wind are incorporated.

When the estimated occurrence wind speed of wind-induced vibration is less than the specified design wind speed, and the estimated response amplitude is also greater than the specified allowable amplitude, suppression of the response is necessary. There are two types of countermeasures: one is aerodynamic and the other is mechanical. The aerodynamic countermeasures intend to modify the aerodynamic characteristics by attaching aerodynamic devices such as fairings, flaps, and deflectors (or corner vanes). Schematics of these elements are shown in Figure 3.3. The mechanical countermeasures often increase damping and sometimes increase the stiffness of the structure. Numerous examples of such countermeasures applied for bridges in Japan are listed in a reference (Fujino et al., 2012).

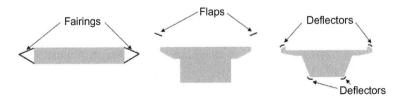

Figure 3.3 Examples of aerodynamic countermeasures.

6 Wind tunnel test and CFD

Wind tunnel tests have been widely used to predict the wind-induced responses of bridges, as well as to estimate wind loading. Because the bridge model scale is much smaller than the actual bridge, it is difficult to satisfy the Reynolds number similitude, so it is usually disregarded, based on the fact that the flow pattern may not change significantly with the different Reynolds numbers if the bridge and its members consist of sharp edges. But attention has to be paid to a structure with curved surface or corner cuts because in such cases, the Reynolds number may change the aerodynamic characteristics significantly. Detailed discussions about the similitude and modeling (Tanaka, 1992), and procedures (Fujino et al., 2012) for wind tunnel test of bridges are provided.

It is important to note that just a small difference in the bridge deck cross-sectional shape, such as modification of the railings, may change the wind-induced response of the bridge greatly. Because it is necessary to have a large-model scale to reproduce the geometric detail and the bridge response to wind is dominantly affected by the response characteristics of the bridge deck, a section model of the bridge deck is often used to check for resistance against dynamic responses. This model is supported with springs so that it represents a dominant response mode of the full bridge. The full bridge model test is also conducted if the three-dimensional (3D) effects along the span cannot be disregarded or the wind effects are so significant that a thorough investigation is necessary.

Computational fluid dynamics (CFD) is a tool to determine the flow using computers. Because of the rapid growth of computing capacity and the development of efficient computation schemes, CFD has become popular in many fields of fluid dynamics. However, the separate flows around a structure are complicated, and generally it is very difficult to obtain a quantitative prediction of the aerodynamic force and response using CFD. To overcome this problem, extensive studies have been and are currently being performed on utilizing CFD in the field of bridge aerodynamics. For instance, a streamline box girder was analyzed using an elaborate numerical model where even railings were reproduced (Sarwar et al., 2008), and the obtained steady and unsteady aerodynamic coefficients agreed well with experimental results. In another example, a numerically less demanding model was used to obtain the coefficients which also agreed reasonably well with experimental results (Nieto et al., 2015). Although it may be still difficult to use CFD for the final estimation of the bridge response to wind, the results by CFD are already used at the first stage of wind-resistant design where the general cross-sectional shape of a deck is chosen.

7 Vortex-induced vibration and its countermeasures

Vortex-induced vibration has sometimes been observed in actual bridges. Two such examples are briefly explained next.

The Tokyo Bay Aqua-Line Bridge (Fujino and Yoshida, 2002; Fujino et al., 2012) is a 10-span continuous steel box-girder bridge, and its longest span is 180 m. Vortex-induced vibration was observed in a full-bridge-model wind tunnel test, but the decision to install tuned mass dampers (TMDs), which increase damping and decrease or suppress vortex-induced vibration, was made after monitoring the bridge's behavior during construction, because there were still a few years to go before the entire road was to be opened. The observed vortex-induced vibration of the actual bridge had a maximum amplitude of 54 cm at a wind speed of 16–17 m/s. The observed turbulence intensity was between 4% and 6%, and this small level of turbulence seemed to contribute to the large response amplitude. The TMDs that are designed to suppress the first and second modes were installed inside the box girder, and the response of those modes decreased significantly. Also, small vertical continuous plates attached outside the railings were installed to reduce the third- and fourth-mode response by modifying the aerodynamic characteristics.

The Great Belt East suspension bridge in Denmark (Larsen et al., 2000; Frandsen, 2001) has main span of 1624 m with 535 m side spans. During the final phases of deck erection and surfacing of the roadway, vortex-induced vibration began to be observed. The estimated maximum response amplitude was about 0.35 m at a wind speed of 8 m/s. The vibration was also observed in a wind tunnel test for the final design, and after that, the decision to install a countermeasure was made because there was uncertainty regarding the full-scale structural damping and test results. As the countermeasure, guide vanes with 2-m widths were installed along at the lower side panel joints of the box girder with an opening of 0.6 m. No harmful response was observed after the installation of the guide vanes. Extensive analysis of the full-scale data and comparison with the wind tunnel test results were made.

Towers of cable-supported bridges are also susceptible to vortex-induced vibration and galloping, particularly when they are freestanding at the erection stage. Examples of countermeasures adopted in Japan are given by Fujino et al. (2012).

8   Verification of buffeting analysis based on field measurements

It is always important to compare the estimated responses and full-scale measurements in order to verify the design procedures. There are several examples of such measurements (e.g., Holmes, 1975; Brownjohn et al., 1994; Larose et al., 1998; Miyata et al., 2002; Macdonald, 2003; Xu and Zhu, 2005; Bakht et al., 2013). A few of them are briefly introduced next.

The wind-induced buffeting responses of the Akashi-Kaikyo suspension bridge were measured during two typhoons (Miyata et al., 2002). The mean of the along-wind direction response agreed well with the analysis that was conducted in the design. However, the fluctuating component of the response was generally less than the specified value in the design specification based on buffeting analysis.

Xu and Zhu (2005) made a comparison between the buffeting response of full-scale health monitoring data of the Tsing Ma suspension bridge in Hong Kong and their elaborate analysis, where the responses under skew wind (i.e., not perpendicular to the bridge axis) were also considered. The agreements were reasonably good.

More recently, the results of 10 years of full-scale monitoring data for the Confederation Bridge in Canada are published (Bakht et al., 2013). The Confederation Bridge is a 13-km-long precast concrete bridge comprised of 43 spans of 250 m. The accuracy of the design wind speed and dynamic characteristics such as natural frequencies were confirmed.

9   Wind-induced vibrations of stay cables

The span length of cable-stayed bridges has become longer over time. Accordingly, their stay-cables have also lengthened. Stay cables are very flexible and low-damping, and they are more prone to wind-induced vibration with longer lengths. Extensive studies have been conducted (FHWA, 2007; Caetano, 2007; Fujino et al., 2012). Among the wind-induced vibrations of stay cables, rain-wind vibration has been clearly noticed by engineers since the mid-1980s, and it is now a common practice to prevent the occurrence of such vibration by installing dampers, modifying the cable surface, or both.

On the other hand, the possibility of wind-induced vibration without rain conditions at relatively high wind speeds has been pointed out, and much research has been conducted on this topic. A possible explanation of the cause is the change of aerodynamic force on cables around the critical Reynolds number range (Jakobsen et al., 2012; Raeesi et al., 2014), and the vibration due to this mechanism is called *dry inclined cable galloping*. Another factor that may be related to this response is the axial flow that forms on the near-leeward side of the cable (Matsumoto et al., 2010). There may be some different causes for the wind-induced vibration of dry stay cables, and experimental (Kimura et al., 2009; Katsuchi and Yamada, 2009) and numerical (Yeo and Jones, 2011) studies have been carried out in order to clarify the mechanism.

Related to the wind-induced vibrations of stay cables, attention must be paid to any strong excitation that may occur if the cables are located in parallel. An example of studies was conducted recently on parallel and unparallel cylinders to clarify their response characteristics (Kim and Kim, 2014).

10   Conclusions

The wind loads and dynamic wind effects on bridges have been briefly introduced in this chapter. For a long-span bridge, it is a necessary and important step to confirm safety against and resistance to strong wind. In order to make the prediction more accurate, research has been conducted where buffeting and flutter analyses are conveniently utilized to understand these phenomena, based on wind tunnel experiments, CFD analyses, and field measurements. Cables are among the bridge elements most vulnerable to wind action, and recent research results on these have been discussed as well.

References

American Society of Civil Engineers (ASCE), 2013. ASCE/SEI 7-10—Minimum Design Loads for Buildings and Other Structures. ASCE, Reston, VA.

Bakht, B., King, J.P.C., Bartlett, F.M., 2013. Wind-induced response and loads for the Confederation Bridge, Part I: On-site monitoring data. Wind Struct. 16 (4), 373–391.

Brownjohn, J.M.W., Bocciolone, M., Curami, A., Falco, M., Zasso, A., 1994. Humber Bridge full-scale measurement campaigns 1990–1991. J. Wind Eng. Ind. Aerod. 52, 185–218.

Caetano, E.S., 2007. Cable Vibrations in Cable-Stayed Bridges (Structural Engineering Documents 9). IABSE, Zürich, Switzerland.

Davenport, A.G., 1962. The response of slender, line-like structure to a gusty wind. P. I. Civil Eng. 20, 389–408.

European Committee for Standardization, 2010. EN 1991-1-4:2005+A1, Eurocode 1: Actions on Structures—Part 1–4: General Actions–Wind Actions. CEN, Brussels, Belgium.

Federal Highway Administration (FHWA), 2007. Wind-Induced Vibrations of Stay Cables. US Department of Transportation. Publication No. FHWA-HRT-05-083.

Frandsen, J.B., 2001. Simultaneous pressures and accelerations measured full-scale on the Great Belt East suspension bridge. J. Wind Eng. Ind. Aerod. 89, 95–129.

Fujino, Y., Yoshida, Y., 2002. Wind-induced vibration and control of trans-Tokyo Bay crossing bridge. J. Struct. Mech. 128 (8), 1012–1025.

Fujino, Y., Kimura, K., Tanaka, H. (Eds.), 2012. Wind-Resistant Design of Bridges in Japan—Developments and Practices. Springer, Tokyo.

Ge, Y.J., 2008. Recent development of bridge aerodynamics in China. J. Wind Eng. Ind. Aerod. 96, 736–768.

Highways Agency, UK, 2001. Design Manual for Roads and Bridges, Vol. 1: Highway Structures, Approval Procedures and General Design, Section 3: General Design, Part 3: BD 49/01, Design Rules for Aerodynamic Effects on Bridges.

Holmes, J.D., 1975. Prediction of the response of a cable stayed bridge to turbulence. In: Eaton, K.J. (Ed.),. Proceedings of the Fourth International Conference on Wind Effects on Buildings and Structures. 8–12 September 1975, Heathrow. Cambridge University Press, Cambridge, pp. 187–197.

Holmes, J.D., 2007. Wind Loading of Structures, second ed. Taylor & Francis, Oxon, UK.

International Standard Organization (ISO), 2009. ISO 4354—Wind Actions on Structures, second ed. ISO, Geneva, Switzerland.

Jakobsen, J.B., Andersen, T.L., Macdonald, J.H.G., Nikitas, N., Larose, G.L., Savage, M.G., McAuliffe, B.R., 2012. Wind-induced response and excitation characteristics of an inclined cable model in the critical Reynolds number range. J. Wind Eng. Ind. Aerod. 110, 100–112.

Jurado, J.A., Hernández, S., Nieto, F., Mosquera, A., 2011. Bridge Aeroelasticity: Sensitivity Analysis and Optimal Design. WIT Press, Southampton, UK.

Katsuchi, H., Yamada, H., 2009. Wind-tunnel study on dry-galloping of indented-surface stay cable. In: Proceedings of the 11th Americas Conference on Wind Engineering. 11ACWE, June 22–26, 2009, San Juan, Puerto Rico.

Kim, S., Kim, H.K., 2014. Wake galloping phenomena between two parallel/unparallel cylinders. Wind Struct. 18 (5), 511–528.

Kimura, K., Kato, K., Kubo, Y., Oh-hashi, Y., 2009. An aeroelastic wind tunnel test of an inclined circular cylinder. In: Proceedings of the Eighth International Symposium on Cable Dynamics. ISCD 2009, September 20–23, 2009, Paris, pp. 143–150.

Kubo, Y., Hirata, K., Mikawa, K., 1992. Mechanism of aerodynamic vibration of shallow bridge girder sections. J. Wind Eng. Ind. Aerod., 41–44. 1297–1308.

Larose, G.L., Johnson, R., Damsgaard, A., 1998. Field measurements of a 1210-m span suspension bridge during erection. In: Proceedings of IABSE Symposium. September 2–4, 1998, Kobe, pp. 217–222. IABSE Rep. No. 79.

Larsen, A., Esdahl, S., Andersen, J.E., Vejrum, T., 2000. Storebaelt suspension bridge: vortex shedding excitation and mitigation by guide vanes. J. Wind Eng. Ind. Aerod. 88, 283–296.

Macdonald, J.H.G., 2003. Evaluation of buffeting predictions of a cable-stayed bridge from full-scale measurement. J. Wind Eng. Ind. Aerod. 91, 1465–1483.

Matsumoto, M., Yagi, T., Hatsuda, H., Shima, T., Tanaka, M., Naito, H., 2010. Dry galloping characteristics and its mechanism of inclined/yawed cables. J. Wind Eng. Ind. Aerod. 98, 317–327.

Miyata, T., Yamada, H., Katsuchi, H., Kitagawa, M., 2002. Full-scale measurement of Akashi–Kaikyo Bridge during typhoon. J. Wind Eng. Ind. Aerod. 90, 1517–1527.

Nieto, F., Owen, J.S., Hargreaves, D.M., Hernández, S., 2015. Bridge deck flutter derivatives: efficient numerical evaluation exploiting their interdependence. J. Wind Eng. Ind. Aerod. 136, 138–150.

Raeesi, A., Cheng, S., Ting, D.S., 2014. A two-degree-of-freedom aeroelastic model for the vibration of dry cylindrical body along unsteady air flow andits application to aerodynamic response of dry inclined cables. J. Wind Eng. Ind. Aerod. 130, 108–124.

Sarwar, M.W., Ishihara, T., Shimada, Yamasaki, Y., Ikeda, T., 2008. Prediction of aerodynamic characteristics of a box girder bridge section using the LES turbulence model. J. Wind Eng. Ind. Aerod. 96, 1895–1911.

Simiu, E., Miyata, T., 2006. Design of Buildings and Bridges for Wind: A Practical Guide for ASCE-7 Standard Users and Designers of Special Structures. John Wiley & Sons, Hoboken, NJ.

Simiu, E., Scanlan, R.H., 1996. Wind Effects on Structures, third ed. John Wiley & Sons, New York.

Sockel, E. (Ed.), 1994. Wind-Excited Vibrations of Structures. Springer-Verlag, Vienna.

Stathopoulos, T. (Ed.), 2007. Wind Effects on Buildings and Design of Wind-Sensitive Structures. Springer, Wien and New York.

Tamura, Y., Kareem, A. (Eds.), 2013. Advanced Structural Wind Engineering. Springer, Tokyo.

Tanaka, H., 1992. Similitude and Modelling in Bridge Aerodynamics. In: Larsen, A. (Ed.), Aerodynamics of Large Bridges. A.A. Balkema, Rotterdam, pp. 83–94.

Xu, Y.L., 2013. Wind Effects on Cable-Supported Bridges. John Wiley & Sons, Singapore.

Xu, Y.L., Zhu, L.D., 2005. Buffeting response of long-span cable-supported bridges under skew winds, Part 2: case study. J. Sound Vib. 281, 675–697.

Yeo, D., Jones, N.P., 2011. Computational study on aerodynamic mitigation of wind-induced, large-amplitude vibrations of stay cables with strakes. J. Wind Eng. Ind. Aerod. 99, 389–399.

Fatigue and fracture

4

Pipinato A.[1], Brühwiler E.[2]
[1]AP&P, Technical Director, Italy
[2]Ecole Polytechnique Fédérale de Lausanne (EPFL), Lausanne, Switzerland

1 Introduction

Fatigue consists of the localized alternating repetitions of concentrated stress cycles in a structure induced by the external application of loads such as vehicles, winds, waves, and temperature. These elements, when below of the structural capacity of the structure, could induce fracture, and over a long period, eventually cause total collapse. Bridges are strategic components of a transportation network mostly at the limit of their traffic capacity, due to overloading or simply for a high number of load vehicles repetitions. ASCE Committee on Fatigue and Fracture Reliability (1982a, 1982b, 1982c, 1982d) reported that 80%–90% of failures in steel structures are related to fatigue and fracture; moreover, also concrete bridges could be affected by fatigue failures (Chen et al. 2011). Fatigue damage could lead to very dangerous incidents: for example, Figure 4.1 shows a train derailment on a fracture-critical truss bridge that severed several members but did not collapse. These data are also confirmed by Byers et al. (1997). The factors behind these failures have been discussed by a number of researchers, including Brühwiler et al. (1990), Kulak (1992), Åkesson (1994), Pipinato (2008, 2010), Stephens et al. (2001); the most relevant of these deal

Figure 4.1 A train derailment on a fracture-critical truss bridge that severed several members but did not cause the bridge to collapse.

Innovative Bridge Design Handbook. http://dx.doi.org/10.1016/B978-0-12-800058-8.00004-9

with geometric imperfections, such as the inclination or deflection of structural elements, and entail the so-called secondary stresses that are difficult to take into account in fatigue safety verifications. However, vibrations, transverse horizontal forces, internal constraints, and localized and diffused defects such as corrosion damage are also causes of fatigue damage (Byers et al., 1997); furthermore, the presence of several joints, detail sizes, and various materials in the same bridge structure lead to different types of fatigue resistance.

The most relevant issue of technical and scientific interest concerns how to extend the service duration of existing bridges and how to improve these structures for higher loads, as it is not realistic or economic to consider reconstruction for all service bridges. A relevant question arises in this context: what is old? The notion of a *design working life* (subsequently called "design service duration") is defined in Eurocode (EN 1990, 2010) as the stipulated period during which a structure or part of it is to be used for its intended purpose with anticipated maintenance but without major repair being necessary. This concept is strictly related to that of maintenance, defined as a set of activities performed during the service duration of the structure in order to enable it to fulfill the requirements for reliability. According to EN 1990 (2010), the intended service duration at the design of a new bridge is generally defined as 100 years, even if for particular cases, the National Annexes and codes could provide other values, which would be longer for strategic and monumental structures and shorter for minor structures. In any case, the bridge shall be in service for several generations, and in no way, the bridge shall be demolished at the end of its intended service duration but regularly, the use of the bridge and its further service duration shall be updated to comply with user demands. The consequent question is: how can a bridge be designed to last for more than 100 years of service?

Therefore, engineering methods aiming to verify the structural reliability of existing structures (also-called *assessment procedures*) are necessary; however, there are not many existing codes in this field. Only some international guidelines are focused on this point, as explored in the rest of this chapter. Verification of the reliability of an existing structure aims at producing evidence that it will function safely over a specified service duration. In evaluating the reliability of an existing structure, the following points should be considered:

- Application of a risk-based approach while respecting commonly accepted safety levels as defined in codes. Uncertainty is reduced by using nondestructive testing, monitoring, and detailed structural analysis.
- Defining risk acceptance criteria, which requires the consideration of different items, such as redundancy, structural importance in the pertaining network, inspection level, and accessibility for inspection of bridge members.
- Defining adequate safety goals for acceptance criteria, which requires the reliability analysis of the structure; setting up target reliability values; performing verifications based on calibrated safety factors.

Two main engineering methods are currently applied in the examination of fatigue safety of existing bridges. The first is the traditional *S-N* curve method, in which the relationship between the constant-amplitude stress range, *S*, and the number of cycles to failure, *N*, is determined by appropriate fatigue experiments and described by a curve. The Palmgren-Miner linear damage hypothesis, also called "Miner's rule" (Miner, 1945), extends this approach to variable-amplitude loadings.

The second method is the fracture mechanics approach, which is dedicated to describe the crack initiation and growth in consideration of the stress field at the crack tip. In general, the two approaches are applied sequentially, with the S-N curve method used at the bridge design stage or for the evaluation of the fatigue endurance, and the fracture mechanics approach used for more refined crack-based evaluation of remaining fatigue endurance or effective decision making on inspection and maintenance strategies (Chryssanthopoulos and Righiniotis, 2006; Ye et al., 2014). The most common application of this latter approach is the linear elastic fracture mechanics (LEFM) method (Cheung and Lib, 2003): in this case, the Paris' law (Paris et al., 1961), the most common LEFM-based crack growth model, is used. The Paris' law is described as

$$\frac{da}{dN} = C \cdot \Delta K^m, \tag{1}$$

where a is the crack size, N is the number of stress cycles, C and m are fatigue growth parameters, and ΔK is the stress intensity range. According to LEFM theory (Cheung and Lib, 2003), ΔK can be estimated as

$$\Delta K = F(a, Y) \cdot \Delta \sigma \cdot \sqrt{\pi \cdot a}, \tag{2}$$

where $\Delta \sigma$ is the tensile stress range, $F(a,Y)$ is the geometry function taking into account possible stress concentrations, and Y represents a vector of random variables, such as the stress concentration coefficient and the dimensions of the specimen under consideration. In the case of welded details, the geometry function is expressed as the product of four separate factors (Tsiatas and Palmquist, 1999, Cheung and Lib, 2003):

$$F(a, Y) = F_g \cdot F_w \cdot F_s \cdot F_e, \tag{3}$$

where F_e, F_s, F_w and F_g are crack shape, free surface, finite width, and stress gradient correction, respectively (Tsiatas and Palmquist, 1999; Cheung and Lib, 2003):

$$F_e = \frac{1}{\int_0^{\frac{\pi}{2}} \sqrt{1 - \frac{c(a)^2 - a^2}{c(a)^2} \sin^2(\vartheta) d\vartheta}} \tag{4}$$

$$F_s = 1.211 - 0.186 \sqrt{\frac{a}{c(a)}} \tag{5}$$

$$F_w = \sqrt{\sec \frac{\pi a}{2 t_f}} \tag{6}$$

$$F_g = \frac{-3.539 \ln \dfrac{z}{t_f} + 1.981 \ln \dfrac{t_{cp}}{t_f} + 5.798}{1 + 6.789 \left(\dfrac{a}{t_f}\right)^{0.4348}}.$$ (7)

In the previous expressions, z is the weld leg size, t_f is the flange thickness, t_{cp} is the cover plate thickness, a is the crack depth, b is half the flange width of the girder, c is half the crack length as a function of crack depth, and ϑ is the angle for an elliptical crack. The relation between c and a is given by $c(a) = 3.549 \, a^{1.133}$.

Hence, the crack propagation law can be written as

$$\frac{da}{dN} = C \left[F(a, Y) \cdot \Delta\sigma \cdot \sqrt{\pi \cdot a} \right]^m.$$ (8)

According to the LEFM approach, the estimation of the crack growth amplitude versus the passage of time, considering the loading history and the estimated traffic flow, gives an accurate method to analyze the remaining service life of a structure.

2 Structural redundancy and safety

2.1 Structural redundancy

Redundancy can be defined as an exceedance of what is necessary or normal. The Federal Highway Administration (FHWA, 2012) carefully analyzed three types of structural redundancy in bridges: load-path, structural, and internal redundancy. A member is considered load-path redundant if an alternative and sufficient load path is determined to exist: this is the case for parallel girders, for example, but the existence of a redundant member is not sufficient. The absence of a failed member and the new load path also should be considered to determine if in this case, the remaining member is able to resist the superimposed loading condition. In the second case, a member is considered structurally redundant if its boundary conditions or supports are such that failure of the member merely changes the boundary or support conditions but does not result in the collapse of the superstructure. In the third case, alternative load paths exist in the same member (for example, multiple plies of a riveted steel member).

2.2 Principles of structural safety

A structure is considered safe if the design has accurately minimized possible economic loss and has ensured to protect people during its whole service duration. The first systematic study of the matter was made by Freudenthal (1945) publishing the first paper on the safety of structures. Safety is strictly related with the concept of *reliability*, referring to the probability that failure will not occur or that a specified criterion will not be exceeded. As exemplification, to design a structural system, the value of the maximal load parameter or the carrying capacity of the structure as expressed by the load parameter value in the limit situation (the ultimate resistance) raises the following safety question: how much higher than the maximal load parameter (the action effect) calculated with a deterministic procedure should the ultimate load value be in the carrying capacity model for the engineer to guarantee that there is either no risk or an extremely small and acceptable risk that a failure will occur? The difference between the two values is called the *safety margin* (Ditlevsen and Madsen, 2005).

The safety documentation for a structure has earlier often been based on the ratio between a calculated carrying capacity R (resistance) and a corresponding loading action effect S (stress). This ratio $N = R/S$ is the safety factor. Since $N > 1$ if and only if $R > S$, the statement $N > 1$ proves that the structure corresponds to a point in the safe set, while $N \leq 1$ says that the structure corresponds to a point in the failure set. In a probabilistic formulation, the safety factor is a random variable $(N = R/S)$, where R and S are random variables corresponding to the chosen resistance definition. The probability that the structure is not failing is, then,

$$P(N > 1) = P(R > S). \tag{9}$$

Unlike the safety factor, this probability does not vary with respect to the definition of R.

Of course, it is required that all considered resistance definitions with respect to a given limit state and corresponding action effects are defined in the same probability space. Let us assume that R and S are mutually independent and distributed according to the normal distribution with parameters (μ_R, σ_R) and (μ_S, σ_S), respectively (with μ = mean value, σ^2 = variance). Then,

$$P(N > 1) = P(S - R < 0) = \Phi\left(\frac{\mu_R - \mu_S}{\sqrt{\sigma_R^2 + \sigma_S^2}}\right), \tag{10}$$

where ϕ is the distribution function of the standardized normal distribution (Ditlevsen and Madsen, 2005).

2.3 Inspection and monitoring

2.3.1 Inspection

Field inspections are necessary for fatigue damage detection, given the implicit uncertainties of key elements related to fatigue like loading history analysis and future traffic estimation. Concurring causes of the constituent material deterioration like corrosion, cracking, and other kinds of damage can affect the structural safety of a bridge. Therefore, all of the elements that directly affect the performance of the bridge, including the footing, substructure, deck, and superstructure, must be periodically inspected or monitored: inspection is the primary nondestructive evaluation method used to evaluate the condition of bridges. Periodic inspections are conducted at defined intervals to survey the actual condition of bridges: the periodicity and inspection methods are commonly regulated by the bridge owners or by code rules. One of the most advanced process is included in the National Bridge Inspection Standards (NBIS, 2004), developed by FHWA: this is composed by a manual guide, and all data from the inspections are then migrated into a bridge inventory, a national database in which all bridges are catalogued and inspection data reported. Outside US, every managing agency has developed a different standard, also in the same nation, which is not a very useful engineering method. Other methods adopted for the fatigue examination of existing bridges include dynamic testing, radiographic inspection, electric inspection, sonic and ultrasonic methods, acoustic emission methods, and dye penetration to detect fatigue cracks. Regular field inspection combined with an accurate testing program could also provide managing authorities with accurate monitoring of cracks and a more precise prediction of the remaining service duration.

2.3.2 Monitoring

Simple monitoring has been replaced nowadays with structural health monitoring (SHM), which is a long-term process that includes an integration of structural analytical skills, bridge design process, construction issues, management, and inspection procedures. These procedures provide accurate information with the support of online SHM systems, which are controlled directly by infrastructural agencies. SHM systems incorporate the use of advanced technologies in sensing, data acquisition, computing, communication, data, and information and communication (ICT) technologies.

3 Codes and standards

3.1 EN 1993-1-9

Eurocode part 3, section 1-9 (EN 1993-1-9), is the European standard that gives methods for the structural fatigue design of members, connections, and joints of steel structures subjected to variable loadings. The fatigue design and verification provided in this code is derived from fatigue tests of common structural details, including the effects of geometrical and manufacturing imperfections, material production, and execution (for example, the effects of tolerances and residual stresses from welding). Materials should conform to the toughness requirements of EN 1993-1-10 (2009). The fatigue strengths provided in this code apply to structures operating under normal atmospheric conditions and with sufficient corrosion protection and regular maintenance, but do not provide any information on different environmental conditions; for example, the effect of seawater corrosion and microstructural damage from high temperatures ($>150°C$) is not covered. Two methods are provided for fatigue safety verification: the damage-tolerant method and the safe-life method. The first should provide an acceptable reliability that a structure will perform satisfactorily over its service duration, provided that a prescribed inspection and maintenance regime for detecting and correcting fatigue damage is implemented throughout the design life of the structure.

The second method should provide an acceptable level of reliability that a structure will perform satisfactorily for its intended service duration without the need for regular in-service inspection for fatigue damage. The methods presented in EN 1993-1-9 describe fatigue resistance in terms of fatigue strength curves for standard details applicable to nominal stresses, and discuss weld configurations applicable to geometric stresses.

The National Annex may give the choice of the verification method, definitions of classes of consequences, and numerical values for γ_{Mf}. Recommended values for γ_{Mf} are given in Table 4.1. Concerning stresses from fatigue actions, modeling for nominal stresses should take into account all action effects, including distortions, and should be based on linear elastic analysis of members and connections. Provided that the stresses due to external loading applied to members between joints are taken into account, the effects from secondary moments due to the stiffness of the connection are considered by adopting a k1-factor (as given in the code tables, for circular hollow sections under in-plane loading and for rectangular hollow sections under in-plane loading). The relevant stresses are nominal direct stresses σ and nominal shear stresses τ, while the relevant stresses in welds (Figure 4.2) are normal stresses σ_{wf} transverse to the axis of the weld and shear stresses τ_{wf} longitudinal to the axis of the weld, for which two separate checks should be performed.

Table 4.1 Recommended values for partial factors γ_{Mf} for fatigue strength (Table 3.1, EN 1993-1-9)

Assessment method	Consequence of failure	
	Low consequence	High consequence
Damage tolerant	1,00	1,15
Safe life	1,15	1,35

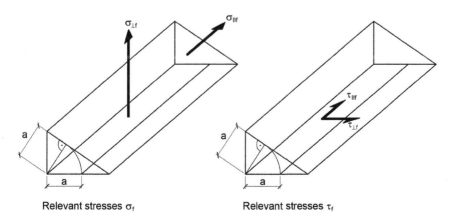

Relevant stresses σ_f Relevant stresses τ_f

Figure 4.2 Relevant stresses in the fillet welds.

Concerning calculation, stresses should be calculated at the serviceability limit state and at the site of potential fatigue initiation. Effects producing stress concentrations at the investigated structural detail should be generally accounted for by using a stress concentration factor (SCF): these values are provided in the code for every detail category. When using geometric (hot spot) stress methods for specific details included in the code, the stresses should be calculated as reported in the code. The fatigue design should be carried out using *nominal* stress ranges for such details as plain members and mechanically fastened joints, welded built-up sections, transverse butt welds, weld attachments and stiffeners, load-carrying welded joints, hollow sections, lattice girder node joints, and orthotropic decks (open and closed stringers), top flange to web junction of runway beams; *modified nominal* stress ranges, in which consistent changes of section occur close to the initiation site that are not included in the code; and *geometric* stress ranges, where high stress gradients occur close to a weld toe. The design value of the stress range to be used for the fatigue design should be the stress ranges $\gamma_{Ff} \Delta\sigma_{E,2}$ corresponding to $N_C = 2 \times 10^6$ cycles. The design value of nominal stress ranges $\gamma_{Ff} \Delta\sigma_{E,2}$ and $\gamma_{Ff} \Delta\tau_{E,2}$ should be determined as follows:

$$\gamma_{Ff} \Delta\sigma_{E,2} = \lambda_1 \times \lambda_2 \times \lambda_i \times \ldots \times \lambda_n \times \Delta\sigma \left(\gamma_{Ff} Q_k \right) \tag{11}$$

$$\gamma_{Ff} \Delta\tau_{E,2} = \lambda_1 \times \lambda_2 \times \lambda_i \times \ldots \times \lambda_n \times \Delta\tau \left(\gamma_{Ff} Q_k \right), \tag{12}$$

where $\Delta\sigma \left(\gamma_{Ff} Q_k \right)$, $\Delta\tau \left(\gamma_{Ff} Q_k \right)$ is the stress range caused by the fatigue loading specified in EN 1991-2 (2010), and λ_i are damage-equivalent factors depending on the spectra as specified in EN 1993. The other cases are reported and specified in the code. The fatigue strength for nominal stress ranges is represented by a series of $(\log \Delta\sigma_R) - (\log N)$ curves and $(\log \Delta\tau_R) - (\log N)$ curves (*S-N* curves), which correspond to typical detail categories. Each detail category is designated by a number that represents, in N/mm^2, the reference value $\Delta\sigma_C$ and $\Delta\tau_C$ for the fatigue strength at 2 million of cycles. For constant amplitude nominal stresses, fatigue strengths can be obtained by:

$$\Delta\sigma_R^m N_R = \Delta\sigma_C^m 2 \times 10^6, \tag{13}$$

with $m = 3$ for $N \leq 5 \times 10^6$, and

$$\Delta\tau_R^m N_R = \Delta\tau_C^m 2 \times 10^6, \tag{14}$$

with $m = 5$ for $N \leq 10^8$;

while

$$\Delta\sigma_D = (2/5)^{1/3} \times \Delta\sigma_C = 0,737 \Delta\sigma_C \tag{15}$$

is the constant amplitude fatigue limit and

$$\Delta\tau_L = (2/100)^{1/5} \times \Delta\tau_C = 0,457 \Delta\tau_C \tag{16}$$

is the cutoff limit.

Finally, for nominal stress spectra with stress ranges above and below the constant amplitude fatigue limit $\Delta\sigma_D$, the fatigue strength should be based on the extended fatigue strength curves as follows:

$$\Delta\sigma_R{}^m N_R = \Delta\sigma_C{}^m 2 \times 10^6, \tag{17}$$

with m = 3 for $N \leq 5 \times 10^6$, and

$$\Delta\sigma_R{}^m N_R = \Delta\sigma_D{}^m 5 \times 10^6, \tag{18}$$

with m = 5 for $5 \times 10^6 \leq N \leq 10^8$;
while

$$\Delta\sigma_L = (5/100)^{1/5} \times \Delta\sigma_D = 0,549\,\Delta\sigma_D \tag{19}$$

is the cutoff limit. The S-N curves have a slope of 3 for up to 5 million cycles where the corresponding stress range is the constant amplitude fatigue limit (CAFL) for that curve. From 5 million cycles to 100 million cycles, a slope of 5 is used. Fatigue strength curves for direct and shear stress ranges are shown in Figures 4.3 and 4.4, respectively: test data used to determine the appropriate detail category for a particular constructional detail refer to the value of the stress range $\Delta\sigma_C$ corresponding to a value of $N_C = 2$ million cycles calculated for a 75% confidence level of 95% probability of survival for log N, taking into account the standard deviation and the sample size and residual stress effects. The number of data points (not lower than 10) was considered in the statistical analysis, as reported in Annex D of EN 1990 (2010). Moreover, the National Annex of European countries may permit the verification of a fatigue strength category for a particular application, in accordance with the aforementioned procedure. Concerning fatigue verification, nominal, modified nominal, and geometric stress ranges due to frequent loads $\psi_1 Q_k$ (see EN 1990, 2010), it should be verified that under fatigue loading:

$$\gamma_{Ff}\,\Delta\sigma_{E,2}/(\Delta\sigma_C/\gamma_{Mf}) \leq 1 \tag{20}$$

$$\gamma_{Ff}\,\Delta\tau\sigma_{E,2}/(\Delta\tau_C/\gamma_{Mf}) \leq 1 \tag{21}$$

Unless otherwise stated in the fatigue strength categories, for combined stress ranges, it should be verified that

$$\left(\frac{\gamma_{Ff}\,\Delta\sigma_{E,2}}{\Delta\sigma_C/\gamma_{Mf}}\right)^3 + \left(\frac{\gamma_{Ff}\,\Delta\tau_{E,2}}{\Delta\tau_C/\gamma_{Mf}}\right)^5 \leq 1,0. \tag{22}$$

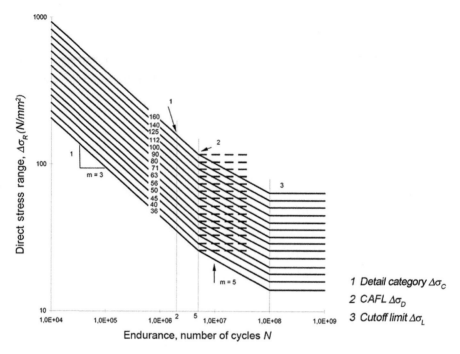

Figure 4.3 Fatigue strength curves for direct stress ranges (EN 1993-1-9).

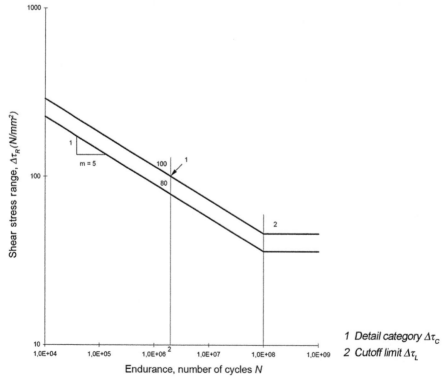

Figure 4.4 Fatigue strength curves for shear stress ranges (EN 1993-1-9).

3.2 North American practice

North American standards include American Association of State Highway and Transportation Officials (AASHTO, 2013), American Institute of Steel Construction (AISC, 2011), American Welding Society (AWS, 2010), and American Railway Engineers Association (AREA) documents. Welded and bolted details for bridges and buildings are designed with reference to the nominal stress range rather than the local "concentrated" stress at the weld detail. Fatigue design is carried out adopting service loads. Usually, nominal stress in the members can be easily calculated without great error. It is a standard practice in fatigue design to separate the details into categories with similar fatigue resistance in terms of nominal stress. Each category detail has an associated *S-N* curve. The *S-N* curves for steel details in the AASHTO (2013), AISC (2011), AWS (2010), and AREA provisions are presented for various detail categories in order of decreasing fatigue strength. These *S-N* curves are based on a lower bound of a large number of full-scale fatigue test data with a 97.5% survival limit. Generally, the slope of the regression line fitting to the test data is typically in the range 2.9–3.1 (Dexter and Fisher, 1996; Dexter and Fisher, 2000). Therefore, in the AISC and AASHTO codes as well as in Eurocode, the slopes have been standardized at 3.0. The fatigue threshold and CAFLs for each category are marked as horizontal dashed lines. When constant-amplitude tests are performed at stress ranges below the CAFLs, noticeable cracking should not occur. The number of cycles associated with the CAFLs is whatever number of cycles corresponds to that stress range on the *S-N* curve for that category or class of detail. The CAFL occurs at an increasing number of cycles for lower fatigue categories or classes. Different details, which share a common *S-N* curve (or category) in the finite-life regime, have different CAFLs (Dexter and Fisher, 2000). The AASHTO Load and Resistance Factor Design (LRFD) Specifications define eight detail categories for fatigue: A, B, B', C, C', D, E, and E'. Figure 4.5 shows the fatigue-resistance curves given in the LRFD Specifications. The plot shows stress ranges on the vertical axis and number of cycles on the horizontal axis for the various categories. Both axes are logarithmic representations. Over some portion of the range, each detail category is a straight line with a constant slope equal to 3. Beyond a certain point, which varies depending on the detail category, the fatigue-resistance line is horizontal.

3.3 S-N curve comparison

In order to compare EN 1993-1-9 (2005) with North American codes, test results of correspondent categories are needed. A first comparison includes riveted details of different experimental investigations and structural types (Adamson and Kulak, 1995; Baker and Kulak, 1982; Mang and Bucak, 1993; Åkesson, 1994; Fisher et al., 1987; Forsberg, 1993; Abe, 1989; Al-Emrani, 2000; Rabemanantso and Hirt, 1984; Brühwiler et al., 1990; Out et al., 1984; Reemsnyder, 1975; Helmerich et al., 1997; Di Battista and Kulak, 1997; Xiulin et al., 1996; ATLSS, 1993): these have been analyzed and compared with Eurocode Category C=63, AASHTO, and

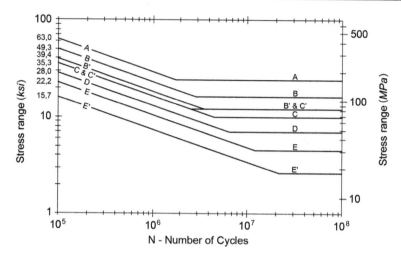

Figure 4.5 Nominal stress *S-N* curves used in AASHTO, AISC, AWS, and AREMA specifications.

AREA (Figure 4.6). The second comparison deals with the specific category of riveted shear details tests data (Stadelmann, 1984; Brühwiler et al., 1990; Pipinato, 2008) compared to Eurocode Category C = 100 and the AASHTO standards (Figure 4.7); finally, the specific category in transverse connection plates due to distortion-induced fatigue strength (Fisher et al., 1990) has been analyzed and compared with Eurocode Category C = 80 and with the AASHTO standards (Figure 4.8). Considering the first and second comparison, failures often occur over the design curves: this could help with understanding that more accurate subdivision of precise details in their specific category works better than grouping details into common categories. This problem relates to the design stage, where deep analysis of the structural details is needed to divide each substructure into its specific category, in order to avoid excessive material and to build lightweight structures. Moreover, this also implies a more efficient detailed study of the examination of existing bridges, as a precise choice of category detail implies a more accurate estimation of the remaining service duration of the considered structure. Concerning AASHTO Category C, the *S-N* curve could be a reasonable way to show the distortion-induced fatigue cracking at the ends of transverse connection plates. Both the AASHTO and Eurocode *S-N* curves seem to equally fit the test data, even though some failures appear at the limit in the high-cycle region. For all the tests and comparisons considered, a change of the nature of the current AASHTO fatigue curves from a linear to a bilinear slope would increase the effort required to calculate the fatigue endurance period.

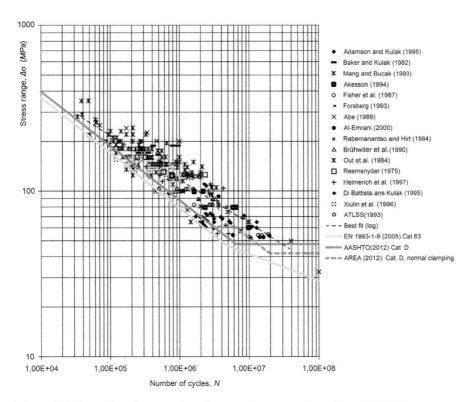

Figure 4.6 Comparison between riveted connection test results and EUROCODE versus AASHTO and AREA provisions.

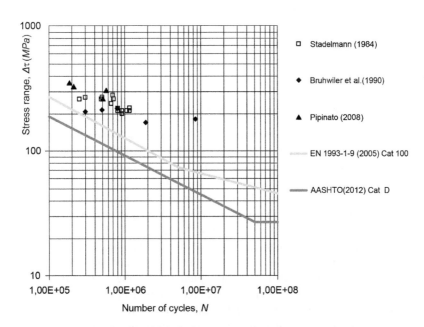

Figure 4.7 Comparison between shear riveted connection test experimental results and EUROCODE versus AASHTO and AREA provisions.

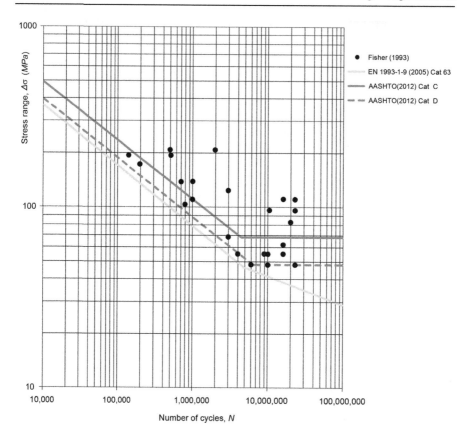

Figure 4.8 Comparison between transverse connection plate test experimental results and EUROCODE versus AASHTO provisions.

3.4   Recent code background and prestandard studies

In the examination of existing structures, as code provisions for new structures do not cover this issue, bridge engineers should cover this gap using guidelines or specific procedures provided by the managing agencies. The most comprehensive recent document on the matter in the United States is "Fatigue Evaluation of Steel Bridges," prepared by the National Cooperative Highway Research Program (NCHRP, 2012); this report summarizes the results of the research effort undertaken as part of NCHRP Project 12-81. This project has a focus on Section 7 dealing with "Fatigue Evaluation of Steel Bridges". Items identified as in need of improvement include utilizing a reliability-based approach to investigate fatigue behavior and aid bridge owners in making appropriate operational decisions; guidance on the evaluation of retrofit and repair details used to stop fatigue cracking; and guidance for the evaluation of distortion-induced fatigue cracks (NCHRP, 2012). To address these needs, a number of analytical and experimental studies were performed: the analytical studies were used to examine various aspects that influence the fatigue behavior. These topics ranged from truck-loading effects on bridge structures, to fatigue resistance–related factors that affect the predicted fatigue duration.

Both analytical and experimental methods were used to further develop an understanding of distortion-induced deformations and the structural behavior of various retrofit details used to improve a bridge suffering from distortion-induced fatigue cracking. Moreover, early in the study, it was decided that it would be beneficial to perform a series of experimental tests to study the influence of tack welds on riveted joints (NCHRP, 2012). The European document that could be compared to is "Assessment of Existing Steel Structures: Recommendations for Estimation of Remaining Fatigue Life" (Kuhn et al., 2008): the document provides background and support for the implementation, harmonization, and further development of the Eurocodes. It has been prepared to provide technical insight on the existing steel structures, how they could be analyzed and how the remaining fatigue endurance period could be estimated. It may be used as a main source of support to further harmonize design rules across different materials and develop the Eurocodes. The European Convention for Constructional Steelwork (ECCS) has initiated the development of this report in the frame of the cooperation between the European Commission (Joint Research Centre) and the ECCS on the further evolution of the Eurocodes. It is, therefore, published as a joint JRC-ECCS report. The aims of these recommendations are the following: (i) to present a stepwise procedure, which can be generally used for the examination of existing structures and steel bridges; (ii) to illustrate all factors to be considered about resistance and to describe ways to get more detailed information on these factors; (iii) to illustrate the remedial measures that can be chosen after fatigue verification showing insufficient fatigue safety and fatigue endurance period; and (iv) to present examples explaining the use of the proposed assessment procedure. The procedure incorporates four phases, differentiating the first phase of investigation from the deeper analysis in which experts are involved (see Figure 4.9).

In 2011, the Swiss Society of Engineers and Architects (SIA) published a series of standards for existing structures. The standard entitled "Existing Structures—Bases for Examination and Interventions" (Brühwiler et al., 2012) specifies the principles, the terminology, and the appropriate methodology for dealing with existing structures. This standard is complemented by a series of standards that treat specific items regarding "actions on existing structures," "existing concrete, steel, composite, timber, and masonry structures," and "geotechnical and seismic aspects of existing structures." These standards provide effective ways to address issues such as higher live loads, accidental actions, or the restoration and improvement of the durability of existing structures. In particular, the following typical challenges concerning fatigue are addressed: if the structural safety for higher live loads can be verified, fatigue safety, the remaining fatigue endurance period (of fatigue-vulnerable structures such as bridges), and serviceability become predominant issues requiring advanced analysis methods; SIA standard 269/1, "Actions on Existing Structures," states that for fatigue safety verification, one must consider correction factors such as past and planned future road or rail traffic and favorable load-carrying effects due to curbs, parapets, road pavement, and railway track (for example, continuous rails on short-span

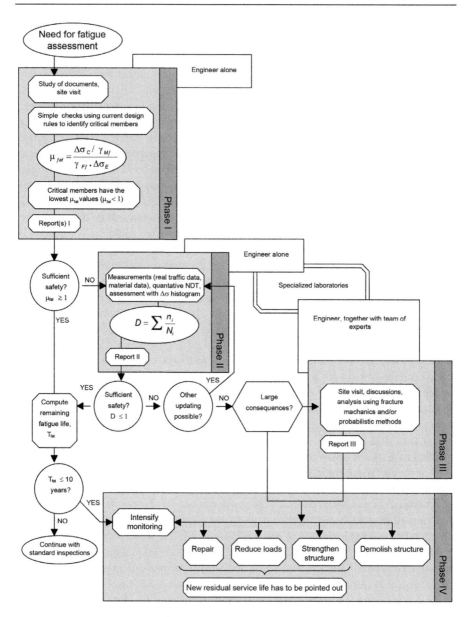

Figure 4.9 Assessment procedure for existing steel structures and recommendations for the estimation of remaining fatigue life (Kuhn et al., 2008).

bridges) to determine fatigue action effects. SIA standard 269/2, "Existing Concrete Structures," covers the fatigue resistance of steel reinforcement, and SIA standard 269/3, "Existing Steel Structures," gives provisions regarding the ultimate resistance (including stability) and fatigue resistance (*S-N* curves) for riveted connections and structural elements.

4   Fatigue and fracture resistance of steel and concrete bridges

4.1   Fatigue

Fatigue is the most common cause of reported damage to steel bridges (ASCE Committee on Fatigue and Fracture Reliability, 1982a), and it is also an issue with concrete bridges (Chen et al., 2011). However, while the fatigue design of steel bridges is clearly discussed in codes and standards, the same does not hold for the examination of existing bridges. Moreover, while steel bridges built in the last two decades have had no significant problems with fatigue and fracture (and should not in the future), bridges designed before the introduction of modern specifications will continue to be susceptible to the development of fatigue cracks and to fracture. In order to avoid stress concentrations and induced deformations in specific key points of bridge structures and control fatigue and avoid fracture, detailed rules have been found to be the most important part of fatigue and fracture design and examination procedures. While for steel bridges a wide amount of information has been given, the following discussion will deal with concrete bridges, which are typically less prone to fatigue than welded steel and aluminum structures.

Plain concrete under high force-controlled compression or tension fatigue loading exhibits strongly increasing strains within a first brief period of fatigue endurance (Dyduch et al., 1994; Cornelissen and Reinhardt, 1984; Schläfli and Brühwiler, 1998), followed by a phase of steady, but only slightly increasing, strains. During the last phase, strains again increase significantly before the specimen fractures (Figure 4.10). The apparent modulus of elasticity of concrete decreases significantly during the test, mainly due to crack formation on a microscopic level. Under uniaxial compression, the concrete matrix shows extensive microcracking during this last period of time. An increasing number of cracks appear parallel to the loading direction on the outer surface of the specimen with subsequent failure. Concrete behavior under tension fatigue loading is also dominated by crack propagation; early age microcracks in the cement matrix and at the interface between aggregates and the cement matrix propagate steadily and perpendicular to the loading direction until the specimen shows one discrete crack. Concrete subjected to tension-compression stress reversals deteriorates more rapidly, which is explained by the interaction of differently oriented microcracks due to compression and tension loading (Weigler and Rings, 1987; Cornelissen and Reinhardt, 1982). Fatigue behavior of steel reinforcement bars can be divided into a crack initiation phase, a steady crack propagation phase, and final fracture with little deformation of the remaining rebar section. Crack initiation on a ribbed, high-yield steel bar usually starts at the root of a rib, which typically is the location of stress concentration. Welds, the curvature of bent bars, and corrosion favor crack initiation and lead to low fatigue strength. The fatigue behavior of reinforced concrete elements is also characterized by progressive deterioration of the bond between reinforcement and concrete. Larger cracks and a smaller contribution of concrete in tension between the cracks result in larger deflection. Failure normally occurs due to fatigue fracture of steel rebars; another failure mechanism may be spalling of

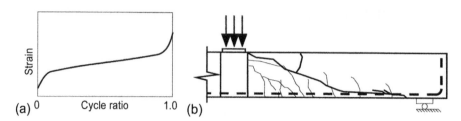

Figure 4.10 (a) Diagram of idealized strain versus cycle ratio for compression fatigue of plain concrete; (b) shear crack pattern due to fatigue loading of a beam without shear reinforcement (Frey and Thurlimann, 1983).

concrete in the compression zone, but then it is possible that the basic ductility criterion for the reinforced concrete section is not fulfilled or the concrete strength is too low (and thus not respecting type 2 verification of structural safety at ultimate static resistance). In fact, fatigue tests show that even overreinforced beams (i.e., concrete compression failure under static loading) fail due to reinforcement fatigue fracture when subjected to fatigue loading.

Beams (without transverse reinforcement, hence not respecting basic rules of good detailing in structural concrete) subjected to predominant shear develop a shear crack pattern after the first few cycles when deformation increases only slightly (Figure 4.10). Subsequently, a critical shear crack appears that crosses the bending cracks. The rather large width of this crack does not allow any stress transfer; as a result, the beam fails due to fatigue of the compression strut (upper flange) (Frey and Thurlimann, 1983). Beams with shear reinforcement show fatigue failure of stirrups, accompanied by spalling of surrounding concrete; failure is ductile. Fatigue tests of scaled deck slabs have shown a punching shear failure mode, and moving wheel loads leading to stress reversals are more detrimental to fatigue strength than stationary pulsating loads (Sonoda and Horikawa, 1982; Perdikaris and Beim, 1988).

4.2 Fracture

Fracture may be defined as rupture in tension or rapid propagation of a crack, leading to large deformation, loss of function or serviceability of the structural element, complete separation of the component (Anderson, 1995). Even if prevention should be focused on fatigue, however, for structural components which are not subjected to significant cyclic loading, fracture could still possibly occur without prior fatigue crack growth. In general, fracture toughness (K_{IC}) has been found to decrease with increasing yield strength of a material, suggesting an inverse relationship between the two properties (Dexter and Fisher, 2000; Crooker and Lange, 1970); moreover, $K_{IC,s}$ values are largely accepted for steel, while $K_{IC,c}$ for concrete has little meaning. However, fracture toughness is more complex than this simple relationship since steels with similar strength levels can have widely varying levels of fracture toughness (Dexter and Fisher, 2000). Steel exhibits a transition from brittle to ductile fracture behavior as the temperature increases. For example, Figure 4.11 shows a plot of the energy required to fracture Charpy V-notch impact test specimens of A588 structural steel at various temperatures. These results are typical for ordinary hot-rolled structural steel. The transition phenomenon shown in Figure 4.11 is a result of changes in the underlying microstructural fracture mode (Dexter and Fisher, 2000). Codes and standards don't provide any specific verification procedures in order to check for fracture in constituent materials: for example, according to Eurocode (EN 1993-2 2006), the material should have the required toughness to prevent brittle fracture within the intended design service duration of the structure, and no further checks against brittle fracture need to be made if the conditions given in EN 1993-1-10 (2009) are met for the lowest service temperature. The National Annex may specify additional requirements depending on the plate thickness. Three main types of fracture with different behavior can be addressed. The first is brittle fracture, which is associated with cleavage of individual grains on selected crystallographic planes. This type of fracture occurs at the lower end of the temperature range, although the brittle behavior can persist up to the boiling point of water in some materials with low toughness. This part of the temperature range is called the *lower shelf* because the minimum toughness is fairly constant up to the transition temperature. Brittle fracture may be analyzed with LEFM because the extent of plastic deformation at the crack tip is generally negligible. The second type, ductile fracture, is associated with a process of void initiation, growth, and coalescence on a microstructural scale, a process requiring considerable energy. This higher end of the temperature range is referred to as the *upper shelf* because the toughness levels off and is essentially constant for higher temperatures. Ductile fracture is also called *fibrous fracture* due to the fibrous appearance of the fracture surface, or *shear fracture* due to the usually large slanted shear lips on the fracture surface (Dexter and Fisher, 2000). The third type is transition-range fracture, which occurs at temperatures between the lower shelf and the upper shelf and is associated with a mixture of cleavage and fibrous fracture on a microstructural scale. Because of the mixture of micromechanisms, transition-range fracture is characterized by extremely large variability.

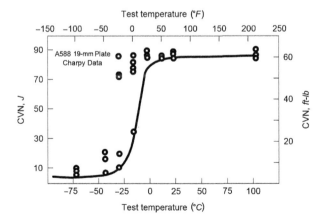

Figure 4.11 Charpy energy transition curve for A588 Grade 50 (350 MPa yield strength) structural steel.

5   Traffic loading and action effects on bridge elements

Traffic running on bridges produces a stress spectrum that may cause fatigue damage. This stress spectrum depends on the geometry of the vehicles, the axle loads, the vehicle spacing, the composition of the traffic, and its dynamic effects. For simplicity, only fatigue loading in Eurocode is reported here: five fatigue load models of vertical forces are defined and given in EN 1991-2 (2010) for road traffic. The use of the various fatigue load models is defined in EN 1992 to EN 1999, and further information is given next: fatigue load models 1, 2, and 3 are intended to be used to determine the maximum and minimum stresses resulting from the possible load arrangements on the bridge of any of these models; in many cases, only the algebraic difference between these stresses is used in EN 1992 to EN 1999. Fatigue load models 4 and 5 are intended to be used to determine stress range spectra resulting from the passage of trucks on the bridge. Fatigue load models 1 and 2 are intended to be used to check whether the fatigue endurance period may be considered unlimited when a constant stress amplitude fatigue limit is given. Therefore, they are appropriate for steel constructions and may be inappropriate for other materials.

Fatigue load model 1 is generally conservative and covers multilane effects automatically. Fatigue load model 2 is more accurate than fatigue load model 1 when the simultaneous presence of several trucks on the bridge can be disregarded for fatigue verification. If that is not the case, it should be used only if it is supplemented by additional data. The National Annex may give the conditions of use of fatigue load models 1 and 2. Fatigue load models 3, 4, and 5 are intended to be used for estimation of fatigue endurance period by referring to fatigue strength curves defined in EN 1992 to EN 1999. They should not be used to check whether fatigue endurance can be considered unlimited. For this reason, they are not numerically comparable to fatigue load models 1 and 2. Fatigue load model 3 may also be used for the direct verification of designs by simplified methods in which the influence of the annual traffic volume and of some bridge dimensions is taken into account by a material-dependent adjustment factor λ_e. Fatigue load model 4 is more accurate than fatigue load model 3 for a variety of bridges and for the traffic when the simultaneous presence of several trucks on the bridge can be disregarded. If that is not the case, it should be used only if it is supplemented by additional data, as specified or defined in the National Annex. Fatigue load model 5 is the most general model, using actual traffic data. A resume of road load fatigue model is depicted in Figure 4.12. A traffic category on a bridge should be defined, for fatigue verification at least, in terms of the following:

- The number of slow lanes
- The number N_{obs} of heavy vehicles (maximum gross vehicle weight of more than 100 kN), observed or estimated, per year and for a slow lane (i.e., a traffic lane used predominantly by trucks)

Indicative values for N_{obs} are given in EN 1991-2 (2010) for a slow lane when using fatigue load models 3 and 4: for example, roads and motorways with two or more lanes per direction with high flow rates of trucks implies $N_{obs} = 2 \times 10^6$ per year and for a slow lane, whereas on each fast lane (i.e., a traffic lane used predominantly by cars),

10% of N_{obs} also may be considered. It should be noticed that these tables are not sufficient to characterize the traffic for fatigue verifications. Other parameters may have to be considered, such as percentages of vehicle types, which depend on the traffic type (i.e., parameters defining the distribution of the weight of vehicles or axles of each type). Other specifications, such as the statistical distribution of the transverse location of loads, should be taken into account according to the code notations. Finally, note that dynamic amplification factors are implicitly accounted in the code (for FLM 1 to 4); however, an additional factor should be added in expansion joints and applied to all loads (see EN 1991-2, 2010).

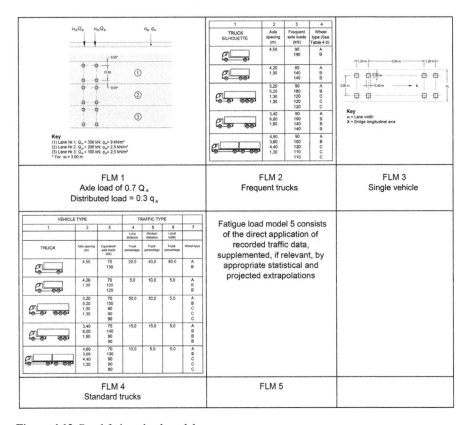

Figure 4.12 Road fatigue load model.

In the specific case of railway bridges, other loading schemes are given. A fatigue damage assessment shall be carried out for all structural elements, which are subjected to alternating stresses. For normal traffic based on characteristic values of Load Model 71 (Figure 4.13), including the dynamic factor Φ, the fatigue safety verification should be carried out on the basis of one of the traffic mixes "standard traffic," "traffic with 250 kN-axles," or "light traffic mix," depending on whether the structure carries mixed traffic, predominantly heavy freight traffic, or lightweight passenger traffic in accordance with the requirements specified. Details of the service trains and traffic mixes considered and the dynamic amplification to be applied are given in annex D of EN 1991-2 (2010). Where the traffic mix does not represent the real traffic (for example, in special situations where a limited number of vehicle types dominate the fatigue loading), an alternative traffic mix should be specified. It should be noted that the dynamic factor Φ is significantly higher than the results from measurements of the dynamic response in bridge elements, as shown by many studies, particularly in the case of ballasted tracks on bridges; hence, the dynamic factor Φ is like an additional partial safety factor which, however, should be considered carefully in the examination of existing bridges (Herwig, 2008; Herwig and Brühwiler, 2011).

6 Common failures

Concerning steel and composite bridges, most failures relate to fatigue cracking from weld defects (Wichtowski, 2013), details with change in section (IIW, 1996; Miki et al., 2003), vibration-induced fatigue cracking in bridge hangers (for example, in the Skellefte River in Sweden as reported by Åkesson, 1991), bridge girders and stringers at timber tie connections (Soudki et al., 1999), diaphragms and cross-bracing connections (Pipinato et al., 2012a), stringer-to-floor-beam connections (Chotickai and Kanchanalai, 2010), elements with coped cut-short flanges (Fisher, 1984), connections between girder splices or welded cover-plates (Kuhn et al., 2008), short diaphragm connections (Pipinato et al., 2009), riveted connections (Brühwiler et al., 1990). Concerning reinforced concrete (r.c.)–steel composite bridges, a wide assessment of different details are given by Leitao et al. (2011), and in particular, for shear studs, some tips can be found in Lee et al. (2005).

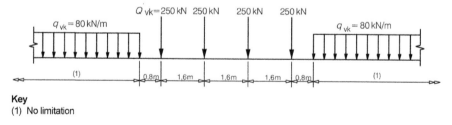

Key
(1) No limitation

Figure 4.13 Railway fatigue load model LM71.

For concrete bridges, a general study could be found in CEB (1988): fatigue damage of reinforced concrete deck slab of road bridges has been identified in Japan in the 1980s after 20 years of service; this damage was due to low concrete strength, inadequate detailing of rebars, and overloaded trucks. Other countries also reported some fatigue damage. Recurrent fatigue-induced problems in r.c. decks have been found in literature (Rodrigues et al., 2013), from the inspection data collected in 40 bridges, in which systematic damages and defects were associated with a particular structural arrangement usually found in two girder-slab r.c. bridges, with cantilever girders at the extremities. For economic reasons and due to the relative simplicity of construction, a great number of these structures can be found along the main roadways analized in Rodrigues et al. (2013). The superstructure is composed of a cast-in-situ concrete deck upon two concrete girders. In some of the investigated bridges, cracks at the midspan were also observed clearly, and in others, evidence of crack repairs in the same region was observed (Figure 4.14).

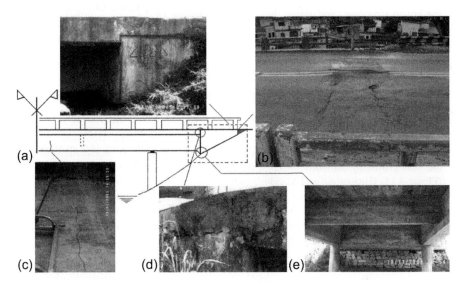

Figure 4.14 (a) Detail of cantilevered deck extremities and inclined embankments; (b) roughness of the access to bridges; (c) cracking at a midspan girder of one of the investigated bridges; (d) damage observed next to the extremities of a cantilevered deck bridge; (e) a rock barrier to restrain embankment slide under deck extremities (Rodrigues et al., 2013).

7   Crack detection, intervention methods, and techniques

7.1   Crack detection

Visual inspection is the most common and useful way to discover fatigue cracks and failures. To further aid observation, two common nondestructive techniques used to expose cracks are dye penetrant and magnetic particle inspection. For the former, a cleaning procedure is first requested to remove contaminants, then a red dye is sprayed on the surface, and finally, the excess is wiped out and a developer is sprayed on, so that the volatile solvent will couple the flaw-entrapped dye penetrant to the powder and speed the penetrant's return to the surface for viewing. With this technique, cracks are shown precisely. Magnetic inspection is used with an electromagnetic yoke, inducing a magnetic field that is eventually disrupted by the crack presence, and then a fine iron filler is sprinkled on the area and is attracted by the magnetic concentration.

In this latter case, rust or paint should be completely removed. Eddy current, ultrasonic testing (either manual or automatic), and time-to-flight-diffraction are other techniques employed for the same purpose.

7.2   Local intervention methods

Local intervention is needed when cracks or defects are evidenced only in a small part of the entire structure, and when these are not loading or distortion-induced fatigue problems. But these flaws should arise from local damage or flaws.

7.2.1   Surface treatment for welded structures

For the case of surface treatment for welded structures, the following solutions can be adopted:

- *Grinding*: The aim is to remove or reduce the size of the weld toe flaws from which fatigue cracks propagate. At the same time, it aims to reduce the local stress concentration effect of the weld profile by smoothly blending the transition between the plate and the weld face (Figure 4.15) (Haagensen and Maddox, 2001).
- *TIG dressing*: Tungsten inert gas (TIG) dressing aims to remove weld toe flaws by remelting the material at the weld toe, reducing the local stress concentration effect of the local weld toe profile by providing a smooth transition between the plate and the weld face (Figure 4.16) (Haagensen and Maddox, 2001).

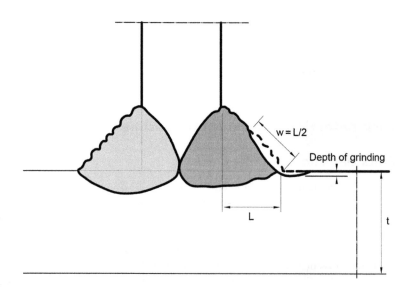

Figure 4.15 The burr grinding technique: the depth of grinding should be 0.5 mm below the bottom of any visible undercut, maximum of 2 mm, or 7% of plate thickness (adapted from Haagensen and Maddox, 2001)

- *Hammer and needle peening:* Compressive residual stresses are induced by repeatedly hammering the weld toe region with a blunt-nosed chisel. It is not applicable to connections with main plate thicknesses of less than 4 mm for steel; for larger areas, needle peening is preferred (Figure 4.17) (Haagensen and Maddox, 2001).
- *Shot peening:* Compressive residual stresses are induced by small spherical media shot bombarding the welding surface (Bandini, 2004).
- *Ultrasonic impact treatment:* This technique involves deformation treatment of the weld toe by a mechanical hammering at a frequency of around 200 Hz superposed by ultrasonic treatment at a frequency of 27 kHz. The objective of this treatment is to introduce beneficial compressive residual stresses at the weld toe by plastic deformation of the surface and to reduce stress concentration by smoothing the weld toe profile (Gunther et al., 2005; Kudryavtsev et al., 2007).

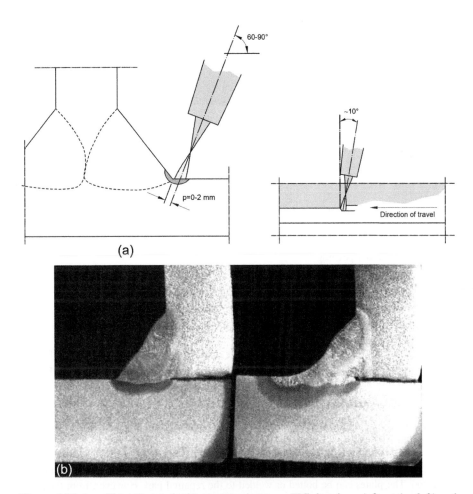

Figure 4.16 A weld toe threated with tungsten inert gas-TIG dressing: a) front (on left) and lateral view (on right) of the application; b) before (on left) and after (on right) the application. (adapted from Haagensen and Maddox, 2001)

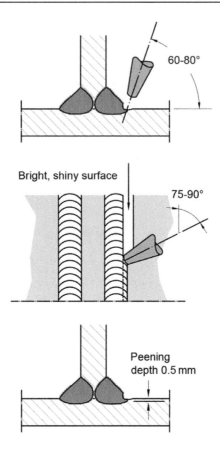

Figure 4.17 Hammer peening

7.2.2 Arresting cracks

Classical solutions for stopping cracks include the following:

- *Stop holes*: The execution of a stop hole is the method most widely used to repair fatigue cracks or to correct details prone to fatigue. It is often used as a temporary measure to stop the propagation of cracks, which might be followed by more extensive repairs. It is rare to implement a remedial action that does not include the drilling of a hole at the top of the crack. In the case of critical details, the hole is often drilled to isolate the detail or to intercept a potential crack before it can propagate far into the root (Connor et al., 2005); the hole dimension could be dimensioned according to Fisher et al. (1990).
- *Cover plating*: Reinforcement plates introduce additional material to the cross section to increase the resisting area. Typically, cracks are repaired with protection plates with this method. The plates can be bolted or welded to the repaired part. However, from the point of view of the fatigue resistance, the bolted connection is the best option because connections with high-strength bolts may be considered as details of Category B (i.e., Category 125 of EC3), while the welded details of Category E (i.e., Category 56 of EC3) or lower. It follows that for permanent repairs, the use of bolted plates is recommended (Connor et al., 2005).

- *Welding*: Welding can be used to repair cracks restoring the continuity of the element; at times, this appears to be the only solution, as the structure to be repaired could not withstand the reduction of a section required by an intervention that uses bolts. However, welding should be done with caution, as it can introduce unfavorable conditions regarding fatigue strength, as discontinuity and residual stresses are inherent in the process (Dexter, 2004).
- *Local heating*: This approach consists of artificially introducing compressive residual stresses to an existing through-thickness crack by local heating near the crack tip. In spot heating, the structure is heated locally, usually with a gas torch, to produce local yielding resulting in compressive thermal stresses. As the locally heated metal cools, it shrinks, causing residual stresses (Jang et al., 2002).

An economic consideration needs to be made here, as the global economy of a new bridge should include maintenance costs, not just construction costs. Considering that repair and any further interventions imply significant costs, an important choice must be made at the design stage. From this perspective, bolted structures offer more advantages in the maintenance stage than welded structures because while a riveted/bolted member will not fail until additional cracks form in one or more additional elements as they are inherently redundant, in a welded structure, one crack could propagate along the whole structural element and can easily cause global failure. For this reason, bolted structures, especially if subjected to cyclic live traffic loadings, are better than welded structures; however, they do require a particular kind of experience in both the design and the construction stages.

7.3   Global interventions

Large interventions are needed when cracks or defects are widely diffused throughout the whole structure, and local retrofits do not produce improvement in the structural behavior. Here are some examples:

- *Load reduction:* An increase in bearing capacity of the structure is obtained through the reduction of permanent loads via the construction of a lighter bridge deck slab. The replacement of the existing slab can appear to be an attractive solution if the high weight of the concrete constituent can be replaced with a system of lower weight. The literature recognizes these main types of system: in situ and precast concrete deck slabs of reduced dimensions using higher-strength concrete, metal gratings with and without fillings, orthotropic steel or aluminum deck slabs, composite fiber-reinforced polymer (FRP) deck slabs, and wooden slabs (Wipf et al., 1993). Deck slabs made of ultra-high-performance fiber-reinforced cement (UHPFRC)–based composites (strengthened with steel rebars) represent a further improvement because of their reduced weight and significantly improved fatigue and ultimate strength (Makita and Brühwiler, 2014a, 2014b), together with FRP or sandwich plate system (SPS), which may also represent suitable solutions, particularly where accelerated reconstruction is required.
- *Composite action:* In the case of simply supported deck systems, composite action could be introduced as a strengthening method to upgrade the loading capacity of the structural system and to reduce hot spot stresses; the structural design is similar to that of new structures; post-installed connectors should be used (Kwon et al., 2007).
- *Cover plating*: The same use of local retrofits could be realized in the whole structure; however, specific interventions could be adopted to gain a larger cross-resistant area, as new angles or members (Figure 4.18).

- *Post-tensioning:* Prestressing can be used as a means to introduce redundancy in critical elements to fracture. It can also be used to repair damaged zones with cracks. In this case, the method can provide the control of the propagation of the crack caused by traffic loads. For example, the total prestressing element of a tensile chord or tensile diagonal, belonging to a lattice, can ensure permanent compression, preventing the crack to propagate to reach the critical size. If this option is not feasible due to the high levels of stress in other parts of the structure, one can choose a partial prestressing; external prestressing by means of high-strength bars

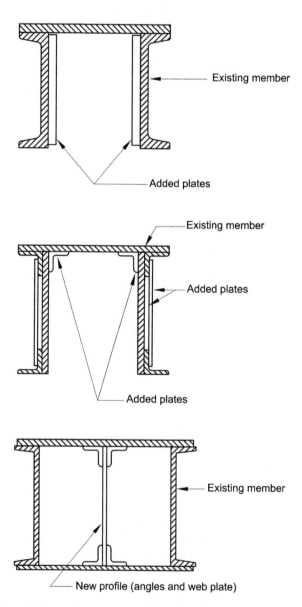

Figure 4.18 Cover plating options

or cables attached to the steel beams has been used as an effective technique for upgrading the load carrying capacity of composite steel-concrete girders (Albrecht and Lenwari 2008). (Figure 4.19).

Figure 4.19 Post-tensioning steel bar application on an existing girder

- *High-strength pretensioned steel plates*: This technique is applicable to beam elements, such as flooring system-bearers of girders and lattice girder bridges. Pretensioned plates are applied to the tension zone.
- *TPSM*: The thermal prestressing method (TPSM) has been proposed for innovative construction of continuous composite girder bridges, as an effective prestressing method to prevent the occurrence of tensile transverse cracks in the concrete deck at the negative bending moment regions (Kim et al., 2010).
- *FRP:* An alternative technique for strengthening steel structures consists of the application of externally bonded FRP sheets, to increase mainly the tensile and flexural capacity of the structural elements. FRP materials have a high strength-weight ratio, do not give rise to problems due to corrosion, and are manageable. However, this technique is not flame resistant and should be shielded from fire.
- *Modification of the static system:* Although not applicable in all cases, modification of static conditions could be applied to improve fatigue resistance, avoiding stress concentrations; for example, by introducing intermediate supports or transforming isostatic spans into hyperstatic systems.

A deeper focus is needed for the strengthening of r.c. bridge deck slabs using UHPFRC: UHPFRC has properties such as high compressive strength (>150–200 MPa) and tensile strength (>7–12 MPa), strain-hardening tensile deformation of 0.1%–0.3%, and very low permeability because of an optimized dense matrix, making the material virtually waterproof (Brühwiler and Denarié, 2013). These properties make UHPFRC suitable for strengthening those parts of structural members that are subjected to mechanically and environmentally severe actions. The tensile behavior of UHPFRC may be effectively improved by arranging steel rebars to create reinforced UHPFRC (R-UHPFRC).

In recent years, the necessity to improve the durability, load -bearing capacity, and fatigue resistance of bridges (primarily deck slabs) is growing due to the increase of traffic loads and volume. Strengthening of concrete bridge deck slabs is efficiently achieved by adding a 30–60-mm-thick layer of UHPFRC (combined with steel rebars) on top of the existing r.c. deck slab without an increase or with only a minor increase in self-weight (thereby avoiding the need to strengthen other structural members like main beams or boxes). The UHPFRC layer increases the fatigue strength and ultimate resistance (in bending and shear) of the deck slab and improves durability due to its waterproofing properties, making the UHPFRC strengthening technology efficient and economical. The technology has been applied since 2004, mostly in Switzerland. A recent large-scale application is the strengthening of a 2.1-km-long highway viaduct the Chillon highway viaduct in Switzerland, implying an increase in fatigue strength of the deck slab (Figure 4.20).

Figure 4.20 Strengthening of the deck slab of the 2.1-km-long Chillon highway viaduct in Switzerland using R-UHPFRC: A view of the viaduct (a), typical section (b),

Figure 4.20 Continued. and R-UHPFRC deck realization (c)

8 Research on fatigue and fracture

In view of a modern structural engineering approach, some changes are required in the common construction of steel and r.c. structures in order to avoid well-known concerns due to material deterioration. For this reason, bolting should be preferred over welding to take advantage of this connection type that features relatively high fatigue strength and rapid construction, and it is also easy to inspect, change, or improve. Moreover, high-strength materials should be used for both steel and r.c. structures, as they are able to sustain higher stresses in hot spots, prolonging their fatigue endurance. New materials are necessary, not only for retrofit interventions on existing structures, but also for new constructions, in order to gain new levels of sustainability, and to reduce economic costs. The most promising areas of research finalized to the aforementioned goals are:

- *Testing*: Testing techniques could be adopted for existing or new structures, at the scale of a single member/component, up to an entire structure. This procedure, helps in finding the most accurate solution for structural retrofit in existing bridges; in new bridges, it is able to anticipate the structural performance of components or entire structures (at the scaled or at the real size), obtaining more sustainable and more efficient design solutions. This method is well established in the area of fatigue since the second half of the eighteenth century, either on a small or a realistic scale. Small-scale specimen tests give longer apparent fatigue lives, either if testing is related to new structures or to FRP-strengthened specimens (Dexter and Fisher, 2000; Pipinato et al., 2012b): therefore, the *S-N* curve must be based on full-size testing of structural components such as girders, cross bracings, and stringer-to-floor connections; moreover, testing on full-scale welded members has indicated that the primary effect of constant-amplitude fatigue loading can be accounted for in the live-load stress range. Relevant experiences with steel bridge testing are shown in Table 4.2 (for small- and full-scale tests of unreinforced steel structures) and Table 4.3 (for small- and full-scale tests of reinforced steel structures).
- *Innovative materials (FRP)*: Some examples of guidelines for the design and construction of externally bonded FRP systems for strengthening of existing metallic structures include the ICE-Institution of civil ENgineers "FRP composites: Life Extension and Strengthening of Metallic Structures" (ICE 2001), the *Construction Industry Research and Information*

Table 4.2 **Full and small scale test of not-reinforced structures**

Author	Year	Experimental data	
Reemsnyder	1975	Type of test	Axial loading costant amplitude test
		Specimens	Riveted gusset plate connections
		Structure	Ore unloading bridge
		Hot spot detail	Tension chord at its connection to a gusset plate
		Note	Five test results have been obtained by newly fabricated specimens
Baker and Kulak	1982	Type of test	Bending and shear test
		Specimens	Built up hanger members
		Structure	Highway bridge
		Hot spot detail	Built up hanger members
Out et al.	1984	Type of test	Bending test
		Specimens	Stringers
		Structure	Railway stringers
		Hot spot detail	Continuous riveted connection between the web and the flange angles
		Note	Not shown results from corroded specimens
Fischer et al.	1987	Type of test	Bending test
		Specimens	Built-up hanger members
		Structure	Railway stringers
		Hot spot detail	Web to flange connection
Bruhwiler et al.	1990	Type of test	Bending test
		Specimens	Built-up plate girders and lattice girders, wrought iron and rolled mild steel
ATLSS	1993	Type of test	Bending test
		Specimens	Flanged angle to web
		Structure	Railway stringers
		Hot spot detail	Web to flange connection
Adamson and Kulak	1995	Type of test	Bending test
		Specimens	Stringers
		Structure	Built-up railway stringers
		Hot spot detail	Horizontal bracing attacchment riveted to the tension flange
Di Battista and Kulak	1995	Type of test	Axial tension
		Specimens	Diagonals
		Structure	Railway truss bridge
		Hot spot detail	Riveted connection of the outstanding legs of these angles to gusset plates
Akesson and Edlund	1996	Type of test	Bending test
		Specimens	Flange angles riveted to web plate
		Structure	Built-up railway stringers
		Hot spot detail	Angle to web connection
Helmerich et al.	1997	Type of test	Bending and axial test
		Specimens	Truss members
		Structure	-
		Hot spot detail	Built-up plate girders

Continued

Table 4.2 Continued

Author	Year	Experimental data	
Matar and Greiner	2006	Type of test	Bending test
		Specimens	Secondary members
		Structure	Railway bridge
		Hot spot detail	Flange to web connection
		Note	Only not corroded specimens were tested
Pipinato	2008	Type of test	Bending test
		Specimens	Full scale girders
		Structure	Railway bridge
		Note	Only not corroded specimens were tested
Pipinato	2008	Type of test	Shear test
		Specimens	Short diaphragm connection
		Structure	Railway bridge
		Hot spot detail	Short diaphragm connection
		Note	Only not corroded specimens were tested

Table 4.3 Full and small scale tests of reinforced structures

Reference	Details	Test type
Bocciarelli, 2009	S275 specimens reinforced on each side with Sika Carbo Dur M614	Normal tension
Iwashita, 2007	Single-lap shear steel specimens	Normal tension
Jones, 2003	5 steel specimens not reinforced and 24 each side reinforced specimens	Normal tension
Monfared, 2008	15 plates of steel reinforcing or not the specimens with FRP, treating the surface with blasting or less, and applying the reinforcements on one or both sides	Normal tension
Zheng, 2006	6 steel specimens with an hoolow at the centre and provided with 2 cracks reiforced with CFRP on one or both side	Normal tension
Colombi, 2003	perforated steel specimens with 2 crack, reinforced with 2 CFRP straps on each side	Normal tension
Taljsten, 2009	5 historical steel specimens perforated, with 2 crack	Normal tension
Liu, 2005	12 steel joints composed by 2 steel plates CFRP reinforced, with normal or HM on each side	Normal tension
Tavakkolizadeh, 2003	5 steel beams hollowed in the flangesand CFRP reinforced on the lower and midspan flange	Bending
Deng, 2007	steel beams CFRP reinforced on the lower and midspan flange	Bending
Bassetti, 1999	truss riveted beams with I section, traditionally reinforced	Bending
Bassetti, 2001	truss riveted beams traditionally reinforced	Bending
	truss riveted beams reinforced with 2 CFRP laminates and 3 CFRP post-tensioned laminates on the lower flange	Bending

Association (CIRIA) Design Guide (Cadei et al., 2004), and the CNR-DT 202/2005 (Italian Research Council, 2005). The benefits of composite strengthening have been applied, for example, in a steel bridge on the London Underground (Moy and Bloodworth, 2007). The benefits of strengthening large cast-iron struts with carbon FRP (CFRP) composites in the London Underground are illustrated in Moy and Lillistone (2006). A state-of-the-art review of FRP strengthened steel structures has been given by Zhao and Zheng (2007). Among these materials, apart from the well-known e-glass, high strength (HS) CFRP, and aramid, high-modulus CFRP (HM CFRP) materials are being widely diffused and have been developed with a tensile modulus that is approximately twice that of steel. We are still far from real improvements in industrial and wide use of new materials, but composite material industries have the right bases to go toward real innovations.

• *Innovative materials (UHPFRC)*: The static behavior of UHPFRC and R-UHPFRC has been investigated by many researchers; however, few findings have been reported so far in the literature. The fatigue behavior of UHPFRC and R-UHPFRC has been investigated by Makita and Brühwiler (2014a, 2014b) with the purpose of using this material for the fatigue strengthening of r.c. bridge deck slabs. Other concurring recent research on this issue includes the local bending tests and punching failure of a ribbed UHPFRC bridge deck (Toutlemonde et al., 2007) and the study of UHPFRC overlays to reduce stresses in orthotropic steel decks (Lamine et al., 2013).

References

Abe, H., 1989. Fatigue strength of corroded steel plates from old railway bridges. In: IABSE Symposium, Lisbon. pp. 205–210.

Adamson, D.E., Kulak, G.L., 1995. Fatigue Test of Riveted Bridge Girders. Structural Engineering report no. 210. Department of Civil Engineering, University of Alberta.

AISC, 2011. Steel Construction Manual, 14th ed. AISC-American Institute for Steel Construction, Chicago.

Åkesson, B., 1991. Older Railway Bridge-Load-Carrying Capacity, Condition and Service Life. Department of Structural Engineering, Chalmers University of Technology, Gothenburg, Sweden.

Åkesson, B., 1994. Fatigue life of riveted railway bridges. Thesis (PhD). Division of Steel and Timber Structures, Chalmers University of Technology, Publ. S94:6, Goteborg, Sweden.

Akesson, B., Edlund, B., 1996. Remaining fatigue life of riveted railway bridges. Der Stahlbau 65 (11), 429–436.

Albrecht, P., Lenwari, A., 2008. Fatigue strength of repaired prestresses composite beams. J. Bridge Eng. 13 (4), 409–417.

Al-Emrani, M., 2000. Two Fatigue-Related Problems in Riveted Railway Bridges. Thesis (Licentiate). Department of Structural Engineering, Chalmers University of Technology, Göteborg, Sweden.

American Association of State Highway and Transportation Officials (AASHTO), 2013. LRFD Bridge Design Specifications, sixth ed. with 2013 Interim Revisions. Washington, DC.

Anderson, T.L., 1995. Fracture Mechanics—Fundamentals and Applications, second ed. CRC Press, Boca Raton, FL.

ASCE Committee on Fatigue and Fracture Reliability, 1982a. Fatigue reliability: introduction. J. Struct. Div. 108 (1), 3–23.

ASCE Committee on Fatigue and Fracture Reliability, 1982b. Fatigue reliability: quality assurance and maintainability. J. Struct. Div. 108 (1), 25–46.

ASCE Committee on Fatigue and Fracture Reliability, 1982c. Fatigue reliability: variable amplitude loading. J. Struct. Div. 108 (1), 47–69.

ASCE Committee on Fatigue and Fracture Reliability, 1982d. Fatigue reliability: development of criteria for design. J. Struct. Div. 108 (1), 71–88.

ATLSS (Center for Advanced Technology for Large Structural Systems), 1993. Assessment of Remaining Capacity and Life of Riveted Bridge Members. Draft Project Report to Canadian National Railways, 1993. ATLSS, Lehigh University, Bethlem, Pennsylvania.

AWS, 2010. D1.5M/D1.5:2010 BRIDGE WELDING CODE. American Welding Society, a Joint Publication of American Association of State Highway and Transportation Officials, Miami.

Baker, K.A., Kulak, G.L., 1982. Fatigue strength of two steel details. Structural Engineering report no. 105, Department of Civil Engineering, University of Alberta.

Bandini, M., 2004. La pallinatura controllata e le sue applicazioni. International Conference on Innovative Solutions, Components and Materials for Transport Industries, Proceedings of the Conference, pp. 122–138, Verona, Italia.

Di Battista, J.D., Kulak, G.L., 1995. Fatigue of Riveted Tension Members. Structural Engineering report no. 211. Department of Civil Engineering, University of Alberta.

Brühwiler, E., Smith, I.F.C., Hirt, M.A., 1990. Fatigue and fracture of riveted bridge members. J. Struct. Eng. 116 (1), 198–214.

Brühwiler, E., Denarié, E., 2013. Rehabilitation and strengthening of concrete structures using ultra-high-performance fibre-reinforced concrete. SEI 23 (4), 450–457.

Brühwiler, E., Vogel, T., Lang, T., Lüch, P., 2012. Swiss standards for existing structures. SEI 22 (2), 275–280.

Byers, W.G., Marley, M.J., Mohammadi, J., Nielsen, R.J., Sarkani, S., 1997. Fatigue reliability reassessment applications: state of the-art paper. J. Struct. Eng. 123 (3), 277–285.

Cadei, J.M.C., Stratford, T.J., Hollaway, L.C., Duckett, W.G., 2004. Strengthening Metallic Structures Using Externally Bonded Fibre-Reinforced Polymers (C595). CIRIA Design Guide. CIRIA, London.

Chen, Y., Ni, J., Zheng, P., Azzam, R., Zhou, Y., Shao, W., 2011. Experimental research on the behavior of high frequency fatigue in concrete. Eng. Fail. Anal. 18, 1848–1857.

Cheung, M.S., Lib, W.C., 2003. Probabilistic fatigue and fracture analyses of steel bridges. Struct. Saf. 23, 245–262.

Chotickai, P., Kanchanalai, T., 2010. Field testing and performance evaluation of a through-plate girder railway bridge. Transport. Res. Rec. 2172, 132–141.

Chryssanthopoulos, M.K., Righiniotis, T.D., 2006. Fatigue reliability of welded steel structures. J. Const. Steel Res. 62 (11), 1199–1209.

Confédération Européene de Billiard (CEB), 1988. Fatigue of Concrete Structures, Bulletin 188. State-of-the-Art Report. Lausanne.

Connor, R.J., Dexter, R., Mahmoud, H., 2005. Inspection and Management of Bridges with Fracture- Critical Details. NCHRP Synthesis 354. Transportation Research Board, Washington, DC.

Cornelissen, H.A.W., Reinhardt, H.W., 1982. Fatigue of plain concrete in uniaxial tension and in alternating tension–compression loading. IABSE Rep. 37, 273–282.

Cornelissen, H.A.W., Reinhardt, H.W., 1984. Uniaxial tensile fatigue failure of concrete under constant-amplitude and programme loading. Mag. Concr. Res. 36 (129), 216–226.

Crooker, T.W., Lange, E.A., 1970. How yield strength and fracture toughness considerations can influence fatigue design procedures for structural steels. Weld. Res. Suppl. 489-s.

Dexter, R.J., Fisher, J.W., 1996. The Effect of Unanticipated Structural Behavior on the Fatigue Reliability of Existing Bridge Structures, Structural Reliability in Bridge Engineering.

In: Frangopol, D.M., Hearn, G., (Eds.), Proceedings of a Workshop at the University of Colorado at Boulder. McGraw Hill, New York, pp. 90–100.

Dexter, R.J., 2004. Sign, signal, and ligth support structures and manual for repair fatigue cracks. In: Third Annual Bridge Workshop: Fatigue and Fracture.

Dexter, R.J., Fisher, W.F., 2000. Fatigue and fracture. In: Chen, W.-F., Duan, L. (Eds.), Bridge Engineering Handbook. CRC Press, Boca Raton.

Ditlevsen, O., Madsen, H.O., 2005. Structural Reliability Methods. John Wiley & Sons, Chichester.

Dyduch, K., Szerszen, M., Desterbecq, J.-F., 1994. Experimental investigation of the fatigue of plain concrete under high compressive loading. Mater. Struct. 27, 505–509.

EN 1990, 2010. Eurocode—Basis of Structural Design. CEN, Brussels.

EN 1991-2, 2010. Actions on Structures—Traffic Loads on Bridges. CEN, Brussels.

EN 1993-1-10, 2009. Eurocode 3: Design of Steel Structures—Parts 1–10: Material Toughness and Through-Thickness Properties. CEN, Brussels.

EN 1993-1-9, 2005. Eurocode 3: Design of Steel Structures—Parts 1–9: Fatigue. CEN, Brussels.

Federal Highway Administration (FHWA), 2012. Steel Bridge Design Handbook: Redundancy. Publication No. FHWA-IF 12-052—Vol. 9.

Fisher, J.W., Mertz, D.R., Zhong, A., 1983. Steel Bridge Members Under Variable Amplitude Long Life Fatigue Loading. Transportation Research Record 267. National Research Council, Washington, D.C.

Fisher, J.W., 1984. Fatigue and Fracture in Steel Bridges. Wiley-Interscience, Hoboken, NJ.

Fisher, J.W., Yen, B.T., Wang, D., 1987. Fatigue and Fracture Evaluation for Rating Riveted Bridges. Transportation Research Record 302, Transportation Research Board, Washington, DC, pp. 25–35.

Fisher, J.W., Yen, B.T., Wang, D., 1990. Fatigue strength of riveted bridge members. J. Struct. Eng. 116 (11), 2968–2981.

Forsberg, B., 1993. Utmattningsha llfasthet hos a¨ldre konstruktionssta °l med korrosionsskador. Master thesis. Division of Steel Structures Lulea University of Technology. ISSN 0349-6023.

Freudenthal, A.M., 1945. Safety of structures. Trans. ASCE 112 (1947), 125–180.

Frey, R., Thurlimann, B., 1983. Ermüdungsversuche an Stahlbetonbalken mit und ohne Schubbewehrung. Institute für Baustatik und Konstruktion Bericht Nr 7801-1, ETH Zurich.

Gunther, H.P., Kuhlmann, U., Dürr, A., 2005. Rehabilitation of welded joints by ultrasonic impact treatment (Uit). In: IABSE Symposium, Lisbon.

Haagensen, P.J., Maddox, S.J., 2001. IIW Recommendations on Post-Weld Improvement of Steel and Aluminium Structures. IIW Commission XIII Working Group 2—Improvement Techniques, Paris.

Helmerich, R., Brandes, K., Herter, J., 1997. Full scale laboratory fatigue tests on riveted railway bridges, evaluation of existing steel and composite bridges. In: IABSE Workshop, Lausanne. pp. 191–200.

Herwig, A., 2008. Reinforced Concrete Bridges Under Increased Railway Traffic Loads— Fatigue Behaviour and Safety Measures. Thèse EPFL No. 4010. EPFL Press, Lausanne.

Herwig, A., Brühwiler, E., 2011. In situ dynamic behavior of a railway bridge girder under fatigue causing traffic loading. In: Proc. ICASP 11—11th Intl. Conf. Appl. Stat. Prob. Civil Eng., Zurich, August 1–4.

ICE-Institution of Civil Engineers 2001. FRP composites: Life Extension and Strengthening of Metallic Structures. In ICE Design and Practice Guide, ISBN: 978-0-7277-3009-1. Thomas Telford, London.

International Institute of Welding (IIW), 1996. Fatigue of Welded Components and Structures. International Institute of Welding—Working Group 5 (WG5): Repair of Fatigue-Loaded Welded Structures. IIW, Villepinte.

Italian Research Council, 2005. Guidelines for the Design and Construction of Externally Bonded FRP Systems for Strengthening Existing Structures. Preliminary Study. Metallic Structures (CNR– DT 202/2005). Italian Research Council, Rome.

Jang, C.D., Song, H.C., Lee, C.H., 2002. Fatigue life extension of a through-thickness crack using local heating. In: International Offshore and Polar Engineering Conf, May. ISOPE edition. Kitakyushu, Japan.

Kim, S.H., Kim, J.H., Ahn, J.A., 2010. Life-cycle cost analysis of a TPSM applied continuous composite girder bridge. Intl. J. Steel Struct. 10 (2), 115–129.

Kudryavtsev, Y., Kleiman, J., Lugovskoy, A., Lobanov, L., Knysh, V., Voitenko, O., Prokopenko, G., 2007. Rehabilitation and Repair of Welded Elements and Structures by Ultrasonic Peening. IIW Document XIII-2076-05. International Institute of Welding, Paris.

Kuhn, B., Lukic´, M., Nussbaumer, A., Gunther, H.-P., Helmerich, R., Herion, S., Kolstein, M.H., Walbridge, S., Androic, B., Dijkstra, O., Bucak, O., 2008. Assessment of Existing Steel Structures: Recommendations for Estimation of Remaining Fatigue Life. Joint report prepared under the JRC-ECCS cooperation agreement for the evolution of Eurocode 3 (program of CEN/TC 250).

Kulak, G., 1992. Discussion of "Fatigue Strength of Riveted Bridge Members" by John W. Fisher, Ben T. Yen, and Dayi Wang (November 1990, Vol. 116, No. 11)". J. Struct. Eng. 118 (8), 2280–2281.

Kwon, G., Hungerford, B., Kayir, H., Schaap, B., Kyu Ju, Y., Klingner, R., Engelhardt, M., 2007. Strengthening Existing Non-Composite Steel Bridge Girders Using Post-Installed Shear Connectors. Report FHWA/TX-07/0-4124-1, Texas Department of Transportation.

Lamine, D., Marchand, P., Gomes, F., Tessier, C., Toutlemonde, F., 2013. Use of UHPFRC overlay to reduce stresses in orthotropic steel decks. J. Construct. Steel Res. 89, 30–41.

Lee, P.G., Shim, C.S., Chang, S.P., 2005. Static and fatigue behavior of large stud shear connectors for steel-concrete composite bridges. J. Construct. Steel Res. 61 (9), 1270–1285.

Leitão, F.N., Silva, J.G.S. da, Andrade, S.A.L. de, Vellasco, P.C.G. da S., Lima, L.R.O. de., 2011. "Fatigue Analysis of Composite (Steel-Concrete) Highway Bridges Subjected to Dynamic Actions of Vehicles". CC 2011 - The Thirteenth International Conference on Civil, Structural and Environmental Engineering Computing. Civil-Comp Press. Chania, Crete, Greece. Vol. 1. pp. 1–17.

Makita, T., Brühwiler, E., 2014a. Tensile fatigue behaviour of ultra-high performance fibre reinforced concrete combined with steel rebars (R-UHPFRC). Intl. J. Fatigue 59, 145–152.

Makita, T., Brühwiler, E., 2014b. Tensile fatigue behaviour of ultra-high performance fibre reinforced concrete. Mat. Struct. 47, 475–491.

Mang, F., Bucak, Ö., 1993. Application of the S-N line concept for the assessment of the remaining fatigue life of old bridge structures. In: Bridge Management 2. Inspection, Maintenance Assessment and Repair. ISBN: 0 7277 1926 2.

Miki, C., Ito, Y., Sasaki, E., 2003. Fatigue and Repair Cases in Steel Bridges. Tokyo Institute of Technology, Department of Civil Engineering, Miki Laboratory, Tokyo.

Miner, M.A., 1945. Cumulative damage in fatigue. J. Appl. Mech. 12 (3), 159–164.

Moy, S.S.J., Bloodworth, A.G., 2007. Strengthening a steel bridge with CFRP composites. Proc. ICE. Struct. Build. 160, 81–93.

Moy, S.S.J., Lillistone, D., 2006. Strengthening cast iron using FRP composites. Proc. ICE. Struct. Build. 159 (6), 309–318.

National Bridge Inspection Standards (NBIS), 2004. U.S. Department of Transportation, Federal Highway Administration (FHWA), Washington, DC.

National Cooperative Highway Research Program (NCHRP), 2012. Fatigue Evaluation of Steel Bridges. National Cooperative Highway Research Program. Report 721/2012, Transportation Research Board, Washington, DC.

Out, J.M.M., Fisher, J.W., Yen, B.T., 1984. Fatigue Strength of Weathered and Deteriorated Riveted Members. Transportation Research Record 950 , vol. 2. 10–20.

Paris, P.C., Gomez, M.P., Anderson, W.E., 1961. A rational analytic theory of fatigue. Trend Eng. 13, 9–14.

Perdikaris, C., Beim, S., 1988. RC bridge decks under pulsating and moving load. J. Struct. Eng. 114 (3), 591–607.

Pipinato, A., 2008. High-Cycle Fatigue Behaviour of Historical Metal Riveted Railway Bridges. PhD thesis, University of Trento, Trento, Italy.

Pipinato, A., 2010. Step-level procedure for remaining fatigue life evaluation of one railway bridge. Baltic J. Road Bridge Eng. 5 (1), 28–37. http://dx.doi.org/10.3846/bjrbe.2010.04.

Pipinato, A., Pellegrino, C., Bursi, O.S., Modena, C., 2009. High-cycle fatigue behavior of riveted connections for railway metal bridges. J. Construct. Steel Res. 65 (12), 2167–2175.

Pipinato, A., Molinari, M., Pellegrino, C., Bursi, O.R., Modena, C., 2011. Fatigue tests on riveted steel elements taken from a railway bridge. Struct. Infrastruct. Eng. 7 (12), 907–920. http://dx.doi.org/10.1080/15732470903099776.

Pipinato, A., Pellegrino, C., Modena, C., 2012a. Assessment procedure and rehabilitation criteria for the riveted railway Adige Bridge. Struct. Infrastruct. Eng.: Main., Manage Life-Cycle Design, Perform. 8 (8), 747–764. http://dx.doi.org/10.1080/15732479.2010.481674.

Pipinato, A., Pellegrino, C., Modena, C., 2012b. Fatigue behavior of steel bridge joints strengthened with FRP laminates. Mod. Appl. Sci. 6 (9), pp. 1–15.

Rabemanantso, H., Hirt, M.A., 1984. Comportement a la Fatigue de Profiles Lamines avec Semelles de Renfort Revetees, ICOM Report 133, Swiss Federal Institute of Technology. Lusanne, Switzerland.

Rodrigues, J.F.S., Casas, J.R., Almeida, P.A.O., 2013. Fatigue-safety assessment of reinforced concrete (RC) bridges: application to the Brazilian highway network. Struct. Infrastruct. Eng.: Main., Manage., Life-Cycle Design Perform. 9 (6), 601–616.

Reemsnyder, H.S., 1975. Fatigue life extension of riveted connections. J. Struct. Div. Proc. Am. Soc. Civ. Eng. 101 (ST12), 2591–2608.

Schläfli, M., Brühwiler, E., 1998. Fatigue of existing reinforced concrete bridge deck slabs. Eng. Struct. 20 (11), 991–998.

Sonoda, K., Horikawa, T., 1982. Fatigue strength of reinforced concrete slabs under moving loads. IABSE Rep. 37, 455–462.

Soudki, K., Rizkalla, S., Uppal, A., 1999. Performance of bridge timber ties under static and fatigue loading. J. Bridge Eng. 4 (4), 263–268.

Stadelmann, W., 1984. Compiling of material tests on wrought iron rivets at EMPA, St. Gallen. Private communication [in German], cited in Bruhwiler, E., Smith, I.F.C., and Hirt, M. 1990. Fatigue and fracture of riveted bridge members. J. Struct. Eng. 116 (1), 198–213.

Stephens, R.I., Fatemi, A., Stephens, R.R., Fuchs, H.O., 2001. Metal Fatigue in Engineering. John Wiley & Sons, New York.

Toutlemonde, F., Renaud, J.-C., Lauvin, L., Brisard, S., Resplendino, J., 2007. Local bending tests and punching failure of a ribbed UHPFRC bridge deck. In: Proc. 6th Intl. Conf. Fracture Mech. Concrete and Concrete Struct. 3, pp. 1481–1489, 2007.

Tsiatas, G., Palmquist, S.M., 1999. Fatigue evaluation of highway bridges. Prob. Eng. Mech. 14, 189–194.

Weigler, H., Rings, K., 1987. Unbewehrter und bewehrter Beton unter Wechselspannungen. Deutscher Ausschuss für Stahlbeton 383, 52.

Wichtowski, B., 2013. Assessment of fatigue and selection of steel on constructions of steel bridges welded according to Eurocode 3. Welding Intl. 27 (5), 345–352.

Wipf, T.J., Klaiber, F.W., Besser, D.M., Laviolette, M.D., 1993. Manual for Evaluation, Rehabilitation and Strengthening of Low Volume Bridges. Iowa Department of Transportation, Ames.

Xiulin, Z., Zhen, L., Yongi, S., Yanman, Y., Zhijiang, S., 1996. Fatigue performance of old bridge steel and the procedures for life prediction with given survivability. Eng. Fract. Mach. 53 (2), 251–262.

Ye, X.W., Su, Y.H., Han, J.P., 2014. A state-of-the-art review on fatigue life assessment of steel bridges. Math. Prob. Eng. 2014, Article ID 956473, pp. 1–13.

Zhao, X.L., Zheng, L., 2007. State-of-the-art review on FRP strengthened steel structures. Eng. Struct. 29, 1808–1923.

Section III

Structural analysis

Bridge structural theory and modeling

5

Pipinato A.
AP&P, Technical Director, Italy

1 Introduction

The bridge engineering and design process involves a number of disciplines, at the basis of which undoubtedly lies the structural analysis theory. This chapter deals with the most interesting aspects of the structural analysis of bridges. Later, the text focuses on finite element method (FEM) theory and its applications to bridges, to emphasize the importance of this design instrument, commonly used by bridge engineers in their everyday applications.

2 Structural theory

Stresses inside a body generated by external excitations (volume and surface forces) can be obtained using equilibrium equations. Three equilibrium equations relate to the six components of the σ_{ij} stress tensor for an infinitesimal element in a static state; in the dynamic case, equations of motion are needed, including second-order derivatives of displacement (with respect to time). Considering the geometrical conditions, strains and displacement could be linked by using strain-displacement equations of kinematics expressing the components of strain ε_{ij} by the displacement components (u_i). The constitutive laws exert a material influence on these mathematical relations. The 15 variables are described and can be connected by 15 equations (3 equilibrium equations, 6 kinematics equations, and 6 constitutive equations). To solve the general problem of solid mechanics, two basic methods are available: the displacement method and the stress method. A combination thereof can also be used. In fact, while these methods could be directly applied for simple elastic problems, the discretization procedure is applied nowadays for complex and irregular structural forms and components, as in bridge engineering. This implies a preference for the use of the so-called FEM model, which can subdivide every complex body into finite small elements, solving every variable in the investigated discretized body with direct or iterative procedures. It is the method most widely adopted by automated software for the solution of real-life structural analysis problems. The systematic method adopted in computational analysis is matrix analysis, a computer code procedure implementing the classical methods used for handmade structural calculations (i.e., the principle of virtual displacement, the minimization of the total potential energy, and the minimization of the total complementary energy). When using matrix analysis to solve issues, two methods are available:

Innovative Bridge Design Handbook. http://dx.doi.org/10.1016/B978-0-12-800058-8.00005-0

the force method (with unknown internal forces) and the displacement method (with unknown displacement). In every case, the solving equations are based on a joint equilibrium and compatibility, giving as a result the stresses, strains, and displacements of every element of the investigated structure. Whenever FEM results are available, human judgment is needed: FEM solutions cannot be used without an expert overview of the reliability of the given output.

2.1 Equilibrium

Newton's First Law of Motion states that an object will remain at rest or in uniform motion in a straight line unless acted upon by an external force. It may be seen as a statement about inertia, that objects will remain in their state of motion unless a force acts to change the motion. This is what we call "equilibrium" state, in which, if the resultant force acting on a body is zero, the particle will remain at rest or will move at a constant velocity. Static is concerned essentially with the case where the particle or body remains at rest. A complete free-body diagram is essential for the solution of problems concerning the equilibrium. In a three-dimensional (3D) case, the conditions of equilibrium require the satisfaction of the following equations of static:

$$\sum F_x = 0 \quad \sum F_y = 0 \quad \sum F_z = 0$$
$$\sum M_x = 0 \quad \sum M_y = 0 \quad \sum M_z = 0$$

(1)

The previous equations state that the sum of all forces acting on a body in any direction must be zero, and the sum of all moments about any axis must be zero. A structure is *statically determinate* when all forces on its members are found by using only equilibrium conditions. If there are more unknowns than available equations of statics, the problem is called *statically indeterminate*. The degree of static indeterminacy is equal to the difference between the number of unknown forces and the number of relevant equilibrium conditions. Any reaction that is in excess of those that can be obtained by statics alone is termed a *redundant*. The number of redundants is, therefore, the same as the degree of indeterminacy. The generalization of the foregoing is analyzed in a free body of volume "V" (Ansel and Saul, 2011).

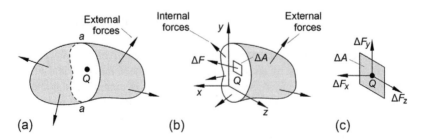

Figure 5.1 Derivation of equations of equilibrium: (a) a loaded body; (b) body with external and internal forces; (c) enlarged area ΔA with force components.

The body with all the appropriate forces, both known and unknown, acting on it is represented in Figure 5.1. An element of area ΔA, located at the internal point Q on the cut surface, is acted on by force ΔF. Put the origin of coordinates at point Q, with x normal and y and z tangent to ΔA, and assume that ΔF is not lying along x, y, or z. Decomposing ΔF into components parallel to x, y, and z (Figure 5.1c), we can define the normal (perpendicular) stress σ_x and the shearing (tangent) stresses τ_{xy} and τ_{xz} as follows:

$$\sigma_x = \lim_{\Delta A \to 0} \frac{\Delta F_x}{\Delta A} = \frac{dF_x}{dA}$$

$$\tau_{xy} = \lim_{\Delta A \to 0} \frac{\Delta F_y}{\Delta A} = \frac{dF_y}{dA} \qquad (2)$$

$$\tau_{xz} = \lim_{\Delta A \to 0} \frac{\Delta F_z}{\Delta A} = \frac{dF_z}{dA}.$$

In the generalized 3D case, distributed forces within a load-carrying member can be represented by a statically equivalent system consisting of a force and a moment vector acting at any arbitrary point (usually the centroid) of a section. These internal force resultants (also called *stress resultants*), exposed by an imaginary cutting plane containing the point through the member, are usually resolved into components that are normal and tangent to the cut section (Figure 5.2). The sense of moments follows the right-hand-screw rule, often represented by double-headed vectors, as shown in the figure. Each component can be associated with one of four modes of force transmission: (i) The axial force P or N tends to lengthen or shorten the member; (ii) the shear forces Vy and Vz tend to shear one part of the member relative to the adjacent part; (iii) the torque or twisting moment T is responsible for twisting the member; (iv) the bending moments My and Mz cause the member to bend.

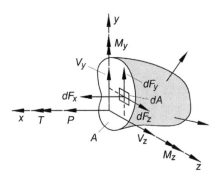

Figure 5.2 Forces and moments on a section of a 3D body.

According to the general case, component of stresses vary from point to point in a stressed body, where stress variations are governed by the conditions of equilibrium and mathematically from the differential equations of equilibrium.

Considering a planar infinitesimal element of sides dx and dy (Figure 5.3), being $\sigma_x, \sigma_y, \sigma_{xy}, \tau_{yx}$ functions of x and y, but independent from z (not varying in thickness),

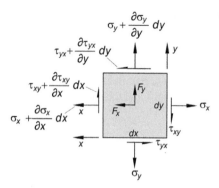

Figure 5.3 3D infinitesimal element with body forces and stresses.

and all the other components zero, this is a typical plane stress situation. In this situation, the variation of stresses (e.g., along the x-axis could be denoted by a truncated Taylor's series with partial derivatives, with σ_x a function of x and y) is

$$\left(\partial\sigma_x + \frac{\partial\sigma_x}{\partial x}\,dx\right). \tag{3}$$

All the other variations are similarly obtained. By the use of Eq. (1), the equilibrium of x forces, $\Sigma F_x = 0$, considering the equilibrium of an element of unit thickness, taking moments of force about the lower-left corner ($\Sigma M_z = 0$), neglecting the triple products involving dx and dy, this reduces to $\tau_{xy} = \tau_{yx}$. In a like manner, it may be shown that $\tau_{yz} = \tau_{zy}$ and $\tau_{xz} = \tau_{zx}$. Consequently:

$$\left(\sigma_x + \frac{\partial\sigma_x}{\partial x}\,dx\right)dy - \sigma_x dy + \left(\tau_{xy} + \frac{\partial\tau_{xy}}{\partial y}\,dy\right)dx - \tau_{xy}dx + F_x dxdy = 0 \tag{4}$$

$$\left(\frac{\partial\sigma_x}{\partial x} + \frac{\partial\tau_{xy}}{\partial y} + F_x\right)dxdy = 0 \tag{5}$$

$$\frac{\partial\sigma_x}{\partial x} + \frac{\partial\tau_{xy}}{\partial y} + F_x = 0$$
$$\frac{\partial\sigma_y}{\partial y} + \frac{\partial\tau_{xy}}{\partial x} + F_y = 0 \tag{6}$$

The 3D case could be similarly obtained giving the following result:

$$\frac{\partial\sigma_x}{\partial x} + \frac{\partial\tau_{xy}}{\partial y} + \frac{\partial\tau_{xz}}{\partial z} + F_x = 0$$
$$\frac{\partial\sigma_y}{\partial y} + \frac{\partial\tau_{xy}}{\partial x} + \frac{\partial\tau_{yz}}{\partial z} + F_y = 0 \tag{7}$$
$$\frac{\partial\sigma_z}{\partial z} + \frac{\partial\tau_{xz}}{\partial x} + \frac{\partial\tau_{yz}}{\partial y} + F_z = 0$$

Or, more synthetically:

$$\frac{\partial \tau_{ij}}{\partial x_j} + F_i = 0, \quad i, j = x, y, z \tag{8}$$

that is denoted also as

$$\tau_{ij,j} + F_i = 0 \tag{9}$$

The problem to find the configuration of the system, subjected to whatever constraints there may be, when all forces are balanced, is solved by the principle of virtual work. Only if the virtual work equality for all, arbitrary, variations of displacement is ensured would the equilibrium be complete (Zienkiewicz and Taylor, 2000).

2.1.1 Numerical method in structural analysis

The minimization of the total potential energy (TPE) is the basic principle of FEM: according to this, the sum of the internal strain energy and external works must be stationary when equilibrium is reached; for elastic problems, the TPE is stationary and minimal. This concept is mathematically expressed as (Zienkiewicz and Taylor, 2000):

$$\frac{\partial \Pi}{\partial a} = \left\{ \begin{array}{c} \frac{\partial \Pi}{\partial a_1} \\ \frac{\partial \Pi}{\partial a_2} \\ \vdots \end{array} \right\} = \mathbf{0}, \tag{10}$$

where a_i represents displacements and Π is the total potential energy ($\Pi = U + W$, and U and W are the total strain energy and the total potential energy, respectively). It is of interest to note that if true equilibrium requires an absolute minimum of total potential energy (Π), a finite element solution by the displacement approach will always provide an approximate (Π) greater than the correct one. Thus, a bound on the value of the total potential energy is always achieved. If the functional Π could be specified, a priori, then the finite element equations could be derived directly by Eq. (10). The Ritz (1909) process of approximation frequently used in elastic analysis uses precisely this approach. The total potential energy expression is formulated, and the displacement pattern is assumed to vary with a finite set of undetermined parameters (Zienkiewicz and Taylor, 2000). According to this interpretation, the static relation between unknown nodal displacements and known external loads is

$$[\mathbf{F}] = \{\mathbf{k}\}[\boldsymbol{\delta}] \tag{11}$$

[F] is the external loads vector, {k} the stiffness matrix, and [δ] the nodal displacement vector. In the force method, {k} is substituted by the force transformation matrix, and **[δ]** by the internal force vector. To exemplify, for a plane truss member (11) become:

$$[\mathbf{F}] = \begin{bmatrix} \bar{f}_{xi} \\ \bar{f}_{yi} \\ \bar{f}_{xj} \\ \bar{f}_{yj} \end{bmatrix} = \frac{EA}{L} \begin{bmatrix} 1 & 0 & -1 & 0 \\ 0 & 0 & 0 & 0 \\ -1 & 0 & 1 & 0 \\ 0 & 0 & 0 & 0 \end{bmatrix} \begin{bmatrix} \bar{u}_{xi} \\ \bar{u}_{yi} \\ \bar{u}_{xj} \\ \bar{u}_{yj} \end{bmatrix} = \{\mathbf{k}\}[\boldsymbol{\delta}] , \qquad (12)$$

where EA/l is the axial stiffness of the truss member.

2.1.2 Influence lines and surfaces

The maximization of stress/strains in a detailed section of a structural member is relevant to bridge structure and affected by moving loads during the whole life of the bridge. For this reason, influence lines (ILs) are used. To define an IL, suppose a generic structure with its boundary condition and an applied load F at the x_1 position in the x-axis (Figure 5.4). Considering the investigated quantity Q (normal tension, shear, bending moment, rotation, displacement, etc.) in the generic section S at the x_s position, this could be defined by

$$Q = Q(x_1, x_s, F) \qquad (13)$$

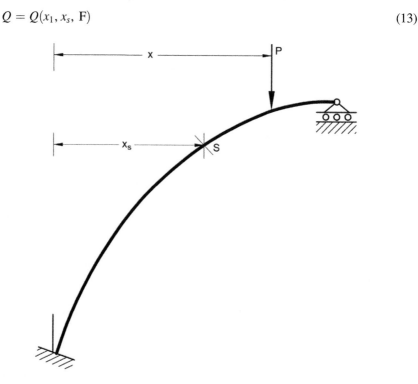

Figure 5.4 IL trivial representation.

The following situations apply: (i) if x_s is variable, solicitation diagrams are represented, describing the precise position of F solicitations and deformations; (ii) if x_1 is variable, solicitation diagrams are represented, describing the various position of F solicitations and deformations at point S. Dimensions of IL include force $\times$ length/force for moments, the pure number for normal forces and shear, length/forces for displacements. According to this definition, ILs could be defined as

$$vs = \int_0^l v_S^F(x_S, x)\,p(x)dx$$

$$ws = \int_0^l w_S^F(x_S, x)\,p(x)dx$$

$$Ms = \int_0^l M_S^F(x_S, x)\,p(x)dx \tag{14}$$

$$Ns = \int_0^l N_S^F(x_S, x)\,p(x)dx$$

$$Ts = \int_0^l T_S^F(x_S, x)\,p(x)dx,$$

where the ILs of v, w, M, N, and T, respectively, are described in the investigated section S along the x-axis under the moving load. By different notation, for a generic variable distributed load $q(x)$, the investigated generic quantity Q is

$$Q = \int_{x_1}^{x_2} q(x)\,\eta\,dx. \tag{15}$$

For an uniform load, $q = \text{cost}$, Eq. (14) becomes

$$Q = q \cdot \Omega, \tag{16}$$

where Ω is the surface area under the IL. In this specific case of a uniform distributed load, the position of the maximum of Q can be found easily as follows (Figure 5.5):

$$d\Omega = \eta_{x1} \cdot dx - \eta_{x2} \cdot dx \tag{17}$$

$$\frac{d\Omega}{dx} = 0 \tag{18}$$

and $\eta_{x1} = \eta_{x2}$. $\tag{19}$

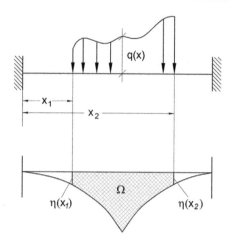

Figure 5.5 IL of a distributed load.

The typical solution of the shear and moment maximization on a continuous bridge is displayed in Figure 5.6.

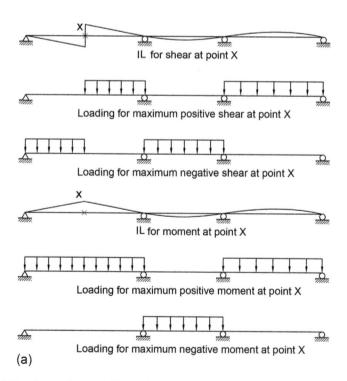

Figure 5.6 ILs of a continuous bridge: (a) shear and moment maximization at point X and the associated critical load diagrams;

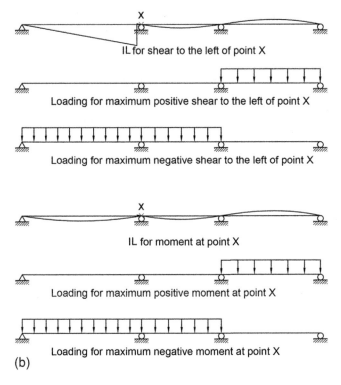

IL for shear to the left of point X

Loading for maximum positive shear to the left of point X

Loading for maximum negative shear to the left of point X

IL for moment at point X

Loading for maximum positive moment at point X

Loading for maximum negative moment at point X

(b)

Figure 5.6 Continued. (b) shear and moment maximization at point X and the associated critical load diagrams.

2.2 Compatibility

The deformation of a continuum is described in strain analysis without any reference to the material property: both the Lagrangian and Eulerian strain tensor are used to describe the deformation; however, in the first, the position and physical properties of the particles are described in terms of the material or referential coordinates and time, and in the second, they are described in terms of the spatial coordinates. In the latter case, this is called the *spatial description* or *Eulerian description;* i.e., the current configuration is taken as the reference configuration. However, strain equations of compatibility for infinitesimal strains could be also written in a different way: given the strain field, it is possible to compute the displacements in the case of *infinitesimal deformations*. For infinitesimal motions, the relation between strain and displacement is

$$\varepsilon_{ij} = \frac{1}{2}\left(\frac{\partial u_i}{\partial x_j} + \frac{\partial u_j}{\partial x_i}\right) \tag{20}$$

Given that there are six strain components (since $\varepsilon_{ij} = \varepsilon_{ji}$), we can determine the three displacement components u_j. First, it should be noted that the displacement field that gives rise to a particular strain field cannot be completely recovered, so the displacements can be determined only if there is some additional information about how much the solid has rotated and translated. Second, the strain-displacement relations can be integrated (a strain field is a symmetric second-order tensor field, but not always vice versa). The strain-displacement relations amount to a system of six scalar differential equations for the three displacement components u_i. To be integrable, the strains must satisfy the compatibility conditions, which may be expressed as

$$\frac{\partial^2 \varepsilon_{ij}}{\partial x_k \partial x_l} + \frac{\partial^2 \varepsilon_{ld}}{\partial x_i \partial x_j} - \frac{\partial^2 \varepsilon_{il}}{\partial x_j \partial x_k} - \frac{\partial^2 \varepsilon_{ik}}{\partial x_i \partial x_l} = 0. \tag{21}$$

Alternatively, expanding this expression in the (1, 2, 3) principal axis notation:

$$\frac{\partial^2 \varepsilon_{11}}{\partial x_2^2} + \frac{\partial^2 \varepsilon_{22}}{\partial x_1^2} - 2\frac{\partial^2 \varepsilon_{12}}{\partial x_1 \partial x_2} = 0$$

$$\frac{\partial^2 \varepsilon_{11}}{\partial x_3^2} + \frac{\partial^2 \varepsilon_{33}}{\partial x_1^2} - 2\frac{\partial^2 \varepsilon_{13}}{\partial x_1 \partial x_3} = 0$$

$$\frac{\partial^2 \varepsilon_{22}}{\partial x_3^2} + \frac{\partial^2 \varepsilon_{33}}{\partial x_2^2} - 2\frac{\partial^2 \varepsilon_{23}}{\partial x_2 \partial x_3} = 0$$

$$\frac{\partial^2 \varepsilon_{11}}{\partial x_2 \partial x_3} - \frac{\partial}{\partial x_1}\left(-\frac{\partial \varepsilon_{23}}{\partial x_1} + \frac{\partial \varepsilon_{31}}{\partial x_2} + \frac{\partial \varepsilon_{12}}{\partial x_3} \right) = 0 \tag{22}$$

$$\frac{\partial^2 \varepsilon_{22}}{\partial x_3 \partial x_1} - \frac{\partial}{\partial x_2}\left(-\frac{\partial \varepsilon_{31}}{\partial x_2} + \frac{\partial \varepsilon_{12}}{\partial x_3} + \frac{\partial \varepsilon_{23}}{\partial x_1} \right) = 0$$

$$\frac{\partial^2 \varepsilon_{33}}{\partial x_1 \partial x_2} - \frac{\partial}{\partial x_3}\left(-\frac{\partial \varepsilon_{12}}{\partial x_3} + \frac{\partial \varepsilon_{23}}{\partial x_1} + \frac{\partial \varepsilon_{31}}{\partial x_2} \right) = 0.$$

All strain fields must satisfy these conditions.

2.3 Constitutive laws

The concepts introduced in this chapter so far, in the framework of nonrelativistic mechanics, are essential to characterize stresses, kinematics, and balance principles. However, they are not capable to distinguish one material from another. As balance laws alone are not able to determine the response (displacement of the body due to an applied force) of a deformable body, they must be helped by additional equations, the constitutive laws, which depend on the material that the body is made of. A constitutive law describes the physical behavior of a specific material under defined conditions of interest.

2.4 Elastic and plastic behavior

The deformations of a body subjected to external actions are principally described by two main behaviors: elastic and plastic behaviors (Figure 5.7). Elastic materials, if subjected to an external force, return to their initial configurations at release; this behavior could be linear or nonlinear, representing the so-called linear elastic and nonlinear elastic laws. Plastic materials, on the other hand, has a final configuration that is different from the initial one at release of the applied force.

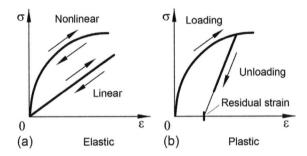

Figure 5.7 Constitutive laws: (a) linear and nonlinear elastic laws; (b) plastic law (stress versus strain plot).

Elastic behavior can be defined in three different ways: (i) the processes in which the original size and shape can be recovered, called *elasticity;* (ii) the processes in which the value of state variables in a given configuration are independent of how it was reached, called elastic; or (iii) a nondissipative process is called an *elastic process.* It can be shown that the Cauchy stress, σ, in an elastic process would depend on the deformation gradient, F; three material unit vectors, D_i; and the state of Cauchy stress in the reference configuration, with σ_R being

$$g\left(\sigma, F, \sigma_R, D_1, D_2, D_3\right) = 0. \tag{23}$$

This is an assumption on how the state variable varies with the motion of the body. Considering definition (ii), it is certain that there is an implicit function that relates to the Cauchy stress and the deformation gradient. Assuming that $\sigma_R = 0$ (the reference configuration is stress free) and that the Cauchy stress is related explicitly to the deformation gradient, g() could be simplified as follows:

$$\sigma = h\left(F, D_1, D_2, D_3\right). \tag{24}$$

Moreover, assuming an isotropic material, the Cauchy stress is

$$\sigma = f\left(F\right). \tag{25}$$

The general relation between the components of the Cauchy stress (σ_{ij}) and deformation gradient (F_{kl}) is linear and takes the following form:

$$\sigma_{ij} = B_{ijkl}\left(F_{kl} - \delta_{kl}\right) = B_{ijkl}F_{kl} - B_{ijkl}\,\delta_{kl}, \tag{26}$$

where B_{ijkl} is the components of a constant fourth-order tensor and δ_{kl} is the Kronecker delta. If Eq. (25) assumes the form

$$\sigma_{ij} = B_{ijkl}\, F_{kl}, \tag{27}$$

the generalization of the Hook principle is described. Whenever the elastic response is not linear, the elastic behavior of the material is not constant; however, deformations remains reversible.

2.4.1 Nonlinear effects

Four sources of nonlinearity could affect the structures: they are the material, geometric, force boundary conditions and displacement boundary conditions. The most relevant for bridge engineers are discussed in the following sections.

Geometric nonlinearity

If nonlinear terms cannot be neglected, or displacements are so large that evident (not infinitesimal) changes in the structure are effected, geometric nonlinearity arises. One of the analytical consequences is the introduction of the geometric stiffness matrix, which takes this effect into consideration. A mathematical explanation could be found in Chen and Lui (1987). However, the most relevant consequences are the following:

- Equilibrium is formulated with respect to the deformed geometry of the structure, and a second-order analysis taking into account second-order effects must be performed: second-order analysis considering the $P - \Delta$ effect (influence of axial forces acting through displacement associated with member chord rotation) and the $P - \delta$ effect (influence of axial forces acting through displacement associated with the member's flexural curvature).
- In the hypothesis of large displacement, the analysis is based on small-strain and small-member deformation, but moderate rotations and large displacement theory (Akkari and Duan, 2000).

An example of geometric nonlinearity behavior of a structure is reported in Figure 5.8.

Figure 5.8 The Olive View Hospital, after the San Fernando, California, earthquake (magnitude 6.7) in 1971.

Material nonlinearity
Steel Figure 5.9, including four phases (elastic, plastic, strain hardening, and soften-ing), represents the stress-strain behavior of structural steel. This is generally described by the following relations:

$$f_s = \begin{cases} E_s \varepsilon_s & 0 \le \varepsilon_s \le \varepsilon_y \\ f_y & \varepsilon_{sy} < \varepsilon_s \le \varepsilon_{sh} \\ f_y + \dfrac{\varepsilon_s - \varepsilon_{sh}}{\varepsilon_{su} - \varepsilon_{sh}}(f_{su} - f_y) & \varepsilon_{sh} < \varepsilon_s \le \varepsilon_{su} \\ f_u \left[1 - \dfrac{\varepsilon_s - \varepsilon_{su}}{\varepsilon_{sb} - \varepsilon_{su}}(f_{su} - f_{sb}) \right] & \varepsilon_{cu} < \varepsilon_s \le \varepsilon_{sb}, \end{cases} \tag{28}$$

where f_s and ε_s is the stress of strain in steel; E_s is the modulus of elasticity of steel; f_y and ε_y is yield stress and strain; ε_{sh} is hardening strain; f_{su} and ε_{su} is maximum stress and corresponding strain; f_{sb} and ε_{sb} are rupture stress and corresponding strain.

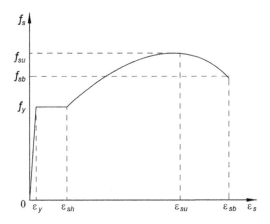

Figure 5.9 Stress-strain graph theoretical curve.

Phenomenological models including nonlinear equations are calibrated on the basis of experimental data. The first model of this type was proposed by Ramberg and Osgood (1943); another noteworthy model came from Menegotto and Pinto (1973). In addition, many other models have been studied in the literature. The Menegotto and Pinto (1973) model is an evolution of the model proposed by Giuffrè and Pinto (1970) and is laid out as follows: for $\varepsilon_s \to 0$, $\sigma_S = E_{S0}\varepsilon_s$; and for $\varepsilon_s \to \infty$, $\sigma_S = E_\infty \varepsilon_s + (E_{S0} - E_\infty)$:

$$\sigma_s = E_\infty \varepsilon_s + \frac{(E_{s0} - E_\infty)\varepsilon_s}{\left[1 + (\varepsilon_s/\varepsilon_0)^R \right]^{1/R}}, \tag{29}$$

where E_{s0} is the initial tangent modulus of the stress-strain curve, E_∞ is the secondary tangent modulus (for large strain), R is the independent parameter that defines the cur-vature; and $\varepsilon_0 = \sigma_0/E_{s0}$ is the strain at the intersection point between the tangent at the origin and the asymptote. This model has some advantages with respect to the implicit Ramberg-Osgood law. The more relevant of these is that each parameter (E_0, E_∞, σ_0,

ε_0, R) in Eq. (28) defines a separate aspect of the curve's geometry, so these can be modified independently (Figure 5.10).

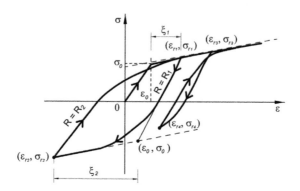

Figure 5.10 The Menegotto and Pinto (1973) model.

Concrete Figure 5.10, including confined and unconfined concrete situations, represent the stress-strain behavior of structural concrete. Analytical models describing the stress-strain model proposed for monotonic loading of confined and unconfined concrete are influenced by the shape of the reinforced concrete (r.c.) section and the transverse reinforcement type and disposition. Phenomenological models including nonlinear equations are calibrated on the basis of experimental data. The more diffused model of this type is that proposed by Mander et al. (1988), based on the assumptions of other studies (mainly William and Warnke, 1975, Schickert and Winkler, 1977, Elwi and Murray, 1979). To determine the confined concrete compressive strength f'$_{cc}$, a constitutive model involving a specified ultimate strength surface for multiaxial compressive stresses is used. The general solution of Mander et al. (1988) is depicted in Figure 5.11, while the model is shown in Figure 5.12, and the effectively confined core for circular and rectangular hoop reinforcement is illustrated in Figure 5.13. According to this model, when the confined core is placed in triaxial compression with equal effective lateral confining stresses f'$_1$ from spirals or circular hoops, the confined compressive strength is

$$f'_{cc} = f'_{c0}\left(-1.254 + 2.254\sqrt{1 + \frac{7.94f'_l}{f'_{c0}}} - 2\frac{f'_l}{f'_{c0}}\right), \tag{30}$$

where f'$_{c0}$ is the unconfined concrete compressive strength; and the effective lateral confining stress on the concrete f'$_1$ is

$$f'_l = 0.5k_e\rho_sf_{yh} \tag{31}$$

Rectangular concrete sections: The rectangular hoops produce two unequal, effective, confining pressures; in the principal x- and y-direction, they are defined as

$$f'_{lx} = k_e\rho_xf_{yh} \tag{32}$$

$$f'_{ly} = k_e \rho_y f_{yh},$$ (33)

where

$$K_e = \frac{\left[1 - \sum_{i=1}^{n} \frac{(w'_i)^2}{6 b_c d_c}\right]\left(1 - \frac{s'}{2 b_c}\right)\left(1 - \frac{s'}{2 d_c}\right)}{(1 - \rho_{cc})}$$ (34)

and

$$\rho_x = \frac{A_{sx}}{s d_c}, \quad \rho_y = \frac{A_{sy}}{s b_c},$$ (35)

While for concrete circular section by circular hoops or spiral

$$f'_l = 0.5 \, k_e \rho_s f_{yh}$$ (36)

$$K_e = \begin{cases} \left(1 - \dfrac{s'}{2 d_s}\right)^2 / (1 - \rho_{cc}) & \text{for circular hoops} \\[2mm] \left(1 - \dfrac{s'}{2 d_s}\right) / (1 - \rho_{cc}) & \text{for circular spirals} \end{cases}$$ (37)

$$\rho_s = \frac{4 A_{sp}}{s d_s}$$ (38)

where K_e is the confinement effectiveness coefficient, f_{yh} is the yield stress of the transverse reinforcement, s' is the clear vertical spacing between transverse reinforcement; s is the center-to-center spacing of the transverse reinforcement; ds is the center-line diameter of the transverse reinforcement; ρ_{cc} is the ratio of the longitudinal reinforcement area to section core area; ρ_s is the ratio of the transverse confining steel volume to the confined concrete core volume; A_{sp} is the bar area of transverse reinforcement; w'_i is the ith clear distance between adjacent longitudinal bars; b_c and d_c are core dimensions to the center lines of the hoop in the x- and y-directions (where

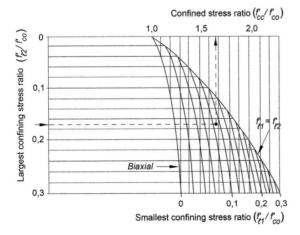

Figure 5.11 Confined strength determination from lateral confining stresses for rectangular sections; Mander et al. (1988).

$b \geq d$), respectively; A_{sx} and A_{sy} are the total area of transverse bars in the x- and y-directions, respectively. A trivial example of the application of this method is given here: Consider a column with an unconfined strength of $f'_{cO} = 30$ MPa and confining stresses [Eqs. (30, 31)], $f'_{lx} = 5,1$ MPa, and $f'_{ly} = 2.7$ MPa, the compressive strength of

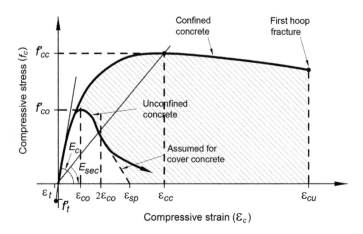

Figure 5.12 Stress-strain model proposed for monotonic loading of confined and unconfined concrete; Mander et al. (1988).

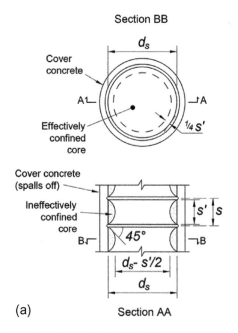

Figure 5.13 Effectively confined core for circular (a)

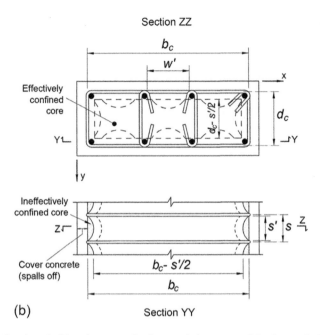

Figure 5.13 Continued. (b) and rectangular hoop reinforcement; Mander et al. (1988).

the confined concrete could be inferred from looking at Figure 5.11, finding $f'_{cc} = 1.65*30 = 49.5$ MPa (dotted line in the figure).

An example of the geometric nonlinearity behavior of a structure is reported in Figure 5.14.

Figure 5.14 The Hanshin Expressway after a magnitude-6.9 earthquake hit Kobe, Japan, in 1995.

3 Structural modeling

3.1 Introduction

Although structural modeling using numerical methods are applied in FEM solutions from many years, we cannot forget the sense and the procedures involved when a FEM procedure is ongoing. The common FEM procedure is well explained by Bathe (2014): a mathematical model, including differential equations describing the geometry, the kinematics, the material law, loadings and boundary conditions, and other elements, describes the investigated physical problem. The FEM solution passes through the identification of finite element types, the mesh density, and the solution parameters. Finally, an assessment of the accuracy of the FEM solution is investigated, and eventually, the mesh is refined, along with some changes to the model.

In the final phase, the results are interpreted, and the analysis could be redone for detailed refinement or for the improvement due to the structure of the final optimization (Figure 5.15). One of the implicit limits of the FEM procedure is that it represents only a simplified description of a physical model, as the most refined mathematical model is not able to reproduce all the information that is present in nature. A further consideration concerns the implicit FEM representation: FEM models are mathematical models and are able to represent physical problems only if they are accurately described. For this reason, a FEM model must be obviously checked for its reliability with an accurate final assessment. One suggested step-by-step solution, which is probably able to simplify engineering doubts during the design process, is the hierarchy of models: in this procedure, a sequence of mathematical models that includes increasingly more complex effects is investigated. For example, a beam structure may be analyzed first with the Bernoulli beam theory, next with the Timoshenko beam theory, then with 2D plane stress theory, and finally with a fully 3D continuum model, including in each case the nonlinear effects (Bathe et al., 1990). Clearly, it should be taken care of the time-consuming procedure that is requested for such a procedure, which may be adopted for rather unknown or complicated structural solutions, or whenever doubts about the interpretation of a complex Fem model are found. Some preliminary conclusions could be drafted as follows:

- The response to be predicted by an FEM analysis is correlated with the mathematical solution adopted.
- The most effective FEM requires a minimal amount of effort, simultaneously providing, with an acceptable margin of error, the most effective answer to the design question.
- Every specific mathematical model is able to illustrate only the effects that it can describe and nothing more (e.g., a beam analysis is not able to predict any further information than that provided by the beam theory).
- Unreliable peak stresses could be found in a FEM model, and it is important to be able to show that these are solely due to the simplifications introduced in the model.

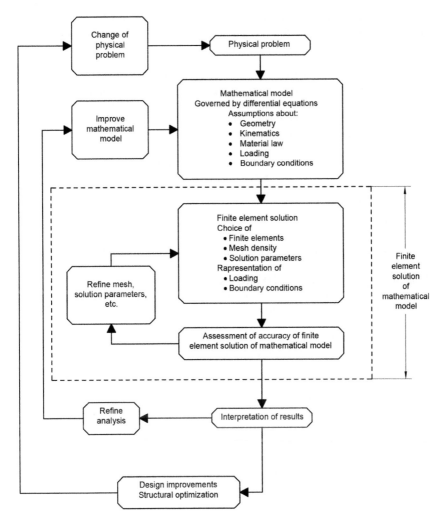

Figure 5.15 The FEM-finite element method process.
(adapted from Bathe, 2014).

3.2 Modeling elements

In order to model a bridge structure, different elements provided in the common FEM software solution are available, including those described in the following sections (HSH, 2007).

3.2.1 1D elements

- *Truss*: A linear member, with a constant section along the axis, subject to compression/tension forces; the degree of freedom (DOF) of a truss is the only axial displacement at the nodes

- *Beam*: Six DOFs at each node (three translations and three rotations) describe the beam element type, which could be subjected to axial and lateral loads and moments; standard beams include axial, bending, and torsional stiffness;
- *Spring/damper*: A longitudinal spring-damper option is a uniaxial tension-compression element with up to three DOFs at each node. Translations in the nodal *x*-, *y*-, and *z*-directions. No bending or torsion is considered. The torsional spring-damper option is a purely rotational element with three degrees of freedom at each node: rotations about the nodal *x*-, *y*-, and *z*-axes. No bending or axial loads are considered; the spring-damper element has no mass, although masses can be added by using the appropriate mass element.
- *Cable*: The cable element is based on the catenary formulation, and it may have a free length that is different from the distance between the end nodes.

3.2.2 2D elements

- *Plane strain:* For modeling very thick structures such as stress analysis through the section of a dam.
- *Plane stress:* For modeling thin 2D sheets subject to in-plane loads.
- *Plate/shell:* For modeling general 3D structures made of relatively thin material. Plate elements in some software may be either thin or thick. Thick plates consider the effects of shear deformation.
- *3D membrane:* For modeling very flexible structures such as draped membranes.
- *Shear panel:* For modeling flat sheets that carry only in-plane shear loads.

3.2.3 3D elements

Tetrahedral: Featuring either 4 or 10 node elements.
Wedge: Featuring either 6 or 15 node elements.
Pyramid: Featuring either 5 or 13 node elements.
Hexahedral: Featuring 8, 16, or 20 node elements

3.2.4 Constraints

- *Rigid link*: Used to rigidly connect nodes. Both translations and rotations may be coupled, selectively. For modeling rigid diaphragms on specific planes (e.g., the floor of a building), automatic tools are available to assign the rigid links, create the master node, and set the required DOFs.
- *Pinned link*: Displays similar behavior as rigid links, except that they couple only the translational DOFs.
- *Master/slave link*: Used to force nodes to share DOFs. For example, the X displacement of one node can be forced to be the same as the X displacement of another node. Master/slave links may be applied in the global Cartesian system or in any user-defined coordinate system (UCS).

A similar identification of finite elements, including their applications, is reported in Table 5.1.

Table 5.1 Identification of FEM Elements (Young and Budynas, 2002)

Element type	Name	Shape	Number of nodes	Applications
Line	Truss		2	Pin-ended bar in tension or compression
	Beam		2	Bending
	Frame		2	Axial, torsional, and bending. With or without load stiffening
Surface	Four-node quadrilateral		4	Plane stress or strain, axisymmetry, shear panel, thin flat plate in bending
	Eight-node quadrilateral		8	Plane stress or strain, thin plate or shell in bending
	Three-node triangular		3	Plane stress or strain, axisymmetry, shear panel, thin flat plate in bending. Prefer quad where possible. Used for transitions of quads
	Six-node triangular		6	Plane stress or strain, axisymmetry, thin plate or shell in bending. Prefer quad where possible. Used for transitions of quads
Surface	Eight-node hexagonal (brick)		8	Solid, thick plate (using midside nodes)
	Six-node pentagonal (wedge)		6	Solid, thick plate (using midside nodes). Used for transition
	Four-node tetrahedron (tet)		4	Solid, thick plate (using midside nodes). Used for transition

Continued

Table 5.1 **Continued**

Element type	Name	Shape	Number of nodes	Applications
Special purpose	Gap		2	Free displacement for prescribed compressive gap
	Hook		2	Free displacement for prescribed compressive gap
	Rigid		Variable	Rigid constraints between nodes

3.3 Modeling methods

Depending on the level of refinement expected, 2D or 3D models are commonly used in FEM models. At first glance, beam models are used, as they provide very useful and fast results, in terms of force and moments (Figure 5.16).

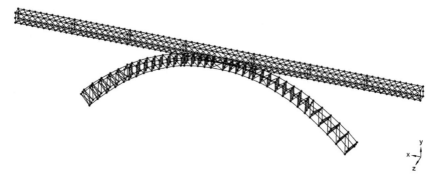

Figure 5.16 Beam model of the Paderno arched railway and road structure (Pipinato, 2010).

The FEM should be built with the aim to anticipate key locations in which principal (axial, shear, torsion, and bending) actions are desired. Concerning refinement, excessively refined meshing could negatively influence the modeling procedure, being a time-wasting action: a proper combination of mesh refining and real component size is the basis of reliable results. The technical judgment at this design stage is a balanced choice, involving time, costs, and accuracy in results, depending on the particular answer that is expected. For instance, while in a static analysis of precast reinforced concrete (p.r.c.) beams, an accurate model of reinforcement is needed, this is not useful in a dynamic analysis, as for this lumped parameter models are enough. Increasing levels of geometric and structural complications requiring detailed FEMs (e.g. composite beam, skewed, and curved bridges) and requiring the local evaluation of stress/strain components increase the complexity of the model itself (Figure 5.17).

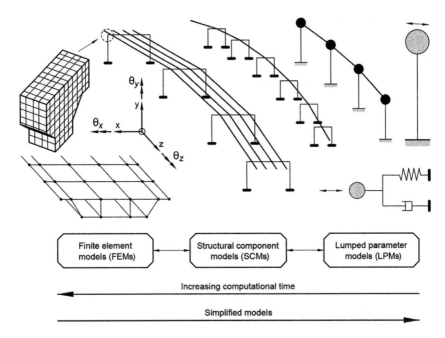

Figure 5.17 FEM modeling methods.

3.4 Materials and cross sections

The material properties are described in the FEM model in order to simulate the aforementioned constitutive laws during the procedure. While most of the structural theories concern homogenous and isotropic materials, and materials as steel are well described by FEM models using the actual section, other materials such as concrete, characterized by a composite nonlinear performance, should be modeled accordingly, particularly when the ultimate behavior is investigated (as this implies the section partialization). For linear constitutive laws, software libraries are commonly used, while for nonlinear models, constitutive laws also could be inferred from structural sampling and testing of specimens. Finally, given the material grade, also theoretical laws could be employed (see the section entitled "Nonlinear Effects," earlier in this chapter).

3.5 Boundaries

The real situation of the boundary condition is expected to be introduced into the model: as far as failing the boundary conditions at this stage, could lead the designer to unexpected fields. To avoid this situation, such boundaries as support conditions, bearings, and expansion joints must be carefully analyzed and designed. Moreover, for simple linear static analysis, boundaries are commonly represented only by fixed/pinned/roller possibilities, avoiding structure/soil interaction, while for dynamic analysis, the structure/soil interaction should be carefully analyzed.

3.6 Modeling strategies

Different strategies are used in order to investigate specific problems of the structure along its service life, including the following:

- *Global models:* These are used for the global static analysis of the structure and for the seismic design.
- *Local models:* Submodels are able to amplify the structural behavior at a higher scale, highlighting specific parameters to be deepened by the use of FEM refinement and FEM hierarchy applications (a sequence of mathematical models that include increasingly more complex effects).
- *Tension and compression models:* To avoid hammering, the tension and compression models are used to capture nonlinear responses for bridges with expansion joints in order to model the nonlinearity of the hinges with cable restrainers. Maximum response quantities from the two models are used for seismic design (Caltrans, 2007). Tension and compression models are able to capture the out-of-phase and in-phase frame movement, respectively.
- *Frame models:* Isolating the structural portion going from an expansion joint to the other, the bridge become always more simple, and gives a sort of upper-bound dynamic behavior; in this situation, seismic characteristics of individual frame responses are controlled by mass of superstructure and stiffness of individual frames. Transverse stand-alone frame models shall assume lumped masses at the columns. Hinge spans shall be modeled as rigid elements with half of their mass lumped at the adjacent column. Effects from the adjacent frames can be obtained by including boundary frames in the model (Caltrans, 2010).
- *Bent models:* A simplified model, including only transverse bent cap together with their columns is used to obtain the maximum solicitation values; in this model, accurate design includes foundation flexibility and a parametrical analysis of different ground motions input versus response.

3.7 Modeling approach

3.7.1 Superstructure

Spine models
In this section, a 3D space frame is modeled in which the superstructure is made of a series of straight-beam elements located along the center line of the superstructure at its center of gravity in the vertical direction. Substructure elements are modeled as beams that are oriented so that their member properties coincide with the 3D orientation of the piers or columns (Figure 5.18).

Grillage models
A 3D space grid of beam elements in which the superstructure is comprised of both longitudinal and transverse beams located at the vertical center of gravity of the superstructure is modeled. Longitudinal members are located at the center of gravity of each

girder line (web and slabs). Transverse beams are intended to model the bridge deck and transverse diaphragms. Substructure elements are also modeled as beams that are oriented so that their member properties coincide with the 3D orientation of the piers or columns (Figure 5.18).

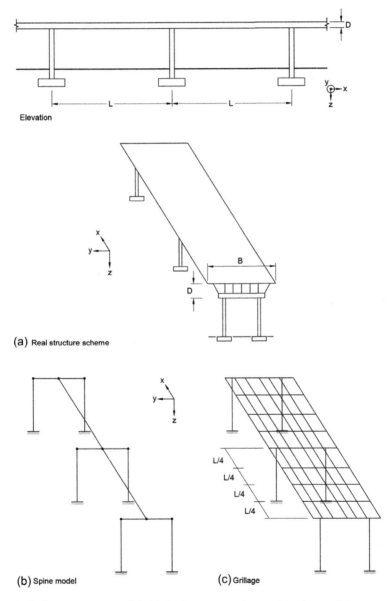

Figure 5.18 Superstructure model: (a) Real structure scheme, (b) spine model, (c) grillage model.

Isotropic and orthotropic plates

Isotropic or orthotropic superstructure differences are defined in Figure 5.19. In this subset, a special attention should be given for the orthotropic deck, which can be solved with the following nonhomogenous differential equation (Huber, 1923):

$$D_x \frac{\partial^4 w}{\partial x^4} + 2H \frac{\partial^4 w}{\partial x^2 \partial y^2} + D_y \frac{\partial^4 w}{\partial y^4} = p\,(x, y), \tag{39}$$

where w is the deflection of the middle surface of the plate at any point (x, y) (Figure 5.20), D_x, D_y, and H are the rigidity coefficients defined by

$$\begin{aligned} D_x &= \frac{E_x\, t^3}{12\left(1 - v_x v_y\right)} \\ D_y &= \frac{E_y\, t^3}{12\left(1 - v_x v_y\right)} \end{aligned} \tag{40}$$

$$2H = 4C + v_y D_x + v_x D_y \tag{41}$$

and p (x,y) is the loading intensity at any point as a function of the coordinates x and y. Consequently, solving equations have been inferred (Girkman, 1959). When modeling an orthotropic deck, a rough model could be built up with a plate element, considering different bending stiffness into the two principal directions. An advanced model, to consider local effects or for the choice of the specific type of rib or for a refined analysis of transverse to rib connection and so on, should be made of plate elements representing the local portion of the substructure.

Bent model

In the hypothesis in which the bridge superstructure could be considered a rigid body under seismic loads, the bent models could be used. Discretizations are reported in Figure 5.21.

Thermal expansion joints

Thermal expansion joints allow movements in long superstructures: these should be modeled as hinges-6 DOF, free to rotate in the longitudinal direction and pin in the transverse direction to represent shear.

3.7.2 Substructure

Substructure modeling should adhere to the following rules of thumb (Figures 5.22A, 22B, 22C):

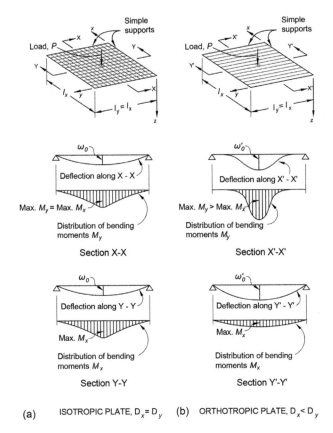

Figure 5.19 Comparison of deflections and bending moment in a square isotropic and a square orthotropic plate.

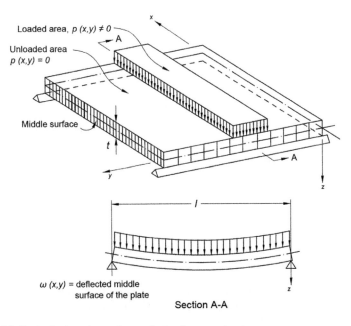

Figure 5.20 Basic designations of an orthotropic superstructure.

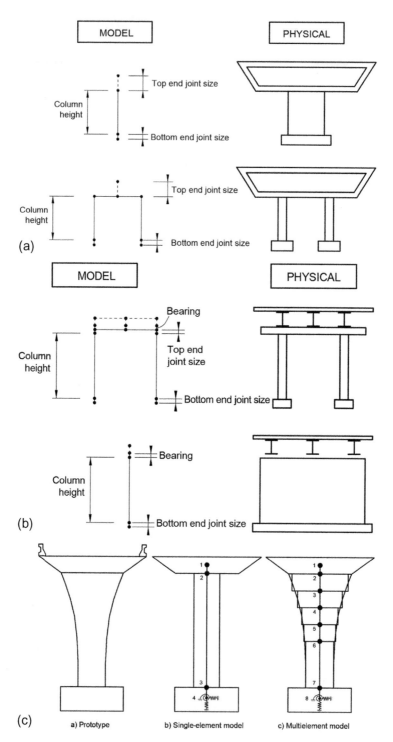

Figure 5.21 Bent models: (a) model discretization for monolithic connection; (b) bearing supported connections for precast concrete girders or steel superstructures on drop cap; (c) single-column bent model.

- *Spring modeling of the foundation node:* The fixed-base connection is generally used in the single-column scheme, while for the multicolumn bents, a pin-base scheme is adopted.
- *Column-bent cap model:* A simplified model including only transverse bent cap together with their columns is used to obtain the maximum level of moment and shear; the design should undergone to the aforementioned rules.

An equivalent fixity model (Figure 5.22D) models the pile shaft (Figure 5.22E) for nonseismic loading. For seismic loading, a soil-spring model (Figure 5.22F) should be considered to capture the soil-structure interaction.

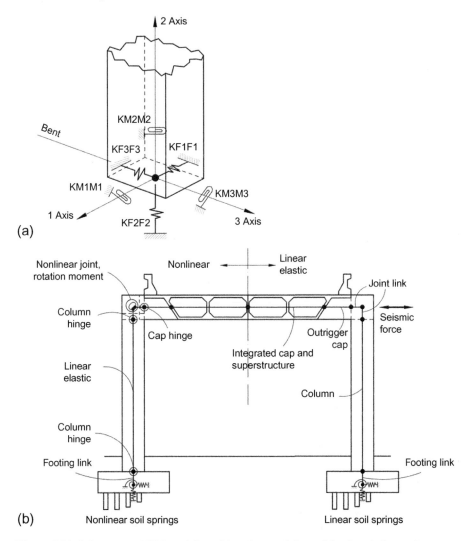

Figure 5.22 Substructure FEM modeling: (a) spring modeling of the foundation node; (b) column-bent cap model;

(Continued)

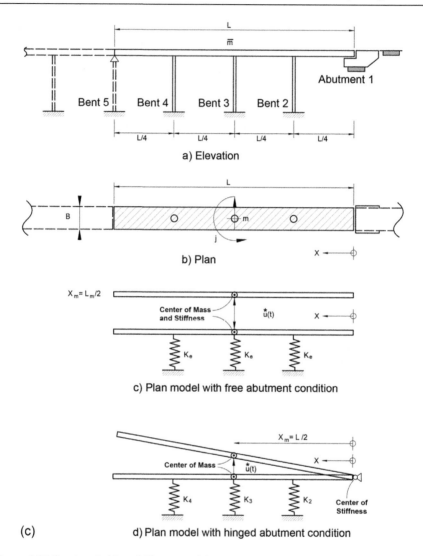

a) Elevation

b) Plan

c) Plan model with free abutment condition

(c)   d) Plan model with hinged abutment condition

Figure 5.22 Continued. (c) multiframe model;

3.8 Modeling by bridge type

3.8.1 R.c. bridges

Slab bridges represent the simplest solution: in this case, an equivalent beam or a grid of equivalent beams model is suggested by the use of equivalent stiffness; otherwise, an adequate number of quadrilateral isotropic plate elements should be modeled, representing the continuous bridge slab, all lying in a plane and connected at a finite number of nodes. In this case, in-plane distortions are not considered, and an accurate analysis should be undergone if the particular details of the bridge require them. In

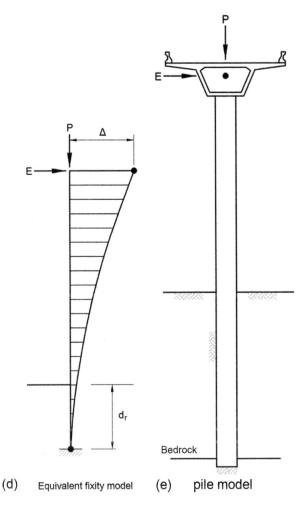

(d)   Equivalent fixity model   (e)   pile model

Figure 5.22 Continued. (d) equivalent fixity model; (e) pile model;

this case, some basic recommendations are (i) regular-shaped plates to be designed with quadrilateral elements, not exceeding 2:1 proportion; if triangular elements are adopted, regular equilateral shapes should be preferred; (ii) avoid discontinuities and irregular subdivision of elements; (iii) bearing locations and pier locations are identified with specific nodes.

In a retrofitting analysis, the fiber-reinforced polymer (FRP) retrofit of r.c. beams or slabs could be modeled with solid elements, with eight nodes and three DOFs at each node, including information on creep, plastic deformation, crushing, and cracking. The special case of curved concrete bridges has three modeling solutions: the spine model, the grillage model, and the 3D FEM model. Depending on the complexity of the structure, it is suggested that the most time-consuming modeling approach be adopted only at the final design stage.

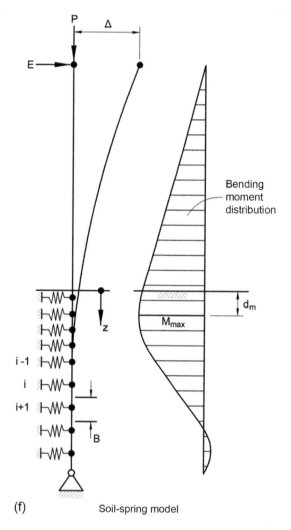

Figure 5.22 Continued. (f) soil-spring model.

3.8.2 Prestressed/post-tensioned concrete bridges

Basic FEM solutions not including prestressing or post-tensioned elements, could be used by introducing applied loading instead of dedicated resisting elements. When prestressing or post-tensioned elements are included in the FEM package, the elements should account for tendon type, immediate loss, elastic shortening, long-term losses, and changes in stress for bending. Different analyses are available, depending on the FEM solution adopted, such as beam type, tendon type, plane stress, and solid type. The beam analysis is the most diffused, including the current beam section and truss elements representing the tendons, including pretensioning forces.

3.8.3 Steel girder bridges

Beam models are sufficient to satisfy the steel girder bridge design at the approximate analysis stage, or to check complex models results with local approximate analysis where the wide amount of elements and the information modeled could give the wrong results. Another approximate way to do this is line-girder analysis: in this case, load distribution factors are used to isolate a single girder from the rest of the superstructure system, evaluating that girder individually. The load distribution factors are determined by approximate formulas for both straight bridges and curved bridges, as reported, for example, by AASHTO (2015), Kim et al. (2007), and Zhang et al. (2005). FEM solutions include:

- *2D grid analysis method*: In this method, the structure is divided into plane grid elements with 3 DOFs at each node (vertical displacement, rotation angles about the longitudinal and transverse axes, or the first derivative of the rotational angle about the longitudinal axis, or both). The element choice, node spacing, and other modeling parameters are often set following simplified guidelines (e.g., AASHTO, 2015).
- *Plate and eccentric beam analysis methods*: The deck is modeled using plate or shell elements, while the girders and cross frames are modeled using beam elements offset from the plate elements to represent the offset of the neutral axis of the girder or cross-frame from the neutral axis of the deck. This approach is covered by AASHTO (2015). The offset length is typically equal to the distance between the centroids of the girder and deck sections. This method is somewhat more refined than the traditional 2D grid method. For this modeling approach, beam element internal forces obtained from this method need to be eccentrically transformed to obtain the composite girder internal forces (bending moment and shear) used in the bridge design.
- *Grid analysis method*: An enhancement of the aforementioned solution, including modeling of cross frames or diaphragms with consideration of shear deformation in addition to flexural deformation, modeling of the warping stiffness of open cross section shapes (such as I-shaped girders), modeling of girder supports, lateral bracing, cross-frames or diaphragms at their physical elevation within the structure.
- *3D FEM analysis methods*: This method includes a computerized structural analysis model where the superstructure is modeled in three dimensions, including girder flanges using line/beam elements or plate/shell/solid type elements; modeling of girder webs using plate/shell/solid type elements; modeling of cross frames or diaphragms using line/beam, truss, or plate/shell/solid type elements (as appropriate); and modeling of the deck using plate/shell/solid elements. This method is dedicated to three main categories of design issues: complicated situations such as severe curvature, skew, or both; unusual framing plans, unusual support/substructure conditions, or other complicating features (AASHTO, NSBA 2011); refined analysis of structure submodels. However, a further complication arises with this solution type, as solicitation parameters are not directly calculated. Instead, the model reports stresses in flanges, webs, and deck elements. If the designer wishes to consider girder solicitations, some type of conversion/integration of the stresses over the depth of the girder cross section will be required. The procedure is long and time consuming, with an increase in the potential for error.

3.8.4 Truss bridges

For truss bridges, an initial 2D model should be sufficient, incorporating the 2D information of the planar truss. In this case, only one side truss is modeled and the vertical loads are applied directly to that. In the final design stage, a 3D model is required, including the two trusses, and the deck and all the structural components are modeled. In this last case, an assemblage of beams, carefully considering the mutual connections, represents the deck: the designer should avoid stress concentrations in connection points, as well-known fatigue prone details are present in truss deck connections (Pipinato, 2012).

3.8.5   Arch bridges

In the case of arch bridges, dedicated FEM solutions are available in the market, including multiphase construction and hangers/cables analysis. The analysis of steel or steel/concrete arch bridges includes the following elements:

- *Arch structure*: For the arch structure, 3D models are necessary, introducing a correct amount of straight-beam elements reproducing the cross-frame arched geometry. The global stability analysis in the two principal planes (arch plane and horizontal plane) is required for the common high compression value in the arch itself. The influence of initial stress in arches should be carefully considered (e.g., an initial tension level in hangers helps to maintain arches in position in deck-through arches). Fatigue analysis of cables and hangers is required. A parametric analysis of cable spacing and geometry, including different arch geometry solutions, could help in reducing weight and optimizing the correct shape of the structure.
- *Deck structure*: Stringers and floor beams, together with the deck permanent loads (e.g. special equipment), should be modeled in a 3D FEM solution to account for the correct stiffness of each component.
- *Hangers*: Hangers could be initially modeled as truss elements, even though special elements are introduced in dedicated FEM solutions.

For masonry arches, general FEM software is used to perform this very simple modeling.

3.8.6   Cable-stayed bridges

For cable-stayed bridges, 3D modeling is recommended, including the following elements:

- *Main girder*: Steel or concrete boxes or composite I-girders are basic solutions; in the first case, the girder could be modeled as a beam at the centroid of its cross section with a longitudinal development, linking the beam to a cable's anchor point by rigid link; if transverse-stiffened beams are used instead of rigid links, bending and shear stiffness along the length of the bridge should be carefully calculated. If composite I-girders are used, the girder can be modeled as a grid of beams.
- *Pylons*: 3D beam elements are usually used to model pylons, changing the cross section shape/geometry/direction according to the vertical development.
- *Cables*: Truss elements are commonly used, except in those cases where sag effect should be accounted for; in this case, appropriate element should be introduced in the model (namely cable element), considering the equivalent Young's modulus;
- *Pylon/girder connection*: According to the specific connection (full separation, rigid connection, vertical support, etc.) the connection is introduced into the 3D model; if a damping system is adopted in the bridge, accurate calibration of this element should be provided, checking for the allowable displacement versus movement.

3.8.7   Suspension bridges

For suspension bridges, 3D modeling is also recommended, including the following items:

- *Main cables and saddles*: If truss elements are used, cables are meshed at the hanger locations; otherwise, specific catenary cable elements are included in specific FEM solutions. Concerning saddles, the modeling and analysis should account for their role in the various erection and exercise phases (moveable/fixed, according to the type of bridge), and considering its relevant role in the whole bridge structure, a 3D submodel should be considered.
- *Hangers*: Truss elements could be used except for the principal rigid connection between the girder and the main cables (at midspan), where beam elements should be preferred, together with a specific submodel of the structure, including solid elements, to account for hot spot stress checks.
- Pylons and Girders: the same procedure described for cable-stayed bridges should be adopted.

4 Research and development

In view of the future development of these relevant structural engineering instruments as FEM software, some improvements are required, as suggested in the following items:

- *Full integration of the modeling procedures:* Output of the analysis should be usable/readable for an extended number of engineers involved in the bridge design and production, facilitating understanding of the model; the close native format of every different software is a commercial understandable choice, however this is a clear limit to the public which cannot easily manage and handle the structural design. To solve this issue, a possibility stands on the introduction of dedicated free software interface, in order to enable everyone to read the output of the Fem.
- *Full integration of FEM and BIM:* Nowadays, construction solutions integrate a wide variety of information on element types, material, geometry, and construction issues. All this information are included in bridge design output that is generated apart from the FEM modeling, and another time, apart from building information modeling (BIM) models; some software solutions are now trying to close the gap among these three phases. However, a great amount of work should be done in this specific field, in order to give bridge engineers a unique software solution integrating all steps of design production.
- *Coded output:* Not all codes and standards are very clear related to the postprocessed FEM structural output content that should be produced for the authority, for the official deposit. However, it seems that software houses haven't very clear that it is not possible to produce code tabulations that are almost impossible to be understood. A coded tabulation should have to be produced in a clear format, leaving to the designer the possibility of getting further specific calculation output (or not), integrating a clear description of the adopted code and standard procedure. This is a relevant issue related to FEM postprocessing and is also very useful for the repeatability of the structural assumptions and overdone calculations.

References

Akkari, M., Duan, L., 2000. Nonlinear analysis of bridge structures. In: Chen, W.-F., Duan, L. (Eds.), Bridge Engineering Handbook. Boca Raton Press, Boca Raton, FL.

American Association of State Highway and Transportation Officials (AASHTO), 2015. AASHTO LRFD Bridge Design Specifications, seventh ed., with 2015 interim revisions American Association of State Highway and Transportation Officials, Washington, DC.

American Association of State Highway and Transportation Officials–National Steel Bridge Alliance (AASHTO, NSBA), 2011. Guidelines for Steel Girder Bridge Analysis. AASHTO/NSBA Steel Bridge Collaboration Task Group—Analysis of Steel Girder

Bridges. American Association of State Highway and Transportation Officials–National Steel Bridge Alliance, Washington, DC.

Ansel, C., Saul, F.K., 2011. Advanced Mechanics of Materials and Applied Elasticity: Analysis of Stress. Prentice Hall, Pearson PLC, London.

Bathe, K.J., 2014. Finite Element Procedures. Monograph, Prentice Hall, Upper Saddle River, NJ.

Bathe, K.J., Lee, N.S., Bucalem, M.L., 1990. On the use of hierarchical models in engineering analysis. Comp. Meth. Appl. Mech. Eng. 82, 5–36.

Caltrans, 2007. Memo to Designers 20-4—Earthquake Retrofit for Bridges. California Dept. of Transportation, Sacramento, CA.

Caltrans, 2010. Seismic Design Criteria, Version 1.6. California Dept. of Transportation, Sacramento, CA.

Caltrans, 2015. Bridge design practice. California Dept. of Transportation, Sacramento, CA.

Chen, W.F., Lui, E.M., 1987. Structural Stability: Theory and Implementation. Elsevier, New York.

Elwi, A.A., Murray, D.W., 1979. A 3D hypoelastic concrete costitutive relationship. J. Eng. Mech. Div. ASCE 105 (4), 623–641.

Girkman, K., 1959. Flachentragwerke, fifth ed. Springer Verlak, Vienna.

Giuffrè, A., Pinto, P.E., 1970. Il comportamento del cemento armato per sollecitazioni cicliche di forte intensità. Giornale del Genio Civile 5, 391–408, Maggio.

HSH, 2007. Strand 7 User Manual. Strand7 Pty Ltd.

Huber, M.T., 1923. Die Theorie der kreuzweise bewehrten Eisenbetonplatten. Der Bauingenieur 4 (1923), 354 392.

Kim, W.S., Laman, J.A., Linzell, D.G., 2007. Live load radial moment distribution for horizontally curved bridges. ASCE J. Bridge Eng. 12 (6), 727–736.

Mander, J.B., Priestley, M.J.N., Park, R., 1988. Theoretical stress-strain model for confined concrete. J. Struct. Eng. 114 (8), 1804.

Menegotto, M., Pinto, P.E., 1973. Method of analysis for cyclically loaded reinforced concrete plane frames including changes in geometry and non-elastic behavior of elements under combined normal force and bending. In: Proc., IABSE Symp. Resistance and Ultimate Deformability of Structures Acted on by Well-Defined Repeated Loads, Intl. Assoc, Bridge Struct. Eng., Libson, Portugal, vol. 13. pp. 15–22.

Pipinato, A., 2010. Structural analysis and fatigue reliability assessment of the Paderno bridge. Practice Period. Struct. Des. Construct. 15 (2), 109–124. http://dx.doi.org/10.1061/(ASCE) SC.1943-5576.0000037.

Pipinato, A., 2012. Assessment procedure and rehabilitation criteria for the riveted railway Adige Bridge. Struct. Infrastruct. Eng. 8 (8), 747–764. http://dx.doi.org/10.1080/15732479. 2010.481674.

Ramberg, W., Osgood, W.R., 1943. Description of Stress-Strain Curves by Three Parameters. Technical Note No. 902. National Advisory Committee For Aeronautics, Washington DC.

Ritz, W., 1909. Uber eine neue Methode zur Losung gewissen Variations Probleme der mathematischen, Physik. J. Reine angew. Math. 135, 1–61.

Schickert, G., Winkler, H., 1977. Results of test concerning strength and strain of concrete subjected to multiaxial compressive stress. Deutcher Ausschuss fur Stahlbeton, Heft 277. Berlin.

Willam, K.J., Warnke, E.P., 1975. Constitutive models for the triaxial behavior of concrete. Proc. Int. Assoc. Bridge Struct. Eng. 19, 1–30.

Young, W.C., Budynas, R.G., 2002. Roark's Formulas for Stress and Strain, seventh ed. McGraw-Hill, New York.

Zhang, H.L., Huang, D.H., Wang, T.L., 2005. Lateral load distribution in curved steel I-girder bridges. J. Bridge Eng. 10 (3), 281–290.

Zienkiewicz, O.C., Taylor, R.I., 2000. The Finite Element Method, fifth ed. Butterworth Heynemann, New York.

Dynamics of bridge structures

6

Chouw N.
Department of Civil and Environmental Engineering, University
of Auckland, Auckland, New Zealand

1 Linear idealization of bridge structures

Bridge structures can be idealized as either a single degree-of-freedom (SDOF) system, a multi-DOF system, or an infinite-DOF system. The equation of motion of each of these models describes the relationship between the loading and the response of the system. As anticipated, the more information that must be obtained, the greater the number of DOFs needed. For an initial and approximate estimate of bridge response, an SDOF system is sufficient.

1.1 SDOF system

Figure 6.1 shows a bridge with two segments under a dynamic loading $P(t)$. As an example, the response of the left segment is considered. For simplicity, it is assumed that the bridge girder is much stiffer than the pier bending stiffness. Hence, a rigid girder is assumed, and the response of the bridge structure to a horizontal dynamic loading $P(t)$ can be described by considering only one DOF, i.e., the horizontal girder response $u(t)$.

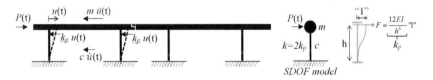

Figure 6.1 A two-segment bridge structure with an assumed fixed base.

The stiffness of the structure of the DOF considered is the force required to cause a unit displacement in the DOF. The rigid girder can be represented by a fixed-base boundary condition (see Figure 6.1, on the right, showing a top boundary condition). The force required to produce unit displacement at the top of one pier is equal to the pier stiffness $k_p = 12EI/h^3$, where EI and h are the bending stiffness and the height of the pier, respectively. The mass m of the SDOF system can be assumed to be the total girder mass and the mass of the top half of the piers. When the bridge girder vibrates to the left, the inertia force $F_I = m\,\ddot{u}(t)$ is initiated and acts in the opposite direction, and each bridge pier will resist the deformation. The resisting force is $F_S = k\,u(t)$. The energy loss during the vibrations (e.g., due to friction at the pier-girder connections) can be described in terms of viscous damping; i.e., the velocity-dependent damping force $F_D = c\dot{u}(t)$, where c is the damping coefficient.

The equation that governs the response of the bridge structure to the load $P(t)$ can be derived from the equilibrium of all forces:

$$m\ddot{u}(t) + c\dot{u}(t) + ku(t) = P(t). \tag{1}$$

In the case of a ground excitation $\ddot{u}_g(t)$, the inertia force F_I is determined by the absolute acceleration $u^t(t)$ (Figure 6.2), while the pier restoring force $F_S(t)$ and the damping force $F_D(t)$ depend on the relative response.

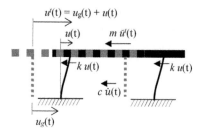

Figure 6.2 SDOF bridge system under an earthquake loading.

Thus, the governing equation is

$$m\ddot{u}^t(t) + c\dot{u}(t) + ku(t) = 0 \ \ or \ \ m\ddot{u}(t) + c\dot{u}(t) + ku(t) = -m\ddot{u}_g(t). \tag{2}$$

1.2 MDOF system

In the previous section, the girder is assumed to be rigid. Consequently, the system can be described by the SDOF $u_1(t)$. Should the girder flexibility be considered, however, depending on the bending stiffness ratio of girder to pier, the rotational DOF $u_2(t)$ and $u_3(t)$ at the pier-girder connections can be influential. The bridge segment is then described by three DOFs (Figure 6.3).

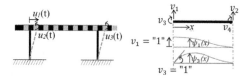

Figure 6.3 MDOF bridge model and member deflection due to unit left nodal deformation.

The stiffness coefficient of each bridge member (i.e., the nodal forces due to unit nodal displacement or rotation) can be determined by a number of approaches. One of these is the principle of virtual deformation, where the following shape functions are considered:

$$\psi_1(x) = 1 - 3\left(\frac{x}{L}\right)^2 + 2\left(\frac{x}{L}\right)^3,$$

$$\psi_2(x) = 3\left(\frac{x}{L}\right)^2 - 2\left(\frac{x}{L}\right)^3,$$

$$\psi_3(x) = x\left(1 - \frac{x}{L}\right)^2 \quad \text{and} \tag{3}$$

$$\psi_4(x) = \frac{x^2}{L}\left(\frac{x}{L} - 1\right).$$

The deflected shape $v(x) = \psi_1(x)v_1 + \psi_2(x)v_2 + \psi_3(x)v_3 + \psi_4(x)v_4$ of the structural member can now be expressed in terms of its nodal displacements v_1 and v_2 and the nodal rotations v_3 and v_4.

With the stiffness coefficient $k_{ij} = \int_0^L EI(x)\psi_i''(x)\psi_j''(x)dx$ (i.e., the force developed at the ith DOF due to a unit deformation at the jth DOF), the stiffness k_e of the structural member of length L with a bending stiffness EI relates the deformation v of each member DOF to the corresponding elastic nodal force f_s:

$$\underbrace{\frac{2EI}{L^3}\begin{bmatrix} 6 & -6 & 3L & 3L \\ -6 & 6 & -3L & -3L \\ 3L & -3L & 2L^2 & L^2 \\ 3L & -3L & L^2 & 2L^2 \end{bmatrix}}_{k_e}\begin{bmatrix} v_1 \\ v_2 \\ v_3 \\ v_4 \end{bmatrix} = \begin{bmatrix} f_{s1} \\ f_{s2} \\ f_{s3} \\ f_{s4} \end{bmatrix}. \tag{4}$$

By transforming the local member coordinate to the global DOF, the system stiffness K of the whole system can be obtained using the direct stiffness approach (see, e.g., Martin, 1966).

$$K = \sum_{e=1}^{n} k_e, \tag{5}$$

where n is the number of structural members.

With regard to the mass matrix of the bridge segment, a lumped-mass model can be assumed; i.e., each structural member will be divided into two segments of $L/2$, with masses concentrated at the segment ends. The mass matrix of the system is, then, a diagonal matrix.

A more realistic model of the inertia forces can be achieved by assuming that the inertia forces initiated along the structural members by unit nodal lateral or angular accelerations are proportional to the corresponding shape function Ψ_i. The mass coefficient of the so-called consistent mass model is $m_{ij} = \int_0^L m(x)\psi_i(x)\psi_j(x)dx$; i.e., the force developed at the ith DOF due to a unit acceleration at the jth DOF. Hence, the mass matrix M_e of the structural member of the length of L with mass per unit length $\bar{m}$ relates the acceleration at each DOF to the corresponding inertial force f_I:

$$\frac{\bar{m}L}{420} \underbrace{\begin{bmatrix} 156 & 54 & 22L & -13L \\ 54 & 156 & 13L & -22L \\ 22L & 13L & 4L^2 & -3L^2 \\ -13L & -22L & -3L^2 & 4L^2 \end{bmatrix}}_{M_e} \begin{bmatrix} \ddot{v}_1 \\ \ddot{v}_2 \\ \ddot{v}_3 \\ \ddot{v}_4 \end{bmatrix} = \begin{bmatrix} f_{I1} \\ f_{I2} \\ f_{I3} \\ f_{I4} \end{bmatrix} \tag{6}$$

and the system mass matrix M can be obtained using the direct stiffness approach (see, e.g., Martin, 1966).

$$M = \sum_{e=1}^{n} M_e \tag{7}$$

where n is the number of structural members.

In order to have the same matrix properties, the Rayleigh damping $C = \lambda M + \mu K$ is often used; i.e., the damping matrix is stiffness and mass matrices are proportional (Clough and Penzien, 1993). By introducing the orthogonality property of the natural modes ϕ_r with the corresponding frequency ω_r (i.e., $\phi_j^T M \phi_i = \phi_j^T K \phi_i = \phi_j^T C \phi_i = 0$), λ and μ can be calculated from two selected damping ratios $\xi_r = \frac{1}{2}\left(\frac{\lambda}{\omega_r} + \mu \omega_r\right)$.

Thus, the governing equation is

$$M\ddot{u}(t) + C\dot{u}(t) + Ku(t) = P(t), \tag{8}$$

where the bold uppercase letters M, C and K denote the mass, damping, and stiffness matrices of the bridge structure, respectively; and the bold lower and uppercase letters u, $\dot{u}$, $\ddot{u}$, and P denote the displacement, velocity, acceleration response, and load vectors, respectively.

1.3 IDOF system

Unit deformations described by the shape functions of Eq. (3) fulfill the differential equation of a flexural beam exactly. For calculating the structural response to a static loading, the stiffness matrix [Eq. (4)] is exact. For analyzing the structural response to dynamic loading, the stiffness and mass matrices of Eq. (5) and Eq. (7), resulting from the member stiffness and mass matrices [Eqs. (4) and (6)], can only describe the dynamic properties of the structure approximately.

The vibration of a structural member with uniformly distributed mass $\bar{m}$, bending stiffness EI, and viscous damping c_t is governed by a partial differential equation (see also Figure 6.4 and Chouw, 1994).

$$\bar{m}\ddot{v} = -EI\, v_{,xxxx} - c_t \dot{v} \tag{9}$$

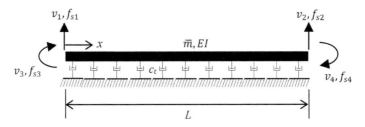

Figure 6.4 Structural member with continuously distributed mass $\bar{m}$, stiffness EI, and damping c_t.

where $(\dot{\;}) = \frac{d}{dt}(\;)$ and $(\;)_{,x} = \frac{d}{dx}(\;)$. Eq. (9) can be transformed into the Fourier or Laplace domain, and an analytically solvable normal differential equation can be obtained. The solution of this equation can be expressed in terms of nodal displacement and rotation $\tilde{v}$, and corresponding force and moment $\tilde{f}_s$. In the following expression, the tilde indicates the variable in the Laplace domain.

The Laplace transformation of Eq. (9) leads to

$$4f^4\tilde{v} = -\tilde{v}_{,xxxx},\tag{10}$$

where $f^4 = \frac{\bar{m}s^2 + c_t s}{4EI}$, and the Laplace parameter s is $\delta + i\,\omega$ and $i = \sqrt{-1}$; $\delta = \frac{8}{T}$; T is the time window considered.

The solution in the Laplace domain is $v = (A\cos f\,x + B\sin f\,x)$ $e^{fx} + (C\cos f\,x + D\sin f\,x)e^{-fx}$, with the four constant values A, B, C, and D being obtained from four boundary conditions involving both ends of the structural member. By relating the nodal deformation to the nodal forces $\tilde{f}_{s1} = -EI\tilde{v}_{,xxx}(x=0,s)$ and $\tilde{f}_{s2} = EI\tilde{v}_{,xxx}(x=L,s)$ and moments $\tilde{f}_{s3} = -EI\tilde{v}_{,xx}(x=0,s)$ and $\tilde{f}_{s4} = EI\tilde{v}_{,xx}(x=L,s)$, the exact member stiffness $\tilde{k}_e$ can be obtained. It is worth mentioning that in the stiffness coefficient, the mass and stiffness component can no longer be separated. Hence, the stiffness $\tilde{k}_e$ is called the *dynamic stiffness* of the structural member:

$$\underbrace{\begin{bmatrix} \tilde{k}_{11} & \tilde{k}_{12} & \tilde{k}_{13} & \tilde{k}_{14} \\ \tilde{k}_{21} & \tilde{k}_{22} & \tilde{k}_{23} & \tilde{k}_{24} \\ \tilde{k}_{31} & \tilde{k}_{32} & \tilde{k}_{33} & \tilde{k}_{34} \\ \tilde{k}_{41} & \tilde{k}_{42} & \tilde{k}_{43} & \tilde{k}_{44} \end{bmatrix}}_{\tilde{k}_e} \begin{bmatrix} \tilde{v}_1 \\ \tilde{v}_2 \\ \tilde{v}_3 \\ \tilde{v}_4 \end{bmatrix} = \begin{bmatrix} \tilde{f}_{s1} \\ \tilde{f}_{s2} \\ \tilde{f}_{s3} \\ \tilde{f}_{s4} \end{bmatrix},\tag{11}$$

where

$$\tilde{k}_{11} = \tilde{k}_{33} = 2\gamma f^2\,(e^{4\alpha} + 2e^{2\alpha}\sin 2\alpha - 1);$$
$$\tilde{k}_{12} = \tilde{k}_{21} = -\tilde{k}_{34} = -\tilde{k}_{43} = \gamma f\,(-e^{4\alpha} + 2e^{2\alpha}\cos 2\alpha - 1);$$

$$\tilde{k}_{13} = \tilde{k}_{31} = 4\gamma f^2 \left\{-e^{3\alpha}(\sin\alpha + \cos\alpha) + e^{\alpha}(-\sin\alpha + \cos\alpha)\right\};$$

$$\tilde{k}_{14} = \tilde{k}_{41} = -\tilde{k}_{23} = -\tilde{k}_{32} = 4\gamma f \left(\sin\alpha(e^{\alpha} - e^{3\alpha})\right);$$

$$\tilde{k}_{22} = \tilde{k}_{44} = \gamma \left(e^{4\alpha} - 2e^{2\alpha}\sin 2\alpha - 1\right);$$

$$\tilde{k}_{24} = \tilde{k}_{42} = 2\gamma \left\{e^{3\alpha}(\sin\alpha - \cos\alpha) + e^{\alpha}(\sin\alpha + \cos\alpha)\right\};$$

$$\alpha = \sqrt{\frac{(ms + c_t)sL^4}{4EI}}; \quad \gamma = \frac{2EIf_t}{e^{4\alpha} - 2e^{\alpha}(1 + 2\sin^2\alpha) + 1}.$$

The dynamic stiffness $\tilde{K}$ of the bridge structure can be obtained using the same direct stiffness approach [Eq. (5)] in the Laplace domain. The equation of motion of the bridge structure with infinite DOF in the Laplace domain is an algebraic equation:

$$\tilde{K}\tilde{u} = \tilde{P}. \tag{12}$$

2 Bridge response to dynamic loading

The equation of motion relates loading to structural response. For a given loading, the responses $u(t)$, $\dot{u}(t)$, and $\ddot{u}(t)$ can obtained from the equation of motion. Depending on loading types and the number of DOFs, the equation can be solved analytically or numerically.

2.1 SDOF system

2.1.1 Harmonic loading

In the case of a harmonic loading [e.g., $P(t) = P_e \sin(\omega_e t)$], the excitation is characterized only by its amplitude $P_e = kA_e$ and frequency $f_e = \omega_e/2\pi$. Because of structural damping c, the free vibration determined solely by the dynamic properties of the structure will die away with the passage of time. The remaining steady-state response is determined by the magnitude and frequency of the load (i.e., A_e, f_e) and the natural frequency and damping ratio of the structure (i.e., f_s and ξ). The damping ratio ξ is c/c_{crit}, where c and $c_{\text{crit}} = 2\sqrt{mk}$ are the actual structural damping and the critical damping that results in monotonic decay of vibration with time, respectively.

The particular solution for Eq. (1) can be assumed as $u(t) = C\sin\omega_e t + D\cos\omega_e t$, and by substituting this solution and its first and second time derivatives into Eq. (1), while equating the sine and cosine terms of both sides of the equation, the coefficients C and D can be obtained. The steady-state response is, then,

$$u(t) = \sqrt{C^2 + D^2} \sin(\omega_e t - \phi), \tag{13}$$

where $\phi = tan^{-1}\left(\frac{-D}{C}\right) = tan^{-1}\left(\frac{2\xi\beta}{1-\beta^2}\right)$, $\quad C = A_{st}\dfrac{1-\beta^2}{\left(1-\beta^2\right)^2 + (2\xi\beta)^2}$, $\quad D = A_{st}\dfrac{-2\xi\beta}{\left(1-\beta^2\right)^2 + (2\xi\beta)^2}$,

$A_{st} = \frac{P_e}{k}$, and $\beta = \frac{f_e}{f_s}$

The amplitude A_{dyn} of the dynamic response is $\sqrt{C^2 + D^2}$. The maximum response ratio $A_{\mathrm{dyn}}/A_{\mathrm{st}}$ shows an amplification or a reduction of the dynamic response relative to the maximum static response A_{st}. In Figure 6.5, the dynamic response amplitude A_{dyn} of a number of SDOF structures due to the same harmonic excitation is displayed relative to the constant static response amplitude A_{st}, as a function of frequency ratio β and damping ratio ξ. A plot of the maximum dynamic response of the SDOF structures to the same dynamic loading is called the *response spectrum* of the harmonic loading (even though it is not a real spectrum), because the maximum response of one structure is independent of the response of another structure, while the spectrum values of a real spectrum (e.g., the Fourier spectrum) are linked. The display of the response is only spectrumlike because the amplitude is shown as a function of the natural undamped frequency f_s of the structures. The spectrumlike display is still useful, though, since it reveals the consequence of the structural frequency f_s relative to the excitation frequency f_e, and the actual damping c relative to the critical damping c_{crit}.

If β is very small, $A_{dyn} \approx A_{st}$; i.e., the structure responds to the dynamic loading in the same way that it responds to a static load. This is the case when the structure is very stiff; i.e., f_s is very high relative to the excitation frequency f_e. The excitation is so slow relative to the speed of the natural vibration of the structure that the structure responds to the excitation as a staticlike loading.

In the case of harmonic ground excitation, the structure moves with the ground with the same amplitude (see top left sketch of the response in Figure 6.5). Since the structure moves like a rigid body, there is no relative deformation along the structure. Consequently, the damping has no effect, and no force will be generated in the structure. Because of this, damage to structures resulting from deformation-related forces can be avoided. Since rigid body–like movement occurs, the whole structure will experience the ground acceleration. This induced acceleration might cause excessive loading for secondary structures attached to the main structure (Lim and Chouw, 2015).

With increasing excitation frequency or with a more flexible structure, the steady-state response becomes larger than the static response. For common civil infrastructure with a damping ratio less than 20%, the influence of damping is less significant. When the excitation frequency coincides with the natural frequency of the structure ($\beta = 1$), the structure is in resonance with the excitation. Without damping, it is only a question of time before the structure will collapse due to the accumulation of induced energy. The response of the strongly excited structure is controlled solely by the damping. Following this observation, structures in earthquake-prone regions should be built with a fundamental frequency that is as far as possible from the dominant frequencies of ground motions predicted for those regions.

With either or both further increases of the excitation frequency or flexibility of the structure, the amplitude of the dynamic response becomes smaller than the static response. The structure responds in the opposite direction to that of the loading (see the sketch on the right side, $\beta > 1$, $A_{dyn} < A_{st}$ in Figure 6.5). The damping plays a less important role than the ratio of the excitation frequency to the structural frequency.

In the case of a very flexible structure or a very fast-moving excitation, the structure hardly responds to the excitation; i.e., A_{dyn} is very small relative to the excitation amplitude A_e. Before the slow-moving structure can react, the load has already moved to the other direction. The load is too fast for the structure to have the opportunity to respond (see the right sketch in Figure 6.5). In the case of a ground excitation, the structure hardly experiences induced acceleration (A_{dyn} «). Consequently, secondary structures (e.g., nonstructural components such as ceilings or cladding piping system) will likely survive the ground motion without difficulty. However, the structure will likely suffer damage due to large relative deformation along the height of the structure.

The advantage of a structure with both a very small β (rigid structure) and a very large β (flexible structure) at the same time can be created by installing a base isolator at the interface between the footing and the supporting ground. The flexibility of the isolator in the lateral direction simulates a flexible system. The system itself is the rigid structure. Hence, when the ground moves, the isolator will be displaced laterally, while the structure performs only rigid body movement (i.e., no deformation in the structure, and thus no generated forces). In the worst-case scenario, the strongly dislocated isolator will be damaged and needs to be replaced prior to seismic motion (e.g., aftershock shaking). The structure itself can survive the earthquake without significant damage.

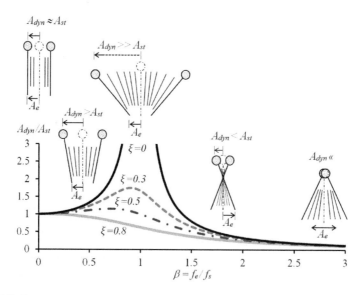

Figure 6.5 Consequence of excitation-to-structural frequency ratio β and damping ratio ξ for the structural response ratio A_{dyn}/A_{st} (A_e= excitation amplitude; A_{dyn} and A_{st}= dynamic and static response of the structure; f_e and f_s= the frequency of the harmonic excitation and the structure, respectively).

Although a steady-state response is unlikely in the case of earthquakes, a long-duration and harmoniclike ground motion is possible; e.g., when a site with soft sediment is excited by the incoming seismic waves, resulting in harmoniclike ground excitation of the ground surface and the structure. The ground motions have the dominant frequency of the sediment, such as that observed in the 1985 Mexico City earthquake. At the Secretaria de Comunicaciones (SCT) station, the harmoniclike earthquake motion lasted more than 2 min. (Chouw, 1994).

2.1.2   Pulse excitation

Should the excitation have a pulselike nature that does not last long, the load can be broken into a sequence of very short impulses. The load duration t_d is short, relative to the fundamental period of the structure, of the order of a few multiples of the fundamental period of the structure. Once the response of the structure to each of these very short impulses has been determined, the response to the total pulselike excitation can be obtained by superimposing the effect of these very short impulses. This consideration of the influence of each impulse on the structure and the superposition of each influence as the total effect of the loading is also called the *Duhamel integral approach* (Figure 6.6).

It is known that an impulse I reflects the change in momentum $\Delta(m\dot{u})$. Consequently, for a structure with a constant mass throughout the time period considered, the impulse causes a sudden change in velocity ($\Delta\dot{u} = I/m$). The effect of this sudden velocity change on an SDOF system, initially at rest with $u(0) = 0$, can be considered as initial velocity $\dot{u}(0) = \frac{I}{m}$ at the time τ of occurrence of the impulse. For an assumed free vibration $u(t) = G\sin(\omega_d t - \alpha)$, the constants G and α can be obtained from introducing the initial conditions $u(0)$ and $\dot{u}(0)$ into the free vibration equation.

With $G = \frac{I}{m\omega_d}$, $\alpha = \frac{\pi}{2}$, the response to the very short impulse is

$$u(t) = \frac{I(\tau)}{m\omega_d} e^{-\xi\omega(t-\tau)} \sin\omega_d(t-\tau), \quad at\, t = \tau, \;\; I(\tau) = P(\tau)\,d\tau, \;\; t > \tau \qquad (14)$$

The total response to the pulse excitation can be obtained by superposition of each impulse response:

$$u(t) = \int_{\tau=0}^{\tau=t} \frac{P(\tau)}{m\omega_d} e^{-\xi\omega(t-\tau)} \sin\omega_d(t-\tau)\,d\tau \qquad (15)$$

Since superposition is applied, only a linear system can be considered; i.e., damage to the structure during the loading cannot be incorporated into the analysis. However, for estimating the impact of the pulse loading on the nonlinear behavior of the structure, an analysis of the linear structure is always useful. It serves as a reference case to reveal the consequence of nonlinear material behavior due to structural damage (i.e., reduction of the structural stiffness resulting from damage).

In near-source earthquake regions, the hypocentral distance of the structure to the source of the earthquake is short (typically within 20 km), and the ground motion may

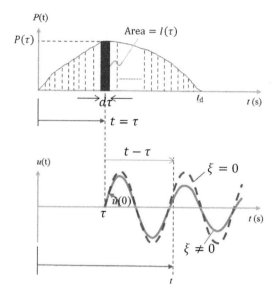

Figure 6.6 Duhamel integral approach.

have strong pulses due to the directivity effects of seismic wave propagation. For simplicity, the ground motion can be described by the strong pulse; i.e., $P(\tau) = -m\ddot{u}_g(\tau)$. Pulselike loadings can also be triggered by underground explosions, such as those due to mining activities. To incorporate possible damage during a strong earthquake, the step-by-step approach described in the next section can be used.

As an example, the response $u(t)$ of the structure to a ground acceleration pulse induced by an underground explosion can be calculated using Eq. (15). It is assumed that the load $P(\tau) = -m\ddot{u}_g(\tau)$ can be described by a half-sine $\ddot{u}_g(t) = \ddot{u}_{go}\sin\pi t$ with a duration t_d of 1 s. For simplicity, the damping effect of this short duration pulse loading can be ignored:

$$u(t) = \int_{\tau=0}^{\tau=t} \frac{\ddot{u}_{go}\sin\pi\tau}{m\omega}\sin\omega(t-\tau)d\tau = \frac{\ddot{u}_{go}}{\omega}\frac{\pi\sin\omega t - \omega\sin\pi t}{\pi^2 - \omega^2}, \quad t \leq 1s \qquad (16)$$

2.1.3 Earthquake loading

Ground motion due to an earthquake has a seemingly random development over time. A closed-form solution as derived in section 2.1.1 for harmonic loading, and utilization of a short pulse excitation using the Duhamel integral approach, as described in section 2.1.2, are not usually an option. The so-called step-by-step approach is commonly used. Since no superposition is performed, nonlinear geometry and material behavior can be incorporated into the analysis. The approach is based on an assumption of the acceleration development within one time step; i.e., it is assumed that the response acceleration one step later relative to the current acceleration is known. Based on this assumption, a number of step-by-step numerical approaches have been developed (see e.g. Figure 6.7).

In the following discussion, the step-by-step approach based on an assumption of constant acceleration development within one time step $\ddot{u}(\tau) = \frac{1}{2}(\ddot{u}_i + \ddot{u}_{i+1})$ is described. With $\frac{c}{m} = 2\xi\omega$ and $\frac{k}{m} = \omega^2$, the equation of motion for a load increment is $m\,\Delta\ddot{u} + c\,\Delta\dot{u} + k\,\Delta u = \Delta P$, which can be rearranged as follows:

$$\Delta\ddot{u} + 2\xi\omega\,\Delta\dot{u} + \omega^2\,\Delta u = \frac{\Delta P}{m} \tag{17}$$

With δ, $\Delta\ddot{u}$ and Δu:

$$\Delta u = \frac{(2\ddot{u}_i + \Delta\ddot{u})\Delta t^2}{4} + \dot{u}_i\,\Delta t$$

and a rearrangement leads to

$$\Delta\ddot{u} = \frac{4(\Delta u - \dot{u}_i\Delta t)}{\Delta t^2} - 2\,\ddot{u}_i \tag{18}$$

With $\Delta\dot{u}$ and Δu:

$$\Delta\dot{u} = \frac{2\Delta u}{\Delta t} - 2\ddot{u}_i. \tag{19}$$

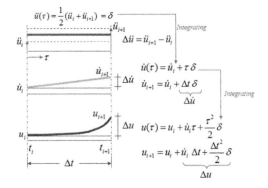

Figure 6.7 Constant acceleration assumption $\ddot{u}(\tau)$ within one time step yielding the velocity $\dot{u}_{i+1}$ and displacement u_{i+1}.

By substituting Eqs. (18) and (19) into Eq. (17), the response increment within one time step can be defined:

$$\Delta u = \frac{\dfrac{4\dot{u}_i}{\Delta t} + 2\ddot{u}_i + 4\xi\omega\dot{u}_i + \dfrac{\Delta P}{m}}{\dfrac{4}{\Delta t^2} + \dfrac{4\xi\omega}{\Delta t} + \omega^2} \tag{20}$$

With the response increment, the response one time step later at $t + \Delta t$, i.e., Δu, u_{i+1}, $\dot{u}_{i+1}$, and $\ddot{u}_{i+1}$, can be calculated.

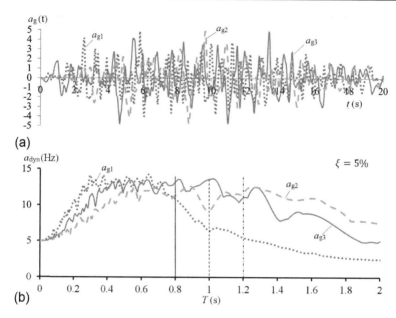

Figure 6.8 Influence of *PGA* and frequency content on the structural response: (a) Time history of ground motions; (b) response spectra of the ground motions.

Since the development of responses in the future cannot be predicted, the assumption of the development is only an estimate. Hence, the time step is relevant, leading to what is referred to as a *conditionally stable computation scheme* (see, e.g., Clough and Penzien, 1993). A large time step, relative to the natural period T of the structure, will lead to an inaccurate result. Time steps that are too small will also lead to inaccuracies. A time step $\Delta t \leq 0.1\,T$ is recommended; i.e., if the structure vibrates in its natural mode, 10 discrete values will be available to describe one natural vibration of the structure appropriately.

Figures 6.8(a) and (b) show the simulated ground accelerations based on JSCE (2000) and their corresponding response spectra, respectively. The maximum response of the structure to each ground motion can be obtained using Eq. (20), and $a_{g1}(t)$, $a_{g2}(t)$ and $a_{g3}(t)$ are ground motions for hard, soft, and medium soil conditions, respectively. In order to reveal the role of the peak ground acceleration (*PGA*), the ground motions considered have the same *PGA* (5 m/s²).

This same *PGA* is evident for all response spectrum values at the zero period in Figure 6.8(b). As is demonstrated in Figure 6.5, a totally rigid structure (i.e., with the natural period $T=0$ s) will move identically with the ground (see top left sketch in Figure 6.5). Hence, it will experience exactly the same acceleration as the ground. Consequently, the maximum response of the structure to the ground motion is the *PGA* of the ground excitation; i.e., the *PGA* is the total acceleration.

Based on a quasi-static approach (i.e., the maximum inertia force F_I is PGA times the mass of the structure), a structure under this loading could be anticipated to have the same maximum force. However, Figure 6.8(b) clearly shows that this is not the case. The reason for this is that the assumption of the quasi-static approach is based solely on one quantity of the loading (i.e., PGA), while both the other significant property of the loading (i.e., its frequency content) and the dynamic property of the structure (i.e., its natural period) are ignored. Even though all ground motions have the same PGA, depending on the period of the structure relative to the dominant frequencies of the loading, the structure will respond differently. Consequently, the maximum response acceleration is not the same for structures with different natural periods. The vertical solid, dashed, and dashed-dotted lines in Figure 6.8(b) indicates the response of a structure with the natural period of 0.8 s, 1 s, and 1.2 s, respectively. While for a structure with the natural period of 0.8 s, the maximum responses due to ground motions of hard- and soft-soil conditions are similar, the excitation of medium-soil condition results in the largest response. In contrast, for a structure with the natural period of 1.2 s, similar maximum responses occur due to excitations of soft- and medium-soil conditions. For a structure with a natural period of 1 s, the excitation of each soil condition evokes very different responses. The largest response occurs when the excitation of medium-soil condition is considered.

These results clearly show that the commonly used approach, which is solely based on PGA, is unreliable. It is advised that the combined effect of the whole dynamic properties of loading and structure should be considered in the analysis of structural responses.

2.2 MDOF system

In contrast to an SDOF system, the equation of motion cannot be solved directly for each DOF. Even if a lumped-mass model is assumed, the damping and stiffness matrices have coupled terms. Hence, the equation of motion [Eq. (8)] needs to be solved simultaneously for all DOFs. For a system with more than two DOFs, a hand calculation will be time consuming. However, this difficulty can be overcome by expressing the total response using the modal response $Y(t)$:

$$u(t) = \sum_{r=1}^{n} \boldsymbol{\Phi}_r Y_r(t) = \boldsymbol{\Phi}\, Y(t), \tag{21}$$

where $\boldsymbol{\Phi}_r$ is the rth mode shape, which does not vary with time.

By substituting Eq. (21) and its time derivatives $\dot{u}(t) = \boldsymbol{\Phi}\,\dot{Y}(t)$ and $\ddot{u}(t) = \boldsymbol{\Phi}\,\ddot{Y}(t)$ into Eq. (8), the equation of motion becomes

$$M\boldsymbol{\Phi}\ddot{Y} + C\boldsymbol{\Phi}\dot{Y} + K\boldsymbol{\Phi}Y = P. \tag{22}$$

The natural frequency of the system can be obtained from its undamped free vibrations; i.e., C and P are not considered. By substituting the assumed solution to be $Y(t) = A\cos\omega t + B\sin\omega t$ and its second time derivative into Eq. (22), the equation becomes $(K - \omega^2 M)\boldsymbol{\Phi}Y = 0$. Since $Y \neq 0$, the frequency equation $(K - \omega^2 M)\boldsymbol{\Phi} = 0$

can be obtained. For nontrivial solutions (i.e., $\boldsymbol{\Phi} \neq \boldsymbol{0}$), det $\left| \boldsymbol{K} - \omega^2 \boldsymbol{M} \right| = 0$ and provides the eigenvalues of the system; i.e., the natural circular frequencies (ω_r, $r = 1, n$) of the structure. By substituting the natural frequency ω_r into the frequency equation, the corresponding natural mode shape $\boldsymbol{\Phi}_r$ can be determined.

Multiply Eq. (22) by any mode $\boldsymbol{\Phi}_i^T$ to obtain:

$$\boldsymbol{\Phi}_i^T \boldsymbol{M} \boldsymbol{\Phi} \ddot{\boldsymbol{Y}} + \boldsymbol{\Phi}_i^T \boldsymbol{C} \boldsymbol{\Phi} \dot{\boldsymbol{Y}} + \boldsymbol{\Phi}_i^T \boldsymbol{K} \boldsymbol{\Phi} \boldsymbol{Y} = \boldsymbol{\Phi}_i^T \boldsymbol{P}, \tag{23}$$

and by decoupling the system matrices using the orthogonality properties of the natural modes of the system, $\boldsymbol{\Phi}_j^T \boldsymbol{M} \boldsymbol{\Phi}_i = \boldsymbol{\Phi}_j^T \boldsymbol{C} \boldsymbol{\Phi}_i = \boldsymbol{\Phi}_j^T \boldsymbol{K} \boldsymbol{\Phi}_i = 0$.

$$\boldsymbol{\Phi}_i^T \boldsymbol{M} \boldsymbol{\Phi}_i \ddot{Y}_i + \boldsymbol{\Phi}_i^T \boldsymbol{C} \boldsymbol{\Phi}_i \dot{Y}_i + \boldsymbol{\Phi}_i^T \boldsymbol{K} \boldsymbol{\Phi}_i Y_i = \boldsymbol{\Phi}_i^T \boldsymbol{P}. \tag{24}$$

Since all system matrices are diagonal matrices, $\boldsymbol{\Phi}_i^T \boldsymbol{M} \boldsymbol{\Phi}_i = M_i$, $\boldsymbol{\Phi}_i^T \boldsymbol{C} \boldsymbol{\Phi}_i = C_i$, $\boldsymbol{\Phi}_i^T \boldsymbol{K} \boldsymbol{\Phi}_i = K_i$ and $\boldsymbol{\Phi}_i^T \boldsymbol{P} = P_i$ are scalar. The coupled equations of motion [Eq. (22)] for n-DOFs are now decoupled into n independent equations of motion, and instead can be solved separately in the manner of SDOF equations. n is the total number of DOFs considered.

$$M_i \ddot{Y}_i(t) + C_i \dot{Y}_i(t) + K_i Y_i(t) = P_i(t), i = 1, n. \tag{25}$$

Once all modal responses $Y_i(t)$ due to $P_i(t)$ are calculated, the total response of the structure can be obtained. To solve Eq. (25), all previously discussed approaches for SDOF systems can be applied (Chouw, 2013b).

2.3 IDOF system

The algebraic equation of motion, Eq. (12), can be solved easily once the load is transformed into the Laplace domain:

$$\widetilde{P}(\delta + i r \, \Delta \omega) = \sum_{r=0}^{N-1} P(n \, \Delta t) \, e^{-\delta (n \, \Delta t)} e - i (r \, \Delta \omega)(n \, \Delta t) \, \Delta t, \tag{26}$$

and the response $\widetilde{u} = \frac{\widetilde{P}}{K}$ can be calculated. A back-transformation of the response $\widetilde{u}$ into the time domain gives

$$u(n \, \Delta t) = \frac{e^{\delta (n \, \Delta t)}}{2 \pi} \sum_{r=0}^{N-1} \widetilde{u}(\delta + i r \, \Delta \omega) \, e^{i (r \, \Delta \omega)(n \, \Delta t)} \, \Delta \omega, \tag{27}$$

where $r = n = 0, \dots, N - 1$; $i = \sqrt{-1}$; $\delta = \frac{8}{T}$; $\Delta t = \frac{T}{N}$; $\Delta \omega = \frac{2\pi}{T}$. The largest circular frequency considered is $\omega_{max} = \frac{N}{2} \Delta \omega$, and N is the number of time steps considered.

For structures with material or geometrical nonlinearities, the nonlinear behavior can be approximated by a sequence of linear behavior; i.e., within each sequence, the structure behaves in a linear manner. The calculation is performed in the Laplace domain while the correction of the result is performed in the time domain (i.e., determination of the unbalanced forces as loading for the subsequent analysis). Details of this Laplace and time domain analysis are given in Chouw (1994, 2002).

3   Influence of supporting soil

In current bridge design, a fixed-base structure is often assumed due to the intrinsic difficulties in modeling coupled soil-foundation-bridge structure response. This design practice of neglecting soil effects is used, although the bridge is always supported by soil, and the significant impact of local soil has been observed in most major earthquakes (Chouw, 1995; Chouw and Hao, 2012).

3.1   Dynamic properties of the soil-structure system

For simplicity, a bridge segment is modeled as an SDOF system. The natural period or frequency of the system can be obtained by observing the free vibration (i.e., vibration behavior after the load that causes the free vibrations is outside the time window under consideration).

Figure 6.9(a) shows a simplified model of a bridge segment [shown in Figure 6.9(b)] with an assumed fixed base; i.e., the influence of the supporting subsoil is ignored as commonly performed in the current design. From the equation of motion for undamped free vibrations [i.e., Eq. (1), with no damping and loading], $m\ddot{u}(t) + ku(t) = 0$, the natural circular frequency ($\omega = \sqrt{\frac{k}{m}}$), and the natural period ($T = \frac{2\pi}{\omega} = 2\pi\sqrt{\frac{m}{k}}$) of the structure can be determined by substituting the mass $m = \frac{u_{st}}{k\,a}$, where u_{st} is the static displacement due to a load that causes an acceleration a of the mass:

$$T = \frac{2\pi}{\sqrt{a}}\sqrt{u_{st}}. \tag{28}$$

Figure 6.9(c) shows the simplified SDOF model of the bridge segment with the subsoil. For simplicity, the mass of the footing is not considered. The influence of the flexibility of the supporting soil is given by the horizontal and rotational springs with stiffness k_x and k_θ, respectively. Here, u, u_x, and u_θ are the static displacement due to a load W induced deformation of the structure, elongation of the horizontal spring, and the rotation of the rotational spring, respectively. The relationship between the load and induced rotation, $k_\theta \frac{u_\theta}{h} = W h$, can be obtained from the base moment. The total static displacement is

$$\bar{u}_{st} = \frac{W}{k} + \frac{W}{k_x} + \frac{Wh^2}{k_\theta}. \tag{29}$$

From Eq. (28), the natural period $\bar{T}$ of the soil-structure system is

$$\bar{T} = \frac{2\pi}{\sqrt{a}}\sqrt{\bar{u}_{st}} = T\sqrt{1 + \frac{k}{k_x}\left(1 + \frac{k_x h^2}{k_\theta}\right)}. \tag{30}$$

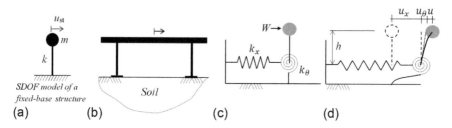

Figure 6.9 Influence of subsoil on the fundamental period of the system. (a) Fixed-base structure and (b) soil-supported structure, (c) simplified SDOF soil-structure system, and (d) its static deformation.

When the structure vibrates, the interaction between the vibrating footing and the supporting soil causes waves in the ground. These waves propagate from the footing-soil interface and transfer part of the vibration energy away. The vibrating footing, therefore, experiences energy loss. This process is termed *radiation damping;* i.e., due to the radiation of waves in the soil. This damping and the soil resistance (i.e., soil stiffness) depend on the frequency of the vibrating footing, the direction of footing vibration, and the dynamic soil properties; e.g., velocities of waves propagating in the soil. The frequency-dependent soil stiffness can be calculated; e.g., using the boundary element method (Chouw, 1994) or from the *Handbook of Impedance Functions* (Sieffert and Cevaer, 1992).

For simplicity, the following static soil stiffness and frequency-independent damping can be used in Eqs. (29) and (30).

For the half-space case shown in Figure 6.10,

$$k_x = \frac{8GR}{2-v} \text{ and } k_\theta = \frac{8GR^3}{3(3-v)}, \tag{31}$$

where R and v are the equivalent radius of the assumed circular footing and Poisson's ratio, respectively.

Surface footing

Half space

Figure 6.10 Surface footing on subsoil.

For a soil layer of thickness H over bedrock with a footing embedment D shown in Figure 6.11,

$$k_x = \frac{8GR}{2-v}\left(1+\frac{R}{2H}\right)\left(1+\frac{2D}{3R}\right)\left(1+\frac{5D}{4H}\right) \text{ and}$$

$$k_\theta = \frac{8GR^3}{3(3-v)}\left(1+\frac{R}{6H}\right)\left(1+2\frac{D}{R}\right)\left(1+0.7\frac{D}{H}\right), \tag{32}$$

where G is the shear modulus of the soil.

$$\bar{\xi} = \xi_s + \frac{\xi}{\left(\frac{\bar{T}}{T}\right)^3}, \tag{33}$$

where $\xi = \frac{c}{c_{crit}}$, $c_{crit} = 2 m \omega$, ξ_s is material and radiation damping of the soil, $\bar{T}$ is the natural period of the soil-structure system, and T is the natural period of the assumed fixed-base structure.

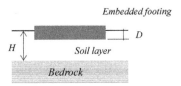

Embedded footing

Soil layer

Bedrock

H   *D*

Figure 6.11 Footing with embedment D on soil layer over bedrock.

Eq. (33) shows that the larger the period ratio $\frac{\bar{T}}{T}$, the less the structural damping will contribute to the system damping. In other words, the more flexible the system is (e.g., due to soft soil), the smaller the effects of structural damping. Details regarding the soil-foundation-structure system are given by Chouw (2013a).

3.2 Effect of spatially varying ground motion

In the case of a long, extended structure (e.g., pipelines or long bridges), the excitation of adjacent bridge supports is normally not the same, since seismic waves need time to travel from one bridge support to the other. Even if the adjacent bridge structures have the same fundamental frequency, relative response will nevertheless will still occur, because spatially nonuniform ground excitation is to be expected due to the effects of wave passage, site response, and coherency loss.

Despite this knowledge, most design specifications still consider spatially uniform ground motions as the design seismic loading. The consequence of this assumption of uniform loading can be seen in severe damage to bridges in almost all major earthquakes in the past (see Figure 6.12).

With an assumption of uniform ground excitation, Eq. (2) can be rewritten with the stiffness $k = k_1 + k_2$, $\xi = \frac{c}{2m\omega}$ and $\omega = \sqrt{\frac{k_1 + k_2}{m}}$ as follows:

$$\ddot{u}(t) + 2\xi\omega\,\dot{u}(t) + \omega^2 u(t) = -\ddot{u}_g(t). \tag{34}$$

Figure 6.13(a) shows the bridge structure under uniform ground excitation. When spatially varying ground motions are considered, it is useful to incorporate the effect of footings. Figure 6.13(b) shows that the left and right footing is of mass m_1 and m_2, respectively. With an assumption of the lumped-mass model, the equation of motion for the three-DOF system is

Figure 6.12 Unseating damage to Llacolen Bridge due to the 2010 Chile earthquake induced relative movements of adjacent segments.

$$\begin{bmatrix} m & 0 & 0 \\ 0 & m_1 & 0 \\ 0 & 0 & m_2 \end{bmatrix}\begin{bmatrix} \ddot{u}^t \\ \ddot{u}_{g1} \\ \ddot{u}_{g2} \end{bmatrix} + \begin{bmatrix} c & 0 & 0 \\ 0 & 0 & 0 \\ 0 & 0 & 0 \end{bmatrix}\begin{bmatrix} \dot{u}^t \\ \dot{u}_{g1} \\ \dot{u}_{g2} \end{bmatrix} + \begin{bmatrix} k_1+k_2 & -k_1 & -k_2 \\ -k_1 & k_1 & 0 \\ -k_2 & 0 & k_2 \end{bmatrix}\begin{bmatrix} u^t \\ u_{g1} \\ u_{g2} \end{bmatrix} = \begin{bmatrix} 0 \\ 0 \\ 0 \end{bmatrix}. \quad (35)$$

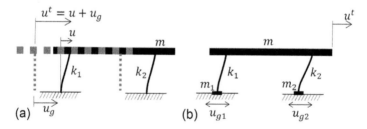

Figure 6.13 Bridge segment under (a) uniform and (b) spatially nonuniform ground motions.

While the quasi-static response of the uniformly excited bridge segment does not occur because of the rigid-body movement of the whole structure, the spatially varying ground motions does cause a quasi-static response since both bridge piers will move differently. By substituting the relationship between dynamic response u and quasi-static response u^{qs},

$$\begin{bmatrix} u^t \\ u_{g1} \\ u_{g2} \end{bmatrix} = \begin{bmatrix} u \\ 0 \\ 0 \end{bmatrix} + \begin{bmatrix} u^{qs} \\ u_{g1} \\ u_{g2} \end{bmatrix}, \quad (36)$$

into Eq. (35), and by ignoring all dynamic components, the quasi-static response can be determined:

$$(k_1 + k_2)u^t - k_1 u_{g1} - k_2 u_{g2} = 0 \text{ and with } u = 0, \ u^{qs} = \frac{k_1 u_{g1} + k_2 u_{g2}}{k_1 + k_2}. \tag{37}$$

For equal pier bending stiffness, $k_1 = k_2 = \frac{k}{2}$, $u^{qs} = \frac{u_{g1} + u_{g2}}{2}$, which indicates an average of ground motions.

From Eq. (35), by ignoring the quasi-static response, the equation of motion for spatially varying ground excitation can be defined:

$$m\ddot{u} + c\dot{u} + (k_1 + k_2)u(t) = k_1 u_{g1} + k_2 u_{g2},$$

or, for the equal pier bending stiffness case, $k_1 = k_2 = \frac{k}{2}$, and

$$\ddot{u} + 2\xi\omega\dot{u} + \omega^2 u = \frac{k}{2m}(u_{g1} + u_{g2}) = \frac{\omega^2}{2}(u_{g1} + u_{g2}). \tag{38}$$

For an MDOF system, the equation for dynamic and quasi-static response can be derived in the same manner.

Eq. (35) becomes

$$\begin{bmatrix} M_{bb} & M_{bs} \\ M_{sb} & M_{ss} \end{bmatrix} \begin{bmatrix} \ddot{u}^t \\ \ddot{u}_g \end{bmatrix} + \begin{bmatrix} C_{bb} & C_{bs} \\ C_{sb} & C_{ss} \end{bmatrix} \begin{bmatrix} \dot{u}^t \\ \dot{u}_g \end{bmatrix} + \begin{bmatrix} K_{bb} & K_{bs} \\ K_{sb} & K_{ss} \end{bmatrix} \begin{bmatrix} u^t \\ u_g \end{bmatrix} = \begin{bmatrix} 0 \\ 0 \end{bmatrix}. \tag{39}$$

Here, the subscripts b and s stand for bridge and soil, respectively.

By ignoring the dynamic response, Eq. (39) gives the quasi-static response

$$u^{qs} = -K_{bb}^{-1} K_{bs} u_g, \tag{40}$$

and the dynamic response can be obtained from the equation of motion:

$$M_{bb} \ddot{u} + C_{bb} \dot{u} + K_{bb} u = P_{eff}, \tag{41}$$

where $P_{eff} = -\left(M_{bb}K_{bb}^{-1}K_{bs} + M_{bs}\right)\ddot{u}_g - \left(C_{bb}K_{bb}^{-1}K_{bs} + C_{bs}\right)\dot{u}_g$.

The total response can be obtained as follows:

$$\begin{bmatrix} u^t \\ u_g \end{bmatrix} = \begin{bmatrix} u \\ 0 \end{bmatrix} + \begin{bmatrix} u^{qs} \\ u_g \end{bmatrix} \tag{42}$$

4 Bridge integrity: consequences of relative response of adjacent bridge structures

Despite the most advanced bridge design specifications, almost all major earthquakes have shown that large relative movements between bridge girders and between bridge decks and abutments can have severe consequences for the structural integrity of a bridge. Normally, the available gap is designed to cope with a relative closing movement due to a large change in temperature. In strong earthquakes, large relative opening and closing displacements may occur. If the relative closing displacement is larger than the available gap, then pounding will take place. To estimate pounding-induced damage, the pounding force needs to be determined (see, e.g., Khatiwada et al., 2014). Should the relative opening response exceed the seat length, then the bridge deck will lose its support and collapse (see Figure 6.12; Chouw, 1995; Chouw and Hao, 2012).

The main causes of relative response are
- Different dynamic properties of adjacent bridge structures
- Noncoherent ground motions at adjacent bridge supports
- Nonuniform soil-structure interaction
- Combined effect of the first three factors

To avoid or minimize relative responses, current design specifications (e.g., CALTRANS, 2010), suggest identical or similar fundamental frequencies of adjacent bridge structures. The fundamental period of the less flexible bridge structure should be at least equal to or larger than 70% of the fundamental period of the more flexible adjacent structure. With equal or similar frequencies, the adjacent structures will then respond approximately in phase. Consequently, the relative response is negligible, and pounding and unseating can be avoided. However, this recommendation of similar frequencies is based on uniform excitation; i.e., all bridge structures will experience the same ground motions at the same time. However, in actuality, this rarely is the case, so instead of reducing relative response, constructing adjacent structures at the same or similar frequencies becomes one of the significant causes of relative response. Thus, the good intentions of most current design specifications could worsen, not improve, bridge performance in many cases. This is because adjacent structures of the same or similar structural frequencies will cause relative responses if spatially nonuniform ground motions occur (Chouw and Hao, 2005; Bi et al., 2010; Li et al., 2013).

Spatially varying ground motions occur mainly due to the effect of
- Wave passage due to the finite speed of the propagating seismic waves
- Site response due to spatially nonuniform soil profiles
- Coherency loss due to refraction and reflection of waves in the wave path

Since the soil along the bridge is never uniform, spatial variation of ground excitations is unavoidable.

Most current design specifications are based on uniform ground excitation. For example, the *New Zealand Transport Agency Bridge Manual* (NZTA, 2004) defines the requirement for the minimum seat length to prevent unseating as

$$SL = 2E + 0.1 \geq 0.4\,m, \tag{43}$$

where E is the relative movement between span and support.

According to the AASHTO LRFD bridge design specification (AASHTO, 2010) for straight bridges, the length of seating required (in meters) is

$$SL = 0.203 + 0.00167\,l + 0.00666H, \tag{44}$$

where l and H are the effective length of the bridge deck to the adjacent expansion joint or to the end of the bridge deck, and the average height of columns supporting the bridge deck from the abutment to the next expansion joint or pier height, respectively.

In contrast to the AASHTO and NZTA specifications, the seat length according to the Japan Road Association (JRA, 2004) is

$$SL = u_{rel} + u_G \geq SL_m, \tag{45}$$

where u_{rel} is the relative displacement of the adjacent structures and u_G describes the relative ground displacement, which depends on the soil strain $u_G = \varepsilon_G L$ and the distance L between the substructures in meters. For hard-, medium-, and soft-soil conditions, the soil strain ε_G has the values of 0.0025, 0.00375, and 0.005, respectively. The minimum value of $SL_m = 0.7 + 0.005\,l$.

To reveal the consequence of the spatial variation of ground motions for the relative opening and closing displacements at an expansion joint, two bridge segments with the same damping ratio of 5% on an assumed half-space with a shear wave velocity of 100 m/s, soil density of 2000 kg/m^3, and Poisson's ratio of 0.33 are considered [Figure 6.14(a)]. The ground motions are stochastically simulated based on the Japanese design spectrum (JDS) for soft-soil conditions [Figure 6.14(b)]. For simplicity, each bridge segment is modeled by an SDOF system. It is assumed that the surface footings have the dimensions of 9 m × 9 m. The bridge structures and subsoil are modeled by finite and boundary elements. While the calculation of the response is performed in the Laplace domain, the unbalanced forces for correcting the response due to pounding and girder separation are determined in the time domain. Details can be found in Chouw (1994, 2002), Chouw and Hao (2006), and Chouw (2008).

To limit the number of influence factors, it is assumed that both bridge structures have the same fixed-base fundamental frequency of 1 Hz, and the effect of spatial variation of the ground excitation is not considered for the moment. The height of the left structure ($h_2 = 9$ m) is kept constant. Since both structures have the same fixed-base fundamental frequency, they experience the same soil-structure interaction (SSI) and uniform ground excitation; i.e., there is no relative displacement between the girders [shown by the bold line in Figure 6.14(c) for $h_1 = h_2 = 9$ m]. Should the bridge piers have different heights, different SSIs will result in relative responses between the girders. While structures with tall piers are mainly controlled by the rotational stiffness of the soil, the response of structures with short piers will likely be determined by the horizontal soil stiffness (see Figure 6.9). Consequently, bridge segments with different slenderness ratios will not respond in phase, even though both structures may have the same fixed-base fundamental frequency. In this example, a height ratio (h_1/h_2) of 0.5 produces a seat length of 12 cm [see Figure 6.14(c), the thin-solid line at 16.96 s]. In contrast, a height ratio of 1.5 produces a seat length of 17.9 cm [see Figure 6.14(c), the dashed line at 17.8 s].

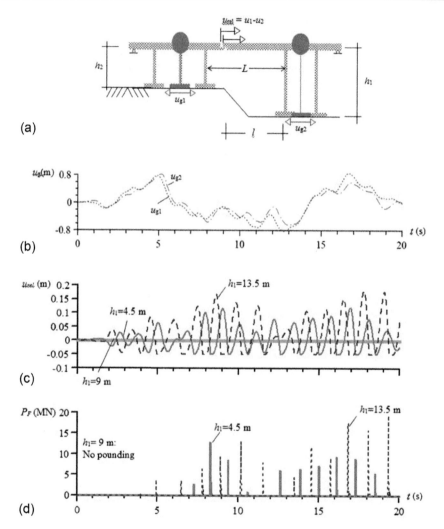

Figure 6.14 Two bridge structures with $h_2 = 9$ m. (a) Simplified double SDOF model, (b) spatially varying ground motion, (c) relative displacement u_{rel} at the joint, and (d) pounding force P_F.

The results clearly show the significance of considering SSI. While using the common assumption of a fixed-base structure, unseating will not take place, but including SSI will result in seat length requirements that cannot be revealed by following a conventional analysis of a fixed-base structure, even if uniform ground excitation can be justified. An assumption of a fixed-base structure clearly may underestimate the damage potential due to unseating and pounding between bridge girders.

When SSI is ignored in the analysis, the influence of the height ratio h_1/h_2 also cannot be revealed. Figure 6.15 shows the consequence of the height ratio h_1/h_2 and SSI for the seat length SL required to prevent unseating, where SL is shown as a function of the frequency ratio f_1/f_2 and the effect of pounding is not considered. The results show that the slender structure with $h_1 = 13.5$ m has the largest required seat length (thin dashed line). In addition, $h_1 = 9$ m results in a large seat length (see bold, solid black line in Figure 6.15). In the frequency ratio range between 0.5 and 1.2, and above 1.8, bridge structures with $h_1 = 13.5$ m need the longest seat length.

The horizontal solid thin line in the figure shows the minimum seat length values SL_m according to the Japanese design specifications. Assuming a distance between bridge piers to be $L = 60$ m, the seat length [according to Eq. (45)] is the shortest (as indicated by the bold, dashed line). The consequence of the spatially varying ground movement for the seat length required can be clearly seen to have a nonzero value at the frequency ratio f_1/f_2 of 1. In contrast to reality, an assumption of constant soil strain for all cases causes a constant quasi-static contribution of 0.3 m. Even though the assumption does not reflect the reality, it is still the most advanced knowledge considered in the current Japanese design specifications. To this author's knowledge, this soil strain factor has not been considered in other specifications. The results show that even if a larger distance ($L = 100$ m) between the substructures is chosen, SL is still smaller than the values calculated using a more realistic numerical soil-structure model (see the bold, solid, black line).

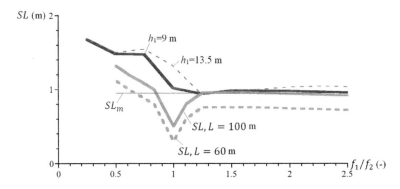

Figure 6.15 Influence of soft-soil JDS spatially nonuniform ground motions on the minimum seat length SL required to prevent unseating, according to JRA.

A large number of physical experiments on the effects of spatial variation of ground motions using multiple shake tables had been performed at the University of Auckland's Centre for Earthquake Engineering Research (UACEER). The ground motions were stochastically simulated based on the New Zealand loadings code (NZS1170.5, 2004) for soft-, medium-, and hard-soil conditions. Field tests on sand (Li, 2013), and fixed-base tests in the laboratory were considered, and these revealed the influence of soil-foundation-bridge structure interaction under spatially varying ground excitations.

Based on the experimental results, the following empirical equation for seat length, SL_g of girders is proposed:

$$SL_g = (5.6f_r - 3.3)d_{ave} + d_{uni}, \tag{46}$$

where d_{ave} and d_{uni} are the maximum relative displacements of fixed-base structures due to uniform ground motion, and the mean of the maximum girder displacements of two adjacent bridge segments with an assumed fixed base, respectively.

At least three ground motions should be considered. The maximum relative displacement between girders under uniform ground motions should consider excitation along two mutually perpendicular directions. Once the average maximum displacement of the bridge structures under the selected ground motions is determined, the seat length can be obtained from Eq. (46). These steps should be repeated for the other ground excitations. The largest seat length possible should be determined to be the seat length required to prevent unseating.

An empirical equation is proposed for the seat length SL_a at the abutments:

$$SL_a = 1.4\,d_{uni}. \tag{47}$$

The background of the development of these empirical equations is given by Li (2013).

To mitigate possible damage to bridge structures due to their relative responses, a number of measures have been developed and applied (e.g., hinge restrainers to prevent excessive relative opening displacements at a joint) so that seat extenders to ensure unseating does not occur through inadequate seat width. Recently, the author has proposed the use of modular expansion joints (MEJs) to cope with relative opening and closing displacements. These would help prevent the unseating and pounding of a bridge girder with adjacent girders or abutments. Details of a MEJ application are given in Chouw and Hao (2008).

Figure 6.16(a) shows a sketch of a modular expansion joint. It consists of a number of intermediate gaps so that in total, the joint can have a gap that is wide enough to cope with the largest relative movements expected at the joint. To determine the largest expected relative movements, fixed-base and field tests have been performed. From the newly developed relative displacement response spectrum at UACEER, the total gap that the joint has to cope with can be determined. Figure 6.16(b) shows the relative displacement response spectrum developed for ground motions according to the Japanese design spectrum (JSCE, 2000).

From Figure 6.16(b), suppose that the left and right bridge structures have the fundamental periods T_1 of 1 s and T_2 of 2.5 s, respectively. Referring to the spectrum, the total gap required of the modular expansion joint can be calculated; in this case, it is 1.4 m.

In the case of a strong earthquake and poor soil, nonlinear soil behavior can be observed (e.g., due to liquefaction, as observed in Kobe and Christchurch earthquakes; see Chouw, 1995; Chouw and Hao, 2012). The nonlinear soil behavior can be described using macro elements with a predefined yield surface (e.g., Chouw and Rincon, 2011). The nonlinear soil-foundation-bridge structure interaction can significantly influence the structural response.

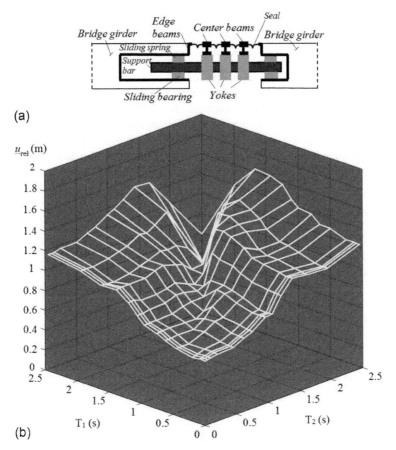

Figure 6.16 Mitigation measure: (a) Modular expansion joint; (b) relative displacement spectrum.

5   Conclusions

Idealization of bridge structures can provide a quick insight into the dynamic behavior of the structure under loading, but with the drawback of limiting the information that otherwise could be obtained by more realistic analyses. Similarly, the idealization of dynamic loadings can be limiting when attempting to estimate the response of the structure, since the result will only be as good as the original correct assumptions of the loading. This chapter has suggested possibilities that may allow for going beyond the conventional analysis of bridge structural response under dynamic loadings.

Possible SDOF modeling, multi-DOF, and infinite DOF descriptions of the bridge structure were introduced. Approaches to solve the governing equations were provided, together with insights into the relationship between loading and structural response characteristics. In particular, the role of peak ground acceleration and the frequency content of the loading in the response of a structure were explored.

In contrast to the conventional consideration of fixed-base structures, the consequences of supporting soil for the dynamic properties of the soil-structure system and for the spatial variation of ground motions were considered and described. The structural response of the soil-structure system, which otherwise cannot be revealed from the assumption of fixed-base structures, was introduced. The relative response resulting from spatially varying ground excitation and spatially nonuniform soil-structure interaction was considered, especially with the aim to estimate its damage potential. Recent developments in mitigating damage development due to the relative response between bridge structures were also described.

Acknowledgments

The author would like to thank the Ministry of Business, Innovation, and Employment through the Natural Hazards Research Platform for the support of the research on "Seismic response of New Zealand bridges" under the Award UoA 3701868.

References

American Association of State Highway and Transportation Officials (AASHTO), 2010. Load and Resistance Factor Design (LRDF) Specification for Highway Bridges. AASHTO, Washington DC.
Bi, K., Hao, H., Chouw, N., 2010. Required separation distance between decks and at abutments of a bridge crossing a canyon site to avoid seismic pounding. Earthqu. Eng. Struct. Dyn. 39 (3), 303–323.
CALTRANS, 2010. Seismic Design Criteria, Version 1.6. November.
Chouw, N., 1994. Analysis of Structural Vibration Considering the Dynamic Transmitting Behaviour of Soil. Technical Report, Ruhr University, Bochum, Germany.
Chouw, N., 1995. Effect of the earthquake on 17th of January 1995 on Kobe. In: Proc. DACH Meeting of the German, Austrian, and Swiss Society for Earthquake Engineering and Structural Dynamics, 135–169.
Chouw, N., 2002. Influence of soil-structure interaction on pounding response of adjacent buildings due to near-source earthquakes. JSCE J. Appl. Mech. 5, 545–553.
Chouw, N., 2008. Unequal soil-structure interaction effect on seismic response of structures. In: Proc. 18th New Zealand Geotechnical Society, Geotechnical Symp. Soil-Structure Interaction—From Rules of Thumb to Reality. University of Auckland, New Zealand, 214–219. September 4–5.
Chouw, N., 2013a. Advanced Structural Dynamics, Civil710 Course Note. University of Auckland, New Zealand.
Chouw, N., 2013b. Structural Dynamics, Civil314 Course Note. University of Auckland, New Zealand.
Chouw, N., Hao, H., 2005. Study of SSI and non-uniform ground motion effect on pounding between bridge girders. Soil Dyn. Earthqu. Eng. 25 (7), 717–728.
Chouw, N., Hao, H., 2006. Comments on bridge girder seating length under current design regulations. JSCE J. Appl. Mech. 9, 691–699.

Chouw, N., Hao, H., 2008. Significance of SSI and non-uniform near-fault ground motions in bridge response. II: Effect on response with modular expansion joint. Eng. Struct. 30, 154–162.

Chouw, N., Hao, H., 2012. Pounding damage to buildings and bridges in the 22 February 2011 Christchurch earthquake. Intl. J. Protect. Struct. 3 (2), 123–139.

Chouw, N., Rincon, E.W., 2011. Numerical simulation of adjacent bridge structures with nonlinear SSI. In: 3rd Intl. Conf. Computational Methods in Structural Dynamics and Earthquake Engineering, Corfu, May 26–28, Greece.

Clough, R.W., Penzien, J., 1993. Dynamics of Structures. McGraw-Hill, Singapore.

Japan Road Association (JRA), 2004. Specification for highway bridges - Part V: Seismic design. 5th ed., Tokyo (in Japanese).

Japan Society of Civil Engineers (JSCE), 2000. Earthquake Resistant Design Codes in Japan. JSCE, Maruzen, Tokyo.

Khatiwada, S., Chouw, N., Butterworth, J.W., 2014. A generic structural pounding model using numerically exact displacement proportional damping. Eng. Struct. 62, 33–41.

Li, B., 2013. Effects of Spatially Varying Ground Motions on Bridge Response. Ph.D. thesis, University of Auckland, Auckland, New Zealand.

Li, B., Bi, K., Chouw, N., Butterworth, J.W., Hao, H., 2013. Effect of abutment excitation on bridge pounding. Eng. Struct. 54, 57–68.

Lim, E., Chouw, N., 2015. Review of approaches for analysing secondary structures in earthquakes and evaluation of floor response spectrum approach. Intl. J. Protect. Struct. 6 (2), 237–261.

Martin, H., 1966. Introduction to Matrix Methods of Structural Analysis. McGraw-Hill, New York.

New Zealand Standards, 2004. NZS1170.5 Structural Design Actions, Part 5: Earthquake Actions, Wellington. ISBN 1-86975-018-7.

New Zealand Transport Agency (NZTA), 2004. Bridge Manual, Section 5: Earthquake-Resistant Design. September.

Sieffert, J.G., Cevaer, F., 1992. Handbook of Impedance Functions. In: Quest Editions Presses Academiques, Nantes.

Risk and reliability in bridges

7

Bhattacharya B.
Indian Institute of Technology, Kharagpur, India

1 Overview

In probability-based design, we explicitly account for uncertainty and variability in
(i) the inputs (loads, etc.), (ii) the properties (strength, stiffness etc.), and (iii) the
(mathematical) model of the system. We then design or assess the system so that it
satisfies its safety and performance objectives with acceptable probabilities for its
intended function under expected service conditions and projected service life. When
a design code is used instead of an explicit probability-based approach, such uncertainties are often accounted for indirectly in the form of partial safety factors (PSFs),
load combination schemes, and other code provisions.

Consider the entire life of a bridge from its conception. It starts with functional
requirements. Suppose that the bridge should be part of the national highway network
at a location over a major river and be able to handle four lanes of unrestricted traffic,
plus shoulders on both sides. The waterway below is used for cargo and passenger
transport. Thus, some clear span and height must be provided. The design life of
the bridge has to be specified.

A concept design follows. Economic considerations play a part here. At some point
in the design cycle, the owner has to argue that the bridge will be worth the expense.
For the given geometry, location, design life, and expected loading, the most economical material (steel, prestressed concrete, reinforced concrete, etc.) and form (box
girder, truss, cable stayed, etc.) are selected. Material properties are required, and a
finite element model of the structure is developed.

Loads have to be obtained. Is the bridge going to be in earthquake prone area? Will
there be tidal waves? Scouring? Barge impact? What kind of wind forces are we
looking at? What are the uncertainties in the loads and what load magnitudes shall
we design for? How do we combine the loads? Do we collect data? How large should
the data set be?

The cycle of design and analysis continues until a final form is obtained. Costs must
be contained, functional requirements have to be met, and safety must be ensured.
How are failure criteria to be defined, for both collapse and functionality? And
how safe is safe enough? Is an explicit dynamic analysis of the bridge structure necessary? Is the finite element model of the structure accurate enough? Does it account
for nonlinearities near failure? Does it represent realistic boundary conditions? What
are the uncertainties in the model?

Construction begins and quality must be maintained. Once construction is complete, the bridge is put into service. Then occasional maintenance (and sometimes

Innovative Bridge Design Handbook. http://dx.doi.org/10.1016/B978-0-12-800058-8.00007-4

major maintenance) need to be performed. Is the bridge deteriorating? Is traffic becoming heavier? How often should the bridge be inspected, and how extensively should it be repaired? Can the bridge be kept closed to traffic, and for how long— for a month, for a day, for 6 h? Should the bridge be made stronger or more durable at the construction stage so that maintenance actions can be minimized? Finally, the bridge may become too unfunctional or too costly to maintain, too risky to operate, or both. It is then demolished, and a new one built in its place. The cycle restarts.

In this chapter, we isolate the key concepts discussed here and treat them systematically.

2 Uncertainty in bridge modeling and assessment

2.1 Probabilistic modeling of uncertain phenomena

In the context of probability theory, the *sample space* is the universal set of all possible events. Probabilities are assigned to an appropriately defined collection of subsets or events (called a *sigma algebra*) of the sample space. A random experiment implies the occurrence of an event. When the outcome of an experiment can be given in numerical terms, then there is a random variable (RV) in hand. Any possible outcome of an RV is called a *realization*. An RV can be either discrete or continuous. If a quantity varies randomly with time, we model it as a stochastic process, which can be viewed as a family of RVs indexed by time. If a quantity varies randomly in space, we model it as a random field, which is the generalization of a stochastic process in two or more dimensions. You should be familiar with the basic concepts of probability theory and RVs and processes; if review is needed, refer to standard texts (Ang and Tang, 1975; Papoulis and Pillai, 2002; Hines et al., 2003) for a refresher.

2.1.1 Common random variables encountered in structural reliability

A RV is governed by its probability laws. The probability law of an RV can be described by any of the following equivalent ways: cumulative distribution function (CDF), probability density function (PDF) for continuous RVs, probability mass function (PMF) for discrete RVs, characteristic function (CF) and moment-generating function (MGF).

Although any nondecreasing function bounded by 0 and 1 can be a candidate probability distribution function for an RV X, only a few models (e.g., normal, Poisson, geometric, and Weibull) are commonly used by the scientific and engineering community. This is because the underlying process appears repeatedly in a wide variety of problems. Deriving models solely from data, without basing them on the underlying physics, is very expensive and often inconclusive as well.

Uniform distribution arises naturally when there is no reason to favor one outcome over another from the sample space, making all sample points equally likely. This distribution also corresponds to the state of maximum Shannon entropy.

The Bernoulli trial refers to a binary outcome: $X = 0$ (often called *failure*) that occurs with probabilty q and $X = 1$ (often called *success*) that occurs with probability p, so that $p + q = 1$. A sequence of independent and identical Bernoulli trials can help model large classes of phenomena of engineering interest. The number of trials before the first success gives rise to the geometric distribution. In general, the number of trials before the rth success gives rise to the Pascal (or negative binomial) distribution. The number of successes in a fixed number of Bernoulli trials, on the other hand, follows the Binomial distribution.

The concept of *mean (or average) return period* (also called *mean recurrence interval*) arises from a sequence of independent and identically distributed (IID) Bernoulli trials. Let success in the Bernoulli trial refer to the occurrence of event A (so that failure means the nonoccurrence of A)—typically a relatively rare phenomenon, like annual rainfall exceeding 50 in., annual maximum wave height exceeding 20 m, annual maximum wind speed exceeding 150 km/h, and earthquake magnitude exceeding 7 on the Richter scale. The time (or, more literally, the number of trials), T, between successive occurrences of event A in a sequence of Bernoulli trials is an RV. T follows the geometric distribution due to the IID nature of the trials. Hence, the mean of T is $1/p$ time units, where p is the probability of occurrence of A in each trial (or time unit). In the continuous time scenario, the time between occurrences is exponentially distributed and the mean occurrence time is the reciprocal of the occurrence rate of the underlying Poisson process.

The Gaussian (or normal) distribution appears as the limiting form for the sum of a number of RVs, subject to certain conditions (Resnick, 1999), and is the most widely used model for continuous RVs. In structural engineering, dead loads are almost exclusively modeled as Gaussian. In fact, in the absence of evidence to the contrary, the Gaussian distribution is the default choice. The exponentiated Gaussian gives the lognormal RV and is popular in the literature of structural reliability, especially for nonnegative quantities. Extreme value theory has given rise to three limiting forms (Galambos, 1987)—Gumbel, Frechet, and Weibull—and are often adopted for the distribution of the maximum (e.g., wind, wave, traffic) or the minimum (e.g., strength) of a process.

The typical loads considered in bridge analysis and design are dead loads, live loads (mostly traffic loads (Wen, 1990; Bhattacharya, 2008; Guzda et al., 2007; Nowak, 1993; Enright et al., 2013)), wind loads, and earthquake loads. Dead loads represent the gravity loads (i.e., self-weight) of various components of the bridge starting from prefabricated elements and cast in situ members to wearing surfaces and fixtures. Depending on the location of the bridge, snow load, wave load, impact load, and other elements may also be considered. NBS 577 (Ellingwood et al., 1980) lists the mean bias and coefficient of variation (COV) and distribution for common material resistance and load RVs, and these are widely adopted in the structural reliability literature.

Uncertainties in material properties, and to a lesser extent those in geometries, lead to uncertainties in strength. Uncertainties in boundary conditions (e.g., the extent of joint fixity) are typically listed under modeling uncertainties.

2.1.2   Common stochastic processes encountered in structural reliability

Various types of stochastic processes appear in the analysis of structural reliability. Mostly, they represent load processes. In some cases, however, strength degradation also needs to be modeled as stochastic processes. Broadly, load processes are either sustained or intermittent. The sustained kind can be further subdivided into (approximately) time-invariant, such as dead loads, and fluctuating, such as occupant live loads. The sustained loads can be modeled as RVs. Intermittent loads, such as seismic loads, are present for a very short period of time compared to the life of the structure. In the limit, intermittent loads can be modeled as pulses with random magnitudes occurring at random instants of time. Wind loads and traffic loads can have both fluctuating components (low-level continuous) and intermittent pulses (storms and heavy trucks). In many cases, it is only the life-time maximum of the fluctuating or intermittent load processes, rather than detailed temporal characteristics, that may be required in structural reliability analysis. In such cases, the said maximum is modeled as a random variable. The corresponding design quantity is a characteristic value of the distribution of the RV, which may be defined as n-year return period value or some other quantile.

Pulse processes, occurring randomly in time with random pulse magnitudes, are particularly suited for modeling the occurrence of heavy trucks, high winds, high waves, and earthquakes on bridges—so long as the within-event variations are not important. If these event variations do need to be considered [e.g., to determine the response history (of the order of a minute) of a bridge due to a strong motion earthquake], then the occurrence can still be modeled as a pulse, but the frequency content, envelope function, and other details of the load time history will also be needed. The Poisson process is the most common model for pulse processes.

The stationary Gaussian process is the most common model for continuous stochastic processes. It can be fully defined in terms of the mean and the covariance functions. Nonstationary and non-Gaussian processes may be created through various transformations and filtering of the stationary Gaussian process (Shinozuka and Sato, 1967; Ghanem and Spanos, 2012; Vanmarcke, 1983).

2.1.3   Types of uncertainty

Uncertainties in engineering problems occur as a result of natural variability, incomplete information, or imperfect knowledge. A traditional classification system for uncertainties is either Type I or Type II. Type I uncertainties (also known as *natural, inherent,* or *aleatory*) cannot be reduced, as they are intrinsically associated with the quantity. Type II uncertainties (also known as *modeling* or *epistemic*), on the other hand, can be reduced with increased information or sophistication. A more modern classification of uncertainty is statistical, parameter, or modeling; these are preferred since they give a greater resolution to the analyst.

Regardless of classification, the mathematical representation of uncertainty must follow the probability laws described here and is generally described by RVs. Apart from the type of distribution, two dimensionless constants are popularly used to describe a RV: the mean bias, B, which is the ratio of the mean to the nominal (or predicted or handbook) value:

$$B = \frac{\mu}{X_n} \tag{1}$$

(the median bias can also be defined similarly), and the COV, which is the normalized standard deviation:

$$V = \frac{\sigma}{\mu}. \tag{2}$$

2.1.4   Statistical uncertainty

Suppose that the mathematical model of a phenomenon requires the use of an RV X with the distribution function F_X. We do not wish to probe further where the uncertainty in X is coming from. We are content to treat X as random and F_X as a black box. We can either use an empirical form for F_X, or assume a parametric form (e.g., normal, Weibull) and obtain its parameters from data. Most RVs used in structural reliability problems represent this kind of uncertainty.

2.1.5   Parameter uncertainty

Suppose that we know from analytical, subjective, or experimental considerations that an RV X follows the distribution g that is governed by a set of parameters, $\underline{\theta}$. However, there may be uncertainties about the exact value of $\underline{\theta}$. In that case, for any fixed value of $\underline{\theta}$, g is the *conditional* distribution of X. If the parameters are now expressed as a vector of random variables, $\underline{\Theta}$, then we can write

$$P\left[X \leq x | \underline{\Theta} = \underline{\theta}\right] = g\left(x; \underline{\theta}\right)$$

$$F_X(x) = P[X \leq x] = \int\limits_{\text{all } \underline{\theta}} g\left(x; \underline{\theta}\right) f_{\underline{\Theta}}\left(\underline{\theta}\right) d\underline{\theta} \tag{3}$$

to obtain the unconditional distribution of X.

Zio and Apostolakis (1996) say that some model uncertainties can in fact be used as parameter uncertainties; e.g., when using a Monte Carlo simulation, a flag can be used to switch some models on or off, or just use a "switch case" kind of parameter that will select one model at a time in repeated simulations. It may also not always be possible to separate model uncertainties for parameter uncertainties—for instance, there may be parameters whose values need to be estimated from available data, but how the estimates themselves are computed may depend on the model chosen in the first place.

Regardless of the classification, probability theory allows us to treat all uncertainties as RVs. For some, though, conditional distributions may be necessary.

2.1.6   Modeling uncertainty

Analysis tools for predicting global response, stress analysis, and other things are commonly deterministic in nature. The model predictions deviate from the actual due to three broad types of reasons: mathematical idealizations, numerical errors, and ignorance. Mathematical idealization includes simplifications such as neglect of nonlinearities; ignorance effectively leads one to neglect a group of variables. The difference between idealization and ignorance is that in the former, one knows what is being left out, while in the latter, one does not. Gallegos and Bonano (1993) named these as mathematical model uncertainty, conceptual uncertainty, and computer code uncertainty, respectively.

All three introduce new RVs into the reliability problem, and hence into the limit state. Ditlevsen (1982) incorporated modeling uncertainty into the limit state equation by transforming the vector of basic variables into another random vector of the same dimension; i.e., by substituting the basic variables $\underline{X}$ with $\underline{V(X)}$.

In the aggregate sense, the model uncertainty, M, in predicting some property or response may be expressed as

$$M = \frac{\text{actual}}{\text{predicted (or nominal)}} \qquad (\text{predicted} \neq 0). \qquad (4)$$

M is an RV because the exact deviation is unknown. Of course, in many situations, actual results may just not be available; e.g., failure pressure of an actual nuclear power plant containment under pressurization due to an actual core meltdown. On such occasions, we can have only competing models, and in some cases, scaled model test results under idealized conditions.

Some modeling uncertainty distributions cannot be estimated from data. They can only be estimated from subjective probabilities given by a group of experts (Lind and Nowak, 1988; Cooke and Goossens, 2008; Keeney and Winterfeldt, 1991).

Examples of treatment of modeling uncertainty

A simple example is the yield strength (Ellingwood et al., 1980), Y, of a steel member, which is commonly modeled as the product of three RVs representing intrinsic and extrinsic uncertainties, and the nominal value, Y_n:

$$Y = B_P B_M B_F Y_n, \qquad (5)$$

where B_P accounts for professional or modeling error, B_M is the material variability, and B_F is the fabrication error. For example, if these three factors are considered to be mutually statistically independent and lognormally distributed, then the yield strength Y also is lognormal.

Nikolaidis and Kaplan (1991) performed a survey of uncertainties in FEA (finite element analysis) in marine and other industries (such as automobile and aerospace). Their conclusions and findings are the following: Depending on the loading case, the mean bias in FEA of containership ranged from 0.9 to 1.4, and the COV from 0.1 to 0.4. For aerospace structures, the stress modeling uncertainty is uniformly distributed with mean 1.0 and COV 0.12. In the automobile industry, FEA underestimates the flexibility of a car body. The error in predicting deflection due to bending or torsion ranges between 10% and 20%. For offshore structures, the uncertainty in modeling member forces has a mean bias between 0.8 and 1.1, with COV between 0.2 and 0.4.

In fatigue strength computation (Fricke and Muller-Schmerl, 1998), the permissible stress range, $\Delta\sigma_P$, in a component may be determined by the use of several adjustment factors on the S-N curve-based reference stress range, $\Delta\sigma_R$, which corresponds to $N = 2\times10^6$:

$$\Delta\sigma_{max} \leq \Delta\sigma_P = \Delta\sigma_R f_n f_m f_R f_t f_s f_w f_c, \tag{6}$$

where f_n takes into account the effect of the stress spectrum (compared to the constant amplitude assumption of the S–N curve); f_m accounts for material type; f_R accounts for the mean-stress effect; f_t accounts for the plate thickness effect; f_s accounts for imperfections; f_w accounts for weld shape improvements; and f_c accounts for corrosive environments. Of these, the factors f_n, f_R and f_t may be considered as representing modeling uncertainty.

Similar approaches have been taken to derive strain-based limit states for nuclear power plant containments. Cherry and Smith (2001) adopted the Hancock model in his fragility analysis of steel containments. They defined the equivalent plastic strain at failure, ε_{fail}, in terms of the fracture ductility, $\varepsilon_{f,uni}$, and four correction factors $(f_1, ..., f_4)$ as follows:

$$\varepsilon_{fail} = \varepsilon_{f,uni} \times f_1 \times f_2 \times f_3 \times f_4 \tag{7}$$

where $f_1 = 1.6 \exp(-3\sigma_m/2\sigma_{von})$, f_2 accounts for material variability, f_3 accounts for model sophistication (i.e., modeling error), and f_4 accounts for random corrosion degradation (reduction of ductility). This failure criterion can be applied locally in conjunction with a finite element analysis.

In seismic design, the target displacement, δ_t, of the control node at the rooftop of a building may be calculated as (Whittaker et al., 1998):

$$\delta_t = C_0 C_1 C_2 C_3 S_a \frac{T_e^2}{4\pi^2}, \tag{8}$$

where C_0 is the modification factor to relate spectral displacement and expected maximum inelastic displacement at the roof; C_1 is the modification factor to relate the expected maximum inelastic displacements to displacements calculated for linear elastic response; C_2 is the modification factor to represent the effects of stiffness degradation, strength deterioration, and pinching on the maximum displacement response; C_3 is the modification factor to represent increased displacements due to dynamic second-order effects; S_a is the response spectrum acceleration at the effective fundamental period and damping ratio of the building; and T_e is the effective fundamental period of the building in the direction under consideration, calculated using the secant stiffness at a base shear force equal to 60% of the yield force. The factors C_1, C_2

and C_3 serve to modify the relation between mean elastic and mean inelastic displacements where the inelastic displacements correspond to those of a bilinear elastic plastic system. The factors C_0 through C_3 may be considered as representing modeling error.

The environmental load effect due to sea-load, wind, ice, and earthquake is (Moan, 1997):

$$S = KC_1C_2...E^{\alpha}, \tag{9}$$

where K is a constant, C_1 is the transfer from environmental conditions to load, C_2 is the transfer from the load to the load effect, E is the characteristic environmental parameter, and α is a constant. E usually follows an extreme value distribution (e.g., Gumbel). K, C_1, C_2, ... are generally random and may be assumed to be lognormal. Each of these transfer functions may be decomposed into a nominal value and a modeling error variable.

In seismic code development, Cornell et al. (2002) separated uncertainty into four parts: β_{DU} (uncertainty in estimating the median demand), β_{DR} (the record to record variability in demand), β_{CU} (the uncertainty in estimating the capacity), and β_{CR} (the randomness in capacity), so that the total uncertainty (in the lognormal standard deviation sense) in demand and capacity are, respectively:

$$\begin{aligned} \beta_D^2 &= \beta_{DU}^2 + \beta_{DR}^2 \\ \beta_C^2 &= \beta_{CU}^2 + \beta_{CR}^2 \end{aligned} \tag{10}$$

β_{DU} and β_{CU} correspond to modeling uncertainty type, while β_{DR} and β_{CR} correspond to the statistical uncertainty type. In a related work, Yun et al. (2002) considered β_{NTH}, the uncertainty in nonlinear time history analysis, and assumed it to be 0.15, 0.20, and 0.25 for 3-, 9-, and 20-story buildings, respectively.

3 Reliability of bridges

For a structure with several critical locations subject to time-dependent loads and possessing time- and space- dependent material properties, the reliability function estimates the probability that the capacity, C, exceeds the demand, Q, at all locations at all times that the structure is in service:

$$\text{Rel}(t) = 1 - P_f(t) = P\left[C^j(\underline{x}, \tau) \geq Q^j(\underline{x}, \tau), \forall \tau \in (0, t), \forall \underline{x} \in \Omega, j \in J\right], \tag{11}$$

where Ω is the set of critical locations of the structure and t is the total time horizon. Both capacity and demand of the structure are generally functions of space and time and constitute a multidimensional stochastic process. Capacity can change randomly due to aging or other time-dependent effects, and can recover due to maintenance operations. Further, there may be several modes of failure j (e.g., shear, flexural, deflection), as indicated by the superscripts to C and Q, associated with any given location. The demand represents the effect of all loads acting simultaneously on the structure (e.g., dead, live, wind) and may be expressed either in load space or in load effect space:

$$Q^j(\underline{x}, \tau) = DL^j(\underline{x}, \tau) + LL^j(\underline{x}, \tau) + WL^j(\underline{x}, \tau) + \ldots \tag{12}$$

The $+$ sign indicates combination, not necessarily superposition, and thus may involve nonlinear effects.

The structural reliability problem in its most general formulation is thus infinite-dimensional both in time and space, which makes it computationally intractable; hence, engineering judgment and various simplifications and restrictions are adopted. For example, if there were only one critical location with only one failure mode, and demand and capacity were time invariant as well, Eq. (11) would boil down to a time-independent element level reliability problem in two RVs:

$$\text{Rel} = P[C_0 - Q_0 > 0], \tag{13}$$

which could easily be computed with the help of the joint PDF of C_0 and Q_0:

$$\left. \begin{aligned} \text{Rel} &= \int_{-\infty}^{\infty} \int_{c=q}^{\infty} f_{C_0, Q_0}(c, q) \, dc \, dq \\ &= \int_{-\infty}^{\infty} [1 - F_{C_0}(q)] f_{Q_0}(q) \, dq \\ &= \int_{-\infty}^{\infty} F_{Q_0}(c) f_{C_0}(c) \, dc \end{aligned} \right\} \text{if } C_0 \text{ is independent of } Q_0 \tag{14}$$

Thus, the way to making the general reliability problem [Eq. (11)] tractable is to identify only the key sets of (i) failure modes, (ii) load combinations, (iii) critical locations, and (iv) temporal statistics of important processes, so that only a manageable number of failure events need to be analyzed and checked against acceptance criteria.

Figure 7.1 shows the important steps involved in structural reliability analysis. The concept of limit states, various solution techniques for the reliability problem, and an introduction to explicit time-dependence into the reliability problem are discussed next.

3.1 Limit states

One of the first steps in a structural reliability analysis is to identify the failure modes (or, more generally, nonperformance modes) of the structure. A *limit state* is the boundary between the safe (or acceptable) and failed (or unacceptable) domains of structural performance in the failure mode under consideration. It is represented with the help of the limit state function (also called the *performance function*), $g(\underline{X})$, in the following manner:

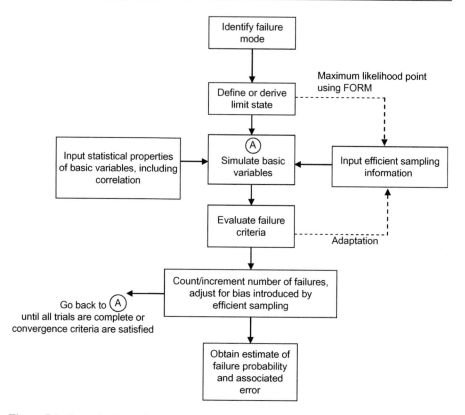

Figure 7.1 General scheme for reliability analysis.

$g\left(\underline{X}\right) < 0$, unacceptable or failed domain

$g\left(\underline{X}\right) \geq 0$, acceptable or safe domain

$$(15)$$

so that $g\left(\underline{X}\right) = 0$ is the limit state equation.

The boundary between the two regions, $g(\underline{X}) = 0$, is called the *limit state equation*. $\underline{X}$ is the set of basic variables that consist of the complete set of quantities used to describe structural performance in the failure mode under consideration. They may include material properties, loads or load effects, environmental parameters, geometric quantities, and modeling uncertainties. Basic variables in a limit state are usually modeled as RVs; however, those with negligible uncertainties may be treated as deterministic.

Limit states may be defined for elements as well as the system (Figure 7.2). The difference between element and system in reliability analysis has less to do with the scale and complexity of the participating component/assembly/substructure, and more to do with the form of the limit state function and whether one needs to undertake Boolean combinations, discussed next.

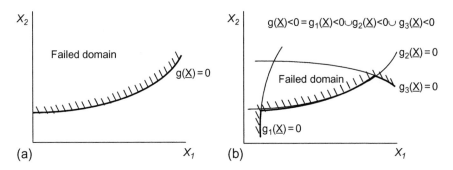

Figure 7.2 Limit state functions (a) for an element, and (b) for a series system. The failure domain is indicated by the hashing.

3.1.1   Structural limit states and load combinations used in bridges

There are broadly two kinds of failure for a structure: *irreversible* and *reversible*. Irreversible failure can be divided into two types:

- *Overload*—e.g., ultimate failure that happens under a single high-loading event. Design codes refer to these as *strength/extreme/accidental limit states*. This type of failure is irreversible in nature. The structure needs to be repaired or replaced after such failure. The consequences of such failure are serious, even catastrophic.
- *Cumulative damage*—e.g., fatigue cracking. It too is irreversible in nature and the structure needs to be repaired or replaced after such failure. The consequences of such failure can be serious. However, this damage proceeds gradually and can be detected through inspection before failure occurs.

Reversible structural damage is temporary in nature and typically has to do with the functional requirements of the bridge (e.g., deflection, vibration). There is no lasting damage, but the structure is not available for the duration of this kind of failure. The consequences of such failure are usually minor. Most serviceability limit states listed in design codes are in this category.

Strength and serviceability limit states can be formulated both as element- and system-level limit states depending on the objective and available information. For each of these, several load combinations need to be evaluated [Eq. (12)]. For example, the American Association of State Highway and Transportation Officials (AASHTO) bridge load and resistance factor design (LRFD) code (AASHTO, 2012) specifies five strength-type load combinations (involving dead loads and various live loads, with or without wind loads), two extreme event load combinations (involving dead loads and reduced live loads with earthquake or ice/collision/flood loads), four service load combinations (involving dead loads, various live loads, and wind loads, among others) and two fatigue load combinations (involving only live loads).

Eq. (11), therefore, simplifies to checking the following groups of limit states one at a time:

overload: $C^j(\underline{x}_i) \geq Q^j(\underline{x}_i; LC_k(t))$

$\vdots$

cumulative: $C^{j'}(\underline{x}_{i'}) \geq Q^{j'}(\underline{x}_{i'}, LC_{k'}(t))$ (16)

$\vdots$

reversible: $C^{j''}(\underline{x}_{i''}) \geq Q^{j''}(\underline{x}_{i''}; LC_{k''})$.

3.1.2 Element-level limit states

If it is possible to define a single differentiable performance function $g(\underline{X})$ of the basic variables for a given failure mode, then we have what is known as an *element reliability problem*. An example is shown in Figure 7.1(a). An element reliability problem is most naturally realized in the case of a single critical cross section of one structural component in a single failure mode (such as flexural failure); in such cases, the function g is commonly derived from analytical/mechanistic modeling. It can be the same function used in a corresponding deterministic analysis, with some or all of the variables now treated as RVs. However, it is entirely possible that the performance function corresponding to the roof displacement of a tall building under wind loading can be derived in the form of a single response surface given in terms of a relevant set of basic variables (obtained from a set of finite element analyses of the structure); in this case, the reliability of excessive roof displacement for the building also will qualify as an element reliability problem.

The simple two-variable linear problem, as in Eq. (13), is an element reliability problem, the basic variables are $\underline{X} = [C_0, Q_0]^T$, and the limit state equation is

$$g(\underline{X}) = C_0 - Q_0. \qquad\qquad (17)$$

Typical modes of failures in bridge structures that give rise to element reliability problems include yielding, crushing, buckling, fatigue failure etc. Element reliability problems are easy to formulate and inexpensive to compute.

Example 1: A small structural design problem (Figure 7.3).

Consider a cable (8-in. diameter) in a suspension bridge made of A36 steel with random yield strength Y (time invariant). Let Y be Weibull distributed with COV 15%. It is a 1-RV problem. No modeling uncertainty is considered. The axial load $q = 1600$ kip and the cross-sectional area $a = 50.3$ in^2 are deterministic. Let cable

Figure 7.3 A cable in tension.

failure be defined as yield of the gross section. Find the failure probability of the cable. The target failure probability is 0.001. Redesign if necessary.

The mean yield strength of A36 steel is 38 ksi. The shape and location parameters of Y are, therefore, $V_Y = 15\% \Rightarrow k = 8$ and $u = \frac{\mu}{\Gamma(1+1/8)} = \frac{38}{.94} = 40.4$ ksi. The failure event is

$$\{\text{Failure}\} = \left\{\frac{q}{a} > Y\right\}, \tag{18}$$

and the probability of failure,

$$P_f = P[\text{failure}] = P\left[Y < \frac{1600\text{kip}}{50.3\,\text{in}^2}\right] \tag{19}$$
$$= P[Y < 31.8] = 1 - e^{-\left(\frac{31.8}{40.4}\right)^8} = 0.14,$$

is solved using the Weibull CDF.

Since it is required that $P[\text{failure}] < .001$, the cable is inadequate. Reliability can be increased in four ways with this problem: increasing the area, reducing the load, increasing the mean strength, and decreasing the variability of strength. Of these actions, the second is not possible without restricting traffic, and the third and fourth would require a different material and possibly be very expensive. Thus, we decided to first try to increase the cross-sectional area.

The revised cross-sectional area can be found by finding the inverse of the CDF at the target P_f:

$$\therefore P\left[Y < \frac{q}{a_{new}}\right] = .001 \Rightarrow 1 - e^{-\left(\frac{q}{a_{new}\,40.4}\right)^8} = .001, \tag{20}$$

which yields

$$a_{new} = \frac{1600}{40.4 \times .4217} = 93.9\,\text{in}^2. \tag{21}$$

Suppose that the resultant diameter (about 11 in.) proves to be impractical. The next option is to try a different grade of steel without changing the diameter. Assume that the distribution of Y_{new} remains Weibull and its COV remains 15%. The approach now is to select a new mean. The target probability of failure remains 0.001:

$$P\left[Y_{new} < \frac{q}{a}\right] = .001, \tag{22}$$

which yields

$$\exp\left[-\left(\frac{31.8}{u_{new}}\right)^8\right] = .999$$
$$\Rightarrow u_{new} = 75.4 \tag{23}$$
$$\Rightarrow \mu_{new} = 75.4\,\Gamma(1+1/8) = 70.9\,\text{ksi}.$$

The new mean strength is acceptable provided this new grade of steel has sufficient ductility, corrosion resistance, and other desirable properties. Otherwise, a totally new design may need to be adopted.

3.1.3 System-level limit states

As should be clear by now, the difference between an element and a system in a reliability analysis context is somewhat arbitrary and largely dependent on the available information and scale of interest. Indeed, a problem of tensile failure of a prismatic rod made of a brittle material that can be treated as a simple element reliability problem from a continuum viewpoint may amount to an intractable system reliability problem from microstructural considerations. For practical purposes, it is mostly the availability of a single, differentiable, and closed-form performance function that separates an element reliability problem from a systems one.

It would be highly desirable, then, to somehow cast the performance of a structural system in terms of a single limit state (perhaps using approximate numerical techniques, such as a response function fit) and thereby take advantage of the speed, elegance, and accuracy of element reliability solution techniques; such formulation unfortunately remains elusive more often than not. Needless to say, structural system failure events are thankfully so rare (and in any case, structural systems can hardly be deemed to constitute a nominally identical sample) that the other alternative—a frequentist interpretation of structural system reliability—is not feasible. The usual systems reliability formulation, therefore, is presented in terms of Boolean combinations of element limit states depending on the logical construct of the system in terms of its components and the definition of failure at the systems level (Bhattacharya et al., 2009). An example is shown in Figure 7.2(b).

If the system failure event can be cast as an intersection of m element failure events (i.e., a classical parallel system), then the system failure probability is

$$\text{Parellel system}: \quad P_{f,sys} = P\left[\bigcap_{i=1}^{m} g_i \leq 0\right], \tag{24}$$

where g_i's are the element limit state surfaces in the basic variable space ($\underline{X}$). For a series system type configuration, the system failure probability is

$$\text{Series system}: \quad P_{f,sys} = P\left[\bigcup_{i=1}^{m} g_i \leq 0\right]. \tag{25}$$

For systems more general than the simple series and parallel organizations, the greatest challenge is to identify the minimal cut sets (at least the dominant ones), particularly in light of the circumstances peculiar to structural systems mentioned previously. A set of elements of a system is a *cut set* if the failure of all members of the cut set causes system failure (Birolini, 1999). A *minimal cut set* is one that if any element is removed from it, the subset no longer remains a cut set.

If the cut sets $C_i, i = 1, \ldots, n_c$ can be identified for the system, the system failure probability becomes

$$\text{Series-parallel system}: \quad P_{f,sys} = P\left[\bigcup_{i=1}^{n_c} C_i\right] = P\left[\bigcup_{i=1}^{n_c}\left\{\bigcap_{j=1}^{n_i} g_{ij} \leq 0\right\}\right], \tag{26}$$

where g_{ij} is the jth limit state in cut set i. Reaching an exact solution of Eq. (26) may be impossible, thus bounds on system reliability, based on marginal events (Cornell, 1967), pairs of joint events (Ditlevsen, 1979) or triplets of joint events (Hohenbichler and Rackwitz, 1983) are available. Cut sets, without regard to an ordering of element failure events, can be determined for elastic–perfectly plastic structures.

The binary nature of elements (being in either failed or safe states), though not always a realistic representation (elements may have multiple failure states), facilitates the use of standard methods such as fault or event trees (or a combination of the two), including their variants to suit the peculiarities of structural systems, to describe system failure in terms of component failure events, and hence to identify minimal cut or minimal path sets of the system. System reliability computation for structures is not straightforward since the component failures are not mutually independent events due to (i) active redundancy in the structure leading to load sharing, (ii) load path dependence in the case of successively applied multiple yet sustained loads, (iii) load redistribution after initial member failures for redundant structures, (iv) nonlinear behavior and nonbrittle failure of the components, (v) failure sequences of different probabilities for the same cut set in a progressive collapse or incremental loading situation, and (vi) possible statistical dependence among the basic variables.

3.2 Computation of reliability

The failure probability corresponding to Eq. (15) is given by the multidimensional integral in the basic variable space:

$$P_f = P\left(g(\underline{X}) < 0\right) = \int_{g(\underline{x}) < 0} f_{\underline{X}}(\underline{x}) d\underline{x}, \tag{27}$$

where $f_X(x)$ is the joint probability density function for $\underline{X}$. The reliability of the structure would then be defined as $Rel = 1 - P_f$. A rather popular measure of reliability in structural reliability literature is the generalized reliability index, β, which is defined as the normal probability inverse: $\beta = \Phi^{(-1)}(1 - P_f)$. Typical values of β range from 2 to 5 for most structural components.

Closed-form solutions to Eq. (27) are generally unavailable. Two different approaches are widely in use: (i) analytic methods based on constrained optimization and normal probability approximations, and (ii) simulation-based algorithms, with or without variation reduction techniques. Both can provide accurate and efficient solutions to the structural reliability problem. The first kind, grouped under first-order reliability methods (FORM), holds a distinct advantage over the simulation-based methods, in that the design points and the sensitivity of each basic variable can be explicitly determined. Furthermore, the reliability index β owes its geometric interpretation to FORM (cf. Eq. (29)).

3.2.1 FORM

FORM calculates the reliability of a system by mapping the failure surface onto the standard normal space and then by approximating it with a tangent hyperplane at the design point (defined as the point on the limit state surface in the standard normal space that is closest to the origin) (Shinozuka, 1983). Provided the limit state surface is well behaved, the solutions obtained by FORM are reasonably close to that obtained by the relatively expensive simulation-based solutions.

The two important steps of FORM are described in detail in the following paragraphs.

1. First, map the basic variables $\underline{X}$ to the independent standard normal space $\underline{Y}$, and hence $g(\underline{X})$ to $g_1(\underline{Y})$. Several mappings are possible, such as (i) Hasofer-Lind (Hasofer and Lind, 1974) or second-moment transformation, which uses information only on the first two moments of each X; (ii) Nataf transformation (Melchers, 1987), which uses marginal distribution of each X and the correlation matrix of the $\underline{X}$ vector; (iii) Rosenblatt transformation (Melchers, 1987), which uses nth order joint distribution information, a special case of which is the so-called full distribution transformation valid when the $\underline{X}$ are mutually independent; (iv) the Rackwitz-Fiessler transformation (Rackwitz and Fiessler, 1978), which converts each X point by point into an equivalent normal U through a marginal distribution and density equivalence, and then the vector $\underline{U}$ into the independent standard normal vector $\underline{Y}$ through a Nataf type transformation.
2. Next, locate on g_1 the point $\underline{y}^*$ closest to the origin:

$$\min F = \underline{y}^T \underline{y}$$
$$\text{subject to } G = g_1\left(\underline{y}\right) = 0. \tag{28}$$

Let the solution to this optimization problem be $\underline{y}^*$ and let β be the distance of this optimal point from the origin. This minimum norm point $\underline{y}^*$, is known as the *checking* or the *design point*. The limit state surface g1 can be approximated by a tangent hyperplane at y^*, yielding the approximate probability of failure as

$$P_f = \Phi\left(-\beta \operatorname{sgn}\left[g_1\left(\underline{0}\right)\right]\right). \tag{29}$$

The signum function determines whether the origin is in the safe domain. The drawback of FORM is that it provides the exact solution only if the original limit state is linear and the basic variables are normally distributed. Otherwise, the extent of error depends on the curvature of the limit state and the method of mapping of $\underline{X}$ to $\underline{Y}$.

After performing a FORM analysis, the design point $\underline{y}^*$ can be transformed back into the basic variable space, yielding the checking point, $\underline{x}^*$, which cannot be obtained from simulation-based solutions. It is implied that if the structural element in question is designed using this combination $\underline{x}^*$, the reliability of the component would be β (within the approximations of FORM). This, in fact, is the basis of load and resistance factor design, discussed subsequently.

The gradient projection method, originally developed by Rosen (1961), is well suited to tackle the constrained nonlinear optimization problem in Eq. (28).

3.2.2 Monte carlo simulations

Except in very special situations, closed-form solution to the structural reliability problem [Eq. (27)] does not exist and numerical approximations are needed. The true probability of failure, P_f,

$$P_f = \int_{\text{all } x} \mathbf{I}[\{\text{Failure}\}] f_{\underline{X}}(\underline{x}) d\underline{x} = \int_{\text{all } u} \mathbf{I}[\{\text{Failure}\}] f_{\underline{U}}(\underline{u}) d\underline{u}, \tag{30}$$

can be estimated using basic (also called *brute-force* or *crude*) Monte Carlo simulations (MCS) in practice as

$$\hat{P}_f = \frac{1}{N} \sum_{i=1}^{N} \mathbf{I}[g(T(\underline{U}_i)) < 0], \tag{31}$$

where a zero-mean normal vector $\underline{U}$ with the same correlation matrix ρ as the basic variables is generated first and then transformed element by element according to the full distribution transformation:

$$T(\underline{u}) = \underline{x} \Rightarrow F_{X_i}(x_i) = \Phi(u_i). \tag{32}$$

The use of the same ρ for $\underline{U}$ as for $\underline{X}$ results in error, but the error is generally small (der Kiureghian and Liu 1986). N is the total number of times the random vector $\underline{U}$ is generated, and $\underline{U}_i$ is the ith realization of the vector. It is well known that the basic Monte Carlo simulation–based estimate of P_f has a relatively slow and inefficient rate of convergence. The COV of the estimate is

$$\hat{V}(\hat{P}_f) = \sqrt{(1 - P_f)/(NP_f)} \approx \sqrt{1/(NP_f)}, \tag{33}$$

which is proportional to $1/\sqrt{N}$ and points to an inefficient relation between sample size and accuracy (and stability) of the estimate. Such limitations of the basic Monte Carlo simulation technique have led to several variance-reducing refinements. Notable among them are Latin hypercube sampling (Ayyub and McCuen, 1995), importance sampling along with its variants (Melchers, 1990; Bjerager, 1988), and subset simulations (Au and Beck, 2001), which, if performed carefully, can significantly reduce the required sampling size. Nevertheless, importance sampling and other variance-reducing techniques should be performed with care, as their results may be very sensitive to the type and the point of maximum likelihood of the sampling distribution, and an improper choice can produce erroneous results (Sen and Bhattacharya, 2015).

3.2.3   System reliability computation

An ordered sequence of failure events from a cut set is variously termed in structural systems reliability analyses, sometimes with subtle differences among them, as failure sequence or failure path. To be specific, a failure sequence under incremental loading accounts for load redistribution after each component failure, while a failure path does not, and leads to different events whenever load redistribution occurs after each successive component failure (Bjerager et al., 1987). The terms *failure mode* and *collapse mode*, unfortunately, have been used in the literature to denote a cut set both with and without regard to ordering of failure events and have led to confusion in some cases. We prefer *collapse mode* to imply a cut set without reference to failure order, and *failure sequence* to imply an ordered sequence from a cut set. A path set is sometimes referred to as a *stable configuration* (Bennet and Ang, 1986), although this approach is rarely taken in structural problems.

Depending on the structural complexity and desired accuracy of the solution, the dominant failure sequences (or collapse modes) can be found in a variety of ways. Some of these involve only a deterministic analysis of the structural system, while others employ a fully probabilistic analysis, and still others use some limited probabilistic information. The assumption of rigid perfectly plastic material behavior is fairly popular in structural system reliability analysis, as it eliminates load history dependence. It is well known that deterministic plastic mechanism analysis can lead to collapse mode identification in case of rigid-plastic framed structures, although the number of modes generated quickly becomes huge (Watwood, 1979; Gorman, 1981). Such deterministic rules have been variously adapted to search for the probabilistically dominant collapse modes by (i) creating linear combinations of those basic mechanisms that have the lowest reliability indices [the beta-unzipping method (Thoft-Christensen and Murotsu, 1986)], (ii) using linear programming (Corotis and Nafday, 1989), (iii) using stochastic programming (Zimmerman et al., 1993), (iv) using genetic algorithms (Shao and Murotsu, 1999) etc.

The probabilistically dominant failure sequences can be searched using truncated enumeration schemes that include the branch and bound method (Thoft-Christensen and Murotsu, 1986) and, importantly, the incremental loading method (Karamchandani, 1987; Moses, 1997). The incremental loading method is particularly useful (and often is the only way out) when component failure is multistate instead of the usual binary (Karamchandani and Cornell, 1992b), material behavior is brittle, semibrittle, or nonlinear instead of ideal plastic (Karamchandani and Cornell, 1992a), and system failure occurs not due to formation of a mechanism, but due to excessive deformation or a specified drop in structural stiffness with regard to specified degrees of freedom (DOF). Nevertheless, one potential drawback of the incremental analysis method is its quasi-static assumption of structural behavior: the load duration needs to be sufficiently long to allow potential redistribution of load effects throughout the system.

3.3 Specifying target reliabilities for design and assessment

It has become increasingly common to express safety requirements, as well as some functionality requirements, in reliability-based formats. A reliability-based approach to design, by accounting for randomness in the different design variables and uncertainties in the mathematical models, provides tools for ensuring that the performance requirements are violated as rarely as considered acceptable. Such an approach comes under the broad classification of performance based design (PBD). In structural engineering, PBD has most enthusiastically been espoused in the seismic engineering community, as evident in SEAOC (SEAOC, 1995), ATC-40 (ATC, 1996), FEMA 273 (FEMA, 1997), FEMA 350 (FEMA, 2000), and other sources.

Mathematically, we go back to Eq. (16) and set a lower limit to the reliability, or equivalently, an upper limit to the failure probability, for each limit state:

$$1 - \text{Rel}(t) = P_f(t) \leq P_f^* = \Phi(-\beta_T). \tag{34}$$

where P_f^* is the maximum permissible failure probability and β_T is the equivalent target reliability index. The cause, reference period, and consequences of violation of different limit states may vary, and if a reliability approach is taken, the target reliability for each limit state must take such differences into account (ISO, 1998; JCSS, 2001; Bhattacharya et al., 2001; Wen, 2001). For example, if the structure gives appropriate warning before collapse, the failure consequences are reduced, and that in turn can reduce the target reliability for that mode (JCSS, 2001; DNV, 1992). Functionality target reliabilities may be developed exclusively from economic considerations. The safety target reliability levels required of a structure (i.e., in strength or ultimate type limit states), on the other hand, cannot be left solely to the discretion of the owner or be derived solely from a minimum total expected cost consideration, since structural collapse causing a large loss of human life, property, or both, may not be acceptable either to the society or the regulators, even if it is an "optimal" solution in some ways. Design codes, therefore, often place a lower limit on the reliability of safety related limit states (Bhattacharya et al., 2001; Galambos, 1992).

3.3.1 Code-specified target reliabilities

Conventional structures that have a history of successful service, such as concrete buildings, highway bridges, and steel vessels, can be deemed sufficiently safe, and their calculated reliability levels may be used as the targets for new structures of the same kind. This, in principle, is done when a new reliability-based code is developed for a given class of structures having a successful history of use and a wide knowledge base about their performance (Ellingwood and Galambos, 1982). The objective is to produce more uniform levels of safety and more optimal structures. ISO 2394 (1998), and later JCSS (2001), proposed three levels of requirements with appropriate degrees of reliability: (i) serviceability (adequate performance under all expected actions), (ii) ultimate (the ability to withstand extreme or frequently repeated actions during construction and anticipated use), and (iii) structural integrity (i.e., progressive collapse in ISO 2394 and robustness in JCSS). Target reliability values were suggested based on the consequences of failure (C) and relative cost of safety measure (S) (JCSS, 2001). In ultimate limit state, permissible P_f ranged from 10^{-3}/year for minor C and large S, to 10^{-5}/year for moderate C and normal S, down to 10^{-6}/year for large C and small S. In serviceability limit state, the maximum annual failure probability ranged from 0.1 (high S) to 0.01 (low S).

The Canadian Standards Association (CSA, 1992) defines two safety classes and one serviceability class (and corresponding annual target reliabilities) for the verification of the safety of offshore structures:

- Safety class 1—Great risk to life or high potential for environmental pollution or damage
- Safety class 2—Small risk to life or low potential for environmental pollution or damage
- Serviceability impaired function and none of the other two safety classes being violated

Det Norske Veritas (DNV, 1992) specifies three types of structural failures for offshore structures and target reliabilities for each corresponding to the seriousness of the consequences of failure. The American Bureau of Shipping (ABS, 1999) identified four levels of failure consequences for various combinations of limit states and component class for the concept mobile offshore base, and assigned target reliabilities for each.

3.3.2   Bridge structures

Ghosn and Moses (1998) suggest three levels of performance to ensure adequate redundancy of bridge structures corresponding to functionality, ultimate, and damaged condition limit states, while Nowak et al. (1997) recommend two different reliability levels for bridge structures corresponding to ultimate and serviceability limit states.

Nowak et al. (1997) recommend a lifetime target component reliability index β of 3.5 and a target system reliability index of 5.5 in ultimate limit states for bridge structures. For serviceability limit states, they recommend a target component (i.e., girder) reliability index of 1.0 in tension and 3.0 in compression. They also compute component reliabilities of different kind of bridges (reinforced concrete, prestressed concrete, and steel built to AASHTO 1992 and BS 5400 specifications) in bending, shear, and serviceability limit states.

Ghosn and Moses (1998) suggest the following reliability requirements to ensure adequate redundancy of a highway bridge structure:

$$\beta_u - \beta_1 \geq 0.85, \ \ \beta_f - \beta_1 \geq 0.25, \ \ \beta_d - \beta_1 \geq -2.7. \tag{35}$$

The subscripts $1, f, u$ and d refer to first member failure, functionality limit state, ultimate state, and damaged condition limit state, respectively.

The design of the Confederation Bridge (in Northumberland, Canada) required that load and resistance factors be calibrated to "a β of 4.0 for ultimate limit states, for a 100 year life" (MacGregor et al., 1997). Sarveswaran and Roberts (1999) chose an acceptable annual failure probability of bridge collapse in UK equal to 2×10^{-5} which corresponded to an FAR of 2 (FAR is discussed in Section 3.3.4).

3.3.3 Loss-based approaches

The risk of an undesirable event is commonly defined as:

$$\text{Risk} = p \times C, \tag{36}$$

where p is the probability of occurrence of the event and C is the consequence of event (lives lost, lost revenue, monetary compensation, lost utility, etc.). Eq. (36) is valid when there is only one level of undesirable consequence. A more general expression would be

$$\text{Risk} = \sum p_i \times C_i. \tag{37}$$

The term *risk* is also used in the public health and actuarial literature in the sense of an individual's probability of death. The definition of risk and what constitutes the consequences of failure depend on whose risk is at stake—the public's, a corporation's, or an individual's. Once the tolerable risk R^* is known, and C can be quantified, the maximum permissible failure probability can be set:

$$P_f^* = \frac{R*}{C}. \tag{38}$$

The actual risk from an activity may be markedly different from the risk perceived by the public. Society's general reaction to hazards of different levels can range from indifference to rationality to dread. If exposure to an activity is voluntary, the acceptable level of risk is generally higher. Involuntary activities, on the other hand, have a much less acceptable risk to an individual. In the absence of proper information about a perceived hazardous activity, the public may have a "dread risk." Appreciating this fact, the maximum tolerable risk suggested in the Netherlands for existing situations is 10^{-5}/person/year, while for new situations, it is 10^{-6}/person/year (Bottelberghs, 1995). However, it needs to be underlined that a society's sense of tolerable risk for a given activity may change with time.

3.3.4 Fatality-based approaches

When the loss from failure is measured in terms of human lives lost, there are several fatality-based approaches to setting target reliabilities. It is nevertheless controversial to put a monetary value on human life. Various agencies and researchers have investigated levels of probability that are acceptable to society for events that cause fatalities, as described next. The acceptable probabilities depend on the nature of the hazard and decrease with the increasing number of fatalities.

As reported in MSC 72/16 (IMO, 2000), the Health and Safety Executive (HSE) in the United Kingdom suggests 10^{-4}/person/year as the limit of fatality risk to members of the general public. A Construction Industry Research and Information Association report (CIRIA, 1977) report developed an empirical formula for setting the annual target failure probability as

$$P_f = \frac{K_s}{n_r} p' \, /yr, \tag{39}$$

where p' is the basic annual probability of death accepted by an individual member of society (its typical value in the UK being 10^{-4}), K_s is the social criterion factor, and n_r is the aversion factor defined as the number of lives involved. K_s accounts for the voluntary nature of hazardous activity (a person may be willing to increase his or her exposure by a factor of K_s) and its typical value is 5. Public aversion to an accident is assumed to be directly proportional to the number of lives involved. However, other nonlinear relations have also been proposed. For example, Allen (1981) offered a somewhat different formula for annual target failure probability that incorporated the nature of warning available for the impending failure:

$$P_f = \frac{A}{W\sqrt{n_r}} 10^{-5} \, /yr, \tag{40}$$

where n_r is the aversion factor as before, A is the activity factor, and W is the warning factor. The factor 10^{-5} in Eq (40) was derived from data on building collapses in Canada. For normal activities, A ranges from 1 (buildings) to 10 (high-exposure edifices like offshore structures) and equals 3.0 for bridges. W ranges from 0.01 (fail-safe conditions) to 1.0 (failure without warning). Note that Eq. (40) uses $\sqrt{n_r}$ rather than n_r in the denominator, implying that the rate of growth in risk aversion decreases with the number of fatalities. Later, ISO (1998) tied the acceptable failure probability to the square of the number of lives involved, signifying perhaps a decrease in the public's sense of tolerable risk in engineered systems.

A somewhat different measure of hazardous activities that accounts for exposure time is the fatal accident rate (FAR). The FAR for an activity is the number of fatalities per 100 million h of exposure to that activity (i.e., 1000 people working 2500 h/year and having working lives of 40 years each):

$$FAR = 10^8 P[F]/T_h, \tag{41}$$

where $P[F]$ is probability of fatality and T_h is the exposure time in person-hours. Typical values of FAR in the UK (Mander and Elms, 1993) range from 5 (chemical processing industry) to 67 (construction industry). FARs for various activities in Japan (Suzuki, 1999) include 0.2 for fires, 4.3 for railway travel, and 46.3 for civil aviation.

4 Reliability-based design codes of bridges

4.1 PSFs

Reliability-based PSF design is intended to ensure a nearly uniform level of reliability across a given category of structural components for a given class of limit state under a particular load combination (Ellingwood, 2000). We approach the topic of optimizing PSFs by noting that any arbitrary point, $\underline{x}^a$, on the limit state surface, by definition, satisfies the following:

$$g\left(\underline{x}^a\right) = 0, \tag{42}$$

We can, for example, choose each member of $\underline{x}^a$ to correspond to a particular quantile of the respective element of the random vector $\underline{X}$, such that Eq. (42) defines a functional relation among these quantiles. By choosing different values for $\underline{x}^a$, we can effectively move the joint density function of $\underline{X}$ with respect to the limit state surface. Clearly, this relative movement in the basic variable space affects the limit state probability. In other words, by specifying a functional relation among quantiles (or some other statistics) of the basic variable $\underline{X}$, we can affect the reliability of the structure.

Extending this idea, a design point $\underline{x}^d$ on the limit state surface can be carefully chosen so that it locates the limit state in the space of basic variables such that a desired target reliability is ensured for the design. The ensuing design equation,

$$g\left(\underline{x}^d\right) = 0, \tag{43}$$

is essentially a relationship among the parameters of the basic variables and gives a minimum requirement type of tool in the hand of the design engineer to ensure target reliability for the design in an indirect manner. Since nominal or characteristic values of basic variables are typically used in design, Eq. (43) may be rewritten as

$$g\left(\frac{x_1^n}{\gamma_1}, \ldots, \frac{x_k^n}{\gamma_k}, \gamma_{k+1}x_{k+1}^n, \ldots, \gamma_m x_m^n\right) \geq 0, \tag{44}$$

where the superscript n indicates the nominal value of the variable. We have partitioned the vector of basic variables into k resistance type and $m-k$ action type quantities. The PSFs, γ_i, are typically greater than 1: for resistance-type variables, they divide the nominal values, while for action type variables, they multiply the nominal values to obtain the design point as follows:

$$
\begin{aligned}
&\text{Resistance PSFs}: \gamma_i = \frac{x_i^n}{x_i^d}, \quad i = 1, \ldots, k \\
&\text{Action PSFs}: \gamma_i = \frac{x_i^d}{x_i^n}, \quad i = k+1, \ldots, m.
\end{aligned}
\tag{45}
$$

If the design equation [Eq. (44)] can be separated into a strength term and a combination of load-effect terms, the following safety checking scheme may be adopted for design:

$$
R_n\left(\frac{S_i^n}{\gamma_i^s}, i = 1, \ldots, k\right) \geq l\left(\sum_{i=1}^{m-k} \gamma_i^q Q_i^n\right),
\tag{46}
$$

where R_n is the the nominal resistance and a function of factored strength parameters, l is the load-effect function, S_i^n is the nominal value of the ith strength/material parameter, γ_i^s is the ith strength/material factor, Q_i^n is the nominal value of the ith load, and γ_i^q is the ith load factor. Note that there is no separate resistance factor multiplying the nominal resistance (as in LRFD), since material PSFs have already been incorporated in computing the strength.

The nominal values generally are fixed by professional practice and thus are inflexible. Some of the m PSFs (often those associated with material properties) can also be fixed in advance. The remaining PSFs can be chosen by the code developer so as to locate the design point, and hence locate the limit state as alluded to previously, and hence achieve a desired reliability for the structure.

4.2 Calibration of PSFs

By normalizing the limit state with the design equation in a two variable problem, the reliability problem can be written as

$$
\begin{aligned}
&\text{Find } \gamma_1^s, \ldots, \gamma_k^s, \gamma_1^q, \ldots, \gamma_{m-k}^q \text{ such that} \\
&P\left[\frac{C}{C^n\left(\gamma_1^s, \ldots, \gamma_k^s\right)} - \frac{Q}{Q^n\left(\gamma_1^q, \ldots, \gamma_{m-k}^q\right)} \leq 0\right] = \Phi(-\beta_T),
\end{aligned}
\tag{47}
$$

where β_T is the target reliability index, C is the random capacity, and C^n is its nominal value. Of course, this is an underdefined problem, and even though some of the PSFs may be fixed in advance as stated previously, it has an infinite number of solutions. Additional considerations are needed to improve the problem definition. Such considerations naturally arise when PSFs are needed to be optimized for a class of structures and are discussed next.

It is common to expect that the design equation be valid for r representative structural components (or groups). Let w_i be the weight (i.e., relative importance or relative frequency) assigned to the ith such component (or group). These r representative components may differ from each other on account of different locations, geometric dimensions, nominal loads, material grades, and other factors. For a given set of PSFs, let the reliability index of the ith group be β_i. Choosing a new set of PSFs gives a new design, a new design point, and consequently, a different reliability index. If there has to be one design equation (i.e., one set of PSFs) for all the r representative components, the deviations of all β_is from β_T must in some sense be minimized. The design equation [Eq. (44)], when using the optimal PSFs obtained this way, can ensure a nearly uniform reliability for the range of components. Several constraints may be introduced to the optimization problem to satisfy engineering and policy considerations (as summarized in Agrawal and Bhattacharya, 2010). Moreover, some PSFs, such as those on material strengths, may be fixed in advance, as stated previously. The PSF optimization exercise takes the following form:

$$\min\left[\sum_{i=1}^{r}w_i\left(\beta_i\left(\gamma_1^q,...,\gamma_{m-k}^q\right)-\beta_T\right)^2\right]\text{ where }\sum_{i=1}^{r}w_i=1$$

$$\text{subject to: }\min(\beta_i)>\beta_T-\Delta\beta,\quad i=1,...,r \tag{48}$$

$$\gamma_i^{\min}\le\gamma_i^q\le\gamma_i^{\max},\ i=1,...,m-k$$

$$\gamma_i^s=m_i,i=1,...,k.$$

The weighted squared error from the target reliability index over all groups is minimized while ensuring that the lowest reliability among all the groups does not drop by more than $\Delta\beta$ below the target. The material PSFs are fixed, while the load PSFs have upper and lower limits.

5 Bridge life cycle cost and optimization

In life cycle cost analysis of bridges, cost to owners ("agency") as well as the public ("users") need to be taken into account (NCHRP, 2003). Agency costs include design, construction, maintenance, repair, and replacement (less salvage value). If failure occurs, then costs may include compensation and cleanup. For users, costs arise from accidents, delays, and detours. Since some costs are fixed (i.e., deterministic) while some are outcome dependent (i.e., random), the total cost (i.e., life cycle cost),

$$C_T=C_I+\sum_{n_i}C_M(t_i)+\sum_{n_j}C_U\left(t_j\right)+C_F, \tag{49}$$

is probabilistic in nature. C_F is either the replacement cost (C_{rep}) at the end of life or the failure cost (C_f) that occurs at some random instant T_f. The maintenance and user costs, C_M and C_U, are also uncertain, as they depend on future loading and aging effects and whether the bridge fails before design life. Hence, the total expected cost can be written as

$$E[C_T]=C_I+\sum_{n_i}E[C_M(t_i)I(t_i)]+\sum_{n_j}E\left[C_U\left(t_j\right)I\left(t_j\right)\right]+\left(1-P_f\right)C_{rep}+P_fC_f.$$

$$\tag{50}$$

The indicator function I verifies whether the bridge has survived up to the indicated time. Discounting of future costs can also be included (Lind, 1993). In a decision making context, the total expected cost is minimized, subject to constraints like available budget and target reliability.

Decisions regarding new design as well as maintenance, therefore, require explicit determination of the bridge's time-dependent reliability function, which is discussed next.

5.1 Time-dependent structural reliability

5.1.1 Descriptors of the TTF

Let T denote the random time to failure (TTF), also known as *failure-free operating time* or *lifetime,* of an item. The reliability function Rel(t) evaluated at time t is the probability that the item survives beyond t:

$$\text{Rel}(t) = P[T > t] = \int_t^\infty f_T(\tau)d\tau, \tag{51}$$

where f_T is the probability density function of T. The hazard function, $h(t)$, which is the conditional density of the TTF, presents the same information differently and can be very useful in revealing unsafe conditions:

$$h(t) = \frac{f_T(t)}{\text{Rel}(t)} \text{ so that } \text{Rel}(t) = \exp\left[-\int_0^t h(\tau)d\tau\right]. \tag{52}$$

Statistics of T are routinely obtained for electronic/electrical components through accelerated testing programs. This is possible because (i) an abundant number of nominally identical specimens can be obtained, (ii) a large amount of test data can be generated in a relatively short time, (iii) tests can be performed in near-actual conditions, (iv) tests are not hazardous and (v) tests are relatively inexpensive.

For civil engineering structures, very seldom are all five of these points satisfied. Also, actual failure data are, thankfully, rare. In the parlance of system reliability, structures constitute active redundant systems with load sharing and dependence— the most difficult type of system to model for reliability analysis.

Nevertheless, time-dependent reliability functions are useful for civil engineering systems not only at the new design stage, but also for scheduling future maintenance, for posting load restrictions, and for managing life cycle costs as already explained. The reliability function is obtained from the mechanics of the problem where time-varying behavior of some of the basic variables is now brought into the picture explicitly.

5.1.2 Capacity and demand vary nonrandomly in time

Without loss of generality, we look at only one critical location and one failure mode of the structure given in Eq. (11). For multiple critical locations and failure modes, the limit state discussed here can be augmented by unions of individual failure events.

At a given location and for a given failure mode, let the capacity and demand vary deterministically in time:

$$C(\tau) = C_0\, d(\tau)$$
$$Q(\tau) = Q_0\, h(\tau),$$
(53)

where C_0 and D_0 are RVs and d, h are nonrandom functions of time, $d > 0, h > 0$. That is, if the process $C(\tau)$ is known at any instant t_1, its value can be known precisely at all other instants of time; likewise for $Q(\tau)$. Due to the nonrandom nature of d and h, the reliability function

$$\mathrm{Rel}(t) = P[C_0\, d(\tau) - Q_0\, h(\tau) > 0, \text{ for all } \tau \in (0, t]]$$
(54)

can be written as

$$\mathrm{Rel}(t) = P\left[C_0 - Q_0 \max_{0 < \tau \leq t} \frac{h(\tau)}{d(\tau)} > 0\right].$$
(55)

d is commonly the aging function. Its form can be derived from the mechanics of damage growth [e.g., corrosion loss (Bhattacharya et al., 2008b) and fatigue crack growth (Kwon and Frangopol, 2011)] and the loading history. Here, $d = 1$ implies that the capacity does not degrade with time, and $h = 1$ implies that the load is sustained in time. This approach still will be valid for several simultaneously occurring loads [cf. Eq. (12)] if one could write

$$Q_0\, h(\tau) = Q_0^{(1)} \cdot h_1(\tau) + Q_0^{(2)} \cdot h_2(\tau) + Q_0^{(3)} \cdot h_3(\tau) + \ldots,$$
(56)

in which h represents nonrandom functions of time and the initial load magnitudes $Q_0^{(i)}$ are RVs.

Example 2: We define a time-dependent problem based on Example 1. The cable is subject to uniform corrosion, causing its radius, whose initial value $r_0 = 4$ in, to deteriorate as $\Delta r(t) = b_1 t^{b_2}$, where $b_1 = 0.1 \mathrm{in/yr}^{b_2}$, $b_2 = 0.9$ are the corrosion law constants. The cross-sectional area thus deteriorates according to $a(t) = \pi(r_0 - \Delta r)^2$.

The cable is made of A36 steel, whose yield strength Y is now assumed to be normally distributed with mean $\mu_Y = 38$ ksi and COV $V_Y = 15\%$. The load, Q_0, is invariant and sustained in time and is now considered a normal RV. Its mean is $\mu_Q = 1000$ kip and the COV is $V_Q = 20\%$. In the context of Example 1, the mean bias of the load is $1000/1600 = 0.625$. The load and capacity are independent.

Compute the reliability and hazard functions. The reliability function [Eq. (55)] for this problem can be simplified as follows:

$$\mathrm{Rel}(t) = P\left[Y - Q_0 \max_{0 < \tau \leq t} \frac{1}{\pi(r_0 - b_1 \tau^{b_2})^2} > 0\right]$$
$$= P\left[Y - Q_0 \frac{1}{\pi(r_0 - b_1 t^{b_2})^2} > 0\right]$$
(57)
$$= P\left[\pi(r_0 - b_1 t^{b_2})^2 Y - Q_0 > 0\right]$$
$$= P[M(t) > 0].$$

Note that due to the monotonically decreasing nature of $d(\tau)$, the limit state is evaluated *only* at the right end point of the interval $(0,t]$. In any other situation, this simplification would be wrong and would lead to dangerous overprediction of reliability.

The margin process M is normally distributed as a linear combination of normals. Its mean and variance at time t are

$$\mu_M(t) = a(t)\mu_Y - \mu_Q$$
$$\sigma_M^2(t) = a^2(t)\sigma_Y^2 + \sigma_Q^2 \tag{58}$$

The reliability function, therefore, can be expressed as the normal CDF:

$$\mathrm{Rel}(t) = \Phi\left(\frac{\mu_M(t)}{\sigma_M(t)}\right). \tag{59}$$

Differentiating the reliability function leads to the hazard function:

$$h(t) = -\frac{\phi\left(\dfrac{\mu_M(t)}{\sigma_M(t)}\right)}{\Phi\left(\dfrac{\mu_M(t)}{\sigma_M(t)}\right)} \frac{\dot{\mu}_M(t)\sigma_M(t) - \mu_M(t)\dot{\sigma}_M(t)}{\sigma_M^2(t)} \tag{60}$$

These two functions are plotted in Figure 7.4. The choice of normal distribution for both RVs in the problem led to the closed-form expressions for reliability and hazard functions given previously. For other distributions, FORM or Monte Carlo simulations may be adopted.

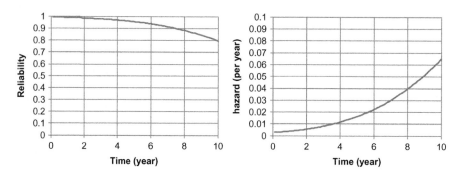

Figure 7.4 Reliability and hazard functions of corroding cable.

5.1.3 Load occurs as a pulsed sequence with random magnitudes

Known number of load pulses and no aging

We first consider the case when C is time invariant [i.e., $d \equiv 1$ in Eq. (53)] and the load occurs as pulses of random magnitude $Q_1, Q_2, \ldots, Q_{n(t)}$, with the number of load pulses n in time t being known. We assume that the loads are IID, that Q_i is independent, and each Q_i has the same distribution F_Q. Further, the loads are independent of the capacity. The reliability function,

$$\text{Rel}(t) = P\left[Q_1 < C_0, Q_2 < C_0, Q_3 < C_0, \ldots, Q_{n(t)} < C_0\right], \tag{61}$$

can be simplified by first conditioning it on an arbitrary value of C_0, and using the IID property of Q_i:

$$\text{Rel}(t|C_0 = c) = \left[F_Q(c)\right]^{n(t)}. \tag{62}$$

The total probability theorem is then applied to yield

$$\text{Rel}(t) = \int_0^\infty \left[F_Q(c)\right]^{n(t)} f_{C_0}(c) \, dc. \tag{63}$$

Q is a poisson pulse process and no aging

A point process $N(t)$ on the line $\mathbb{R}^+ = [0, \infty)$ is a set of randomly occurring points such that (i) any finite interval contains a finite number of points with probability 1, and (ii) the number of points in disjoint intervals is the sum of the individual counts (Kovalenko et al., 1996). The points are commonly designated as arrival times: $T_1, T_2, \ldots, T_i \geq 0$. The interarrival times are $\tau_1 = T_1, \tau_2 = T_2 - T_1, \ldots$, so that $T_n = \tau_1 + \tau_2 + \cdots + \tau_n$. The point process can be described by the joint distribution of (i) the arrival times, (ii) the interarrival times, or (iii) the increments in disjoint intervals. $N(t)$ is a renewal process if the interarrival times are mutually independent and identically distributed. A renewal process is Poisson if the interarrival times are exponentially distributed, or equivalently, if the increments in disjoint intervals are independent.

A Poisson process $N(t)$ is completely defined by its rate of occurrence, λ. The Poisson RV, N_t, with its mean being equal to λt, represents the number of arrivals in the Poisson process $N(t)$ in the interval $(0,t]$.

The joint distribution of the interarrival times $T_1, T_2, \ldots T_n$ given $N(t) = n$ is

$$f_{T_1, T_2, \ldots, T_n | N(t) = n}(t_1, t_2 \cdots t_n) = \begin{cases} \dfrac{n!}{t^n} & 0 < t_1 < \ldots < t_n < t \\ 0, & \text{otherwise} \end{cases}. \tag{64}$$

Now we generalize this situation and consider the loads to occur according to a Poisson pulse process (with rate λ). As before, the magnitude of the pulses are IID and independent of capacity. No aging is considered. Since the number of pulses in time interval $(0,t]$ is random, the reliability function is expressed as

$$\begin{aligned}
\text{Rel}(t) &= \sum_{n=0}^\infty P\left[\bigcap_{i=1}^n Q_i < C_0 | N(t) = n\right] P[N(t) = n] \\
&= \int_{c=0}^\infty \sum_{n=0}^\infty P\left[\bigcap_{i=1}^n Q_i < c | N(t) = n, C_0 = c\right] P[N(t) = n] f_{C_0}(c) \, dc
\end{aligned} \tag{65}$$

By using the algebraic form of the Poisson PMF, the reliability function simplifies to

$$\text{Rel}(t) = \int_0^\infty e^{-\lambda t\left(1 - F_Q(c)\right)} f_{C_0}(c)\,dc \tag{66}$$

Q is a poisson pulse process and structure ages deterministically

We now introduce aging, as in Eq. (53). Figure 7.5 shows a schematic of this situation. Since the loads occur as a Poisson pulse, the occurrence times, T_i, are random in nature, and the individual limit states are evaluated at these random instants of time:

$$\text{Rel}(t) = \sum_{n=0}^\infty P\left[\bigcap_{i=1}^n Q_i < C_0 d(T_i) | N(t) = n\right] P[N(t) = n]. \tag{67}$$

Since these random occurrence times are ordered, $T_1 < T_2 < \ldots < T_i < T_{i+1} < \ldots$, their conditional joint PDF, given that n pulses occurred in $(0,t]$, is $1/t^n$ [cf. Eq. (64)]. The reliability function, conditioned on a fixed value of C_0, then can be written as

$$\text{Rel}(t|C_0 = c) = \sum_{n=0}^\infty \iiint_{\text{all } \tau_i} P\left[\bigcap_{i=1}^n Q_i < cd(\tau_i) | N(t) = n, T_i = \tau_i, T_i < T_j, 1 \le i < j \le n\right] \times$$

$$f_{\underline{T}}\left(\underline{\tau}\right) d\underline{\tau} P[N(t) = n] \tag{68}$$

$$= \sum_{n=0}^\infty \left[\frac{1}{t} \int_{\tau=0}^t F_Q[cd(\tau)]\,d\tau\right]^n P[N(t) = n].$$

By using the algebraic form of the Poisson PMF and removing the conditioning on C_0, the reliability function simplifies to

$$\text{Rel}(t) = \int_0^\infty e^{-\lambda t\left(1 - \frac{1}{t}\int_{\tau=0}^t F_Q[cd(\tau)]\,d\tau\right)} f_{C_0}(c)\,dc. \tag{69}$$

Note that Eq (69) reduces to Eq. (66) when d is identically equal to 1.

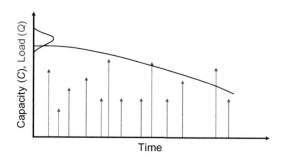

Figure 7.5 Deteriorating capacity and Poisson pulse loads with random magnitudes.

5.1.4 Load and capacity vary randomly in time

This is the most general case, and it constitutes a first passage problem (Ditlevsen and Bjerager, 1986; Lin, 1976). The rate at which the margin process $M(\tau) = C(\tau) - Q(\tau)$ crosses the zero barrier (i.e., enters or leaves the "safe" domain) at an arbitrary time t is given by the joint PDF of the process and its derivative, $\dot{M}$, at that instant:

$$\bar{\nu}_0(t) = \int\limits_{-\infty}^{\infty} |\dot{m}(t)| f_{M(t)\dot{M}(t)}(0, \dot{m}) d\dot{m}. \tag{70}$$

If the margin process is statistically stationary, the passages into the unsafe domain become asymptotically Poisson, so that the reliability function represents the probability of the first passage into the unsafe domain beyond time t:

$$R(t) = (1 - F_T(0))e^{-\bar{\nu}_0 t}, \tag{71}$$

where $F_T(0)$ is the probability that the margin is negative at $t = 0$. In this stationary case, the constant rate of downcrossing (into the unsafe domain) is

$$\nu_0^- = \int\limits_0^{\infty} \dot{m} f_{M\dot{M}}(0, \dot{m}) d\dot{m}. \tag{72}$$

Further, if the margin is stationary Gaussian, it is independent of its derivative at the same instant, and the downcrossing rate becomes

$$\nu_0^- = \frac{\sigma_{\dot{M}}}{\sqrt{2\pi}} \frac{1}{\sigma_M} \phi\left(\frac{\mu_M}{\sigma_M}\right) \text{ if } M \text{ is stationary Gaussian.} \tag{73}$$

5.2 Reliability-based maintenance of bridges

Reliability-based maintenance of nonrepairable systems is preventive in nature, as opposed to corrective maintenance, which is performed to maintain the availability of repairable systems. Consider the reliability function shown in Figure 7.4. If the target reliability is 0.9 and the remaining life is 10 years, then this item becomes unacceptable at around $t_u = 8$ years. Four options are available:

- Replace the item with a new item at t_u.
- Repair the item before t_u (preventive maintenance).
- Make a stronger item, so that no repair becomes necessary.
- Restrict loads.

This section discusses the second option. It can be placed in the context of minimizing the total expected cost [Eq. (50)] subject to such constraints as budget and reliability.

Repair can be either perfect (in which the item is made as new), or partial (only a fraction of the original strength is restored). The question is: How is the reliability function altered due to periodic maintenance? In other words, we are looking to describe the conditional reliability $\text{Rel}(t|M_{0:t})$ given the maintenance plan, M, up to time t. Note that we are still looking into the future when we are trying to predict $\text{Rel}(t|M_{0:t})$, i.e., the analyst's position on the time axis is $t=0$. Thus, the conditional reliability function would still have the essential properties of the unconditional reliability—namely, it is a nonincreasing function that drops from 1 to 0 with time.

Although not generally recommended (but some authors do), one could also add the survival history up to time t and look at the question again. The difference is subtle but important. This would happen if the analyst were placed at some point in time in the future (say at t_0) and asked how the reliability function would behave henceforth. That is, one would estimate the conditional reliability $\text{Rel}(t|M_{0:t}, S_{0:t})$, where S gives the survival information up to time t. The plot of $\text{Rel}(t|M_{0:t}, S_{0:t})$ would no longer behave monotonically but would jump to 1 at each discontinuous point t_0 where the structure is known to have survived. It is easy to show that this jump would happen even in the absence of any maintenance operation, but just because the structure survived up to t_0. $\text{Rel}(t|M_{0:t}, S_{0:t})$ is not a reliability function in the strict sense; rather, it is a piecewise juxtaposition of several reliability functions and must be interpreted cautiously.

To illustrate this point, we assume that only one maintenance operation is performed on the structure, which occurs at time t_R. It is convenient to start with the hazard function. It is altered due to the maintenance operation:

$$h(t) = \begin{cases} h_0(t), t < t_R \\ h_1(t), t \geq t_R \end{cases}. \tag{74}$$

The reliability function [Eq. (52)] then becomes

$$\text{Rel}(t) = \begin{cases} \text{Rel}(t), & t < t_R \\ \text{Rel}(t_R) \exp\left[-\int_{t_R}^{t} h_1(\tau)d\tau\right], t \geq t_R \end{cases}. \tag{75}$$

If perfect repair is undertaken at t_R, then the hazard function undergoes a time shift:

$$\text{Perfect repair at } t_R: \quad h_1(t) = h_0(t - t_R), \ t \geq t_R \tag{76}$$

and the reliability function is repeated as a scaled version of itself:

$$\text{Perfect repair at } t_R: \quad \text{Rel}\left(t|M_{t_R}^{100\%}\right) = \begin{cases} \text{Rel}(t), & t < t_R \\ \text{Rel}(t_R) \cdot \text{Rel}(t - t_R), t \geq t_R \end{cases}. \tag{77}$$

The event $M_{t_R}^{100\%}$ signifies 100% repair at time t_R.

Example 3: Now we repeat Example 2 with 100% repair performed at 5 years (shown by the red lines in Figure 7.6). It is clear that due to the repair, the reliability function stays above 0.9 at the end of the 10-year life as required.

Generalizing, if the repair is imperfect, we start with the second factor in Eq. (75) for $t \geq t_R$ and rewrite it as

$$
\begin{aligned}
\exp\left[-\int_{t_R}^{t} h_1(\tau)d\tau\right] &= \exp\left[-\int_{0}^{t-t_R} h_1(\tau + t_R)d\tau\right]\\
&= \exp\left[-\int_{0}^{t-t_R} h_1'(\tau)d\tau\right]\\
&= \mathrm{Rel}'(t - t_R),
\end{aligned}
\tag{78}
$$

where h_1' is a legitimate hazard function (generally different from h_0 due to the imperfect nature of the repair), and $\mathrm{Rel}'(t)$ is the corresponding reliability function, which is generally different from (and less benign than) $\mathrm{Rel}(t)$. The reliability function due to imperfect repair can then be written as

$$
\text{Imperfect repair at } t_R: \quad \mathrm{Rel}\left(t|M_{t_R}^{\alpha\%}\right) = \begin{cases} \mathrm{Rel}(t), & t < t_R \\ \mathrm{Rel}(t_R) \cdot \mathrm{Rel}'(t - t_R), & t \geq t_R \end{cases}.
\tag{79}
$$

$M_{t_R}^{\alpha\%}$ represents imperfect repair at time t_R, in which the strength is restored to $\alpha\%$ of the initial value. The green lines in Figure 7.6 correspond to $\alpha = 90$. The effect is not as good as perfect repair, as can be expected.

If, in addition, the condition is imposed that the structure is found to survive at t_R, then the conditional reliability starts from 1 at t_R as stated previously, and all past information is erased:

$$
\begin{aligned}
\mathrm{Rel}\left(t|T > t_R, M_{t_R}^{\alpha\%}\right) &= \frac{P\left[T > t|M_{t_R}^{\alpha\%}\right]}{P\left[T > t_R|M_{t_R}^{\alpha\%}\right]}, \quad t \geq t_R\\
&= \frac{\mathrm{Rel}\left(t|M_{t_R}^{\alpha\%}\right)}{\mathrm{Rel}(t_R)}, \quad t \geq t_R\\
&= \mathrm{Rel}'(t - t_R), \quad t \geq t_R.
\end{aligned}
\tag{80}
$$

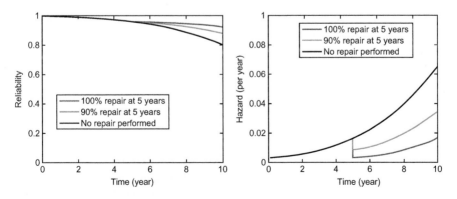

Figure 7.6 Effect of perfect and partial repair on reliability and hazard functions.

6   Load and resistance factor rating methodology

As part of periodic inspection, a bridge may need to be rated for load-carrying capacity. Load rating a bridge gains urgency in the face of changed traffic pattern or due to any change in the health of the bridge. When load-rating a bridge, the best model is the bridge itself. By monitoring the bridge, one can gather in-service traffic and performance data and conduct in-service evaluations. NCHRP Project 12-46 (NCHRP, 2001) led to the development of a reliability based bridge rating Manual (AASHTO 2003) that was consistent with AASHTO's reliability-based LRFD approach for design of new bridges. The method was termed load and resistance factor rating (LRFR), and like LRFD, LRFR specifications were still based on design parameters and non-site-specific data. Nevertheless, they did open the door for using site-specific information to load rate bridges, e.g., by using weigh-in-motion data and obtaining site specific live load factors. The recent AASHTO (2015) Manual for Bridge Evaluation includes the older deterministic allowable stress and load factor rating methodologies, in addition to the modern LRFR approach for condition evaluation of bridges.

The load-rating equation for exisiting bridges takes the following general form (Wang et al., 2011a):

$$RF = \frac{\phi C_n - \gamma_D D_n}{\gamma_L (L_n + I_n)},$$
(81)

where C_n is the nominal capacity; D_n is the nominal dead load; L_n is the nominal live load; I_n is the nominal impact; and ϕ, γ_D, and γ_L are the capacity, dead load, and live load factors, respectively. The rating may be performed at various live load levels—inventory, operating, etc. The factors should ideally be derived from probabilistic considerations.

It may be relatively time consuming and expensive to inspect and instrument every bridge in a jurisdiction's inventory (Bhattacharya et al., 2005). If in-service response from a limited number of sites can be deemed representative of a larger suite of bridges, the rating factors can be optimized for the entire suite of bridges (similar to the principle applied in LRFD and LRFR), and bridge owners may determine the safety of bridges in their inventory using such optimized rating equations. The factors can be adjusted to take care of aging (Bhattacharya et al., 2008a) and system (Wang et al., 2011b) effects.

Summary

Various sources of uncertainty affect a bridge structure during its life: first at the design stage, then during construction, and then throughout its useful life after it has been put into service. These uncertainties are modeled as random variables, random processes or random fields as appropriate. Performance requirements of a bridge are described in terms of limit state functions. The exceedance probabilities of these limit states, i.e., the probabilities of non-performance, need to be kept within acceptable limits. Reliability-based design and maintenance, whether through first principles or by using codes of practice, can ensure compliance. There are various methods of deciding acceptable failure probabilities (or, equivalently, target reliabilities). Once the bridge is put into service, its load characteristics may change and the structure may be subjected to various forms of (generally random) deterioration. Time dependent reliability analyses of an aging bridge, coupled with preventive maintenance, can ensure that reliability does not fall below acceptable limits. A suit of bridges can be rated in service by optimized site-specific partial safety factors.

References

Agrawal, G., Bhattacharya, B., 2010. Optimized partial safety factors for the reliability based design of rectangular prestressed concrete beams. J. Struct. Eng. SERC Madras 37, 263–273.

Allen, D.E., 1981. Criteria for design safety factors and quality assurance expenditure. In: Third Intl. Conf. on Structural Safety and Reliability, Trondheim, Norway. June 1981.

AASHTO, 2003. Guide Manual for Condition Evaluation and Load and Resistance Factor Rating (LRFR) of Highway Bridges. American Assoc. of State Highway and Transportation Officials, Washington, DC.

American Association of State Highway and Transportation Officials (AASHTO), 2012. LRFD Highway Bridge Design Specifications, sixth ed. American Association of State Highway and Transportation Officials, Washington, DC.

AASHTO, 2015. Manual for Bridge Evaluation. American Assoc. of State Highway and Transportation Officials, Washington, DC.

American Bureau of Shipping (ABS), 1999. Draft Mobile Offshore Base Classification Guide. American Bureau of Shipping, Houston, TX.

Ang, A.H.S., Tang, W.H., 1975. Probability Concepts in Engineering Planning and Design. Wiley, New York.

Applied Technology Council (ATC), 1996. ATC-40 Siesmic Evaluation and Retrofit of Existing Concrete Buildings. Applied Technology Council, Redwood City, CA.

Au, S., Beck, J., 2001. Estimation of small failure probabilities in high dimensions by subset simulation. Prob. Eng. Mech. 16, 263–277.

Ayyub, B.M., Mccuen, R.H., 1995. Simulation-based reliability methods. In: Sundararajan, C. (Ed.), Probabilistic Structural Mechanics Handbook: Theory and Industrial Applications. Chapman Hall, New York.

Bennet, R.M., Ang, A.H.-S., 1986. Formulation of structural system reliability. J. Eng. Mech. 112, 1135–1151.

Bhattacharya, B., 2008. The extremal index and the maximum of a dependent stationary pulse load process observed above a high threshold. Struct. Safe. 30, 34–48.

Bhattacharya, B., Basu, R., Ma, K.-T., 2001. Developing target reliability for novel structures: the case of the mobile offshore base. Marine Struct. 14, 37–58.

Bhattacharya, B., Li, D., Chajes, M.J., Hastings, J., 2005. Reliability-based load and resistance factor rating using in-service data. J. Bridge Eng. 10, 530–543.

Bhattacharya, B., Li, D., Chajes, M.J., 2008a. Bridge rating in the presence of strength deterioration and correlation in load process. Struct. Infra. Eng. 4, 237–249.

Bhattacharya, B., Li, D., Chajes, M.J., 2008b. Bridge rating using in-service data in the presence of strength deterioration and correlation in load process. Struct. Infrastruct. Eng. 4, 237–249.

Bhattacharya, B., Lu, Q., Zhong, J., 2009. Reliability of redundant ductile structures with uncertain system failure criteria: a study on a highway steel girder bridge. Sadhana 34, 903–921.

Birolini, A., 1999. Reliability Engineering Theory and Practice. Springer-Verlag, New York.

Bjerager, P., 1988. Probability integration by directional simulation. J. Eng. Mech. 114, 1285.

Bjerager, P., Karamchandani, A., Cornell, C.A., 1987. Failure tree analysis in strucutral system reliability. ICASP 5.

Bottelberghs, P.H., 1995. QRA in the Netherlands. In: Conference on Safety Cases, IBC/DNV. London.

Canadian Standards Association (CSA), 1992. General Requirements, Design Criteria, the Environment, and Loads. A National Standard of Canada, Toronto.

Cherry, J.L., Smith, J.A., 2001. Capacity of steel and concrete containment vessels with corrosion damage, NUREG/CR-6706. USNRC, Washington, DC.

Construction Industry Research and Information Association (CIRIA), 1977. Rationalization of Safety and Serviceability Factors in Structural Codes. CIRIA, London Report No. 63.

Cooke, R.M., Goossens, L.L.H.J., 2008. TU Delft expert judgment data base. Reliab. Eng. Sys. Saf. 93, 657–674.

Cornell, C.A., 1967. Bounds on the reliability of structural systems. J. Struct. Div. ST1, 171–200.

Cornell, C., Jalayer, F., Hamburger, R., Foutch, D., 2002. Probabilistic basis for 2000 SAC federal emergency management agency steel moment FREM guidelines. J. Struct. Eng. 128, 526–533.

Corotis, R.B., Nafday, A.M., 1989. Structural system reliability using linear programming and simulation. J. Struct. Eng. 115, 2435–2447.

der Kiureghian, A., Liu, P.L., 1986. Structural reliability under incomplete probability information. J. Eng. Mech. ASCE 112 (1), 85–104.

Det Norske Veritas (DNV), 1992. Structural Reliability Analysis of Marine Structures, Classification Notes No. 30.6. Hovik, Norway.

Ditlevsen, O., 1979. Narrow reliability bounds for structural systems. J. Struct. Mech. 7, 453–472.

Ditlevsen, O., 1982. Model uncertainty in structural reliability. Struct. Safe. 1, 73–86.

Ditlevsen, O., Bjerager, P., 1986. Methods of structural systems reliability. Struct. Safe. 3, 195–229.

Ellingwood, B.R., 2000. LRFD: implementing structural reliability in professional practice. Eng. Struct. 22, 106–115.

Ellingwood, B.R., Galambos, T.V., 1982. Probability-based criteria for structural design. Struct. Safe. 1, 15–26.

Ellingwood, B.R., Galambos, T.V., Macgregor, J.G., Cornell, C.A., 1980. Development of a Probability Based Load Criterion for American National Standard A58. NBS Special Publication 577, U.S. Department of Commerce, National Bureau of Standards, Washington, DC.

Enright, B., Carey, C., Caprani, C.C., 2013. Microsimulation evaluation of Eurocode load model for American long-span bridges. J. Bridge Eng. 18, 1252–1260.

Federal Emergency Management Agency (FEMA), 1997. FEMA-273 NEHRP Guidelines for the Siesmic Rehabilitation of Buildings. Washington, DC.

Federal Emergency Management Agency (FEMA), 2000. FEMA-350 Recommended Seismic Design Criteria for New Steel Moment-Frame Buildings. Washington, DC.

Fricke, W., Muller-Schmerl, A., 1998. Uncertainty modeling for fatigue strength assessment of welded structures. J. Offshore Mech. Arctic Eng. 120, 97–102.

Galambos, J., 1987. The Asymptotic Theory of Extreme Order Statistics. Krieger, Malabar, FL.

Galambos, T.V., 1992. Design codes. In: Blockley, D. (Ed.), Engineering Safety. McGraw Hill, London.

Gallegos, D.P., Bonano, E.J., 1993. Consideration of uncertainty in the performance assessment of radioactive waste disposal from an international regulatory perspective. Reliability Eng. Sys. Safe. 42, 111–123.

Ghanem, R.G., Spanos, P.D., 2012. Stochastic Finite Elements: A Spectral Approach. Dover, Mineola, New York.

Ghosn, M., Moses, F., 1998. Redundancy in Highway Bridge Superstructures. Report 406, Transportation Research Board, Washington, DC.

Gorman, M.R., 1981. Automatic generation of collapse mode equations. J. Struct. Div. 107, 1350–1354.

Guzda, M., Bhattacharya, B., Mertz, D., 2007. Live load distribution on highway bridges using in-service bridge monitoring system. J. Bridge Eng. 12, 130–134.

Hasofer, A.M., Lind, N.C., 1974. Exact and invariant second-moment code format. J. Eng. Mech. 100, 111–121.

Hines, W.W., Montgomery, D.C., Goldsman, D.M., Borror, C.M., 2003. Probability and Statistics in Engineering. Wiley, Hoboken, NJ.

Hohenbichler, M., Rackwitz, R., 1983. First-order concepts in system reliability. Struct. Safe. 1, 177–188.

International Maritime Organization (IMO), 2000. Formal Safety Assessment. Report No. MSC 72/16. IMO, London.

International Organization for Standardization (ISO), 1998. ISO 2394 General Principles on Reliability for Structures. ISO, Geneva.

Joint Committee on Structural Safety (JCSS), 2001. Probabilistic Model Code, 12th Draft. Retrieved 16 August, 2015, from http://www.jcss.byg.dtu.dk/Publications/ Probabilistic_Model_Code.

Karamchandani, A., 1987. Structural System Reliability Analysis Methods. Reliability of Marine Structures Program. Report No. 83. Department of Civil Engineering, Stanford University, Stanford, CA.

Karamchandani, A., Cornell, C.A., 1992a. An event-to-event strategy for nonlinear analysis of truss structures I. J. Struct. Eng. 118, 895–909.

Karamchandani, A., Cornell, C.A., 1992b. Reliability analysis of truss structures with multistate elements II. J. Struct. Eng. 118, 910–925.

Keeney, R.L., Winterfeldt, D.V., 1991. Eliciting probabilities from experts in complex technical problems. IEEE Trans. Eng. Manage. 38, 191–201.

Kovalenko, I.N., Kuznetsov, N.Y., Shurenkov, V.M., 1996. Models of Random Processes. CRC Press, Boca Raton

Kwon, K., Frangopol, D.M., 2011. Bridge fatigue assessment and management using reliability-based crack growth and probability of detection models. Prob. Eng. Mech 26, 471–480.

Lin, Y.K., 1976. Probabilistic Theory of Structural Dynamics Huntington. Krieger, New York.

Lind, N.C., 1993. Target reliability levels from social indicators. In: 6th Intl. Conf. on Structural Safety and Reliability. Innsbruck, Austria. August 1993.

Lind, N.C., Nowak, A.S., 1988. Pooling expert opinions on probability distributions. J. Eng. Mech. 114, 328–341.

Macgregor, J.G., et al., 1997. Design criteria and load and resistance factors for the Confederation Bridge. Canad. J. Civil Eng. 24, 882–897.

Mander, J., Elms, D., 1993. Quantitative risk assessment of large structural systems. In: Sixth Intl. Conf. on Structural Safety and Reliability. Innsbruck, Austria. August 1993.

Melchers, R.E., 1987. Structural Reliability Analysis and Prediction. Ellis Horwood, Chicester, UK.

Melchers, R.E., 1990. Radial importance sampling for structural reliability. J. Eng. Mech. 116, 189–203.

Moan, T., 1997. Current Trends in the Safety of Offshore Structures. In: 7th ISOPE Conf. Honolulu. May 1997.

Moses, F., 1997. Problems and prospects of reliability-based optimization. Eng. Struct. 19, 293–301.

National Cooperative Highway Research Program (NCHRP), 2001. Calibration of Load Factors for LRFR Bridge Evaluation. Report 454. Transportation Research Board, National Research Council, Washington, DC.

National Cooperative Highway Research Program (NCHRP), 2003. Bridge Life-Cycle Cost Analysis. Transportation Research Board, National Research Council, Washington, DC.

Nikolaidis, E., Kaplan, P., 1991. Uncertainties in Stress Analyses on Marine Structures. SSC-363. Ship Structure Committee, Washington, DC.

Nowak, A.S., 1993. Live load model for highway bridges. Struct. Safe. 13, 53–66.

Nowak, A.S., Szerszen, M.M., Park, C.H., 1997. Target safety levels for bridges. In: 7th Intl. Conf. on Structural Safety and Reliability. Kyoto, Japan.

Papoulis, A., Pillai, S.U., 2002. Probability, Random Variables, and Stochastic Processes. McGraw-Hill, New York.

Rackwitz, R., Fiessler, B., 1978. Structural reliability under combined random load sequences. Comp. Struct. 9, 489–494.

Resnick, S., 1999. A Probability Path. Birkhauser, Boston.

Rosen, J.B., 1961. The gradient projection method for nonlilnear programming. Part II. Nonlinear constraints. J. Soc. Indus. Appl. Math. 9, 514–532.

Sarveswaran, V., Roberts, M., 1999. Reliability analysis of deteriorating structures—the experience and needs of practising engineers. Struct. Safe. 21.

Sen, D., Bhattacharya, B., 2015. On the Pareto optimality of variance reduction simulation techniques in structural reliability. Struct. Safe. 53, 57–74.

Shao, S., Murotsu, Y., 1999. Approach to failure mode analysis of large structures. Prob. Eng. Mech. 14, 169–177.

Shinozuka, M., 1983. Basic analysis of structural safety. J. Struct. Eng. 109, 721–740.

Shinozuka, M., Sato, Y., 1967. Simulation of nonstationary random processes. J. Eng. Mech. Div. 93, 11–40.

Structural Engineers Association of California (SEAOC), 1995. Vision 2000, Performance-Based Seismic Engineering of Buildings. Sacramento, CA.

Suzuki, H., 1999. Safety target of very large floating structure used as a floating airport. In: Third Intl. Workshop on Very Large Floating Structures. Honolulu, HI. September 1999.

Thoft-Christensen, P., Murotsu, Y., 1986. Application of Structural Systems Reliability Theory. Springer-Verlag, Berlin.

Vanmarcke, E., 1983. Random Fields: Analysis and Synthesis. MIT Press, Cambridge, MA.

Wang, N., O'Malley, C., Ellingwood, B.R., Zureick, A.-H., 2011a. Bridge rating using system reliability assessment. I: assessment and verification by load testing. J. Bridge Eng. 16, 854–862.

Wang, N., O'Malley, C., Ellingwood, B.R., Zureick, A.-H., 2011b. Bridge rating using system reliability assessment. II: improvements to bridge rating practices. J. Bridge Eng. 16, 863–871.

Watwood, V.B., 1979. Mechanism generation for limit analysis of frames. J. Struct. Div. 109, 1–15.

Wen, Y.-K., 1990. Structural Load Modeling and Combination for Performance and Safety Evaluation. Elsevier, Amsterdam.

Wen, Y.-K., 2001. Minimum lifecycle cost design under multiple hazards. Reliab. Eng. Sys. Saf. 73, 223–231.

Whittaker, A., Constantinou, M., Tsopelas, P., 1998. Displacement estimates for performance-based seismic design. J. Struct. Eng. 124, 905–912.

Wood, R.H., 1968. The reinforcement of slabs in accordance with a predetermined field of moments. Concr. (London) 2, 69.

Yun, S., Hamburger, R.O., Cornell, C.A., Foutch, D.A., 2002. Seismic performance evaluation of steel moment frames. J. Struct. Eng. 128, 534–545.

Zimmerman, J.J., Ellis, J.H., Corotis, R.B., 1993. Stochastic optimization models for structural reliability analysis. J. Struct. Eng. 119, 223–239.

Zio, E., Apostolakis, G.E., 1996. Two methods for the structures assessment of model uncertainty by experts in performance assessments of radioacive waste repositories. Reliab. Eng. Sys. Saf. 54, 225–241.

Innovative structural typologies

Adriaenssens S.[1], Boegle A.[2]
[1]Princeton University, Princeton, NJ, USA
[2]HafenCity Universität, Hamburg, Germany

1 Introduction: aim and context

The objective of this chapter is to demonstrate how advances in computational form-finding, optimization, and digital fabrication techniques have brought about a new realm of bridge typologies. The engineering design of bridges requires the solution of a complex brief that stipulates economic cost, technical quality, ease of maintenance, durability, site suitability, and esthetic appeal. Solutions that satisfy these criteria are not unique. In the past, engineers identified feasible instances using trial and error, accumulated knowledge, and deductive reasoning. Today, computational tools are available to the bridge designer to aid decision making and steer the design process to novel typology solutions. These tools in the engineer's toolbox are most useful in the preliminary design stage, where they can achieve the most gain in terms of economic and environmental cost (Mueller, 2014).

There are at least five distinct reasons that these new techniques have not been widely embraced. First, form-finding and optimization techniques are perceived as implying a large computational cost, which is undesirable in a preliminary design context. Second, the design solutions generated by these techniques are wrongly thought to entail fragility and lack robustness or redundancy. Third, in contrast with products from the automotive and aerospace industries, the uniqueness of each bridge project excludes the repetition of gains. Fourth, the limitations of overconstrained bridge design codes might not allow unconventional typologies generated by these computational techniques. Finally, with the exception of a few instances, these computational methods are not taught in a traditional undergraduate or graduate civil engineering curriculum and, hence, require expert knowhow, which is unlikely to be available in standard structural design offices.

2 Literature review

This section provides a succinct literature review of numerical form finding, optimization, and computer numerically controlled (CNC) machine techniques that have been integrated into the preliminary design of bridges.

Structural form finding can be defined as a forward process in which parameters are directly controlled to find an "optimal" geometry of a structural system, which is in static equilibrium with one design loading (Adriaenssens et al., 2014). This process is particularly suited for "form-active" and certain "form-passive" bridge typologies that

Innovative Bridge Design Handbook. http://dx.doi.org/10.1016/B978-0-12-800058-8.00008-6

resist external loads, predominantly through axially loaded members or membrane action. The axially loaded members could include the main and the hanger cables of a suspension bridge, the hangers in a bowstring arch bridge, the elements forming an arch, and the struts connecting the bridge deck to a below-deck arch. Membrane stresses, on the other hand, could be taken within the thickness of a structural surface, as in the case of thin shell systems supporting a bridge deck. Initially, the geometry of a structural system is unknown. However, the process may require some arbitrary starting geometry. To steer the form-finding process, the structural designer can manipulate certain parameters, such as (i) the boundary conditions and external loads, (ii) the topology of the model, and (iii) the internal forces and their relationship to the geometry. Once the final shape is found, the numerical model is updated by assigning real physical material and member properties. The rapid feedback of the form-finding program and the interactive designer experience are key to making informed decisions in the preliminary design process. After the form-finding procedure, the designer needs to carry out a rigorous static and dynamic analysis according to relevant bridge design codes.

Several form-finding methods have been used. The force density method (Schcck, 1974) for instance solves the problem of static equilibrium without requiring material properties. Descamps et al. (2011), for example, used this method for generation of the geometry of three-dimensional (3D) systems of suspension bridges and arches. To balance stress levels in the hanger attached to a rigid bowstring, Caron et al. (2009) used the same force density technique. An extended form of this technique was developed by Quaglaroli and Malerba (2013) for flexible bridge decks suspended by cable nets. An alternative computational form-finding approach, the dynamic relaxation method (Day, 1965), incorporates fictitious material stiffness and element properties to solve for equilibrium. Methodologies have been presented based on dynamic relaxation to find the shape of a shallow arch (Halpern and Adriaenssens, 2014), suspension (Segal et al., 2014) and tensegrity (Rhode-Barbarigos et al., 2010) bridges.

Most academic work related to bridge optimization can be found in the domain of economic cost and maintenance optimization (Ayd and Ayvaz, 2013; Hassan et al., 2013); however, this chapter focuses on optimization approaches that affect the bridge topology, not their shape or the size of their elements. In this context, structural optimization is an inverse process in which parameters are indirectly optimized to find the optimal structural layout of a bridge system such that an objective function or fitness criterion is minimized. Fauche et al. (2010) demonstrated the use of topology optimization as a design tool for a thin-shelled bridge structure. Their optimization routine, which was coupled to a finite element analysis, aimed at maximizing compliance and finds its solution using the fixed point iteration method. Briseghella et al. (2013) also proposed a slightly different topology optimization approach for shell bridge design by minimizing compliance using a solid isotropic material with penalization algorithm. Nagase and Skelton (2014) presented a design methodology for tensegrity bridges, which is based on parametric design concepts, fractal geometry, and mass minimization through an iterative linear programming approach. To minimize operational energy, Thrall et al. (2012) employed simulated annealing to optimize the topology of deployable linkage bridges. Rahmatalla and Swan (2003) developed a methodology that optimizes the topology of truss bridges to maximize buckling stability. Using the homogenization method with the objective of maximizing stiffness, Lochner-Aldinger (2011) demonstrated how two-dimensional (2D) bridge topologies could be generated to inform footbridge design using the homogenization method. Islam et al. (2014) used a global optimization algorithm, evolutionary operation (EVOP), to minimize the cost of a network arch bridge by varying geometric shape, rise-to-span ratio, cross-section of arch and hangers, and topology of the hangers.

In traditional bridge construction, techniques to manufacture members rely on material subtraction, deformation (such as bending), and casting methods. Many of these methods predate the arrival of the computer and have recently been adjusted to suit a numerical control process (Schodek et al., 2005). As a result, a person no longer directly operates the machine; computer algorithms do. Other more recent manufacturing processes (such as lasers) completely depend upon computer technologies for both their operation and control. Based on a computer-aided design (CAD) manufacturing layout file, the machine itself prepares a set of commands, reads them, and instructs tools to execute coded movements. There are two benefits of CNC tools for the design and construction of bridges. First, these tools can be more accurate, faster, and more economical than conventional machines for the manufacture of bridge components. Second, they can facilitate the construction of novel components needed for nontraditional bridge typologies (Adriaenssens et al., 2009).

3 3D bridges force-modeled for one loading condition

Throughout history, engineers have been shaping bridge systems (or, more specifically, their structural elements) to follow the flow of forces. Their objective was to obtain a dominant load-bearing behavior with tension or compression, minimizing any shear and avoiding bending. In particular, since the 17th century, when the first interactions between science and building practice became visible, various form-finding methods were developed. Long before the development of numerical methods, these form-finding techniques were based on physical experiments. It was, for example, British architect-philosopher Robert Hooke (1635–1703) who formulated the ideal shape of an arch through a hanging model and its inversion. Experimental form finding was (and still is) considered to be good practice, but a strong restriction remains—namely, that it can be applied for only one load condition.

Thus, the geometry of early force-modeled bridges is carefully formed, but it only works because of a larger self-weight and a comparatively lower life load. One of the pioneers of using physical hanging models for form finding was Spanish architect Antoni Gaudi (1852–1926), known for his virtuously shaped arches and vaults. For example, the crypt in his Güell Chapel and the arches in his Casa Mila are all perfectly shaped for a high self-weight, which is true of the stone material they are constructed from. Connected to his garden designs, he developed a number of bridge designs. Some were realized as viaducts, like the Park Güell (shown in Figure 8.1), while others were unrealized, like the bridge over the Torrent de Pomeret between Sarrià and Sant Gervasi, on the outskirts of Barcelona, Spain.

Figure 8.1 Gaudí's design and construction of a force-modeled viaduct in the Park Güell (https://creativecommons.org/licenses/by-sa/2.0/legalcode; Valerie Hinojosa).

Physical form finding methods were essential for the visionary structures of Italian architect Sergio Musmeci (1926–1981). At a time where numerical methods were not available, he designed prestressed spatial membrane structures based on the idea of minimal surfaces and inverted them into shell structures under compression. Thereby, the Basento Viaduct in Potenza, built in 1969, is said to be his masterpiece (Figure 8.2). A slender continuous shell structure serves as the load-bearing system and experiences uniform but not isotropic compression stresses. The form-finding process started with soap models, but the most promising results were achieved with the second generation of experiments. With this viaduct, Musmeci used a neoprene material; the model was not only more stable than the soap models, but also able to model the nonisotropic behavior typical of a concrete membrane. Last, but not least, a large-scale model (two segments, scale 1:10) made of microconcrete was created in Bergamo's laboratory. This model was essential to optimize the form, to obtain information about the strains and stresses, and to discuss details and questions concerning

Figure 8.2 The form finding process of the Basento Bridge employed several physical techniques, including the pre-stressing of membranes.

the construction process. The resulting bridge is a magnificent, spatial, concrete structure that gets its aesthetics directly from its efficiency.

Another typology, where spatial equilibrium is achieved through geometry, is the circular ring girder bridge, a pioneering work in efficient 3D load-bearing behavior. These bridges have a distinctive, nonlinear layout of their deck trajectory. Traditionally, the plan layout of a bridge is straight or only slightly curved, which has no major effect on the load-bearing behavior. In contrast, the principle of the circular ring girder allows much more design freedom for footbridges.

The general structural principle behind the circular ring girder, shown in Figure 8.3, is that it needs only a single hinged support either on the inner or outer edge, without flipping downward or being stressed with torsion. Similarly, a straight slab strip needs either a support on both sides or, if only supported on one side, a fixed support. Instead, the circular girder transfers the overturning moment—resulting from dead load or any uniform distributed load—along the supported edge into a bending moment along the horizontal axis. This overturning moment can be replaced by a pair of radial distributed forces along the length of the ring girder, one facing the center of the circle, the other pointing in the opposite direction. As in a thin-walled pressure

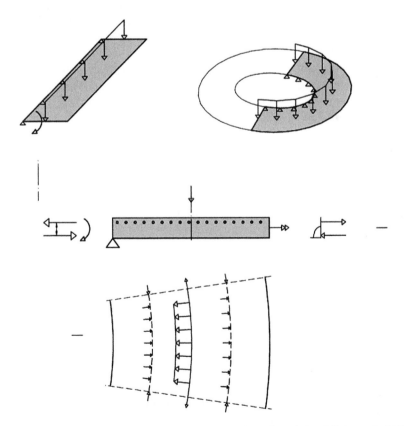

Figure 8.3 Principles behind the structural behavior of the ring girder (Bögle et al., 2003).

vessel, any radial distributed force along a circular line causes a normal force in the longitudinal direction. Thus, if the circular ring girder rests on the inner edge, a line load causes ring tension on the upper side and ring compression on the lower side of the slab, or vice versa for a line support on the outer edge. Any load with geometric affinity may be supported in this manner, but any loads without geometric affinity, such as point loads or unbalanced live loads, cause moments in the girder that require an appropriate stiffness in the horizontal axis.

Circular ring girders are mostly supported by a suspension cable and inclined hangers. The inclination of the hangers introduces other horizontal forces into the bridge deck, creating other compression ring forces in the deck when supported at the inner edge, and other tension ring forces when supported at the outer edge. The spatial load-bearing behavior of the ring girder, under geometric affined loads, is quite descriptive and does not truly require numerical form-finding techniques. However, the spatial geometry of the suspension cable, the exact inclination of the hangers, and their analysis necessitate numerical form-finding techniques. With the aid of numerical methods, this idea can be developed further (Bögle et al., 2003).

The earliest suspension bridge built on this principle is the footbridge over the Rhine-Main-Danube Canal in Kehlheim, Germany (built in 1987 by the architectural firm schlaich bergermann and partner), which features a compact, prestressed concrete cross section that is supported on the inner edge by inclined hangers. These are attached to a suspension cable, and eventually to a mast and foundations. The compact cross section is structurally inefficient, as only the top layer is structurally fully used. Thus, in later designs, the cross section has been transformed into a network of cables and struts subjected to tension or compression, respectively. The Westpark Bridge in Bochum, Germany (built in 2003 by schlaich bergermann and partner; see Figure 8.4) elegantly illustrates this circular girder typology with network. In this case, the S-shaped walkway (66 m long, each half with a radius of 46 m) is supported on the inner perimeter. The flow of forces is expressed in a lightweight deck, to handle the tensile ring forces, and a circular compact steel strut, to take the compression ring forces beneath the deck. This bridge shows an additional novelty: two inclined masts, each inside one of the semicircles, are stabilized only by the main suspension cable cables (see Figure 8.5). The mast tip, the geometry of the main suspension cable, and the hangars are in spatial balance for one load case. This structural arrangement is sufficient for different load cases even without stay cables because the foundations of the masts are placed lower than the anchorages of the cables. Then, each individual load case leads to a new equilibrium geometry, and here, stability is coupled to deformation.

Further optimization will be reached if any overturning moment due to self-weight or any uniformly distributed load is completely avoided. This equilibrium can be achieved if the resulting force of the hangers passes through the gravity center of the deck. In the practical sense, this requires a deck with cantilevers of different heights to connect to the hangers as in the "balcony to the sea" bridge in Saßnitz, Germany (built in 2007 by schlaich bergermann and partner). This 120-m span suspension bridge only acts as a circular ring girder when a lateral pedestrian load is applied, with both a compression ring below and a tension ring on the deck level.

Figure 8.4 The spatial geometry of the circular ring girder, suspension cables, and inclined mast of the West Park Bridge, Bochum, Germany © Nicolas Janberg (www.structurae.de).

Figure 8.5 The geometry of the inclined masts in spatial equilibrium.

Other realized projects have added to the potential of the principle of circular ring girders. For example, the Footbridge Harbor Grimberg in Gelsenkirchen, Germany (built in 2009 by schlaich bergermann and partner) is supported on the outside perimeter of the curve. Its main cables are not anchored on the abutments, but instead are 24 m in front of them (see Figure 8.6). In this case, the bridge has to be designed for torsion and bending as well. The challenge of the form-finding process was to find a stable equilibrium between the anchorage of the main suspension cables, the position and inclination of the mast and hangars, the slenderness of the deck, and the stiffness of the abutment.

According to the principle of inversion, which dates back as far as the 17th century, the inversion of a cable-suspended bridge leads to an arch bridge, including all the structural challenges arising from a thin arch under compression. There are various examples existing, a few of them already exploring spatial values concerning aesthetics and structural behavior. Further, by inverting the spatial main cable of a cable-suspended ring girder, an impressive spatial curve arises. This lightweight solution provides a structural answer to the challenge of bringing an arch in longitudinal

Figure 8.6 Spatial static equilibrium between the different elements of the Footbridge Harbor Grimberg, Gelsenkirchen, Germany (© schlaich bergermann and partner, Michael Zimmermann).

view together with a curve in the plan layout. Initial approaches have been made by schlaich bergermann and partner (e.g., the footbridge over the Rhine-Herne-Canal, Germany, and footbridges in Esslingen, Germany, and Belfast, Ireland). One of the most recent examples is the design of FEHCOR for the Salford Meadows Bridge Competition in Salford, UK, shown in Figure 8.7 (2014). A 130-m-long curved pathway, supported on only one side, will be carried by one spatial arch. The form of the spatial arch follows the efficient flow if forces of a circular curved girder are supported on only one side. This extraordinary form is not the result of pure architectural expression, but instead, the result of structural optimization with an arch only under compression for evenly distributed loads. The proposal goes even further when it entirely exploits the possibilities of contemporary structural steel manufacturing, with its capacity to construct spatial and complex shapes without a significant cost increase.

Figure 8.7 The spatial form of the arch follows the flow of forces of the circular deck girder only supported on one side; competition entry, Salford Meadow Bridge (© FEHCOR).

4 3D bridges, optimized for one or more criteria and composed of surface elements

The advances of CAD manufacturing technologies have introduced a new bridge design vocabulary that is starting to result in new bridge typologies. It is no longer difficult to fabricate designs that involve highly complex, geometrical 3D forms that cannot be described by straight lines and circular arcs. Likewise, the form-finding methods have developed. Originally, numerical form-finding methods could focus on only one parameter and search for the one equilibrium of forces using mainly linear elements. Now, the new numerical methods allow multiobjective optimization. Therefore, the structural focus shifts toward surface elements.

The development of the so-called steel sails is the enhancement of a continuous trough bridge with the webs shaped according to the bending moments. Just like the main cables and hangers in a suspension bridge, these webs experience mainly tensile stress, suggesting that they should be made out of steel. The deck, in compression, is ideally realized in concrete. Dissolving the steel webs more and more leads to elegantly curved sail-like steel plates, suspended from short, reinforced concrete (r.c.) masts. Structurally, this assembly is related to "extradosed bridges," a Japanese variation of multispan cable-stayed bridges with small deflections under heavy loads. Thus, these structures are often used for railway bridges. The load-bearing behavior becomes immediately obvious when flipping its orientation upside down into an inverted strutted frame, as seen in the Felsegg Bridge (designed in 1932 by Robert Maillart), shown in Figure 8.8, or the Ganter Bridge (designed in 1980 by Christian Menn), both in Switzerland. An example of such a bridge with steel sails is the Neckar Rail Bridge (schlaich bergermann and partner), a train bridge in Germany with two main spans of 72 and 78 m and two small side spans (see Figure 8.9).

The idea of using structural steel surface elements aligns with the potential offered by new CNC manufacturing methods (particularly steel-laser-cutting techniques);

Figure 8.8 The shape of the Felsegg Bridge (1932) reflects the inverted form of the Rail Bridge Bad Cannstatt (photo taken by authors).

Figure 8.9 A continuous trough bridge with sidewise webs shaped according to the bending moments, giving the appearance of sails; Rail bridge, Bad Cannstatt, Stuttgart, Germany (competition 1998, completion 2021).

existing typologies can be reinterpreted and constructed more economically and efficiently while new typologies can be envisaged.

The process of fabrication of the Abetxuko Bridge, located in Vitoria, Spain (Pedelta, 2006) across the Zadorra River, illustrates the enormous possibilities available through CAD/CNC techniques (see Figure 8.10). The structural system is a continuous beam is quite simple, with spans of 26 m on its sides and 40 m in the center, but the two trusses make it a true landmark. The organic curved trusses are above the deck and separate the road traffic from the pedestrian walkway on both sides. With its irregular and curved forms, this bridge mirrors a different engineering attitude than the classical one of purity and flow of forces. Still, the dimensions are adjusted according to the structural analysis. The expressive forms were cut, bent, and welded at a local steelyard. Segments were transported to the building site to be finally assembled. Thus, these fabrication techniques allow for divergence from standard geometry.

The last design case study presented here draws on the concepts of the plate-stayed bridge but uses form finding and topology optimization to generate a new bridge typology: the hanging shell bridge, entirely constructed out of surface elements. The Knokke-Heist footbridge in Belgium (designed in 2008 by Ney and Partners), shown in Figure 8.11, was designed unlike traditional structures, where the loads in the longitudinal and transverse directions are decoupled (Adriaenssens et al., 2010). From a topological point of view, the bridge is based on a cutout, curved shape that efficiently carries external loads through membrane action and satisfies the site requirements. The shape of the steel bridge was numerically form-found, like a network of connected springs supported at the abutments and the mast heads. The grid is allowed to relax or "fall" under gravity loads applied at the spring connections. The idea behind the curved structural form is to carry all loads within the steel surface shell without needing additional structural elements (see Figure 8.12).

Figure 8.10 The design of the Abetxuko Bridge across the Zadorra River exploited the new formal possibilities of CNC technologies (Pedelta). (Copyright Ricardo Ferraz, TU Berlin, Fachgebiet Entwurf und Konstruktion – Massivbau).

Figure 8.11 Side view of the Knokke-Heist footbridge in Knokke, Belgium (Ney and Partners); photo credit Jean-Luc Deru.

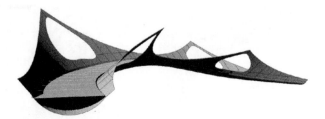

Figure 8.12 The force-modeled bridge shape; photo credit Ney and Partners.

Once the overall shape has been found, the geometry is further refined to comply with the CNC manufacturing constraint of single-curvature steel sheet bending, and then it is numerically optimized to maximize the overall stiffness of the bridge. The latter task presents a typical topology optimization problem that consists of distributing a given amount of material in a design domain subject to load and support conditions, such that the stiffness of the structure is maximized. Figure 8.13(a) shows the optimal thickness distribution in the shell for different values and the mean

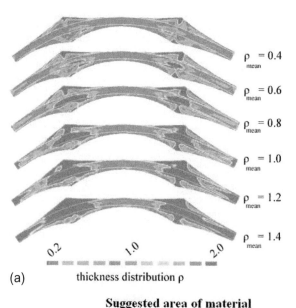

Figure 8.13 (a) Optimal shell thickness distribution; (b) suggestion of location of openings (images from authors).

$\rho_{mean} = 0.4$

$\rho_{mean} = 0.6$

$\rho_{mean} = 0.8$

$\rho_{mean} = 1.0$

$\rho_{mean} = 1.2$

$\rho_{mean} = 1.4$

0.2   1.0   2.0

(a)   thickness distribution ρ

Suggested area of material

Suggested opening   Suggested opening

(b)

thickness ρ_{mean}, which is a measure of the total material volume constraint. Figure 8.13(b) shows a close-up of the results of the topology optimization for the area around the intermediate supports, and also suggests the optimal location for openings and shows where the steel plate thickness must be increased. Topology optimization provided a powerful tool for the preliminary design of this thin-shelled bridge. By combining topology optimization with form finding and CNC manufacturing constraints, a 3D typology that might not have been conceivable in a purely analytical or intuitive fashion was generated.

5 Future prospects and conclusions: role of the designer and the toolbox

Most of this chapter focused on constructed bridges made of steel components. However, many historic bridges that articulate their force flow, such as the Felsegg Bridge and the Basento Viaduct, were made of r.c. and were constructed at a time and place where manual labor was cheap. Over recent years, a gap has developed between the formal opportunities offered by digital additive fabrication techniques and available construction methods for r.c. systems. Current additive methods focus on nonstructural materials, produce small components, and rely on an additive layering approach. Therefore, these approaches not suitable for r.c. bridge systems. To overcome these challenges, recent research progress has developed techniques such as smart dynamic casting, which combines digital fabrication with slipforming (Lloret et al., 2015) and robotic swarm printing (Oxman et al., 2014). However, it is yet to be seen how these techniques will influence the discovery of new r.c. bridge typologies and drive their construction.

It is said that "an engineer is a (wo)man who can do for a dime what any fool can do for a dollar." While it is customary for an engineer to be responsible for achieving a specific technological need for the lowest economic cost, this saying is crippling to both the engineer's creativity and the design's potential. With the available new tools, the role of the bridge designer needs to be emphasized. Traditionally, designers use tools such as back-of-the-envelope hand calculations, physical experiments and tests, design charts, and 2D sketches to develop and advance the preliminary design process. The computational techniques presented in this chapter are new tools in this existing toolbox; they allow for the rapid generation of a large set of design alternatives that fulfill specific requirements. These alternatives might be unusual and surprising to the traditional bridge designer, who is grounded in intuition and accumulated knowledge; yet they present unexplored feasible domains in the design space. These instances can inform the bridge designer of new typologies beyond the existing archives of acceptable systems, developed from 19th- and 20th-century analytical and construction techniques. Prior studies of existing successful designs are important, but the use of previous typologies might exclude the creation of more efficient ones. Instead, this chapter argues that by drawing on a broad body of existing knowledge and utilizing 21st-century computational tools, the designer might uncover a range of novel bridge typologies, waiting to be discovered.

References

Adriaenssens, S., Ney, L., Bodarwe, E., Dister, V., 2009. Centner footbridge bridges the gap between structural design and digital fabrication. Steel Construct. 2 (1), 33–35.

Adriaenssens, S., Devoldere, S., Ney, L., Strauven, I., 2010. Shaping Forces: Laurent Ney. Bozar Books - A+ Editions, Brussels.

Adriaenssens, S., Block, P., Veenendaal, D., Williams, C., 2014. Shell Structures for Architecture: Form Finding and Optimization. Routledge, Abingdon, UK.

Ayd, Z., Ayvaz, Y., 2013. Overall cost optimization of a prestressed concrete bridge using genetic algorithm. J. Civil Eng. 17 (4), 769–776.

Bögle, A., Schmal, P., Flagge, I., 2003. Leicht Weit—Light Structures, Jörg Schlaich, Rudolf Bergermann. Prestel, Munich.

Briseghella, B., Fenu, L., Feng, Y., Mazzarolo, E., 2013. Topology optimization of bridges supported by a concrete shell. Struct. Eng. Intl. 3, 285–294.

Caron, J., Julich, S., Baverel, 2009. Self-stressed bowstring footbridge in FRP. Comp. Struct. 89 (3), 489–496.

Day, A., 1965. The Engineer, vol. 29. 218–221.

Descamps, B., Coelho, R.F., Ney, L., Bouillard, P., 2011. Multicriteria optimization of lightweight bridge structures with a constrained force density method. Comp. Struct. 89, 277–284.

Fauche, E., Adriaenssens, S., Prevost, J., 2010. Tpology optimization of a thin shell structure. J. Intl. Assoc. Shell Spatial Struct. 51 (2), 153–160.

Halpern, A., Adriaenssens, S., 2014. In-plane optimization of truss arch footbridges using stability and serviceability objective functions. Struct. Multidisc. Optim. 51 (4), 971–985.

Hassan, M., Nassef, A., El Damatty, A., 2013. Optimal design of semi-fan cable-stayed bridges. Canad. J. Civil Eng. 40, 285–297.

Islam, N., Rana, S., Ahsan, R., Ghana, S., 2014. An optimized design of network arch bridge using global optimization algorithm. Adv. Struct. Eng. 17 (2), 197–210.

Lloret, E., et al., 2015. Complex concrete structures: merging existing casting techniques with digitak fabrication. Comput. Aided Des. 60, 40–49.

Lochner-Aldinger, I., 2011. Formfindung und Optimierung im Brückenbau. Altair University, Bonn.

Mueller, C., 2014. Ph.D. thesis: Computational Exploration of the Structural Design Space. Massachusetts Institute of Technology (MIT), Cambridge, MA.

Nagase, K., Skelton, R.E., 2014. Minimal mass design of tensegrity structures. In: SPIE Smart Structures and Materials + Nondestructive Evaluation and Health Monitoring (pp. 90610W–90610W). International Society for Optics and Photonics.

Oxman, N., et al., 2014. Towards robotic swarm printing. Arch. Des. 108–115. May–June.

Quagliaroli, M., Malerba, P., 2013. Flexible bridge decks suspended by cable nets: a constrained form finding approach. Intl. J. Solids Struct. 50 (14–15), 2340–2352.

Rahmatalla, S., Swan, C., 2003. Form finding of sparse structures with continuum topology optimization. J. Struct. Eng. 129 (12), 1707–1716.

Rhode-Barbarigos, K., Bel Hadj, N., Motro, R., IFC, S., 2010. Designing tensegrity modules for pedestrain bridges. Eng. Struct. 32 (4), 1158–1167.

Scheck, H., 1974. The force density method for the form-finding and computation of general networks. Comp. Meth. Appl. Mech. Eng. 3, 115–134.

Schodek, D., Bechthold, M., Griggs, K., Kao, K.M., Steinberg, M., 2005. Digital Design and Manufacturing: Cad/CAM Applications in Architecture and Design. John Wiley & Sons, Hoboken, NJ.

Segal, E., Rhode-Barbarigos, L., Adriaenssens, S., Filomeno Coelho, R., 2015. Multi-objective optimization of polyester-rope and steel-rope suspended footbridges. Eng. Struct. 99, 559–567.

Thrall, A., et al., 2012. Structural optimization of deploying steel pantographs. ASCE J. Comput. Civ. Eng. http://dx.doi.org/10.1061/(ASCE)CP.1943-5487.0000272.

Section IV

Bridge design based on construction material type

Reinforced and prestressed concrete bridges

Balázs G.L., Farkas G., Kovács T.
Budapest University of Technology and Economics, Budapest, Hungary

1 Types of reinforced concrete bridges

The type of reinforced or prestressed concrete bridge deck depends mainly on the functional requirements, the structural form, and the main span length of the construction. Precast or cast in situ reinforced concrete (r.c.) bridge decks can be practically applied for all structural types, like arch, cable-stayed, extradosed, and even suspension bridges with a majority of girder bridges. In this chapter, we mainly discuss simply supported and continuous girder bridges, which differ from each other mainly by their cross section. The structural depths of r.c. girders largely depend on the selected cross section.

Solid slabs with rectangular cross sections are suitable for spans of up to 15 m. If self-weight is reduced by using side cantilevers, the spans from 18 to 20 m are particularly economic (Figure 9.1). The main benefits of this type of cross section are the relatively high torsional rigidity, the smaller structural depth than beams, and the easily fixable reinforcement. These types of bridge decks require prestressed solid slabs, which can be economically used due to these benefits for spans of 15–23 m owing to having less formwork than beams, and a lesser quantity of earthwork in the approach embankments. The disadvantages of this construction are the greater quantities of reinforcement and concrete beams, which consequently leads to greater self-weight of the deck.

Voided slabs with rectangular cross sections or with side cantilevers are used for spans longer than about 20 m in order to reduce the self-weight (Figure 9.2). This type of cross section can be economic up to 25 m (with constant depth) or up to 35 m (with variable depth). Voids are produced by excluding the concrete by appropriate materials (such as cardboard or expanded polystyrene) and are located near mid-depth, causing minimum inertia reduction. To avoid lifting of void formers due to the upward pressure of fresh concrete, particular attention is required during casting.

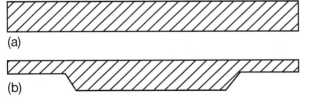

Figure 9.1 Solid slabs:
(a) with constant thickness
(b) with side cantilevers.

Innovative Bridge Design Handbook. http://dx.doi.org/10.1016/B978-0-12-800058-8.00009-8

Figure 9.2 Voided slab.

Ribbed slabs can be used for spans between 20 and 40 m (Figure 9.3). The reduction in self-weight of this type of bridge deck is significant compared to solid slabs. For simply supported spans with constant depth, the typical span/depth ratio is between 22 and 25. This value may be about 30 to 35 for continuous prestressed concrete structures. For continuous r.c. bridge decks with variable depth, the span/depth ratio can be 40 at the middle of the span and 25 over intermediate supports. The disadvantage of this type of cross section is the deep section at the ribs. The construction requires relatively complicated formwork and special construction techniques.

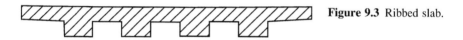

Figure 9.3 Ribbed slab.

A *cast-in-situ beam and slab system* is a development of the ribbed slab that is used for longer spans (Figure 9.4). For simply supported spans with posttensioned beams, the range of use is 30–50 m.

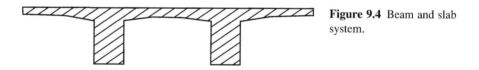

Figure 9.4 Beam and slab system.

Beam and slab systems with precast elements are most often used for simply supported spans (Figure 9.5). The continuity of the deck can be achieved by additional ordinary reinforcement over the piers or by posttensioning. It provides an economic solution with pretensioned beams for spans up to 30 m and up to 50 m in the case of posttensioning. Many different cross sections can be used for precast beams. The most usual one is an I-shaped section, placed at 0.6 m to about 4.0 m apart with a cast in situ concrete slab. The formwork of the slab is often created by a series of thin, precast concrete slab elements. The span/depth ratio of 18 is usual for decks with beams 2–3 m apart. A span/depth ratio of 25 is possible for continuous bridge decks.

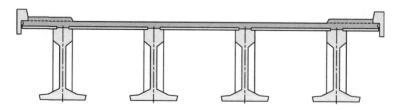

Figure 9.5 Beam and slab system with precast elements.

Box girder bridge decks with single or multiple cells are necessary for spans longer than 80 m (Figure 9.6). The longest span that can be constructed with a box girder cross section is about 300 m. The main benefit of these structures is significant torsional rigidity. For r.c. decks of constant depth, the span/depth ratio is normally within the range of 14–30. The optimum span/depth ratio for constant depths is between 18 and 22. For spans that exceed 60 m, it is structurally and economically favorable to apply longitudinally varied depths. For spans in excess of 150 m, variable depth is essential. The span/depth ratio at the piers in this case can be between 15 and 22, and the span/depth ratio at mid-span will be in the range of 35 to 22 for decks simply supported on the piers, and between 40 to 45 for decks embedded in the heads of piers.

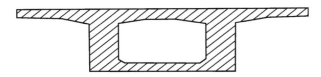

Figure 9.6 Box girder cross section.

2 Prestressing in bridges

The cross section of a bridge deck subjected to bending moments will carry the load by development of internal compressive and tensile stresses. In r.c. elements, cracks will be formed in zones where high tensile stresses develop. An efficient possibility to reduce, or even avoid, these tensile stresses is the use of prestressing. The prestressing technology is widely applied in bridges, especially when modern construction methods (e.g., bridge decks built by large precast elements, incremental launching, or other cantilever constructions).

2.1 Principle of prestressing

The principle of prestressing is to artificially create stress distribution in the structure before loading, which will contribute to balance the external loads. One way to produce the suitable amount of stress distribution is to apply compressive force to the structural element. This can be achieved by use of high-strength steel tendons that are stressed before loading of the structure (Figure 9.7). The anchorage forces at the ends of the tendons provide compression in the concrete, straight tendons with eccentricity giving an additional bending effect, while the use of curved tendons can reduce axial, bending, and shear effects due to external loads.

In addition to reducing the external forces in the case of curved tendons and increasing the rigidity by delayed cracking of the r.c. element, prestressing reduces deflections under service conditions.

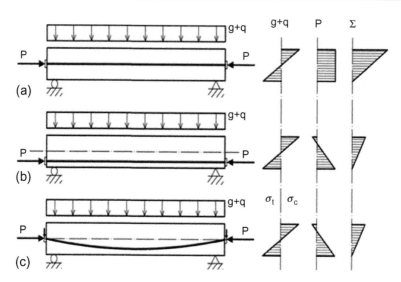

Figure 9.7 Stresses in a mid-span cross section due to external loads and prestressing forces: (a) centric prestressing along the full length (b) constant eccentricity prestressing along the full length (c) varying eccentricity prestresssing along the length.

2.2 Prestressing systems

The main systems of prestressing are *pretensioning* and *posttensioning*.

In the case of pretensioning, the prestressing tendons are tensioned before the casting of concrete. The prestress will be released from the temporary anchorages and transferred to the element after the concrete hardens. The transmission of the prestressing force to the structural element is generally ensured by bonding over the interface of prestressing tendons and concrete. The common types of prestressing steels are eccentrically placed straight strands with nominal diameters of 13 or 15 mm. This technology is mainly used to produce precast bridge girder beam elements in precast plants.

Posttensioning is the commonly used technology in long-span cast-in-situ r. c. bridge constructions. In this case, the prestressing tendons are tensioned after the concrete hardens. Special-end anchorages are fixed to the concrete, which transfer the prestressing forces to the structural element. Prestressing can be provided by prestressing wires, strands, cables, or bars. Cables usually consist of 3 to 55 strands as a function of the prestressing force. The layout of tendons for posttensioning can be selected so as to have the optimal effect of prestressing.

The tendon must be freely movable in the element to be tensioned after hardening the concrete. To achieve this, the tendons are enclosed in metal or plastic ducts. The ducts are generally filled by cement grout to improve corrosion protection and provide bonding between tendons and the surrounding concrete (Figure 9.8).

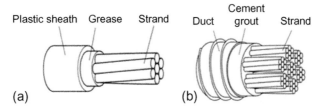

Figure 9.8 (a) A 7-wire strand protected by plastic sheath and (b) a cable formed by strands grouted in a duct.

In the case of conventional (internal) prestress, the cables are embedded in the concrete part of the cross section. For the construction of posttensioned bridge decks, the use of external prestressing is more and more frequent as part of up-to-date construction technologies. In the case of external prestressing, the tendons are outside the concrete cross section. A polygonal layout of tendons is favorable in order to follow the internal forces due to external loads. (Figure 9.9).

Figure 9.9 An example of external prestressing applied in both the longitudinal and transverse directions with unbonded tendons.

Posttensioning of bridge girders can be provided by bonded or unbonded tendons. The bonded situation is considered if ducts are cement grouted and the tendons are in direct contact with the cement grout. In the case of bonded tendons, the strain variations due to live loads are equal both in the tendon and in the surrounding concrete at the same level of cross section. Cracks are well distributed along the length of the structural element.

External prestressing or tendons in plastic sheaths represent unbonded behavior. In principle, there is no friction between the tendon and concrete, so there is no local increase of strain in prestressing steel. Consequently, cracks will be concentrated with wide openings without ordinary reinforcement (Figure 9.10).

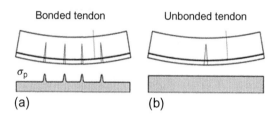

Figure 9.10 Stresses and cracks with bonded (a) and unbonded (b) tendons.

2.3 Detailing rules

In posttensioning systems, special anchorage devices are used to transfer the prestressing forces to the concrete (Figure 9.11). The so-called active anchorage allows the application of tensioning of the tendon, and the passive anchorage fixes the end of the tendon without tensioning. The so-called coupling device connects the end of one of the tensioned tendons to another tendon that will be tensioned in a second phase. Ducts for posttensioning systems connected to the anchorages form a channel inside the concrete element for the installation of tendons and provide an interface suitable for the transfer of bond stresses in the case of bonded tendons (using cement grout) for internal prestressing. For external prestressing, the duct can be placed inside or outside the concrete element, including single or multiple strands that are covered by grease, allowing the longitudinal movement of strands without developing bond stresses (unbonded prestressing). Posttensioning systems are developed and produced by qualified specialist companies, which can be used for bridges.

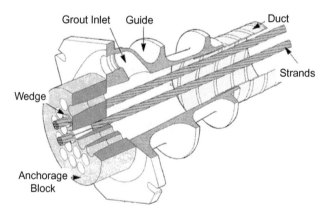

Figure 9.11 System of active Freyssinet anchorage for posttensioning.

For external prestressing, deviating devices must be placed between the tendon and the structural element to insure the required layout of the cable (see Figure 9.9). These devices are designed to transfer the cable deviation forces to the structure. The minimum radii of curvature for tendons must meet the requirement that the maximum tensile stress in the curvature complies with the tensile strength requirement. This minimum curvature must be declared by the tendon supplier in the prestressing system documentation.

2.4 Losses and time depending effects to prestressing forces

The prestressing force applied to the structure decreases along the tendon length, and also with time.

Immediate losses of prestress occurring during stressing are the following:

- Losses due to elastic shortening of concrete develop when tendons cannot be tensioned at the same time. The loss in tendon stress corresponds to the elastic deformation of concrete during prestressing. The range of the average loss for bridge decks is about 25 MPa, which is not very much compared to the level of prestressing.
- Losses due to friction between the prestressing steel and the duct in case of posttensioning at a distance x from the stressing anchorage depend on the stress of the tendon at the anchorage, σ_{po}, the coefficient of friction between the prestressing steel and the duct, μ, the sum of the angular deviation in radian from a distance x to the anchorage, α, and the unintentional angular deviation per unit length of the tendon k. The stress in the tendon at a distance x from the anchorage is given by the following formula:

$$\sigma_{px} = \sigma_{po}\, e^{-\mu(\alpha + kx)} \approx \sigma_{po}\, \mu(\alpha + kx).$$

- Values of μ and k are obtained by experiments and are given in the system documentation. The coefficient of friction for bare strands or wires over steel ducts is between 0.25 and 0.30, for seven wire strands over plastic ducts is in the range of 0.15–0.20, while for individually greased strands in plastic sheaths is 0.05–0.07.
- Losses due to draw-in of prestressing anchoring wedges occurs before the wedges are fully gripped into the surface of prestressing tendons at the anchorage during prestressing. This displacement causes a reduction of the prestress of the tendons in the anchorage zone. The loss of prestress can be calculated as a function of the draw-in value divided by the overall length of the tendon. Therefore, this type of loss is considerable for short tendons and almost negligible for long tendons.

Time-dependent losses of prestress due to the shrinkage and creep of concrete as well as relaxation of prestressing tendons occur during the whole lifetime of the structure. The national and international codes give different expressions for the time-dependent losses, but the fundamental expression is

$$\Delta\sigma_p = E_p\,\varepsilon_{cs} + E_p\,\varphi\,\sigma_{cp}/E_c + \varkappa\sigma_p,$$

where

E_p is the modulus of elasticity of prestressing steel,
E_c is the modulus of elasticity of concrete,
ε_{cs} is the shrinkage of concrete,
φ is the creep coefficient of concrete,
σ_{cp} is the compressive stress of the concrete from quasi-permanent load at level of tendons,
$\varkappa$ is the percentage ratio of relaxation loss relative to tendon stress, and
σ_p is the stress in the tendon.

The value of the time-dependent losses for bridge constructions is between 10% and 15% of the initial prestress.

2.5 Effective values of the prestressing force

The effective value of the prestressing force, P(t), at time t is

$$P(t) = A_p \sigma_p(t),$$

where A_p is the cross sectional area of the prestressing reinforcement, and $\sigma_p(t)$ is the effecive stress in the tendon at time t which is equal to the initial prestress reduced by the losses of prestress:

$$\sigma_p(t) = \sigma_{po} - \Delta \sigma_p(t).$$

The maximum value of the initial prestress cannot exceed the minimum of $k_1 f_{pk}$ or $k_2 f_{p0\cdot1k}$, where f_{pk} and $f_{p0\cdot1k}$ are the characteristic tensile strength and the characteristic 0.1% proof-stress of the prestressing steel, respectively. The values of k_1 and k_2 may be given in national and international codes; the recommended values are 0.8 and 0.9, respectively.

2.6 Effects of prestressing

Two basic methods can be used in order to take into account the effects of pre-stressing in the structure. The effects of prestressing can be considered *internal forces* by introducing a normal force, $N_p = P\cos\alpha$, a shear force, $V_p = P\sin\alpha$, and a bending moment, $M_p = eP\cos\alpha$, in each concrete cross section, where P is the pre-stressing force, e is its eccentricity, and α is the angle between the prestressing rein-forcement and the neutral axis of the structural element in the considered cross section. These isostatic forces can also produce additional hyperstatic effects in the case of hyperstatic structures.

Prestressing effects can be considered *external loads* by introducing forces cre-ated by the prestressing onto the concrete element. These forces represent the anchorage forces, the distributed normal forces due to the tendon curvature perpen-dicular to the tendon P(x)/R(x), and the distributed friction force parallel to the ten-don dP(x)/dx. If the tendon layout has a quadratic parabolic shape or a tendon shape with constant curvature, the effect of prestressing as external forces is shown in Figure 9.12.

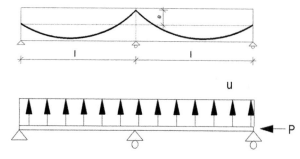

Figure 9.12 Effects of prestressing as external forces in case of quadratic parabolic tendon layout.

The intensity of the uniformly distributed equivalent external load perpendicular to the neutral axis of the element is given in this case by

$$u = 8Pf/l^2.$$

This formula enables one to determine the necessary value of the prestressing force, which can be directly balanced as part of a uniformly distributed external force q in uncracked elastic state:

$$P_{nec} = ql^2/8f.$$

3 Design of reinforced and prestressed concrete bridge decks

3.1 Conceptual design

The main objective of the conceptual design is to find the optimal structural form of the bridge to satisfy the needs of the client as well as meet the aesthetic, economic, and social aspects of the project by the various alternatives. The design procedure at this stage will be based on a relatively simple analysis to determine the main dimensions of the primary members of the structure to compare possible solutions, which can differ in the following ways:

- The structural system, including the longitudinal configuration and the, corresponding distribution of spans
- The construction materials, reinforced or prestressed concrete, use of normal, high performance or lightweight concrete, etc.
- The type and the dimensions of the cross section, slab, beam, box girder, etc.
- The erection technique, precast system, or cast-in-situ concrete, including the definition of the main steps of the construction sequences

All these elements are correlated, which is why a good conceptual design must be based on the intuition, the knowledge, and the experience of the engineers responsible for the project.

3.2 Structural modeling and analysis

Structural analysis consists of evaluating the response of the bridge to the external effects. For structural analysis, the structure has to be idealized by suitable models. An r.c. structure consists of combination of structural elements, like beam, column, slab, shell, etc. The response of the global structure (for example, the distribution of the internal forces due to external loads) is determined using analytical or numerical methods. However, the idealizations of the behavior used for the analysis of the structural elements are described next.

In the current design codes for structural analysis of reinforced or prestressed concrete structures, four methods are proposed:

- *Linear elastic analysis* for the determination of action effects for both the serviceability and the ultimate limit states may be used assuming linear stress-strain relationships for concrete and steel with uncracked cross sections. The results are realistic, with assumptions that actions are low.
- *Linear elastic analysis with limited redistribution* for analysis of structural members in ULS can be used for continuous beam or slab decks predominantly subjected to bending when the ratio of the length of adjacent spans is between 0.5 and 2 (Figure 9.13). Redistribution of bending moments can be applied only if sufficient rotation capacity of the considered sections of the structural member is provided. Requirements to check the rotation capacity are given in national and international codes. The reinforcement of the cross sections are determined according to the redistributed bending moments.

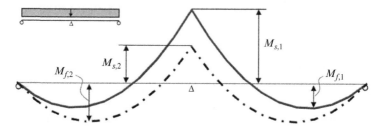

Figure 9.13 Example for redistribution of bending moments for a continuous bridge deck.

- *Theory of plasticity* should be only used to check at ULS conditions of r.c. bridges. In this case, the sufficient deformation capacity of the critical regions of the structure corresponding to the envisaged plastic mechanisms must be ensured. According to European Standard EN 1992-2, the required rotation capacity in the region of plastic hinges for beams, slabs, or frames, generally may be considered to be ensured if the area of reinforcement of a cross section fulfills both of the following:

$$x_u/d \leq 0.30 \text{ for concrete strength classes} \leq C50/60$$

$$x_u/d \leq 0.23 \text{ for concrete strength classes} \geq C55/67$$

where x_u is the neutral axis depth in ULS, and d is the effective depth of the cross section.
- *Nonlinear modeling* for dimensioning of r.c. bridges may be used under the conditions that equilibrium and compatibility of the structure are satisfied and for materials accurate nonlinear behavior is applied. The model must properly cover all failure modes, like due to axial forces, bending, shear, etc. The resistance should be evaluated in incremental steps. The process should be continued until the structure reaches its ultimate capacity.

4 Methods of construction

Construction of bridge decks with precast prestressed concrete beams connected by in situ concrete deck slab is economical for spans up to 50 m, mainly for structures such as long viaducts, where a large number of beams is required.

In the case of in situ concrete construction, the use of classical scaffolding to support the formwork is particularly suitable for bridges built over land if the ground can provide a suitable foundation and the structure is neither too high nor too long (Figure 9.14).

Figure 9.14 Scaffolding of the middle part of the railway bridge at Zalalövő, Hungary.

The use of launching girders is a particular way to support the formwork of an incrementally concreted bridge deck, which requires the use of a special movable girder supported from the previously completed part of the structure.

Prestressed concrete bridge decks built by balanced cantilever construction method can be used for spans from about 50 m up to 300 m. This method involves assembling the elements of the deck by building outward from either side of the piers symmetrically. Each segment of the construction is prestressed to the previously completed part of the structure (Figure 9.15).

Figure 9.15 Construction of the highway viaduct at Köröshegy, Hungary.

In the case of in situ construction, each segment is cast in situ using a formwork usually suspended from a steel frame supported by the previously cast segment. In order to limit its weight and avoid problems with deformation during construction, the length of each segment is limited to between 3 and 5 m. The length of the segment at the piers is approximately twice the length of a subsequent segment. The typical length of the segments in the use of precast elements is between 2 and 4 m, depending upon their depth and width, to limit the bending moment and to avoid a large amount of prestress at one location. The treatment of the joints between segments is an important factor of precast construction. The mean types of joints are coupled joints, mortared joints, and in situ concrete joints. The most common are coupled joints, when joints are filled usually by epoxy resin.

The principle of construction by the incremental launching method is to concrete the deck on the ground in a succession of segments, located at one or both ends of the bridge. When a segment is completed, it is prestressed to the previously completed part of the structure. Then the whole assembled structure will be advanced toward its final position to clear the casting area for the construction of the next segment. This method is suitable to built long bridge decks with various spans. The range of spans varies between 15 and 20 m for slab decks, and 40 and 70 m for box girders. The length of each segment can be between 12 m and the full length of the span. During construction, each section of the structure will be subjected to a considerable bending moment, which varies in function with the current position of the section. For this reason, it is beneficial to provide a uniform prestress during launching to the sections of intermediate sections of the bridge. Although in the first span, where the hogging moments are mostly high, eccentric prestress is favorable. The most common solution to reduce the high hogging moment in the first span is to use a launching nose (Figure 9.16).

Figure 9.16 Construction of the main part of the 1670 m length railway viaduct at Zalalövő.

The launching noises are usually a necessary light and rigid steel construction with a length equal to 60% or 80% of the span. The easiest solution to reduce the excessive moments during construction is to use temporary intermediate supports placed midway between piers. Another possibility to reduce hogging moments is to apply a mast located approximately a span-length behind the end of the deck to support cable stays when launching. For the longest spans, cable stays may be combined with launching nose as well.

5 Design example

In this section, a simplified structural analysis of the main longitudinal girder for static loads according to the Eurocode practice is presented for a monolithic beam and slab-type superstructure.

5.1 Basic design data

5.1.1 Geometry

The longitudinal axis of the deck is assumed to be straight in plan and perpendicular to the planes of supports.

The main dimensions are as follows (Figs. 9.17 and 9.18):
Span: $L = 19.00$ m
Total height of main girder: $h = 1.4$ m (at girder axis)
Carriageway width: $w = 8.0$ m
Thickness of deck slab: $v = 220$ mm (at the symmetry axis).
Pavement structure: 4 cm wearing layer
                                         6 cm binding layer
                                         4 cm protective layer
                                         1 cm waterproofing
                                         20–25 cm r.c. deck slab.

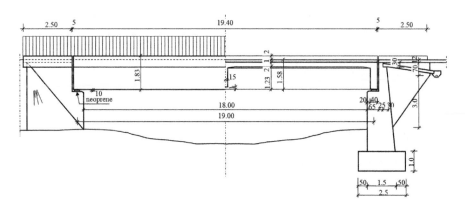

Figure 9.17 Longitudinal section and side view.

5.1.2 Design codes

Here are the relevant design codes:

EN 1990   Eurocode—Basis of structural design
EN 1991-2   Eurocode 1—Actions on structures. Traffic loads on bridges
EN 1992-1-1 Eurocode 2—Design of concrete structures. General rules and rules for buildings
EN 1992-2   Eurocode 2—Design of concrete structures. Concrete bridges

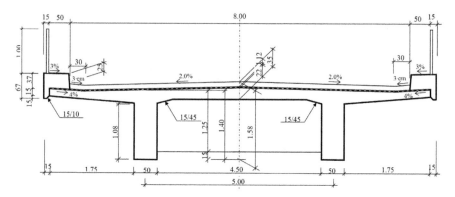

Figure 9.18 Cross section at midspan.

5.1.3 Material properties

The material properties are as follows:

Concrete	C35/45
Characteristic value of compressive strength, f_{ck} (N/mm^2)	35
Mean value of tensile strength, f_{ctm} (N/mm^2)	3.2
Mean value of the modulus of elasticity, E_{cm} (N/mm^2)	34,000
Long-term modulus of elasticity, $E_{c.eff} = E_{cm}/(1+\phi)$ (ϕ: final creep coefficient)	
Ultimate strain, ε_{cu} [‰]	3.5
Strength reduction factor for bridges	$\alpha = 0.85$
Partial factor for concrete:	$\gamma_c = 1.5.$
Reinforcing steel	B500B
Characteristic value of yield strength, f_{yk} (N/mm^2)	500
Characteristic value of the elongation at maximum load, ε_{uk} (‰)	No limit
Modulus of elasticity, E_s (N/mm^2)	200,000
Partial factor for steel:	$\gamma_s = 1.15$

5.1.4 Actions

For simplification, only self-weight of the superstructure is considered as permanent action.

Self-weight

Specific weights:	Concrete:	25.0 kN/m^3
	Asphalt:	24.0 kN/m^3
	Waterproofing:	0.25 kN/m^2
	Safety barrier:	0.50 kN/m

The self-weight of the superstructure is calculated for the half of the cross section (for one longitudinal girder) as follows:

- Self-weight of structural parts (load-bearing structure): $g_1 = 40.28$ kN/m
- Self-weight of non-structural parts (kerb, pavement, barrier, equipment): $g_2 = 19.22$ kN/m
- Total self-weight: $g = g_1 + g_2 = 40.28 + 19.22 = \mathbf{59.50}$ kN/m.

Variable actions

As typical variable actions on bridge decks, vertical and horizontal traffic loads, wind, and temperature actions will be discussed here. However, for simplification, only vertical traffic loads (i.e., Load Model 1/LM1) will be considered in these calculations.

Traffic loads Number of notional lanes ($6\,m \leq w = 8.0\,m$): 2
   Width of

Traffic lanes:	$w_l = 3.0\,m$
Remaining area	$w_r = 2.0\,m$

Vertical traffic loads For simplification, only LM1 is considered. Characteristic values of LM1 ($\alpha_{Qi}Q_{ik}$, $\alpha_{qi}q_{ik}$, and $\alpha_{qr}q_{rk}$) are shown in Table 9.1 and Figure 9.19.

Table 9.1 Characteristic Values of LM1

Lane	Tandem System (TS)	UDL
	Axle weight, Q_{ik} (kN)	q_{ik} (or q_{rk}) (kN/m²)
Lane 1	300	9.0
Lane 2	200	2.5
Remaining area (q_{rk})	0	2.5

For simplification, the tandem system on each lane is replaced by a one-axle load of equal weight ($L > 10\,m$); and the values of the adjustment factors are set as $\alpha_{Qi} = \alpha_{qi} = \alpha_{qr} = 1.0$.

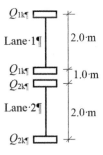

Figure 9.19 Axle positions.

Traffic loads on footways and cycle tracks: $q_{fk} = 0.0\,kN/m^2$ (no footway and cycle track in this case).

Horizontal forces Centrifugal force is disregarded due to the straight longitudinal axis of the deck.

   Braking and acceleration force ($Q_{\ell k}$)

The $Q_{\ell k}$ force acts at the top level of the pavement in the longitudinal axis of the carriageway:

$$Q_{\ell k} = 0.6\alpha_{Q1}(2Q_{1k}) + 0.10\alpha_{q1}q_{1k}wL = 497\,\text{kN (but } 180\alpha_{Q1} = 180\,\text{kN} \leq Q_{\ell k} \leq 900\,\text{kN)}$$

Wind action (F_{wk}) For simplification, the vertical and longitudinal (horizontal) components of wind are disregarded.

The height of the reference area, d_{tot}, as a function of parapet type, as well as the calculation of the reference area, $A_{ref,x}$, and the associated force coefficient, $c_{f,x}$, are seen in Figure 9.20.

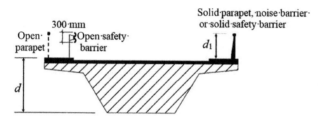

Type of parapet	On one side	On both sides
Open parapet or open safety barrier	$d + 0.3\,\text{m}$	$d + 0.6\,\text{m}$
Solid parapet or solid safety barrier	$d + d_1$	$d + 2d_1$
Open parapet and open safety barrier	$d + 0.6\,\text{m}$	$d + 1.2\,\text{m}$

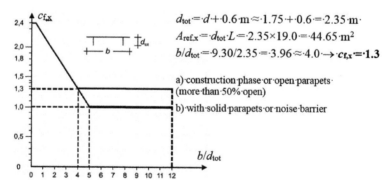

$d_{tot} = d + 0.6\,\text{m} \approx 1.75 + 0.6 = 2.35\,\text{m}$

$A_{ref.x} = d_{tot} \cdot L = 2.35 \times 19.0 = 44.65\,\text{m}^2$

$b/d_{tot} = 9.30/2.35 = 3.96 \approx 4.0 \rightarrow c_{f,x} = 1.3$

a) construction phase or open parapets (more than 50% open)

b) with solid parapets or noise barrier

Figure 9.20 Definition of reference area, $A_{ref,x}$, for a horizontal wind component.

The horizontal component (horizontal wind force) uniformly distributed along the superstructure length acts perpendicularly to the longitudinal axis of the superstructure:

$$F_{wk,x} = \frac{1}{2}\rho v_b^2 C_x A_{ref,x} = 0.5 \times 1.25\,[\text{kg/m}^3] \times (20\,[\text{m/s}])^2 \times 2.34 \times 44.65$$

$$= 26.1\,\text{kN,}$$

where the wind load factor, C_x, is calculated assuming usual suburban terrain (terrain category II) and a distance $z=5$ m of the superstructure from the ground level, which resulted in a terrain roughness factor $c_e(z)=1.8$, as follows:

$$C_x = c_{f,x}\, c_e(z) = 1.3 \times 1.8 = 2.34$$

The basic wind velocity, v_b, is a nationally determined parameter (NDP); here, $v_b=20$ m/s is taken.

Temperature action (T_k) Bridge temperatures as a function of air temperatures (NDP) are determined as follows:

- Minimum bridge temperature (for -15 °C minimum air temperature): $T_{e,min}=-7$ °C
- Maximum bridge temperature (for 35 °C maximum air temperature): $T_{e,max}=37$ °C
- Assumed initial temperature: $T_0=+10$ °C

Uniform temperature component The uniform temperature component is as follows:

- Maximum contraction: $\Delta T_{N,con}=T_0 - T_{e,min}=10-(-7)=17$ °C
- Maximum expansion: $\Delta T_{N,exp}=T_{e,max} - T_0=37 - 10=27$ °C

Uneven (linear) temperature component (in vertical plane) Temperature differences between the bottom and top fibers of the main girder (for 150-mm thick pavement) are as follows:

- Top face warmer: $\Delta T_{M,heat}=15\ °C \times 0.5=7.5$ °C
- Bottom face warmer: $\Delta T_{M,cool}=8\ °C \times 1.0=8$ °C

Simultaneity of temperature components The simultaneity of the uniform and linear temperature components should be assumed to be as follows:

$$T_k = \max \begin{cases} \Delta T_{N,con}\left(\text{or}\,\Delta T_{N,exp}\right) + 0.75\Delta T_{M,cool}\left(\text{or}\,\Delta T_{M,heat}\right) \\ 0.35\,\Delta T_{N,con}\left(\text{or}\,\Delta T_{N,exp}\right) + \Delta T_{M,cool}\left(\text{or}\,\Delta T_{M,heat}\right) \end{cases}$$

5.1.5   Combination of actions

For simplification, only self-weight and LM1 vertical traffic load are considered in the calculations given here.

Partial and combination factors

Partial factors for permanent actions (NDP) are as follows:

- $\gamma_{G,inf}=1.00$ if favorable
- $\gamma_{G,sup}=1.35$ if unfavorable
- $\xi=0.85$ reduction factor for unfavorable permanent actions

Partial factors for variable actions (NDP) are as follows:

- Traffic load: $\gamma_Q=1.35$
- Other variable actions: $\gamma_Q=1.5$

ψ factors (NDP) for traffic loads are as follows:

- UDL: $\psi_{0,q}=\psi_{0,r}=0.4$; $\psi_{1,q}=\psi_{1,r}=0.3$; $\psi_{2,q}=\psi_{2r}=0.0$
- TS: $\psi_{0,Q}=0.75$; $\psi_{1,Q}=0.6$; $\psi_{2,Q}=0.0$

Combination of traffic loads with other actions

For simplification, only group gr1a of traffic loads (including LM1 + loads on footways and cycle tracks) is considered here.

For ULS verifications, actions will be combined according to the alternative combinations as follows:

$$E_{Ed} = \max \begin{cases} \gamma_{G,\sup}E_G + \gamma_Q\left(\psi_{0,q}E_q + \psi_{0,Q}E_Q\right) \\ \xi\gamma_{G,\sup}E_G + \gamma_Q\left(E_q + E_Q\right) \end{cases}$$

Combinations of actions for SLS are as follows:

- Characteristic combination: $E_{car}=E_G + (E_q+E_Q)$
- Frequent combination: $E_{fr}=E_G + (\psi_{1,q}E_q+\psi_{1,Q}\, E_Q)$
- Quasi-permanent combination: $E_{qp}=E_G + (\psi_{2,q}E_q+\psi_{2,Q}\, E_Q)$

5.2 Calculation of internal forces

5.2.1 Influence line in the transverse direction

For simplification, a linear influence line is assumed (Figure 9.21). Reduction of traffic loads due to LM1 to one longitudinal girder using the one-axle model for the concentrated vehicle load:

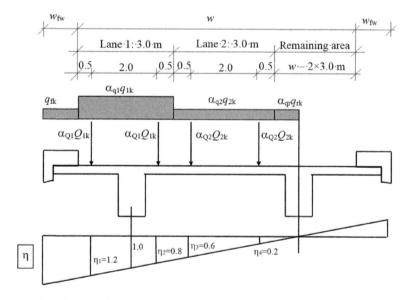

Figure 9.21 Reduction of traffic loads to one longitudinal girder.

Reduction of UDL to one girder (influence area below the ith notional lane: $A_\eta^1 = 3.0\,\text{m}$; $A_\eta^2 = 1.2\,\text{m}$; $A_\eta^r = 0.025\,\text{m}$; $A_\eta^{fw} = 0.89\,\text{m}$):

$$q_{red} = \alpha_{q1}\, q_{1k}A_\eta^1 + \alpha_{q2}\, q_{2k}A_\eta^2 + \alpha_{qr}\, q_{rk}A_\eta^r + q_{fk}A_\eta^{fw}$$

$$= 1.0 \times 9.0 \times 3.0 + 1.0 \times 2.5 \times 1.2 + 1.0 \times 2.5 \times 0.025 + 0.0 \times 0.89$$

$$= 30.06\,\text{kN/m}$$

Reduction of TS (one-axle concentrated vehicle load) to one girder:

$$Q_{red} = \alpha_{Q1}Q_{1k}(\eta_1 + \eta_2) + \alpha_{Q2}Q_{2k}(\eta_3 + \eta_4)$$

$$= 1.0 \times 300 \times (1.2 + 0.8) + 1.0 \times 200 \times (0.6 + 0.2) = 760\,\text{kN}$$

5.2.2 Bending moments

For design purposes, bending moments are calculated only at cross section K. The load arrangement (of self-weight and reduced traffic loads) is shown in Figure 9.22.

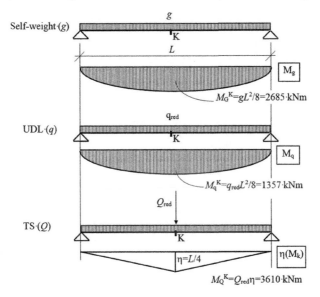

Figure 9.22 Load arrangement resulting in a maximum bending moment in the longitudinal girder.

The design bending moment is calculated as follows:

$$M_{Ed}^K = \max \begin{cases} \gamma_{G,sup}M_G^K + \gamma_Q\left(\psi_{0,q}M_q^K + \psi_{0,Q}M_Q^K\right) \\ \xi\gamma_{G,sup}M_G^K + \gamma_Q\left(M_q^K + M_Q^K\right) \end{cases}$$

$$= \max \begin{cases} 1.35 \times 2685 + 1.35(0.4 \times 1357 + 0.75 \times 3610) \\ 0.85 \times 1.35 \times 2685 + 1.35(1357 + 3610) \end{cases} = 9786\,\text{kNm}$$

The frequent value of bending moment is as follows:

$$M_{fr}{}^{K} = M_{G}{}^{K} + \left(\psi_{1,q}M_{q}{}^{K} + \psi_{1,Q}M_{Q}{}^{K}\right) = 2685 + (0.3 \times 1357 + 0.6 \times 3610)$$
$$= 5258\,kNm$$

The quasi-permanent value of bending moment is as follows:

$$M_{qp}{}^{K} = M_{G}{}^{K} + \left(\psi_{2,q}M_{q}{}^{K} + \psi_{2,Q}M_{Q}{}^{K}\right) = 2685 + (0.0 \times 1357 + 0.0 \times 3610)$$
$$= 2685\,kNm$$

5.2.3 Shear forces

For design purposes, design shear force is calculated at cross sections A and A′. The load arrangement (of self-weight and reduced traffic loads) is shown in Figure 9.23.

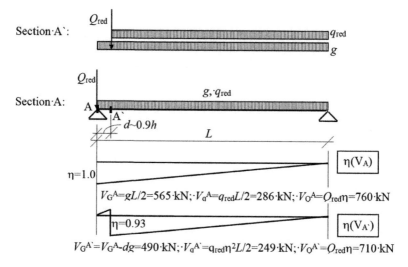

Figure 9.23 Load arrangement resulting in the maximum shear force in the longitudinal girder.

The design shear forces are figured as follows:

$$V_{Ed}{}^{A} = \max \begin{cases} \gamma_{G,sup}V_{G}^{A} + \gamma_{Q}\left(\psi_{0,q}V_{q}^{A} + \psi_{0,Q}V_{Q}^{A}\right) \\ \xi\gamma_{G,sup}V_{G}^{A} + \gamma_{Q}\left(V_{q}^{A} + V_{Q}^{A}\right) \end{cases}$$

$$= \begin{cases} 1.35 \times 565 + 1.35(0.4 \times 286 + 0.75 \times 760) \\ 0.85 \times 1.35 \times 565 + 1.35(286 + 760) \end{cases} = 2175\,kN$$

$$V_{Ed}{}^{A'} = \max \begin{cases} \gamma_{G,sup} V_G^{A'} + \gamma_Q \left(\psi_{0,q} V_q^{A'} + \psi_{0,Q} V_Q^{A'} \right) \\ \xi \gamma_{G,sup} V_G^{A'} + \gamma_Q \left(V_q^{A'} + V_Q^{A'} \right) \end{cases}$$

$$= \begin{cases} 1.35 \times 490 + 1.35(0.4 \times 249 + 0.75 \times 710) \\ 0.85 \times 1.35 \times 490 + 1.35(249 + 710) \end{cases} = 1956 \, kN$$

5.3 ULS

5.3.1 Effective width of flange

The effective width of flange (Figure 9.24) can be determined as follows:

- on the outer side of the beam (cantilever side): $b_{eff,o} = \min(0.2l_c + 0.1 \, l_0; \, l_c; \, 0.2 \, l_0) = \min(0.2 \times 1.75 + 0.1 \times 19.0; \, 1.75; \, 0.2 \times 19.0) = 1.75 \, m$
- on the inner side of the beam (toward the symmetry axis): $b_{eff,i} = \min(0.2l_r/2 + 0.1 \, l_0; \, l_r/2; \, 0.2 \, l_0) = \min(0.2 \times 2.25 + 0.1 \times 19.0; \, 2.25; \, 0.2 \times 19.0) = 2.25 \, m$, where $l_0 = 19 \, m$ is the distance along the beam between sections with zero bending moment (here equal to the span); $l_c = 1.75 \, m$ is the length of the cantilevered deck slab, and $l_r = 4.5 \, m$ is the clear distance between the longitudinal beams.

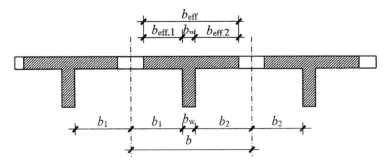

Figure 9.24 Effective width of flange.

The total effective width of flange is as follows:

$$b_{eff} = b_w + b_{eff,o} + b_{eff,i} = 0.5 + 1.75 + 2.25 = 4.5 \, m$$

5.3.2 Design for flexure

For simplification, only cross section K is sized here.

Applied σ-ε diagrams (Figure 9.25):

Estimation of the effective depth (d) is done assuming the following parameters:

- Concrete cover: $c = 40 \, mm$ (corresponding to exposure class XD3)
- Design increase of cover: $\Delta c_{dev} = 10 \, mm$
- Diameter of longitudinal bars: $\phi = 36 \, mm$ (arranged in three rows)
- Diameter of stirrups: $\phi_k = 16 \, mm$

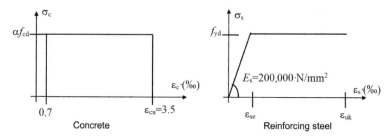

Figure 9.25 Stress-strain diagrams for materials.

- The following equation is used:

$$d \approx h - (c + \phi_k + 2.5\phi + \Delta c_{dev}) = 1400 - (4.0 + 1.6 + 2.5 \times 3.6 + 1.0)$$
$$= 1244\,mm.$$

Calculation of the depth of compression zone (x_c) is done as follows:

$$M_{Ed}{}^K = x_c b_{eff} \alpha f_{cd}(d - x_c/2)$$
$$\Rightarrow 9786\,kNm = x_c \times 4500 \times 0.85 \times (35/1.5) \times (1244 - x_c/2)$$
$$\Rightarrow x_c = 92\,mm\,(remains\,in\,flange)$$

Strain at the level of longitudinal bars is as follows:

$$\varepsilon_s = \varepsilon_{cu}(d - 1.25x_c)/(1.25x_c) = 0.035(1244 - 1.25 \times 92)/(1.25 \times 92)$$
$$= 0.0346 >$$
$$> \varepsilon_{se} = f_{yd}/E_s = 0.0022\,(yields)$$

Required amount of longitudinal tension reinforcement is as follows:

$$A_s = x_c b_{eff} \alpha f_{cd}/f_{yd} = 92 \times 4500 \times 0.85 \times (35/1.5)/(500/1.15)$$
$$= 18784\,mm^2\,(19\phi36 \to 3 \times 6 + 1\,pieces\,of\,bars)$$

Minimum amount of tension reinforcement is as follows:

$$A_{s,min} = \max \begin{cases} 0.26 \dfrac{f_{ctm}}{f_{yk}} b_w d \\ 0.0013 b_w d \end{cases}$$

$$= \max \begin{cases} 0.26 \dfrac{3.2}{500} 500 \times 1244 = 1035\,mm^2 << A_s\,(OK) \\ 0.0013 \times 500 \times 1244 \end{cases}$$

Bars shall be spaced according to the relevant detailing rules, and then the bending resistance of the section must be recalculated on the basis of the provided amount and position of reinforcement.

5.3.3 Design for shear

For simplification, the resistance of compression struts is verified at cross section A and the shear reinforcement is sized only at cross section A'. As shear reinforcement, only vertical stirrups are applied ($\alpha = 90°$).

Design for shear is carried out based on the variable strut inclination method as described next (Figure 9.26).

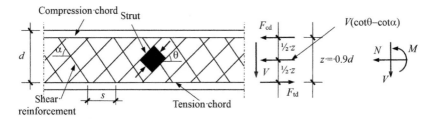

Figure 9.26 Calculation model for shear.

The data necessary to calculate shear resistance are as follows:
Strength reduction factor:

$$v = 0.6(1 - f_{ck}/250) = 0.52$$

Size effect factor:

$$k = \min\left(1 + \sqrt{\frac{200}{d\,[\mathrm{mm}]}};\ 2.0\right) = \min\left(1 + \sqrt{\frac{200}{1244}};\ 2.0\right) = 1.40$$

Longitudinal steel ratio (assuming that 50% of A_{sl} required at midspan (cross section K) is fully anchored behind the supports):

$$\rho_l = \min\left(A_{sl}/(b_w d);\ 0.02\right) = \min\left(0.5 \times 18784/(500 \times 1244);\ 0.02\right) = 0.015$$

Assumption of the compression strut inclination (angle between compression strut and longitudinal axis of the beam):

$$\cot\theta = 1.3 \text{ corresponding to } \theta = 37.5° \ (1.0 \le \cot\theta \le 2.5 \text{ condition is fulfilled})$$

Verification of compression struts (maximum shear force at cross section A):

$$V_{Rd}, \max = 1{,}0 \, \nu f_{cd} b_w 0.9 \, d \frac{\cot\theta}{1+\cot^2\theta}$$

$$= 1.0 \times 0.52 \times (35/1.5) \times 500 \times 0.9 \times 1244 \times \frac{1.3}{1+1.3^2}$$

$$= 3257 \, \text{kN} \geq V_{Ed}{}^A = 2175 \, \text{kN (OK)}$$

Check whether design shear reinforcement is necessary by solving the following:

$$V_{Rd,c} = \left[0{,}18/\gamma_c k \left(100 \, \rho_l f_{ck} \left[\text{N/mm}^2\right]\right)^{1/3}\right] b_w d$$

$$= \left[0.18/1.5 \times 1.4 (100 \times 0.015 \times 35)^{1/3}\right] 500 \times 1244$$

$$= 392 \, \text{kN} < V_{Ed}{}^{A'} = 1956 \, \text{kN},$$

with shear reinforcement being required.

Required amount of shear reinforcement (from condition $V_{Rd,s} = V_{Ed}{}^{A'}$):

$$A_{sw}/s = \frac{V_{Ed}^{A'}}{0.9 d f_{yd} \cot\theta} = \frac{1956}{0.9 \times 1244 \times 500 \times 1.3}$$

$$= 3091 \, \text{mm}^2/\text{m} \left(\phi 16/125 \text{ vertical stirrups}, \, (A_{sw}/s)_{prov} = 3217 \, \text{mm}^2/\text{m}\right)$$

Minimum amount of shear reinforcement:

$$\rho_{w,\min} = 0.08 \frac{\sqrt{f_{ck}}}{f_{yk}} = 0.08 \frac{\sqrt{35}}{500} = 0.00095 << \rho_{w,prov} = (A_{sw}/s)_{prov}/b_w = \frac{3217}{500}$$

$$= 0.0064 \, \text{(OK)}$$

Stirrups shall be spaced according to the relevant detailing rules, and then the shear resistance of the section must be calculated on the basis of the provided shear reinforcement.

5.4 SLS

5.4.1 Crack control

Crack control is carried out by calculating crack width on the frequent level of actions (NDP). For simplification, only cross section K is analyzed.

Applied crack width limit: $w_{lim} = 0.3$ mm (for exposure class XD3)

Cross-sectional data necessary for crack-width calculation (omitting calculation details) is calculated as follows:

Final value of creep coefficient: $\phi_c = 1.65$ (RH = 80%, 28 days of concrete age at initial loading)

Effective modulus of elasticity:

$$E_{c,\,eff} = E_{cm}/(1+\phi_c) = 34000/(1+1.65) = 12853\,\text{N/mm}^2$$

Depths of the neutral axis and moments of inertia are calculated assuming the following:

Uncracked (I) stage:

For short-term loading:	$y_{I,0} = 422$ mm,	$I_{I,0} = 0.33$ m^4
For long-term loading:	$y_{I,t} = 510$ mm,	$I_{I,t} = 0.45$ m^4

Cracked (II) stage:

For short-term loading:	$y_{II,0} = 237$ mm,	$I_{II,0} = 0.15$ m^4
For long-term loading:	$y_{II,t} = 386$ mm,	$I_{II,t} = 0.33$ m^4
Effective depth of the outer row of longitudinal bars:		$d_{so} = 1326$ mm.

Division of the frequent value of bending moment into short-term and long-term parts considers the following:

Long-term part: $M_{fr,t} = M_G^K = 2685$ kNm,

Short-term part: $M_{fr,0} = M_{fr}^K - M_{fr,t} = 5258 - 2685 = 2573$ kNm

Cracking moment is calculated as follows:

$$M_{cr} = f_{ctm}\frac{I_{I,0}}{h - y_{I,0}} = 3.2\frac{0.33 \times 10^{12}}{1400 - 422} = 1078\,\text{kNm}$$

Calculation of steel stresses is calculated as follows:
From the cracking moment at the outer bar ($\alpha_0 = E_s/E_{cm} = 200000/34000 = 5.9$):

$$\sigma_{sr} = \alpha_0\frac{M_{cr}}{I_{II,0}}\left(d_{so} - y_{II,0}\right) = 5.9\frac{1078 \times 10^6}{0.15 \times 10^{12}}(1326 - 237) = 47\,\text{N/mm}^2.$$

From the quasi-permanent value of moment at the outer bar:
($\alpha_t = E_s/E_{c,eff} = 200000/12853 = 15.6$):

$$\sigma_{s,qp} = \alpha_t\frac{M_{qp}}{I_{II,t}}\left(d_{so} - y_{II,t}\right) = 15.6\frac{2685 \times 10^6}{0.33 \times 10^{12}}(1326 - 386) = 121\,\text{N/mm}^2.$$

From the frequent value of moment at the outer bar:

$$\sigma_s = \alpha_t\frac{M_{fr,t}}{I_{II,t}}\left(d_{so} - y_{II,t}\right) + \alpha_0\frac{M_{fr,0}}{I_{II,0}}\left(d_{so} - y_{II,0}\right)$$

$$= 15.6\frac{2685 \times 10^6}{0.33 \times 10^{12}}(1326 - 386) + 5.9\frac{2573 \times 10^6}{0.15 \times 10^{12}}(1326 - 237) = 232\,\text{N/mm}^2$$

Effective tension area and corresponding steel ratio (Figure 9.27) are calculated as follows:

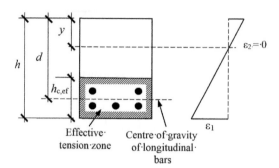

Figure 9.27 Effective tension area.

Effective tension area:

$$h_{c,ef} = \min\left[2.5(h-d-\Delta c_{dev});\ (h-y_{II,t})/3,\ h/2\right]$$
$$=365\,\text{mm}$$

$$A_{c,eff} = b_w h_{c,ef} = 500 \times 365 = 182500\,\text{mm}^2$$

Effective steel ratio:

$$\rho_{s,eff} = A_s/A_{c,eff} = 18784/182500 = 0.103$$

Calculating the difference of average strain in steel (ε_{sm}) and average strain in concrete (ε_{cm}) (for simultaneous long-term and short-term loading: $k_t = 0.5$):

$$\varepsilon_{sm} - \varepsilon_{cm} = \max\left[\frac{\sigma_s - k_t\dfrac{f_{ctm}}{\rho_{s,eff}}\left(1+\alpha_0\rho_{s,eff}\right)}{E_s},\ 0.6\frac{\sigma_s}{E_s}\right]$$

$$= \max\left[\frac{232 - 0.5\dfrac{3.2}{0.103}(1+5.9\times0.103)}{200000},\ \frac{0.6\times232}{200000}\right] = 0.00104$$

Calculation of maximum crack spacing (for ribbed bars: $k_1 = 0.8$; for bending: $k_2 = 0.5$):

$$s_{r,max} = 3.4c + 0.425\,k_1\,k_2\,\phi/\rho_{s,eff} = 3.4 \times 40 + 0.425 \times 0.8 \times 0.5 \times 36/0.103$$
$$= 171\,\text{mm}$$

Calculation of crack width:

$$w_k = s_{r,max}\,(\varepsilon_{sm} - \varepsilon_{cm}) = 171 \times 0.00104 = 0.18\,\text{mm} < w_{lim} = 0.3\,\text{mm}\,(\text{OK})$$

5.4.2 Deflection control

With regard to appearance and drainage, the longitudinal beams are generally designed with a camber equal to the deflection due to self-weight. For simplification, the effect of cracking on deflections is assessed by the use of distribution coefficient ζ, which allows for tension stiffening and enables an interpolation between the uncracked and the cracked state of the structure as follows.

Distribution coefficient allowing for tension stiffening (for sustained and repeated loading: $\beta = 0.5$; for σ_{sr} and σ_s see section 9.5.4.1) is calculated as follows:

$$\text{For quasi} - \text{permanent level of actions}: \zeta_{qp} = 1 - \beta\left(\frac{\sigma_{sr}}{\sigma_{s,qp}}\right)^2 = 1 - 0.5\left(\frac{47}{121}\right)^2$$

$$= 0.92$$

$$\text{For frequent level of actions}: \zeta_{fr} = 1 - \beta\left(\frac{\sigma_{sr}}{\sigma_{s,fr}}\right)^2 = 1 - 0.5\left(\frac{47}{232}\right)^2 = 0.98$$

Deflection control at midspan (cross section K) to avoid unacceptable appearance of the structure (quasi-permanent level of actions) is calculated as follows:

$$e_{qp}^K = \frac{5}{48}\frac{M_{qp}L^2}{E_{c,\,eff}}\left(\frac{1 - \zeta_{qp}}{I_{I,t}} + \frac{\zeta_{qp}}{I_{II,t}}\right)$$

$$= \frac{5}{48}\frac{2685 \times 10^6 \times (19.0 \times 10^3)^2}{12853}\left(\frac{1 - 0.92}{0.45 \times 10^{12}} + \frac{0.92}{0.33 \times 10^{12}}\right) = 23.6\,\text{mm}$$

$$\text{verification condition}: \quad e_{qp}^K = 23.6\,\text{mm} \leq \frac{L}{500} = 38\,\text{mm (OK)}$$

Deflection control at midspan (cross section K) to avoid user discomfort (assuming a camber at midspan, $e_0^K = -e_{qp}^K$) is calculated as follows:
Deflection from UDL (q):

$$e_q^K = \frac{5}{48}\frac{M_q^K L^2}{E_{cm}}\left(\frac{1 - \zeta_{fr}}{I_{I,0}} + \frac{\zeta_{fr}}{I_{II,0}}\right)$$

$$= \frac{5}{48}\frac{1357 \times 10^6 (19 \times 10^3)^2}{34000}\left(\frac{1 - 0.98}{0.33 \times 10^{12}} + \frac{0.98}{0.15 \times 10^{12}}\right) = 10.1\,\text{mm}$$

Deflection from TS (Q):

$$e_Q^K = \frac{1}{12}\frac{M_Q^K L^2}{E_{cm}}\left(\frac{1 - \zeta_{fr}}{I_{I,0}} + \frac{\zeta_{fr}}{I_{II,0}}\right)$$

$$= \frac{1}{12}\frac{3610 \times 10^6 (19 \times 10^3)^2}{34000}\left(\frac{1 - 0.98}{0.33 \times 10^{12}} + \frac{0.98}{0.15 \times 10^{12}}\right) = 21.4\,\text{mm}$$

Deflection from frequent value of traffic load:

$$e_{q+Q, fr,}{}^{K} = \psi_{1,q} e_q{}^{K} + \psi_{1,Q} e_Q{}^{K} = 0.3 \times 10.1 + 0.6 \times 21.4 = 15.9 \, \text{mm}$$

Verification condition:

$$e_{fr}{}^{K} = e_0{}^{K} + e_{qp}{}^{K} + e_{q+Q, fr}{}^{K} = -23.6 + 23.6 + 15.9 = 15.9 \, \text{mm} \leq \frac{L}{400}$$
$$= 47.5 \, \text{mm} \, (\text{OK})$$

6 Research and development

6.1 Shell pedestrian bridge in Madrid

Two pedestrian bridges (Matadero and Invenadero Bridges) with a concrete shell cover have been constructed in Madrid on the banks of the Manzanares River (Corres et al., 2012) (Figure 9.28). These pedestrian bridges are excellent examples of creativity, optimal use of material, and ability to create an extraordinary appearance. The purpose of these special bridges was to establish communication between downtown Madrid and its surroundings. The structural solution consists of an r.c. arch-vault with suspended composite deck spanning 43.5 m and 7.7 m rise. The deck is suspended by means of two series of 8.1-mm-diameter ties every 0.6 m at both sides.

Figure 9.28 Shell pedestrian bridge in Madrid, Spain
(courtesy of Hugo Corres, Fhecor, Madrid, Spain).

6.2   Large-span arch bridge, Colorado

With a main span of 323 m, the Hoover Dam Bypass Bridge (also known as the Mike O'Callaghan–Pat Tillman Memorial Bridge) is the fourth-longest single-span concrete arch bridge in the world (Figure 9.29). Each half-arch rib is made up of 26 cast-in-place sections, with construction starting from the canyon walls and a closure pour that locks the two halves together. Approximately 6880 m^3 of concrete of 69 MPa strength is cast into the arches. The outer dimensions of each hollow-arch rib are 6 m wide by 4.26 m long. Structural steel struts connect the arches at each column and are covered with pre-cast concrete panels. The largest struts weigh nearly 40 tons. Each of the 440 3-m-tall concrete segments was precast off-site and erected to form the pier columns. The precast columns are 90 m tall. The structural steel tub girders were fabricated off-site and placed with cableway cranes. The temporary cable stay tower and support system for erection of the arch incorporated more than 600,000 m of cable-stayed strand. The bridge design satisfies objectives for both architecture and performance.

Figure 9.29 Hoover Dam Bridge
(photo by Balázs).

6.3   Lightweight concrete for bridges (Stolma bridge, Norway)

The Stolma Bridge in Norway, a lightweight aggregate concrete structure built in 2000, had the longest span of this type of bridge (another example of which is shown in Figure 9.30), with a main span of 301 m (total length 467 m). The concrete grade was LC 60, with density of 1930 kg/m^3 (fib, 2000).

6.4   UHPC bridge, Sherbrooke, Canada

The first large-span (60-m) reactive powder concrete (RPC) or ultra-high-performance concrete (UHPC) pedestrian bridge was erected in Sherbrooke, Canada, in 1997 (Figure 9.31). It consists of 6 pieces of 10-m-long match-cast elements with two posttensioned bottom arches and posttensioned inclined diagonals. The structure was completed by external posttensioning (Figure 9.31) (Aitcin, 2014).

Figure 9.30 Lightweight aggregate concrete bridge currently under construction in Norway (fib, 2000).

Selection of materials was done by the University of Sherbrooke under the supervision of Professor Pierre-Claude Aitcin. The deck is 30 mm thick and posttensioned both longitudinally and transversally. Posttensioned diagonals connect the deck to the two bottom arches made of stainless steel tubes filled with RPC, with 2 mm thickness and 3.2 m length. The RPC contained a relatively high amount of modified CEM II type cement with low hydration heat, silica fume and crushed quartz, sand, superplasticizer, and water with a low water-cement ratio. The elements were steam cured. The RPC reached an average strength of 199 MPa with a standard deviation of 9.5 MPa, the modulus of elasticity was 48 000 MPa, and the modulus of rupture was 40 MPa (Aitcin, 2014).

Figure 9.31 View of the Passarelle of Sherbrooke bridge
(photo by Balázs).

6.5 Seonyugyo bridge, Seoul, South Korea

A 120-m-span UHPFRC pedestrian bridge known as the Seonyugyo Bridge or Rainbow Bridge (Figure 9.32) was erected in Seoul, South Korea, to mark the occasion of the 100 years of diplomatic relations between South Korea and France. It was designed by Rudy Ricciotti. The main arch is made of Ductal®, with a high percentage of steel fibers.

Figure 9.32 Seonyugyo Bridge, a 120-m-span UHPC pedestrian bridge in Seoul, South Korea (photo by Balázs).

6.6 MuCEM footbridge, Marseille, France

The MuCEM footbridge (Figure 9.33) has a particular role to connect the Museum of European and Mediterranean Civilization (MuCEM) to Fort Saint-Jean in Marseille. It is a very elegant solution to bridge the gap between the two constructions, featuring a highly elevated structure. The footbridge is constructed of precast segments of Ductal, each being 4.60 m long, created from a single mold, and including high steel fiber dosages. The precast elements are posttensioned together.

Figure 9.33 The MuCEM footbridge, a UHPFRC structure between the Museum of European and Mediterranean Civilization and Fort Saint-Jean in Marseille (photo by Balázs).

The Association Francaise de Genie Civil developed recommendations for the design of UHPFFRC structural elements (AFGC, 2007, 2013).

6.7 Tomai expressway, Shizuoka, Japan

The new Tomai Expressway between Tokyo and Kyoto is constructed parallel to the first Tokai Expressway to avoid traffic congestion. The Tomai Expressway is the most heavily used road operated by the Central Nippon Expressway, with some sections seeing more than 100,000 vehicles a day.

Earthquake resistance has been one of the main design aspects of the freeway viaduct in the vicinity of Shizuoka. A special solution has been developed for the web, which consists of steel tubes cast with concrete. This structure (Figure 9.34) reduced not only weight, but also transparency for the superstructure.

Figure 9.34 Viaduct on the Tomei freeway between Tokyo and Kyoto at Shizuoka (photo by Balázs).

6.8 Butterfly web bridge, Terasako Choucho bridge, Japan

The butterfly web bridge style is named after the shape of the prefabricated web of 150 mm thickness. The web is prestressed along one of the diagonals, which is subjected to tension with 15.2-mm-diameter strands of indented surface. The 80-MPa design strength concrete includes short steel fibers. The web does not contain non-prestressed reinforcement.

One of the reasons for this special web is the intention to reduce the weight of the structure, and hence increase earthquake resistance. The bridge is constructed by using the balanced cantilever method. By using butterfly webs, not only the earthquake resistance but also sustainability improves because less concrete is required than for an ordinary concrete web box girder.

The Terasako Choucho Bridge (Figures 9.35 and 9.36) has been constructed with butterfly webs. It received the Tanaka Award from the Japan Society of Civil

Engineers in 2013 (Figure 9.36). It is a 10-span continuous butterfly web bridge with a length of 712.5 m, spans of 58.6 m + 87.5 m + 7 × 73.5 m + 49.2 m and a width of 9.26 m.

Figure 9.35 Side view of Terasako Choucho Bridge with the butterfly web design, in Miyazaki, Japan
(courtesy of Akio Kasuga, Sumitomo Mitsui Construction, Japan).

Figure 9.36 Prestressing of main girder by the Terasako Choucho Bridge
(courtesy of Akio Kasuga, Sumitomo Mitsui Construction).

References

AFGC (Association Francaise de Genie Civil), 2013. Ultra-High-Performance Fiber-Reinforced Concrete: Recommendations, revised ed. AFGC Publication, Paris.

AFGC (Association Francaise de Genie Civil), 2007. Concrete Design for a Given Structure Service Life: State of the Art and Guide for the Implementation of a Predictive Performance Approach Based upon Durability Indicators. AFGC Publication, Paris.

Aïtcin, P.-C., 2014. The first UHSC structure, 15 years later. In: Proceedings of All Russian Conf. on Concrete and Reinforced Concrete. 12-16 May, 2014, Moscow, vol. 7, pp. 7–22.

Corres, H.-P., Diestre, S., León, J., Pérez, A., Sánches, J., Cruz, C., 2012. New materials and construction techniques in bridges and building design. In: Fardis, M.N. (Ed.), Innovative Materials and Techniques in Concrete Construction. Springer Science and Business Media, Dordrecht, Netherlands, pp. 17–41. http://dx.doi.org/10.1007/978-94-007-1997-2_22.

fib, 2000. Lightweight Aggregate Concrete: Part 1 (Guide) - Recommended Extensions to Model Code 90; Part 2 (Technical Report) - Identification of Research Needs; Part 3 (State-of-the-art Report) - Application of Lightweight Aggregate Concrete, May 2000, 118 pages.

Steel and composite bridges

10

Pipinato A.[1], De Miranda M.[2]
[1]AP&P, Technical Director, Italy
[2]Studio de Miranda Associati, Technical Director, Italy

1 Introduction

Steel bridges represent a design solution, able to handle all the necessary span alternatives. Advantages of this solution relate to the smallest foundations; the industrialization of the construction process; an easy dismantling and reuse procedure; greater control of members, substructures, and connections; a rapid construction phase; and a lighter solution if compared to classical r.c. bridges. For small and medium-size spans, girders and trusses are the most common solutions. Commercial (up to 30-m spans) or plate girders are the widely diffused solutions until 100-m spans, with a superstructure made of an r.c. deck acting compositely (Figures 10.1 and 10.2). Over 100m of span length, box girders are more convenient, for their improved torsional stiffness and aerodynamic behavior (Figures 10.3 and 10.4). For longer spans (more than 250 m), orthotropic deck solutions resting on a box girder are more practical. Also, truss bridges are employed especially for railway bridges, and could represent a practical and economical solution (Figure 10.5).

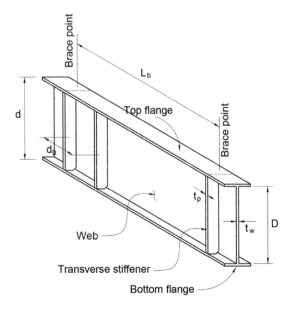

Figure 10.1 Girder components.

Innovative Bridge Design Handbook. http://dx.doi.org/10.1016/B978-0-12-800058-8.00010-4

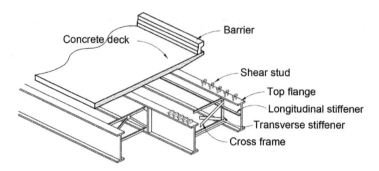

Figure 10.2 Steel-r.c. composite girder components.

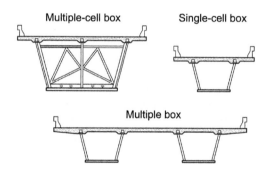

Figure 10.3 Box-girder component types.

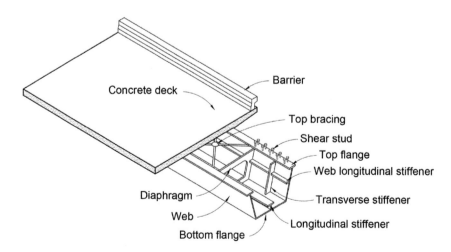

Figure 10.4 Steel-r.c. composite box-girder components.

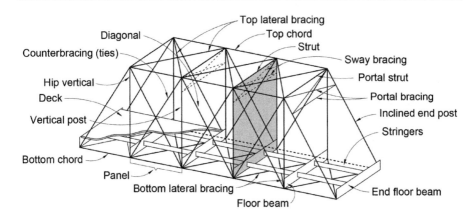

Figure 10.5 Truss main components.

2 Design

2.1 Steel bridges

A bridge in which the entire superstructure is made of steel can be referred to as a steel bridge, marking a difference with the steel-concrete composite bridge, where the deck is an r.c. slab: in "pure" steel bridges, the deck is also a steel structure. The weight of a steel deck is around one-third the weight of a concrete deck, but its cost is around twice as much. This has three consequences:

- Steel bridges are usually more expensive than composite bridges up to spans of around 100 m. In fact for small–medium spans, a steel deck is more costly than a concrete deck, and the gain in weight of the main girders due to the lighter structure is not compensated, while this occurs in movable bridges, which usually have a steel deck.
- A completely steel bridge is more subject to fatigue than a composite one because the stresses induced by self-weight are less, and, in turn, those induced by live loads increase, as do stress ranges.
- Steel decks become convenient to use on the longest spans due to the reduced self-weight; the importance of this increases proportionally with the bridge size.

For smaller and medium spans, a steel deck can present the advantage of allowing full inspection in any part, as opposed to an r.c. or post-tensioned (PT) concrete slabs, where inspection of rebars or cables is difficult. A steel bridge can be designed by using any of the classic bridge types, listed here in order of increasing span:

- *Plate girder type*—Typically with two girders and side cantilevers for spans up to around 120 m.
- *Box girder type*—For curved alignments or longest spans; the shape can be rectangular or trapezoidal.
- *Truss girder*—For medium span bridges, especially used for railway lines.
- *Arch type*—Or bowstring.
- *Cable-supported bridge*—cable-stayed or suspension structures.

A particular type of steel deck is composed of three elements:

* A deck plate
* A set of longitudinal stiffeners
* A system of transverse girders – A set of longitudinal girders (two or more)

The orthogonal mesh and the different stiffness of this double-supporting system is the origin of the name of this type of deck: the *orthotropic plate deck*. The deck plate, if dimensioned only with regard to strength, could have thickness of 10–12 mm. In any event, a minimum thickness of 14 mm is required these days in order to meet the fatigue requirements for road bridges of normal traffic. The longitudinal stiffeners have the scope of supporting the deck plate and can be typically flat plates, angles, tees, or trapezoidal channels. The latter type, although involving higher fabrication costs, presents greater torsional stiffness and has been popular in Europe in recent decades. Spacing of open stiffeners is of the order of 300 mm, while for closed sections, the spacing is twice that value. Their size and depth is proportional to the transverse girder spacing. Transverse girders have the scope of supporting the longitudinal ribs and typically have an inverted T-section, which acts compositely with the deck plate to form an asymmetrical I-section. The spacing of transverse girders can vary from 2.5 to 6 m, a spacing of around 4 m being typical in Europe.

2.2 Composite bridges

2.2.1 General

Composite bridges are more and more popular around the world since they combine some advantages of steel bridges with some key factors of concrete bridges. In fact, a composite bridge presents the following:

* A steel main structure that is much easier to erect if compared to the construction of a concrete girder
* A light structure, which imposes smaller loads on piers and foundations, allowing for economy
* A concrete slab, which is cheaper and easier to build than a steel orthotropic deck, and also presents two other advantages:
    * A higher mass, which induces fewer vibrations, noise, and dynamic loads on the supporting structure
    * A top surface that allows for easy paving with traditional methods, while weak points of the orthotropic deck consists of the difficulty of realizing a strong binding, the delicate execution, and some concern about durability of its paving

Aside from these advantages, a composite deck has the following drawbacks:

* Longitudinal tension forces can cause cracks in the slab, and the link with steel causes tensile stresses due to restrained concrete shrinkage.
* With respect to a steel deck, the weight increases, which is a disadvantage for the longest spans
* With respect to a concrete girder, the steel structure is usually more expensive in terms of material costs

In any case, composite bridges, if well designed, have shown to be competitive with concrete bridges in all small–medium spans and competitive with steel bridges in spans up to 120 m.

2.2.2 Typical structures

The main structure of a composite bridge is formed by the following elements: (i) main longitudinal girders; (ii) transverse diaphragms or transverse girders; and (iii) concrete slab. These elements are comprised as follows:

- For spans up to around 70 m, main girders are typically plate I-girders, with two of them for widths up to around 12 m. Two main girders can be maintained even for larger widths if a central stringer is provided. For longer spans, the structure is typically a box girder of constraint or variable depth. The cross-section shape is usually rectangular, but the trapezoidal shape, even if more complicated to fabricate, often gives the advantage of a more attractive and slender appearance, as well as a bottom flange of smaller width, which gives economic benefits.
- Transverse diaphragms in smaller girders are commonly realized with simple I-beams. For longer spans, a trussed structure is more appropriate. For box girders, plated diaphragms are more suited to small boxes, while trussed diaphragms fit well into larger structures.
- Concrete slabs can be cast in place, cast over prefabricated slabs, reinforced by steel joists, or made of full-thickness, precast elements. Construction time decreases from cast in place to prefabricated.

2.2.3 Composite cable-stayed bridges

In self-anchored cable-stayed bridges, composite decks are very competitive for spans ranging between 200 to 500 m. This mainly occurs for the following structural systems:

- Self-anchored three span cable-stayed bridge
- Two planes of stay cables
- Plated girders of small depth both in the longitudinal and transverse directions
- Concrete slab built in partially or fully precast elements

Such a cable-stayed bridge will present a concrete slab fully compressed in two directions:

- Longitudinally, by the horizontal component of stay cables
- Transversally, the slab being the top flange of the transverse girders simply supported by the stay cables' planes

This biaxial compressive stress is advantageous for strength and durability. The global self-weight of such a structure is much less than the weight of a full concrete deck, requiring fewer stay cables, and compensating for the extra cost that the steel structure implies; but construction can be easier and quicker. Further, if compared with the full steel deck, the increased mass of a composite deck brings significant benefits in terms of reduction of vibration, both of deck and of stay cables, and higher aerodynamic stability for larger spans.

2.2.4   Erection

Erection methods of composite bridges typically include the following:

- *Erecting the steel structure from the ground, by using cranes and temporary steel piers;* This is the cheaper and simpler way for smalls spans with small height from ground.
- *Longitudinal launching;* This is the preferred method for continuous girders, with regular alignment, constant depth, and at least three spans. It is very convenient for girders high over the ground, such as over a deep valley, a river, or a sea strait; by using a long launching nose, this method can be used even for one girder.
- *Building by balanced progressive cantilever;* This is still an option mainly for major bridges; in this case, entire segments, transported by barges or trucks, can be lifted and jointed to the erected structure.
- *Erecting by rotation;* This is convenient when erection can be done on the sides of the obstacle to be crossed: river or motorway.
- *Other special types of erection,* such as transversal launching or full transportation by barges of entire girders, are also possible.

3   Product specifications

Steel is an iron-carbon alloy characterized by specific percentages of the constituent components. Structural steel have a carbon content between 0.1 and 0.3%: the carbon component has the task of improving the strength, but at the same time it reduces the ductility and weldability of the base material. Different strength types of structural steels are commonly used for bridge structures. Depending on the reference nations, different codes should be used in the design phase: in this chapter, North American and European practice are presented. Designs are based on minimum properties such as those shown in Table 10.1, including ASTM A709 (2010a) for North America and Eurocodes for Europe. Additional special requirements are often provided in other codes that are available in Europe, such as National Annexes, and the United States, such as American Association of State Highway and Transportation Officials (AASHTO) standards; these standards mainly differ in notch toughness and weldability requirements.

3.1   Codes

ASTM A709 (2010a) specification covers carbon and high-strength, low-alloy steel structural shapes, plates, and bars and quenched and tempered alloy steel for structural plates intended for use in bridges. Seven grades are available in four yield strength levels as depicted in Table 10.1. The nominal values of material properties given in the EN 1993-1-1 (2005) code should be adopted as characteristic values in design calculations. EN 1993-1-1 (2005) covers the design of steel structures fabricated of steel material conforming to the four steel grades listed in Table 10.1.

Table 10.1 Structural Steel Materials According to ASTM and Eurocode

Standard	Designation	Product Categories	Nominal Thickness (mm/in)	f_y (MPa)	f_u (MPa)
ASTM709	36	Plates, shapes, bars	t < 101,6/2,5	250	400
ASTM709	50	Plates, shapes, bars, sheet piles	t < 101,6/2,5	345	450
ASTM709	50S	Shapes	t < 101,6/2,5	345	450
ASTM709	50W	Plates, shapes, bars	t < 101,6/2,5	345	450
ASTM709	HPS 50W	Plates	t < 101,6/2,5	345	482
ASTM709	HPS 70W	Plates	t < 101,6/2,5	485	586
ASTM709	HPS 100W	Plates	t < 101,6/2,5	690	690
EN10025-2	S235	Hot-rolled members	t < 40/1,57	235	360
EN10025-2	S275	Hot-rolled members	t < 40/1,57	275	430
EN10025-2	S355	Hot-rolled members	t < 40/1,57	355	510
EN10025-2	S450	Hot-rolled members	t < 40/1,57	440	550

3.2 Stress-strain behavior

Both the ASTM A370 (2010b) and Eurocode EN 6892-1 (2009) defines the testing requirements to determine the tensile strength of steel products. The test method requires the determination of the yield strength, tensile strength, and percent elongation for each test. A stress-strain curve can be measured by graphically or digitally recording the load and elongation of an extensometer during the duration of the test. The elastic modulus or Young's modulus for steel is the slope of the elastic portion of the stress-strain curve. It is conservatively taken as E = 200.000 MPa (29,000 ksi) for structural calculations for all structural steels used in bridge construction.

3.3 Hardness

Indentation resistance is the hardness property of steel materials. Measurable with a wide variety of testing methods, including the Brinell, Vickers, and Rockwell methods, it is not a direct test but an indirect measure of tensile and ductility properties. Hardness testing is commonly used to assess the residual properties of structural steel that has been exposed to fire (FHWA, 2012).

3.4 Ductility

A minimum ductility is required for steel, that should be expressed in terms of limits for the following:

- The ratio f_u/f_y of the specified minimum ultimate tensile strength f_u to the specified minimum yield strength f_y
- The elongation at failure on a specific gauge length

- The ultimate strain ε_u, corresponding to the ultimate strength f_u

The material ductility is required both by ASTM A709 (2010a) and Eurocode EN 1993-1 (2005): however, the material ductility does not automatically translate into structural ductility. The design choice involving connection types, section transitions, bracings, etc., can lead to steel member fails in a brittle mode. To provide structural ductility, the steel must have a sufficient strain-hardening capability to increase the local net section strength sufficiently to allow the gross section to reach yield before rupture occurs at the net section.

3.5 Fracture toughness

The material should have the required material toughness to prevent brittle fracture within the intended design working life of the structure. No further checks against brittle fracture need to be made if the conditions given in codes are met: for example, those of EN 1993-1-10 (2010) give the maximum permissible element thickness appropriate to a steel grade, its toughness quality in terms of K_V-value, the reference stress level (σ_{Ed}), and the reference temperature T_{Ed}. According to a well-based scientific procedure, the linear elastic fracture mechanics (LEFM) approach is the way to predict brittle fracture in bridges, or generally in steel structural components. A measure of fracture toughness could be realized with the Charpy V-notch test (K_V-value).

3.6 Fatigue resistance

A comprehensive overview of fatigue resistance is provided in Chapter 4.

3.7 Strength property variability

Members' property variability is an inherent consequence of the steel manufacturing and is considered in the resistance factors of both Eurocode (EN 1993-1-1, 2005) and in load and resistance factor design (LRFD) specifications (AASHTO, 2013).

3.8 Residual stresses

Residual stress is a permanent state of stress in a structure that in itself is in equilibrium and is independent of any applied action. These stresses can result from the rolling processes, cutting processes, welding shrinkage, or lack of fit between members or from any loading event that causes part of the structure to yield. Distortion during fabrication is the direct consequence of residual stresses. In order to avoid this problem, mandatory tolerances must be specified during the design stage.

3.9 Durability

According to Eurocode (EN 1993-2, 2006), to ensure durability, bridges and their components may be designed to minimize damage or be protected from excessive deformation, deterioration, fatigue, and accidental actions that are expected during the working life of the designed structure. Structural parts of a bridge to which guardrails or parapets are connected should be designed to ensure that plastic deformations of the guardrails or parapets can occur without damaging the structure. The possibility of the safe replacement of any replaceable components of a bridge should be verified as a transient design situation. Permanent connections of structural parts of the bridge should be made with preloaded bolts of specific category connections. Alternatively, closely fitted bolts, rivets, or welding may be used to prevent slipping. Joints where the transmission of forces occurs purely by contact may be used where justified by fatigue assessments.

3.10 Robustness and structural integrity

According to Eurocode (EN 1993-2, 2006), the design of the bridge should ensure that when the damage of a component due to accidental action occurs, the remaining structure can sustain at least the accidental load combination with reasonable means. The National Annex may define components that are subjected to accidental design situations and also details for the assessments. Examples of such components are hangers, cables, and bearings. The effects of corrosion or fatigue of components and material should be taken into account by appropriate detailing (see also EN 1993-1-9 and EN 1993-1-10).

4 Structural connections

4.1 Bolted connections

EN 1993-1-8 (2005) integrates the general part of EN 1993-1-1 (2005) dealing with verification procedures and requirements for bolted connections. The different classes of bolts, with diameters measured in 12, 14, 16, 18, 20, 22, 24, 27, and 30 mm, can be separated into classes 4.6, 5.6, 6.8, 8.8, and 10.9. For each class, the yield strength f_{yb} and the ultimate strength f_{ub} are given. In the construction of bridges, the last two classes are more diffused. Only bolt assemblies of Classes 8.8 and 10.9 may be used as preloaded bolts with controlled tightening. The reference standard for these bolts in Europe is EN 14399-1 (2005). Bolts with controlled tightening are very sensitive to differences in manufacturing and lubrication.

European regulations on bolts with controlled tightening have the aim to ensure that, with a given torque, the required preload is obtained with a good reliability and sufficient safety margins to avoid excessive tightening of the screw and consequent plastic deformation. For this reason, a test method to verify the suitability of the components in controlled tightening is included in the Eurocode. The adopted safety factors are given in EN 1993-1-1 (2005) and EN 1993-2 (2006), respectively, for general rules for buildings and bridges. The main safety factors are summarized as follows:

$\gamma_{M0} = 1,05$, strength of gross cross sections
$\gamma_{M1} = 1,25$ strength of net sections at the position of bolts
$\gamma_{M2} = 1,25$ strength of the bolts
$\gamma_{M2'} = $ strength of the contact plates;
$\gamma_{M3} = 1,25$ sliding resistance at the ultimate limit state (ULS)
$\gamma_{M7} = 1,10$ preload of high resistance bolts.

Diameters and characteristics of bolts in U.S. code (AASHTO, 2013) are different from those in the Eurocode. The standard in the United States for the design of bolted connection (AISC, 2010a) is mainly included in ASTM A325M (2013), ASTM A490M (2012), and related standards. Screws, nuts, and washers are described in AISC (2010a) and in ASTM A325M (2013) specifications. The diameters are 15.88, 19.05, 22.23, 28.58, 31.75, 34.93, and 38.10 mm (the smallest diameters approximately correspond to the European 16-, 20-, 22-, 27-, and 30-mm specifications). ASTM A325M and ASTM A490M classes are similar to European classes 8.8 and 10.9, respectively. Under U.S. code, manufacturers' certifications shall be sufficient proof of compliance with the code standard. The use of high-strength bolts is described in RCSC (2009). High-strength bolts are classified in this document according to the strength of the material as follows:

Group A: ASTM A325, A325M, F1852, A354 Grade BC, and A449
Group B: ASTM A490, A490M, F2280, and A354 Grade BD.

4.2 Riveted connections

Eurocode 3 EN 1993-1-8 (2005) integrates the general part of EN 1993-1-1 (2005) dealing with verification procedures and requirements for riveted connections. The material properties, dimensions, and tolerances of steel rivets should comply with the requirements given in 1.2.6 Reference Standards, Group 6 of the National Annex. Minimum and maximum spacing and end, and edge distances for rivets are the same for bolts and are given in the same Eurocode. Riveted connections should be designed to transfer shear forces, so if tension exists, the design tensile force $F_{t,Ed}$ should not exceed the design tension resistance $F_{t,Rd}$ given in the code. The standard (AISC, 2010a) in the United States do not cover the design of riveted connection. In the evaluation of existing bridges, rivets shall be assumed to be ASTM A502, Grade 1, unless a higher grade is established through documentation or testing (AISC, 2010a), and the same code suggests that because removal and testing of rivets is difficult, assuming the lowest rivet strength grade simplifies the investigation.

4.3 Welded connections

Submerged arc welding (SAW) is probably the most widely used process for welding bridge web-to-flange fillet welds and inline butt welds in thick plate to make up flange and web lengths. A continuous wire via a contact tip forms a molten pool. The weld pool is submerged by flux fed from a hopper. The flux immediately covering the molten weld pool melts, forming a slag and protecting the weld during solidification; surplus flux is collected and recycled. This process is mainly automatic or robot-assisted. Metal-active gas welding (MAG), process is the most widely used manually controlled process for factory fabrication work; it is sometimes known as *semiautomatic* or *carbon dioxide welding*. When the shielding gas used is an inert argon or nonreactive carbon dioxide the process is named metal inert gas (MIG). The manual metal arc welding (MMA) process remains the most versatile of all welding processes, but its use in the modern workshop is limited.

Eurocode 3 EN 1993-1-8 (2005) integrates the general part of EN 1993-1-1 (2005), which deals with verification procedures and requirements for welded connections: the provisions apply to weldable structural steels conforming to EN 1993-1-1 and to material thicknesses of 4 mm and over. The provisions also apply to joints in which the mechanical properties of the weld metal are compatible with those of the parent metal. For welds in thinner material, refer to EN 1993 part 1.3; and for welds in structural hollow sections in material thicknesses of 2.5 mm and over, guidance is given in section 7 of EN 1993-1-8 (2005). Welds subjected to fatigue should also satisfy the principles given in EN 1993-1-9. The specified yield strength, ultimate tensile strength, elongation at failure, and minimum Charpy V-notch energy value of the filler metal should be equivalent to or better than that specified for the parent material. A fillet weld with an effective length of less than 30 mm or less than 6 times its throat thickness (whichever is larger) should not be designed to carry loads. The effective throat thickness a of a fillet weld should be taken as the height of the largest triangle (with equal or unequal legs) that can be inscribed within the fusion faces and the weld surface, measured perpendicular to the outer side of this triangle. The effective throat thickness of a fillet weld should not be less than 3 mm. In determining the design resistance of a deep penetration fillet weld, consider its additional throat thickness, provided that preliminary tests show that the required penetration can consistently be achieved (Figure 10.6). A uniform distribution of stress is assumed on the throat section of the weld, leading to the normal stresses and shear stresses (shown in Figure 10.7) as follows: $\sigma_\perp$ is the normal stress perpendicular to the throat, $\sigma_\parallel$ is the normal stress parallel to the axis of the weld, $\tau_\perp$ is the shear stress (in the plane of the throat) perpendicular to the axis of the weld, and $\tau_\parallel$ is the shear stress (in the plane of the throat) parallel to the axis of the weld. The design resistance of the fillet weld will be sufficient if the following are both satisfied:

$$\left[\sigma\perp^2 + 3\left(\tau\perp^2 + \tau\|^2\right)\right]^{0,5} \leq f_u/(\beta_w\gamma_{M2}) \text{ and } \sigma\perp \leq 0.9f_u/\gamma_{M2}, \tag{1}$$

where f_u is the nominal ultimate tensile strength of the weaker part joined, and β_w is the appropriate correlation factor taken from the code (which varies from 0.8 to 1).

Finally, welds between parts with different material strength grades should be designed using the properties of the material with the lower strength grade. The design resistance of a full penetration butt weld should be taken as being equal to the design resistance of the weaker of the parts connected, provided that the weld is made with a

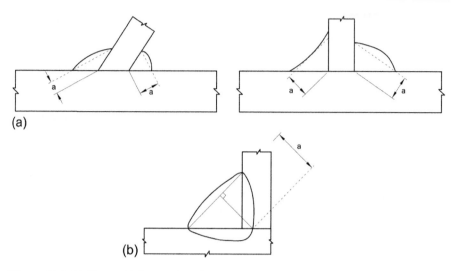

Figure 10.6 (a) Throat thickness of a fillet weld: (b) throat thickness of a deep penetration fillet weld.

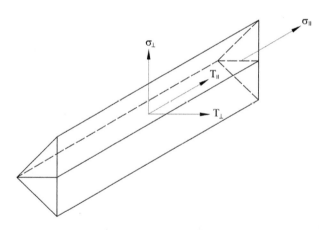

Figure 10.7 Stresses on the throat section of a fillet weld.

suitable consumable that will produce all-weld tensile specimens with both a minimum yield strength and a minimum tensile strength not less than those specified for the parent metal. The design resistance of a partial penetration butt weld should be determined using the method for a deep penetration fillet weld given in the code. The throat thickness of a partial penetration butt weld should not be greater than the depth of penetration that can be consistently achieved. The design resistance of a T-butt joint, consisting of a pair of partial penetration butt welds reinforced by superimposed fillet welds, may be determined as for a full penetration butt weld if the total nominal throat thickness, exclusive of the unwelded gap, is not less than the thickness t of the part forming the stem of the tee joint, provided that the unwelded gap is not more than $(t/5)$ or 3 mm, whichever is less. In lap joints, the design resistance of a fillet weld should be reduced by multiplying it by a reduction factor β_{Lw} to allow for the effects of nonuniform distribution of stress along its length.

The standard in the United States for the design of welded connection (AISC 2010a) is mainly included in AWS (2010) and related standards. The selection of weld type [complete-joint-penetration (CJP) groove weld versus fillet versus partial-joint-penetration (PJP) groove weld] depends on base connection geometry (butt versus T or corner), in addition to required strength and other issues discussed in the code. Consideration of notch effects and the ability to evaluate with NDE may be appropriate for cyclically loaded joints or joints expected to deform plastically (AISC 2010b).

4.4 Connection choice

Transport needs are commonly a constraint parameter that induces engineers to select the appropriate construction phase and consequently the connections design. In general, bolt connections are cheaper than welding, do not require skilled labor, are not difficult to inspect, and finally can be applied faster. However, it is not attractive in appearance, and when a good-looking structure is required, bolting is not generally permitted. On the other hand, welded joints require expensive and improved skill labor, and of course, repair is less fast than it is for bolted connections.

5 Steel bridge analysis

5.1 Structural modeling

Analysis should be based upon calculation models of the structure that are appropriate for the limit state under consideration. The calculation model and basic assumptions for the calculations should reflect the structural behavior at the relevant limit state with appropriate accuracy and reflect the anticipated type of behavior of the cross sections, members, joints, and bearings. The method used for the analysis should be consistent with the design assumptions. For the structural modeling and basic assumptions for bridge components, accurate details are given in the codes and standards adopted. Dealing with Eurocode, for the structural modelling and basic assumptions see EN 1993-2 (2006), while for the design of plated components and cables see also EN 1993-1-5 (2007) and EN 1993-1-11 (2007). For US codes, FHWA (2012) is a comprehensive reference.

The effects of the behavior of the joints on the distribution of internal forces and moments within a structure, and on the overall deformations of the structure, they may generally be taken into account where significant (such as in the case of semi-continuous joints). To identify whether the effects of joint behavior on the analysis need to be considered, the following distinctions may be made between three joint models if the Eurocode procedure is adopted (i.e., EN 1993-1-8):

- *Simple*, in which the joint may be assumed not to transmit bending moments
- *Continuous*, in which the behavior of the joint may be assumed to have no effect on the analysis
- *Semicontinuous*, in which the behavior of the joint needs to be taken into account in the analysis

These three models are classified as nominally pinned, rigid and semi-rigid connections. The requirements of the various types of joints are given in EN 1993-1-8 (2005). Ground-structure interaction should be taken into account, considering the deformation characteristics of the supports where significant. For example, EN 1997-1 (2005) gives guidance for the calculation of soil-structure interaction.

Concerning the global analysis of the structure, the internal forces and moments may generally be determined using either of the following:

- First-order analysis, using the initial geometry of the structure
- Second-order analysis, taking into account the influence of the deformation of the structure

The effects of the deformed geometry (second-order effects) should be considered if they significantly increase the action effects or modify the structural behavior. First-order analysis may be used for the structure if the increase of the relevant internal forces or moments or any other change of structural behavior caused by deformations can be disregarded. This condition may be assumed to be fulfilled if the following criterion is satisfied:

$$\alpha_{cr} = F_{cr}/F_{ed} \geq 10, \text{ for elastic analysis}$$

$$\alpha_{cr} = F_{cr}/F_{ed} \geq 15, \text{ for plastic analysis,}$$

where α_{cr} is the factor by which the design loading would have to be increased to cause elastic instability in a global mode, F_{Ed} is the design loading on the structure, and F_{cr} is the elastic critical buckling load for global instability mode based on initial elastic stiffness.

The bridges and components may be checked with first-order theory if the following criteria are satisfied for each section. Elastic analysis should be used to determine the internal forces and moments for all persistent and transient design situations. The National Annex may give guidance to enable the user to determine when a plastic global analysis may be used for accidental design situations. Concerning the possible presence of imperfections in the structure, appropriate allowances should be incorporated into the structural analysis to cover the effects of imperfections, including residual stresses and geometrical imperfections such as lack of verticality, lack of straightness, lack of flatness, lack of fit, and any minor eccentricities in joints of the unloaded structure. Equivalent geometric imperfections should be used, with values that reflect the possible effects of all type of imperfections unless these effects are included in the resistance formulas for member design. The following imperfections should be taken into account: (i) global imperfections for frames and bracing systems and (ii) local imperfections for individual members. The internal forces and moments may be determined using either elastic global analysis or plastic global analysis. In the first case, elastic global analysis should be based on the assumption that the stress-strain behavior of the material is linear, regardless of the stress level; plastic global analysis allows for the effects of material nonlinearity in calculating the action effects of a structural system.

5.2 Verification for static loading in ULS

According to EN 1993-2 (2006), the partial factors $\gamma_M = R_k / R_d$ shall be applied to the various characteristic values of resistance. Neglecting general information on gross sections, shear lag effects, effective properties of cross section with class 3 webs, and Class 1 or 2 flanges, precise information about the effects of local buckling for class 4 cross sections are given in EN 1993-2 (2006). In this case, the effects of local buckling should be considered using one of the following two methods specified in EN 1993-1-5 (2007) (i) effective cross section properties of class 4 sections in accordance with EN 1993-1-5 (2007) section 4; or (ii) limiting the stress level to achieve cross section properties in accordance with EN 1993-1-5 (2007), section 10. Also, for tension members, the general rules of EN 1993-1-1 (2005) apply. For compression members, the design resistance of cross sections for uniform compression $N_{c,Rd}$ should be determined as follows:

Without local buckling:

$$N_{c,Rd} = A f_y/\gamma_{M0} \text{ for class 1, 2 and 3 cross sections.} \tag{2}$$

With local buckling:

$$N_{c,Rd} = A_{eff} f_y/\gamma_{M0} \text{ for class 4 cross sections or} \tag{3}$$

$$N_{c,Rd} = A \sigma_{limit}/\gamma_{M0} \text{ for stress limits.} \tag{4}$$

where $\sigma_{limit} = \rho_x f_y$ is the limiting stress of the weakest part of the cross section in compression (see EN 1993-1-5 (2007)).

Concerning bending moment, the design resistance for bending about the major axis should be determined as follows:

Without local buckling:

$$M_{c,Rd} = W_{pl} f_y/\gamma_{M0} \text{ for class 1, 2 cross sections} \tag{5}$$

$$M_{c,Rd} = W_{el, min} f_y/\gamma_{M0} \text{ for class 3 cross sections.} \tag{6}$$

With local buckling:

$$M_{c,Rd} = W_{eff, min} f_y/\gamma_{M0} \text{ for class 4 cross sections or} \tag{7}$$

$$M_{c,Rd} = W_{el, min} \sigma_{limit}/\gamma_{M0} \text{ for stress limits,} \tag{8}$$

where $W_{el,min}$ and $W_{eff,min}$ are the elastic moduli that correspond to the fiber with the maximum elastic stress, and σ_{limit} is the limiting stress of the weakest part of the cross section in compression. No more details than those provided in EN 1993-1-1 (2005) and EN 1993-1-5 (2007) are needed for shear. While torsional and distortional effects should be taken into account for members subjected to torsion. The effects of transverse stiffness in the cross section or of diaphragms that are built into reduce distortional deformations may be taken into account by considering an appropriate elastic model that is subject to the combined effect of bending, torsion, and distortion. Distortional effects in the members may be disregarded where the effects from distortion, due to the transverse bending stiffness in the cross section or diaphragm action, do not exceed 10% of the bending effects. Diaphragms should be designed to take into account the action effects resulting from their load distributing effect. The interaction among bending, axial load, shear, and transverse loads may be determined using either interaction methods or interaction of stresses using the yielding criterion (EN 1993-1-5, 2007). Other combinations and specific cases are illustrated in EN 1993-1-1 (2005).

5.3   Verification for earthquake loading

The required provisions are included in EN 1998-2 (2005), and applies to the earthquake resisting system of bridges designed by an equivalent linear method taking into account a ductile or limited ductile behavior of the structure. Also for bridges provided with isolating devices and for verifications on the basis of results of nonlinear analysis, EN 1998-2 (2005) shall be applied.

5.4   Verification of SLS

The following serviceability criteria should be met according to EN 1993-2 (2006):

- Restriction to elastic behavior in order to limit excessive yielding, deviations from the intended geometry by residual deflections, and excessive deformations
- Limitation of deflections and curvature in order to prevent unwanted dynamic impacts due to traffic (combination of deflection and natural frequency limitations), infringement of required clearances, cracking of surfacing layers, damage of drainage
- Limitation of natural frequencies in order to exclude vibrations due to traffic or wind that are unacceptable to pedestrians or passengers in cars using the bridge, limit fatigue damages caused by resonance, and limit excessive noise emission
- Restriction of plate slenderness, in order to limit: excessive rippling of plates; breathing of plates; reduction of stiffness due to plate buckling, resulting in an increase of deflection, see EN 1993-1-5 (2007)
- Improved durability by appropriate detailing to reduce corrosion and excessive wear
- Ease of maintenance and repair, to ensure: accessibility of structural parts for maintenance and inspection, renewal of corrosion protection and asphaltic pavements; replacement of bearings, anchors, cables, expansion joints with minimum disruption to the use of the structure

5.5   Verification associated with durability

The most relevant indications concerning design for durability in steel bridge design relates to the fatigue endurance, as provided in EN 1993-1-9(2005). Moreover, EN 1993-2 (2006) provides further insights relating to the specific argument, concerning the following:

- Structural detailing for orthotropic steel decks
- Material
- Fabrication conforming to EN 1090

Finally, according to EN 1993-2 (2006), components that cannot be designed with sufficient reliability to achieve the total design working life of the bridge should be replaceable. These may include:

- Stays, cables, hangers
- Bearings
- Expansion joints
- Drainage devices
- Guardrails, parapets
- Asphalt layer and other surface protection
- Wind shields
- Noise barriers

6 Composite bridge analysis

6.1 Introduction

Steel members in association with a concrete deck in a bridge structure are often abbreviated under the general nomenclature of composite bridges. In fact, this bridge type has been used throughout the world, mainly in the I-girder and box-girder shapes. In this subsection, these two types will be illustrated, only from a structural and code-based point of view. Although this is not the unique application of composite structures. Other applications deals with: composite steel concrete foundations associating steel beam to concrete piles; towers or special transfer modules at lower cable anchorages in long span bridges; special structures for tunnels; Concrete Filled Steel Tube (CFST), an interesting application for long span arches (Pipinato and Modena, 2010).

A typical I girder composite section is shown in Figure 10.2. Steel I-section or box girder may be a rolled or built-up plated section consisting of top and bottom flange plates welded to a web plate. Hot-rolled steel beams are applicable to shorter-span bridges, and plate girders to longer-span bridges (about 40–90 m). If connectors are not provided, so that the r.c. deck is simply supported by the deck, the composite action is not provided by this structure, while a steel section that acts with the concrete deck to resist flexure is a composite section. Connecting device details are provided in EN 1993-1-8 (2005) for requirements for fasteners and welding consumables, while for headed stud shear connectors, refer to EN 13918 (2008).

The various structural members included in a composite steel-r.c. structure should be designed according to the following considerations: the web mainly provides shear strength for the girder and is commonly taken as 1/16–1/18 of the girder span. The web thickness should be as small as the buckling resistance allows, while the web height could be also adopted where a variable cross section enhance the material savings. Longitudinal and transverse stiffeners are usually designed in order to increase flexure resistance of the web-controlling lateral web deflection and preventing bending and buckling, respectively, and shear resistance in correspondence of supports and concentrated loads. The bending strength is provided by flanges that have been designed according to the specific code requirements provided in the erection site. The general advantage of r.c. composite bridges lies principally on the very slender, aesthetically pleasant shape due to the optimal combination of the high tensile strength of the structural steel, the high compressive strength of concrete, and the high durability of normal r.c. decks due to restrictive crack width limitation (Hansville and Sedlacek, 2010). A typical cross section includes those illustrated in Figure 10.8.

6.2 Structural modeling

The structural model and basic assumptions shall be chosen in accordance with EN 1990 (2005) and reflect the anticipated behavior of the cross sections, members, joints, and bearings. Where the structural behavior is essentially that of a reinforced or prestressed concrete structure, with only a few composite members, global analysis should be generally in accordance with EN 1992-2 (2005). Analysis of composite structures should be in accordance with EN 1994-2 (2006). Concerning joint modeling, the effects of the behavior of the joints on the distribution of internal forces and moments

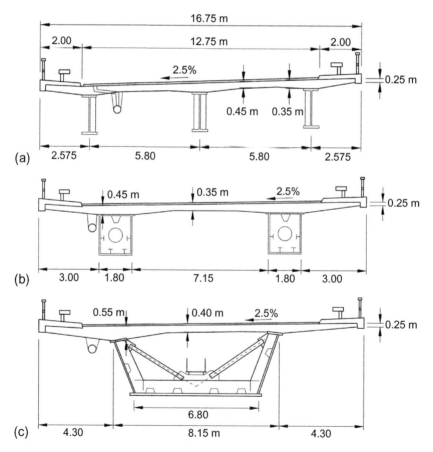

Figure 10.8 Typical cross section of r.c. composite bridges: (a) plate girder bridge with three rolled or welded built-up main girders (b) cross section with two separated box girders (c) box girder.

within a structure, as well as on the overall deformations of the structure, may generally be ignored, but where such effects are significant (such as in the case of semicontinuous joints) they should be taken into account; see EN 1993-1-8 (2005). To identify whether the effects of joint behavior on the analysis need to be considered, see the definitions given before for simple, continuous and semicontinuous joints.

Ground-structure interaction should be taken into account as discussed in EN 1994-2 (2006). The structural stability is to be taken into account, and the action effects may generally be determined using either first-order analysis, using the initial geometry of the structure; or second-order analysis, taking into account the influence of the deformation of the structure. The effects of the deformed geometry (second-order effects) shall be considered if they significantly increase the action effects or modify the structural behavior. Equivalent geometric imperfections should be used with values that reflect the possible effects of system imperfections and also member imperfections unless these effects are included in the resistance formulas; the imperfections and design transverse forces for stabilizing transverse frames should be calculated in accordance with EN 1993-2 (2006).

6.3   Verification for static loading in ULS

Composite beams should be checked for (EN 1994-2, 2006):

− resistance of cross-sections (see 6.2 and 6.3)
− resistance to lateral-torsional buckling (see 6.4)
− resistance to shear buckling and transverse forces applied to webs (see 6.2.2 and 6.5)
− shear connections (see 6.6)
− resistance to fatigue (see 6.8).

6.4   Verification for earthquake loading

The most relevant requirements are illustrated in EN 1998-2 (2005).

6.5   Verification of SLS

A structure with composite members shall be designed and constructed such that all relevant serviceability limit states (SLS) are satisfied according to the principles of Section 3.4 of EN 1990 (2005). Calculation of stresses for beams at the serviceability limit state (SLS) shall take into account the following effects, as needed:

• Shear lag
• Creep and shrinkage of concrete
• Cracking of concrete and tension stiffening of concrete
• Sequence of construction
• Increased flexibility resulting from significant incomplete interaction due to slip of shear connection
• Inelastic behavior of steel and reinforcement, if any
• Torsional and distortional warping, if any

Moreover, deflections and vibrations are checked according to EN 1990 (2005), EN 1993-2 (2006), and EN 1991-2 (2003).

6.6   Verification associated with durability

The relevant provisions given in EN 1990, EN 1992, and EN 1993 should be followed. Detailing of the shear connection should be in accordance with EN 1994-2 (2006). The corrosion protection of the steel flange should extend into the steel-concrete interface at least 50 mm.

7   Truss bridges analysis

A particular type of bridge is the truss, which is typically all made of steel. Trusses are assumed to be pin-jointed where the straight-force components meet. This assumption means that members of the truss (chords, verticals, and diagonals) will act only in tension or compression. A more complex analysis is required where rigid joints impose significant bending loads upon the elements, as in a Vierendeel truss. The large amount of truss bridges built worldwide is well known, probably for the ability to distribute the forces in the structure assuming different geometric configurations. Modern materials and fabrication methods (e.g., automated welding), the specific use of the bridge (roadway, railway, etc.), and such other data as the lane numbers and the traffic category, influence the truss typology choice.

7.1   Truss typologies

A wide number of truss types have been developed, each with a special use. Many variations on these common schemes could be found in literature; however, this typological presentation could help find the best design solution. In Table 10.2, the most diffused truss types are listed and shown.

7.2   Analysis methods

The most common analysis method includes force member methods (FMMs), based on assuming that the truss joints are frictionless pins. This means that so long as loads are applied to the joints and not along the member length, the two forces acting on each member act along its axis. However, this scheme rarely works in real members, as the physical pins are never really friction-free, so secondary bending are effectively present in members. When riveted, bolted, or welded connections start to be used, a common construction method to reduce eccentricities or to compensate bending involves the alignment of the working line of members into each node. Two variations of the method are used: the method of sections and the method of joints (see e.g. Krenk and Høgsberg, 2013). However, these methods are scholarly based solutions for simple structures that are statically determinate; and for more complex structure the computer methods are preferred.

Table 10.2 Truss Typologies

Designation	Geometric Scheme	When First Used	Typical Length	Comments
Pratt		1844	9–75 m	Diagonals in tension, verticals in compression, except for hip verticals adjacent to inclined end post.
Baltimore (petit)		1871	75–180 m	A: with substruts, B: with subties
Warren		1848	15–120 m	Triangular in outline, the diagonals carry both compressive and tensile forces. An original Warren truss has equilateral triangles.
Pratt half-hip		Late 19th century	9–45 m	A Pratt with inclined end posts that do not horizontally extend the length of a full panel.
Pennsylvania (petit)		1875	75–180 m	A: Parker with substruts, B: Parker with subties
Warren		Mid-19th century	15–120 m	Diagonals carry both compressive and tensile forces; verticals serve as bracing for triangular web system.
Truss leg bedstead		Late 19th century	9–30 m	A Pratt with vertical end posts embedded in their foundations.
Lenticular-parabolic		1878	5–110 m	A Pratt with top and bottom chords parabolicy curved over the entire length.
Double intersection Warren		Mid-19th century	23–120 m	Structure is indeterminate; members act in both compression and tension; two triangular web systems are superimposed upon each other with or without verticals.

(Continued)

Table 10.2 Continued

Designation	Geometric Scheme	When First Used	Typical Length	Comments
Parker		Mid- to late 19th century	12–75 m	A Pratt with a polygonal top chord
Greiner		1894	23–75 m	Pratt truss with the diagonals replaced by an inverted bowstring truss.
Pegram		1887	45–195 m	A hybrid between the Warren and Parker trusses; upper chords are all of equal length.
Howe		1840	9–45 m	Diagonals in compression, verticals in tension (wood, verticals of metal).
Camelback		Late 19th century	30–90 m	A Parker with polygonal top chord of exactly five slopes.
Double intersection Pratt		1847	21–90 m	An inclined end-post Pratt with diagonals that extend across two panels.
Post		1865	30–90 m	A hybrid between the Warren and the Double intersection Pratt.
Bowstring arch-truss		1840	15–40 m	A tied arch with diagonals serving as bracing and verticals supporting the deck.
Camelback		Late 19th century	30–150 m	A: Pennsylvania truss with a polygonal top chord of exaclty five slopes, B: Same as A, with horzontal struts
Schwelder		Late 19th century	30–90 m	A double-intersection Pratt positioned in the center of a Parker.
Bollman		1852	23–30 m	Verticals in compression, diagonals in tension; diagonals run from end posts to every panel point.

Waddell A-truss		Late 19th century	8–23 m	Expanded version of the king post truss, usually made of metal.
Kellogg		Late 19th century	23–30 m	A variation on the Pratt with additional diagonals running from upper chord panel points to the center of the lower chords.
K-truss		Early 20th century	60–240 m	Takes the name from the particular shape remembering K members.
Fink		1851	23–45 m	Verticals in compression; diagonals in tension; longest diagonals run from end posts to center panel points.
Wichert		1932	122–305 m	Identified by a characteristic pin connected support system over the piers; truss is continuous over piers.
Stearns		1890	15–60 m	Simplification of fink truss with verticals omitted at alternative panel points

8 Research and development

Toward a more rational and sustainable construction industry, and considering research and development fields of interest, steel bridges could be discussed as a paramount in the framework of innovative constructions, growing every year with the increasing content of innovations and futuristic solutions. However, increasing R&D actions are needed, and the most promising areas of exploration could defined be as follows:

- *High-strength steels:* More cost effective solution, stronger, lighter, and even more resistant to weather, corrosion, fatigue, and exceptional loads steel bridges are required in the construction market. For this reason, high-strength solutions should be deepened and researched; the U.S. Federal Highway Administration (FHWA) reported that high-strength solutions in steel bridges were found to provide lifetime cost savings of up to 18% and weigh 28% less than traditional steel bridge design materials (FHWA, 2002); however, research in this area is not as concentrated and diffused as necessary.

- *Steel protection technologies*: A design life requirement of 120 years is often required in modern bridges, and the performance of the protective system is a critical factor. Furthermore, reductions in the number of repainting cycles have become significant in the evaluation of whole life costs. There has been a widely held view that most steel bridges require frequent attention to maintain the original protective coating system. In reality, coating lifetimes have progressively increased from 12 and 15 years to 20 and 25 years. From continued developments in coating technology, modern high-performance coating systems may be expected to achieve lives so as not having the first major maintenance for more than 30 years.
- *Weathering steel*: Weathering steel is a low-maintenance solution that is exploited less than would be useful. Although it is not an optimal solution for all environmental conditions (e.g., marine locations and highly contaminated sites with deicing salt or SO_2 industrial fumes), there are at least three motivations to prefer the weathering steel alternative: (i) low maintenance with periodic inspection and cleaning; (ii) the cost benefit avoiding painting (initially and over the whole life cycle); and (iii) the managing authority recently prefers the use of nude weathering steel to painted solutions, as mature weathering steel bridges blends well with surrounding protected landscape environments. However, some inherent problems due to the superficial debris or to the uncorrected maintenance of the superficial patina (high-pressure water washing should be avoided) could accelerate the steel decay. Innovative solutions are requested to reduce these problems. Furthermore, the availability on the market of a wide variety of steel components should be increased.
- *Innovative/optimized structural shape*: Combining new materials, with innovative structural and aesthetically pleasant shape solutions, unexpected advances could be reached. The principle goals should be: increaseing cost savings, accelerating construction, enabling life longer solutions. To support these goals, computational and modeling researches, industrial interests and supports are requested.

References

American Association of State Highway and Transportation Officials (AASHTO), 2013. LRFD Bridge Design Specifications, sixth edition with 2013 interim revisions, AASHTO, Washington, DC.

American Institute of Steel Construction (AISC), 2010a. Specification for Structural Steel Buildings. N. 360-10, Chicago.

American Institute of Steel Construction (AISC), 2010b. Commentary on the Specification for Structural Steel Buildings. N. 360-10, Chicago.

American Society for Testing and Materials (ASTM), 2010a. Standard Specification for Structural Steel for Bridges. A 709/A709M, ASTM, West Conshohocken, PA.

American Society for Testing and Materials (ASTM), 2010b. Standard Test Methods and Definitions for Mechanical Testing of Steel Products. A370 - 10a, ASTM, West Conshohocken, PA.

American Society for Testing and Materials (ASTM), 2012. Standard Specification for High-Strength Steel Bolts, Classes 10.9 and 10.9.3, for Structural Steel Joints (Metric). 490M-12, West Conshohocken, PA.

American Society for Testing and Materials (ASTM) A325M, 2013. Standard Specification for Structural Bolts, Steel, Heat Treated 830 MPa Minimum Tensile Strength (Metric). ASTM International, West Conshohocken, PA.

American Society for Testing and Materials (ASTM) A490M, 2013. Standard Specification for High-Strength Steel Bolts, Classes 10.9 and 10.9.3, for Structural Steel Joints (Metric). ASTM International, West Conshohocken, PA.

American Welding Society (AWS), 2010. AISC Specification Referenced Structural Welding Code—Steel Standard Code AWS D1.1/D1.1M. AWS, Miami, FL.

EN 13918, 2008. Welding. Studs and Ceramic Ferrules for Arc Stud Welding. Comité Européen de Normalisation (CEN), Brussels.

EN 14399-1, 2005. High-Strength Structural Bolting Assemblies for Preloading—Part 1: General Requirements. Comité Européen de Normalisation (CEN), Brussels.

EN 1990, 2005. Eurocode - Basis of Structural Design. Comité Europeén de Normalisation (CEN), Brussels.

EN 1991-2, 2005. Eurocode 1: Actions on Structures - Part 2: Traffic Loads on Bridges. Comité Europeén de Normalisation (CEN), Brussels.

EN 1992-2, 2005. Eurocode 2: Eurocode 2 - Design of Concrete Structures - Concrete Bridges - Design and Detailing Rules. Comité Europeén de Normalisation (CEN), Brussels.

EN 1993-1-1, 2005. Eurocode 3: Design of steel structures—Part 1-1: General Rules and Rules for Buildings. Comité Européen de Normalisation (CEN), Brussels.

EN 1993-1-10, 2005. Eurocode 3: Design of Steel Structures—Part 1–10: Material Toughness and Through-Thickness Properties. Comité Européen de Normalisation (CEN), Brussels.

EN 1993-1-5, 2007. Eurocode 3 - Design of Steel Structures - Part 1-5: Plated Structural Elements. Comité Europeén de Normalisation (CEN), Brussels.

EN 1993-1-8, 2005. Eurocode 3: Design of Steel Structures—Part 1–8: Design of Joints. Comité Européen de Normalisation (CEN), Brussels.

EN 1993-1-9, 2005. Eurocode 3: Design of Steel Structures—Part 1–9: Fatigue. Comité Européen de Normalisation (CEN), Brussels.

EN 1993-1-10, 2005. Eurocode 3: Design of Steel Structures—Part 1-10: Material Toughness and Through-Thickness Properties. Comité Europeén de Normalisation (CEN), Brussels.

EN 1993-2, 2006. Eurocode 3: Design of Steel Structures—Part 2: Steel Bridges. Comité Européen de Normalisation (CEN), Brussels.

EN 1994-2, 2006. Eurocode 4 - Design of Composite Steen and Concrete Structures - Part 2: General Rules and Rules for Bridges. Comité Europeén de Normalisation (CEN), Brussels.

EN 1997-1, 2005. Eurocode 7: Geotechnical Design—Part 1: General Rules. Comité Européen de Normalisation (CEN), Brussels.

EN 1998-2, 2005. Eurocode 8: Design of Structures for Earthquake Resistance—Part 2: Bridges. Comité Européen de Normalisation (CEN), Brussels.

EN 6892-1, 2009. Metallic Materials. Tensile Testing. Method of Test at Ambient Temperature. Comité Europeén de Normalisation (CEN), Brussels.

Federal Highway Administration (FHWA), 2002. High Performance Steel Designers' Guide. US Federal Highway Safety Administration, Washington.

Federal Highway Administration (FHWA), 2012. Steel Bridge Design Handbook: Bridge Steels and Their Mechanical Properties. Publication No. FHWA-IF 12-052 Vol. 1, Washington.

Hansville, G., Sedlacek, G., 2010. Steel and composite bridges in Germany. In: State of the Art. Proceedings of the 7th Japanese–German Bridge Colloquium Osaka 2007.

Krenk, S., Høgsberg, J., 2013. Statics and Mechanics of Structures. Springer. http://dx.doi.org/10.1007/978-94-007-6113-1 2.

Pipinato, A., Modena, C., 2010. Il progetto dei ponti ad arco CFST. Strade e Autostrade, n. 6/2010, Edizioni Edi-Cem, Milano.

Research Council on Structural Connections (RCSC), 2009. Specification for Structural Joints Using High-Strength Bolts. Chicago.

Timber bridges

Malo K.A.
Department of Structural Engineering, Norwegian University
of Science and Technology (NTNU), Trondheim, Norway

1 Wood used in bridges

1.1 Introduction

Throughout the history of societies in the Middle East and Mediterranean, as well as in China, several quite large timber bridges were built. The first bridge built by humans, more than 10000–15000 BC, was probably a structure made of timber logs spanning over a waterway. Later, the Romans built timber bridges to ease transport; in particular, one of those bridges, known as Caesar's Bridge (55 BC), is well documented by the Italian architect Andrea Palladio (1508–1580).

Palladio was also among the first to extend the design of timber bridges into trusses, making longer spans possible. In the 19th century, many timber bridges were built all over the world, using many variants of trusses and arches; an overview of popular structural systems, together with comprehensive information on nearly all relevant topics regarding timber bridges may be found in Ritter (1990). At the end of the 19th century, steel bridges became popular, and from the early 20th century on, reinforced concrete (r.c.) become available as bridge material and both r.c. and steel largely replaced wood for building bridges. Although the use of timber for bridges was low throughout most of the 20th century, interest in this material has grown over the last few decades. Patents for glued and laminated timber (glulam) by the German carpenter K. F. O. Hetzer (1846–1911) made it possible to build large structural members out of very small pieces, which could easily be made straight as well as curved. Modern wood preservation techniques and the growing need for sustainable building materials have resulted in a renewed interest in timber bridges.

Wood is a renewable resource that is easily available in the inhabited parts of the world. Most climatic zones have at least a few tree species that may be used for structural purposes. Older trees are harvested and replaced by young trees, transforming carbon dioxide, water, and small amounts of nutrients from the earth into a structural material via solar energy. The material production by photosynthesis is a natural process, necessary for all life on Earth. When old trees die, the material is broken down and restored by natural and sustainable processes. A piece of wood stores a large amount of carbon dioxide, which makes wood an environmentally friendly material. It is, in fact, the only structural material that has a positive effect on global warming caused by greenhouse gases. Depending on the type, 1 kg of wood can contain about 1.7 kg of carbon dioxide and store it for as long as the wood is used (for instance, in a bridge).

Innovative Bridge Design Handbook. http://dx.doi.org/10.1016/B978-0-12-800058-8.00011-6

Decay and disintegration of wood mainly occur due to activities from fungi, insects, and bacteria. The disintegration of wood is strongly dependent on the moisture content (MC), temperature, and surrounding conditions. By controlling the surrounding conditions, the deterioration of wood can be postponed for a very long time; well-known examples of this use are the stave churches in the Nordic countries, many of which are still standing after nearly a thousand years. On the other hand, the ingredients in wood might easily be returned to nature by exposing the material to the natural conditions on the ground. A rough rule of thumb is that wooden structures with moisture content below 20% (by weight) are not prone to decay.

2   Wood as structural material

2.1   Structure of wood

Wood used for structural purposes may broadly be divided into two groups: conifer and deciduous, which are more commonly called *softwood* and *hardwood*. Hardwood trees typically have broad leaves, but solely in the growing season (losing their leaves in winter), while the conifer species have needle-shaped leaves throughout the year. The terms *softwood* and *hardwood* are somewhat misleading in a general sense, but for the species most commonly used for engineered structures, they have some relevance. Most engineered timber bridges are made of conifer wood (softwood), as explored in the next sections.

Wood is a fibrous, strongly anisotropic material on multiple scales. A useful conceptual model is a composite structure consisting of a bundle of lightly glued, thin-walled drinking straws, conceptually illustrated in Figure 11.1. The tubular structure can be observed for a piece of wood by the human eye, especially by use of a simple magnifying glass. The cell walls in the tubes consist of wood-fiber layers having inclined orientations relative to each other. The tubes have different sizes but are mainly oriented in the direction of the stem and are held together in a matrix of mainly lignin, an organic polymer. Transport of nutrients and water takes place in the tubes, and in order to serve the branches as well, the trunk also has some tubes (known as *rays*) in the radial directions. The structure of the branches is quite similar to the trunk itself; i.e., the branches also have pith, an annual ring structure, and bark. The connection between the branches and the trunk is known as the *knot*. The knots give rise to some disturbance in the annual ring and tubular structure; some of the tubes are spliced to the branch with tubes, while others have some deviations relative to the longitudinal axes of the stem in order to pass the branching connection. Knots in sawn timber give discontinuities in the fibers, as well as fiber inclinations (grain deviations).

Usually the growth of trees depends on the season, and each year creates an additional seasonal ring to the cross section of the stem, commonly denoted as annual rings that typically are a few millimeters thick. Rapid growth gives large tubular cells with small cell thickness, and consequently (on an average volume basis) less load-carrying capacity in the wood material, since it is the fiber material in the cell walls that gives strength to the material.

The growth along the perimeter produces new wooden fibers by forming wooden cells and strawlike structures held together with a matrix material (gluelike substrate).

From the pith in the center of the stem, the annual rings form concentric circles in the stem; hence, the natural coordinate system for a piece of wood is cylindrical with its origin in the pith. The direction along the stem is called *longitudinal (L)*, meaning along grains or fibers, while the outward direction from the pith is called *radial (R)*, and the direction along the perimeter is called *tangential (T)* (see Figure 11.2). However, as most pieces of wood are sawn with rectangular shapes and the exact location in the stem is unknown, only longitudinal and transversal directions are used in design calculations. The transversal properties are weighted averages of the R and T properties, which in reality might differ quite significantly. It is common to indicate the L direction (i.e., the stem direction) with a subscript 0 that indicates zero degree angles relative to the grain (fiber) direction, and the transverse direction with subscript 90 (degrees).

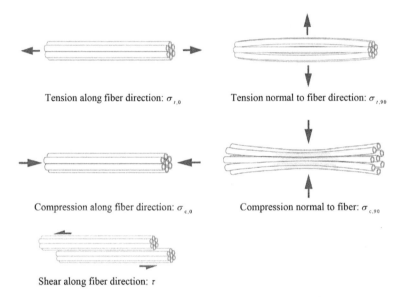

Tension along fiber direction: $\sigma_{t,0}$ Tension normal to fiber direction: $\sigma_{t,90}$

Compression along fiber direction: $\sigma_{c,0}$ Compression normal to fiber: $\sigma_{c,90}$

Shear along fiber direction: τ

Figure 11.1 Conceptual tube structure of wood subjected to typical stress situations with commonly used symbols for stresses.

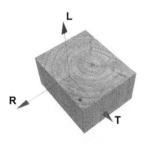

Figure 11.2 Annual rings, pith, and natural material axes at a point for wood.

2.2 Mechanical properties of wood

For anisotropic materials like wood, it is usually necessary to relate all stresses in the material to the material axes in order to evaluate the loading capacity (see Figure 11.1). As already explained, practical design of load-carrying structures is based on strength and stiffness in the grain direction (transversely isotropic). Furthermore, linear material models based on Claude-Louis Navier's hypothesis of plane deformation and Robert Hooke's linear relation between stress and strain are used.

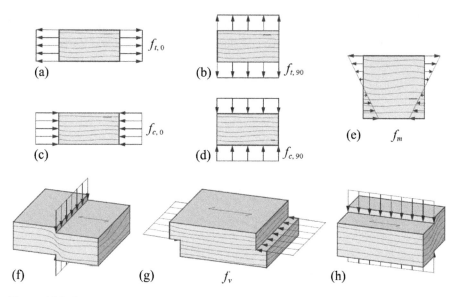

Figure 11.3 Stress exposure and associated strength (drawing: K. Bell).

Notations and terms used in the following discussion are those adopted from the European standards for wood materials and design of timber structures (see EN 1995-1-1:2004/A1:2008). For design verification, stresses are denoted by σ (normal stress) and τ (shear stress), and for normal stresses, indices t and c denote tensile and compression, respectively, while indices 0 and 90 indicate the angle between stress direction and fiber direction (e.g., $\sigma_{t,0}$). The stresses are compared to the corresponding strength, denoted by f (e.g., $f_{t,90}$). Note that in general, no stress criterion is available for combined stresses in wood the way that von Mises applies to steel, but certain stress combinations are considered to have significant interactions which are taken into account. Figure 11.3 shows the usual configurations for strength evaluations, using the same subscript system as for stresses described previously. Although the shear stresses occur with the same value both in the transversal (Figure 11.3f) and longitudinal (Figure 11.3g) directions, the shear strength is much less along the fibers. Consequently, the shear strength f_v corresponds to the situation shown in Figure 11.3g. Some structural details might be exposed to "peeling" loads, see Figure 11.3h, and in those special cases, it might be necessary to evaluate these stresses and compare them to the corresponding strength (rolling shear strength).

Numerous experiments have showed that the bending strength is greater than that obtained by simple linear models using tensile or compressive strength at extreme fiber as the strength criterion, and it is well known that with bending, considerable nonlinear stress redistribution occurs. However, this effect is accounted for by the introduction of special design rules for bending. The stresses are computed by linear models (see Figure 11.3e), but are compared to a specific nominal bending strength denoted as f_m, which in principle accounts for the nonlinear stress distribution.

The wood material is produced by nature, and humans cannot do much to control the production of it. The natural variation in properties is large; to get more unified mechanical and physical properties, sorting procedures are necessary. Many sorting strategies are used, based either on visual grading by the human eye or machine grading using some sort of correlation between strength, stiffness, density, and visual properties. The mechanical properties for use in structural design calculations are based on statistical distributions and statistical measures like the mean and characteristic (5% fraction) values.

The properties for evaluating the strength and stiffness for wood are specified in standards. In Europe, valid standards are currently EN 338:2009 for solid wood and EN 14080:2013 for glulam. For other locations or species the Wood Handbook (Forest Products Laboratory, 2010) is useful. Some properties of Norway spruce, one of the most used species for structural construction in Europe, are given in Table 11.1 for the European grade C24. The C in the grade notation C24 stands for the conifer species (with D standing for deciduous), while 24 indicates a characteristic bending strength of 24 MPa. For glued laminated members (glulam), the notation GL is used in the European codes, and some of these properties are listed in Table 11.1. Further properties and grade classes are given in EN 338 (solid wood) and EN 14080 (glulam).

Table 11.1 Material Properties of Softwood According to EN 338 and EN14080

Symbols	Strength Properties (MPa)						Stiffness Properties (MPa)				
	$f_{m,k}$	$f_{t,0k}$	$f_{t,90k}$	$f_{c,0k}$	$f_{c,90k}$	$f_{v,k}$	$E_{0,mean}$	$E_{0,05}$	$E_{90,mean}$	G_{mean}	$G_{0,05}$
C24	24	14	0.4	21	2.5	4.0	11,000	7400	370	690	460
GL30h	30	24	0.5	30	2.5	3.5	13,600	11,300	300	650	540

3 Design of timber components

3.1 Loads on timber bridges

In this discussion, the previously mentioned Eurocodes will be used as the model design code. They are based on the limit state concept used in conjunction with a partial factor method. EN 1990 Basis of Structural Design (EN 1990:2002, commonly denoted Eurocode 0) states how the fundamental limit states shall be verified by design. For each of the two fundamental limit states, the ultimate limit state (ULS) and serviceability limit state (SLS), several scenarios are defined. Eurocode 0 has guidelines as to how the actions in each scenario shall be combined by the use of partial factors giving a design value, indicated by subscript d, of the combined effects of the actions. The actions are given by the EN 1991-x series of standards, like EN 1991-2 for traffic loading and EN 1991-1-4 for wind action. Outside Europe other design codes apply, for instance for USA see (AASHTO, 2015). Note that for timber bridges, additional evaluations might be necessary for self-weight due to effects of moisture and use of preservatives. Furthermore, moisture might give considerable dimensional changes which need to be considered, as do temperature effects.

3.2 Design values

Eurocode 5 Part 1-1, "Design of Timber Structures" (EN 1995-1-1:2004/A1:2008) and Part 2, "Bridges" (EN 1995-2:2004), describe the principles and requirements for safety, serviceability, and durability of timber bridges. The mechanical behavior of wooden materials shows considerable time and moisture dependencies. Long duration of loading (DOL) significantly decreases the measureable strength of the material; this effect is accounted for in modern design codes. Wood is also a hygroscopic material; i.e., water is exchanged with the surroundings. Increased moisture content (MC) leads in general to a decrease in strength and stiffness properties. In air, the MC and exchange of water are dependent on the relative humidity (RH). Most material properties of wood are related to standardized climatic condition (RH 65% and 20 °C), leading to approximately 12% MC. Furthermore, standardized DOL is used in order to have a common reference for determination of mechanical properties.

The effects from DOL and MC cannot be neglected in the design of timber structures and are taken into account in a simplified manner, through the use of a modification factor k_{mod}, which is dependent on the climatic conditions (i.e., MC) and the DOL, applicable to the timber structure during its design life. The climatic conditions are characterized into three service classes, each of which is related to the expected MC during a given design life EN 1995-1-1:2004/A1:2008. Service class 2 may be applied to timber bridges where the timber parts are properly covered and not exposed directly to rain and water, while in all other cases, service class 3 should be used for timber bridges.

The DOL effect is included in design by characterizing the typical load duration into classes, e.g., self-weight is permanent loading and wind is instantaneous. Traffic loading on bridges is normally assumed to be short-term loading. The design value for a strength property is then calculated by

$$R_d = k_{\mathrm{mod}} \frac{R_k}{\gamma_m} \tag{1}$$

Recommended values for k_{mod} and the material factor γ_m are stated in EN 1995-1-1:2004/A1:2008 and EN 1991-2. The partial factor for material properties γ_m depends on the type of wood-based product, as well as on the design problem at hand. All the safety and strength properties are based on the use of the characteristic (5%) value R_k (denoted by the use of subscript k or 05), while serviceability issues like deformation and vibration use the mean values of the material properties (subscript *mean*).

3.3   Design strength for structural timber members

Some design formulas essential for timber bridges are presented in the following, but it should be emphasized that these represent only a subset, and references are made to EN 1995-1-1:2004/A1:2008 and EN 1991-2 for more comprehensive information. In the following discussion, it is assumed that the axis along the structural member is denoted x, while y and z are the principal axes of the cross section. Furthermore, it is assumed that bending about the strong axis is about the y-axis.

3.3.1   Bending and axial actions

The design strength in Eurocode 5 (EN 1995-1-1:2004/A1:2008 and EN 1995-2:2004) is formulated on the basis of linear elastic methods combined with the use of various factors k_{xx} where the subscript is dependent on the physical effect that it applies to. The factors k_{xx} account for effects neglected by the simplified and linear elastic calculations. For timber members having stresses mainly in the direction of the longitudinal material axes, the following requirements apply;

For bending and axial tension:

$$\frac{\sigma_{t,0,d}}{f_{t,0,d}} + \frac{\sigma_{m,y,d}}{f_{m,y,d}} + k_m\frac{\sigma_{m,z,d}}{f_{m,z,d}} \leq 1 \quad \text{and} \quad \frac{\sigma_{t,0,d}}{f_{t,0,d}} + k_m\frac{\sigma_{m,y,d}}{f_{m,y,d}} + \frac{\sigma_{m,z,d}}{f_{m,z,d}} \leq 1 \tag{2}$$

For rectangular cross-sectional shapes, the bending stress redistribution shape factor k_m can be set equal to 0.7, while it should be set to 1.0 for other cross sections (EN 1995-1-1:2004/A1:2008).

For combined bending and axial compression of members prone to buckling:

$$\frac{\sigma_{c,0,d}}{k_{c,y}f_{c,0,d}} + \frac{\sigma_{m,y,d}}{f_{m,y,d}} + k_m\frac{\sigma_{m,z,d}}{f_{m,z,d}} \leq 1 \quad \text{and} \quad \frac{\sigma_{c,0,d}}{k_{c,z}f_{c,0,d}} + k_m\frac{\sigma_{m,y,d}}{f_{m,y,d}} + \frac{\sigma_{m,z,d}}{f_{m,z,d}} \leq 1 \tag{3}$$

Here, the buckling effect is brought into the design formulas by use of the factors $k_{c,y}$ and $k_{c,z}$ where subscript c indicates compression, and y or z relates to buckling about the y-axis or z-axis, respectively. The buckling factor $k_{c,i}$ is defined as

$$k_{c,i} = 1 \Big/ \left(k_i + \sqrt{k_i^2 - \lambda_{rel,i}^2}\right) \quad \text{and} \quad k_i = 0.5\left[1 + \beta_c\left(\lambda_{rel,i} - 0.3\right) + \lambda_{rel,i}^2\right] \tag{4}$$

where y or z replace subscript i. The member slenderness λ_i enters the expressions through a material scaled relative slenderness defined as

$$\lambda_{rel,i} = \frac{\lambda_i}{\pi}\sqrt{\frac{f_{c,0,k}}{E_{0,05}}} \qquad (5)$$

The factor β_c reflects the fact that highly industrialized products, like glulam and LVL (Laminated Veneer Lumber), generally have smaller geometrical imperfections, so $\beta_c = 0.2$ for solid timber and $\beta_c = 0.1$ for glulam and LVL.

Members subjected to bending about the strong axis shall also be checked for lateral-torsional instability by

$$\frac{\sigma_{m,d}}{k_{crit}f_{md}} \leq 1 \quad \text{and} \quad \left(\frac{\sigma_{m,d}}{k_{crit}f_{md}}\right)^2 + \frac{\sigma_{c,0,d}}{k_{c,z}f_{c,0,d}} \leq 1 \qquad (6)$$

The latter expression of Eq. (6) takes into account the possible interaction of lateral-torsional instability and weak axis buckling. The reduction factor due to lateral-torsional instability k_{crit} is determined by the simplified expression

$$k_{crit} = \begin{cases} 1 & \text{for } \lambda_{rel,m} \leq 0.75 \\ 1.56 - 0.75\lambda_{rel,m} & \text{for } 0.75 < \lambda_{rel,m} \leq 1.4 \\ 1/\lambda_{rel,m}^2 & \text{for } \lambda_{rel,m} > 1.4 \end{cases} \quad \text{where} \quad \lambda_{rel,m} = \sqrt{\frac{f_{m,k}}{\sigma_{m,crit}}}.$$
$$(7)$$

The critical bending stress level is determined by classical theory for lateral-torsional instability for elastic members. For timber, warping of cross sections can usually be neglected, leading to

$$\sigma_{m,crit} = \frac{M_{y,crit}}{W_y} = \frac{\pi\sqrt{E_{0,05}I_zG_{0,05}I_{tor}}}{l_{ef}W_y} \qquad (8)$$

where W_y is the section modulus about the strong y-axis, I_z is the second moment of area about the weak axis, and I_{tor} is the torsional moment of area. The effective length of the structural members is denoted l_{ef}; and the ratio of l_{ef}/l is usually in the range 0.5 to 1.0, where l is the actual length of the member.

3.3.2 Shear action

The shear strength along the grain is quite low for most wood species, and for high beams, this might limit the utilization of the timber member. The design requirement is

$$\frac{\tau_d}{k_vf_{vd}} \leq 1 \qquad (9)$$

The shear stress along and normal to grain ($\tau_{zxd} = \tau_{xzd}$) is calculated by

$$\tau_d = \frac{V_{zd}S_y}{I_yb_{ef}} = \frac{3V_{zd}}{2h_{ef}b_{ef}} \qquad (10)$$

The latter expression in Eq. (10) is only valid for rectangular cross sections. The introduction of an effective width b_{ef} is meant to account for the risk of cracking due to wetting and drying; it is defined as $b_{ef} = k_{crack}b$, where k_{crack} represents the amount of noncracked material. The effective height h_{ef} will only be smaller than the height of the cross section h in cases where some material is locally removed, as in connections and notches. If a notch leads to a combination of tension normal to grain and shear stresses, a critical stress concentration may occur, and this situation is accounted for by a correction factor k_v for the strength [see Eq. (9)], which in such a case will be less than 1.0. In other cases k_v equals unity.

3.3.3 Local effects

Stresses may be transferred between wooden members by compressive contact stresses between mating surfaces, or by use of additional elements like metallic fasteners. Contact stresses on inclined surfaces should be related to the material axes of the wood or checked by simplified rules offered by the codes (see, e.g., EN 1995-1-1:2004/A1:2008). For contact stresses normal to grain on limited surfaces relative to the member size, an increase in capacity is achieved and can be utilized in design calculations.

The use of fasteners normally requires removal of material due to drilling of holes, cutting of grooves, or similar action. The removal of material reduces the effective load-bearing cross section, which must be taken into account, and it is especially important in cases with tensile stresses.

3.3.4 Curved and tapered members

Special rules apply for curved glulam members, taking into account the reduction in strength due to bending of lamellas during production, and the occurrence of tensile stresses normal to grain due to straightening bending moments. In many cases interaction of stress components may be the design case. For wooden members with tapered cross sections, the stresses at the surface with inclination relative to the grain direction will have a multi-axial stress state, which need special consideration. Most design codes have guidelines for the handling of these effects (see e.g., EN 1995-1-1:2004/A1:2008).

3.4 Structural modeling

It is generally sufficient to use linear elastic models in order to distribute the effects of the actions in a wooden structure. Care must be taken regarding the effect of DOL since creep effects may influence the force distribution within a structure, especially in cases where different materials are combined. Somewhat simplified, it can be stated that for calculations in ULS, the *characteristic* values of the material properties are used, while for SLS, the *mean* values are used. It may be necessary to make further evaluations in cases where second-order deformations affect the internal distribution of forces.

Wood is a strongly anisotropic material and cannot be adequately represented by isotropic material models. While the E/G ratio is about 2.6 for structural metals, it is roughly 16 for wood. Consequently, general isotropic models requiring two parameters as input (E and G, or E and Poisson ratio) are deemed to fail for wood. The most used material model for three-dimensional (3D) finite elements is the transverse isotropic linear elastic model, neglecting the difference between the tangential and radial directions, but including the difference between transverse and longitudinal directions. In this case, the material axes of the wooden elements have to be represented correctly.

For the overall behavior of beamlike structural members, good results are usually achieved by the use of simple beam elements, provided the shear deformations are included (e.g., use of Timoshenko beam elements).

4   Design of connections

4.1   Connectors

Metallic fasteners made of steel with grades ranging from 4.6 to 8.8 (ISO 898-1:2013) are mostly used. For modern timber bridges, the rod-type connections (dowels, bolts, and screws) are most popular. The fasteners are either axially or shear loaded. Herein, only the shear-loaded dowel-type connection will be discussed. It should be noted that for timber bridges, all metallic parts should have adequate protection against corrosion. Stainless steel dowels are widely used for noncovered bridges, but zinc-coating (hot-dipping) is also quite common.

4.2   Dowel type connection

A dowel is a smooth rod cut in appropriate lengths and is very similar to a bolt, but lacks the threads, nut, and head. The dowel cannot transfer forces in the direction of its own rod axis; otherwise, the nominal capacity of dowels and bolts is similar. The most effective dowel-type connection is achieved by the use of slotted-in steel-plates where the capacity of the dowel is balanced with the capacity of the wooden layers between the plates. A conceptual model of a dowel type connection is visualized on the left side in Figure 11.4, while on the right, a similar joint from a bridge is depicted.

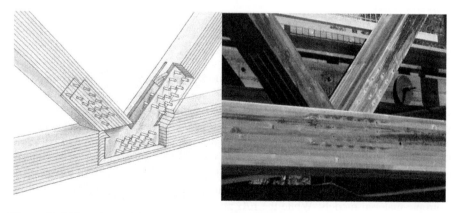

Figure 11.4 Dowel-type joint with slotted-in steel plates.

Typically, a shear-loaded connection will transmit forces from one structural member to one or more fasteners, which in turn will transfer the forces to the receiving structural member. This leads to three natural steps in the design of timber connection using connectors: evaluation of the capacity of the transmitting member, the receiving member, and finally, the capacity of the transferring elements (e.g., the fasteners).

There are several possible failure modes, as illustrated in Figure 11.5. Failure mode 1 is due to the limited embedding strength or capacity of the fasteners; design considerations usually aim at this failure mode since this is the most ductile type of failure. Failure mode 2 includes splitting along a row of fasteners in the grain direction; this failure is minimized by adequate spacing in the fiber direction and end/edge distances. In addition, a reduced computational capacity is used depending on the spacing and the number of fasteners on rows parallel to the grain. Failure mode 3 should be avoided by using proper end distance. Failure mode 4 is a block shear failure that may occur in connections with steel plates and numerous and dense groups of fasteners. Failure mode 5 is a tension failure in the net section and may often govern the design capacity. Failure mode 6 is splitting due to tension normal to grain, a load exposure that always should be minimized. However, in many cases, a force component normal grain occurs and a splitting check should be performed.

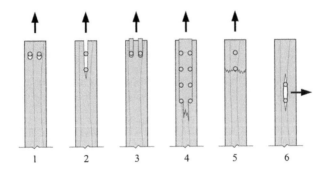

Figure 11.5 Basic failure modes for wood in steel-to-wood dowel type connections.

4.3   Design expressions for dowel-type connection with multiple slotted-in plates

The theory for the capacity of connections using shear-loaded rod-type connections is usually based on work done by Johansen (**YEAR**). The theory is based on the assumptions of rigid plasticity, where the crushing or embedding strength of the wood as well as the yielding of the rods exhibit perfect rigid plastic behavior. A set of possible plastic failure mechanisms is shown in Figure 11.6 for wood to steel connections. For multiple slotted-in steel-plates in a structural wooden member, only failure mechanisms (c), (d), and (e) in Figure 11.6 are relevant for the external (outermost) shear planes in a connection, while (j/l) or (m) will govern the internal shear planes.

The capacity expressions for the external shear planes (per shear plane and connector) are given by Eq. (11) where t_1 is the thickness of the external (outer) wooden layer, d is the diameter of the dowel, $f_{h,k}$ is the characteristic embedding strength of the wood, and $M_{y,Rk}$ is the characteristic bending strength of the connector.

$$F_{v,Rk} = \min \begin{cases} f_{h,k} \cdot t_1 \cdot d & \text{(c)} \\ f_{h,k} \cdot t_1 \cdot d \left[\sqrt{2 \cdot \dfrac{4M_{y,Rk}}{f_{h,k} \cdot d \cdot t_1^2} - 1} \right] & \text{(d)} \\ 2.3 \cdot \sqrt{\cdot M_{y,Rk} \cdot f_{h,k} \cdot d} & \text{(e)} \end{cases} \tag{11}$$

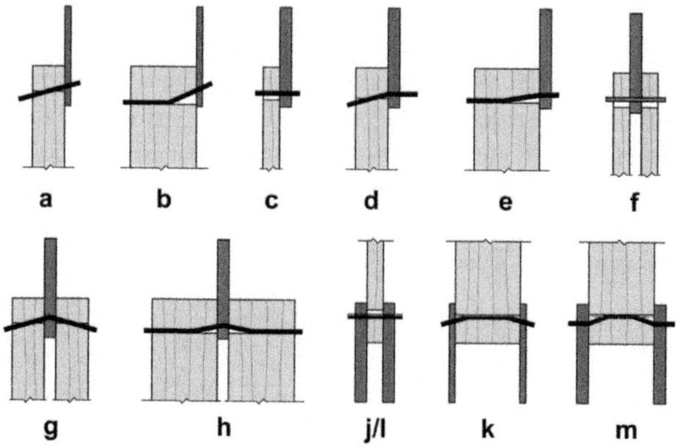

Figure 11.6 Basic failure modes of fasteners for steel-to-wood dowel-type connections.

For the internal shear planes (i.e., all shear planes between the outer steel plates), the capacity per shear plane and connector is expressed in Eq. (12). Note that t_2 is the thickness of the inner wooden layer. Capacity formulations like Eqs. (11) and (12) are often denoted European Yield Models (EYMs), and more on this may be found in EN 1995-1-1:2004/A1:2008.

$$F_{v,Rk} = \min \begin{cases} 0.5 \cdot f_{h,k} \cdot t_2 \cdot d & \text{(j/l)} \\ 2.3 \cdot \sqrt{\cdot M_{y,Rk} \cdot f_{h,k} \cdot d} & \text{(m)} \end{cases} \tag{12}$$

5 Design of modern timber bridges

5.1 Building elements

5.1.1 Glulam

Most timber bridges built today use glulam, which is a stack of parallel solid wood lamellas with a thin layer of glue in between, brought together into a single statical element by applied external pressure during the curing of the glue. Several glulam stacks can be glued together side by side, a process usually referred to as *block-gluing*. In this way, a wide range of cross-sectional sizes can be made, ranging from the size of solid timber to several square meters. The individual lamella is finger jointed and therefore can be made continuous in any practical length. Usually, the size of a glulam component is limited by transportation obstacles from the factory to the building site, e.g. height under bridges and road curvature.

5.1.2 Stress-laminated decks

The stress-laminated deck system has become popular due to its light weight and high lateral stiffness. In principle, it can be made continuous to any width or length. It consists of parallel wooden lamellas placed with their flatwise faces side by side, but displaced lengthwise relative to each other, spreading the joints to avoid weak sections (see Figure 11.7a). The joints in the lamellas' longitudinal direction are simple butt joints, with just one end facing the other. Depending of the size of the bridge and the loading, lamellas of both solid timber and glulam beams are used.

The lamellas are pressed together by prestressing rods made of high-strength steel and usually placed an equal distance apart. The design of the prestressing system is governed by the need for minimum friction to avoid vertical slip due to concentrated wheel loads (as illustrated in Figure 11.7c), as well as possible occurrence of gaps between the lamellas, which may result in a too-soft deck (see Figure 11.7b). Most design codes for timber bridges have guidelines for the density of butt joints and necessary prestressing force.

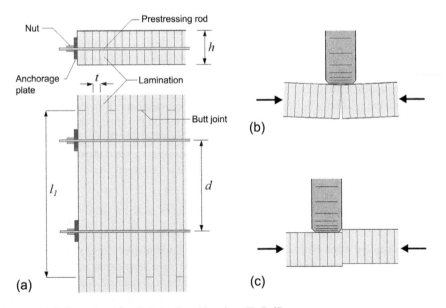

Figure 11.7 Stress-laminated deck plate (drawing: K. Bell).

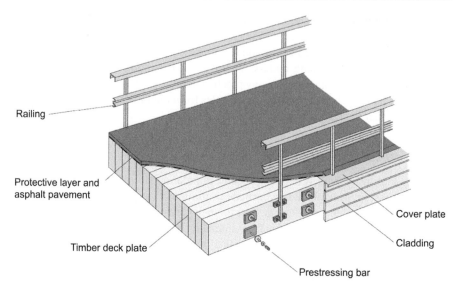

Railing

Protective layer and
asphalt pavement

Cover plate

Timber deck plate

Cladding

Prestressing bar

Figure 11.8 Bridge deck using prestressed laminations.
(Reproduced with permission from Svenskt Trä, Swedish Wood, 2011).

5.1.3   Other materials

A timber bridge usually contains other materials, such as steel and concrete and materials for protection against water access to the wood materials. In addition, r.c. is used in
the abutments and sometimes as decking. Steel is used in the fasteners of the timber
joints, in hangers used in arches, and sometimes as tension members in truss work.
In combination with stress-laminated timber decks, steel crossbeams are often preferred due to higher stiffness, less height, and smaller volume, leading to a more slender
appearance of the bridge. Concrete decks can either be designed as a separate plate or in
composite action with timber members. In the latter case, concrete will be in compression, while the tension will be handled by the timber members. Two different layouts
have been used: either distributed shear connectors leading to almost continuous shear
force transfer between the parts, or concentrated connections between the concrete
plate and the timber structure at the timber joints connecting the concrete slab directly
to the slotted-in steel connector plates without contact between wood and concrete.

5.2   Structural systems

5.2.1   Beams and slabs

Short bridges are often built as simple beam-type glulam structures, either as simply
supported single-span or multiple-span bridges. The main beams span in the lengthwise
direction, and in most cases, crossbeams on top, with small spacing, form the transversal bearing. A top wearing layer of concrete or wooden planks are usually added. An
alternative to crossbeams is to use a concrete plate on top of the main beams, with shear

connectors in between forming a composite system. However, in some cases, it is preferred to avoid composite action between the wooden structure and the concrete slab due to differences in expected creep and temperature behavior. The bridge depicted in Figure 11.9 has no composite action between the concrete top layer and the timber trusses. A slab-type wooden bridge is often produced by using stress-laminated decks; see Figure 11.7 for the layout and Figure 11.8 for a simple application.

5.2.2 Trusses

Trusses in modern bridges are mostly made of glulam members. The truss can be beneath (Figure 11.9) or above the carriageway (Figure 11.10). The choice depends on the available free height under the carriageway and aesthetic, economic, and durability considerations. The trusses are prefabricated in as large pieces as possible, the size of which is commonly limited by transport regulations and road obstacles. Splices in the chords are usually placed at locations suitable for assembling the separate parts on site. All the inclusive splices of the connections are of the slotted-in steel plate and dowel types (see Figure 11.4).

Figure 11.9 Kjøllsaeter Bridge, Norway, whose length is 158 m, features 6 spans, the longest of which is 45 m (photo: Norwegian Public Roads Administration).

5.2.3 Arches

Arches are often used in the design of timber bridges; they may have massive cross sections (see Figure 11.12) or, for longer spans, the arches may be formed by trusses (see Figure 11.11). The use of a truss arch is beneficial to allow handling of the considerable moment actions originating from the loads transferred through vertical hangers. A structural feature of the arch is large horizontal thrusts at the footing; these can be accommodated by the use of heavy foundations, which was the chosen solution for the Tynset Bridge in Norway (depicted in Figure 11.11). For shorter bridges, a tension tie can be a better and cheaper solution, and this has been used for the Fretheim Bridge (Figure 11.12), a bowstring bridge in Flåm, Norway. The double tension tie and the chosen detail at the footing of this bridge are shown in Figure 11.13.

Figure 11.10 Flisa Bridge, whose length is 196 m, has three spans, the longest of which is 70 m (photo K. A. Malo).

Figure 11.11 Tynset Bridge in Norway, whose length is 124 m, has three spans, the longest of which is 70 m (photo K.A. Malo).

Figure 11.12 Fretheim Bridge in Flåm, Norway (span 38 m; photo: Sweco Norway AS).

Figure 11.13 Footing detail at Fretheim Bridge (photo: Sweco Norway AS).

The bridges shown in Figure 11.10 through Figure 11.13 all have stress-laminated timber decks. This type of deck is light and can allow for smaller dimensions in other parts of the bridges, as well as reduced foundation costs. Existing foundations (of an old bridge) can often be reused. This was the case for Flisa Bridge (Figure 11.10), where a one-lane steel bridge was replaced by a timber bridge with a pedestrian lane as well as two road lanes.

6   Design verifications of timber bridges

6.1   Structural information

Some important points concerning an actual timber bridge deign are presented in this section. The design specifications are for the Fretheim Bridge in Flåm, Norway, depicted in Figure 11.12 and Figure 11.13. The bridge is a three-hinged bowstring bridge with arches of glulam, tension ties made of steel rods, and a stress-laminated timber deck made of solid timber where the layout is as shown in Figure 11.7a. The hangers are fastened to the arches and to transversal steel crossbeams beneath the timber deck. The arches are slightly slanted inward, with a ratio of 9/100. The span of the bridge is 37.9 m, and the radius of curvature of the circular arches is 35.2 m.

There is no horizontal wind truss between the two arches, and the arches are clamped sideways at the supports. The horizontal stabilization of the arches is increased by replacing the hangers closest to the support with rigid U-shaped steel frames fastened to the deck, which in turn transfer the horizontal forces to the supports. The design load combinations are stated here without further explanation.

6.2 Verification of arch in ULS

The structural system is treated as being symmetrical about the center hinge, and only the left part is shown on the structural system drawing in Figure 11.14. The cross section of the glulam arch member has a width and height of 800 mm and 1000 mm, respectively. The height is gradually reduced to 800 mm in the vicinity of the hinges.

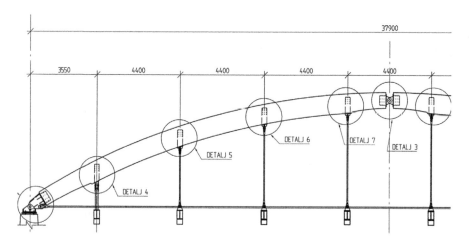

Figure 11.14 Structural system of Fretheim arch bridge (design and drawing: Sweco Norway AS).

The actual loading on the bridge is governed by its location and expected use, together with guidelines from local authorities and requirements from the bridge owner. Furthermore, the loading and load combinations should be in accordance with the current design regulations (e.g., as stated in EN 1990:2002 and EN 1991-2 for Europe). Although the structural system is quite simple, many load combinations need to be investigated, and the results lead to many possible combinations of design actions. The arches are subjected to high compressive forces with moment action about both axes, which vary along the arch. In order to exemplify the use of the timber design verification, this example will use a severely simplified approach by considering just the values stated in Table 11.2, which are based on the original design calculations (Sweco Norway, 2005).

Table 11.2 Design Values of Actions (Load Factors Included)

Symbol	Meaning	Value
N	Axial compressive force	2000 kN
M_y	Moment about y-axes (strong axes)	1500 kNm
M_z	Moment about z-axes (weak axes)	70 kNm
l_{ky}	Buckling length about y-axes (i.e., in the z-x plane)	23 m
l_{kz}	Buckling length about z-axes (i.e., in the y-x plane)	28 m
l_{ef}	Effective length, lateral-torsional instability	15 m

It is assumed that the wooden arches are produced according to EN 14080:2013, fulfilling the requirements for the glulam class GL30h; hence, the properties stated in Table 11.1 are used in the calculations. Furthermore, the traffic loading is treated as short-term loading; and provided that the wooden members are protected against direct water exposure, the modification factor for material strength $k_{mod} = 0.9$ (EN 1995-1-1:2004/A1:2008). The material factor for glulam members is set to $\gamma_m = 1.25$ (EN 1995-1-1:2004/A1:2008), and finally the design strength values are determined by use of the characteristic values given in Table 11.1 modified according to Eq. (1).

The design strength with respect to compression and bending about both cross-sectional axes can be evaluated by using Eq. (3). First, it is assumed that the buckling takes place about the y-axis and that the compression is combined with full moment action about the y-axis and reduced moment about z-axes. The slenderness about the y-axis becomes $\lambda_y = l_{ky} / \sqrt{I_y/A} = 79.7$, and the relative slenderness can be evaluated by Eq. (5):

$$\lambda_{rel,y} = \frac{\lambda_y}{\pi} \sqrt{\frac{f_{c,0,k}}{E_{0,05}}} = \frac{79.7}{\pi} \sqrt{\frac{30}{11300}} = 1.31 \tag{13}$$

Next, the buckling parameter $k_{c,y}$ is evaluated by use of Eq. (4), resulting in $k_{c,y} = 0.52$. The stresses are determined by use of common linear elastic relationships. The final step is to evaluate the interaction of compression and bending by use of the left-hand expression of Eq. (3), which reads

$$\frac{\sigma_{c,0,d}}{k_{c,y} f_{c,0,d}} + \frac{\sigma_{m,y,d}}{f_{m,y,d}} + k_m \frac{\sigma_{m,z,d}}{f_{m,z,d}} = 0.26 + 0.57 + 0.02 = 0.85. \tag{14}$$

By comparison of the terms, it is obvious that in this assumed failure mode, the effect of bending about the strong axis dominates the utilization of the member.

Next, it is assumed that buckling takes place about the weak axis and the following buckling parameters result: $\lambda_z = 121.2$, $\lambda_{rel,z} = 1.99$, $k_{c,z} = 0.24$. The right-side expression of Eq. (3) becomes

$$\frac{\sigma_{c,0,d}}{k_{c,z} f_{c,0,d}} + k_m \frac{\sigma_{m,y,d}}{f_{m,y,d}} + \frac{\sigma_{m,z,d}}{f_{m,z,d}} = 0.56 + 0.40 + 0.03 = 0.99. \tag{15}$$

In this case, buckling about the weak axis is the dominating effect and the member is fully utilized.

The buckling lengths in this example have been determined by the use of linearized instability analyses, quantifying the critical axial force P_{cr}. Moreover, the buckling length then has been estimated from the simplified relation $l_k = \pi \sqrt{EI / P_{cr}}$, where P_{cr} is the axial force at the buckling load level.

6.3 Verification of a dowel connection in ULS

A dowel-type connection transferring the force from the hanger to an arch is depicted in Figure 11.15. The four slotted-in steel plates are extended outside the arch, and through a hole in each plate, a common pin has been installed to create a hinge beneath the arch to avoid any moment action on the dowel connection. It is essential to design the groups of fasteners with no eccentricity, as eccentricities will cause unequal force distribution on the dowels and lead to a more expensive connection.

The detail denoted as "Detalj 5" in the Fretheim Bridge drawing (Figure 11.14) is used here as an example of the calculations. This connection is located 11 m from the center point of the arch and is shown in Figure 11.16. It differs slightly from the design shown in Figure 11.15.

Figure 11.15 A dowel-type connection between hanger and arch (photo K.A. Malo).

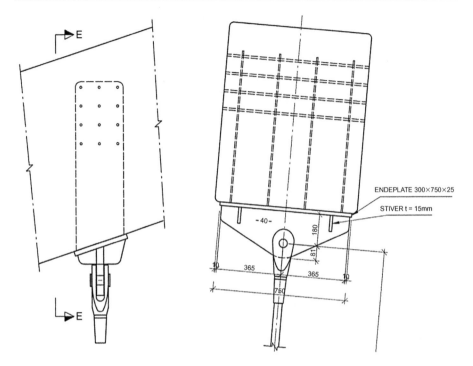

Figure 11.16 Dowel-type connection between hanger and arch (Detalj 5 in Figure 11.14); side view (left); cross section (right), design and drawing: Sweco Norway AS.

The input parameters for the calculation are four steel plates of thickness that are 8 mm thick, with 9-mm slots in the wood. The thickness of the external wood layer is 100 mm, while the internal layers are 188 mm thick. The 12 dowels have diameters of 12 mm and characteristic bending strength $M_{y,Rk} = 67152$ Nmm; also, they are installed in a regular pattern with spacing of 100 mm in both directions. The embedding strength of glulam is dependent on the wood density, dowel diameter d, and the angle α between the grain and force directions, and is evaluated by

$$f_{h,\alpha,k} = \frac{f_{h,0,k}}{k_{90} \sin^2 \alpha + \cos^2 \alpha} \tag{16}$$

where $f_{h,0,k} = 31.0$ MPa and is the basic embedding strength along the grain, and parameter $k_{90} = 1.53$ for 12-mm dowels in softwood.

6.3.1 Transfer of forces from steel plates to wood

The hangers are vertical and the angle between the arch and the vertical force (at this location) is 71.8 degrees, leading to $f_{h,\alpha=71.8,k} = f_{h,k} = 21.0$ MPa. The characteristic load-bearing capacity of a single dowel is the sum of the individual shear planes acting on the dowel. The shear planes may have different capacities, but they should be compatible with respect to deformation at the ultimate load. Here, the two external shear planes will be identical, due to the symmetry of the layout in the cross section. This is also the case for the six internal shear planes, but the capacities of the internal and external shear planes might very well be different unless they all have the same failure mode.

The capacity of an external shear plane is determined by use of Eq. (11) and an evaluation gives

$$
F_{v,Rk,ext} = \min \left\{ \begin{array}{ll} f_{h,k} \cdot t_1 \cdot d = 25188\,\text{N} & \text{(c)} \\ f_{h,k} \cdot t_1 \cdot d \left[\sqrt{2 \cdot \dfrac{4M_{y,Rk}}{f_{h,k} \cdot d \cdot t_1^2} - 1} \right] = 11371\,\text{N} & \text{(d)} \\ 2.3 \cdot \sqrt{\cdot M_{y,Rk} \cdot f_{h,k} \cdot d} = 9459\,\text{N} & \text{(e)} \end{array} \right\}. \quad (17)
$$

The capacity of an external shear plane is determined by using Eq. (12):

$$
F_{v,Rk,int} = \min \left\{ \begin{array}{ll} 0.5 \cdot f_{h,k} \cdot t_2 \cdot d = 23677\,\text{N} & \text{(j/l)} \\ 2.3 \cdot \sqrt{\cdot M_{y,Rk} \cdot f_{h,k} \cdot d} = 9459\,\text{N} & \text{(m)} \end{array} \right\}. \quad (18)
$$

In this connection, the dowels are very slender compared to the thickness of the wooden layers surrounding the steel plates, and the failure mode is governed by the bending of the dowels. It turns out to give the same failure mode for both external and internal shear planes. The capacity for a single dowel then becomes

$$
R_{v,Rd} = R_{v,Rk} \frac{k_{\text{mod}}}{\gamma_m} = \left(\sum_{int} F_{v,Rk,int} + \sum_{ext} F_{v,Rk,ext} \right) \frac{k_{\text{mod}}}{\gamma_m}
$$
$$
= (2 \cdot 9459 + 6 \cdot 9459) \frac{0.9}{1.3} = 52390\,\text{N}. \quad (19)
$$

It is common during the calculation of connections to set material factor $\gamma_m = 1.3$. In this case, the dowels are equally loaded and capacity of the dowel connection is therefore

$$
R_{d1} = 12 \cdot 52.390 = 629\,\text{kN} \quad (20)
$$

6.3.2 Splitting along dowel rows caused by force parallel to grain

A complete verification also requires splitting control of the glulam member. This is performed by evaluating the design capacity for splitting along the fibers due to several dowels on a row, or splitting due to tension normal to grains caused by the force on the group of fasteners. The capacity of a single dowel [given in Eq. (19)] represents the capacity where force and grain directions have a 71.8° angle deviation. For evaluating the risk of splitting due to several dowels on a row, the capacity of a single dowel with the force along the grain is eevaluated, and thereafter, the capacity of the rows are determined.

Letting $\alpha = 0$ in Eq. (16), the embedment strength becomes $f_{h,k} = f_{h,0,k} = 31$ MPa. Evaluation of Eqs. (17) and (18) gives $F_{v,R,k} = 11501$ N; and consequently, by use of Eq. (19), the design capacity of a single dowel along the grain becomes $R_{v,Rd} = 63697$ N. The increased risk of splitting along a row of dowels depends on the number of dowels and the spacing between them and is taken into account by use of a reduced average strength or a reduced effective number (n_{ef}) for dowels on a row (EN 1995-1-1:2004/A1:2008). In the dowel connection example shown in Figure 11.16, the dowels are not placed on rows along the fiber direction. However, the deviation from alignment along the grain is too small to satisfy the spacing requirements between rows normal to the grain direction, and consequently three dowels in a row ($n = 3$) are used to verify splitting strength. The effective number on a row is determined by

$$n_{ef} = \min \begin{cases} n \\ n^{0.9} \sqrt[4]{\dfrac{a_1}{13d}} = 2.44, \end{cases} \tag{21}$$

where a_1 is the spacing in the grain direction; in this instance, it equals 105 mm. In this case, all four rows are equally stressed, and the capacity along grain becomes

$$R_{0d} = 4 \cdot 2.44 \cdot 63.697 = 622 \, \text{kN}. \tag{22}$$

R_{0d} shall be compared to the force component in the fiber direction, which means that the capacity for vertical hanger force is

$$R_{d2} = R_{0d}/\cos 71.8 = 1990 \, \text{kN}. \tag{23}$$

6.3.3 Splitting along grain caused by tensile force normal to grain

This type of failure may occur on the rear side of a group of fasteners. The current failure criterion in EN 1995-1-1:2004/A1:2008 is based upon simplification of fracture mechanic models; it reads:

$$F_{90,Rk} = 14b \sqrt{\dfrac{h_e}{1 - \dfrac{h_e}{h}}} = 14 \cdot 800 \sqrt{\dfrac{900}{1 - \dfrac{900}{1000}}} = 1.063 \cdot 10^6 \, \text{N}. \tag{24}$$

An increase in the effective height h_e will give increased capacity; and therefore, these types of connections should always be located on the rear side of wooden members (relative to the force direction). In this case, a suitable distance to the rear surface of the member is 50–100 mm. For convienience, 100 mm is used here. The design capacity with respect to tensile failure of the normal grain for a shear force caused by the normal component of the force in the hanger is obtained by

$$F_{90,Rd} = F_{90,Rk} {}^{k_{\text{mod}}/\gamma_m} = 736 \, \text{kN}. \tag{25}$$

The design criterion in this case is that the shear force on either side of the connection should be less than $F_{90,Rd}$. The two shear force components shall be determined such that the sum of them equals the normal component of the external force. In principle, the distribution of the shear forces shall be determined from a static analysis, but doing this is not necessary in this case since $F_{90,Rd}/\sin 71.8 = 775 \, \text{kN}$, which is greater than the force that can be transferred through the dowels. This failure mode, therefore, will not govern the design, regardless of the distribution of the shear forces.

It can be concluded that the maximum force that can be transferred in the hanger is limited by the capacity of the dowel connectors $R_{d1} = 629 \, \text{kN}$.

7   Design and durability

Good designers of timber structures follow either of two simple rules:

- Keep water out of the structure.
- If you cannot keep water out, make sure that it can easily get out again.

It is obvious that the durability of wooden bridges is governed by the design. Numerous historical examples (e.g., Sétra, 2007) have showed that timber bridge design resulting in a too high level of humidity in the wood has led to fungus attacks, which cause the most serious damage. The most important objective of timber bridge design, therefore, is to avoid excessive humidity in the wood. This is not new knowledge, of course. Italian architect André Palladio (1508–1580) published an architectural treatise recommending that if timber bridges were built, they should at least be covered. And in fact, many of the durable timber bridges in Switzerland and the United States are covered by a complete roof (Pierce, 2005). But these bridges are mainly pedestrian bridges or made for vehicles that are very different from today's 20-m-long trucks. On a rainy day, a modern truck-train at high speed will create a considerable blast wave that will throw up a large spray of water and bring it into the bridge structure. In such a situation, the roof in fact may reduce the ability of rapid drying and lead to high moisture in the structure. Hence, the covering of bridges by a roof is a doubtful approach for modern road bridges. On the other hand, bridges with a deck on top that protects the supporting structure from weathering have demonstrated a better state of preservation than those where the deck was between or below the carrying structure (Kropf, 1996). To obtain good durability, design of the details is essential (Sétra, 2007). More on recent findings related to durability of timber bridges can be found in Kleppe (2010) and Pousette and Sandberg (2010).

The design of a bridge depends on many factors, including topography, required waterway clearance, load, and appearance. Decisions made at an early planning stage can have a decisive influence on the long-term behavior of the structure. The less the structure protects itself, the more effort must be invested in protecting individual endangered parts. Much of this can be resolved at the drawing table, assuming that the design engineer is responsive to the needs and limits of the construction material and keeps in mind that sun exposure and high temperatures might also damage wood (Kropf, 1996). The difference in change of volume due to unequal moisture distribution through a wooden member, together with very low strength normal to the grain, can cause longitudinal checks to develop into large cracks. Large cracks in connection areas may reduce the strength of both connections and members. By combining good detail design with supplemental measures (e.g., cover, water-repellent surface coating, and chemical treatment where needed—but only there), it is possible to equip weather-exposed wooden structures for a service life comparable to other construction materials, and still maintain the advantage of wood as an ecological material without disposal problems (Kropf, 1996).

References

AASHTO-LRFD Bridge Design Specifications, 7th edition with 2015 revisions. American Association of State Highway Transportation Officials, Washington, DC, 1,960 p. 2014.

Forest Products Laboratory. Wood handbook - Wood as an engineering material. General Technical Report FPL-GTR-190. Madison, WI: U.S. Department of Agriculture, Forest Service, Forest Products Laboratory: 508 p. 2010.

International Organization for Standardization: INTERNATIONAL STANDARD ISO 898-1:2013. Mechanical properties of fasteners made of carbon steel and alloy steel. Part 1: Bolts, screws and studs with specified property classes — Coarse thread and fine pitch thread.

Johansen, K.W., 1949. Theory of Timber Connections, vol. 9. IABSE, Zürich, Switzerland, pp. 341–348.

Kleppe, O., 2010. Durability of Norwegian timber bridges. In: Proc. Intl. Conf. Timber Bridges (ITCB2010 and Tapir Academic Press), ISBN 978-82-519-2680-5, pp. 157–168.

Kropf, F.W., 1996. Durability and detail design—the result of 15 years of systematic improvements. Paper presented at the National Conference on Wood Transportation Structures, Madison, Wisconsin. http://www.fs.fed.us/eng/bridges/documents/desinpln/Durability_And_Detail_Design.pdf.

Pierce, P.C. Brungraber, R.L. Lichtenstein, A. Sabol, S. *Covered Bridge Manual*, FHWA-HRT-04-098. McLean, VA: Federal Highway Administration. 327 p. 2005.

Pousette A. and Sandberg K., 2010. Outdoor tests on beams and columns. In: Proc. Intl. Conf. Timber Bridges (ITCB2010 and Tapir Academic Press), ISBN 978-82-519-2680-5, pp. 169–178.

Ritter, M.A., 1990. Timber Bridges. Design, Construction, Inspection, and Maintenance. EM 7700-8 Forest Service, U.S. Department of Agriculture, Washington, DC.

Sétra (Service d'études sur les transports, les routes et leurs aménagements, France), 2007. Timber Bridges. How to Ensure their Durability. Technical Guide. © 2007 Sétra - Reference: 0743A- ISRN: EQ-SETRA–07-ED40–FR+ENG, http://www.setra.equipement.gouv.fr.

Sweco Norway, 2005. Drawings and Documentation of Fretheim Bridge, Flåm, Norway. Lillehammer, Norway, December 19, 2005.

Swedish Wood, 2011: Design of timber structures. Swedish Forest Industries Federation, ISBN 978-91-637-0055-2, Stockholm, Sweden.

Masonry bridges

12

Pipinato A.
AP&P, Technical Director, Italy

1 Structural theory of masonry structures

In this chapter, the structural theory of masonry structures is introduced, as well as the history and technology of masonry structures.

1.1 History of masonry structures

The history of masonry structures began as a spontaneous process of construction mainly related to simple walls built with stone or caked mud, with mud-smeared mortar to increase stability and to make the edifice watertight. Stone was preferred to brick in many situations depending on the geographical location and the availability of quarries. An increase in the use of masonry began when quarry capacity and stone workmanship became more prevalent, and on the other hand when fires built brick began to be used. Another fundamental development was the introduction of using lime instead of mortar in construction. After buildings in Mesopotamia were erected with stone and natural, sun-dried brick, and later the Egyptian pyramids were constructed, the Greeks built lime and marble constructions of a superior class. Still later, in the 1st century B.C., Romans introduced a number of refined masonry constructions featuring masonry arches and walls of imposing size, aqueducts, palaces, and churches of an unforgettable beauty and impressive durability. Another step forward in masonry construction took place in medieval times, when masonry was developed at an highly sophisticated level, mainly in Europe but also in the Islamic world. The Industrial Revolution, which began in the mid-18th century, fostered further advances in masonry, as quarry and working machines were developed, together with a strong impulse to find advanced mechanical solutions, and the widespread use of Portland cement mortar increased the strength and durability of masonry buildings and bridges. The 19th century saw a strong change in the use of masonry, as reinforced concrete (r.c.) and steel structures developed rapidly in order to meet the growing demand for taller buildings. Finally, during the twentieth century, innovative solutions to increase the use of masonry arose, including high-strength mortar, steel-reinforced masonry, and industrialized lighter-masonry blocks. Together with the history of masonry construction, bridges developed similar innovations; however, they are mostly no longer used, and have been replaced completely by steel and concrete bridges. For this reason, most of this discussion should be seen as an assessment study on existing structures, rather than a design chapter. The main masonry bridges built in the last several centuries are listed in Table 12.1.

Innovative Bridge Design Handbook. http://dx.doi.org/10.1016/B978-0-12-800058-8.00012-8

Table 12.1 Main masonry bridges in the world.

Reference photo	Name	Place	Nation	Main span (total lenght)	Year of construction
	Pont de la Libération	Villeneuve-sur-Lot	France	96 (315)	1919
	Syratalviadukt	Plauen	Germany	90 (295)	1905
	Longmen Bridge	Luoyang	China	90 (295)	1961

Table 12.1 Continued

Reference photo	Name	Place	Nation	Main span (total lenght)	Year of construction
	Solkan Bridge	Nova Gorica	Slovenia	85 (278)	1906
	Adolphe Bridge	Luxembourg City	Luxembourg	84 (275)	1904

Continued

Table 12.1 Continued

Reference photo	Name	Place	Nation	Main span (total lenght)	Year of construction
	Pont de Montanges (Pont-des-Pierres)	Valserine river	France	80 (262)	1910
	Viaduc de la Roizonne	La Mure	France	79 (260)	1928

1.2   Theory of masonry structures

The term *masonry* can be defined as an assemblage of classified stones, bricks, or both, which is often put together with mortar. The geometrical shape of the elementary stone could be squared and well fitted, or unworked units just placed one on top of another to shape the form of the structure. Mortar usually has been used to fill interstices, and it decays over time. The stability of this conventional structure is ensured by compaction under gravity of these elements; the main state of tension relates to compression, and only a low amount of tension can be resisted.

An indirect parameter to determine the strength of stone is the height at which a prismatic column could be theoretically be built before crushing at its base due to its own weight (Heyman, 1995), and this can be predicted easily: for example, a medium sandstone might have a unit weight of 20 kN/mc and a crushing stress of 40,000 kN/mq, and dividing one number by the other returns the maximum height of the column as 2 km. According to this observation, and belonging to other extensive past studies (e.g. Villarceau 1854) Heyman (1995) suggested to limit the nominal stress to 1/10 of the crushing stress of the material.

However, observing real cases of existing constructions throughout history is fundamental to understand how these structures works: for example, from studies of Beauvais Cathedral by Benouville (1891), he observed that the maximum stress of that church was not greater than 1.3 N/mm^2, and if compared to the crushing stress, the safety factor was found to be more than 30. Observing other masonry structures, Heyman (1997) stated that the main portions of the load-bearing structure of a church will be working at 1/100 of the crushing stress, and infill panels or walls that carry little more than their own weight may be subjected to a background stress as low as 1/1000 of the potential of the material. So the safety factor against crushing that is implicit in these statistics makes their rough derivation unimportant. This observation obviously does not consider lateral horizontal forces.

According to Heyman, the three most fundamental assumptions that apply to masonry structures are:

- Masonry has no tensile strength.
- Stresses are so low that masonry has an effectively unlimited compressive strength.
- Sliding does not occur.

Concerning the first assumption, one could consider that as individuals, stone may be strong in tension, but the mortar between stones is weak. The second one will be approximately correct if average stresses are in question, even if stress concentration that could arise in common masonry structures should lead to failure only locally (i.e., splitting or surface spalling). As for the third assumption, even if there were evidence of the slippage of individual stones, generally the masonry structure retains its shape well; only a very small compressive prestress is all that is necessary to avoid the dangers of slippage and general loss of cohesion.

Structure in general also could be analyzed considering the three main structural criteria of strength, stiffness, and stability. The structure must be strong enough to carry whatever loads are imposed, including its own weight; at the same time, it must not deflect unduly; and it must not develop large unstable displacements, whatever local or overall. If these three criteria are satisfied, then the designer can run through a checklist of secondary limit states to make sure that the structure is otherwise serviceable. Concerning masonry structures, Heyman (1997) observed a paradox: strength and stiffness do not lie in the foreground of masonry design, nevertheless is the third criterion of stability that is more relevant for masonry. For example, considering a semicircular arch structure carrying a given load P and its own weight, stresses are low and the deflections negligible, and both will remain so as the value of P is increased; however, a certain value of P destroys the structure's stability, and a point is reached at which the structural forces can no longer be contained within the arch, so stresses remain low, but an instable mechanism of collapse takes place.

1.3 History and technology of masonry arches

The first masonry bridges had only modest span lengths, with partially or completely underground foundations that used the land as an abutment. With the advent of Roman bridges, robust and elegant new constructions were developed; preference were on odd number of arches, and on cicrcle arch profiles arc profile circle . The construction of these periods reached longer spans, to use the land ground instead of the riverbed, to erect foundations on a soil with superior mechanical properties and avoiding at the same time avoiding the undermining danger. Of course this solution was applied whatever the total length could be covered with a single span. In the Pont-Saint-Martin in the Aosta Valley, Italy (Figure 12.1a), which has a span of 31.4 m, the low vault is made of big blocks of cube-shaped stone (90 cm wide) of five rings parallel, the space between the rings is filled in conglomerate. A typical structure of Roman masonry arches was the aqueduct, generally built with one, two, or three levels of rounded arches; For example, the Pont du Gard (Figure 12.1b), which carried water to Nimes, has three rows of arches whose width decreased as they moved upward: the lights of the higher arches are 4.4 m, while in the lower orders, they range from 15.5 to 24.4 m. In the early Middle Ages, the construction of bridges stops almost completely because of the lack of commercial movement. In constructions during this period in Sicily and Spain, a strong influence came from Eastern culture (lancet vaults, often with polycentric profiles). In France, however, different forms were used; for example, in the bridge of Avignon (Figure 12.1e), built in the 12th century, the arch approximates an arch of parabola, with 20–25 m span length and thickness of 70 cm.

During the same period in Italy, a trend started toward slimness and bold proportions: the Ponte Vecchio in Florence and that of Castelvecchio in Verona (Figure 12.1c) were examples of this design; the latter has the greater arch of 48.7 m, and the three brick arches are lowered, while piles are mixed masonry brick and stone. Another noticeable bridge of this period was the Devil bridge (1321–1341), with a 45-m span (Figure 12.1.d).

This evolution led to a decrease in the ratio of the thickness of the arch to the span length; the materials were often coming from other constructions, even if well-worked, the mortars were of improved quality if compared to the previous period.

At the end of the eighteenth century, the École Nationale des Ponts et Chaussées in Paris introduced many technical innovations: shoulder-piles were abandoned, and arches were designed and built directly over the full river level. The Concordia Bridge

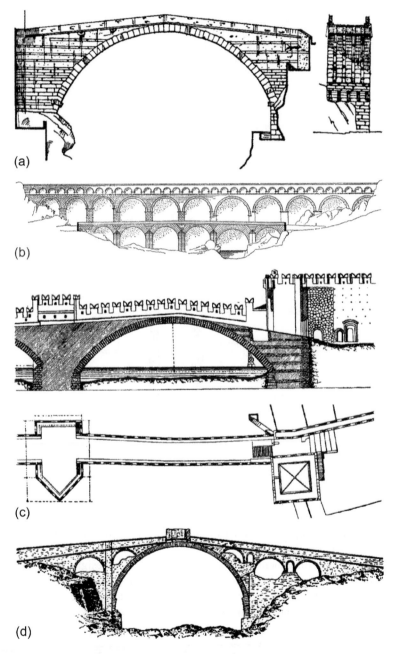

Figure 12.1 (a) Pont Saint Martin, (b) Pont Du Gard, (c) Castelvecchio bridge (lateral and plan view), (d) Pont du Diable (Ceret),

(Continued)

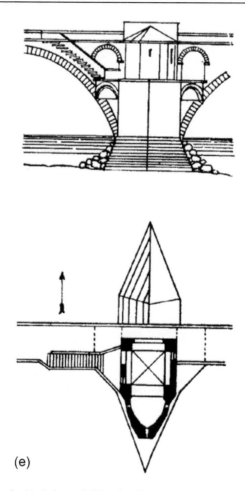

(e)

Figure 12.1 Continued. (e) Avignon bridge details.

in Paris is a typical example of this new structural shape (Figure 12.2). A new design philosophy was developed, in which the materials were selected considering their characteristics, the walls were built with great care and with suitable binders, and design loadings and consequent stresses in foundations were designed to be uniformly distributed, with adequate design and detail solutions. In addition, greater importance was placed on the phase of camber and disarmament, which proceeds from the abutment to the key at midspan, as soon as mortars have cured sufficiently.

Figure 12.2 Concordia Bridge, Paris.

The bridge over the Dora Riparia river in Turin is a wonderful example of construction stone cutting. The arch intradox is circular, with a 45-m span and 5.5 m of sag, and the abutment is built with larger size blocks to be more stable.

Major rail projects were built from Séjourné using the circular arch. On the Castres-Montauban rail line, the Lavaur Bridge (erected in 1884; Figure 12.3), with a span of 61.5 m; the Castelet Bridge (1883) with a span of 41.2 m; and the Antoinette bridge at Vielmur (1884), with a 50-m span are works of interest due to their slenderness and the width of the arches and the slender shoulders and vaults.

Figure 12.3 Bridge across the Agout River, Lavaur.

The railway bridges constructed in the second half of the eighteenth century generally were viaducts, as they were built as a sequence of many spans with small arches (Figure 12.4). This turns out to be the most effective solution when you need to move on a plan of very irregular morphology. The oldest bridges with tall columns usually had several arches, and the lower ones tended to be smaller, narrowing and becoming lower. However, you can find some bridges with high columns and a single row of

Figure 12.4 Railway viaduct, Lockwood.

arches of modest span, such as the Lockwood viaduct (Figure 12.4), or multi-span solutions with a central large span (e.g. the Wiesen Viaduct, Figure 12.5).

In Figures 12.6 and 12.7 the constitutive elements of a typical masonry arch bridge are described.

Figure 12.5 Railway viaduct, Wiesen.

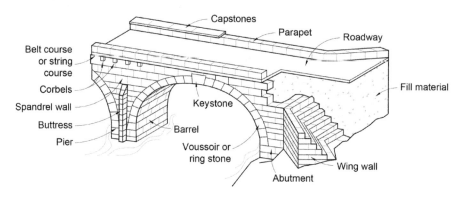

Figure 12.6 Constitutive elements of a masonry bridge.

The sizing of these elements was entrusted to empirical work relations that provided a geometric scaling (Corradi, 1998). Element sizing were not always supported by theoretical formulations, but often based on the builders' past experience.

The shoulders of the bridges were massive pieces of masonry designed to balance the arch and anchor the bridge to the two sides. The shape of the shoulders is usually trapezoidal. The masonry, brick, or square blocks generally have horizontal courses; in the abutment area, to avoid sliding in the joints due to the high thrust of the arc with respect to the weight of the overlying masonry, inclined stone slabs or other special joint pieces were often used. In the technical manuals of the time, several formulas for sizing can be found, which did not take into account the mechanical characteristics of

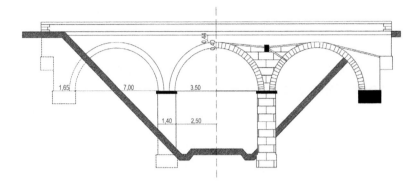

Figure 12.7 Longitudinal arched bridge typical section.

the masonry and the soil; the most simple and widely used provide the width of the shoulder as follows (in meters):

$$s = 0.005h + 0.2c + c/f*(0.10 + 0.005c), \tag{1}$$

where h is the height of the shoulder from the floor to the foundation plan, c is the arch span, and f the sag. Other formulas also take into account the thickness and height of the arch, as well as other specific parameters.

   The columns had rectangular horizontal sections; river bridges can be equipped with protections to improve the hydraulic behavior and to avoid undermining. Intermediate decks were often placed in the viaducts, to limit the buckling of high columns. The vestments of columns had slopes of 1/20–1/25 in the lateral faces and 1/10–1/16 in the frontal ones; the slope can be linear or logarithmic.

   For the sizing of columns, various empirical formulas were proposed that provide the width of the columns in the plant s' function of parameters such as the height h' of columns from the foundation plan to the arches, the light or rope arch c, the thickness of the arch; the Colombo proposes to adopt the greater of these two values (in meters):

$$s' = 0.2h' + 0.6 \tag{2a}$$

$$s' = 0.125c \tag{2b}$$

The shoulder-pile has a larger cross section and is strengthened by pilasters, with a considerable slope; they were included in the multispan bridges because, during the construction of arches, they acted in contrast to the horizontal, unbalanced thrusts. In addition, they protected against the accidental collapse of an arch. As for aesthetics, the pile shoulder breaks up the monotony of similar-looking piles.

   The vaults (or arches) were generally used in the barrel and plant straight. The intrados profiles are circular arches, and although they do not represent the optimal shape in relation to the distribution of the loads, they do meet the requirements of construction simplicity (tracing, camber). Three types are generally built: semi-circular, lowered arch (circular, semielliptical, or polycentric), and acute or ogive arch. In the viaduct, the semi-circular arch is more frequent; the acute arc is suitable in cases where heavy mid-span loads are present, however it is weak to heavy loads on the abutments; the lowered arch is necessary only for lower bridges crossing a riverbed.

2 Assessment of the load-carrying capacity of arch masonry bridges

2.1 Historical methods

Past studies, such as La Hire (1695, 1712) and Couplet (1729–1730), involved theoretical analysis and experimental activities on the line of thrust, and especially on the arch collapse. Then Gregory (1697) deepened the shape of the catenary arch as the most appropriate solution to carry its own shape. Heyman (1982) looked into this last observation in more detail, concluding that it could be interpreted as defining the lower-bound theorem of plasticity. Navier (1833) introduced the middle third rule which was applied to masonry arches by Rankine (1898), such that the line of thrust was constrained to lie within the middle third of the arch in order to avoid tensile stress. Barlow (1846) and Fuller (1875) worked on graphical solutions on the line of thrust. Castigliano (1879) used the theorem of minimum strain energy to develop elastic methods, and for masonry arches, he performed an iterative analysis quite similar to modern, nonlinear finite element model (FEM) analysis (no tension/nonlinear procedure, cutting out tension portion of the arch). Further works developed by Rankine (1898) analyzed the role of backings in masonry arches and recorded the first geometrical/mechanical observations including that strong backings are particularly likely for semicircular or elliptical arches which would probably otherwise have been unstable during constructions; or to give the greatest possible security to a hidrostatic arch, the backings ought to be built of solid rubber masonry up to the level of the crown of the extrados, and the importance of squared side-joints in backings to avoid failures.

2.2 Recent methods

The first consistent studies on arch bridges were performed by Pippard and Chitty (1951), Heyman (1969, 1972, 1976, 1980, 1982), Withey (1982) and Tellet (1983): these studies made significant contributions to our body of knowledge using elastic methods and collapse mechanisms. Pippard also looked into assessment methods: the arch behaved elastically until the first hinge or crack was formed. Then it failed with a four-hinged mechanism. He saw that after the first hinge occurred, there was a significant amount of reserve of strength in the arch before it collapsed. The elastic method enhanced the preliminary estimation of masonry arches: a two-pinned arch with horizontal forces keeping the arch in place is the basic assumption. Alfred John Sutton Pippard Pippard's (1948) approach used the partial derivative of the strain energy, U, with respect to a force that is equal to the displacement in the direction of the force (Castigliano's theorems; Castigliano 1879); the ring was treated as a two-pinned parabolic arch with a secant variation of $I = I_0 \sec\alpha$, where I_0 is the second moment of area at the crown. The axial thrust and shearing force terms in the strain energy equation were ignored in this theory. Hence, the strain energy was assumed to be totally dependent upon the flexural response of the arch.

Therefore, the limiting value of the point load at the crown derived by Pippard would be given by

$$W = \frac{\dfrac{256 f_c h d}{L} - 128 \rho L h \left(\dfrac{1}{21} + \dfrac{h+d}{4a} - \dfrac{a}{28d} \right)}{\left(\dfrac{25}{a} + \dfrac{42}{d} \right)}.$$

The following conditions are applied to this simple solution:

- The arch is assumed to be parabolic, with a span-to-rise ratio of 4.
- The arch is assumed to be pinned at the abutments (i.e., it is a two-hinged arch).
- The dispersal of loading applied at the surface of the fill was assumed to occur only in the transverse direction, with a 45-degree load spread angle.
- Pippard considered the case of a single point load applied at the midspan; the effective width of the arch was taken as twice the fill thickness at the crown (b = 2 h).
- The fill was assumed to have no structural strength and to only impose vertical loads on the arch.

The fill was assumed to be of the same density as the arch ring (i.e., 22 kN/m3); the limiting compressive stress was taken to be $f_c = 1.40$ N/mm^2, and the limiting tensile stress was taken to be $ft = 0.7$ N/mm^2. The expression was then modified by the Military Engineering Experimental Establishment (MEXE) in the form of a nomograph and is currently recommended by the UK Department of Transport in its departmental standard (Department of Transport, 2001). The MEXE method is a long-established system of assessing masonry arch load-carrying capacity. It has been subject to review in recent years, and some shortcomings have been identified. There is now a growing consensus that the current version of MEXE overestimates the load-carrying capacity of short span bridges, but for spans over 12 m, it becomes increasingly conservative. This method was based on the two-hinged elastic analysis by Pippard, which was then calibrated with both field and laboratory tests in the 1930s (Oliveira et al., 2010).

The method was most predominately used in World War II as a way to quickly classify the load-carrying capacity of older masonry arch bridges. However, since that time, the MEXE method has still been used as a way to load-rate masonry arch bridges. The modified axle load (MAL) depends equally on the arch and backfill thickness, although the ring thickness has significantly more influence on the arch behavior than does the backfill. The modification factors are introduced without taking account of the arch geometry; the backfill depth, ring thickness, and even the mortar thickness could have differing influences on arches with different geometries. The method comprises of the primary calculation of

$$Modified\ axle\ load = \frac{740(d+h)^2}{L^{1.3}} F_{sr} F_p F_m F_j F_{cm},$$

where d = thickness of arch barrel adjacent to the keystone (m); h = average depth of fill at the quarter points of the transverse road profile between the road surface and the arch barrel at the crown, including road surfacing (m); L = span (m); Fsr is the span/rise factor; Fp is the profile factor; Fm is the material factor; Fj is the joint factor; and Fcm is the condition based, to be determined on site.

More details about and limitations of this basic formula and its development can be found in Department of Transport (2001).

2.3 Empirical rules

Numerous methods have been employed during the last several decades to assess the load-carrying capacity of masonry arch bridges, using tools ranging from easy-to-use geometrical rules to the most sophisticated finite element software. The first and simplest approach deals with empirical rules, based on geometrical relations, coming from proportions in the construction of arch components (span, rise and thickness, width and height of piers, etc.). There are not always mechanical confirmation of these rules; however, they have been extensively applied in the past with real-life arches. A summary of these methods is reported in Table 12.2.

2.4 Classic solution

A classic solution of arch bridges relies mainly on Heyman's theory (Heyman, 1997, 1982, 1996), which simplifies the calculation of the ultimate load of a masonry arch by making the following assumptions: (i) masonry units have an infinite compressive strength, (ii) masonry units behave as a rigid body, (iii) joints transmit no tension, and (iv) masonry units do not slide at the joints. As a result of these assumptions, the bounding theorems of plasticity can be applied to determine the ultimate load of a masonry arch. Plasticity theories incorporate two theorems: (i) an upper bound, or mechanism solution; and (ii) a lower bound, or equilibrium solution.

The mechanism method is based on upper-bound plastic analysis (Heyman, 1982): masonry arch collapse loads can be determined by analyzing the arch as a mechanism

Table 12.2 Empirical rules for crown arch thickness (MEXE, 1952).

Date	Author	Deep arch	Shallow arch
15th century	Alberti	$t=s/10$	-
1714	Agutier, $s>10$ cm	$t=0.32+s/15$	-
1777	Perronet	$t=0.325+0.0035\,s$	$t=0.325+0.0694\rho$
1809	Gauthey, $s<16$ m	$t=0.33+s/48$	-
1809	Gauthey, $16<s<32$ m	$t=s/24$	-
1809	Gauthey, $s>32$ m	$t=0.37+s/48$	-
1809	Sganzin	$t=0.325+0.3472\,s$	-
1845	Dejardin	$t=0.30+0.045\,s$	$t=0.30+0.025\,s$
1854	L'Eveille	$t=0.333+0.033\,s$	$t=0.33+0.033\,s^{1/2}$
1862	Rankine	$t=0.19R^{1/2}$	-
1870	Dupuit	$t=0.20\,s^{1/2}$	$t=0.15\,s^{1/2}$
1885	Croizette-Desnoyers	$t=0.15+0.20\rho^{1/2}$	-
1885	Lesquiller	$t=0.10+0.20\,s^{1/2}$	$t=0.10+0.20\,s^{1/2}$
1914	Sejourne	$t=0.15+0.15\,s^{1/2}$	-

$s=$span; $R=$radius of the circle passing through the crown and intrados springing; $\rho=$curvature radius

instead of an elastic structure. The effects of hinges on the collapse load of masonry arches were analyzed, suggesting that the possible hinge point locations should be identified. Then the forces and stresses in the indeterminate structure should be calculated. Next, the locations of the hinges should be adjusted based on the calculations and the process continued iteratively until the location of the hinges stabilizes (Livesley, 1978).

The thrust analysis is based on lower-bound plastic analysis: Heyman recommended determining the smallest possible arch thickness in which the thrust line with assumed hinge locations would fit. That thickness was then compared to the actual arch thickness. He considered the ratio of the two values to be the geometric factor of safety for the arch. From his work, he developed what he called the "quick analysis" method. It is based on an arch with inputs of dimensionless parameters and a point load P. The equation is based on a failure occurring with hinges at each of the springings, under the live load, and at the crown.

2.5   FEM analysis

Three-dimensional (3D) nonlinear FEM analysis offers the opportunity to model the entire structures, checking for structural behavior of a well-defined and precise structure. All loads affecting the structure could be modeled.

3   Analysis, repair, and strengthening

Bridge analysis, repair options, and strengthening for existing bridges are reported in detail in this section.

3.1   Material modeling

Masonry is defined as a structural material made by the assemblage of natural (stones) or artificial (bricks) elements, with or without mortar, suitable for the realization of the bearing elements of a construction. The difficulty of modeling masonry depends on the following factors:

- Masonry is a discrete material (blocks and mortar) in which the dimension of the single constituting element is large compared to the dimensions of the structural element.
- The geometry, origin, and placing of the blocks can vary considerably.
- Blocks are stiffer than mortar.
- The mortar thickness is limited (compared to the block dimensions).
- Stiffness of the vertical joints is remarkably smaller than the one of the horizontal joints.

The physical-chemical and mechanical parameters in the interaction between the stone units and the mortar joints depend on the factors described next.

Properties of the stone elements, such as:

- Compression and tension strength with monoaxial and pluriaxial stresses

- Elasticity module, Poisson coefficient, ductility, and creep
- Waterproof and superficial (roughness) characteristics
- Chemical agent resistance
- Volume variation for humidity, temperature, and chemical reaction
- Weight, shape, and dimension of the holes

Properties of the mortar, such as:

- Compression strength and behavior under pluriaxial stresses
- Elasticity module, Poisson coefficient, ductility, creep, and adhesive force
- Workmanship, plasticity, and capacity of detaining water

Construction formality, such as:

- Geometry and placing of the stone elements
- Filling of the joints at the head
- Ratio of the joint thickness and dimensions of the stone elements
- Hand-made construction and consequent lack of uniformity of the layers

Actually, if some monoaxial tests are carried out separately on the constituting masonry elements (mortar and blocks), the typical qualitative behavior shows good compression strength and very poor tensile strength. But while the stone has a nearly linear behavior, larger elastic module, and brittle failure, the mortar shows a nonlinear behavior, larger elastic module and certain ductility.

Depending on the desired level of accuracy and simplicity, the following methods could be used:

- *Detailed micromodeling:* The block and the mortar in the joints are represented by continuum models, while the interface unit-mortar is represented by discontinuous elements. The Young model, the Poisson coefficient, and the inelastic properties of the units and the mortar are taken into account.
- *Simplified micromodeling:* The blocks are represented by continuum elements whereas the behavior of the mortar joints and unit-mortar interface is lumped in discontinuous elements. Poisson coefficient and the inelastic properties of the unit and the mortar are neglected.
- *Macromodeling:* Blocks, mortar, and interface unit-mortar are represented as a continuum homogenization theories have been developed in order to derive the global behavior of masonry from the behavior of the constitutive materials (block and mortar).

This physical-mathematic abstraction (i.e., transforming the reality into a scheme governed by mathematically treatable laws) can appear arbitrary when dealing with masonry. In reality, each material is provided with a microstructure and the assimilation to a continuum implies an operation of stress average on a suitable reference volume. The masonry material, realized through the assemblage of two components, shows a constitutive bond characterized by a nonlinear law and intermediate compression strength to each single component. The limit of the linear behavior coincides with the beginning of the partialization of the cross section. Therefore, micromodeling is necessary to better understand the local behavior of masonry structures; macromodeling is applicable when the structure is composed of walls of sufficient dimensions so that the stresses along the length of the element are uniform. This type of modeling is preferable when accuracy and efficiency are both required. The other two important aspects related to the material in the analysis and behavior of masonry are the size effect (unit size versus structural size) and the influence of the material parameters on the numerical analysis.

3.2   Structural modeling

Another complex topic in masonry is the choice of a suitable model representing the structure. According to the hypothesis of homogeneous material, the following model types can be distinguished:

- *With lumped masses:* This is a rough approximation of the geometry of the structure, but it can be sufficient in order to determinate the structural dynamic response (if the non-linearity of the material and the resultants effects of the real geometry of the structure are included). Obviously, this type of model cannot be used to predict the local or global collapse mechanisms or the damage levels of the single structural components.
- *With beams and columns:* This defines in greater detail the behavior of the system than the previous item. It is possible to determine the sequential formation of the collapse mechanisms both statically and dynamically.
- *Macro elements:* This considers the structure as a whole of wall panels, each of which is a recognizable and complete part of the building. It can also coincide with an identifiable part of it in architectonical and functional terms (for example: the façade, the apse, or the chapels); usually, it is formed by more panels and horizontal elements connected to each other so that they represent a unitary constructive part, even if it is joined and not independent from the whole of the construction.

Concerning the FEM element types, models that can be distinguished according to the following details:

2D or 3D alternatives: two-dimensional (2D) or three-dimensional approaches are available, adopting one-dimensional frame, two dimensional shell, or three dimensional brick elements or a combination thereof. Shell elements produce faster and more controllable models because of the presence of a smaller number of joints compared to the brick elements. On the contrary, the model with brick elements allows the visualization of the stress evolution inside the structure. Notwithstanding, the results gained in the two analysis types are similar, both in terms of structural strains and stress distribution.

Meshing: by an increasing of elementary elements, the result reliability is strongly influenced by convergence problem solution; therefore, using a dense mesh could not be the better option. The most appropriate mesh dimension derives from the engineering judgement, taking into account also the dimension of the investigated structure, and the expected results sophistication.

3.3   Damage classification in masonry bridges

This section presents some details on masonry bridge damage classification. One clear and innovative way was described in Sustainable bridges (2007). In particular, the general bridge structure damage classification is reported in Figure 12.8; the

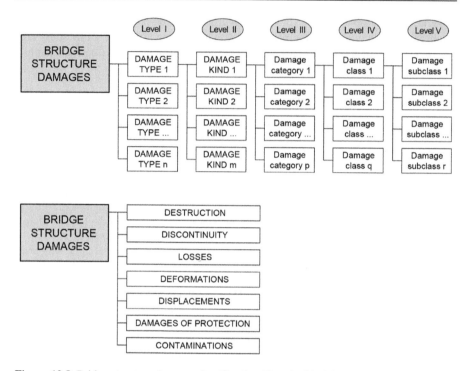

Figure 12.8 Bridge structure damage classification (Sustainable bridges, 2007).

classification according to the damage localization is depicted in Figure 12.9; the classification for damage discontinuity is illustrated in Figure 12.10; the classification for losses type is highlighted in Figure 12.11; the classification according to the deformation type is reported in Figure 12.12; the classification according to the displacement type is reported in Figure 12.13; finally, Figures 12.14 and 12.15 deals with the description of the overall damage causes, and with the contaminations type respectively.

3.4 Common damages in masonry arch bridges

Structural defects normally fall into the following categories:

- Construction
- Long-term loading
- Transient loading
- Environmental

A combination of all of the above types of defect can usually be found in existing masonry bridges. Modern traffic loads, heavier than those in the past, could induce an older bridge to exhibit serious problems, while well-maintained masonry arches not subjected to heavy loads are probably among the most durable constructions.

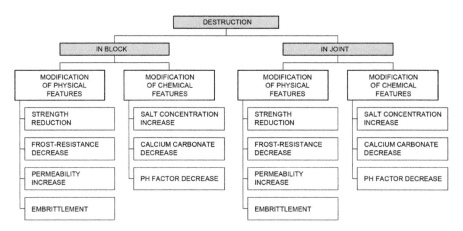

Figure 12.9 Classification according to the damage localization (Sustainable bridges, 2007).

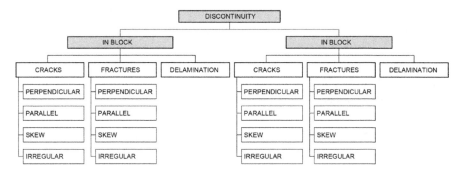

Figure 12.10 Classification for damage discontinuity (Sustainable bridges, 2007).

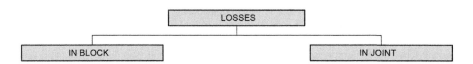

Figure 12.11 Classification for losses type (Sustainable bridges, 2007).

3.4.1 Scour of foundations

One of the most common causes of collapse for masonry arch bridges is scour of foundations, expecially for shallow foundations , being more sensitive than deep foundations. However, this damage type is hidden in the river bed, some element make the damage worse, such as: an increase in flow speed in the river (e.g., for environmental causes) and a local disturbance of the flow due to the design of the piers. Scour problems can be avoided by adding deep foundations linked to the existing structure.

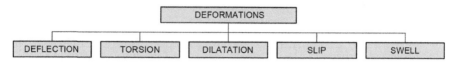

Figure 12.12 Classification for damage deformations type (Sustainable bridges, 2007).

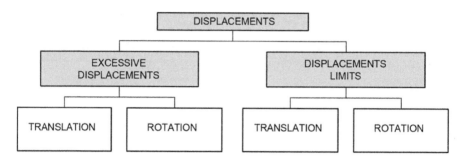

Figure 12.13 Classification according to the displacement type (Sustainable bridges, 2007).

3.4.2   Arch ring issues

The arch ring of a masonry arch can be affected by a wide variety of elements, including the following:

- *Splitting beneath the spandrel walls:* Spandrel walls are employed in order to stiff the arch ring at its edges; the typical failure here is cracks induced by shear stresses in the ring for traffic loads.
- *Abutment movements:* Foundation lack of capacity to substain dead and live loads, is the principal cause of this defect, and they also can produce hinge cracks that need to be repaired; the presence of three well-defined hinges in an arch may allow it to articulate under service loads, resulting in loss of mortar.
- *Spandrel walls:* These walls are the masonry component of arch bridges most exposed to the environmental cyclic action; as a result, lacks of units, or local small rotations could be found.
- *Filling material:* An accurate waterproofing or an efficient drainage system could avoid long-term water saturation of infill material; if not provided, care should be taken to prevent water stagnation, as it could also increase lateral pressure on the spandrel walls.
- *Natural stone:* Stone masonry was largely adopted by the Romans, and structures built in this way have lasted for thousands of years, reaching medium-span size (i.e., about 150 m); although this material is no longer used, these historic bridges stand as landmarks and probably represent the longest-lasting (if most expensive) construction solution.
- *Salt crystallization:* White efflorescence is often the visible part of this defect; it could be concentrated on the top layers or deeply in the masonry, inducing massive decay in this latter case; water or sand brushing solves the problem only for superficial contaminations.
- *Air pollutants:* Especially in urban areas, or in the industrial and marine environment, air pollutants can lead to superficial color changes; and sometimes (in rare cases) damages are enclosed in the masonry.

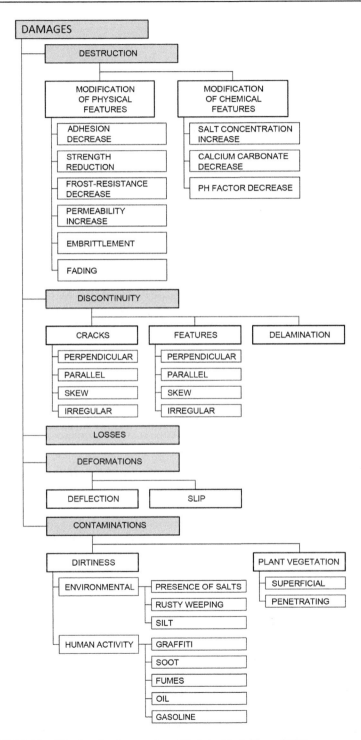

Figure 12.14 Classification for damage type (Sustainable bridges, 2007).

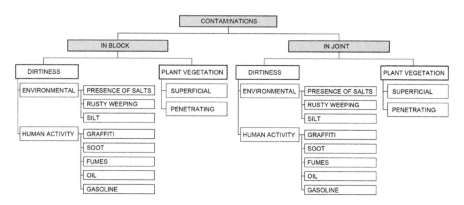

Figure 12.15 Classification of damage according to the contaminations. (Sustainable bridges, 2007).

- *Freezing/thawing:* If it freezes, wet stone can flake, break off, and wash away again when the ice melts. Cycles of freezing/thawing could change completely the structure of bridge components; unit replacement, considering also the best replacing material and mortar, is the solution, together with superficial treatment of the whole masonry.
- *Plant growth:* These are usual "inhabitants" of masonries, however, a short period presence could not influence the structural behavior, a long-term presence could shorten the life of the structure itself.
- *Load traffic:* Increasing loads on masonry arches may or may not be a structural issue; external signs like visible cracks in key positions such as in the spandrel walls or beneath the arch ring should provide warnings that the bridge should be assessed for its ability to handle actual traffic conditions, and eventually, retrofits should be instituted.

3.5 Structural intervention techniques for masonry arch bridges

3.5.1 Identification of defects

Visual inspection is considered to be sufficient as a first step when analyzing existing arches. The presence of a crack or settlement in parapets could be a sign of the abutment movement, and longitudinal cracks in the arch barrel could indicate spandrel wall detachment. However, some defects can be discovered only with nondestructive testing (NDT).

3.5.2 Structural intervention

Pressure pointing and grouting

Even if it is considered a possible way to reduce voids, fill cracks, and improve the condition of the arch, grouting of the contained ground above and behind the arch should be carefully evaluated, as the distribution of the grouted mass could change the structural behavior of the whole arch in a negative way.

Tie bars

The use of tie bars, a traditional technique widely used in masonry buildings, should be carefully applied with masonry arch bridges: e.g., passing a bar through the full section of the arch to restrain spandrel movements could lead to cracked regions in the nearest of the end plates of the bars.

Rebuilding spandrel/wing walls

The simplest solution if the roadway could be closed: the excavation behind the wall and refilling, incorporating a reinforced earth system, avoiding excessive pressure against the spandrel walls.

Saddling

Saddling is a common repair technique in which the fill is removed so that the top surface of the arch barrel is exposed. An r.c. saddle is then put in place over the original barrel. Saddles are typically 150–200 mm thick and made of relatively weak concrete. The nature of the bridge and its behavior is changed, so lack of stresses within the arch could be evident.

Concrete slabs

A concrete slab realized on the existing deck is able to reduce local loadings. It can be helpful in reducing drainage problems and easing lateral pressure on walls.

Underpinning

Underpinning includes installing a new foundation for the bridge, excavating material from beneath the foundations, and replacing it with concrete beams or slabs. Using deep foundations as piles could enhance the behavior of the structure, providing a safer ground interface.

Partial reconstruction

When arch ring damage is extensive, the only real resource is to rebuild either the entire structure or a part of it.

Maintenance

Routine maintenance consists of the following:

- Keeping the road surface maintained, checking that the waterproofing is in good condition, and minimizing dynamic loading from traffic due to overloading and fast braking/accelerating movements
- Removing vegetation on the structure
- Repairing of lateral guardrails
- Repairing deteriorated mortar regions

These four areas of maintenance involve modest expense compared with what may have to be done to fix problems resulting from neglect.

4   Structural assessment and retrofit

Masonry bridge assessment and retrofit applied in the field are commonly the most important lessons to deep the bridge engineering practice. Two case studies, are presented in the following sections, the first dealing with structural assessment, the second with a bridge retrofit.

4.1   Structural assessment: case study

The case study discussed herein has been described in detail in Sustainable bridges (2007). It concerns a bridge located in Poland, about 30 km from the city of Wrocław. The bridge, built in 1875, is a masonry arch structure with spandrel walls. The basic geometric dimensions are presented in Figure 12.16. The arch is barrel-shaped, and the plan shape is rectangular; the span horizontal clearance is 9.93 m, the width 8.55 m, the vertical clearance 5.84 m, and the arch radius 4.97 m. The constituent material is brick, backfill material is unknown, and the brick dimensions is $6.5 \times 12 \times 25$ cm; the joint thickness is $1 \div 1.5$ cm, and the brick strength and joint strength are unknown. The structure experiences local rail with very low traffic; the available formal documentation about the bridge were an inventory card (1965) and sketch drawing (1953).

The bridge has defects typical of masonry structures, including an increase in salt concentration, deterioration and loss of material, and longitudinal cracks. Loss of bricks and joints on both spandrel walls of the bridge was filled with concrete and new bricks (Figures 12.17, 12.18). The displacement measurements were carried out by means of three independent systems:

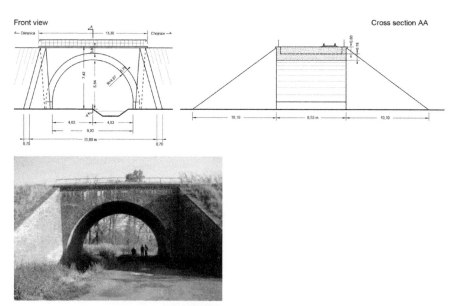

Figure 12.16 Side view, cross section, and photo of the case study (Sustainable Bridges, 2007).

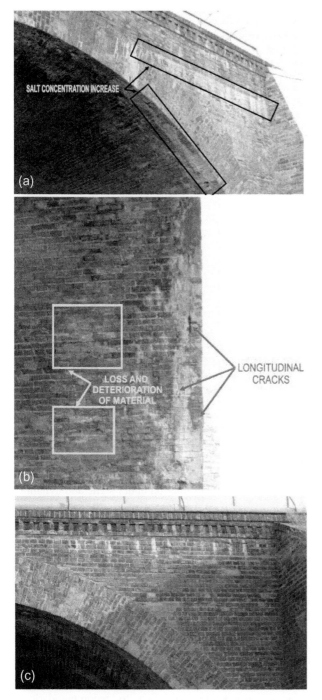

Figure 12.17 (a) Salt concentration increase; (b) material deterioration, loss of material and cracks; (c) filled losses of masonry (adapted from Sustainable bridges 2007).

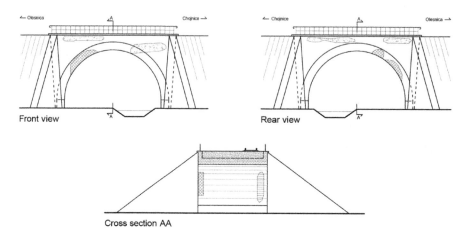

Figure 12.18 Damage localization (adapted from Sustainable bridges 2007).

- *Laser measurements* below the axis of the track in the middle of the span (L1)
- *Microradar measurements* from two different radar positions in five points of the middle cross sections (R1–R5) and in two points in quarter-point sections (R6, R7)
- 3. *Linear variable differential transformer (LVDT) measurements* in three points in the middle of the span (D1–D3)

In addition, accelerations of selected points (A1–A4) were monitored.

The configuration of the measurement points for all measuring systems are shown in Figure 12.19.

For the load tests, the Polish railway provided one two-boggie engine with three axles in each bogie, with axle loads equal to 200 kN.

The aim of the test was the measurement of deformation under static and dynamic loads

- *Laser displacement measurements:* The reaction of the middle of the arch was much higher than the reaction of the quarter points. So the backfill and the ballast distribute the load very well. During this testing session, only velocities were measured, so only the relative load distribution was estimated.
- *Microradar displacement measurements:* Two different positions of the radar were applied: A and B. For displacement measurements of points R1–R5 (Figure 12.19) located along the transverse profile, the lateral position (A) of the radar against the bridge was chosen. For displacement measurements of the points R6 and R7 located along the longitudinal profile under the track axis, the radar was located under the bridge (B).
- *LVDT displacement measurements:* LVDT gauges were located in points D1–D3 (Figure 12.19) along the transverse profile a half-meter from the crown cross section.

The radar measurements were carried out according to recommendations included in Sustainable bridges (2007). The aim of the test was to measure masonry

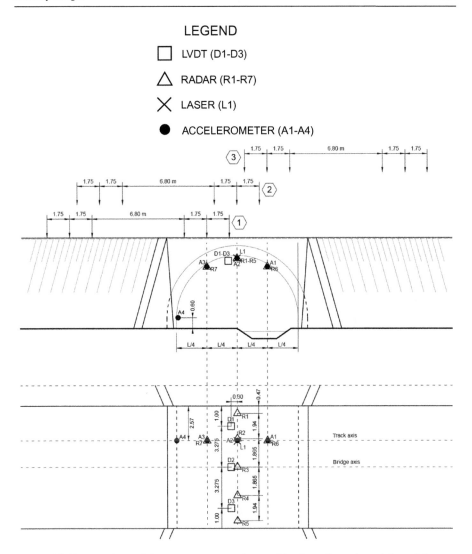

Figure 12.19 Locations of the measurement points and load configurations (adapted from Sustainable bridges 2007).

elements' thicknesses (arch barrel, abutment, and wing walls), detect voids or structural anomalies in masonry elements and backfill, and evaluate moisture or water content. Most results are given in form of radar-grams representing profiles of the structural elements perpendicular to their surfaces (e.g. Figure 12.20).

Darker areas of the radar-grams indicate anomalies in material that can include wet areas, boundaries between masonry and backfill or ballast, and brick layer bond,

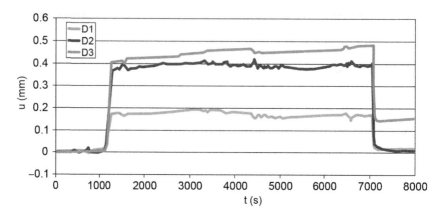

Figure 12.20 Exemplary results for LVDT measurements—displacements of points D1–D3 for loading in 1/4 of the span.

with or without cracks. Radar antennas of different frequencies (with different penetration depths) have been used to estimate the thickness of the walls. Because of high attenuation in the inner masonry structure, the measurements have not produced satisfying results for thickness estimation.

Radar measurements have been successfully applied to investigate the moisture distribution in the masonry. These results have been verified by coring and through geoelectrical measurement.

Crossing the brick layers at the vertical profile, time undulation of the reflection bands of approximately 0.5 ns correspond well with the material changes of brick and mortar between the brick layers. The general structure of the brickwork based on brick layers of different orientation of the bricks (stretcher and header course) is already visible at the small time variation of the reflected signal on the surface. These changes between bricks and mortar are less visible at the horizontal profiles, probably caused by a smearing effect of wave propagating to the depth by the antenna movement along the brick layers.

As a result, it could be observed that the concrete cover for the purpose of draining the arch is visible from the top of the bridge with 500 MHz and additionally with the 900 MHz under the ceiling of the arch, and a wall thickness of the abutment of approximately 2 m is expected to be derived from the end of the concrete cover reflection. Other details are reported in Sustainable bridges (2007).

Another test performed was the electrical conductivity tests, with the aim of detecting voids, moisture/water content, testing/comparison of the NDT technique.

4.3 Structural intervention: case study

The report on the following case study has been synthetized from Paeglītis and Paeglītis (2000). The bridge over the Venta River was built in 1874 (Figures 12.21–12.25), spanning 164 m with 17 arches. The material used for the reconstruction

Figure 12.21 Bridge over Venta River in Kuldiga, built in 1874, after the recent restoration.

Figure 12.22 Bridge over Venta River in Kuldiga, built in 1874: on site inspection.

was chosen in accordance with tests developed at the Riga Technical University labs. The bridge consists of two parts—the 133-year-old initial part and the 81-year-old restored part. Several deteriorations were uncovered in 2006, so a general intervention was chosen as the response. In particular, masonry units had crumbled away in piers, corrosion of surfaces and joint material decay were diffused in the structure, and in the deck, the water drainage was not working due to damage, leading also to water filtration in bricks. Lime-based mortar bricks resulted to be the best construction components, allowing the water to migrate in masonry. On the contrary, cement-based mortar for joints with soft bricks revealed a lot of water inside the brick, that was the principle cause of damages due to the freezing/thawing cycles.

The Stone Conservation and Restoration Center of Riga Technical University was asked to research bricks and mortar: all bricks analyzed, and coming from the existing construction, presented a high level of porosity, while the tests on the existing r.c. components, has found to be of considerably lower quality. Chemical tests of salt content in bricks, r.c., dolomite stone, and the old dolomite grout indicated very low levels of salt soluble in water. The pH level in r.c. that was found variable between 8.5 and 9.3, depending on the position of the

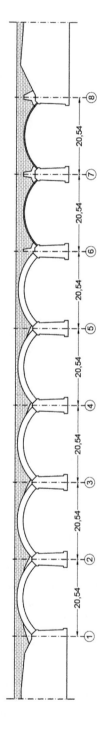

Figure 12.23 Longitudinal view of the bridge over Venta River in Kuldiga, built in 1874 (adapted from Paeglitis and Paeglitis 2000).

Figure 12.24 Retrofit yard of the bridge over Venta River (adapted from Paeglītis and Paeglītis 2000): view of the arches reinforcement.

Figure 12.25 Retrofit yard of the bridge over Venta River (adapted from Paeglītis and Paeglītis 2000): waterproofing of the arches.

investigation. After the first assessment phase, supported by on-site investigation on the costituent materials, the retrofit yard started: In the first phase, the road surface was dismantled and the filling of arches taken out. The bridge was found to be in good condition, and the r.c. surface did not revealed noticeable damage or cracks. As the traffic load capacity was not good enough, a steel framework for arches 6 and 7 was built. The existing waterproofing and protection layers were restored, and the arches were finally filled using draining soil. A new r.c. slab resting on the filling was built to redistribute loadings and to prevent waterproofing premature damages. All structural elements were finally restored with an external layer washing.

References

Barlow, W.H., 1846. On Arches. Min. Proc. ICE 5, London, 439–474.

Benouville, L., 1891. Étude sur la Cathédrale de Beauvais. In: De Badout, M.A. (Ed.), In: Encyclopédie d´Architecture, 40 serie, vol, 3. pp. 52-54, Librairies imprimeries réunies, Paris.

Castigliano, A., 1879. Théorie de l'équilibre des systèmes élastiques et ses applications1-2, Front Cover. Alberto Castigliano. Auguste Frederic Negro Editeur, Turin.

Corradi, M., 1998. Empirical methods for the construction of masonry arch bridges in the 19th century. In: Sinopoli, A. (Ed.), Arch Bridges: History, Analysis, Assessment, Maintenance, and Repair. Balkema, Rotterdam.

Couplet, P., 1729. De la poussee des voutes. In: Histoire de l'Academie Royale des Sciences. p. 79.

Couplet, P., 1730. De la poussee des voutes. In: Histoire de l'Academie Royale des Sciences. p. 117.

Department of Transport, 2001. The assessment of highway bridges and structures, design manual BD 21/01. In: Design Manual for Roads and Bridges, vol. 3, Section 4. Department of Transport Highways Agency, London.

Fuller, G., 1875. Curve of equilibrium for a rigid arch under vertical forces. Proc. I.C.E. 40.

Gregory, D., 1697. Catenaria. Philosophical Transactions, No. 231, p. 637.

Heyman, J., 1969. The safety of masonry arches. Int. J. Mech. Sci 11, 363–385.

Heyman, J., 1972. Coulomb's Memoir on Statics: An Essay in the History of Civil Engineering. Cambridge University Press, Cambridge.

Heyman, J., 1876. Copulet's Engineering Memoirs, 1726-33. In: Hall, A.R., Smith, N. (Eds.), History of Technology. Mansell, London.

Heyman, J., 1980. The estimation of the strength of masonry arches. Proc. Inst. Civ. Eng. 69, 921–937, Part 2.

Heyman, J., 1982. The Masonry Arch. Ellis Horwood, Chichester.

Heyman, J., 1997. The Stone Skeleton, Structural Engineering of Masonry Architecture. Cambridge University Press, 172 p, ISBN 9780521629638. Cambridge.

Heyman, J., 1982. The Masonry Arch. Ellis Horwood, Chichester, UK.

Heyman, J., 1996. Arches, Vaults, and Buttresses: Masonry Structures, and Their Engineering, 418 pp., Collected studies CS546 series. Variorum Edition, Aldershot.

La Hire, Philippe de. Traité de mécanique, on explique tout ce qui est nécessaire dans la pratique des Arts, et les propriétés des corps pesants lesquelles ont eu plus grand usage dans la Physique. Paris: imprimerie Royale.

La Hire, P. de, 1712. Sur la eonstruction des voûtes dans les édifiees. Mémoires de l'Académie Royale des Sciences, Année. 1712, Paris 1731, pp. 69–77.

Livesley, R.K.A., 1978. Limit analysis of structures formed from rigid blocks. Intl. J Num. Meth. Eng. 12, 1853–1871.

Navier, M., 1833. Résumé des leçons de mécanique données à l'École polytechnique. Carilian-Goeury et V. Dalmont, Paris 1841, p. 491.

Oliveira, D.V., Lourenco, P.B., Lemos, C., 2010. Geometric issues and ultimate load capacity of masonry arch bridges from the northwest Iberian Peninsula. Eng. Struct. 32 (12), 3955–3965.

Paeglitis, A., Paeglitis, A., 2000. Restoration of masonry arch bridge over Venta River in Kuldiga. In: Proceedings of the XXVII Baltic Road Conference, Riga, Latvia.

Pippard, A.J.S., 1948. The approximate estimation of safe loads on masonry bridges. In: Civil Engineer in War, Vol 1. ICE-institution of Civil Engineers, London, pp. 365–372.

Pippard, A.J.S., Chitty, L., 1951. Study of the Voussoir Arch, National Building Studies. Research Paper 11, HMSO, London.

Rankine, W.J.M., 1898. A manual for civil engineering. Charles Griffin & Co London, article 295, 429–432.

Sustainable bridges, 2007. Possibilities of unification of bridge condition evaluation Background document SB3.3. Sustainable Bridges—Assessment for Future Traffic Demands and Longer Lives. Research Programme Developed Under the 6th Framework EU Programme— Priority 6 Sustainable Development Global Change and Ecosystems Integrated Project. Contract number: TIP3-CT-2003-001653.

Tellet, J., 1983. A review of the literature on brickwork arches. In: Proc 8th int. Symp. on Load Bearing Brickwork, Tech Sec. 2-Structures, Building Materials Section. Britisch Ceramic Society, London.

Villarceau, Y., 1854. On the arch bridge construction. Academie des Sciences of Paris, Imprimerie Imperiale, Paris, p. 325.

Whitey, K., 1982. Assessment of the Masonry Arch Bridge Work Pap. WP/B/28/82 Transport and Road Research Laboratory, Crowthorne, Berks (unpublished).

Section V

Bridge design based on geometry

Arch bridges

Schanack F.[1], Ramos O.R.[2]
[1]Universidad Austral de Chile, Valdivia, Chile
[2]University of Cantabria, Santander, Spain

13

1 Introduction

Arch bridges are generally classified into three types: (i) deck arch bridges (with arches below the deck; Figure 13.1), (ii) through arch bridges (those with arches above the deck, generally tied arches; Figure 13.2), and (iii) half-through arch bridges (where parts of the arches are below and others above the deck; Figure 13.3).

Since the last one is a combination of the former two, this chapter only discusses deck arch bridges and tied arch bridges.

Figure 13.1 Deck arch bridge.

Figure 13.2 Through arch bridge.

Figure 13.3 Half-through arch bridge.

Ideally, an arch is only subject to compressive forces, free of any bending moments and shear forces. In such a case, the arch is extremely efficient because every part of its cross sections is subject to the same stress; it is also very economical because it can be made of cheap materials that resist great compressive forces, such as concrete. In order to obtain an ideal arch free of bending moments, the arch axis must coincide with the line of thrust caused by all loads acting on it.

Innovative Bridge Design Handbook. http://dx.doi.org/10.1016/B978-0-12-800058-8.00013-X

Nevertheless, in arch bridges, there is no single line of thrust, and sometimes there is more than one arch axis. Moving loads on the bridge cause changes in the line of thrust. Time-dependent material behaviors, such as creep, shrinkage, and elastic deformations under load and thermal expansion, cause the arch shape to change. A geometrical difference between the arch axis and the line of thrust causes a bending moment that is proportional to the arch compression force; therefore, it can become important quickly. And even if the arch shape would ideally fit the line of thrust, care must be taken to prevent arch buckling.

Consequently, the structural difficulty in the design of all arch bridges is, on the one hand, the minimization of the misalignment of the arch axis and the line of thrust, and on the other hand, the provision of sufficient bending and buckling resistance.

Another important issue regarding arch bridges is how to carry loads to the ground—not only the vertical loads but, especially, the thrust of the arch. Deck arches require a very competent foundation (generally made of rock, with a high bearing capacity) to support the high vertical and horizontal forces transmitted by the arch. This is the natural concept of an arch bridge, also known as a *true arch*. However, most of the through and half-through arches are tied arches; that is, the horizontal thrust of the arch is tied by the deck, so the foundation of the arches only need to support vertical reactions.

2 Deck arch bridges

2.1 Types and classification

In order to design a deck arch bridge, two requirements must be met: (i) enough clearance under the deck to place the arch and (ii) foundation materials with competent bearing capacity to resist the thrust. This type of bridge can be classified according to various criteria, such as arch form, the material used to build the arch, the relation between the arch and the deck, the restraint conditions, and the arch cross section.

Regarding the arch form, the following types can be distinguished:

- *Rounded or semicircular arch.* This form was extensively used by the ancient Romans. There are superb examples of arch bridges built by the Romans using the rounded arch: the Pont du Gard in the south of France, the Segovia Aqueduct in Spain, and probably the most magnificent Roman bridge of all, the Alcántara Bridge, also in Spain.
- *The pointed arch.* The use of this kind of arch started its development in the Middle Ages, with the expansion of the gothic style. A concentrated load on top of the key of this arch type is needed to maintain its structural efficiency. Nowadays, this form is relatively commonly used in bridges for high-speed railways (HSRs). Several examples can be found in Germany and Spain (like Deza bridge or Rombachtal bridge).
- *Low-rise arch.* A circular arc is used in this style of bridge instead of a complete semicircle. The 18th-century French engineer Jean Perronet built many arch bridges of this kind (like the first Pont de la Concorde, for example). Nowadays, this type is frequently used.
- *Parabolic arch.* This form is structurally very efficient, given that the parabola is the inverse funicular curve for a uniform load.

- *Several circular arcs.* This shape is an attempt to approximate the parabolic arch by chaining circular segments turning into a more easily constructible geometry.
- *Polygonal arch.* When there are few columns connecting the arch and the deck, the load is not uniformly distributed enough, so a polygonal form is a better approximation of the inverse funicular curve.

The following kinds of material can be used to make arches:

- *Stone arches.* All arch bridges built before the 19th century were made of stone. Since that time, this material has been replaced by steel and concrete. Nevertheless, stone arches are still constructed sometimes in China, such as the New Danhe Bridge, currently the stone arch bridge with the longest span (Feng, 2010). Featuring a span of 146 m and a deck width of 24.8 m, it opened in 2000.
- *Concrete arches.* The first concrete arches were built in the late 19th century. These days, concrete is the material used most often to construct arches. The current world record for the span of a concrete deck arch bridge is held by the Wanxian Yangtze River Bridge in China. Opened in 1997, this bridge has a 420-m-long span. The 390-m-span KRK-1 Bridge in Croatia is also impressive. When the 445-m-span Beipanjiang Railway Bridge in China is completed (scheduled for 2016), it will become the new world record holder in this category.
- *Metal arches.* The first cast-iron arches were built in the early 19th century. Soon, however, the cast iron was replaced by steel, which became the most used material for arches until the mid-20th century. Today, many arches are still built with steel, but not as many as those made of concrete. The current world record of steel deck arch is the New River Gorge Bridge, when opened in 1977 in West Virginia, with a 518-m-long span.
- *Concrete-filled steel tubular (CFST) arches.* This is a relatively recent type of deck arch. The current world record for CFST arches is the Zhijinghe River Bridge in the Three Gorges area of China, built in 2009 with a 430-m span.

Regarding the relation between the deck and the arch, two types can be distinguished:

- *Close-spandrel arches.* This type of arch was used in ancient stone arch bridges.
- *Open-spandrel arches.* Modern concrete and steel deck arch bridges belong in this category. The connection between the deck and the arch is made by means of piers.

Regarding the restraint conditions of the arch, there are three types:

- *Fixed arch.* This is the type most often used nowadays. It is statically indeterminate, and due to its fixed condition, it is subject to internal stress due to thermal or time-dependent actions.
- *Two-hinged arch.* It has pinned connections at both springings. It is also statically indeterminate, and its supports are free of stress due to thermal or time-dependent actions.
- *Three-hinged arch.* This type has an additional third hinge at the crown, so it is statically determinate. It is, therefore, free of stress due to thermal or time-dependent actions, but, on the other hand, it has the largest deflections of the three types.

Finally, regarding the arch cross section, the following is true:

- In concrete arches, there are solid cross sections or cell cross sections, which can be with a single or several cells.
- In steel arches, the cross section can be a truss, a solid web girder, or a tube filled with concrete.

2.2   Review of recent innovative deck arch bridges

2.2.1   Steel truss arches

In 2009, the Daninghe Bridge over the Daning River in Wuxia, China, opened to traffic (Figure 13.4). This steel truss deck arch spans 400 m by means of a 10-m constant deep truss. It has three arch ribs. The top and bottom chords of each rib are of a steel box section (1.5 m deep and 1.0 m wide). The spandrel columns are also steel frame structures corresponding to the three arch ribs. The 24.5m-wide deck is supported by 1.7-m-deep continuous steel-concrete composite girders with a span length of 27 m. The chosen erection procedure for this bridge was the cable-stayed cantilever method.

Figure 13.4 Daninghe Bridge, Wuxia, China.

2.2.2   CFST arches

CFST arch bridges have been increasingly used since the 1990s, mostly in China. This type of bridge combines the advantages of concrete and steel arch bridges, so it has become a good alternative for deck arch bridges spanning 250 m and longer.

CFST members have greater strength than concrete ones and greater stiffness than steel tubes used by themselves. Steel tubes give confinement to the concrete core, increasing its compression capacity; on the other hand, concrete allows the steel tubes to resist local buckling. In addition, CFST arches have advantages from the point of view of the erection procedure: namely, CFST members have less self-weight than concrete ones, so they can be put up easily. Then the concrete filling of the tubes can be made without falsework or formwork.

One of the most outstanding bridges of this type is the Zhijinghe Bridge, built in 2009 over the expressway from Yichang to Enshi, Hubei, China (Figure 13.5). It consists of a 430-m-long span CFST deck arch with a rise of 78 m and a rise-to-span ratio of 1:5.5. The arch has two CFST ribs, 4 m wide each, with a distance between them of 13 m. The depth of the arch varies from 13 m at springing to 6.5 m at crown. The spandrel columns are made of reinforced concrete (r.c.). The deck is supported by continuous, prestressed concrete girders spanning 21 m. The arch was constructed using the cantilever method.

Figure 13.5 Zhijinghe Bridge, Hubei, China.

2.2.3 HPC and UHPC concrete arches

Arches are subjected to very high compression forces. Due to that fact, high-performance concrete (HPC), or even ultra-high performance concrete (UHPC), is highly suitable for concrete deck arch bridges. The use of HPC or UHPC allows smaller cross sections for the arch, so lighter self-weight structures can be achieved, easing the erection procedure and lessening the thrust at springing.

HPC is commonly used in big deck arch bridges. Los Tilos Bridge, built in 2004 in the Canary Islands, Spain (Figure 13.6), is a 255-m-span deck arch bridge erected by the cantilever truss method. Both the arch and the spandrel columns are made of 75 MPa concrete.

Figure 13.6 Los Tilos Bridge, in the Canary Islands.

The Contreras Reservoir Bridge, built in 2010 in Cuestas de Contreras, Spain (Figure 13.7), is an HSR bridge. It consists of a 70-MPa concrete arch, spanning 261 m and erected by the cable-stayed cantilever method.

Figure 13.7 Contreras Reservoir Bridge, Spain.

Another example of this type of bridge is the Colorado River Bridge, constructed in 2010 on the Hoover Dam Bypass, on the border of Arizona and Nevada in the United States (Figure 13.8). This bridge has a 70-MPa deck arch that spans 323 m, and it was built by the cable-stayed cantilever method.

Figure 13.8 Colorado River Bridge, on the Hoover Dam Bypass.

Currently, other two HSR bridges are under construction in Spain: the Alcántara Reservoir Bridge (Figure 13.9) and the Almonte River Bridge (Figure 13.10), whose completion is expected in 2015. Both are deck arches erected by the cable-stayed cantilever method. The Alcántara Reservoir Arch spans 324 m using 70 MPa concrete whereas the span of the Almonte River arch reaches 384 m with 80 MPa concrete.

Figure 13.9 Alcántara Reservoir Bridge, in Spain.

Figure 13.10 Almonte River Bridge, in Spain.

To date, there are few examples of deck arch bridges constructed with UHPC, as this is a very new material and the technology associated with it is still in its infancy.

2.2.4 HSR arch bridges

Deck arch bridges for HSR have to face several challenges in comparison with highway bridges. First, the traffic loads (and therefore the dead-to-live load ratios) are higher. This makes it more difficult to design an accurate inverse funicular curve for the arch, and consequently, higher bending moments are expected. The dynamic effect caused by the passage of the trains is another issue that must be handled. In addition, the continuous passage of trains increases fatigue concerns. Finally, the limitations for service deflections and accelerations pose a very pressing problem.

In recent years, some remarkable deck arch bridges for HSR have been constructed or are currently under construction. The three Spanish HPC arch bridges mentioned previously deserve to be highlighted; but apart from these three magnificent examples, another two are currently being constructed in China. The Nanpanjiang Railway

Bridge in Qiubei, Yunnan, China (Figure 13.11), is planned to be completed in 2015; it is a concrete deck arch spanning 416 m. The rise of the arch is 99 m, which leads to a rise-to-span ratio of 1:4.2. The arch rib cross section has three cells within a single box with a constant depth of 8.5 m and a variable width of 28 m at springing and 18 at crown. Double-column framed piers are used to support the deck. CFST are used as a skeleton during cable-stayed cantilever method construction and before the casting of concrete (Pérez-Fadón and Sánchez Ramírez, 2015).

Figure 13.11 Nanpanjiang Railway Bridge, in Yunnan, China.

The Beipanjiang Railway Bridge in Qinlong, Guizhou, China (Figure 13.12), expected to be finished in 2016, is of the same type as the Nanpanjiang Railway Bridge. This bridge will be not only the world's highest railway bridge (at 283 m), but also the world's longest concrete arch ever built (with a 445-m span). The rise of the arch is 100 m, with a rise-to-span ratio of 1:4.45. The arch rib cross section has three cells within a single box with a constant depth of 9 m and a variable width of 28 m at the springing and 18 at the crown. As in the Nanpanjiang Railway Bridge, double-columned framed piers support the deck and CFST is used as a skeleton via the cable-stayed cantilever method.

Figure 13.12 Beipanjiang Railway Bridge, Guizhou, China.

2.2.5 Innovative erection procedures

The use of steel scaffolding as a reinforcement for the concrete arch was developed by the Austrian engineer Josef Melan in the late 19th to early 20th centuries. One of the most outstanding examples of this procedure is the Martin Gil Railway Bridge, built in 1942 in Zamora, Spain (Figure 13.13), a 210-m-span deck arch bridge. This technique was abandoned because it is expensive to use such a large amount of embedded steel scaffolding. However, since the 1990s, this technique has become popular in China by means of the replacement of the steel scaffolding by CFST as embedded scaffolding. The steel tubular truss arch is erected by the cantilever method, and the concrete is then poured into the steel tubular chords. Using CFST as embedded scaffolding makes this construction method very economical and increases the possible span of concrete arch bridges.

Figure 13.13 Martin Gil Railway Bridge, Zamora, Spain.

The longest concrete arch bridge in the world (at least until the completion of the Beipanjiang Railway Bridge, expected in 2016, as mentioned earlier), the Wanxian Yangtze River Bridge (Figure 13.14), was erected by this method. It spans 420 m with a 24-m-wide deck. The arch rib is a three-cell rectangular box that is 7 m deep and 16 m wide.

Figure 13.14 Wanxian Yangtze River Bridge

Another innovative erection procedure for long-span concrete arch bridges is the partial cantilever method. Using this method, two partial half arches are erected by the cantilever method, and the rest (i.e., the crown) is lifted into its final position.

2.3   General design recommendations and common details

In the range of short to medium-length spans, both concrete and steel can be used equally. Based on economic factors, concrete deck arch bridges are more commonly used in this range of span length, taking advantage of the efficient behavior of the concrete under compression forces. For large spans (300 m or longer), steel arches have traditionally been dominant in the construction of arch bridges. In this case, the low self-weight of the steel sections in comparison with the concrete ones represents an important advantage during the erection. Furthermore, truss steel arch bridges, for example, can be fabricated in small, lightweight segments that can be easily transported to remote locations. Composite steel–concrete arches are also an interesting solution to use for medium-length to long spans in order to take advantage of the properties of both materials. CFST technology is a good example of this composite. The use of concrete (reinforced or prestressed) for long-span arch bridges, despite the fact that it has a more complex construction, may sometimes be interesting in order to achieve an adequate global stiffness (in the case of HSR bridges, for example), with a slenderer cross section; in addition, the dead load of the arch in relation to the live load allows the line of thrust to stay close to the arch axis, achieving an optimized shape-behavior.

The rise-to-span parameter controls the behavior of many structural variables. The most common rise-to-span ratio of an arch is between 1:5 and 1:6. Shallow arches (with a rise-to-span ratio close to 1:8 to 1:10) offer an aesthetic appearance, but they also increase the horizontal thrust on the foundations and the compression forces in the arch.

From an academic point of view, there were three ways to carry out the springing line connection: fixed, hinged, and three-hinged (springing line and crown hinges). Historically, there have been more bridges with fixed and two-hinged arches than three-hinged arches (because of their high flexibility). Nowadays, most arches are fixed in order to increase the stiffness of the structure and to reduce the installation and maintenance costs of the hinges. Current finite element method (FEM) techniques allow proper analysis of the response of fixed arches to secondary and thermal effects.

The more closely the shape of the arch axis matches the theoretical line of thrust, the better will be the response of the arch in terms of bending moment reduction. Due to the nature of the loads (permanent and variable), it is not possible to design a perfect inverse funicular arch shape for all cases. In concrete arch bridges, the self-weight of the arch is big enough in compare to the other loads, so it is very common to determine the shape of the arch axis as the inverse funicular curve of dead load (generally a parabolic curve). In steel arch bridges, sometimes the shape of the arch axis matches the thrust line of the dead load plus half of the live load. Regarding this issue, it is interesting to point out the influence of the spacing between the piers on top of the arch. In a classical road bridge, with 8–10 piers on top of the arch, the line of thrust has the classical appearance of a smooth curve (Figure 13.15).

Figure 13.15 La Regenta Bridge, Asturias, Spain.

Sometimes the number of piers is smaller (six or less); this happens, for example, in arch bridges with a span shorter than 100 m and distances between piers of 15–20 m, or in longer arches that have pier distances of about 40–60 m. In such cases, the inverse funicular curve is quite similar to a polygonal line. On some occasions, the designer sacrifices the "structural truth" in order to avoid an inadequate aesthetic perception of the bridge, but in other cases, [for example, if the loads transmitted by the piers are heavier, as occurs in HSR bridges like the Jauto Bridge in Spain (Ramos et al., 2014)], it is necessary to design a nonsmooth shape of the center line of the arch, matching the actual thrust line of the design loads (Figure 13.16).

Figure 13.16 Jauto Bridge, Spain.

An extreme example of this can be studied when the number of piers on top of the arch is only 1. Such is the case of the viaduct over Rio Deza, an HSR arch bridge (Figure 13.17), where the same pattern of the span deck length (75 m) is maintained throughout the whole viaduct. Here, the most appropriate shape is an ogee arch that also serves as the point of fixity for horizontal actions affecting the deck (mainly the braking forces of the train and the friction development into the pot bearing on the piers). The total rise of the arch is 96 m, and the span is 131.50 m long. The shape of the arch is the compromise inverse funicular curve generated by the self-weight of the arch and the load induced by the deck (Pardo de Vera et al., 2011).

Figure 13.17 Deza Bridge in Galicia, Spain.

In concrete arch bridges, the box section is the most typical cross section due to its high efficiency. In the case of decks wider than 10 m, usually bicellular box sections are designed. For arches shorter than 140–160 m, or for very wide decks (more than 20–25 m), sometimes the arch is comprised of two individual box sections (Figure 13.18).

Figure 13.18 Ceceja Bridge, Cantabria. Spain.

In steel arch bridges, the truss section was widely used during the last years of the 19th century and well into the 20th century, such as the New River Gorge Bridge, built in 1977 (Figure 13.19).

During the second half of the 20th century, many solid rib steel arches were also designed, mainly with box sections (Pantaleón and Ramos, 2015). In this case, it is more common to design two parallel braced arches (Figure 13.20), but unique box section solutions also have been constructed (Figure 13.21).

Figure 13.19 New River Gorge Bridge.

Figure 13.20 Rainbow Bridge, Niagara Falls. Canada.

Figure 13.21 Ricobayo Dam Arch Bridge in Zamora, Spain.

In medium-length to long arch spans, most of the solid rib cross sections (both concrete and steel) have variable depths (i.e., larger at springing line and decreasing toward the crown). For roadways, the depth–span ratio at the springing line usually ranges from 1:50 to 1:60, and the depth–span ratio at the crown usually is from 1:60 to 1:90. In the case of truss cross sections, the depth–span ratio is bigger, from 1:25 to 1:50.

2.4 Selected structural problems

Today, FEM gives an adequate understanding of the behavior of arch bridges, including nonlinear phenomena, secondary effects, and local buckling. However, it is very important to design a calculation model that properly represents all bridge elements (arch, foundations, piers, deck, and their connections).

There are multiple issues related to the design of an arch bridge that deserve particular attention, including the following:

- *Influence of the compression shortening of the arch in the inverse funicular shape.* In order to allow the correct behavior of the arch under permanent loads, it is convenient to introduce a horizontal force into the crown of the arch before closing it, counterbalancing its deformation due to the compression shortening.
- *Secondary and thermal effects.* In fixed concrete arches, it is very important to carry out an exhaustive analysis of the secondary effects and thermal effects to evaluate properly the redistribution of axial forces and bending moments along the arch.
- *Nonlinear effects.* Both in-plane and out-of-plane buckling must be controlled. The response of the arch against out-of-plane buckling is highly influenced by the bracing system between the arches. The connection between arch, piers, and deck is important, too.
- *Deflection control under live loads.* Especially in HSR arch bridges, it is very important to control the deflection under nonsymmetric live load cases.
- *Aeroelastic instability.* In large arch bridges, it is convenient to carry out wind tunnel tests in order to dismiss any kind of aeroelastic instability. Some problems have also appeared in arches during the erection stage.
- *Connection details.* There are many structural details in both concrete and steel arches that deserve special analysis, such as the pier–arch connection, the joint weld between members of steel truss arches, the congestion of the reinforcement in the arch springing area, etc.

2.5 Construction methods

Several methods have been developed for the construction of deck arch bridges: the scaffolding method, the cantilever method, the swing method, and the embedded scaffolding method. These are described in the next sections.

2.5.1 Scaffolding method

The scaffolding method was the first erection procedure used for arch construction, and all stone arch bridges were built via that method. For concrete arch bridges, that was the most commonly used method until the mid-20th century.

The scaffolding is usually made of timber or steel. A careful camber design of the scaffolding must be carried out in order to counteract the deflection caused by the self-weight of the arch. In order to save costs, the scaffolding should be easy to assemble and reuse.

Nowadays, this method is competitive only for short span bridges because its cost increases significantly with the span length (Figure 13.22).

Figure 13.22 The scaffolding method in progress on the Jauto Bridge.

2.5.2   Cantilever method

Since the mid-20th century, the cantilever method has been used most often for the construction of deck arch bridges. The method consists of constructing both halves of the arch rib, with each springing to the crown, and then performing the final closure at the crown. As each half arch cannot bear itself, a temporary bearing structure is needed. There are two types of bearing structures: the cable-stayed cantilever and the cantilever truss.

For the cable-stayed cantilever method (or pylon cantilever method), the temporary supporting structures are either pylons, the adjacent piers and stayed cables anchored to the ground, the pier foundations, or adjacent approach bridges (Figure 13.23). This erection procedure has been used for steel, concrete, and CFST deck arch bridges. The lifting equipment for the arch members ranges from cranes on the ribs to cable cranes spanning the whole bridge. The member also can be lifted from the ground or barges on the water surface. For concrete deck arch bridges, a formwork traveler can be used for in-place casting.

The truss cantilever method uses temporary diagonal members between spandrel columns and horizontal cables or the deck as the upper tension chord of the temporary truss cantilever beam (García-Arango et al., 2009). In order to install the arch members, the same procedures as in the cable-stayed cantilever method can be used. The truss cantilever method is most commonly used in the erection of concrete deck arch bridges (Figure 13.24).

Figure 13.23 Cable-stayed cantilever method for the Colorado River Bridge.

Figure 13.24 Truss cantilever method in progress on La Regenta Bridge.

2.5.3 Swing method

The swing method consists of the construction of each half of the arch in a vertical or horizontal position. Then, if the half arch is built vertically, it will be rotated around the springing from a high position to the closure. Figure 13.25 shows the vertical downward swing method in action. On the other hand, if the half arch is built horizontally it will be rotated around the springing from the low position up to the closure; this is the vertical lift-up swing method.

2.5.4 Embedded scaffolding method

The embedded scaffolding method is used for concrete deck arch bridges. It was developed by Austrian engineer Josef Melan in the late 19th century, and then it was improved during the first half of the 20th century. This method involves the erection of a light steel truss that serves as scaffolding for the casting of the concrete. It was abandoned in the 1950s due to the increasing cost of the embedded truss scaffolding as the span length increased.

Figure 13.25 Vertical downward swing method being used on the Deza Bridge.

Since the 1990s, this erection procedure has had a renaissance in China with the replacement of embedded truss scaffolding with embedded CFST scaffolding. Due to the lower cost of the CFST members, the method is competitive again. First, the steel tubular truss arch is erected. Then the tubes are filled with concrete. After that, the concrete cross section of the arch is cast in place, with the CFST arch serving as scaffolding.

2.6 Research

One of the major areas of research concerning deck arch bridges deals with steel-concrete composite arches. As is well known, one of the main problems of long-deck, concrete arch bridges is the increase in self-weight. Due to heavy self-weight, the thrust at the springing is very high, and the erection procedure becomes difficult and expensive. Therefore, concrete deck arch bridges are less competitive when long spans are needed.

Since 2003, a series of studies have been conducted at the Fuzhou University in China in order to reduce the self-weight of concrete arches. Several steel-concrete composite cross sections for arches have been tested. The idea is to keep the upper and bottom r.c. slabs, but to use steel instead of concrete for the web. For the steel web corrugated plates, plain plates and tubular trussed have been investigated. A 30% saving of self-weight has been achieved in comparison with common r.c. cross sections.

3 Tied arch bridges

3.1 History

In the literature, tied arch bridges are also called *bowstring arches, langer beam bridges, Nielsen bridges, Lohse girder, network arches* and so on. The first tied arches were actually truss bridges with a curved upper chord. The first drawing of such a bridge appeared about 1482, in a manuscript of Leonardo Da Vinci (Figure 13.26).

In 1796, American engineer Robert Fulton published a design of a bridge with a type of truss beam that he called "bowstring." His idea is based on an arch where concentrated loads are uniformly distributed by a truss and the lateral abutments are replaced by a tie between the arch ends.

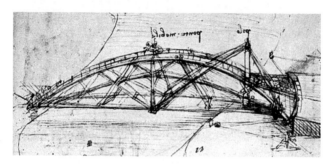

Figure 13.26 Drawing of a tied arch bridge by Leonardo Da Vinci, c.1482.

In the 19th century, many arch bridges with ties between its ends have been built. There are, for example, hundreds of bridges over Erie Canal, in New York, constructed between 1850 and 1870 (Griggs, 2002), which were based on an 1841 patent by American civil engineer Squire Whipple. At the same time, the lenticular girder, also known as *Pauli girder* or *fish-belly girder,* was developed. These girders are essentially trusses with a curved top and a curved bottom chord, but about 1860, engineers began using trusses where only the top chord was curved, so that the bottom chord could handle traffic directly.

In 1872, German engineer Hermann Lohse built the first lenticular girder with only vertical members between the chords. Today, the term *Lohse girder* is sometimes used to refer to tied arches with vertical hangers. In 1883, the Austrian engineer Josef Langer built the first tied arch bridge with vertical hangers. The Langer beam has a steel beam with constant depth, which is stiffened by a top arch.

The idea of tied arches with inclined hangers was returned to and built upon by Octavius F. Nielsen in 1925. Due to necessary simplifications for the analysis, Nielsen arch bridges were limited to inclined hangers that were not intersecting. They had an extraordinary slenderness and reached spans of up to 145 m (Bridge of Castelmoron, France, 1993). In 1955, the norwegian engineer and profesor Per Tveit started working on Nielsen bridges using new analysis methods that allowed him to calculate and develop tied arch bridges with multiple intersecting hangers, which he called "network arch bridges" (Tveit, 1966). (In Japan, this type of structure is called *Nielsen-Lohse girder,* while in China, the term *X-style arch bridge* is used). The first network arch bridges were built in 1963.

To date, the network arch is the latest development in tied arch bridges. It has allowed the construction of the the world's slenderest arch bridges. For example, the Brandangersund Bridge finished in 2010, has a 40-cm-deep deck supported by two 711-mm steel tube arches in a 220-m span (with a slenderness ratio of 1:198). Today, it is usually sufficient to differentiate two types of tied arch bridges: those with vertical hangers and network arch bridges.

3.2   General design recommendations and common details

The choice between a network arch and a tied arch with vertical hangers depends on the general design concept of the bridge. Vertical hangers usually require more material but are often easier to build. Today, network arches are used with increasing frequency but still most new tied arches have vertical hangers (Figure 13.27).

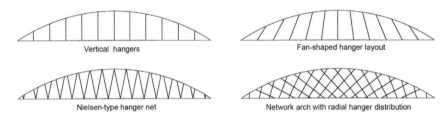

Figure 13.27 Hanger layouts in tied arch bridges.

In the majority of tied arch bridges, the arch rise measures between 14% and 17% of the span. The extreme values are a 14% and 23% (Nakai et al., 1995). Higher arches give lower axial forces in arches and ties, but they are often less aesthetic.

In network arches, the recommended arch shape is circular in order to reduce construction costs; tied arches with vertical hangers are generally parabolic in order to reduce bending moments under permanent loads. The arch slenderness in network arches is between 1:128 and 1:309, while the recommended value ranges from 1:150 to 1:250. When the hangers are vertical, the arch slenderness is between 1:70 and 1:160, with an average of 1:100.

Today, the arches are made almost exclusively of steel. Their cross sections are usually rectangular boxes, but tubes and in some cases rolled profiles also have been used (in network arches, the use of H-shape profiles is recommended).

Most tied arch bridges have a top wind bracing in order to reduce transverse arch bending and increase lateral buckling resistance. This bracing preferably is a symmetrical truss, such as the rhombic truss, the K-truss, or a truss with struts and crossing diagonals, or a Vierendeel beam.

Some tied arch bridges have transversely inclined arches, one toward the other, like the two handles of a basket (known as a *basket-handle arch*; Figure 13.28). This improves the aesthetics and the collaboration of both arches for unsymmetrical loads, seismic actions, and wind loads. But basket-handle arches have major disadvantages as well, such as additional transverse axial forces in the deck and additional bridge width that is needed to keep the arches off the traffic gauge and bigger transverse bending moments in the wider deck.

The number of hangers depends on the distance between them. In network arches, the distance increases with the span, being between 2.0 and 4.0 m equidistant along the arch. The hanger inclination should vary along the bridge, which can be obtained by the radial hanger arrangement where the hanger-arch angle is constant, between 55 and 60 degrees (Schanack and Brunn, 2009). Vertical hangers are usually placed

Figure 13.28 Example of a basket-handle arch on the Villa Maria Bridge, in Villa Maria, Argentina.

at the transverse deck beams, which are about 4 to 5 m apart. The hangers are steel rods, steel plates, wire cables, or parallel strand cables; special attention must be paid to the fatigue resistance of their connections.

The deck of network arch bridges is subject to great tensile force and transverse bending moments. In narrow bridges, it can be a simple r.c. slab with longitudinal pre-stressing cables. In wide bridges, a composite deck with transverse steel beams is more economical because of its lighter self-weight.

In tied arches with vertical hangers, the deck is subject to additional large longitudinal bending moments. Therefore, it requires steel stiffening girders in the arch planes, which usually are I-beams or have a rectangular box section (Fiedler, 2005). In very large span bridges, even orthotropic steel decks are used in order to reduce self-weight.

3.3   Review of recent innovative tied arch bridges

Tied arches, together with trusses, have traditionally been used for railway bridges when long spans are required. In HSR lines, the allowed bridge deflection is very limited. Therefore, tied arches for HSR bridges incorporate innovative design features to increase the stiffness of the structure. In the Mornas Viaduct (erected in France in 1999 and featuring a span of 121.4 m) and the Xinkaihe Bridge (built in China in 2005, with a span of 140 m), this was achieved by double arches. In other bridges, such as the Rego Das Lamas viaduct (built in Spain in 2012, with a span of 80 m), the Yichang Railway Bridge (built in China in 2008, with a span of 275 m) or the Donzére-Mondragon Viaduct (built in France in 1999, with a span of 110 m), the deck is a box girder. The network arch itself is very stiff, but for HSR traffic, the arches are less slender than usual. For example, in the Huihe River Bridge of the Beijing-Shanghai railway line (in China in 2011, with a span of 96 m) and in the East Lake Bridge of the Wuhan-Guangzhou railway line (in China in 2010, with a span of 112 m) the arches are made of two CFSTs each (Hu et al., 2014).

From a global point of view, CFST arches can still be considered innovative, although the first CFST arch was built in 1936 and there are currently more than

150 CFST tied arches in China. This technique combines the easy erectability of steel arches with the benefit of using concrete (poured after arch assembly) for handling compressive forces. Outside of China, very few CFST tied arch bridges have been built; one example is the Ravenna Viaduct in Nebraska, built in 2005, with a 53-m span).

In the first half of the 20th century, concrete has been used very frequently for arch bridges, but was then replaced by steel due to high labor costs for the arch formwork. There are very few recent bridges with concrete arches, like the Norridgewock Bridge in Maine (built in 2011, with a span of 91.44 m), the Depot Street Bridge in Oregon (built in 2006, with a span of 100 m) and the Third Millennium Bridge in Zaragoza, Spain (built in 2008, with a span of 216 m). The concrete arches of theses bridges were cast in place on scaffolding. A very innovative technology was used for the 7th Street Bridge in Ft. Worth, TX (built in 2013, with a span of 49.68 m). It has six spans and a total of 12 identical concrete network arches, which were precast off site (Figure 13.29).

Figure 13.29 Precasting and rotation of the arches, 7th Street Bridge, Fort Worth, TX.

3.4 Selected structural problems

Arch buckling can occur both in the plane of the arch and out of it. The in-plane buckling resistance of network arches usually does not dictate the design. In contrast, an arch with vertical hangers has a buckling length of about 50% of the arch length, which requires enlarging the cross section to obtain the same buckling resistance as in a network arch. Out-of-plane buckling is very similar for both bridge types. Here, the buckling length of unbraced arches is about 80% to 90% of the span length. If the width-to-length ratio of the bridge is smaller than about 1:5, it is usually more economical to add a wind bracing than to enlarge the arch cross section.

The recommended buckling assessment is a second-order FEM analysis with an appropriate equivalent geometric imperfection (Schanack, 2009). It is the standard today that tied arch bridges are calculated by three-dimensional (3D) FEM models using beam elements for arches, hangers, and girders, as well as shell elements for the deck slab. The hanger elements should account for nonlinear effects, such as cable sag and resistance to tension.

The seismic performance of tied arches with wind bracing is generally good. They have a high stiffness-to-mass ratio and only the arch springings require a localized strengthening. Externally, the tied arch behaves much as a simply support beam bridge does. It can easily be equipped with seismic isolators, pendulum bearings, and damping devices; however, very few bridges are built with these devices.

In some tied arch bridges with vertical hangers but inclined arches, the phenomenon of wind- and rain-induced hanger vibration has been observed. During design, special measures to avoid these vibrations must be taken if the natural hanger vibration frequency under permanent loads is smaller than 9 Hz. In network arch bridges, the hangers should be connected at their intersections to avoid any harmful vibrations.

3.5   Erection methods

The most usual construction method is the erection on scaffolding just in the final bridge position. First, the deck is built. Second, the arches are erected, lying on temporary scaffolding towers. Third, the hangers are installed. After that, the bridge can be lifted from the scaffolding by prestressing of the hangers; or the scaffolding can be lowered slowly and the bridge itself stresses the hangers.

Another common method is the construction of a steel skeleton consisting of ties, arches, hangers, wind bracing, and transverse deck beams. This skeleton is erected next to the final position of the bridge, whether on the riverbanks, on the road or track, or parallel to an existing bridge. Then, it is lifted at its ends and is transported longitudinally or transversely, or is rotated toward the final bridge position (Figure 13.30). Afterward, the deck slab is casted and the remaining construction tasks are executed.

Figure 13.30 Transporting the steel skeleton of the Rego Das Lamas Viaduct.

3.6 Research

Since tied arch bridges have been used for 150 years, much of the basic research about them has already been done. However, new analysis methods, materials, construction methods, and application areas continue to attract and require the attention of researchers. For example, arch buckling assessment requires the consideration of safe equivalent geometric imperfections. Consequently, recent research focuses on the shape and magnitude of equivalent geometric imperfections for arch buckling analysis. The use of network arches for railway bridges produces increased fatigue action on the hanger connections. Therefore, another research topic is the fatigue detailing of different hanger types for railway bridges. A further research field is dedicated to construction methods for network arch bridges because optimal network arches do not have a steel skeleton that can be transported as a whole; but concrete decks are too heavy for most transportation methods. Finally, research activities aim to introduce new materials and material combinations for arches, hangers, and ties to achieve economic improvement.

References

Feng, M., 2010. Recent development of arch bridges in China. In: Arch'10–6th Intl. Conf. Arch Bridges. pp. 9–21.

Fiedler, E., 2005. Der Stabbogen bei stählernen Straßenbrücken—Entwicklungstendenzen der letzten 50 Jahre. Stahlbau 74, 96–107. 281–294.

García-Arango, I., Fernández-Nespral, C., Pantaleón, M.J., Ramos, O.R., 2009. Ampliación del viaducto del Pintor Fierros (Arco de La Regenta). Revista de Obras Públicas. 3495.

Griggs Jr., F.E., 2002. Squire Whipple—father of iron bridges. J. Bridge Eng. 7 (3), 146–155.

Hu, N., Dai, G.-L., Yan, B., 2014. Recent development of design and construction of medium and long-span high-speed railway bridges in China. Eng. Struct. 74, 233–241.

Nakai, H., et al., 1995. Proposition of methods for checking the ultimate strength of arch ribs in steel Nielsen-Lohse bridges. Stahlbau 64, 129–137.

Pantaleón, M.J., Ramos, O.R., 2015. Los puentes arco metálicos modernos. Revista de Obras Públicas 162 (3561), 49–64.

Pardo de Vera, I., Pantaleón, M.J., Ramos, O.R., Ortega, G., Martínez, J.M., 2011. Viaductos sobre Río Deza y Anzo 2. Hormigón y Acero 62 (259), 61–74.

Pérez-Fadón, S., Sánchez Ramírez, J.J., 2015. Los arcos más grandes del mundo. Revista de Obras Públicas 162 (3562), 37–50.

Ramos, O.R., Martínez, J.M., De Vena, J., Pantaleón, M.J., 2014. Viaduct over Jauto River. In: 37th Madrid IABSE Symp, pp. 2140–2147.

Schanack, F., 2009. Berechnung der Knicklast in Bogenebene von Netzwerkbögen. Bautechnik 86 (5), 249–255.

Schanack, F., Brunn, B., 2009. Netzgenerierung von Netzwerkbogenbrücken. Stahlbau 78 (7), 477–483.

Tveit, P., 1966. Design of network arches. Struct. Eng. 44 (7), 247–259.

Girders

14

Bharil R.K.
AECOM, Orange, CA, USA

1 Introduction

Girder bridges are the most natural and simplest form of bridging between two points. Chances are very high that today, a bridge engineer will learn how to design a bridge girder before any other bridge type. The use of girder as a natural bridging element is abundantly evident in nature, such as a fallen tree trunk over a stream or rock formations over eroded soil—all providing both people and animals dry and safe access across an obstacle. The use of girders as a manmade bridging element probably evolved as an outdoor extension of an indoor dwelling's floor or roofing system.

The span length and the site conditions often dictate the type of bridge that can be feasible at a given site. There are physical and economic limitations, and the bridge selection process often starts by considering a simple culvert, progressing to a slab or girder system, and ultimately evolving to truss and other more complex systems if and when needed. Figure 14.1 shows the commonly used and economical span ranges of various bridge types. Keep in mind that there are often exceptions to the recommended bridge type selection driven by aesthetic demand, special site conditions, environmental regulations, political influence, and many other factors.

Originally, the girder selection relied on time-proven depth-to-span ratios that controlled deflections and served the function of carrying the load. The most commonly used span-to-depth ratios for various popular bridging elements are described in Figure 14.2. The primary function of these ratios is to control live load deflections and vibrations; however, modern innovation is constantly pushing these ratios toward more lean and efficient systems.

As mathematical formulas evolved, moments and shear were added to the beam equation, and factors of safety were used to guard against uncertainties in building materials and prevalent loads. As the girder shape evolved from untreated tree trunks, swan timber, and cut stones to steel, the material properties begin to play a greater role in its selection. As the analysis and design methodologies evolved, the girder bridges became more complex from simple rectangular beams to fabricated or rolled shapes, concrete with steel reinforcement, concrete with prestressing strands, and various other complex structural systems such as stringer–floor beams and box girders.

Today's girder bridges consist of the elements described next.

The primary structural elements are as follows:

- Girders—Transfer load to substructure elements (the primary focus of this chapter)

Innovative Bridge Design Handbook. http://dx.doi.org/10.1016/B978-0-12-800058-8.00014-1

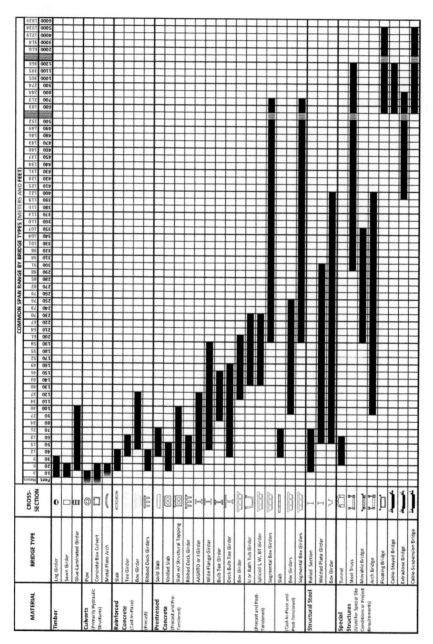

Figure 14.1 Common span range, by bridge type. Compiled in part from California Department of Transportation (2014) and Washington State Department of Transportation (2014).

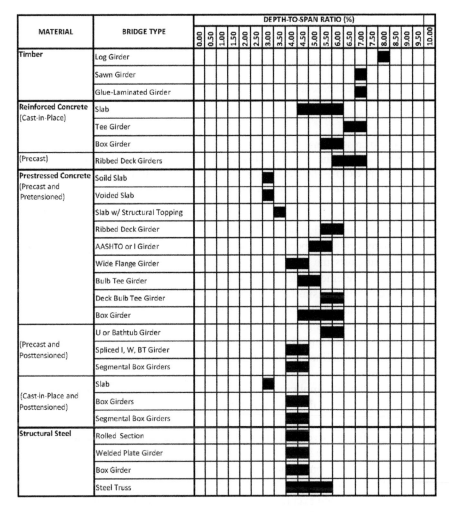

Figure 14.2 Common bridge girder depth-to-span ratios. Compiled in part from California Department of Transportation (2014) and Washington State Department of Transportation (2014).

- Deck—Provides a riding surface and transfers external loads to stringers or girders
- Stringers—Transfer load from slab to floor beams (not always present)
- Floor beams—Transfer load from stringers to girders (not always present)

The secondary structural elements are as follows:

- Diaphragms—Provide stability to girders during construction (often eliminated due to their initial cost)
- Barriers and railings—Serve as a traffic safety element and confine external loads to the designated riding surface

- Bearings—Transfer loads to substructure elements while providing for necessary superstructure rotation and translation
- Joints—Allow movements of superstructure segments to thermal, shrinkage, and seismic demands (use sparingly to reduce maintenance cost)

The substructure elements are as follows:

- Abutments, wingwalls, and approach slabs—Connect the bridge structure to the roadway embankment
- Pier caps and cross-beams—Transfer loads to columns or piles
- Piers, bents, and columns—Transfer loads to foundation interface elements
- Footing and pile cap—Transfer loads to soil/rock strata or other foundation elements
- Piles, shafts, and caissons—Transfer loads to final soil/rock strata via bearing, friction, or both

Before discussing bridge design based on geometry, let's define a bridge girder correctly. Many bridge inspection manuals define a *girder* as a longitudinal bridge element that supports the deck slab carrying external loads and transmits the load to a substructure element such as bearings or abutment/pier cap or cross beam. A *stinger* is defined as a similar longitudinal element that transmits loads to another superstructure element (such as a floor beam) and is typically a part of a more elaborate bridge type such as a truss or a cable-supported system. Other names such as *beam* and *joist* are also interchangeably used but do not necessarily mean the bridge girder. This chapter defines a girder bridge as a bridge whose primary load-carrying members are girders.

Due to its inherent simplicity, the girder bridge is the most common form. The bridging of two distant points by joining them by a straight line is not only intuitive, but also a very efficient form of overall space planning. For example, with girders, there is no sacrificing of vertical clearance below to accommodate an arch springing line or deck truss, no constraining above to accommodate the lateral bracings of a through truss, and no complicated geometry of overhead cables or tied arches. It requires relatively simpler formwork or erection procedures and is often a first choice. However, a girder bridge eventually loses out to other complex forms as new geometry constraints begin to play key roles, spans become longer, or construction access becomes difficult. Such limitations are further given in Figure 14.1 and are described in detail in the following sections.

2 Planning

All successful design begins with a solid plan. Before a bridge project is conceived, the project need, data collection, funding procurement, project delivery methods, and other key steps must be identified, as shown in Figure 14.3.

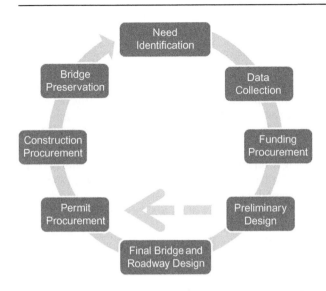

Figure 14.3 Typical bridge project planning cycle— From inception to preservation.

2.1 Project need identification

The planning for a bridge begins with identification of the project—for example, a route alignment that needs to cross an obstacle, followed by data collection and preliminary studies to identify the type of bridge needed and amount of funds needed to plan, design, and build this bridge. Traffic demands and freight mobility needs typically dictate a new crossing, and the condition and capacity of an existing bridge lead to rehabilitation or replacement alternatives. The bridge type selection process depends heavily upon site and funding constraints, and to some extent on the preferences of the bridge owners. Generally, the first preference is a girder bridge unless other more complex bridge options seem to make sense. It is safe to say that a bridge will use girders unless decided otherwise for specific reasons.

2.2 Data collection and preliminary design

Once a project need is identified, the next step is to collect some preliminary site data to develop some viable options. Topographic surveys are needed after a field reconnaissance so that a route can be laid out and an approximate bridge size can be determined. The advent of newer and faster surveying techniques such as three-dimensional (3D) laser scanning (also known as LiDAR) can allow the collection of a vast amount of survey data in days, instead of weeks, to an accuracy of $\frac{1}{4}$ in. (about 1 mm). The use of scanning has been found to be very beneficial in verifying existing as-built plans needed for a bridge widening or rehabilitation project. Some level of preliminary design is essential to identify the ballpark cost, to begin the permitting process, and to plan the funding, design, and construction tasks. Preliminary design, described in detail in section 2.4, is one of the most important tasks.

2.3 Funding procurement

Obtaining funding (whether public or private) is one of the most critical steps in making any project a reality, and it requires a significant amount of planning and effort. Public (i.e., government) funding is the most common funding used for bridges; however, private funding is becoming increasingly popular, as the cost of a bridge project can be directly charged to the public in the form of a toll or user fee. Public-private-partnership (PPP, also sometimes P3) is changing how some very large public projects throughout the world can be funded and maintained, usually in a very short time frame. Typically, bridge cost estimates at the funding level are very generic and use a conventional cost-per–square foot (or meter) of the bridge footprint for estimating purposes. Since the cost-per–square foot (or meter) of various bridge types can vary a great deal, and delays, planning, and procurement costs can consume a lot of money very early on, early consideration needs to be given to better bridge costing to secure an accurate level of financing. It is customary to use 150% to 300% of the bridge's square foot (or meter) cost during the planning stage to arrive at the overall funding needs.

2.4 Project development, delivery, and execution

The project delivery method, whether it is a conventional design-bid-build, design-build (DB), engineer-procure-construct (EPC), PPP, or construction manager–general-contractor (or construction at risk), can have an impact on the bridge planning process, but this decision is often deferred. In the past, the amount of project funding precluded certain type of project delivery methods. For example, it used to be considered that DB procurement should be used only for projects costing over US $20 million, and PPP would be worth the additional effort for projects exceeding US $100 million. However, such boundaries no longer apply these days, and the procurement methods have become more of a comfort level of the bridge owners than money and complexity of the project. The procurement method plays a much greater role in the final design of bridges and delivery of the bridge project. By using DB delivery, the bridge owner is no longer limited by the size of the available workforce. Often, outside consultants are hired as the project manager and construction manager to facilitate such deliveries. Another method, known as *construction at risk* or *general contractor - construction manager (GCCM)*–is where a contractor is retained earlier in the design process, which helps to sync both the design and construction together so that the cost of the project is more certain. An early determination of probable project delivery methods can be very helpful in the planning process and very cost effective for the bridge owners as well.

The other steps of the bridge project cycle, such as construction and preservation, are described in later sections of this chapter.

3 Preliminary bridge design

The process of determining a bridge type, size, and location is commonly known as *preliminary bridge design,* but it is also referred to by other names, such as *project study report, advance planning study, type selection process,* or *type size and location (TS&L)* study, depending on the naming preferences of the bridge owners. Preliminary design is often more detailed than the feasibility or planning studies, and it has a great bearing on the final design and project costs.

All successful bridge projects owe a great deal to the bridge type selection process since this selection basically seals the bridge's fate, whether it is an iconic or just an ordinary bridge. Will it be perceived as aesthetically pleasing or an eyesore for decades to come? Will it meet the construction budget or blow a hole in it? Will it create traffic nightmares during its construction, or will most not even notice it is being built? The list goes on.

It is difficult to understate how important this early design process is, and how much impact it has on almost every aspect of a bridge project. Most engineers with some training can design a bridge once it has been sized up, but the process of successful type selection requires years of experience in bridge engineering, understanding of the multidisciplinary nature of bridge projects, and consideration of funding, construction, inspection, maintenance, hydraulics, traffic, highway geometrics, and many other constraints, as described in this section. Even though a girder bridge may clearly be the best option in most locations, there are many suboption choices that must be made for this bridge type that require additional thought. A thorough type selection process should include the following elements:

- Site constraints—Topography, utilities, traffic, right-of-way, geometry
- Function—The facility serves a function to carry or cross stream, railroad, highways, canals, navigational waterways
- Span length—Total length and lengths of individual spans, limitations, and types
- Substructure—Caps, columns, walls, footings, piles, shafts
- Seismic considerations—Seismic zone, stiffness ratio, balanced spans
- Material selection—Constraints, cost effectiveness, availability
- Aesthetics—Form and function, requirements, public input, local influence
- Environmental considerations—Sustainability, permitting, construction constraints
- Schedule—Fabrication, delivery, construction sequencing, and in-service deadlines
- Cost—Funding, cost effectiveness, life cycle cost comparative analysis (LCCA)

3.1 Site constraints

A topographic map showing existing utilities, right-of-way boundaries, topographic contours, hydraulic boundaries, photographic layers, and other site features is essential to correctly lay out a bridge. Site constraints can play a major role in determining the feasible bridge types and typically dictate how the bridge can be built successfully. For example, a deep ravine may make the placement of falsework very expensive and require launching girders from the banks and therefore, can limit the bridge selection to prefabricated concrete or steel plate girders. Underground utilities may require that the bridge footings are placed only at certain locations, which may affect the span layout. Limited or difficult construction access to the site may limit the use of large cranes and the length of a prefabricated girder that can be delivered to the site. A very active railroad overcrossing may dictate an entirely different type of span that will limit the falsework placement and construction closure of tracks (BNSF Railway and Union Pacific Railway, 2007). Actually, site constraints make each and every bridge unique even though most casual viewers of bridges think they all look alike.

3.2 Function

Besides the main traffic on the bridge deck, what needs to be carried across a bridge may dictate the most efficient type to use. For example, deflection and pedestrian comfort may require a transit bridge to have a certain level of stiffness and may preclude certain types of flexible spans. If a bridge will receive a combination of railway and highway traffic, it may require a double-deck system that is better suited as deck truss, and the combination of pedestrian and truck loading on a long-span bridge may need some modification of standard bridge codes to ensure that the bridge is not overdesigned due to rather unrealistic load combinations. Unusual combinations, such as waterway canal and highway traffic, also have been used sometimes, and these situations require paying careful attention to the design loads and the bridge's performance criteria.

3.3 Span length

Nothing has a more direct influence on the bridge type than the span length. The placement of piers is typically based on the site constraints and often decides the span arrangement. There are some basic guidelines that are difficult to bypass when it comes to selecting the bridge type, based on span. For example, it will difficult to justify erecting a cable-stayed bridge with a span of 150 feet (50 m) where a girder bridge is better suited. Addition of superficial elements without providing a function does not fare well in bridge design. Figure 14.1 illustrates what bridge types are generally suited for which span ranges. For a long crossing, the span arrangements can play a critical role in minimizing the overall project cost. The selection of the number of spans and span lengths requires some consideration of the basic principles of engineering economics. The optimal bridge project cost can be achieved (at least in theory) when the cost of superstructure is almost equal to the cost of substructure, which means balancing individual spans. For the example shown in Figure 14.4, the most cost-effective individual span for this bridge crossing will be about 180 feet

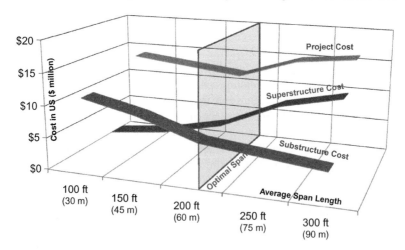

Figure 14.4 Optimizing bridge project cost by balancing superstructure and substructure.

(55 m). There are other factors as well, such as construction risks of a deep foundation, availability of erection equipment, and ease of access and delivery of materials to the site, which can play a big role in the final selection.

3.4 Substructure

The type of substructure can play a significant role in overall bridge type selection. For example, if the foundation soil is relatively poor, requiring a deep foundation, then it may be worth using larger spans and fewer piers; however, increased loading and dead weight due to larger span may affect the foundation design adversely, particularly in a high-seismicity region. From an economic point of view alone, the most preferred footing type is spread footing, followed by driven piles and then drilled shafts (also called *caissons* or *concrete in drilled holes*). The connection of pier column to superstructure also plays a major role in the transfer of superstructure forces to the foundation, which can affect the cost of the foundation. Substructure design in a highly seismic area requires an entirely different set of considerations, as compared to substructures located in a relatively low seismic region. The proportion of substructure with respect to superstructure, bent arrangement, and column heights cannot only affect the overall aesthetics; it can also have a large impact on the seismic design. Typically, substructure thickness when viewed from the bridge elevation profile should be similar to or smaller than the thickness of the substructures. Fortunately, most common substructure configurations work well for girder bridges.

3.5 Seismic considerations

In high seismic zones, the type selection process can have a substantial impact, and in some cases, it is better to size the substructure, span arrangements, joint locations, column width, and length for seismic forces in the preliminary design phase than to shift such responsibilities to the final design phase. Experience shows that planning for balanced structural stiffness pays off later. A detailed discussion on this topic can be found in the *Caltrans Seismic Design Criteria* (California Department of Transportation, 2013). Uneven stiffness attracts additional seismic forces to substructure elements and can be costly to mitigate in the final design phase. Seismic design should be used to plan for joint spacing, span lengths, and the number and length of bends and columns early on.

3.6 Material selection

The material selection goes hand in hand with span lengths. Steel offers a lot of flexibility in span length and curvature, but its use is often restricted by the location of the nearest certified fabrication shops and long-term maintenance considerations such as painting. Cast-in-place reinforced concrete (r.c.) has its limitations due to span length, and prestressed (pretensioned or posttensioned) concrete seems to be a very popular choice of material. The use of structural timber (sawn or glue-laminated) is often limited by the availability of harvestable forests and fabricated timber treatment facilities. Use of masonry was prevalent a few centuries ago but is almost nonexistent today for bridges. Structural composite [i.e., fiber-reinforced polymer (FRP)] bridges were used in special applications as bridge decks due to their light weight; however, their use as a common bridge material is still limited due to their cost and availability (Iyer & Bharil, Testing and Evaluation of FRP deck system... 2004). Concrete, steel, and timber, in that order, make up the vast majority of modern girder bridges.

3.7 Aesthetics of girder bridges

There is an enormous amount of research on the aesthetics of bridges, and an ordinary girder bridge design can also benefit from the very same aesthetic principles applied to the design of signature bridges. Some simple rules specifically applicable to girder bridges are as follows:

- Form follows function, so do not add extraneous items to a bridge since they typically do not work well.
- Keep all horizontal lines (e.g., girders, railing, and deck) continuous and smooth flowing, if they must break (such as at piers), incorporate vertical features of functional elements.
- Pay special attention to span ratios of adjacent spans; keep them as constant as possible, and if they must change, keep the rate of change uniform.
- Pair spans in odd numbers if possible. Three spans look better than two or four. Place end spans with a slightly smaller span to help create balance. This will also aid in the total design later (Figure 14.5).
- Abutments are often designed separately and often do match the main structure elements. For example, uneven heights of two abutment walls may help with the design, but they are a distraction when viewed together.
- When abutments are flanked by retaining walls, match the bottom of the bridge barrier rails with the coping of the retaining wall barrier. This detail is often left to non-bridge engineers and can be very noticeable. The incongruity at this location can be a distraction from the smooth horizontal lines of the bridge spans.
- Use mild vertical slopes and flares to accentuate the column shape that supports the heavy girder superstructure. For example, thickening at the tops of columns appears to be correct logically, although it may not be needed.
- If girder thickness must change between the spans, use gradual variations; but it is best not to use variable depths at all.
- Traffic barriers and safety railings provide ample opportunities to bring life to an otherwise plain girder superstructure. Try to set them so they do not add to the visual perception of girder thickness.
- Nothing makes a statement like a slender superstructure over proportionally sized piers. So do not use anything thicker than necessary; instead, invest in better materials and employ other schemes to increase the serviceability of the structure.
- Visualize your bridge design in actual settings to see how it will look. Also, get bridge architects involved, use computer animation and 3D visualization to iron out the kinks, see how it will look when illuminated at night, to the public, and to local businesses and residents from afar. Use the most contemporary tools you can get to refine the structural features early (Figures 14.8 and 14.9).
- Human imperfections can be magnified in a very long straight line, such as with railings. It is better to break up these lines (e.g., by incorporating a light post pedestal) so such imperfections are less distracting.

Figure 14.5 Aesthetic elements of Sandifer Memorial Bridge over the Spokane River in Washington State. This design included varying spans of timber, steel, and concrete. Photograph courtesy of CES, Inc.

3.8 Environmental considerations

A bridge project can come to a dead halt due to an environmental permitting issue. The key to avoiding this problem is to start early: and identify permitting issues as soon as possible and work with regulatory agencies to see what can be realistically permitted and what will take a Herculean effort. For example, a short construction window for a bridge over a fish-bearing stream may limit you to only prefabricated girder types. Floodways also affect how long the bridge opening needs to be so that rise in backwater can be limited. Scour the potential and possibility of meandering channels, which may preclude spread footings. In addition, wetlands and sensitive cultural resources may require a special type of bridge construction (e.g., temporary construction platforms) that may exclude certain types of bridges, requiring a lot of falsework (Figure 14.7).

3.9 Schedule

Most projects are very schedule-driven, which can often influence the bridge selection. If the lead time for fabrication is too long, cast-in-place options may be preferred. If the incentives for opening to traffic early are great enough, fabricated options may take precedence. If the in-water work window is too small, the bridge span may be made much longer to keep the piers off the natural water boundaries of a river. A preliminary schedule for a bridge project should be made to analyze if the funding, design, construction, and permitting timelines are realistic.

3.10 Cost

Some bridge types inherently cost more. It is important to consider the overall and life cycle costs, rather than just the per square foot (or meter) costs, to compare the real benefits of certain bridge types. For example, in a heavily traveled freeway, steel spans may result in less overall cost when the maintenance of traffic costs and risk of delay costs due to accidents are taken into account. A weathering steel option may offset the cost of repainting the bridge every 20 to 40 years. The following factors should be kept in mind with regard to bridge costs (Calfornia Department of Transportation, 2014; Washington State Department of Transportation, 2014):

- Factors that can result in a project being in the lower end of the cost range include short and simple spans, low structure heights, no special environmental constraints, very large project (i.e., mass and repeat structural quanitities), no aesthetic issues, dry conditions, square bridge (no skew), urban location (i.e., easy access), seat abutments (with low height), spread footing, bridge site closed to general traffic, and no staged construction.
- Factors that can result in a project being in the higher end of the cost range include long spans, high structure height, environmental constraints, small project (i.e., no repeat element or small quantities), aesthetic issues, wet conditions (i.e., cofferdams required), skewed bridges, remote location, tall cantilever abutments, deep foundation (i.e., piles, drilled shafts, micropiles).
- Some factors that can have a very high impact on the cost (from 25% as much as 150%) are unique urban conditions requiring more than two construction stages and very small widening (less than 15 feet or 5 m). However, although architectural and aesthetic requirements on bridges may seem to suggest higher costs, they actually do not cost much.

The bridge cost ranges (based on deck square footage or meter as shown in Figure 14.6) are calculated using "Bridge Costs Only," as defined by the U.S. Federal Highway Administration for the western United States in 2014. These costs reflect the bridge cost only; they do not include items such as time-related overhead, mobilization, existing bridge removal, bridge approach slabs, abutment slope paving, soundwalls, retaining walls, or unusually large wing walls.

4 Final design

Once a girder type is selected and preliminary design is carried out, the final design process is relatively straightforward. There are plenty of resources available, including software, literature, guidelines, examples, and codes, to help in the design development process. Constructability evaluation (that is, designing backward with construction in mind during various phases of design) is highly recommended to reduce project risk. The following elements govern the design development process:

- Design criteria—Codes, specifications, guidelines
- Material properties—Steel, cast-in-place concrete, precast concrete, timber, and FRP
- Loading type—Highway, railroad, transit, pedestrian, and utility
- Design considerations—Structural analysis, seismic, software applications
- Detailing practice—Standards, computer-aided design and drafting (CADD), automation
- Construction specifications—General specifications and special provisions
- Construction cost—Engineer's estimate of cost of hard bid
- Construction schedule—Engineer's estimate of construction working days

4.1 Design criteria

Unlike buildings, bridges are almost always designed for public use, are subject to public scrutiny due to funding sources, and require that the safety of the traveling public remains paramount. There are numerous specifications and codes governing bridge

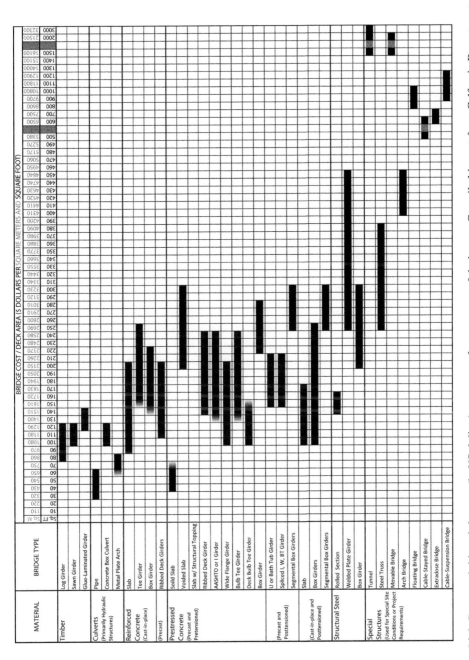

Figure 14.6 Common bridge girder cost (in US dollars) per square foot or meter of deck area. Compiled in part from Calfornia Department of Transportation (2014); Washington State Department of Transportation (2014).

design that are often regulated by the bridge owners or the jurisdiction where the bridge is located. For example, in the United States, almost every 1–2 years, new American Association of State Highway and Transportation Officials (AASHTO) specifications (AASHTO, 2014) are adopted and supplemented by additional special publications by various state agencies, which become the governing design criteria for all bridges under the aforementioned state jurisdiction. The American Railway Engineering and Maintenance-of-Way Association (AREMA) publishes a guide (AREMA, 2014) every year and becomes the code for designing all heavy railroad bridges since it is developed by all Class 1 railroad companies. Modern bridge design is based on probabilistic analysis, which is geared to provide an acceptable probability of failure for all elements instead of the previous factor of safety-based design. Most codes oversimplify the bridge girder design, and therefore, it is important to understand how bridge codes were evolved when encountering an unusual condition that may not be covered in the codes. Codes are often years behind the latest technology and can cover only common conditions. As bridge design gets more complicated, understanding the intent of code becomes more important, and only experienced and well-informed engineers can make full use of innovations when comes to materials, loading, and construction.

4.2 Material properties

Use of high-strength and unique materials is advancing faster than the bridge codes or specifications. For example, yield strength of steel and compressive strength of concrete is much higher today than it has ever been. It is also worth keeping in mind that many of the material properties, such as concrete strength, can depend heavily on the local region and cannot be adopted easily outside the region. For example, the superb quality of coarse aggregates in the Pacific Northwest of the United States can help easily produce a prestressed concrete strength of over 10,000 psi (70 MPa), which is not readily possible elsewhere.

4.3 Loading type

The type of traffic that a bridge carries can and should heavily influence bridge design. Highway loading is relatively lighter than railway loading and may require that the girders be placed much farther apart. The use of two or even three girders is often discouraged due to lack of redundancy in fewer lines of girders. Special loads, such as airplanes, barge impacts, heavy ice, frequent mining trucks, special permit trucks, and transit systems, require special consideration, but they are not covered in most codes. Incidence of load impact damage on highway bridge overpasses has been well documented and should be considered in the design of critical bridges.

4.4 Utilities

In urban areas, utility relocation can have a huge impact on the bridge design, and construction schedule and can be extremely costly if not planned correctly. Utilities should be surveyed and potholed (i.e., water jetting to determine the exact location of buried pipes) early in the design phase so that they can be avoided if possible. Overhead utilities (such as high-voltage power) can interfere in girder erections and crane movements and wet utilities (i.e., water and sewer) are heavy and need to be accounted for in the design. Inflammable utilities (oil and gas) and electrical utilities should be handled with extreme care, and fiber optic lines are difficult to splice. Utilities mounted on bridges should be given special design considerations in the final design phase to ensure that the essential services are not disrupted (Bharil et al., 2003).

4.5 Design considerations

Depending on the materials and loadings, design considerations vary. Unique bridge design guidelines exist for a wide variety of bridges; however, they should be verified to include concerns of the stakeholders and to comply with the jurisdictions. For large projects, the number of stakeholders and jurisdictions can be large and may have conflicting interests that must be handled carefully. Questions in the following categories should be asked to arrive at the final design considerations:

- *Bridge location and ownership:* Where is the bridge located? Who owns the bridge now, and who will maintain it? Typically, this is a major factor in deciding which design codes and manuals will be used. For example, if the bridge is located in the City of Los Angeles, the Caltrans design codes will apply, but various city codes will also have to be followed.
- *Bridge funding:* Who is funding the bridge? The type of funding may impose some additional conditions and design considerations that must be included. For example, if the project funding includes federal funds, the bridge project may require more rigid review and oversight,while purely local funding may allow more leeway on the project design. Some funding may limit the bridge cost and type, and some may preclude certain types of bridges or materials.
- *Bridge traffic:* What will be carried on the bridge? The type of traffic, such as bikes and pedestrians, heavy rail, light rail, vehicular traffic, water, airplanes, oil/gas, and mining trucks will determine the technical and functional design criteria for the bridge. For example, a bridge carrying heavy rail traffic over a highway will require the bridge to be designed by AREMA codes (AREMA, 2014) and *Railroad Grade Separation Guidelines* (BNSF Railway and Union Pacific Railway, 2007) but also satisfy the state design manuals for highway safety features for pier and abutments located underneath.

4.6 Detailing practices

Use of CADD is widespread in design, and its use in construction [e.g., geographic information system (GIS) and building information modeling (BIM)], will become a reality for bridge design in the upcoming years. Tools and programs are available to make better use of the available technology, and bridges form an integral part of the overall project CADD package. CADD for bridges is less regulated than it is for roadways, and most states and agencies allow some flexibility, but others do not. Use of CADD is also widespread at the engineering design level, and most young graduates already have some level of CADD training and knowledge of popular platforms such as AutoCAD and MicroStation. There are some good design practices for bridges, and due to the close interfaces with other disciplines, these practices should be followed closely to accommodate changes and to make key bridge information, such as the bridge foundation footprint, readily available to other disciplines such as utilities.

4.7 Construction specifications

Most people do not realize that in the order of precedence, certain specifications can supersede the design plans and cause a great deal of confusion, schedule delays, claims, and budget overruns. For example, Most agencies have standard construction specifications (published almost yearly) that are typically modified by special provisions written for unique bridge elements such as special bearings may conflict with the design drawings and create anticipated change orders. The process of construction specification writing for bridges should be done by seasoned professionals and should be reviewed to balance the risks to both the owner and contractors.

4.8 Construction cost estimates and schedule

The engineer's estimate of the probable construction cost and likely construction schedule complete the construction bid package on a bridge project. Again, a good estimate of the cost and schedule will determine if there will surprises at the day of the bid (tender) opening and whether there is enough money and time allocated to make it realistic. At this stage, a more detailed estimate of probable construction cost based on actual quantities and the prevailing bid cost of the various items will be required. It is customary to allocate some percentage (10% to 25%, depending on the complexities of the project) of the estimated construction cost to contingency and administration cost. The total project cost can be almost double the hard bid construction cost when all costs (from project planning to construction closeout) are taken into account.

5 Construction

A wide variety of methods are used in girder fabrication, erection, and casting. In addition, bidding, award, and delivery methods affect the construction methods. The construction procurement can also have a huge impact on the schedule and cost. In addition to conventional design-bid-build procurement, alternate project delivery (APD) methods such as DB, EPC, general contractor–construction-manager (GCCM) and PPP are typically used for large bridges and often for major highway segments involving scores of bridges. Schedule saving is a primary goal achieved in APD methods, where after a certain level of preliminary design, the project can be designed, built, operated, and financed by an APD contractor (typically a consortium of contractors, designers, and financiers and concessionaires). APD methods have been used for many years in other industries; however, its application in public infrastructure has increased manifold in the last decade and will continue to do so.

Construction of girder bridges requires attention to the following:

- Bid/tender advertisement, selection, award, and execution—Bid advertisement and contractor selection

- Preconstruction and mobilization—Construction schedule, progress payments, equipments, shop drawings, material certifications, preconstruction conference, and mobilization
- Removal and demolition—Often used in staged construction and can be tricky
- Delivery and erection—Site safety, cranes, launching, etc.
- Resident engineering and construction methods—Cast-in-place construction using false-work, maintenance of traffic; precast and prefabricated construction using fabrication, on-site casting, shop drawings, shipping, and storage
- Project closeout—Record of materials, as-built drawings, final acceptance

Figure 14.7 Wishkah River Bridge over sensitive wetlands in Washington State. Existing bridge (left) as traffic detour and work platform (right).
Courtesy of CES, Inc.

In general, the construction of bridges requires special expertise and should be per-formed by experienced construction personnel. The risk mitigation of unexpected bridge failure is many times costlier than similar risks involved in roadway construc-tion, for example. One of the biggest causes of unbalanced bids is the allocation of construction risks between the owners and contractors. The cost and schedule penal-ties to amend construction problems are much higher, and some construction risks (e.g., uncertainly of foundation) cannot be easily mitigated. It is beneficial to perform thorough constructability reviews by construction personnel at various phases of the design (e.g., preliminary design, intermediate design, and final design) than to handle them during the construction.

6 Preservation

Modern bridges must be designed to allow easy access to preservation activities such as inspection, maintenance, and common repair. Some of design considerations for bridges for preservation activities are described in the next sections.

6.1 Provide Arm's reach inspection access

Provide ladders, maintenance walkways, and access doors for full manual access throughout the bridge elements. Use current specifications of under-bridge inspection trucks (UBITs) to reach various parts of the bridge without rope-assisted climbing. For example, providing a gap of at least 7 feet (2 m) between two adjacent bridges will facilitate the UBIT arm to reach under the bridge (Washington State Department of Transportation, 2014). Not providing these essential amenities during the design phase will only increase the cost of future preservation activities multifold.

6.2   Design for rope-assisted inspection

If full manual climbing or bucket truck access cannot be provided due to cost or other site constraints, access for rope-assisted climbing must be provided. Such access can be easily facilitated by providing rope anchorage points (predrilled holes or brackets) at selected locations. Providing lifeline cables along the girders during original construction is not expensive and often leads to easier and less expensive rope access operations in the future. Keep in mind that if the access is too difficult, the member is likely not to be inspected often or properly. More detailed rope-assisted work requirements have been issued by two organizations: Society of Professional Rope Access Technicians (SPRAT) and International Rope Access Trade Association (IRATA) who are dedicated to such rope access work.

6.3   Design to account for maintenance

All bridges require routine and special repairs over time. For example, bearings and joints need replacement, drains get plugged, and the deck may eventually need an overlay. The designer should provide jacking locations for bearing replacement or repositioning, the joints style should allow for gland replacement, drains should have a cleanout pipe, and the deck should have enough clear cover on the top-reinforcing steel to allow for future scarification (up to 1 in. or 25 mm) and enough structural capacity to absorb the added weight of 2–3 in. (50–75 mm) of overlay (future wearing surface). Other future maintenance items include accounting for stream bed scour and aggradation, ability to clear river debris and ice accumulation underneath, accounting for reduction of vertical clearance due to surface overlay underneath, and added cover and protection for corrosion due to soil and climate. Most of these items are common sense, but they can be easily neglected when designing a brand new bridge with the idea of maximizing cost savings.

6.4   Consider the life-cycle cost of bridge

There is too much focus on the upfront cost of a construction project; the life cycle cost of bridges is often ignored. New bridges are expected to last 75–100 years, and repair, retrofit, and rehabilitation typically can extend the life to 20–40 years. During the expected life of the bridge, maintenance expenses such as painting, deck overlays, joint replacements, and scour mitigation can really add up. For example, choosing between a replacement versus repair, retrofit, and rehabilitation should be based on the life cycle cost, not just upfront costs. The general rule indicates that if the cost of rehabilitation approaches 50% of replacement, extreme caution should be taken before embarking on rehabilitation. The life of an aging bridge can be increased by reducing the number of lanes or changing the type of traffic (banning trucks, for example), and a rigid deck overlay can reduce impact loading while extending the life of the deck by 20 or more years. A simple economic analysis of the bridge (using a present worth or sensitivity analysis) can provide enough information to provide a valuable comparison of various feasible alternatives.

Once a girder bridge is built, it needs to be preserved, which includes the following required elements:

- Inspection and testing—Types of inspection include crack inspection and fracture-critical and fatigue-prone details, testing, and instrumentation. An inspection interval of every two years is common.
- Load rating, posting, and overloads—Once a load rating has been performed, it needs to be updated to reflect the condition of the bridge. Special permits for bridge use by trucks need to be reviewed, evaluated, and permitted.
- Bridge maintenance and management—Preventative maintenance activities such as instrumentation, testing, repairs, overlays, widening, strengthening, scour, and seismic retrofit can play a key role in keeping girder bridges functioning. The use of bridge management practices can prioritize maintenance and repair funding.
- Rehabilitation—This typically involves a major upgrade to the structural capacity and may also involve retrofits (with no change in capacity), widening, and strengthening. It typically uses inspection findings and a bridge management system to prioritize work.
- Seismic—Vulnerability evaluation, prioritization, and seismic retrofitting to prevent collapse during a designated seismic event.

7 Innovation

Innovation has been generally slow in the field of bridge engineering due to the heavy emphasis placed on public safety and the relatively long process of adapting changes to the existing code and practices by bridge owners. Girders were very quick to evolve at an early stage, but many significant innovations, such as wide-flange prestressed supergirder or high-strength welded plate girders, took a long time to develop compared to other industries. Many trends still in their infancy today may one day become the norm. Innovation will continue to change how girder bridges are planned, selected, designed, constructed, monitored, and maintained. Some of these innovations are described as follows:

7.1 Predominance of APD procurement

APD procurement of megaprojects costing billions and encompassing hundreds of bridges to be built in short duration, will boost an unprecedented level of innovation in all sectors of girder bridges. APD may lead to mass girder production, new and rapid fabrication techniques, longer girder spans, use of high-performance materials, advent of special shipping trucks, heavier erection cranes, new girder shapes (Figure 14.8), and integration of GIS/BIM technology into CADD design drawings, and use of drones for remote data collection and construction monitoring. In addition, shifting of maintenance responsibilities to the private sector will lead to innovation in bridge health monitoring, jointless and low-maintenance bridges, and designing for preservation.

Figure 14.8 Steel delta girders of I-90 George V. Voinovich Bridge (Innerbelt DB Project), Cleveland, Ohio.
Courtesy of AECOM, Trumbull-Great Lakes-Ruhlin Joint Venture, and Ohio Department of Transportation.

7.2 High-performance materials

The general trend is the increase in the strength and versatility of cast-in-place and precast concrete. The purpose of using higher-strength materials is not always increasing capacity, but rather increasing structural service performance in terms of durability, imperviousness, and chemical resistivity. Similarly, high-performance structural steel continues to break new ground in terms of tension strength, ductility, and corrosive resistivity (Figure 14.8). Use of new coating techniques in bridges is also evolving to make the cost of a steel bridge as a girder bridge very competitive with precast and cast-in-place concrete bridges.

7.3 Use of structural composites

Although the use of structural composites has reduced in bridges, primarily due to cost and lack of technical expertise, the promise of developing lighter weight with greater strength remains. The use of structural composites in bridges, particularly in everyday girder bridges, from new construction to retrofit, has come a long way from the early testing and instrumentation phases (Iyer and Bharil, 2004).

7.4 Automatic bridge health monitoring

This area is evolving fast from its experimental stages to real life applications. The advancement in durable and less expensive instrumentation techniques, and remote monitoring via the Internet and wireless technology, will enable bridge engineers to understand and track bridge deterioration better, and will help bridge owners to minimize damage during catastrophic events such as earthquakes.

7.5 Improved girder fabrication and shipping lengths

Girder fabrication and shipping lengths are improving as the trucking and shipping industry applies more modern technology to maneuver tight radii and constraints. Precambering of precast girders allows for vertical clearance below, the casting cycle of concrete girders is much shorter, and steel girder fabrication is mostly automated. It is not uncommon to see a single 150-foot (45-m) precast concrete girder being shipped today, while only 120 feet (35 m) was the norm in the past (Figure 14.9). Megaprojects are also contribute to this trend to meet constant demand for increasing efficiency and maneuvering larger spans to avoid traffic closures and to minimize environmental permitting constraints.

7.6 Longer jointless bridges

Deck joints, either due to seismic, thermal, or shrinkage demands, are a maintenance headache and generally expensive. Elimination of these joints can entail a tedious design and approval process that most bridge engineers prefer to avoid. Secondary effects can be substantial and should be avoided to circumvent new maintenance problems. In the future, more of these expensive joints will be eliminated, as jointless bridges of 1000 feet (300 m) long or longer become more commonplace.

7.7 Better girder erection procedures

Modern erection methods and use of supercranes have increased the lifting and launching weight and girder length capabilities and reduced the construction turnaround schedule. These innovations can dramatically increase the use of prefabricated girders.

7.8 Highly efficient girder shapes

Partnerships between precasters/fabricators and bridge engineers, researchers, suppliers, and the trucking/shipping industry have prompted the refinement of the precast

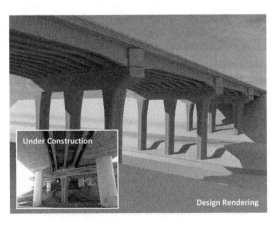

Figure 14.9 The 907-foot (276-m)-long US-101 Petaluma River Bridge in Sonoma County, California, is one of the longest spliced prestressed concrete girder bridges in the United States.
Courtesy of the Bridge Designer, AECOM.

concrete girder shape from a box-shaped beam to highly refined wide-flange, bulb-tees, and supergirders capable of spanning up to 240 feet (75 m). As the demand for longer clear spans increases, the impetus to push the girder shapes beyond what is deemed possible today will continue.

7.9 Hybrid girders

Combining various materials and shapes will push the limits of spans and load-carrying capacity well beyond the classic marriage of concrete deck over steel beams. Use of new materials such as structural composites, time-proven materials such as steel, and glue-laminated timber can yield very high strength-to-weight ratios and can be cost effective as well (Figures 14.5 and 14.8).

7.10 Improved design codes

Nothing can have a greater impact than making refinements in current design codes, and effecting changes in the bridge owners' mindsets to push the limits of girder bridges. The codes were designed to keep things simple so that complex analyses were not required for everyday bridge designs. Code does allow the use of more detailed analysis, but with a caveat that the designer will be left to defend his or her work. Sophisticated software can be used to create a very efficient design, but it may not be easy for the results to pass the reviews of bridge engineering peers. Change does not come easy in this old-fashioned industry, which has served the public well for hundreds of years. The risk of litigation also keeps much innovation at bay, but the question remains: Are consultants, contractors, funding agencies, and the bridge owners willing to take advantage of newfound knowledge, particularly when the established codes may be ambiguous (or just silent) on those topics?

8 Conclusions

There is no doubt that the girder bridge remains the most popular form of bridge worldwide. Given this popularity, innovations will continue despite the mediocrity of mundane girder form (Figure 14.10). Girder bridges will continue to break records in terms of span length, material strength, and longevity, resulting in a more slender, clutter-free system. Innovation comes from the bridge engineering community, comprised of designers, fabricators, constructors, and bridge owners who constantly demand more, care a bit more about the impact of their work products, and challenge themselves to improve both the engineering processes and design codes to move bridge engineering practices forward.

We see and drive over girder bridges almost daily, and the bridge engineering community thus has the opportunity to make a real impact on everyday people. An ugly bridge will remain an eyesore to millions throughout its life (which can last hundreds of years), but a beautiful, elegant, and well-constructed bridge will not only provide safe travel over time but also will complement its setting. It is very easy to get lost in the everyday practice of designing bridges and not look back and consider if we might have done a little better. If we can all promise only one thing to ourselves today, I truly hope that will be that as bridge engineers, we will never design an ugly bridge again.

Figure 14.10 Prestressed concrete box girders span the SR-91/I-15 Interchange of 91 Corridor Improvement DB Project in Southern California.
Courtesy of AECOM and Atkinson-Walsh Joint Venture.

References

American Association of State Highway and Transportation Officials (AASHTO), 2014. AASHTO LRFD Bridge Design Specifications. seventh ed. Washington DC.

American Railway Engineering and Maintenance-of-Way Association (AREMA), 2014. AREMA Manual of Railway Engineering, Structures, vol. 2. Lanham, Maryland.

Bharil, R.K., et al., 2001. Guidelines for bridge water pipe installations. Proceedings from ASCE Pipeline 2001 Conference, San Diego, CA.

BNSF Railway and Union Pacific Railway, 2007. Guidelines for Railroad Grade Separation Project.

California Department of Transportation, Caltrans Publications, 2013. Seismic Design Criteria, SDC Version 1.7, and Memos to Designers.

California Department of Transportation, 2014. Caltrans Comparative Bridge Costs.

Iyer, S.L., Bharil, R.K., 2004. Testing and evaluation of the FRP deck system for the Douglas County in Washington State. In: SAMPE Spring Conference Proceedings, Long Beach, CA.

Washington State Department of Transportation, WSDOT Publications, 2014. Bridge Design Manual (LRFD), Version M 23-50.13, and Washington State Bridge Inspection Manual. Version M 36-64.03.

Long-span bridges 15

De Miranda M.
Studio de Miranda Associati, Technical Director, Italy

1 Introduction

1.1 Concepts and problems of long-span bridges

The definition of the term *long-span bridge* derives from the context and the historical epoch, in terms of the limits reached at that time by the builders of bridges as large span. In Roman times, the maximum spans were in the order of a few tens of meters. At the start of the Industrial Revolution, with the first railways and roads for vehicles, long spans were in the order of 150 m.

In the twentieth century, with the construction of bridges with spans exceeding 1000 m, and up to nearly 2000 m in this century, *large span* generally means a span over 300–500 m. However, spans of these lengths present problems that are mainly linked to the method of construction, aerodynamic stability, and the effect of self-weight on the bridge's static load. In fact, large span structures can be seen today as structures in which the so-called scaling law is dominant.

Scaling law was described as early as 1638 by Galileo Galilei in the *Discorsi* (Galileo, 1638); it expresses the circumstance where, upon increase of the geometrical dimensions of an object (even if the shape does not vary), the stress to which the object is subjected due to its weight increases.

A cube with side 1 and specific weight γ is stressed at the base with a tension of

$$\gamma \cdot 1^3/1^2 = \gamma 1,$$

which expresses a stress that is directly proportional to 1, by a factor of γ.

It is for this reason that the elephant has a more massive bone structure and much wider feet than the gazelle: that is, when considering a geometric scale ratio of 1:10, if the shape remains unvaried, the pressure on the feet and the ground would be 10 times higher in the case of the larger animal. Nature has adapted by changing the animal's shape and limiting its size.

In the same way, if the same shape and material is maintained in the structures, as the dimensions increase, the stress increases proportionally. Then a larger amount of structural material is required, which in turn would lead to an increase in weight. This implies that there is a limit in the dimension of structure that is proportional to the ratio between the resistance and specific weight of the structural materials.

Innovative Bridge Design Handbook. http://dx.doi.org/10.1016/B978-0-12-800058-8.00015-3

For a vertical tension rod of uniform cross section A, length l, and unit weight γ, the maximum stress will be

$$\sigma_{MAX} = \frac{\gamma \cdot l \cdot A}{A} = \gamma \cdot l \rightarrow l = \frac{\sigma_{MAX}}{\gamma} \rightarrow l_{MAX} = \max\left(\frac{\sigma}{\gamma}\right).$$

For a strong steel, this means

$$L_{MAX} = \frac{\sigma_{ULT}}{\gamma} = \left(1.9E6\frac{kN}{m^2}/78.5\frac{kN}{m^3}\right) \cong 24,200\,m.$$

This gives the stress-weight ratio σ/γ a suggestive physical meaning.

For a parabolic rope that has span L, sag f, and section A and is made of the same material, with a uniform weight that is a good approximation if f/L < 1/8 (as is usually the case with bridges), the maximum tension force results:

$$T_{MAX} \cong (\gamma \cdot A) \cdot \frac{L^2}{8 \cdot f} \cdot \left(1 + \left(\frac{4f}{L}\right)^2\right)^{0.5}.$$

The second term, for a typical value if f/L = 1/10, is 1.08. Therefore:

$$\sigma_{MAX} = \frac{T_{MAX}}{A} = \left(\gamma \cdot \frac{L^2}{8 \cdot f}\right) \cdot 1.08 = \gamma \cdot \frac{L}{0.1 \cdot 8} \cdot 1.08 = 1.35\,\gamma L,$$

and

$$L_{MAX} = \frac{\sigma}{\gamma} \cdot 0.74 = 17,900\,m,$$

which is only 26% less than the vertical rod.

For a realistic allowable stress for suspension bridge cables of 650 MPa, we get:

$$\sigma/\gamma = 650\,MPa/78.5\,kN/m^3 = 8280\,m$$

and results in the following:

$$L_{MAX} = 6127\,m.$$

Considering, then, that the load supported by the rope must include the weight of the bridge deck, hangers, and the live load, and assuming that the ratio k between the total load and the self-weight of the cable must be of the order of 1.80, we get:

$$L'_{MAX} = L_{MAX}/k = 3400\,m.$$

It can be seen from the previous examples that large-span structures must have, from the point of view of theoretical feasibility, the following features:

- To be built, as much as possible, using materials with a high σ/γ ratio; therefore, today, high-tensile steel is used.

- As the elements subjected to traction are not subjected to phenomena of instability, they can be used with the maximum tension and have the minimum weight for the force transmitted, so they are highly efficient.
- As a result, the large-span structures are formed mainly from high-tensile steel tension elements (that is, cables).

However, from the point of view of feasibility, a large-span structure must have a structural shape that allows it to be built safely, even when it has large dimensions.

The types of structures that have these features and thus are the most suitable for large spans are suspension bridges and cable stayed bridges.

1.2 Historical evolution of long-span bridges

It is interesting to give a brief history of these bridge types in order to highlight their origin, as well as present and future developments.

1.2.1 Suspension bridge

The suspension bridge has the ancient origins. Crossing a body of flowing water or a ravine using textile ropes anchored to the ends was the first archetype of a suspension bridge. The first data regarding these can be traced back to the fifth century, from Asia and Latin America.

These structures were lightweight and flexible, and they put the following two concepts into practice at the base:

- To suspend the loads on one or more ropes anchored to the ground and configured as a catenary
- To realize the structure by laying a carrying cable from one side of the obstacle to the other and successively equipping it with secondary structures and elements.

The culture of textile rope bridges never developed in the Western world.

The first rope bridge, with a layout that is known today as cable stayed, dates back to 1615, when the Italian polymath and inventor Fausto Veranzio demonstrated an idea for military bridging supported by inclined cables in a book of inventions called *Machinae Novae* (Figure 15.1). (Veranzio, 1968). The cables were formed of iron bar chains.

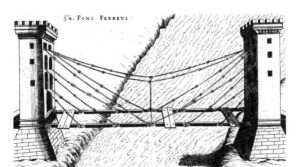

Figure 15.1 The Faustus Verantius bridge, first idea of cable-stayed bridge. It is not sure that it was actually built, although it's possible.

It is not known if the bridge was actually constructed, but it remains the first example of a bridge supported by metal ropes.

The first iron chain suspension bridges were built in Europe in 1730–1740 in Prussia and in the Brittany region of France. The modern suspension bridge was developed as a response of nineteenth-century engineers to the requirements that the Industrial Revolution imposed in terms of new roads, railways, and crossings. It therefore had to provide suitable tensile strength and stiffness, especially regarding the heavy and concentrated moving loads of trains. The stiffness of the structure was brought about by a stiffening beam lattice in the first bridges, which distributed the localized loads onto the ropes in such a way as to guarantee reduced deformations and angular distortions of the road surface.

The first experiences with such bridges were in the United States in the early nineteenth century, by James Finley in Pennsylvania. These were pioneering ventures, hindered by several accidents, collapses, and immediate reconstructions; but these problems quickly increased the existing store of technical expertise.

In Europe, Finley's concept was released by Thomas Pope in *A Treatise on Bridge Architecture* and was developed in England starting with the Dryburgh Abbey Bridge, with a span of 79.30 m, built in 1818. This bridge collapsed six months after its construction; it was then rebuilt and stiffened with stays (Figure 15.2). This was the first integration of the two structural systems.

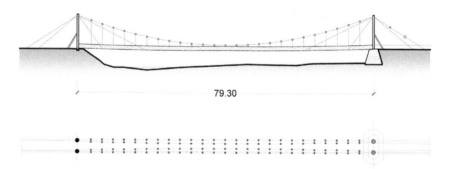

Figure 15.2 The Dryburgh bridge, first suspension bridge in Europe, successively stiffened by stay-cables.

Other bridges followed, with increasing spans and varying technical details, in England, France, and Italy, reaching a span of 280 m with the Fribourg Bridge in 1835. This was followed by the Lewiston-Queenston Bridge in the United States in 1850 and the Roebling Bridges at Niagara Falls, New York, in 1870 with 380 m (Roebling, 1854) and the Brooklyn Bridge in 1883, with a 486-m span (Figure 15.3).

Spans increased fourfold in the twentieth century. They exceeded 1000 m in 1931 with the George Washington Bridge (L = 1067 m) designed by Othmar Amman and Leon Moisseiff, and they continued to increase even after the Tacoma Bridge collapsed due to wind and other not fully known aeroelastic forces. Later bridges that even exceeded this length include the Verrazano Narrows Bridge (L = 1298 m), built in 1964, the Humber Bridge (L = 1410 m) in England, erected in 1981; and the Storebaelt Bridge (L = 1624 m) in Denmark, (Figure 15.4) and the Akashi Kaikyo

Figure 15.3 The Brooklyn bridge, in New York, first long-span suspension bridge in America.

Bridge in Japan (L = 1991 m), both built in 1998. These latter two bridges represented two completely different concepts. The Japanese bridge, with its deep and weighting lattice-stiffening girder, represented the continuation and evolution of the American way after the Tacoma event. The Danish bridge, designed by Danish engineers and built by an international pool of companies, has its slender, light, and streamlined deck that represented the evolution of the European concept of the suspension bridge, with a box-girder deck with an aerodynamically efficient shape.

Figure 15.4 (a) The longest span in the word today, the Akashi Kaykio Bridge, Japan. (b) The longest span in western countries: the Storebaelt East Bridge.

1.2.2 Cable stayed bridges

After several pioneering ventures at the end of the nineteenth century, and the canal bridge of Tamul by Eduardo Torroja in the 1920s, the development of modern cable stayed bridges began in Europe, immediately after World War II. The first cable stayed bridges had very high, rigid girders and very widely spaced stays.

The static pattern recalled the concept of a girder resting on quite distant elastic supports, made from the stays. Examples of this type are the Donzère Mondragon Bridge in France (with a main span of 81 m), designed by Albert Caquot in 1952; the Stromsund Bridge in Sweden (182 m), designed by Franz Dischinger, opened in 1956; the Theodore-Heuss Bridge (260 m), in Mainz, Germany (Figure 15.5); and the Knie Brücke Bridge in Düsseldorf, Germany, designed by Fritz Leonhardt (Leonhardt et al., 1969) and built in 1958; and the Wye Bridge in Scotland, with just two couples of stay cables for each tower.

These bridges were fashioned completely of steel, except for the Tempul reinforced concrete canal bridge by Torroja and the prestressed concrete bridges with PC cables of Riccardo Morandi.

Figure 15.5 Theodor-Heuss bridge, Germany, stiff deck in steel plated girders, harp shaped stay cables, 260 m main span.

The first bridges with a central suspension and box girder with high torsional rigidity were developed by Hellmut Homberg, including the Rhine Bridge in Bonn, erected in 1967 (Figure 15.6).

Figure 15.6 Rhine Bridge in Bonn, Germany, box girder steel deck, central stay cables, harp-fan layout.

The requirement of a torsionally stiff deck girder took the consequence of deck with high flexural stiffness, and in turn the structural behavior corresponded to a girder on elastic supports, well distributed in this case due to the short spacing among the stay cables.

The setting for the development of a new conception of cable stayed bridges was an international design competition (Figure 15.7), and studies by Fabrizio de Miranda (de Miranda, 1971, 1980), which can be resumed by the following concepts discussed next.

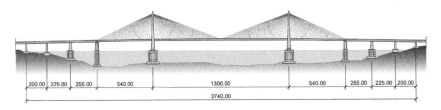

Figure 15.7 Design of a stay cable bridge for the Messina Strait Crossing, 1969, Ref. 5, Design Competition, 1300 m central span, A-shaped towers, box girder deck, cross-tie cables designed for stiffening the cable system.

By interrupting the bridge deck with articulations in the stay-coupling points, the structural layout can be seen as the large lattice girder type Gerber, a Gerber girder is a statically determinate continuous beam (Figure 15.8a) or that of an overturned trussed arch (Figure 15.8b). In both these diagrams, the diagonals (stays) reach the bridge deck starting from the top of the towers.

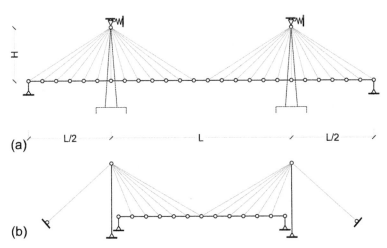

Figure 15.8 Structural systems of truss-like concept of cable-stayed bridge: statically determinate systems obtained by neglecting the deck stiffness. The equilibrium is fully insured only by axial forces in cables and deck. Two typical systems can be defined: (a) The lattice girder of Gerber type; (b) An overturned trussed arch.

The static behavior of these structures is characterized by the prevalent "normal-force" status in various areas, determined by the axial stress in the stays and in the bridge deck, while the "flexure" status in the girder becomes almost secondary if the spacing of the stays to the bridge deck are closer to each other. Densifying of the stays also simplifies the construction details relative to said couplings.

These innovative concepts, together with others expressed in the first design for the Messina Bridge (1968), as well as in the design and construction of the Paranà (1970) and Rande (1973) bridges (de Miranda, 1971; Baglietto et al., 1976; de Miranda et al., 1979), started the development of a new generation of cable stayed bridges. These structures were characterized by the following features:

- Slender and streamlined bridge decks
- Closer stays
- A "lattice" structure, whereas the main resisting system can be idealized by a statically determinate system obtained by neglecting the low flexural stiffness of deck (that is, by ideally placing hinges at the intersection of deck cables)
- Continuous suspension of the deck by stay cables for all bridge lengths
- Towers laid out with inclined legs in an A shape, as proposed in the Messina Bridge

By further developing and improving these concepts, modern cable stayed bridges have today surpassed 1100 m of free span.

Comprehensive discussions of suspension and cable stayed bridges can be found in a number of books (Podonly and Scalzi, 1976; de Miranda, 1980; Leonhardt, 1982; Gimsing, 1983; Walter et al., 1999).

2   Cable stayed bridges

2.1   Structural principles and concepts

Basically, a cable stayed bridge is a deck structure suspended by inclined cables. This suspension system can be realized in many different forms, and this leads to various structural concepts, described in the next sections.

2.1.1   Suspension system

As already mentioned, a cable stayed bridge can be interpreted in two different ways:

- As a girder on elastic supports, given by the stay cables system
- As a trusslike structure; that is, a bridge in which the equilibrium is guaranteed by a system made of elements such as stay cables, decks, and pylons, arranged in a triangulated layout and subjected to axial forces

The first concept corresponded with the idea of replacing some support points of a continuous girder with a series of supports made of the stays. These supports were at a distance of several meters from each other and required very high and heavy girders, making assembly difficult.

2.1.2 Deck slenderness

The modern concept of the cable stayed bridge has replaced the discrete distribution of the suspension points with widespread distribution, reducing the distance between the stays. In this way, the decks could be made more slender, as bending stiffness was no longer necessary to guarantee resistance and stability to the structural system, but just axial stiffness. The thickness of the deck was no longer related to the length of the central span; rather, it depended mainly on the width of the deck itself.

For example, the Rande Bridge (Figure 15.9), designed in 1970 and completed in 1977, had a slenderness (i.e., a ratio between central span and thickness of the deck) equal to 400 m/2 m = 200, when the largest cable stayed bridge realized up to that moment, with spans of the order of 300 m had girder depth of 3–4 m, with a slenderness ratio of less than 100.

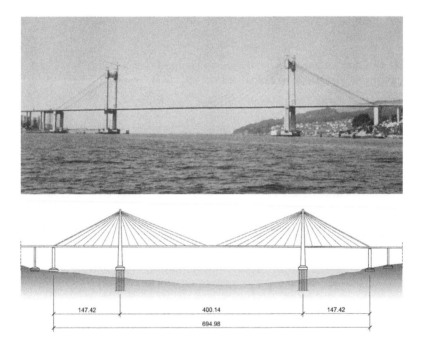

Figure 15.9 The Rande Bridge, Spain, 400 m of main span, longest span at time of design and construction, 1970–1977, first composite deck cable-stayed bridge, first application of multi-strand system for stay-cables.

2.2 Structural systems

On the basis of internal restraints and restraints attached to the ground, cable stayed bridges can be realized according to two different structural systems:

- Earth-anchored system
- Self-anchored system

These will be described in the next sections.

2.2.1   Earth-anchored

In the Earth-anchored bridge, the horizontal forces induced by the horizontal components of the stays are transferred to the Earth. The resulting structure is therefore a true and proper "tensile structure," as the deck girder is subject to traction and the only elements compressed are the pylons (Figures 15.10 and 15.11).

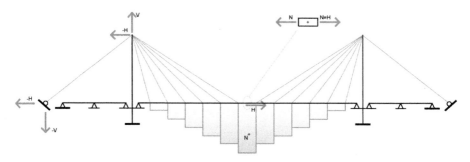

Figure 15.10 Earth anchored system: the horizontal thrust is equilibrated by the deck, in tension, and by the foundation blocks.

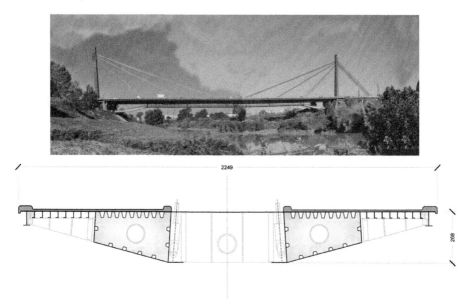

Figure 15.11 The Indiano Bridge, Ref. 6, 200 m of main span, 1968–1976, first earth-anchored cable stayed bridge, first twin-deck bridge.

The structural system is, therefore, similar as that of classical suspension bridges, with the following relative advantages and disadvantages:

- The deck, in traction, is subject to positive effects of the second order, which reduces their bending moments.
- High horizontal forces must be transmitted to the Earth, and this constitutes a technical problem that is sometimes difficult, and always expensive, to solve.

2.2.2 Self-anchored

In the self-anchored bridge, the horizontal components of the stay tensions are balanced by the deck girder, which is therefore subjected to compression (Figures 15.12 and 15.13).

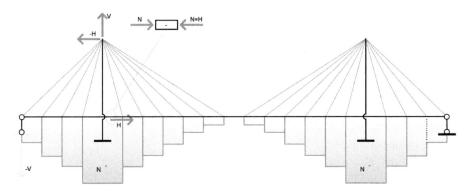

Figure 15.12 Self-anchored system: the deck only, in compression, equilibrates the horizontal thrust.

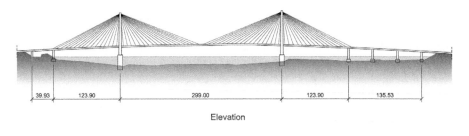

Elevation

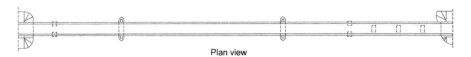

Plan view

Figure 15.13 Pasco-Kennewick Bridge, Ref. 29: first long-span, slender concrete deck, cable-stayed bridge in America, self-anchored. The horizontal thrust is equilibrated in a really economical way by the concrete deck.

This compression induces negative effects of the second order, as the bending moments tend to increase. But the great advantage of the self-anchored bridge consists in the absence of horizontal reactions to the Earth due to the effect of vertical loads acting on the structure.

Moreover, as we will see later in this chapter, this system allows for the realization of the construction using the progressive symmetrical cantilevers method (i.e., one of the most simple and effective construction methods), which has been the main reason of the success of the cable stayed bridge in the medium- to large-span field.

2.3 Cable configuration

The positioning of the cables in cable stayed bridges can follow two different basic configurations:

- Fan (Figure 15.14)
- Harp (Figure 15.15)

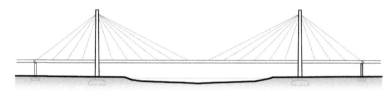

Figure 15.14 Fan configuration: forces are transferred by a clear truss behaviour, minimum weight, maximum stiffness.

Figure 15.15 Harp configuration: high stiffness of deck or towers are required for transferring of forces by bending; equal inclination of the stay cables leads to very ordered visual image.

In the fan layout, the stays are anchored to the upper end of the pylons and branch in various ways toward the deck. In this way, the main structure is effectively a lattice, formed from a series of triangular links, in which axial action prevails.

If articulations should occur or be formed in the nodes, equilibrium would be guaranteed by the main lattice system. The result is a highly efficient structure, whose structural dimensions can be set at minimum values.

In the harp layout, the stays are anchored along the pylons and are parallel. The bending moments in the pylon toward the calibration of the tensions in the stays can be eliminated for permanent loads. This is not possible for mobile loads, however, and the pylons and the deck are greatly stressed by bending.

In reality, these moments can be eliminated with the introduction of intermediate supports in the back spans (Figure 15.16). Otherwise, the deck or girder or both are

Figure 15.16 Harp configuration with side piers; main forces are transferred by truss behaviour: deck and towers can be slender.

subjected to the combined action of compression and bending moment and require greater stiffness and thickness than the case showing the fan configuration.

However, the fan layout leads to a high concentration of stress at the top of the pylons. This consequently creates difficult technical problems that must be solved. Therefore, in many cases, it is preferred to distribute the upper anchorages of the stays, placing them at short distance on the upper pylon.

The result is an intermediate configuration between the two described previously. This is called *harp-fan,* and it maintains the advantages of the fan structure to the extent that the distribution of the upper anchorages is compact (Figure 15.17).

Two planes of stays are typically envisioned in the transversal plane, which suspend the deck at the two transversal ends. In this case, the deck can be very slender and simple from a construction point of view and does not require great torsional stiffness.

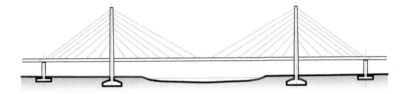

Figure 15.17 Fan-Harp or Semi-Fan configuration: anchorages on top of pylon, shared on a certain length, can be simply detailed.

Several planes of stays can be realized for very wide decks. By envisioning box-shaped girders under torsion, the deck can be suspended via just one central plane of stays (Figure 15.18).

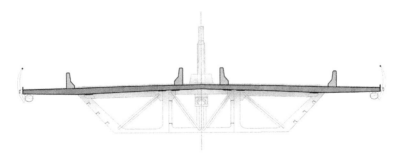

Figure 15.18 Box girders cross sections for central suspension of cable-stayed decks.

In this case, high torsional stiffness is typically accompanied by significant bending stiffness, and as a consequence, bending moments are high due to the effect of the mobile loads. The result is more complex and often heavier decks than in the first case. The requirement of high torsional stiffness calls for a great girder depth and higher aerodynamic drag, which makes this configuration less adapt for large-span bridges.

For long spans, the best layout comprises two planes of stay cables anchored at the top of A-shaped pylons in order to get the maximum torsional stiffness of the deck and thus increase its aerodynamic stability.

2.4   Structural elements

The main structural elements of cable stayed bridges are as follows:

*   Decks
*   Pylons
*   Stay cables

These will be discussed next.

2.4.1   Decks

Decks can be realized entirely in steel, entirely in concrete, or as a steel-concrete composite structure. Generally, the concrete solution is the most convenient for spans up to approximately 250 m. The composite structure can be used successfully in all spans up to about 600 m, but it adapts well to spans from 200 to 500 m (i.e., crossings of large rivers). The steel solution is the most expensive, but it is also the most suitable for bridges with spans exceeding approximately 500 m (that is, for long-span bridges).

Bridge deck (Figure 15.9a) with two planes of stays are realized effectively with two lateral girders, a series of cross-members with pitches varying from 3.50 to 7.00 m, and one slab in a concrete or orthotropic deck. Aerodynamic fairings, improving aerodynamical stability, and reducing drag, can be required for larger spans or higher winds (Figure 15.19b and c, and see de Miranda and Bartoli, 2001). The decks of bridges with a unique central plane of stays are always realized with a central box girder; two lateral overhangs are usually envisioned with variable length from 2 to 8 m. Bridge decks, mainly for longer spans, must be streamlined, with minimum depth and good aerodynamic properties (Figure 15.19b). This not only reduces the wind drag and increases the deck flutter stability, but also reduces vibration amplitudes in the deck so it can withstand vibrations of the stay cables.

2.4.2   Towers

Like decks, towers can be realized in steel, concrete, or in a steel-concrete composite. Typical configurations are illustrated in Figure 15.20.

The main technical problems are related to the upper part of the pylons, where very high vertical loads must be transferred in a limited space to the tower shaft and the horizontal components of the stay cables have to be equilibrated. Towers are not only a fundamental structural element, but they become the main aesthetical element in a cable stayed bridge. For this reason, their design is a difficult, challenging task of integrations of structural/engineering statements and aesthetical/architectural aspects.

2.4.3   Stay cables

Stay cables are the main and more special elements in this type of bridges. Their behavior, and mainly their axial stiffness due to the sag effect, are nonlinear. The nonlinear axial stiffness can be taken into account in an effective engineering form by the equivalent Ernst modulus (Figure 15.21).

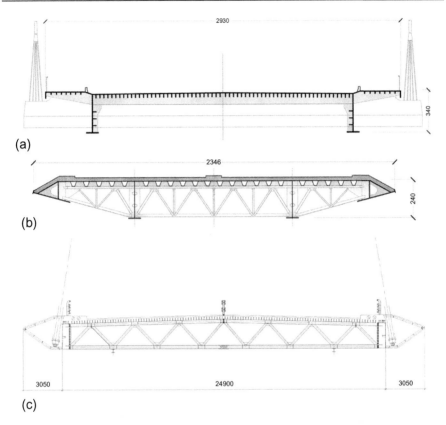

Figure 15.19 Typical cross sections for side suspension of deck: (a) Kniebrucke; (b) Rande Bridge; (c) Higuamo Bridge.

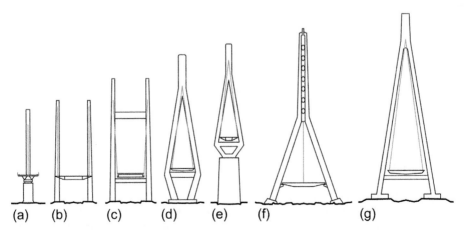

Figure 15.20 Types of towers for cable stayed bridges.

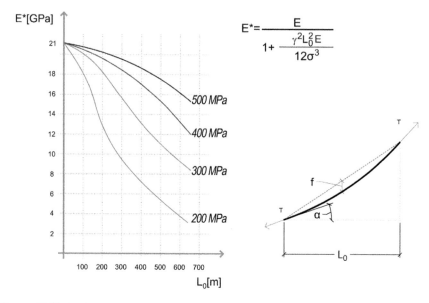

Figure 15.21 Equivalent Elastic Modulus for stay cables.

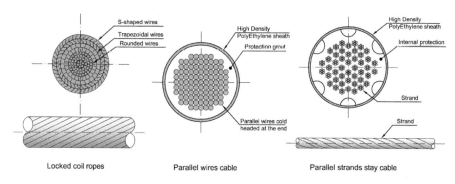

Figure 15.22 Typical cross sections of stay cables.

The following types of cable are mainly used (Figure 15.22):

- Locked coil rope
- Parallel wire cables
- Parallel strand cables

Solid bars and twisted ropes are used less frequently today.

The locked coil rope system was used in the first German cable stayed bridges, and it is still used today, especially for bridges with small and medium spans. The advantages of prefabrication and the consequent high executive quality are balanced by the difficulties regarding transport and installation of these very long elements (which, for larger bridges, may weigh a great deal).

Cables with parallel wires are very stiff and have high resistance to fatigue and low aerodynamic resistance and therefore, except for the difficulty of installation of the large prefabricated elements, they are suitable for bridges with large spans. Cables with parallel strands, in which the strands are installed on site one after the other, are currently the most popular system just by virtue of their easy installation, which requires light and easy tensioning in the cantilevered construction.

The design of stay cables is influenced by four main aspects: strength, fatigue, durability, and aerodynamic stability, listed in order of increasing severity. In fact, strength aspects are well addressed, knowledge is sufficient, and the codes seem to cover all aspects. Fatigue, although there is more uncertainty, is a clear issue from a design point of view, even if the uncertainties of aerodynamic aspects must be taken into account. The durability of stay cables, related to frequent lack of proper inspection and maintenance, is an important issue that requires more research.

Although much knowledge has been acquired over the last decades, the aerodynamic stability of stay cables still presents some degree of uncertainty. This instability is basically due to direct aerodynamic sources, such as:

- Von Karman vortices
- Wake galloping of closely spaced cables
- Buffeting, induced by wind turbulence
- Galloping of inclined cable, or ice accumulation
- Rain/wind induced vibration

In addition, there can be dynamic sources, like cable excitation due to deck/towers vibration from wind or traffic.

The aerodynamic causes depend on the wind actions on the cables. The parameters involved are:

- Wind speed V
- Cable diameter D
- Cable damping c, ρ
- Cable unit mass m

Their relative influence can be studied by looking at the dynamic equilibrium equation:

$$\ddot{y} \cdot m + \dot{y} \cdot c + y \cdot k = F(V(t), D),$$

where

  $y =$ transverse displacement of cable
  $k =$ stiffness of cable, inversely proportional to its length: $k \, \alpha \, L^{-1}$
  $c =$ damping.
  Also, it can be seen that the response y to wind action F are directly proportional to V, D, and L, and inversely proportional to c and m.

The nondimensional Scruton number takes most of these factors into account:

$$Sc = \frac{m \cdot \left(\dfrac{c}{c_{CR}}\right)}{\rho \cdot D^2}.$$

In order to avoid cable instability, the following empirical/experimental criteria were proposed:

- Von Karman vortices usually induce small oscillations and the inherent damping of stay cables results sufficient.
- Wind/rain oscillations can be kept small enough if
  $Sc \geq 10$ for smooth cable surfaces
  $Sc \geq 5$ for cable surfaces with helical ribs or protuberances that can prevent the stabilization of rain rivulets on the cable

According to this criteria, for Sc lower than 5, an additional damping system should be provided in most stay cables, independently of their length. Since the vibration of cables that are shorter than approximately 100 m is rare, the Sc criteria should be used only above this length threshold.

- Wake galloping of closed-spaced cables and dry galloping of inclined cables occurs (PTI, 2007) only above a critical speed of

$$V'_{CR} = (25 \div 80) \cdot f \cdot D \cdot \sqrt{Sc}.$$

Further investigations (FHA, 2007) showed that this statement is too conservative for real stay cables, and that, if Sc is greater than 3 and if the criteria for wind/rain vibration are fulfilled, no risk of dry galloping occurs.

The last cause of vibration is the forced oscillation of the cable ends. This effect does not depend on aerodynamic effects on cables and is often more difficult to control.

In this case, there are two control criteria:

- To increase damping of the cable
- To tune the cable frequencies in order to avoid the range of forcing frequencies

The increase of damping, for this and other instabilities, can be achieved in various ways:

- By installing internal dampers between the cable and the protection pipe, which can be made by high-damping elastomers and viscous or friction dampers
- By installing tuned mass dampers on the stay cable
- By installing external hydraulic/oil/viscous fluid dampers

The change (typically an increase) in the cable frequencies can be achieved by means of cross-cables, or cross-ties, or "aiguilles", interconnecting the stay cables in various ways.

The idea of introducing cross-cables connecting the main stay cables was first proposed by Fabrizio de Miranda, who patented the system (de Miranda, 1969) in the previously mentioned design of a cable stayed bridge for the Messina Strait Crossing.

The purpose of the cross-cables was mainly to reduce the sag effect of the longest cables in order to increase their stiffness. These cables can accomplish this very well, but also intuitively, to reduce their tendency to move and vibrate.

Later, the first experiences of cross-ties with the purpose of stabilizing vibrating cables occurred with the Stormsund Bridge in Denmark in 1971 and later in Japan. And more recently, cross-ties have been used in many large bridges which, after their opening, presented excessive cable vibrations, like the Dames Point Bridge and the Pont de Normandy.

Cross-ties segment the free length of the stay cables, in the cable planes, increasing their first vibration frequencies and adding damping to the system due to interference due to the connected cables vibrating at different frequencies. Research on optimal cross-tie configurations is in progress, but cross-ties have already proved to be effective, and they are the most efficient way of counteracting the vibration of cables in very long-span bridges.

2.5 Analysis and design

The analysis of a cable stayed bridge is divided into three different phases:

- Equilibrium conditions for permanent loads
- Construction phases
- Analysis of the structure in service

The first phase defines the forces to be applied to the stay cables at the end of construction such that:

- The deck will have design geometry and present a design distribution of bending moments, usually as uniform as possible.
- The tower will stay vertical; i.e., in equilibrium under the action of the horizontal components of the stay cables from side and central spans. (Figure 15.23).

This phase defines the initial state of the bridge.

The second phase defines the forces to apply to the stay cables and the precambers to apply to the deck so that:

- The deck under construction will be in a safe condition: i.e., checking of the bending moments in deck and towers and forces in stay cables.
- The final force distribution in the stay cables and the final bending moment distribution in deck and towers will be that defined in the initial-state analysis.
- The deck profile will be the design geometry at the end of all construction phases.

The third phase is the elastic, linear, and nonlinear analysis for all in-service load conditions: live load, wind, earthquake, temperature, etc. Although FEM methods are used for the final checks, the design phase can utilize equilibrium handmade analysis and the result of useful closed-form formulations.

Bending moments in the deck girder (Figure 15.24) are inversely proportional to the deck stiffness since they are related to the imposed deformation of the cable system; for a three-span fan-shaped cable stayed bridge, they were calculated in closed

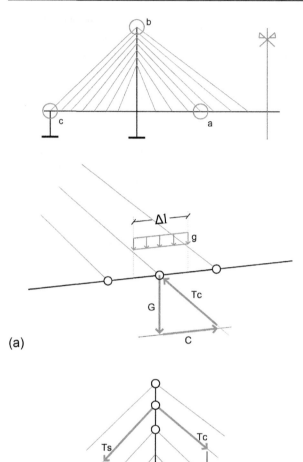

(a)

Figure 15.23 Equilibrium conditions and pre-design fundamental equations of stay-cables of a three-span self-anchored cable stayed bridge. (a) Stay cable-deck equilibrium: define design forces in main span stay-cable. (b) Horizontal equilibrium of side-spans and central span stay cables: define the side spans cables. (c) Global half-central span rotational equilibrium: define design forces in tie-down and anchor cables.

(b)

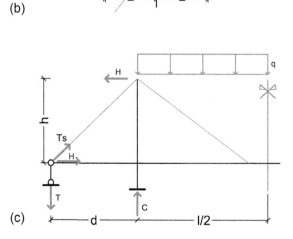

(c)

form by a differential equation (de Miranda, 1980); the maximum value at midspan can be estimated by the following equation:

$$M(L/2) = 0.165 * q * \sqrt{4 \cdot E \cdot J \cdot d \cdot \Delta},$$

where:

$q =$ live load per unit length
$E =$ elastic modulus of deck
$J =$ deck moment of inertia
$\Delta =$ spacing of the stay cables
$d =$ maximum flexibility of the cable system for a unit-concentrated load, given by

$$\frac{N_b^2 \cdot s_b}{E_b^* \cdot A_b} + \frac{N_m^2 \cdot s_m}{E_m^* \cdot A_m} + \frac{N_g^2 \cdot s_g}{E_g \cdot A_g},$$

where:

$N =$ element force
$s =$ element length
$E^* =$ Ernst modulus
$A =$ element area
Cable index $m =$ midspan cable
$b =$ back cable
$g =$ deck girder

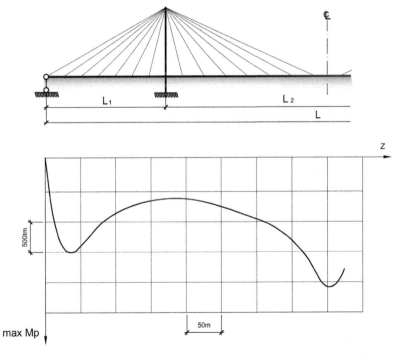

Figure 15.24 Typical bending moment diagram in deck of three spans cable-stayed bridge with slender deck and continuous suspension.

The circular frequency of the first vertical mode of the deck can be estimated in a very synthetic manner still, for a three-span bridge and neglecting the deck stiffness as reasonable for long-span bridges (Wyatt, 1991). If C is a function of geometric ratio, and for $L_{SIDE}/L_{MAIN} = 0.36$ and $H/L_{MAIN} = 0.22$, $C = 1.3$. If h is the pylon height above the deck, σ is the average stress in cables for permanent loads, and g is gravity acceleration, the following results:

$$\omega^2 = C \cdot E \cdot g \cdot h / \left(\sigma \cdot L^2_{MAIN} \right).$$

Cable stayed bridges have a very wide variety of structural systems, shapes, and technologies that make this bridge type conducive to strong innovations and development all over the world. In the last 25 years, the main span lengths have increased by a factor of 2. Remarkable long-span cable stayed bridges in the last 15 years have been the Pont du Normandy, France (built in 1995, 856 m), Tatara, Japan (built in 1998, 890 m), Sutong, China (built in 2008, 1088 m) and Stonecutter, Hong Kong (2009, 1018 m).

The longest span of a cable stayed bridge today belongs to the Vladivostok Bridge, at 1104 m, which was built in 43 months and completed in July 2012 (Figure 15.25; also see SKMost, 2012).

Figure 15.25 Vladivostok bridge, the presently longest span cable stayed bridge.

2.6 Construction methods

The following procedures are typically adopted:

* Installation with provisional supports, which includes:
  * Installation of the deck on temporary supports, with possible longitudinal launching
  * Erection or construction of the towers
  * Installation and tensioning of the stays

This procedure is necessarily adopted for Earth-anchored type decks, but it also can be used for all bridge types, provided there is easy access for the deck propping. Therefore, it is not suitable for long-span bridges.

* Installation by progressive cantilever (Figure 15.26), which includes:
  * Realization of the towers

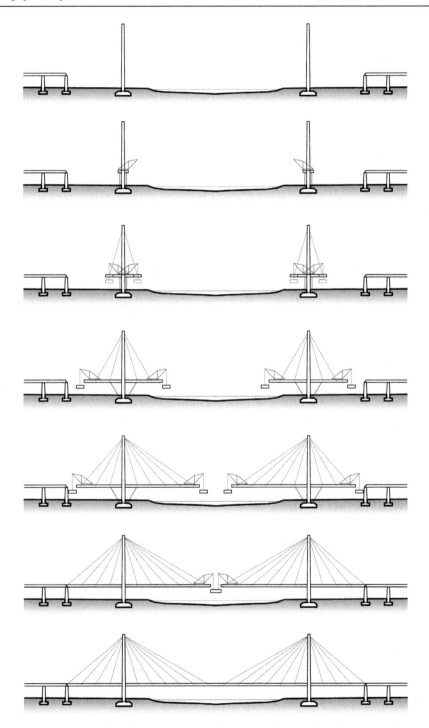

Figure 15.26 Typical erection procedure by balanced symmetrical cantilevering of a CSB. A temporary bracing system by means of bottom counter-stays is shown.

- Lifting off the ground and installation (or assembly on site) of the segments of deck, pro-
  ceeding symmetrically from the piers
- Progressive installation of the stays, in parallel with the assembly of the segments
- Key joints between the two half-bridges

This procedure is adopted when it is not possible (or inconvenient) to install provisional supports. It is the typical procedure for self-anchored cable stayed bridges,
and is also suitable for long-span bridges.

A detailed description of the construction of two bridges by cantilever method are
given by M.de Miranda (2003, 2008).

- Longitudinal launching of the entire bridge, which includes:
  - Preassembly on deck scaffolding, tower and stays of each half-bridge on the access
    abutments
  - Longitudinal translation of the two half-bridges to the final position and subsequent clo-
    sure in the key
- Launching by rotation, which includes:
  - Preassembly on deck scaffolding, antenna, and stays on each half-bridge perpendicular to
    the alignment of the definitive deck
  - Rotation of about 90° of each half-bridge around a vertical axis, coinciding with the axis
    of the mast, to reach the final position and the subsequent closing in key

The procedures outlined here, described with reference to typical three-span bridges with two towers, are generally also applied to bridges with a single antenna or
multiple spans.

For bridges with span over 200 m, the progressive cantilever is typically the most
affordable system, so it is generally used. For long-span bridges of the earth-anchored
type, a progressive assembly of the segments of deck is possible, starting from the
center line of the central span and proceeding to the towers.

3   Suspension bridges

3.1   Static principles and structural form

As stated previously, in a suspension bridge, the deck is sustained by means of vertical
or subvertical cables and by one or more parabolic main cables supported by vertical
pylons. The following points are key:

- The main cables take the profile of the funicular curve of the loads applied to them. The
  funicular curve of the cables self-weight is a catenary.
- The funicular curve of the weight of deck is a second-order parabola. The actual cable profile
  is a curve that will stay between these two, and also, for small sag/span ratios, stays very
  close to the parabola.
- The suspenders, or hangers, simply transmit the load from the deck to the main cables.
- The deck has the simple function of transferring self-weight and live load to the hangers.
  Therefore, if its function were limited to this purpose, it can be very light and slender.

The main cables, being shaped like a funicular curve, generally have low stiffness for localized loads, related mainly to second-order effects. Therefore, excessive slenderness, and in turn excessive flexibility of the deck, lead to large displacements under localized loads, as well as aerodynamic instability. For this reason, the deck structure also has the function of stiffening the entire structure, and it is assigned the correct stiffness in the design phase.

3.1.1 Self-anchored versus earth-anchored

Similar to the cable stayed bridges, the main cables of a suspension bridge can be anchored to the ground or to the deck girder. The first case is the classical configuration, which has the following advantages:

- The construction can be realized without intermediate supports, since the main cables can be installed before the deck and this can be erected by suspending its segments to the cables.
- The second-order effect given by the tensile force in the main cables increases system stiffness and reduces the bending moments in the deck.

The disadvantages of Earth anchoring occurs in the difficulty of anchoring very large horizontal forces, which amount to hundreds of thousands of tons and are usually applied meters above the strong layers of soil. The self-anchored suspension bridge removes this last difficulty, transmitting only vertical loads to the soil.

Inversely, construction is more difficult—at least for long-span bridges—since the deck must be present when the cables are installed. Therefore, the deck must be erected on temporary supports. Furthermore, the positive second-order effects due to the tension in main cables are fully compensated by the negative second-order effect of the compression force on the deck. The latter forces are typically very high, meaning the deck structure must be strengthened.

For long-span bridges, the Earth-anchored system is usually more convenient.

3.1.2 Cable layout

The main cables can be arranged in many configurations, according to the dimensions and number of the spans to be crossed (Figure 15.27). The typical sag/span ratio is in the order of $1/8 \div 1/10$; a deeper profile gives economy of cable steel, while a tight profile gives greater stiffness and fewer deck bending moments.

In order to reduce system flexibility for asymmetrical load conditions (i.e., those giving maximum displacement), a link between the main cables and the deck at midspan is very useful: the horizontal movement of cables is prevented and, in turn, the vertical displacements are reduced.

The hangers are usually vertical. However, it is possible to incline them in order to form a trussed layout. In this way, they tend to behave like shear-resistant structures in which the inclined cables, pretensioned by the deck self-weight, act as the diagonal of an ideal truss structure.

The positive result of this is a stiffer structure. The disadvantage, however, is an increase in the stress range in the hangers, which increases fatigue.

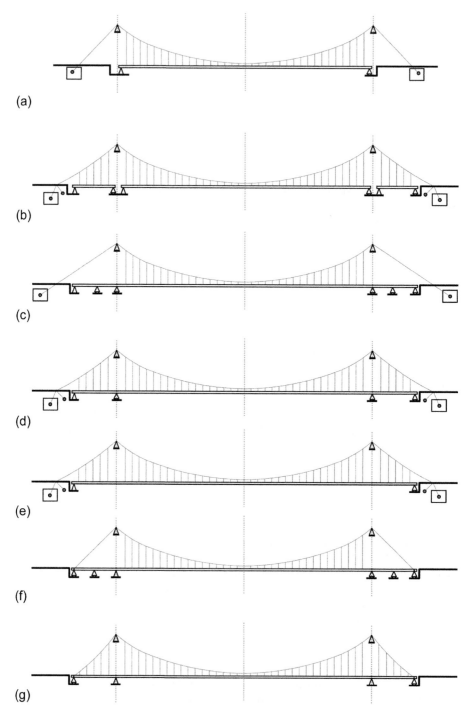

Figure 15.27 Different types of suspension bridge configurations: (a) Simple span; (b) Three spans simply supported; (c) Continuous deck, side spans earth supported; (d) Continuous three spans, fully cable supported; (e) Continuous suspension between anchor blocks; (f) Self-anchored central span cables supported; (g) Self-anchored three spans cable supported.

3.1.3 Deck

The deck structure is usually of an orthotropic plate type; for smaller spans, it can be a concrete slab. Deck girders basically come in three types:

- Truss structure
- Plate girder
- Box girder

The trusslike girder was chosen by American engineers when building bridges until 1970 in order to give great stiffness to the whole structure while maintaining relatively low aerodynamic drag.

For the longest spans, the weight of steel is higher than for other systems.

The longest suspension bridge in the world today, the Akashi Kaikyo (Figure 15.28; also see Kashima, 1998), has a truss-stiffened girder.

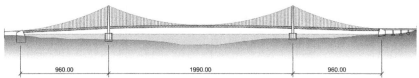

Figure 15.28 Akashi Kaikyo bridge, with deep but transparent truss stiffening girder, simply supported spans. Layout and erection phases.

Plate girders allow the lighter and simplest structural system. But its disadvantages are that its aerodynamic behavior is worse than that of a streamlined box girder, relatively low flexural stiffness and, mainly, very low torsional stiffness. Nevertheless, the use of aerodynamic fairings can improve their aerodynamic performance, and a bottom bracing can improve torsional stiffness.

Box girders, which have a streamlined-aerodynamic profile, apart from a relatively high fabrication cost, present many advantages, just like the deck of suspension bridges, as follows:

- Flexural and torsional stiffness are high.
- Aerodynamic drag is low.
- Aerodynamic properties of the cross section, related to the flutter stability, are good.

However, the aerodynamic stability for a classical box girder suspension bridge depends on its first-mode torsional frequencies, and therefore on its main span length. For very long spans (i.e., > 1600 m), or for very high design wind speed, the stability of a single box girder would not be good enough.

Splitting the deck into two streamlined box girders increases flutter stability greatly. This circumstance was observed in the wind tunnel tests of the Indiano Bridge in Florence, with a double box-girder deck, at the National Physical Laboratory (NPL), London, in 1970, and later formalized by Richardson (1984).

The higher stability of decks with central openings, however, was already out looked by Farquharson (1950–1958) in the wind tunnel tests in 1950, and put in practice in the Mackinac Bridge in Michigan by David Steinman in 1965.

3.1.4 Pylons and anchor blocks

The pylons of suspension bridges (Figure 15.29) can be steel structures, as is usually the case for U.S. and Japanese bridges, or concrete structures, like the Storebaelt Bridge. Concrete structures are typically more economical, at least in areas of low seismicity and with good soil conditions.

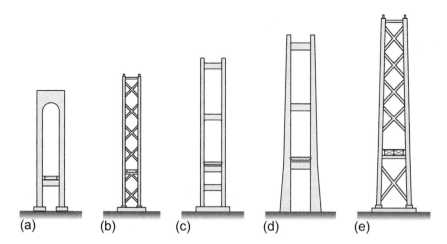

Figure 15.29 Typical layouts of suspension bridge towers, in size ascending order. (a) simple frame; (b) multiple frame; (c) trussed; (d) stain lined frame; (e) stain lined truss.

The tops of pylons have to accommodate the cable saddles, where practically all the load of the half bridge is concentrated. The anchor blocks, always set in concrete, have the purpose of transmitting both vertical and horizontal forces to the soil, transferred by the cable anchorages. They are massive structures, which contribute to much of the total cost of the bridge, as well as lengthening construction time.

3.2 Analysis: special aspects

The global analysis includes static and dynamic analysis, linear and nonlinear aero-elastic checks, and examining the stability of the tower cables and deck. This analysis includes the local stress and fatigue checks of the elements of the deck, as well as checking the local forces on cables, transverse forces at saddles and hanger clamps, and bending moments localized in the vicinity of saddles.

3.2.1 Analysis for vertical loads

The classic theme in the analysis of suspension bridges consists of determining the state of deformation and stress of the cables and the stiffening girder of the bridge deck, taking into account the interaction between them. Starting in the early nine-teenth century, basically, three theories have been developed: first was the theory of Rankine (1869), which was simple but only approximate, then the first-order theory (elastic theory), by Ritter, Levy, and Melan, which leads to correct results at the first order for self-anchored bridges, and precautionary but acceptable for bridges with very stiff girders. Finally, the second-order theory (i.e., deflection the-ory), expressed by Josef Melan (1888) and firstly applied by Lev Moisseiff and Lienhard (1933), which made it possible to obtain accurate results for bridges with deformable decks and, in essence, made it possible to realize the modern long-span suspension bridges.

Geometry and forces for permanent loads

We typically assume that the stiffening girder appears devoid of bending moments at the end of construction, and therefore, the shape of the cable corresponds to the funic-ular of applied permanent loads. This is actually achieved by proper erection procedures.

The geometry of the cable in the central span, with distance L between the ends of the cable and sag f, is defined by the following function:

$$y(x) = M_o(x)/H = \text{ordinate axis of the cable,}$$

where

$$H = M_o(L/2)/f$$

$= $ horizontal component of the cable force for permanent loads, and

$M_o(x) = $ moment of permanent external loads on a beam in simple support of span l.

In the case of the permanent p load uniformly distributed, on a beam of span L, we get:

$$y = 4f/L^2 \cdot (L-x) \cdot x$$

and:

$$H = pL^2/8f.$$

Geometry and loads for moving loads

The differential equation of vertical equilibrium of deck, in the theory of second-order, can be written in the following form:

$$p = EJ\frac{d^4y}{dx^2} - h \cdot \frac{d^2y}{dx^2} - (H+h)\frac{d^2y}{dx^2},$$

where h represents the variation of the horizontal component of the force of the cable due to the movable load.

The last term of the equation, which depends on the variation of load applied to the beam because of the tension in the cable produced by the change of geometry, is neglected in the first-order theory, which then provides acceptable results when:

$$(H+h) \cdot \frac{d^2y}{dx^2} \text{ is negligible compared to the term}: h \cdot \frac{d^2y}{dx^2},$$

which happens to girders with very stiff decks.

This equation can be solved through different methods for succesive iterations, and the bending moment in deck girder can then be expressed by

$$M = M' - hy, \quad (\text{first} - \text{order})$$

$$M = M' - hy - (H+h)v, \quad (\text{second} - \text{order})$$

where M' represents the moment of moving loads on a simply supported beam of span l.

It may be noted that in a generic section of the deck, the bending moment is equal to the product between the horizontal component of the cable force and the displacement between the actual deformed configuration of the cable and the configuration that it would assume in the absence of a deck; that is, the funicular of the external forces (Figure 15.30).

In the first approximation, if we consider, instead of the actual deflected profile, the configuration that the cables assume for the action of permanent loads, we obtain the results of the theory of first order.

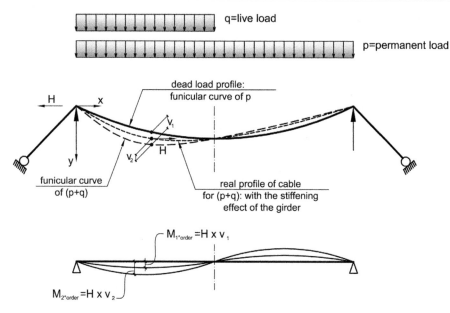

Figure 15.30 Cable profile: (a) Permanent loads (p) shape: funicular curve of p; (b) Real cable profile after loading live load (q), with stiffening effect by girder: total load = p+q; (c) Funicular curve of cable for p+q.

Predimensioning

For the analysis of the final suspension bridges, numerical procedures are adopted in nonlinear regimes of large displacement, for both the analysis of the construction phases and computing initial geometries, and then for the analysis of the bridge in service. However, for the first dimensioning calculation, it is possible to use the following approximate expressions, derived in part from the theory of linearized second-order.

- Geometry and balance of the cable for permanent loads (Pugsley, 1968):

  - Development length of the cable:
  $$l = L\left(1 + \frac{8}{3}\left(\frac{f}{L}\right)^2 - \frac{32}{5}\left(\frac{f}{L}\right)^4\right)$$

  - Horizontal component of the cable force:
  $$H = \frac{\rho L^2}{8f}$$

  - Max. cable force:
  $$T = H\left(1 + 16f^2/L^2\right)^{1/2}$$

  - Variation of the cable length
  $$\Delta l = \frac{Hl}{AE}\left(1 + \frac{16f^2}{3\,L^2}\right)$$

  - Vertical displacement of the center line due to Δl:
  $$v = \frac{\Delta l}{\dfrac{16\,f}{15 - L}\left(5 - \dfrac{24f^2}{L^2}\right)}$$

- Vertical displacement at the center line for the horizontal displacements ΔL of the heads of the towers, neglecting the stiffness of the deck:

$$v' = \frac{\Delta L \left(15 - 40\frac{f^2}{L^2} + 288\frac{f^4}{L^4}\right)}{16\frac{f}{L}\left(5 - 24\frac{f^2}{L^2}\right)}$$

- Variation of cable force and vertical displacement at the load section, due to the action of a concentrated load P applied at abscissa x, neglecting the stiffness of the deck:

$$\Delta H = \frac{3Pl}{4f} \cdot k(1-k)$$
$$v = 4/3 \cdot f \cdot \Delta H/(H - \Delta H) \cdot (3k^2 - 3k + 1),$$
$$with:$$
$$k = x/l$$

- The bending moment for maximum traffic load distributed over a stretch ($0 < x < a$) for a simply supported girder of deck:

$$M_{MAX} \cong 0.161 \cdot p \cdot L^2 \cdot \left(4/\alpha \cdot EJf/wL^4\right)^{1/2}$$

in which is the equivalent stiffness of the cables system, and can be approximated as

$$\alpha \cong 1100 \, \text{kN/m}^2.$$

3.2.2 Analysis for horizontal loads

The great amount of slenderness of the deck in the horizontal plane for large-span suspension bridges (L / B = 40–60) have significant second-order geometric effects on the calculation of displacements and bending moments in the horizontal plane of the deck.

In short, the supporting cables, which are connected to the deck by means of the hangers, follow the movement of the deck to the action of the wind arranging the hangers on inclined planes, thus absorbing part of the horizontal actions and transferring them to the top of the towers (Figure 15.31).

The deck is then subject to directly applied horizontal actions and to the opposite reactions provided by the hangers and transmitted to the cables; by this mechanism, the resulting bending moments and displacements are significantly less than with those calculated with the theory of the first order, and a huge transfer of horizontal transverse force from deck to the top of the towers occurs.

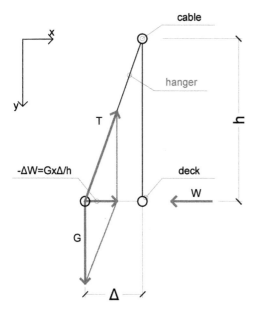

Figure 15.31 Pendulum effect of hanger and restraining force, windward directed, responsible of a substantial reduction of horizontal bending moment in deck.

3.3   Methods of construction

In classic suspension bridges, the operational sequence must pass through the following stages:

- Construction of towers and mooring blocks
- Formation of the supporting cables and installation of hangers
- Installation of the girder deck

The installation of the deck is performed by lifting the structural elements of the deck (or panels of the truss segments or whole segments) from the sea (or river, or from the ground, depending on the environment) by a crane positioned on the cables or on the already-erected deck.

In bridges with trussed decks, it is normal to start from the towers and proceed symmetrically toward the middle of the central span and toward the end of moorings (Figure 15.32b). In bridges with box-girder decks, the segments are assembled initially starting from the center line and proceed symmetrically to the towers to reduce the risk of flutter during construction (Figure 15.32a).

After lifting, the segments are connected temporarily with devices designed to allow mutual rotation during the erection of the adjacent segments, but also to guarantee the necessary stability due to the dynamic effects of the wind. A detailed description of the erection phases, as well as of problem solved in building a suspension bridge, is given by de Miranda and Petrequin (1998).

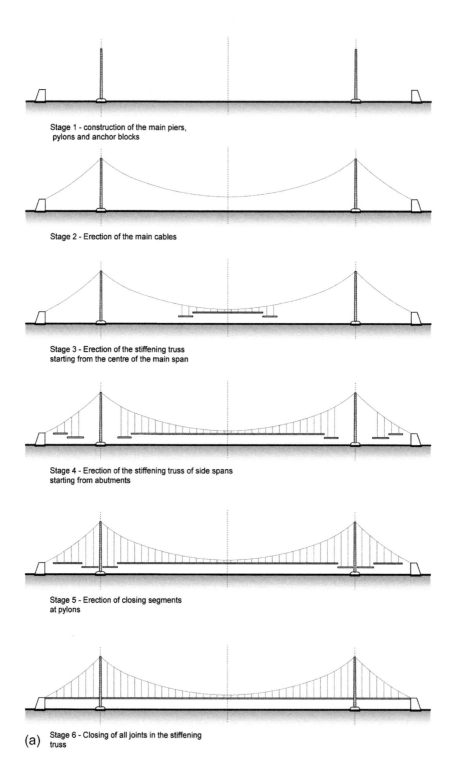

Stage 1 - construction of the main piers, pylons and anchor blocks

Stage 2 - Erection of the main cables

Stage 3 - Erection of the stiffening truss starting from the centre of the main span

Stage 4 - Erection of the stiffening truss of side spans starting from abutments

Stage 5 - Erection of closing segments at pylons

(a) Stage 6 - Closing of all joints in the stiffening truss

Figure 15.32 (a) Erection sequence of typical box-girder suspension bridge, by starting from mid-span in order to minimize the risk of flutter during construction.

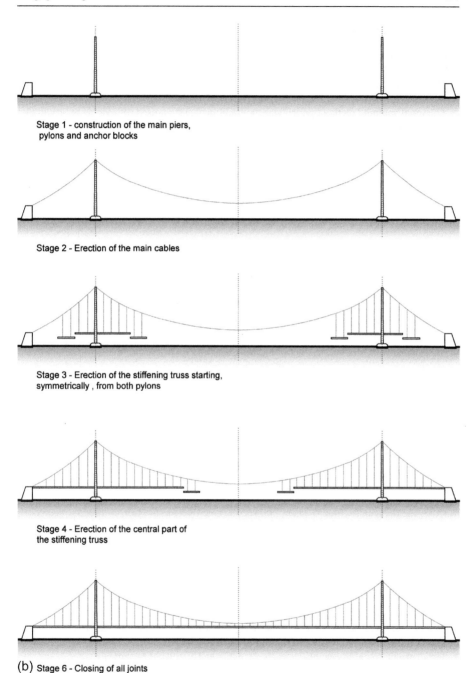

Stage 1 - construction of the main piers,
pylons and anchor blocks

Stage 2 - Erection of the main cables

Stage 3 - Erection of the stiffening truss starting,
symmetrically , from both pylons

Stage 4 - Erection of the central part of
the stiffening truss

(b) Stage 6 - Closing of all joints

Figure 15.32 Continued. (b) Erection sequence of deck starting from pylons.

3.4   Technology of main cables and hangers

Supporting cables are always made of strong steel in parallel wires and are galvanized, with diameters of 5.2–5.7 mm. The following two methods are distinguished by the mode of formation of the cable systems:

- Aerial spinning method, which was the traditional system used for over a century in the construction of suspension bridges, which consists of cable assembly on site, working with individual wires (Figure 15.33)

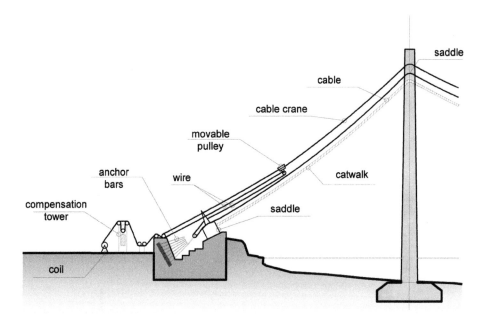

Figure 15.33 Traditional erection method of cables, by wires spinning.

- With bundles of prefabricated strands of parallel wires, the system that allows (theoretically) greater independence from environmental conditions and a higher execution speed.

The realization of the cables typically comprises the following steps:

1. Installation of walkways (i.e., catwalks) made of stranded steel wire, wooden sleepers, and network security, arranged approximately 1.00–1.50 m below the axis of the cable
2. Mounting of a catwalk bracing and stabilizing system consisting of cross-cables and hangers and of transverse walkways; installation of saddles
3. Installation of a cable car, placed over the cable and stabilized by catwalk
4. Spinning of the wires (or strands) from an anchoring end block to the other block and formation of the cable through tiling of the wires, according to the geometry of the project
5. Progressive compaction of the cables
6. Installation of hanger clamps
7. Installation of the surface protection system, consisting typically of a painting and a bandage with galvanized wire with small diameter (3–4 mm)

The cables of the longest suspension bridges that exist today have diameters of 0.82 m (Storebaelt, Denmark) and 1.12 m (Akashi Kaikyo, Japan).

Hangers are made with wire ropes or spiral cables or locked coil ropes or parallel wire ropes, always in galvanized steel and protected by sheaths, often in high-density polyethylene. The clamps that support and anchor the hangers to the main cables are typically made of cast steel, with bolt anchors.

The anchorages to the deck are necessarily of the hinge type for the hangers next to the mooring blocks, where the longitudinal rotation of the hangers is at its maximum, while they may be of the rigid type, with possible adjustment rings, for the intermediate hangers.

3.5 Aerodynamic stability

3.5.1 Deformable structures

For deformable bridge structures, the dynamic effects of the wind must be considered. In particular, these effects are mainly the following:

- Dynamic amplification of the structural response of the turbulent component of the wind and the impulsive action of periodic gusts (buffeting)
- Dynamic action induced by the detachment of vortex wakes (Von Karman vortices)
- Actions induced by aeroelastic instability, that can formally be divided into the following areas:
- Divergence for pure torsion
- Flutter for pure bending (galloping)
- Flutter for pure torsion (stall-flutter)
- Flutter for coupling of bending and torsion (classical flutter).

For spans typically longer than 200 m, the low vibration frequencies determine the need to verify the conditions of aerodynamic stability, briefly summarized here along with some verification criteria.

Von karman vortex
The alternating vortex street wake downwind of the deck induces aerodynamic pulsing actions which, when resonating with the frequency of vibration of the deck, can induce bending and torsional oscillations. The critical speed, for which the vortex shedding occurs at the same frequency of the deck, is equal to

$$V_{cr} = n_r \cdot d / S_t (m/s),$$

where: n_r = natural frequency of mode r in the plane normal to the wind direction (Hz) with:

b = effective width of the deck (m)
d = height of the deck (m)
S_t = Strouhal number, $\cong 0.08$ for $b/d \geq 10$

$0.08 < S_T < 0.15$ for $10 < b/d < 5$

$S_t \cong 0.15$ for $b/d \leq 5$

Vertical oscillations can occur if

$V_{cr} \leq 1.2x\, V_m,$

where $V_m =$ average characteristic wind speed (average of $10'$) (m / s).

In such a case, the aerodynamic loads applied to the structure, the amplitude of the oscillations, and the stress induced must be evaluated.

The maximum flexural displacement is given, in first approximation, by the following expression:

$$y_{MAX} = \frac{b^{1/2} \cdot d^{5/2} \cdot \rho}{4 \cdot m \cdot \delta_s},$$

with values generally overestimated with respect to the actual displacements, especially in the case of continuous, very long bridge decks.

The extent of the structural response to the wind action depends on the aerodynamic shape of the cross section of the deck, atmospheric turbulence (and therefore orography and the height above the ground), and the actual aerodynamic damping.

Torsional-flexural flutter

The torsional-flexural flutter phenomenon consists, in synthesis and with some simplification, of coupled oscillations in bending and torsion of the deck, fed and amplified by the action of the wind. Instability occurs when the wind speed has the effect of reducing the torsional frequency (which is decreased by the aerodynamic torque) to the same value as the bending frequency.

The critical speed for flutter, provided the ratio between flexural and torsional frequencies are far from unity, results in a first approximation by a modified Selberg formula:

$$V_f = k_f \cdot 3.7 \left(1 - \eta_B/\eta_T\right) \left(m \cdot r/\rho b^3\right)^{\frac{1}{4}} \cdot b \cdot \eta_T (m/s)$$

where

$\eta_B =$ first bending frequency (Hz)
$\eta_T =$ first torsional frequency (Hz)
$m =$ mass unit (kg/m)
$\rho =$ density of air (kg/m^3)
$b =$ width of deck (m)
$r =$ polar radius of inertia of the section center line (m)

$k_f =$ shape coefficient equal to unity for flat plate or aerodynamically well profiled sections, and lower up to 0.2 for not streamlined sections.

For safety, it should be:

$V_f \geq 1.5 \cdot V_m,$

in which the multiplier 1.5 takes into account the possible increase of the wind speed for short periods and a safety margin, and V_m is the average characteristic speed for a period of 10 min at the height of the deck.

For sections that are not aerodinamically profiled (that is, for "bluff" sections), V_f is lower than the relative value at profiled sections, and is evaluated based on tests run on similar sections in the wind tunnel.

It's interesting to note that the aerodynamic torque, responsable for the reduction of the torsional frequency and in turn of the critical speed, is due to the forward shift of the lift force in a simple wing airfoil that is proportional to the profile width or chord. But in the case of a twin airfoil, the halved airfoil chord more or less also halves the aerodynamic torque and then increases flutter speed. The following points are important:

- In several suspension bridges, severe oscillations have occurred due to wind as a result of Von Karman vortices and buffeting. In some cases, remedial actions (usually aerodynamic devices like winglets or fairings) are taken to reduce or eliminate them. However, the oscillations for flutter, which rarely occur, can have catastrophic results.
- The aerodynamic stability is, therefore, a primary issue in the design of suspension bridges.
- The conditions for aerodynamic stability are more important in the construction phase than during service, since the structure is incomplete and more flexible during construction. However, the design speed can be considered during the construction phase, related to a shorter window time and then to a shorter return period that is less than the desired life of service.
- On bridges of significant length (300 m), the oscillations due to Von Karman vortices can be triggered even on relatively high eigenmodes, unlike what occurs in girder bridges.
- The action of cross-winds on deck over long spans has a mostly dynamic character: the effects of buffeting, which are often prevalent, must be added to the effects of the uniform component of the wind pressure.

4   Limits of long-span bridges

As was stated at the beginning of this chapter, long spans are generally intended for bridges for which the structural weight is overwhelming, and it becomes necessary to adopt a structural system based on lightweight, strong cables. Effective discussions of the limits and optimal structural systems of long-span bridges were proposed several decades ago by Steinman (1922), Stüssi (1954), and Gimsing (1983).

The choice of the structural system depends on the following three aspects, as stated previously:

- Optimization of weight and cost of the structural material
- Guarantee of aerodynamic stability during service and during construction
- Feasibility of the construction system

From the point of view of the quantity of material used, for spans of between 500 and 1500 m, cable stayed bridges are more convenient than suspension bridges of the same span (Figure 15.34; also see de Miranda, 1971).

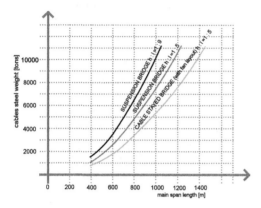

Figure 15.34 Comparison of weight of cables for suspension and cable-stayed bridges. A cost comparison will also include the increase of tower weight and deck weight in cable-stayed bridge, and the large cost of anchor blocks in suspension bridge.

It involves, however, towers higher by 60–70% and, in the self-anchored layout, requires the realization of large cantilevers in the construction phase, which are sensitive to the effects of the wind. They are, in fact, the construction aspects and aerodynamic stability during construction which until now have favored the suspension bridge for longer spans.

Currently, the longest spans of the various types of bridges are:

- Cable stayed bridge, wing deck: L = 1104 m (in Vladivostok, Russia erected in 2012)
- Cable stayed bridge, twin deck: L = 1018 m (Stonecutter, Hong Kong, erected in 2009)
- Suspension bridge with wing deck: L = 1650 m (Xihoumen Bridge, China, 2009)
- Suspension bridge with trussed stiffening girder: L = 1991 m (Akashi Kaikyo, Japan, erected in 1998).

These bridges represent the culmination of a long and gradual evolution (Figure 15.35) of the structural systems and cross sections of traditional decks, where it seems that the limits of free spans have been achieved, at least for suspension bridges. To overcome these limitations, and mainly to increase the aerodynamic stability of even longer spans, various solutions have been proposed along the last decades, such as the following, which appear promising:

- Decks with central openings or with multiple box girders to improve aerodynamic stability;
- Cross bracings between cables and decks in suspension bridges
- Mono-cable suspension systems
- Cross-tie in cable stayed bridges
- Use of aerodynamic fixed or active control
- Systems of cable stayed bridges that are partially earth-anchored
- Mixed suspension systems: cable stayed and suspension, net systems

The effective design implementation and integration of these solutions, together with the use of high performance materials should help to overcome the current limit of 2000 m and achieve progressively even larger spans.

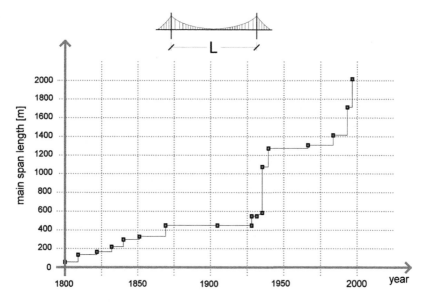

Figure 15.35 Increase in maximum span length of suspension bridges along the time.

Anyway, most of the abovementioned solutions introduce erection complications, and the effects of wind, temperature, and heavy weights of structural elements increase with the bridge dimension. Then, to overcome spans still longer than those achieved so far, these problems must be solved by the construction methods, as well as by all those activities for which it is difficult to extrapolate a real assessment of the operational difficulties to the new size.

Looking at the graph that shows the progress in the free spans of the bridges over the past two centuries, we note that progress has not been continuous and regular, and that when growth was sudden and too rapid, it also abruptly stopped. In 1940, the cause of this discontinuation was the collapse of the bridge in Tacoma, Washington, which led to a season of projects characterized by great caution and conservatism, and certainly excessive in the light of current knowledge. This incident is not derived from errors or chance, but rather from the lack of perception (or even of the lack of the relevant scientific knowledge at that time) of a technical problem: the aeroelastic stability of the unstreamlined sections.

It is interesting to note that, despite the great advances in wind engineering over the last 70 years, stimulated by that incident, only recently has a realistic explanation (Larsen, 2000) on the aerodynamic and aeroelastic mechanism that led to the collapse of the Tacoma bridge been found.

Ultimately, it seems reasonable to say that the future progress of long-span bridges must follow solutions that do not deviate too far conceptually from those already tested and implemented and that they should not stray too far from the previous ones regarding size and spans, creating and developing both continuous and gradual progress.

Innovative Bridge Design Handbook

References

Baglietto, E., Casirati, M., Castoldi, A., de Miranda, F., Roberto Sammartino, R., 1976. Modelli matematici e strutturali del comportamento statico e dinamico dei ponti. Zarate—Brazo Largo Costruzioni Metalliche 4, 151–161.

de Miranda, F., 1969. Stay cables anchored to the deck by means of stiffening additional cables. Patent 7368 A/69–Genova, October, 7 (in Italian).

de Miranda, F., 1971. Il ponte strallato—soluzione attuale del problema delle grandi luci (The cable stayed bridge—an actual solution for long spans bridges). Costruzioni Metalliche 1. and Ponti a struttura d'acciaio, Collana Tecnico Scientifica, ITALSIDER (then CISIA), Milano.

de Miranda, F., 1980. Ponti strallati di grande luce. Zanichelli, Bologna.

de Miranda, M., 2001. Il ponte sul rio Higuamo. Il Giornale dell'Ingegnere n. 5 –Milano, 15 Marzo 2008.

de Miranda, M., 2003. Cable-stayed bridge over the Guamà River, Brazil. Struct. Eng. Int. 3, 171–173.

de Miranda, M., Bartoli, G., 2001. Aerodynamic optimization of decks of cable stayed bridges. In: IABSE, International Association of Bridges and Structural Engineers, Symposium in Kobe, 1998. "Long-Span and High-Rise Structures," vol. 79, pp. 143–148.

de Miranda, M., Petrequin, M., 1998. Storebaelt Bridge, Aspects du Montage et de la Realisation, Construction Metallique, 2-2000, Costruzioni Metalliche n. 6, 1998. In: Beyond the Limits of Erection Activities, IABSE Conference, Kobe, 1998.

de Miranda, F., Leone, A., Passaro, A., 1979. Il ponte strallato sullo Stretto di Rande presso Vigo (Spagna) della Autopistas del Atlantico. Costruzioni Metalliche 2, 55–62

Farquharson, F.B., 1952. Aerodynamic stability of suspension bridges with Special Reference to the Tacoma Narrows Bridge. Bulletin n 116, part III, June. University of Washington, Engineering Experimental Station.

Federal Highway Administration (FHA), 2007. Wind Induced Vibrations in Stay Cables. Report FHWA-RD-05-083, FHA, Washington, DC.

Galileo, G., 1638. Discorsi e Dimostrazioni Matematiche, Intorno a due Nuove Scienze Attenenti Alla Meccanica e i Movimenti Locali.

Gimsing, N., 1983. Cable-Supported Bridges. John Wiley & Sons, New York.

Kashima, S., 1998. Technical Advance in the Honshu-Shikoku Bridges. In: IABSE, Kobe Conference Proceedings, pp. 23–34.

Leonhardt, F., 1982. Brucken, Bridges. Deusche Verlags-Anstalt. Stuttgart, Germany.

Leonhardt, L., Andra, W., Wintergest, L., 1969. Kniebrücke – Dusseldorf – Ed. Tamms-Beyer – Beton–Verlag.

Mélan, J., 1888. Theorie der eisernen Bogenbrücken und der Hängebücken. In: Handbuch der Ingenieuwissenschaten, second ed. Wilhelm Engelmann, Leipzig, Germany, vol. 2, Part 4.

Moisseiff, L., Lienhard, F., 1933. Suspension bridges under the action of lateral forces. Transaction of the American Society of Civil Engineers, n. 98, 1080–1095; 1096-1141.

Podonly, W., Scalzi, J., 1976. Construction and Design of Cable Stayed Bridges. John Wiley & Sons, New York.

Post-Tensioning Institute (PTI), 2007. Recommendations on Stay Cables. Phoenix, AZ.

Pugsley, A.G., 1968. The Theory of Suspension Bridges. Edward Arnold, London.

Rankine, W.J.M., 1869. A Manual of Applied Mechanics. Charles Griffin, London.

Richardson, J.R., 1984. The influence of aerodynamic stability on the design of bridges. In: 12th IABSE Congress, Vancouver, BC, Canada.

Roebling, J.A., 1854. Final Report on the Niagara Railway Suspension Bridge. Lee, Mann and Co., Rochester, N.Y.

SK Most, 2012. Construction of the Bridge Crossing to Russky Island over the Eastern Bosphorus Strait in Vladivostok. http://www.skmost.com.

Steinman, D.B., 1922. A Practical Treatise of Suspension Bridges. John Wiley & Sons, London.

Stussi, F., 1954. Das Problem der Grossen Spannweite. VSB Verlag, Zurich.

Veranzio, F., 1968. Machinae Novae–Ferro Edizioni. Milano via Brera, 6.

Walter, R., Houriet, B., Isler, W., Maia, P., Klein, J.F., 1999. Cable Stayed Bridges. Thomas Telford, London.

Wyatt, T.A., 1991. The dynamic behaviour of cable stayed bridges: fundamental and parametric studies. In: Ito, M., Fujino, Y., Miyata, T., Narita, N. (Eds.), Cable Stayed Bridges - Recent Developments and their Future. Proceedings of a Seminar, Yokohama, Japan, 10–11 December 1991. Elsevier, pp. 151–170.

Section VI

Special topics

Integral bridges

Dicleli M.
Middle East Technical University, Ankara, Turkey

16

1 Introduction

Bridges are traditionally built with expansion joints at the ends to allow for longitudinal displacements of the superstructure due to temperature variations. Thus, most conventional bridges possess expansion joints and bearings, which are expensive in their materials and installation. Furthermore, expansion joints may allow water, salt, and deicing chemicals to penetrate them and cause extensive deterioration to the bearings, substructure, and superstructure components. Consequently, for many years, they have caused considerable maintenance problems for transportation agencies (Wolde et al., 1988a, 1988b; Burke, 1988, 1990; Steiger, 1993). Elimination of expansion joints in bridges may reduce the construction costs, overcome many of the maintenance problems, and increase the stability and durability of the bridges. These economic and functional advantages are generally recognized by bridge engineers (Dicleli, 2000a, 2000b) leading to the concept of integral construction or integral bridge. The lack of expansion joints in integral bridges results in reduced repair and maintenance costs throughout the service life of the bridge. In addition, when used as part of highways or railways, integral bridges enhance the comfort of travelers due to lack of expansion joints and provide better lateral rigidity to breaking loads minimizing the likelihood of rail buckling in continuous railway construction due to the smaller lateral displacement of the bridge. Moreover, modern integral bridges are known to have performed well in recent earthquakes due to their monolithic construction (Erhan and Dicleli, 2014).

2 Historical background

The use of integral bridges began thousands of years ago in the shape of masonry arches, and today there are many of similar arches survived for more than hundred years (Hambly, 1997). The construction of reinforced concrete arch bridges in North America began in the early decades of the 20th century. Bridge engineers began eliminating the deck joints at piers and abutments after the moment distribution method was first developed in the early 1930s (Cross, 1932), which allowed engineers to analyze statically indeterminate structures such as rigid frame bridges. By the mid-20th century, concrete rigid frame bridges became a standard type of construction for many departments of transportation (Burke, 1993). Ohio and Oregon were the first states to use integral bridges in the 1930s, and Illinois and Iowa followed in the 1940s.

Innovative Bridge Design Handbook. http://dx.doi.org/10.1016/B978-0-12-800058-8.00016-5

Washington State had its first integral bridge in the 1960s, and New York State began building integral bridges in the late 1970s.

By 1980, 30 states were using integral abutment bridges as a standard form of construction. In the past few decades, engineers started to notice the benefits of integral bridges over jointed bridges in terms of their superior stability and serviceability and lower maintenance demand. Consequently, most bridge engineers focused their attention on the design and construction practice of integral bridges. Nowadays, most departments of transportation in the United States, Canada, and Europe consider the integral bridge construction as a standard form of construction.

3   Modern integral bridges

Modern integral bridges are single- or multiple-span bridges with a continuous deck and a flexible movement system composed primarily of abutments supported on a single row of piles (Hambly, 1997; Chen, 1997; Dicleli, 2000a, 2000b). A typical slab-on-girder, integral bridge is shown in Figure 16.1. The details of a typical integral bridge are shown in Figure 16.2. In these types of bridges, the road surfaces are continuous from one approach embankment to the other, and the abutments (and occasionally the piers) are cast integrally with the girders and the deck slab. A flexible abutment with a single row of piles is essential to allow for the longitudinal bridge movements due to temperature variations, shrinkage, and creep. The most common type of piles used at the abutments of integral bridges is steel H-piles. Cycle control joints are provided at the ends of the approach slabs to accommodate the longitudinal movements of the bridge.

Figure 16.1 Single-span integral bridge with a slab-on-steel-girder deck (Blackmud Creek Bridge, Edmonton, Canada).

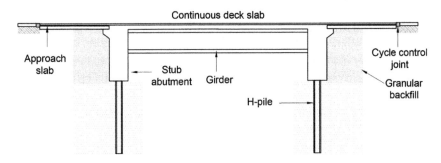

Figure 16.2 Details of a typical single-span integral bridge.

4   Thermal effects on integral bridges

Daily and seasonal temperature changes result in the imposition of cyclic horizontal displacements on the continuous bridge deck of integral bridges, and thus on the abutments, backfill soil, steel H-piles at the abutments, and cycle control joints at the ends of the approach slabs. The magnitude of these temperature-induced cyclic displacements is a function of the temperature difference and the length of the bridge. Thermal-induced cyclic displacements are especially important for the performance of the steel H-piles at the abutments and the abutments themselves. In the following subsections, the effect of cyclic thermal displacements on the steel H-piles and abutments are discussed.

4.1   Thermal effects on integral bridge piles

Daily and seasonal temperature changes result in the imposition of cyclic horizontal displacements on the continuous deck of integral bridges, and thus on the steel H-piles at the abutments. As the length of integral bridges grows, the temperature-induced displacements in the steel H-piles may become larger as well. Consequently, the piles may experience deformations beyond their elastic limit. The ability of steel H-piles to accommodate such large displacements is an important factor that affects the maximum length of an integral bridge.

The displacement capacity of steel members, including steel H-piles at the abutments of integral bridges, is affected by their buckling instability. Instability in steel structural members includes local buckling of the plates forming the cross section of the member, as well as lateral-torsional and global buckling of the steel member. Local buckling instability in steel H-piles may occur in either the flange, the web, or both, depending on the width-to-thickness ratios of the flange and web plates. Lateral torsional buckling, which occurs when steel members are subjected to bending about their strong axis, is critical for steel sections with relatively narrow flanges and is not much concern for steel H-piles that have wider flanges. Furthermore, as the steel H-piles in integral bridges are laterally supported by the surrounding soil, the lateral- torsional or global buckling instability need not be considered. Thus, the local buckling is the only instability type that will be considered when determining the displacement capacity of steel H-piles. The width-to-thickness ratios of the flanges and the web for steel H-piles must be limited to prevent local buckling. Many researchers worked out limits for the width-to-thickness ratios of web and flange to prevent local buckling effects, and hence to ensure ductile behavior of steel members. Most of this research has been implemented in design codes such as the American Institute of Steel Construction (AISC) Load and Resistance Factor Design (LRFD) manual for steel structures (AISC, 2010). This manual divides steel sections into three categories based on their ability to reach a certain compressive stress level and deform without experiencing local buckling problems. These are compact sections, non-compact sections, and sections with slender plate elements (web and flange). Compact sections are capable of developing full plastic flexural capacity. Noncompact sections cannot develop full plastic capacity, but they are capable of developing yield stress in compression elements. The third category covers steel sections with slender plate elements that experience local buckling before the yield stress is achieved. The dividing lines between these three categories are defined by slenderness parameters λ_p and λ_r, which define the limiting width-to-thickness ratios for compact and noncompact sections, respectively. For compact sections, the width-to-thickness ratios for the web and flange are smaller than λ_p. For noncompact sections, they are larger than λ_p but smaller than λ_r; and for slender sections, they are larger than λ_r. Table 16.1 displays the expressions for λ_p and λ_r for web and flange under monotonic loading.

Table 16.1 Limiting Width-to-Thickness Ratios for Compression Elements in Steel Sections (AISC, 2010)

Width-Thickness Ratio	Limiting Width-Thickness Ratios	
	λ_p	λ_r
b_f/t_f	$\dfrac{163}{\sqrt{F_Y}}$	$\dfrac{250}{\sqrt{F_Y}}$
d_w/t_w	For $P_U/\varphi_b P_Y \leq 0.125$, $\dfrac{640}{\sqrt{F_Y}}\left(1-\dfrac{2.75 P_U}{\varphi_b P_Y}\right)$ For $P_U/\varphi_b P_Y \geq 0.125$, $\dfrac{500}{\sqrt{F_Y}}\left(2.33-\dfrac{P_U}{\varphi_b P_Y}\right) \leq \dfrac{665}{\sqrt{F_Y}}$	$\dfrac{665}{\sqrt{F_Y}}$

In Table 16.1, b_f is the flange width; d_w is the clear height of the web plate between flanges; t_f and t_w are the flange and web thickness, respectively; P_u and P_y are the required and yield axial forces; and F_y is the yield stress in ksi. Under cyclic thermal movements, the steel H-piles are expected to reach their plastic capacity. Thus, it is recommended that compact steel HP sections should be used to avoid local buckling instability.

Fatigue of steel H-piles is another important problem that should be considered in bridge design. Steel H-piles supporting the abutments of integral bridges are subjected to cyclic loading due to temperature changes. In the summer, the superstructure expands and pushes the abutment and the piles toward the backfill, and in the winter, the superstructure pulls the abutment and the piles in the opposite direction. Consequently, the piles may be subjected to one dominant cyclic displacement each year due to seasonal temperature changes (Girton et al., 1989). Additionally, the piles may be subjected to numerous small cyclic displacements due to daily and/or weekly temperature fluctuations (Girton et al., 1989). The magnitude of these temperature-induced cyclic displacements is a function of the temperature difference and the length of the structure. As the length of the integral bridges grows, the temperature-induced cyclic displacements in steel H-piles may become larger as well. As a result, the piles may experience cyclic deformations beyond their elastic limit. This may result in the reduction of their service life due to low-cycle fatigue effects.

Examination of the strain versus time records of instrumented steel H-piles for two integral abutment bridges in the state of Iowa (Girton et al., 1989) revealed that both bridges exhibited one large strain cycle per year due to seasonal temperature changes, and about 52 small strain cycles per year due to weekly temperature fluctuations, as illustrated schematically in Figure 16.3. Moreover, the field test records demonstrated that the strain amplitude of the small cycles in the piles supporting the abutments fall within the 20%–40% range of the strain amplitude from the large cycles, as shown in Figure 16.3.

It is noteworthy that the net difference between the seasonal and reference (construction) temperatures may be disparate in the summer and winter based on the climatic conditions of the area where the bridge is located. Therefore, the amplitudes of

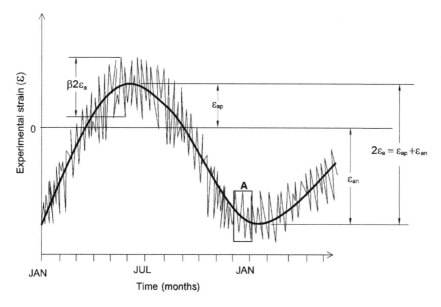

Figure 16.3 General experimental strain versus time for integral bridge piles.

the positive (ε_{ap}) and negative (ε_{an}) strain cycles corresponding to the summer and winter may not be equal, as shown in Figure 16.3. However, as the range of strain amplitudes (rather than the strain amplitude itself) defines the extent of fatigue damage, the positive and negative strain amplitudes may be assumed to be equal.

4.2   Thermal effects in integral bridge abutments

The earth pressure that is exerted on the abutment by backfill soil depends on the extent of movement of the abutment. An integral bridge will experience elongation and contraction due to temperature variations during its service life. Thus, the earth pressure at the abutments should be considered in correlation with temperature variation. A very small displacement of the bridge away from the backfill soil can cause the development of active earth pressure conditions (Barker et al., 1991). Therefore, when the bridge contracts due to a decrease in temperature, active earth pressure will develop behind the abutment. At rest, earth pressure behind the abutment is assumed when there is no thermal movement. When the bridge elongates due to an increase in temperature, the intensity of the earth pressure behind the abutment depends on the magnitude of the bridge displacement toward the backfill soil. Thus, the actual earth pressure coefficient, K, may change between at rest, K_o, and passive, K_P, earth pressure coefficients, depending on the amount of displacement.

5   Conditions for integral bridge construction and recommendations

5.1   Length of the bridge

It is intended that currently, when the overall length of a bridge is less than 150 m, an integral bridge design will be considered. For structure lengths greater than 150 m, the consent of the governing bridge authority (the owner) should be obtained. In considering the movement requirements, due consideration should be given to the place and types of joint, joint seal, backfill and approach slab details, and construction temperatures. The limitation placed on the total length of the structure is mainly a function of local soil properties, seasonal temperature variations, resistance of abutments to longitudinal movements, and the type of superstructure being considered.

The integral bridge length limits, as governed by the low-cycle fatigue performance of steel H-piles and flexural strength of the abutments, are given next (Dicleli and Albhaisi, 2004, 2005).

Integral bridge length limits as governed by low-cycle fatigue performance of the piles:

$$L = \frac{2}{\gamma_T \alpha_T \Delta T} \left[\frac{M_y (\lambda l_c)^2}{6 E_p I_p} \left(1 + \frac{M_y}{M_p}\right) + \frac{0.0085 (\lambda l_c)^2}{6 d_p} \left(2 - \frac{M_y}{M_p} - \left(\frac{M_y}{M_p}\right)^2\right) \right], \qquad (1)$$

where l_c is the critical length of the pile, defined as the length beyond which the pile's top displacement and rotations have practically no effect. This is defined as

$$l_c = 4 \sqrt[4]{\frac{E_p I_p}{k_h}}, \tag{2}$$

where for clay:

$$k_h = \frac{9C_u}{2.5\varepsilon_{50}} \text{ (soft to medium − stiff clay)} \tag{3a}$$

$$k_h = \frac{9C_u}{4.0\varepsilon_{50}} \text{ (stiff clay)} \tag{3b}$$

In the absence of geotechnical data, for soft, medium, medium-stiff, and stiff clay, corresponding values of $C_u = 20$, 40, 80, and 120 kPa (Bowles, 1996) and $\varepsilon_{50} = 0.02$, 0.01, 0.0065, and 0.0050 (Evans, 1982) shall be used.

For sand:

$$k_h = kx \tag{4}$$

$$x = H + 8d_p \tag{5}$$

In the absence of geotechnical data, for loose, medium, medium-dense, and dense sand, corresponding values of $k = 2000$, 6000, 12,000, and 18,000 kN/m^3 shall be used.

In most cases, the subsoil is composed of layers of soils with different stiffness properties. Here, soil stiffness shall be averaged over the top $10 \times b_p$ length of the pile and used in Eqs. (3) and (4), where b_p is the width of the pile perpendicular to the direction of the movement.

The λ values are given in Table 16.2 as follows;

Table 16.2 λ Values for Different Soil and Abutment-Pile Connection Types

Soil	Abutment-Pile Connection	Strong Axis	Weak Axis
Clay	Fixed	0.5	0.55
	Pinned	1.15	1.40
Sand	Fixed	0.65	0.75
	Pinned	1.1	1.40

In lieu of adequate structural and geotechnical data at the bridge site, integral bridge length limits for different pile sizes are given in Table 16.3.

Table 16.3 Maximum Length Limits for Steel and Concrete Integral Bridges Based on the Pile's Low-Cycle Fatigue Performance

Pile Size	Steel Bridges		Concrete Bridges	
	Moderate Climate	Cold Climate	Moderate Climate	Cold Climate
	L (m)	L (m)	L (m)	L (m)
HP310 x 125	220	145	320	265
HP310 x 110	205	135	300	250
HP250 x 85	160	110	240	195
HP200 x 63	125	80	180	150

Integral bridge length limits as governed by flexural strength of abutments are described next.

Abutments have adequate shear strength to accommodate thermal-induced shear forces, and hence, the shear strength of the abutments does not govern the integral bridge length limits. Although the flexural strength of the abutments generally does not govern the bridge length limits either, for abutments taller than 4 m, the following equations may be used to calculate the length limits of integral bridges based on the flexural strength of the abutments (Dicleli, 2005):

For clay:

$$L = \frac{2H}{\alpha_T \alpha \Delta T m^{\frac{1}{n}}} \left[\left(\frac{1}{\alpha_E \gamma S} \right) \left(M_r - n_p \left(M_p + \frac{M_p}{\lambda_v l_c} (H - h_D) \right) \right) \right.$$

$$\left. \left(\frac{6}{(h_D + 2H)(H - h_D)^2} \right) - K_0 \right]^{\frac{1}{n}} \tag{6}$$

For sand:

$$L = \frac{2H}{\alpha_T \alpha \Delta T m^{\frac{1}{n}}} \left[\left(\frac{1}{\alpha_E \gamma S} \right) \left(M_r - n_p M_p \left(1 + (H - h_D) \sqrt[4]{\frac{k(H + 8d_p)}{2 E_p I_p}} \right) \right) \right.$$

$$\left. \left(\frac{6}{(h_D + 2H)(H - h_D)^2} \right) - K_0 \right]^{\frac{1}{n}} \tag{7}$$

where $\lambda_v = 0.30$, m = 10 and n = 0.33 for compacted backfill, and m = 28 and n = 0.56 for uncompacted backfill (Dicleli, 2005).

5.2 Superstructure type

The types of structures used with integral bridges include:

- Steel girders with concrete deck
- Prestressed concrete girders with concrete deck
- Prestressed concrete box girders with concrete deck

Posttensioned construction is not suitable for integral design, as the abutments supported on a single row of piles may not be able to accommodate the lateral forces exerted during the posttensioning process.

5.3 Geometry of the bridge

The geometry of the structure should be considered when deciding the feasibility of integral bridge design. Owing to the nonuniform distribution of loads and difficulties in establishing the movement and its direction, structures with skew greater than 35 degrees or where an angle subtended by a 30-m arc along the length of the structure is greater than 5 degrees are not considered suitable for integral designs. Skews greater than 20 degrees but not exceeding 35 degrees may be considered if a rigorous analysis is carried out to account for the skew effects. In carrying the analysis for skew, the effects such as torsion, unequal load distribution, lateral translation, pile deflection in both the longitudinal and transverse directions, and increase in the length of the abutment exposed to soil pressure shall be considered.

5.4 Abutments and wing-walls

It is recommended that abutment height and wing-wall length shall be limited to 4 m and 7 m, respectively. The abutment should be kept as short as possible to reduce the soil pressure; however, the minimum penetration required for frost protection should be provided. The frost penetration requirement can be reduced to minimize abutment height by providing insulation at the bottom of the abutment. It is recommended to have abutments of equal height at the bridge ends. A difference in abutment heights causes unbalanced lateral load resulting in side-sway, which should be considered in the design by balancing the earth pressure, which is consistent with the direction of side-sway, at the abutments. This procedure requires an iterative process which may result in a pressure ranging from active to at rest on the taller abutment and at rest to passive on the shorter abutment. Wing-walls parallel to the roadway, carried by the structure, shall be used, and their size should be minimized to allow the substructure to move with minimum resistance.

5.5 Multiple-span integral bridges

The spans and the articulation at the supports of multispan structures should be selected such that equal movement would occur at each end of the structure. The deck diaphragms may either be integral with the piers, made fixed in the lateral direction, or move laterally, as appropriate. The piers should be flexible and supported on flexible foundations if made integral with the deck diaphragms.

5.6   Foundation soil conditions

Subsoil condition is an important consideration in the feasibility of the integral arrangement of a structure. The primary criterion is the need to support abutments on relatively flexible piles. Therefore, where load-bearing strata is near the surface, or where the use of short piles (i.e., less than 5 m in length) or caissons is planned, the site is not considered suitable for integral bridges. Where piles are driven in dense and stiff soils, preaugured holes filled with loose sand shall be provided to reduce resistance to lateral movement. Where soil is susceptible to liquefaction, slip failure, sloughing, or boiling, the use of integral arrangement should be avoided.

6   Construction methods of integral bridges

Construction considerations and sequence shall be given on the drawings to specify the following requirements:

- The abutments, including wing-walls, shall be constructed first to bearing seat elevation.
- The girders shall be placed on a support that allows rotation and deflection of the girders due to self-weight and dead weight of the deck. A 20-mm-thick natural rubber sheet is generally adequate to accommodate the rotations of the girders.
- The deck and the portion of the abutment above bearing seat elevation shall be cast integrally with the girders.
- The deck and the abutment to the bearing seat level shall be poured in sequence so that the structure becomes integral, with no residual stresses. This may require careful consideration of the concrete-pouring sequence and use of retarder. The ends of the deck and the abutments shall be placed last unless concrete can be retarded sufficiently to allow the placement from one end to the other in a single pour.
- The stability and the integrity of the structure shall be maintained at all stages of construction.
- Backfill shall not be placed behind the abutments until the deck has reached 75% of its specified strength.
- Backfill shall be placed simultaneously behind both abutments, keeping the height of the backfill approximately the same. At no time shall the difference in the heights of the backfills be greater than 500 mm.

7   Design of integral bridges

The design procedure defined in this section is applicable to integral bridges with slab-on-steel or prestressed-concrete girder deck.

7.1   Construction stages, loads, and load combinations

The construction of an integral bridge is done in stages. Therefore, the bridge must be analyzed at each construction stage to ensure that the structure has adequate capacity to sustain the applied loads particular to the stage under consideration.

Two construction stages are considered for the design of slab-on-prestressed-concrete-girder integral bridges. The loads applied at each construction stage are listed in Table 16.4. In the first stage, the slab concrete is assumed to be wet. Accordingly, the prestressed-concrete girders alone resist the applied loads. The structure is analyzed for the effects of prestressing force, dead weight of the girders, weight of wet concrete slab, and weight of the diaphragms. In the second stage, the bridge is assumed to be in service. Full composite action is considered between the slab, girders, and abutments. The effects of superimposed dead loads, ballast weight, temperature variation, soil pressure, and live load are considered in this stage.

Three stages are considered for the design of slab-on-steel-girder integral bridges. The loads applied at each stage are listed in Table 16.5 In the first stage, the naked steel girders are assumed to be fully assembled and supported on the abutments and piers, but the slab concrete is assumed to be wet. Therefore, the steel girders alone resist the applied loads. The loads due to the wet concrete slab, diaphragms, and self-weight of the steel girders are considered in this stage. In stage two, the steel girders are assumed to be composite with the concrete slab. However, the modulus of elasticity of the concrete slab is assumed to be one-third of its actual final value to consider the effect of creep due to superimposed dead loads. In addition to the loads applied in the first stage, ballast load (in the case of railway bridges) and superimposed dead loads are considered during this stage. In the final stage, the bridge is assumed to be in service. Full composite action is assumed between the slab, girders, and abutments. The effects of temperature variation, soil pressure, and live load are considered in this final stage.

Table 16.4 Summary of Stage Loading for Slab-on-Prestressed-Concrete-Girder Deck Integral Bridges

Stage #	Stage Name	Load ID	Load Description
1	Simply supported beams	1	Own weight of girder
		2	Pretensioning
		3	Weight of wet concrete slab, diaphragms, and abutment
2	Composite structure	4	Superimposed dead load
		5	Asphalt/ballast weight
		6	Long-term prestress losses
		7	Highway/railway live loading at fatigue limit state (FLS)
		8	Similar to load 7, but at serviceability limit state (SLS)
		9	Similar to load 7, but at ultimate limit state (ULS)
		10	Thermal load due to longitudinal expansion
		11	Thermal load due to longitudinal contraction
		12	Passive earth pressure
		13	At-rest earth pressure
		14	Active earth pressure

Table 16.5 **Summary of Stage Loading for Slab-on-Steel-Girder Deck Integral Bridges**

Stage #	Stage Name	Load ID	Load Description
1	Naked steel beam	1	Own weight of steel beam and diaphragms
		2	Weight of wet concrete slab, diaphragms and abutment
2	Composite structure (3n)– superstructure only	3	Superimposed dead load
		4	Asphalt/ballast weight
3	Composite structure (n)	5	Same as load 3
		6	Same as load 4
		7	Highway/railway live loading at FLS
		8	Similar to load 7, but at SLS
		9	Similar to load 7, but at ULS
		10	Thermal load due to longitudinal expansion
		11	Thermal load due to longitudinal contraction
		12	Passive earth pressure
		13	At-rest earth pressure
		14	Active earth pressure

7.2 Modelling integral bridges for analysis under gravitational loads

For the analysis of integral bridges subjected to gravitational loads, a separate structure model is proposed for each construction stage. The proposed structure models are subject to the following assumptions:

- The analysis of bridges having slab-on-girder type deck is reduced to the consideration of one beam and an effective width of the slab for the purpose of gravity load analysis. Accordingly, the abutments are idealized to have a tributary width equal to that of the slab. Similarly, the number of piers and piles per tributary width is calculated, and their stiffness is lumped to obtain a single pier or pile element for analysis purposes.
- The effect of frictional forces between the approach slab and soil, as well as between the wing-walls and soil, resulting from movements due to temperature variations is ignored.
- An equivalent pile length or Winkler spring model is assumed in the structural model.
- The live load applied on the structure is proportioned to one girder, considering the transverse distribution of the live load effects.

Figure 16.4 illustrates a typical two-span, prestressed-concrete-girder integral bridge and its two-dimensional (2D) structure model for construction stage 1. The naked girder alone is considered in the structure model, assuming that the concrete is not hardened. Accordingly, the composite action between the girder and slab and the continuity between the girders of adjacent spans and at the deck-abutment joints are ignored. The bridge is modeled considering each span as a simply supported beam. The structure is analyzed for stage 1 loads as given in Table 16.4. The resulting internal forces (stresses) are then kept so that they can be superimposed on the ones resulting from the loads to be applied in stage 2.

For slab-on-steel-girder integral bridges, the steel beams are fully assembled to form a continuous beam before the slab is cast in construction stage 1. Therefore, the hinge shown at the middle support of the structure model depicted in Figure 16.4 is removed for this type of integral bridges. Furthermore, if steel columns are rigidly connected to the steel girders, the middle simple-support shown in Figure 16.4 is replaced by a column element rigidly connected to the beam. Each span is idealized using 2D beam elements. The structure is then analyzed for stage 1 loads tabulated in Table 16.5. The resulting internal forces (stresses) are stored so that they can be superimposed on the ones resulting from the loads to be applied in subsequent stages.

Figure 16.5 illustrates the structure model for the rest of the construction stages for the same bridge shown in Figure 16.4. The structure model shall be used for the analysis of both steel and prestressed concrete, slab-on-girder, integral bridges. The bridge is idealized as a plane frame considering only one girder and an effective width of slab. Full continuity at the intermediate supports and at the abutment-deck connection joints is considered assuming that the concrete is fully hardened. The idealized

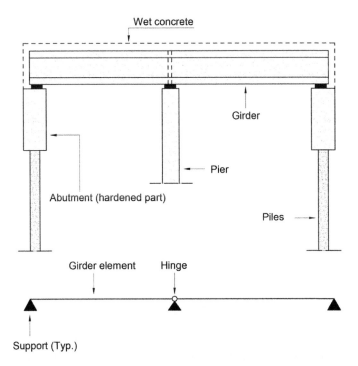

Figure 16.4 Typical two-span integral bridge and analytical model for construction stage I.

abutment and pier members are connected to the deck nodes by abutment-deck or pier-deck connection elements. The pile member is connected to the abutment member by a pile-abutment connection element. If the connections between the abutment and deck, as well as pile and abutment, are normally assumed to be rigid, then the connection elements may be removed from the model.

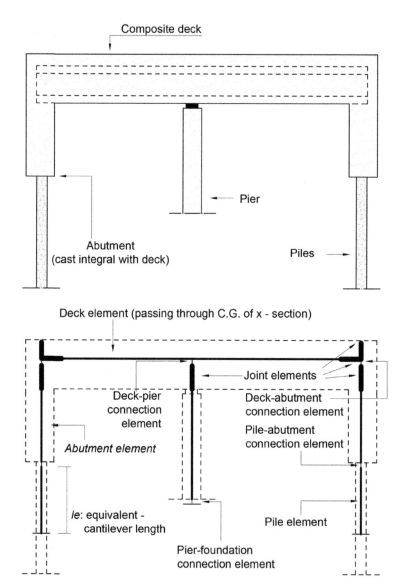

Figure 16.5 Typical two-span integral bridge and analytical model for the final construction stage.

7.3   Thermal variations and associated soil-bridge interaction

The earth pressure coefficient is a function of the displacement or rotation of the earth retaining structure. An integral bridge will experience elongation and contraction due to temperature variations during its service life. Thus, the earth pressure at the abutments should be considered in correlation with temperature variation. A very small displacement of the bridge away from the backfill soil can cause the development of active earth pressure conditions. Therefore, when the bridge contracts due to a decrease in temperature, active earth pressure will be developed behind the abutment. At-rest earth pressure behind the abutment is assumed when there is no thermal movement. When the bridge elongates due to an increase in temperature, the intensity of the earth pressure behind the abutment depends on the magnitude of the bridge displacement toward the backfill. The actual earth pressure coefficient, K, may change between at-rest (K_O) and passive (K_P) earth pressure coefficients depending on the amount of displacement. The dependency of the earth presure coefficient on the thermal displacement of the bridge shall be taken into consideration in the design of the bridge using the following equation:

$$K = \frac{2K_o + \alpha\,\delta T\,L\varphi}{2 + \dfrac{LH^2 S\gamma_s \varphi}{2E_g\left(A_g + nA_s\right)}} \leq K_P, \tag{8}$$

where φ is the rate of the variation of the earth pressure coefficient between at-rest and passive states. In lieu of geotechnical data for the properties of granular soil (backfill), $\varphi = 24\ \mathrm{m}^{-1}$. For bridges with unequal abutment heights, Eq. (8) shall be used cautiously by considering unequal movements at both ends of the bridge due to the difference in abutment height.

7.4   Live load distribution in integral bridges

The maximum live load effect in a bridge is based on the position of the truck both in the longitudinal and transverse directions, the number of loaded design lanes, and the probability of the presence of multiple loaded design lanes. To calculate the maximum live load effects in an integral bridge, the position of the truck in the longitudinal direction, as well as both the position and the number of trucks in the transverse direction, need to be considered in a three-dimensional (3D) finite element model of the bridge. Furthermore, in the estimation of live load effects, the probability of the presence of multiple loaded design lanes need to be taken into consideration by using the multiple presence factors defined in design specifications such as AASHTO (2010). Although, using 3D finite element models to determine live load effects in bridge components is possible due to the readily available computational tools in design offices, using such complicated methods throughout the design process is tedious and time consuming. Therefore, most bridge engineers use simplified 2D structural models and live load distribution factors (LLDFs) readily available in bridge design specifications to determine live load effects in bridge girders. LLDFs have been used in bridge design since the 1930s.

LLDFs are needed for the composite interior and exterior girders of integral bridges for the loading cases where only a single design lane is loaded [fatigue limit state (FLS)] and two or more design lanes are loaded [serviceability limit state (SLS) and ultimate limit state (ULS)]. For this purpose, first the maximum live load effects (moment and shear) from 3D analyses for the composite girders are calculated as the summation of the maximum effects in the girder element and within the tributary width of the slab at the same location along the bridge. For the case where two or more design lanes are loaded, the transverse-loading case producing the maximum girder live load effect after multiplying by the multiple presence factors is used to obtain the LLDFs. These are then calculated as the ratio of the maximum live load effects obtained from 3D analyses to those obtained from 2D analyses under a single truck load. In the calculation of LLDFs, the AASHTO HL-93 truck is used (AASHTO, 2010).

In this section, LLDFs are provided for the interior and exterior girders of commonly used slab-on-girder integral bridges. Using these, it is possible to obtain the actual 3D live load effects in bridge girders by multiplying the response from a 2D model of a bridge represented by a single girder over a tributary width of girder spacing by appropriate LLDFs. Followings are the equations that may be used to calculate the LLDFs for the interior and exterior girder moment and shear for the cases where one design lane is loaded by trucks or two or more design lanes are loaded by trucks.

Interior Girder Moment—Two or More Design Lanes Loaded:

$$LLDE_{IAB} = \frac{S^{0.82}}{500L^{0.06}} \tag{9}$$

Interior Girder Moment—One Design Lane Loaded:

$$LLDE_{IAB} = \frac{3S^{0.72}}{500L^{0.13}} \tag{10}$$

Interior Girder Shear—Two or More Design Lanes Loaded:

$$LLDE_{IAB} = 0.2 + \frac{S}{3600} - \left(\frac{S}{10700}\right)^{2.0} \tag{11}$$

Interior Girder Shear—One Design Lane Loaded:

$$LLDE_{IAB} = 0.36 + \frac{S}{7600} \tag{12}$$

Exterior Girder Moment—Two or More Design Lanes Loaded:

$$LLDE_{IAB} = \frac{L^{0.09} S^{0.53} t_s^{0.06}}{80 K_g^{0.04}} \left(0.5 + \frac{d_e}{5000}\right) \tag{13}$$

Exterior Girder Moment—One Design Lane Loaded:

$$LLDE_{IAB} = \frac{L^{0.06} S^{0.45}}{18 t_s^{0.02} K_g^{0.04}} \left(0.4 + \frac{d_e}{6000}\right) \tag{14}$$

Exterior Girder Shear—Two or More Design Lanes Loaded:

$$LLDE_{IAB} = \frac{L^{0.10} S^{0.43} t_s^{0.03}}{14 K_g^{0.07}} \left(0.4 + \frac{d_e}{3000}\right) \tag{15}$$

Exterior Girder Shear—One Design Lane Loaded:

$$LLDE_{IAB} = \frac{2L^{0.05}S^{0.34}}{15t_s^{0.01}K_g^{0.04}} \left(0.5 + \frac{d_e}{3000}\right) \tag{16}$$

In these equations, S is girder spacing, L is span length, t_s is slab thickness, d_e is cantilever length measured from the centroid of the exterior girder up to the face of the barrier wall length, and K_g is a parameter representing the longitudinal stiffness of the composite slab-on-girder section of the bridge expressed as

$$K_g = n\left(I + Ae_g^2\right)$$

where n is the ratio of the modulus of elasticity of the girder material to that of the slab material, I is the moment of inertia of the girder, A is the cross-sectional area of the girder, and e_g is the distance between the centers of gravity of the girder and the slab.

7.5   Design for seismic loads

Compared to conventional jointed bridges, integral bridges perform better during an earthquake due to the fixity and restraint at the abutments. However, caution should be exercised in the design of substructures to minimize damage in the event of an earthquake. The nonlinear time history analysis method, considering the passive backfill resistance behind the abutments, is more appropriate for the seismic design of integral bridges. In lieu of nonlinear time history analyses, the maximum earth pressure acting on the abutment in the longitudinal direction shall be assumed to be equal to the maximum longitudinal earthquake force transferred from the superstructure to the abutment. To minimize abutment damage, the abutment should be designed to resist the passive pressure being mobilized by the backfill, which should be greater than the maximum estimated longitudinal earthquake force transferred to the abutment. When longitudinal seismic forces are also resisted by piers or columns, it is necessary to estimate the stiffness of the components in order to compute the proportion of earthquake load transferred to the abutment. The wing-walls should be treated similarly for transverse seismic forces. The capacity of piles in both directions should be checked to resist the earthquake forces. It may be necessary in some cases to batter the piles sufficiently in the transverse direction, to adequately transfer the earthquake forces, or to provide stability in the transverse direction.

8   Important considerations in integral bridge design

8.1   Superstructure

The bridge deck components are designed assuming a continuous frame action at the joints linking the bridge deck to the abutments. A connection detail consistent with the degree of continuity assumed at the joints shall be provided. A typical reinforcement detail that provides full continuity at the deck-abutment joints is illustrated in Figure 16.6. The effect of temperature variation and axial compression in the steel and prestressed girders due to backfill soil pressure is considered in the design.

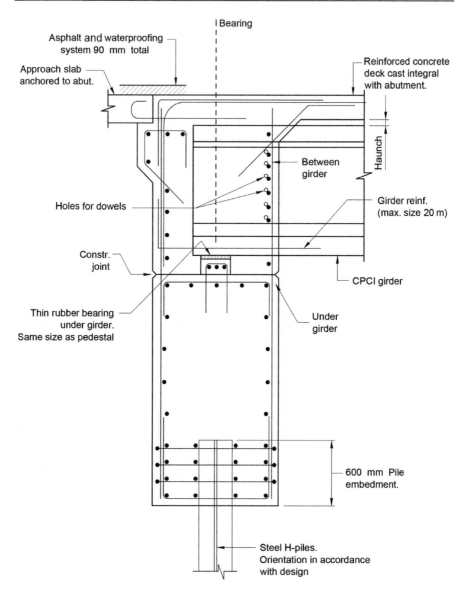

Figure 16.6 Typical reinforcement details of deck-abutment-pile joints.

8.2   Abutments, wing-walls, and approach slab

The abutment shall be connected monolithically to the deck, as shown in Figure 16.6, to avoid any expansion joint. The abutment height shall be restricted to the minimum practical value to reduce the soil pressure and to limit the weight, which moves with the deck. However, the minimum penetration required for frost protection shall be provided. The frost penetration requirement can be reduced to minimize abutment height by providing insulation at the bottom of the abutment. It is recommended that abutments at both sides of the bridge be of equal height since a difference in abutment heights causes unbalanced lateral load, resulting in side-sway. Additionally, the soil under the approach slab shall be sloped to reduce the height of the soil behind the abutment. This practice is also useful in preventing the compaction of the soil behind the abutment wall due to rail or highway traffic. It also reduces the resistance of frictional forces between the soil and the approach slab to bridge movement.

Turnback wing-walls parallel to the railway, carried by the structure, shall be preferably used. Their size shall be minimized to allow the substructure to move with minimum resistance.

The approach slab shall be built integral with the abutment to prevent water penetration. An expansion joint shall be provided at the end of the approach slab, as shown in Figure 16.7. The approach slab shall be designed as a simply supported structure spanning over the backfill behind the abutment to prevent the compaction of backfill material.

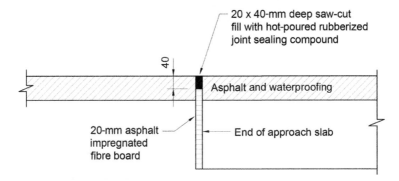

Figure 16.7 Typical joint detail at the end of the approach slab.

8.3 Piles at abutments

A single row of piles shall be used to support the abutments. The design of piles may be carried out using the equivalent cantilever method as a beam-column with a fixed base at a certain distance below the ground surface or using a Winkler soil model. A pin connection is recommended between the pile top and abutment to allow free rotation of the pile top about an axis perpendicular to the bridge longitudinal direction. If the connection is designed as fixed, plastic bending moments may be produced at the pile top due to thermal movements and the effect of live load. The maximum integral bridge length shall be estimated considering low-cycle fatigue effects due to thermal movements.

If the pile-supporting system utilizes the frictional forces between the piles and the soil, consideration shall be given to the effect of lateral displacement of the piles on the frictional resistance. As the piles will be moving laterally with temperature variations, a gap may be produced between the disturbed soil and the pile. This may result in considerable decrease of the frictional resistance of the piles. Therefore, the piles should be designed using the effective frictional pile length reduced by pile displacements.

If the piles are driven into stiff soils, their longitudinal displacement may somehow be restrained. Predrilled oversize holes filled with loose send may be provided to reduce the resistance to lateral movements. A typical diagram of such an arrangement is provided in Figure 16.8.

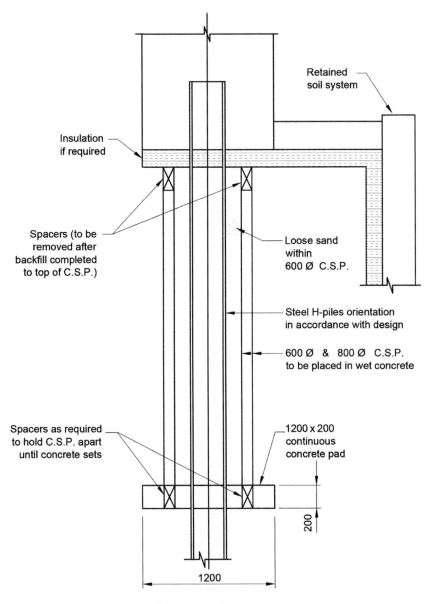

Figure 16.8 Drawing of typical pile in stiff soil.

8.4 Bearings, piers, and foundations

The pier is expected to deflect and rock on its foundation when the structure contracts or expands due to temperature variation. Elastomeric bearings of adequate thickness may be used to reduce the flexibility demand of the pier. The bearings are designed to accommodate the movements of the bridge and to support vertical loads coexisting with rotation of the deck. The pier footing is designed as narrow as possible in the longitudinal direction of the bridge to allow partial rotation of the pier at its base. If the footing is supported on piles, the pile group is designed to allow some rotation of the footing.

References

American Association of State Highway and Transportation Officials (AASHTO), 2010. LRFD Bridge Design Specifications, fourth ed. Washington, DC.

American Institute of Steel Construction (AISC), 2010. Load and Resistance Factor Design (LRFD) Manual for Steel Structures.

Barker, R.M., Duncan, J.M.K., Rojiani, K.B., Ooi, P.S.K., Kim, S.G., 1991. Manuals for the Design of Bridge Foundations, NCHRP Report 343. Transportation Research Board, National Research Council, Washington, DC.

Bowles, J.E., 1996. Foundation Analysis and Design, fifth ed. McGraw-Hill, New York.

Burke Jr., M.P., 1988. Bridge deck joints. In: NCHRP Synthesis of Highway Practice, No. 141. Transportation Research Board, National Research Council, Washington, DC.

Burke Jr., M.P., 1990. Integral bridge design is on the rise. AISC Mod. Steel Const. 30 (4), 9–11.

Burke Jr., M.P., 1993. Bridge deck joints. In: Integral bridges: attributes and limitations. Transportation research record, no. 1393. Transportation Research Board, National Research Council, Washington, DC, pp. 1–8.

Chen, Y., 1997. Assessment on pile effective lengths and their effects on design—I. Assessment. Comp. Struct. 62 (2), 265–286.

Cross, H., 1932. Analysis of continuous frames by distributing fixed-end moment. Trans. Am. Soc. Civil Eng. 96, 1–156.

Dicleli, M., 2000a. A rational design approach for prestressed-concrete-girder integral bridges. Eng. Struct. 22 (3), 230–245.

Dicleli, M., 2000b. Simplified model for computer-aided analysis of integral bridges. J. Bridge Eng. 5 (3), 240–248.

Dicleli, M., 2005. Integral abutment-backfill behavior on sand soil-pushover analysis approach. J. Bridge Eng. 10 (3), 354–364.

Dicleli, M., Albhaisi, S.M., 2004. Effect of cyclic thermal loading on the performance of steel H-piles in integral bridges with stub-abutments. J. Constr. Steel Res. 60 (2), 161–182.

Dicleli, M., Albhaisi, S.M., 2005. Analytical formulation of maximum length limits for integral bridges built on cohesive soils. Can. J. Civil Eng. 32 (5), 726–738.

Erhan, S., Dicleli, M., 2014. Comparative assessment of the seismic performance of integral and conventional bridges with respect to the differences at the abutments. Bull. Earthq. Eng., Springer, 13 (2), pp. 653–677.

Evans, L.T., 1982. Simplified Analysis of Laterally Loaded Piles. Ph.D. thesis, University of California, Berkeley, CA. page 211.

Girton, D.D., Hawkinson, T.R., Greimann, L.F., Bergenson, K., Ndon, U., Abendorth, R.E., 1989. Validation of design recommendations for integral piles. Ames, Iowa: Iowa Department of Transportation, Project HR-292.

Hambly, E.C., 1997. Integral bridges. Proc. Instit. Civil Eng. Transport 123 (1), 30–38.

Steiger, D.J., 1993. Jointless bridges provide fuel for controversy. Roads and Bridges 31 (11), 48–54.

Wolde-Tinsae, A.M., Klinger, J.E., Mullangi, R., 1988a. Bridge Deck Joint Rehabilitation or Retrofitting. Final Report, Department of Civil Engineering, Maryland University, College Park, MD.

Wolde-Tinsae, A.M., Klinger, J.E., White, E.J., 1988b. Performance of jointless bridges. J. Perf. Constr. Facil. 2 (2), 111–128.

Movable bridges

17

Reiner S., Humpf K.
Leonhardt, Andrä und Partner Beratende Ingenieure VBI AG, Stuttgart, Germany

1 Introduction

1.1 General

One of the great beneficiaries of globalization is the transport sector, especially maritime transport. With cost between the Far East and Europe of about $2 for a DVD player and $30 for a television set, even the longest way pays off! This has led to an explosionlike increase of container traffic (e.g., between 2004 and 2005 in Shanghai by 24%, in Dubai by 17%, and in Hamburg by 17%) (BMVBS, n.d). Consequently, the number and size of container ships has increased permanently (Figure 17.1).

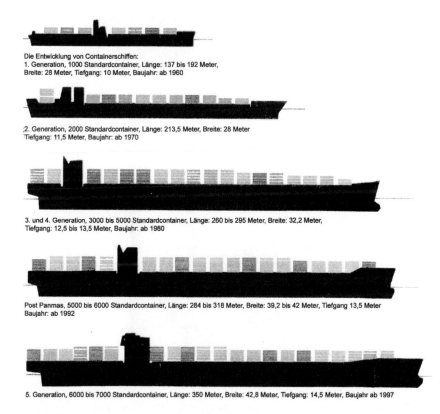

Die Entwicklung von Containerschiffen:
1. Generation, 1000 Standardcontainer, Länge: 137 bis 192 Meter,
Breite: 28 Meter, Tiefgang: 10 Meter, Baujahr: ab 1960

2. Generation, 2000 Standardcontainer, Länge: 213,5 Meter, Breite: 28 Meter
Tiefgang: 11,5 Meter, Baujahr: ab 1970

3. und 4. Generation, 3000 bis 5000 Standardcontainer, Länge: 260 bis 295 Meter, Breite: 32,2 Meter,
Tiefgang: 12,5 bis 13,5 Meter, Baujahr: ab 1980

Post Panmas, 5000 bis 6000 Standardcontainer, Länge: 284 bis 318 Meter, Breite: 39,2 bis 42 Meter, Tiefgang 13,5 Meter
Baujahr: ab 1992

5. Generation, 6000 bis 7000 Standardcontainer, Länge: 350 Meter, Breite: 42,8 Meter, Tiefgang: 14,5 Meter, Baujahr ab 1997

Figure 17.1 Development of container ships.

Innovative Bridge Design Handbook. http://dx.doi.org/10.1016/B978-0-12-800058-8.00017-7

In places with sufficient space for long-ramp bridges, normally high-level bridges are built (Figure 17.2). In places with restricted space, road bridges may still be built as high-level bridges, but railway bridges as low-level movable bridges (Figure 17.3). Because in many ports high-level bridges are unfeasible due to the very restricted space, movable bridges have experienced a veritable renaissance during the last decades.

Figure 17.2 High-level bridge for road and railway traffic: The Zárate-Brazo Largo Bridges across the Paraná River, Argentina (Leonhardt et al., 1979).

Figure 17.3 A high-level bridge for long-distance road traffic and a low-level bridge for local road and railway traffic: the Strelasund Crossing at Stralsund, Germany (Kleinhanß and Saul, 2007).

1.2 Short description of movable bridge types

1.2.1 Lift bridges

Lift bridges are suitable for great spans, but their clearance is limited by the lift towers, which have a great impact on the environment, even when the bridge is closed (Figure 17.4). The cables linking the bridge and the counterweights may suffer from significant wear.

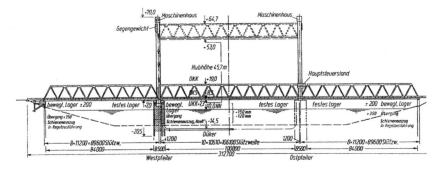

Figure 17.4 Kattwyk lift bridge at Hamburg, Germany (Rüster, 1974).

1.2.2 Swing bridges

Swing bridges are also suitable for great spans and do not limit the clearance. The biggest bridge of this type crosses the Suez Canal at El Ferdan, Egypt, with a free span of about 300 m (Figure 17.5). (Binder, Pfeiffer, Weyer 2001)

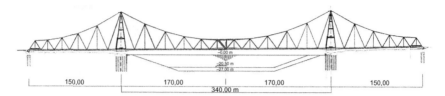

Figure 17.5 Swing bridge across the Suez Canal at El Ferdan, Egypt (Binder et al., 2001).

The disadvantages of swing bridges include the following:

- When opened, they occupy the embankment over a length of about their main span.
- Due to geometrical reasons, it is impossible to have separate bridges for railways and highways in close vicinity.

1.2.3 Bascule bridges

Bascule bridges may have a single flap or two flaps and are also adequate for long spans without limiting the clearance. The connection between the two flaps may transmit shear forces only, or shear forces and bending moments.

For great heights above the water, the counterweight may be attached to the rear arm as a pendulum (Figure 17.6), for reduced heights it has to be integrated with it (described further in section 4 of this chapter).

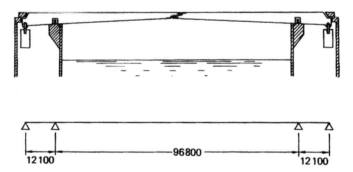

Figure 17.6 Sample of a bascule bridge with hang-on counterweight: Bridge across the Bay of Cadiz, Spain (Freudenberg, 1971).

1.2.4 Balance beam bridges (draw bridges)

Drawbridges, the precursors of bascule bridges, are most probably the oldest type of movable bridge (Figure 17.7). Compared to bascule bridges, they have the advantage of rather simple piers and a high architectural potential (Figure 17.8), but the disadvantage that they permit only rather reduced spans.

Figure 17.7 Vincent van Gogh: Langlois Bridge at Arles, France. Courtesy of Rheinisches Bildarchiv Köln.

Figure 17.8 Diffené Bridge at Mannheim, Germany (Freudenberg, 1989).

2 Example of lift bridge: the Guaiba River Bridge with concrete towers at Porto Allegre, Brazil (1954–1960)

2.1 General information

The Guaiba River Bridge (Leonhardt and Andrä, 1963) has a total length of 5665 m. It consists of the following:

- A 2013-m-long access bridge and a flyover, linking roads parallel to the river with the road crossing it
- The bridge across the Guaiba River, with a total length of 777 m. Its main span, with a clear span of 50 m and a clearance of 40 m, is designed as a lift bridge
- The 344-m-long bridge across the Furado Grande River

- The 774-m-long bridge across the Saco Alamôa Bay
- The bridge across the Jacui River, with a total length of 1757 m and main openings of 50 x 20 m

With the exception of the lift bridge, the entire bridge is from prestressed concrete, with regular spans of 43 m above water and 21.5 m over land.

2.2   The lift bridge

2.2.1   General information

The lift bridge has a free span of 50 m and a clearance above the low-water level of 13.5 m when in service, and 40 m when opened. The lifting height, therefore, is 26.5 m.

It consists of the bridge deck, a steel bridge with orthotropic plate, and four rounded towers made of reinforced concrete (r.c.), which hoist (and hide) the concrete counterweights and machinery. Due to the graceful design of these towers, the often ugly appearance of lift bridges is avoided (Figure 17.9).

Figure 17.9 View of the lift bridge: (a) under service, (b) opened to a major ship.

2.2.2 Bridge deck

The bridge deck has a span of 55.8 m and hoists

1. a four-lane roadway	16.00 m
2. the walkways 2 x 1.15 m	2.30 m
	18.30 m

The distance of the main girders is 13 m and the two cantilevers are 2.65 m long (Figure 17.10).

The orthotropic deck consists of the deck plate, with a thickness of 12 mm; the bulb-shaped longitudinal ribs with a distance of 310 mm and a depth of 160 mm; the narrowly spaced (d = 1.65 m) cross-girders with a depth of 640 mm corresponding to 1/20 of their span; the 60-mm-thick asphalt layer.

The main girders have a depth—as the approach viaducts—of 2.64 m corresponding to 1/21 of their span. They are stiffened by vertical stiffeners on the outside only, and therefore, they are an early application of the tension field theory. The weight of the steel structure is 381 tons, corresponding to 360 kg/m², and the total weight of the bridge deck is 540 tons.

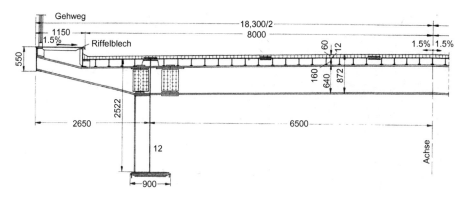

Figure 17.10 Section of the bridge deck.

2.2.3 Tower and piers

The towers and piers have a total height of 48.2 m above the lowest water level. They consist of the following (Figure 17.11):

- Four freestanding towers, with a distance of 51.8 m in the longitudinal direction and 18.6 m in the transverse direction and a height of 35 m. They have overall dimensions of 4 x 4 m and are rounded on their outer faces. Their walls parallel to the bridge axis are 300 mm thick, and the other walls are 250 mm thick. The towers surround the counterweight Ø 3 m x 6 m of heavyweight concrete.
- The 11.6-m-high piers connecting the towers underneath the bridge deck. They have two walls with a distance of 3.0 m and a thickness of 250 mm.
- Pile caps with dimensions of 28.4 m x 4.9 m x 2.0.
- A total of 66 driven piles Ø 0,52 m per pier, system Franki.

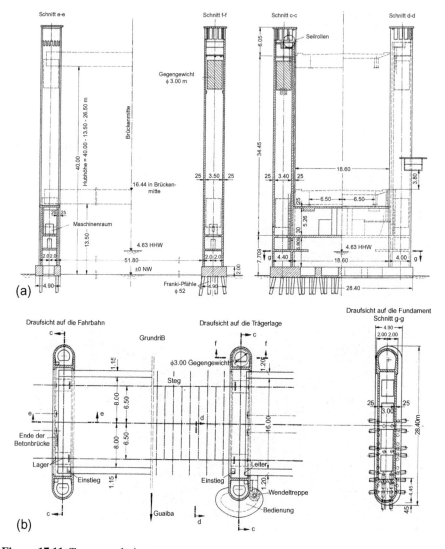

Figure 17.11 Towers and piers.

2.2.4 Mechanical installations

The wheels for turning around the cables, which connect the bridge deck and the counterweights, and the entire machinery at the top are also included in the towers. These features improve the aesthetical appearance of the bridge substantially.

The hoisting and lowering of the bridge deck are controlled from a cabin on the outside of one of the towers (Figure 17.9a).

3 Swing bridges

3.1 The prestressed concrete bridge across the Shatt-Al-Arab, Iraq (1972–1978)

3.1.1 General

The prestressed concrete bridge across the Shatt-Al-Arab (Seifried and Wittfoth, 1979) consists of the following (Figure 17.12):

- The western section, with a total length of 331.75 m
- The eastern section, with a total length of 430.15 m
- A viaduct linking the main bridge to Sindibad Island.

The center part of the western section is a swing bridge with a total length of 67 m, providing space for two shipping canals of 23 m each. The main bridge has regular spans of 46.9 m and a width of 21 m; the viaduct has regular spans of 28 m and a width of 10.75 m.

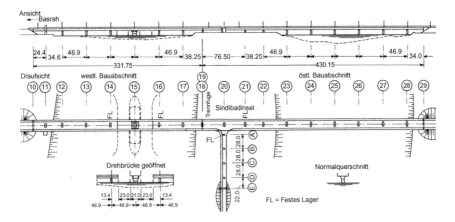

Figure 17.12 General layout.

The entire bridge deck, including the swing bridge, is made of prestressed concrete. The main bridge was built by incremental launching, with a unit length of 15.63 m corresponding to 1/3 of the regular span. With respect to this construction procedure, its depth is 3.65 m, corresponding to 1/12.8 of the regular span.

3.1.2 The swing bridge

Bridge deck
The swing bridge has two cantilevers of 33.5 m each (Figure 17.13a). The cross section consists of the following:

- A trapezoidal box girder with a width of 7 m at its bottom and 10.5 m at the top
- Two 5.25-m-wide cantilevers.

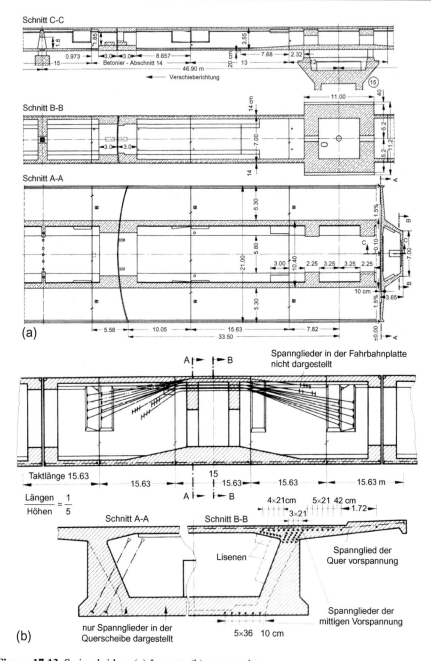

Figure 17.13 Swing bridge: (a) Layout, (b) prestressing.

The bridge deck is prestressed in the longitudinal and the transverse directions (Figure 17.13b). For the launching, continuity tendons were introduced at both bridge ends, which were cut after the bridge had reached its final position.

Main pier

The main pier (Figure 17.14a) consists of the following:

- A solid pier table square 12.8 m
- A hollow shaft square 6.5 m, with a wall thickness of 1.0 m
- A 2.0-m-thick pile cap
- 16 drilled piles Ø 2 m, with a length of about 40 m

The pier table is heavily prestressed (Figure 17.14b). The bridge deck rests at the pier on a turning circle with a radius of 10 m (Figure 17.14c).

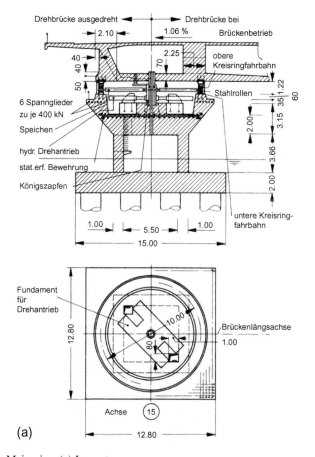

(a)

Figure 17.14 Main pier: (a) Layout,

(Continued)

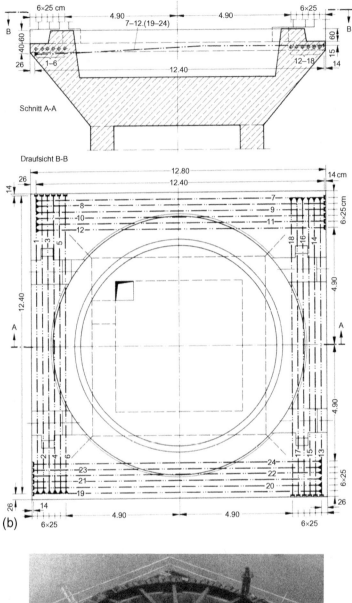

Schnitt A-A

Draufsicht B-B

(b)

(c)

Figure 17.14 Continued. (b) prestressing of the pier table, (c) turning circle.

3.1.3 Joint to the fixed part

The swing bridge is locked to the fixed part by locking devices that can be retracted to facilitate the opening of the bridge (Figure 17.15, top). The circular expansion joint is open, with a gap of 30 mm (Figure 17.15, bottom).

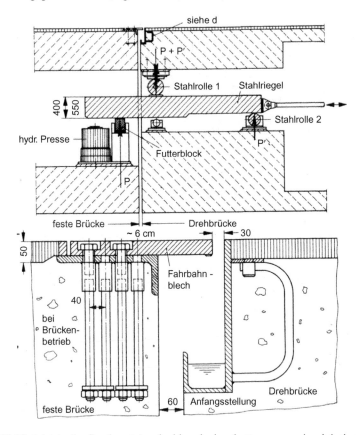

Figure 17.15 Joint to the fixed part: top locking device, bottom expansion jointing.

3.2 Cable-stayed bridge in the port of Barcelona, Spain

3.2.1 Introduction

The bridge in the port of was the first of a growing number of movable bridges built during the last decade the ports of in Spain with the aim of adapting these ports to the needs of modern ship traffic.

The tender design called for a double flap bascule bridge with a free span of 85 m (Figure 17.16). This span is small for the design ship—20,000 dwt, L = 250 m, W = 35 m, sailing at 2.2 m/s—and left the main piers, founded on piles, in the water. Hence, they would be exposed to impact from ships.

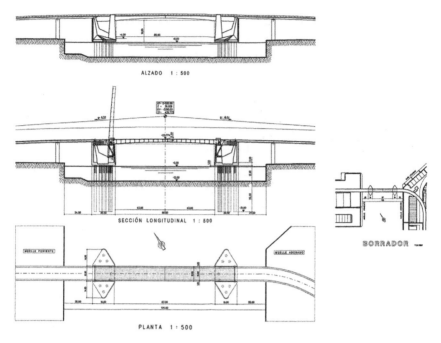

Figure 17.16 Bascule Bridge of tender design.

LAP (Leonhardt und Andrä, 1997) prepared with a group of Spanish contractors an alternative design as a swing bridge. The main aim of this design was to avoid the expensive piers in the water, thereby increasing the safety of navigation.

Unfortunately, this alternative was not selected for the ultimate construction. Nevertheless, for ease of grammar we use the verb forms corresponding to a built bridge.

3.2.2 Description of the design

Main structural system

The main structure of the swing bridge is a cable-stayed bridge with spans of 180 m and 75 m and a single tower with 4 legs (Figure 17.17). The effective span lengths are reduced by cantilevers of the approach viaducts to 159 m and 68 m, respectively. In order to have the permanent loads centered with respect to the axis of the towers, the steel composite deck of the main span is counteracted by a concrete side span.

In the longitudinal direction, the cables are anchored at regular intervals of 17 m in the transverse direction at both borders of the deck.

Bridge deck

The bridge deck consists of a two lane, 10 m wide roadway and two 2,25 m wide walk-ways and cable anchorage zones, yielding a total width of 14,50 m.

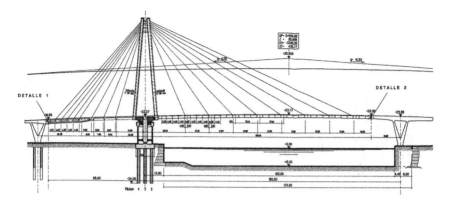

Figure 17.17 Layout of the alternative.

The steel composite bridge deck (Figure 17.18a) is built up from

- The two 2.3-m-deep main girders with a distance of 12.3 m.
- 2.10-m-deep cross girders spaced 4.25 m apart.
- The 225-mm-thick roadway slab with an 80-mm-thick asphalt layer; in the walkway and cable anchorage zone, the slab thickness is 500 mm.
- Concrete cable anchorages at the outside of the main girders.

The bridge deck in the side span, close to the pier, made of prestressed concrete, is a plate-beam structure with a depth of 2.3 m (Figure 17.18b). It consists of

- The main girders, with an outer distance of 12.3 m and a width of 1.75 m to 2.0 m
- Cross girders with a spacing of 8.5 m
- A 600-mthick slab.

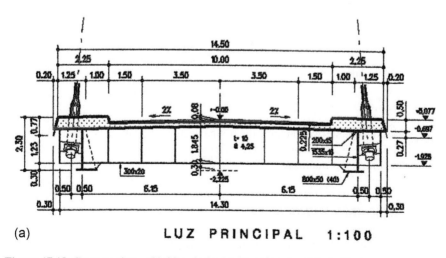

(a)   **LUZ PRINCIPAL   1:100**

Figure 17.18 Cross sections of bridge deck: (a) At main span, (b) at side span,

(Continued)

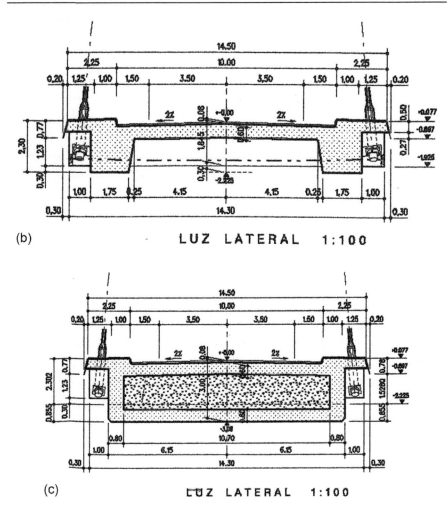

(b)   LUZ LATERAL   1:100

(c)   LUZ LATERAL   1:100

Figure 17.18 Continued. (c) at counterweight.

In the side-span, counterweight area, the bridge deck is a box girder with outer dimensions of 12.3 m x ~3.2 m (Figure 17.18c). The box is filled with heavyweight concrete to counteract the long main span.

Tower and pier (Figure 17.19)
The steel tower has a height of 67.7 m above the bridge deck. It consists of four legs with outer dimensions of 1.5 m x 2 m and is stiffened by four cross girders at their top. The tower is supported by a solid part of the bridge deck that is prestressed in both the longitudinal and transverse directions.

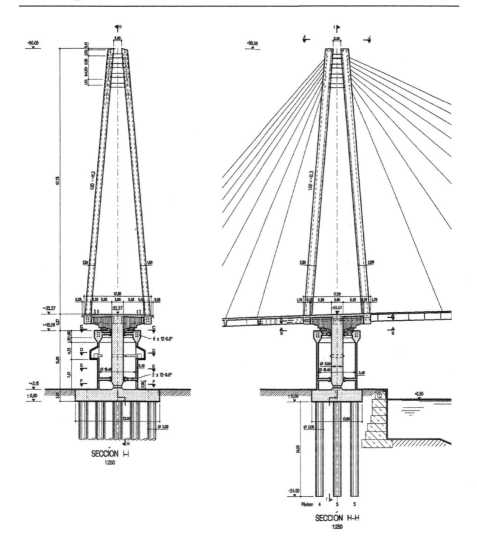

Figure 17.19 Towers and piers.

The circular pier has an outer diameter of 11.2 m and a wall thickness of 400 mm, which is thickened to 2 m at the top, where it hoists the turning table with a diameter of 10.8 m. The pier is founded on 14 24-m-long drilled piles Ø 2 m and a 3-m-thick pile cap.

Mechanical equipment
The bridge deck is turned around a pivot of 3 m, which takes horizontal forces only, by two hydraulic cylinders (Figure 17.20).

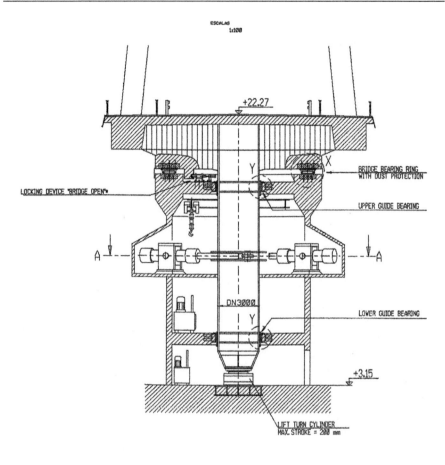

SECTION A-A

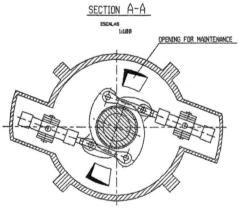

SWING BRIDGE DRIVE

Figure 17.20 Mechanical equipment.

3.2.3   Construction

The bridge was assembled parallel to the embankment. Later, it was turned into its service position.

3.3   Railroad bridge across the Sungai Prai River, Malaysia (2008–2013)

3.3.1   Introduction

A double track electrified railway line between Padang Besan and Ipoh crosses the Sungai Prai River in Malaysia from west to east over a railway bridge designed by LAP (Leonhardt and Andrä 2008) with two spans of 45 m. LAP prepared for a contractor responsible for the construction of the complete rail link the concept and tender design, including the mechanical and electrical elements.

3.3.2   Description of the design

The main structure consists of two balanced spans composed of a steel grid supporting the concrete slab and of two steel sails at the edges as main load-carrying members. The deck width is 11 m, and together with the steel sails of 1 m each, the overall width is 13 m (Figure 17.21). Cross beams of 1.40 m depth at 4 m distance carry the traffic loads to the main girders.

The main girders consist of a plate girder of variable depth (between 2 and 12 m), a V-shaped strut in the axis of the central pier, and two openings in the web of the plate

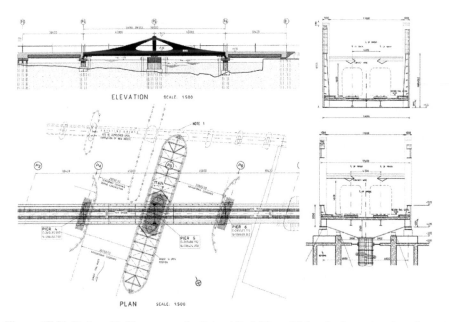

Figure 17.21 Railroad bridge across the Sungai Prai River, Malaysia: Layout and sections.

girder. All stiffeners of the plate girders are located at the inner side, providing a smooth outside face of the structure.

Supply of electricity for the railway can be held independent of the structure, with typical posts and contact wire supports as throughout the railway link. Below the center pier, all turning equipment and rotational bearings are located in a hollow pier partially below water.

The center pivot shaft will provide vertical and horizontal support during the swinging operation. A hydraulic lift/turn cylinder allows to transmit the torque for turning and to lower the structure on bearings for the railway service situation. The pivot shaft is free from loads under service conditions.

In the service position, the bridge is locked with wedge-shaped end locks, and a rail locking device is engaged to provide continuity of the rail. In the open position parallel to the river, the bridge is protected by a guidance steel structure that keeps the footprint of the bridge in open position free from navigation.

4   Bascule bridge: the New Galata Bridge with twin double flaps at Istanbul, Turkey (1985–1993)

4.1   Introduction

The New Galata Bridge across the Golden Horn (Saul et al., 1992) in Istanbul, Turkey, links the quarters of Eminönü and Karaköy directly on the former site of a steel floating bridge built in 1912.

The 477.45-m-long and 42-m-wide bridge (Figures 17.22 and 17.23) consists basically of the following elements:

- A center bascule bridge with a clearance of 80 m and the corresponding bascule bridge piers
- Double deck approach bridges with 8 spans of 22.3 m each, with road and light railway traffic on the upper deck and shops, restaurants, and similar structures on the lower deck
- Abutments.

Between these 2 x 3 elements and between the bascule bridge piers and their piles, buffer bearings are provided.

Due to a water depth of up to 40 m and poor soil of another 40 m, the bridge is founded on driven or drilled hollow steel piles, with a diameter of 2 m, a wall thickness of 20 mm and cathodic corrosion protection.

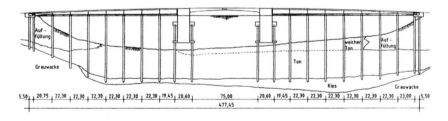

Figure 17.22 General layout of the New Galata Bridge.

Figure 17.23 The New Galata Bridge, nearly finished.

4.2 Design

4.2.1 Bascule bridge

The free span of 80 m and a total width of 42 m render the bascule bridge the world's largest (Figure 17.24 and 17.25). The total length of the flaps (54.5 m each) is divided by the axis of rotation into two cantilevers of 42.8 m and 11.7 m.

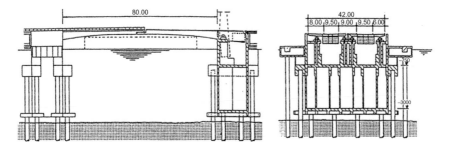

Figure 17.24 General arrangement of the bascule bridge.

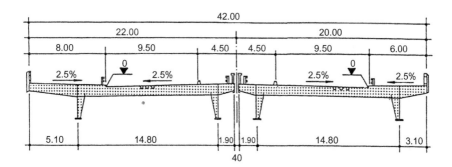

Figure 17.25 Cross-section of the bascule bridge.

In the design of the bascule bridge piers, two contradictory requirements had to be fulfilled: For the absorption of ship impact, they had to be stiff, but for earthquakes, they needed to be flexible. This could be achieved by a pier going down to the seabed and founded on 12 piles, which are fixed to the pier between -13 m and $-7,5$ m and elastically supported at -32 m (Figure 17.26). In order to avoid an overloading of the pile or the addition of piles, the piers are made hollow. In spite of being exposed to a water pressure of up to 35 tons/m^2, the pier walls are not waterproofed; rather, they are reinforced for a crack width of $w_{95} = 0.2$ mm.

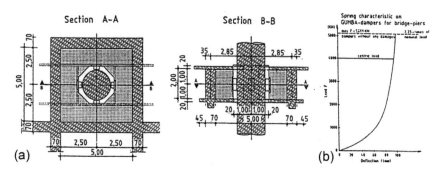

Figure 17.26 Bearings at -32 m: (a) Layout, (b) load-displacement diagram.

4.2.2 Approach bridges

Structural design
Both decks of the approach bridges have four T-beams with a constant depth of about 1.2 m and a width of 3 m, enlarged to 4 m at the piers (Figure 17.27). The prestressing consists transversaly of 4 Ø 0,6 in. St 1570/1770 per linear metre, and longitudinally, of 9 tendons 15 Ø 0.6 in. St 1570/1770 each per beam.

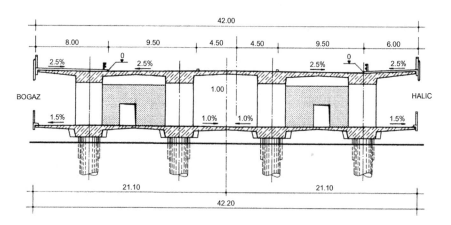

Figure 17.27 Approach bridge.

Bearings

Bearings for vertical loads are needed at the bridge ends and the main piers only due to the longitudinal elasticity of the piles. In order to keep them out of the splash water zone, they support the upper deck only, so the end walls are tension walls. The displacement of these bearings has been sized generously in order to avoid a dripping-down of the end spans in case of an unforeseen strong longitudinal earthquake.

Bearing for transverse forces are also at the abutment and the main piers only; they are designed as Teflon sliding bearings. Longitudinal forces are absorbed at both ends of the approach bridges by buffer bearings, which are working under compression only. In order to avoid bending of the walls, these bearings are at both deck levels. They consist of rubber disks and have a pronounced hysteresis (Figure 17.28).

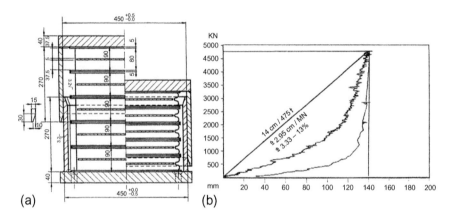

Figure 17.28 Longitudinal buffer bearings: (a) Design, (b) load-displacement diagram.

4.2.3 Piles

In order to reduce the masses involved in an earthquake and to save costs, the pile shafts are designed as hollow steel pipes, with an outer diameter of 2 m and a wall thickness of 20 mm only, with steel quality of St 52-3.

The piles of the bascule bridge piers are filled with tremie concrete B 35 and reinforced in the upper parts. The design of these piles as composite columns proved that shear connectors were needed at both ends only.

4.3 Special aspects of dimensioning

4.3.1 Ship impact

The bridge had to be designed to withstand the head-on impact of a 8000-dwt ship sailing at 2.5 m/s. The corresponding impact force is, according to the Nordic Road Council Regulations for Ship Impact,

$$P_{[kN]} = 500 \cdot \sqrt{dwt} = 500 \cdot \sqrt{8000} = 45.000 \, kN.$$

As a consequence of an eventual ship impact, the loss of buoyancy of the upper or lower part of the pier due to breaching also had to be considered. As the bascule bridge could not be designed against ship impact, of course, two worst-case scenarios were investigated (Figure 17.29)

*   Formation of a hinge in front of the pier
*   Loss of a flap between this hinge and the center

These scenarios led to neither a loss of the other flap nor the rear arm with the counterweight.

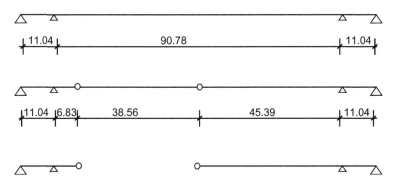

Figure 17.29 Worst-case scenarios.

4.3.2   Earthquake analysis

For the check of the structure's safety during earthquakes, two methods were used. In a first step, a response-spectrum analysis was performed, assuming that the six elements of the bridge are completely independent in the longitudinal direction. In order to determine the displacements of bearings and jointings and the forces acting on the buffers, a time-history analysis was performed next.

Response-spectrum analysis

A response-spectrum analysis was performed for closed flaps, opened flaps, and for construction stages. It was done with a spectrum given in the tender documents and with the spectrum according to an American Association of State Highway and Transportation Officials (AASHTO, Guide Specifications for Seismic Design of Highway Bridges. Washington 1983.) earthquake code that yields substantially higher accelerations for the governing, rather low frequencies (Figure 17.30). The response-modification factor was assumed to be 1.0 for the spectrum according to the tender documents and 3.0 for the spectrum according to AASHTO.

Under the first spectrum, the bridge behaved in a completely elastic manner. That means that the safety against yield is 1.0 at the maximum stressed point of the maximum stressed pile.

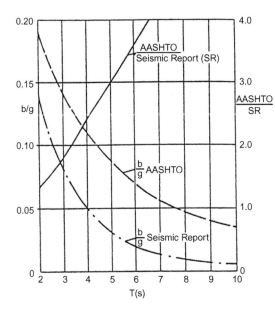

Figure 17.30 Peak acceleration versus natural frequency.

Time-history analysis

The velocity of the surface (Love) waves may be assumed to be 3 km/s, whereas the governing eigenfrequency of the bridge is in the range of 0.25 per second. An earthquake, hence, moves along the bridge in $470/3000 = 0.15$ s, which is substantially less than the period of eigenvibration $t = 1/0.25 = 4$ s. Therefore, it was assumed that the bridge would accelerate uniformly over its entire length, which means no phase difference was considered.

Acceleration diagrams

For the time-history analysis, six acceleration diagrams have been generated that are compatible with the energy content of the response spectrum (Figure 17.31a).

Investigated systems

Corresponding to the progress of design, and especially of the buffers, different connections between the main elements of the bridge were assumed [e.g., elastic springs and springs with a gap for the displacements under service conditions, friction, assumed and real hysteresis of the buffers (Figure 17.31b)].

Results

The results were given graphically. For example, the displacements between the abutment and the approach bridge and the reactions of the corresponding buffers are given in Figure 17.31c and 17.31d.

The design was prepared by LAP and Temel Mühendislik, of Istanbul, Turkey, jointly. The main contractor was a joint venture of STFA, of Istanbul, and Thyssen Engineering GmbH, of Essen, Germany.

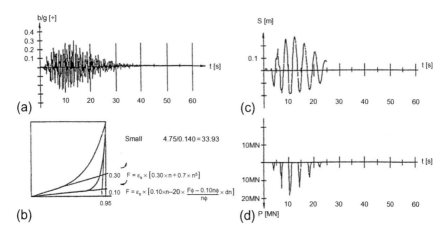

(a)

(b)

(c)

(d)

Figure 17.31 Time-history analysis: (a) Acceleration diagram, (b) analytical description of buffers, (c) deformation of buffer 1, approximately symmetric, (d) forces in buffer 1, pronouncedly nonsymmetric: upward friction only, downward friction and buffer force.

5 Double Balanced Beam Bridge (DBBB)—design proposal

5.1 Design concept

This section discusses the design for a DBBB proposed by Saul and Humpf (2007). So far, balance beam bridges have been built as single-span bridges. Due to the articulation of the balance beam, this system takes permanent loads only. In DBBBs, the joint at the center would have to transmit under live loads the bending moment of a single-span beam. If, instead, the rotation of the balance beam is blocked by a second bearing, the staying system also participates in handling the live loads. This allows balance beam bridges to be built with two flaps, thereby doubling their span range. This solution is advantageous in areas where the piers of a bascule bridge have to be built in water or groundwater.

In more detail, we make use of the fact that for cinematic reasons, the balance beam has to have an eccentricity toward land. With an additional bearing with eccentricity toward the water – which can take compression only and is automatically activated when lowering the flaps (Figure 17.32) – the live loads can also be taken by the balance beam and the pylon, and thereby, the moments of the bridge deck – especially at the center – are substantially reduced.

5.2 Comparison of section forces

5.2.1 System and loads

The free span is 80 m and the bridge width 12 m. The permanent load, including surfacing, is 5 kN/m^2, and the equivalent live load was also 5 kN/m^2. Only the live load over the full main span is considered.

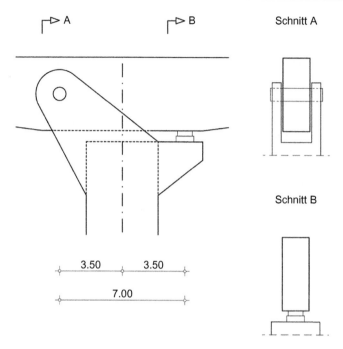

Figure 17.32 Bearing at the top of the tower of a DBBB.

5.2.2 DBBB

The static system for a DBBB is as follows:

- For permanent loads, a span and a cantilever of 22 m each
- For live loads, a continuous beam with spans of 22–44–22 m and elastic, intermediate supports.

The tensile rod is inclined by 1:3, and the distance of the counterweight from the axis of rotation is 80% of that of the rod. The stiffness of the balance beam and the tower are the five-fold of that of the bridge deck. The governing bending moments and reaction forces are given in Figure 17.33.

5.2.3 Double bascule bridge

The static system for a double bascule bridge is as follows:

- For permanent loads, a span and a cantilever of 11 m and 45 m, respectively
- For live loads, a continuous beam with span of 11–90–11 m.

The stiffness at the axis of rotation is five times that at the center. The governing bending moments and reaction forces are given in Figure 17.34.

5.2.4 Comparison of DBBB and the double bascule bridge

s. Figures 17.33c and 17.34.c

- The maximum bending moment of the bridge deck of the DBBB is about 15% of that of the double bascule bridge only. This reduces the construction depth and lowers the gradient.
- At the center joint, the live load moments of both bridge types are virtually the same.
- The counterweight of the DBBB (3165 kN) corresponds to 45% of that of the double bascule bridge (7140 kN) only, due to the longer lever arm.
- The reaction force of the rotation bearing of the DBBB (6900 kN) corresponds to 40% of that of the double bascule bridge (17,450 kN) only.
- The governing moments of the balance beam (72,100 kNm) and the pylon (43,000 kNm) of the DBBB corresponds to 70% and 40% of the maximum moment of the double bascule bridge (107,600 kNm).
- 6. The bending moments acting on the foundation of the DBBB (42,940 kNm) and the double bascule bridge (46,850 kNm) are basically the same.

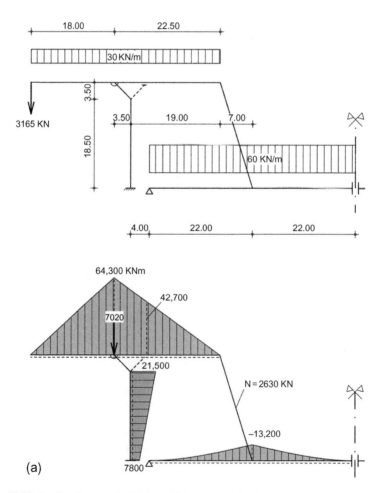

(a)

Figure 17.33 Section forces of a DBBB: (a) Permanent loads,

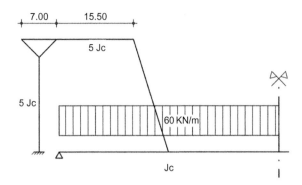

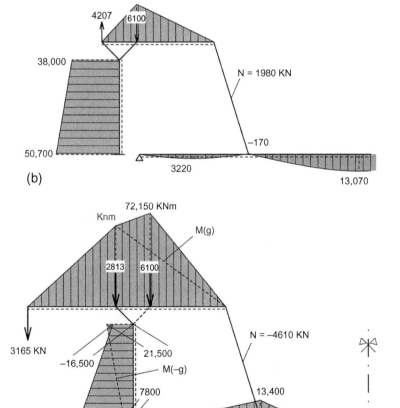

Figure 17.33 Continued. (b) traffic, (c) permanent loads and traffic.

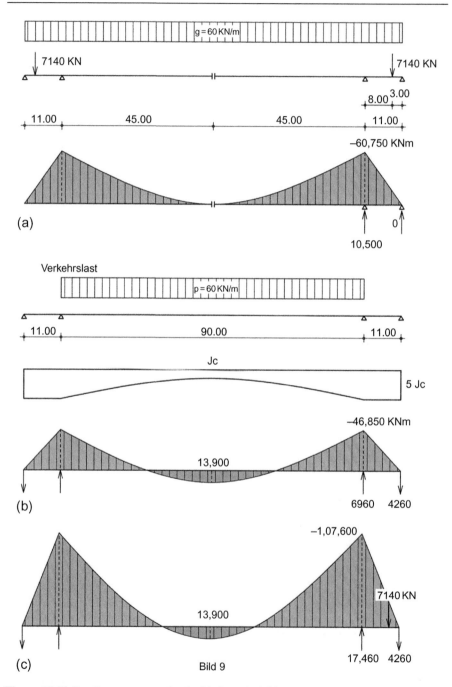

(a)

(b)

(c)

Bild 9

Figure 17.34 Bending moments of a double bascule bridge: (a) Permanent loads, (b) traffic, (c) permanent loads and traffic.

5.2.5   Comparison of DBBB with the single bascule beam bridge

The live load moments of the single bascule beam bridge are that of a beam with a span of 44 m (that is, 13,200 kNm).

The comparison shows the following:

- The governing moments of the bridge deck are basically the same.
- The governing force of the rotation bearing is that under permanent loads (6900 kN). The live load (-4200 kN) reduces it but does not invert it.
- The force of the tensile rod is about 70% bigger.
- The governing moment of the balance beam is increased by about 12% from 64,300 kNm to 72,100 kNm.
- The governing moment at the base of the tower is increased from 7800 kNm to 42,900 kNm.

This is not a problem, however, due to the large dimensions of the tower.

5.3   Summary

The presented innovative system of a DBBB allows taking all loads by a simple blockage of the rotational axis. Compared to a double flap bascule bridge, it allows a substantial reduction of the cost of the bascule bridge pier, especially when situated in water, poor soil, or both, and of the construction depth of the bridge deck. Compared to a single balance beam bridge, it allows to double the span. The increased normal force of the tensile rod and bending moment at the bottom of the tower may be absorbed without serious problems.

References

Binder, B., Pfeiffer, M., Weyer, U., 2001. Die El-Ferdan Brücke-Konstruktion, Fertigung und Montage der größten Drehbrücke der Welt über den Suez-Kanal (The El-Ferdan Bridge design, shop fabrication, and erection of the biggest swing bridge worldwide). Stahlbau 70, 231–244.

BMVBS, n.d. Zukunft Elbe, eine Initiative für Norddeutschland (Future of Elbe River, an initiative for northern Germany). Undated folder.

Freudenberg, G., 1971. Die größte Doppelklappbrücke der Welt über die Bucht von Cadiz (The worldwide biggest double flap bascule brideg across the Cadiz Bay). Steel 463–472.

Freudenberg, G., 1989. Zwei neue Waagebalken-Klappbrücken in Mannheim (Two new scale beam bascule bridges at Mannheim). Stahlbau 58, 35–43.

Kleinhanß, K., Saul, R., 2007. The Second Strelasund Crossing—a modern cable-stayed bridge in close vicinity to the World Cultural Heritage, Stralsund, Germany. Struct. Eng. Intl., 30–34.

Leonhardt, F., Andrä, W., 1963. Die Guaiba-Brücken bei Porto Alegre, Brasilien (The Guaiba Bridges at Porto Alegre, Brazil). Beton Stahlbetonbau 58, 273–279.

Leonhardt, Andrä, Partner GmbH, Stuttgart, July 1997. Puerto de Barcelona, Puente entre Muelle de Poniente y Muelle Adosado. Puente Principal, Alternativa: Sistema giratorio atirantado (Port of Barcelona, bridge between Muelle de Poniente and Muelle Adosado. Main bridge, cable-stayed swing bridge alternative). Stuttgart.

Leonhardt, Andrä, Partner GmbH, June 2008. Design documents for the Sungai Prai River Bridge, Malaysia. Stuttgart.

Leonhardt, F., Zellner, W., Saul, R., 1979. Zwei Schrägkabelbrücken für Eisenbahn und Straßenverkehr über den Rio Paraná/Argentinien (Two cable-stayed bridges for railway and highway traffic across the Paraná River, Argentina). Der Stahlbau 48, 225–236, 272–277.

Rüster, R., 1974. Die Kattwyk Hubbrücke in Hamburg, eine vollständig geschweißte Fachwerkbrücke (The Kattwyk lift bridge at Hamburg, a completely welded truss girder bridge). Stahlbau 43, 257–267.

Saul, R., et al., 1992. Die neue Galata Brücke in Istanbul (The new Galata Bridge at Istanbul). Bauingenieur 67, 433–444 and 1993;68:43–51.

Saul, R., Humpf, K., 2007. Doppelwaagebalkenbrücke – Vorschlag für einen innovativen Klappbrückentyp (Double balance beam bridge – Proposal of a new type of bascule bridge). Stahlbau 76, 559–564.

Seifried, G., Wittfoth, H., 1979. Die Brücke über den Shatt-al-Arab in Basrah (Irak) [The bridge across the Shatt-al-Arab in Basrah (Iraq)]. Beton Stahlbetonbau 74, 77–85.

Highway bridges

Martin B.T.
Modjeski and Masters, Inc., Poughkeepsie, NY

18

1 Introduction

In a report called "The Three Mentalities of Successful Bridge Design," F. de Miranda (1991) stated that a successful bridge design must address three areas: (1) creative and aesthetic, (2) analytical, and (3) technical and practical. The lack of any of these mentalities leads to a less-than-successful design. In today's "team" approach, it is fairly easy to achieve the first two of these mentalities, but the last mentality is the one that is often the most troubling. Without an in-depth familiarity with economical construction alternatives, the selected bridge type, though innovative in its technical aspects, will not be practical.

2 Practical considerations for selection of a highway bridge type

The selection of a bridge type for a given site is driven by many variables, and there is no single correct solution to the problem. For any given span length, there are always many bridge types that can satisfy the design objectives of the project. The type of bridge that is selected can be driven by such variables as the availability and cost of certain materials, the skill set of the local labor force, and the experience of local contractors. It is entirely possible that for a given set of constraints, the bridge type that is preferred in one jurisdiction or country will be entirely different from the one selected in another.

2.1 Selecting a bridge type

2.1.1 Geometric demands of the roadway

Quite often, the type of bridge selected will be driven by the geometric alignment of the approach roadways. The vertical and horizontal geometry of the bridge approaches are driven by such factors as desired vertical clearances and roadside elements/roadway facilities either up-station or down-station of the bridge. For example, if vertical clearance beneath the proposed structure is critical and the elevation of the approach roadways cannot be raised, a shallow, beam-type bridge might be in order. Or if the approach roadways demand the use of a curved structure and the site is such that aesthetic considerations are important, then a continuous curved box-girder bridge might

Innovative Bridge Design Handbook. http://dx.doi.org/10.1016/B978-0-12-800058-8.00018-9

be used due to its pleasing lines and great torsional resistance. If the purpose of the bridge is to cross a navigable channel where a large amount of vertical and horizontal clearance is needed, an entirely different bridge type would be used rather than a bridge serving as an elevated roadway in an urban area.

2.1.2   Utilization requirements

The utilization demands on a bridge will play a major role in its final configuration. The bridge must not only have a sufficient number of lanes to carry today's traffic volume, but also the demands projected for the future. Will the bridge carry pedestrians? Will it also have bicycle pathways? Is there a need to separate opposing traffic lanes with a median barrier? Is there a need to separate sidewalks/bicycle pathways from the roadway with a barrier? The answers to all of these questions will drive the ultimate width of the bridge. Another consideration is future expansion. Are there conditions in the geographic area that might lead to an increase in traffic volume, but there is a high degree of uncertainty? If that is the case, then bridge types with the capability of future widening, such as multiple-girder bridges, should be considered.

2.1.3   Surface site conditions

The terrain of the site will play a major role in selection of a bridge type. Bridges over wide canyons with inaccessible side slopes will require long-span bridge types such as arch spans. Depending on the type of vessel traffic, bridges over navigable waterways may require large spans and high vertical clearances such as truss span, cable stayed bridges, or suspension bridges. If the approach constraints will not allow long, high approaches, then a moveable bridge might be required. Does the bridge cross a flood plain? If so, the elevation of the bottom of structure, span lengths, and impacts to flow during flood season will have to be taken into consideration. All of these factors will play a major role in the selection of bridge type.

2.1.4   Subsurface site conditions

The subsurface condition at a proposed bridge site will play a major role in the selection of a bridge type. Basic questions have to be asked, such as the following:

- Will the soil conditions allow spread footings or will piles be required?
- Can drilled shafts be used, and are there economic advantages to using them?
- Are the soil types such that future settlement might occur? If so, the bridge type selected must be able to accommodate such movement.
- Does the site have a potential for seismic activity? If it does, the resulting configuration of the substructure may drive the superstructure loads and vice versa.
- Are foundation conditions such that high lateral loads can be resisted? For example, if geometric requirements and utilization requirements result in a the need for a long-span bridge, can the subsurface conditions allow the large loads that would result later at the anchorages of a suspension bridge, or would it be best to use a cable stayed bridge that results primarily in vertical loads?

2.1.5   Construction considerations

Erection and construction processes often dictate the type of bridge that is selected to be built. As stated previously, the type of bridge that is selected is often driven by such elements as the availability and cost of certain material, the skill set of the local labor force, and the experience of local contractors. For example, if there is a preponderance of steel fabricators in the area of the bridge site but no precast concrete fabricators nearby, then one might lean toward using precast concrete. If the local workforce does not have the skill set required for steel erection, then bridge types that maximize the use of cast-in-place concrete might be the best solution. Cast-in-place concrete is well suited for grade separation structures with limited restrictions under them in locations of the world where the workforce can build false work quickly and cheaply.

The time allotted by the owner for construction can also drive the bridge type. If time is short, then maximizing the use of precast elements might be in order. This would lead to the potential use of segmental concrete bridges or some of the newer accelerated bridge construction techniques.

It is always recommended to use bid histories for previous jobs in the selection of bridge types. In some parts of the world, such information does not exist; therefore, the construction advantages of one bridge type comprised of one material or the other is difficult to discern. In this case, there is a definite advantage to bid alternative designs. This typically is cost-effective only for large projects.

2.1.6   Project delivery system

In the recent past, most projects in the United States have used a design-bid-build project delivery system. Even more recently, the United States, like most other countries of the world, has started using a design-build delivery system. This is actually a return to the system used in the early years of bridge building in the United States (Barker and Puckett, 2007). During the great bridge-building era of the 19th century, an owner would express an interest in having a bridge built at a particular location and then solicit proposals from engineers not only for the design, but also for the construction. In many cases, the engineer would recognize a need and then present the concept to the affected parties. All services, in the areas of both design and construction, were the responsibility of one entity.

The design-bid-build approach was meant to provide a quality product while also providing a system of checks and balances between the designer/owner and the contractor. As is often the case, the problem with design-bid-build is not the concept, but its execution. Often, problems that develop during construction result in an attempt to assign blame rather than seek a practical solution that increases the financial risks of all parties.

Because design-build more clearly defines lines of responsibility, this delivery system is being used more and more in the United States. That being said, the successful application of this delivery system is dependent on a knowledgeable owner that has staff capable of judging the quality of work provided. This delivery system seems to work best on large bridge projects, though it is being applied in some jurisdictions to smaller bridge projects that are consolidated into a single contract.

2.1.7   Regulatory requirements

Almost every bridge design project in the world has to comply with the regulatory requirements of the jurisdiction in which the bridge is built. These requirements can have a profound impact on the bridge type that is selected and the location of the bridge. There are many environmental regulations and agencies with which coordination must take place, all of which require permits with stipulations that will drive the design process. It is essential that the engineer be knowledgeable about these agencies, regulations, and permits prior to the beginning of the design process.

Local and regional politics also have to enter into the bridge selection process. Often, national, state, or local officials have made commitments to their constituents that must be understood and included by the designer. In some cases, such political drivers override many of the engineering-driven criteria.

2.1.8   Aesthetics

It should be the desire of every engineer to design a bridge that is aesthetically pleasing. That being said, in the opinion of this writer, every effort should be made to design a pleasing functional bridge that is visually appealing but one should not begin with the desire to achieve "uniqueness" at all cost nor should structural efficiency be abandoned for the sake of appearance. A detailed discussion of the aesthetics of bridge design is discussed by others in this volume

3   Bridge types

On the basis of this site criteria, a general idea of the required span length can be established and studies can be performed to determine the most desirable bridge type to be used at the site. There are various publications that can assist in the selection of bridge type as a function of span, but span length alone is not the determining factor. The other factors mentioned here must also be taken into consideration. The following sections include a discussion of the bridge types that are most often used for different span lengths and is based primarily on the information contained in the Pennsylvania Department of Transportation Design Manual, Part 4, Page A.2 (PennDOT, 2012) and data collected by Barker and Puckett (2007).

3.1   Short-span bridges

Short span bridge types (i.e., span lengths up to 15 m) include single-unit or multiunit culverts, concrete slab bridges, precast I-beam bridges, and rolled I-beam bridges.

3.1.1   Culverts

Most often used to provide passage through roadway embankment for small streams, drainage channels, pedestrians, livestock, and, in some cases, vehicles, culverts far outnumber bridges in the United States. The National Bridge Inventory in the United States lists culverts as bridges only if the span exceeds 6.5 m. Almost 20%

of all bridges in the United States are classified as culverts. Culverts take many structural forms and are comprised of many materials ranging from concrete steel, aluminum, and thermoplastics.

3.1.2 Slab-span bridges

Slab-span bridges are simple and cost-effective for spans up to 12 m. They can be built on false work or precast and shipped to the site. They can be used as simple spans or, if a topping slab and reinforcement is added, they can be made continuous over intermediate bents, thereby limiting the number of joints. The spans can be extended to approximately 15 m if prestressing is used. These structure types are shallow and project a simple, slender appearance. Maintenance is rather low except where transverse joints are used.

3.1.3 T-beam

Traditionally, these bridges have been built on scaffolding and poured in place. They are generally economical for spans of 10–20 m. Formwork can be rather complex for the bridge if built in the field. It was a workhorse bridge throughout the middle of the 20th century but has been pretty much replaced by prestressed slab bridges and precast box beam bridges. Should a designer choose to use this bridge type, careful attention should be paid to reinforcement for crack control, as well as clearance above waterways as the underside collects debris, resulting in potential damage to the stems of the T-beam. The bridge has a neat clean appearance, with the exception of the underside. Maintenance costs are low except in the case where transverse deck joints are used.

3.1.4 Wooden beams

Most often used for secondary roads where truck traffic volume is low, wooden beam bridges remain a primary bridge type for rural locations. The bridges are used for spans up to 15 m and usually have a wood pile substructure. With the exception of elements coming in direct contact with pedestrians the components of the bridge are chemically treated for preservation. Main members are usually pre-cut and drilled prior to chemical treatment and installation. The bridge can accommodate a spiked wooden deck, concrete deck, or a combination of wood planking and asphalt. Because of the propensity of the bents and abutments to catch debris when over water, the substructure units are built parallel to the stream. These bridges can be visually appealing in the right environment though it doesn't lend itself to urban environments.

3.1.5 Precast concrete Box beams

Fast becoming a mainstay in the short-span bridge market, the precast box beam bridge can have spread boxes or adjacent boxes. This bridge type is typically used for spans of 10–45 m. It is not advised to use the top of the boxes as the riding surface due to the uneven riding surface that results from variable camber between boxes. The boxes are often transversely posttensioned with grouted shear keys between the boxes. The riding surface is often comprised of an asphalt topping or a concrete slab. It is common for differential movement between the boxes to result in cracking of the asphalt or concrete topping along the joint between the boxes. This is often alleviated by using a highly reinforced concrete deck. The boxes can be used as simple spans or made continuous by pouring concrete between the ends of the boxes. The bridge has an appearance similar to that of the T-beam bridge, except that in the case of adjoining boxes, a smooth underside results. This bridge, like all concrete bridges, requires little maintenance, except where transverse deck joints are used.

3.1.6   Precast concrete I-beams

Competitive with steel girders for spans of 10–45 m, precast, prestressed concrete I-beams have many of the same advantages and disadvantages as the precast concrete box beams. In most cases, the beams are designed as noncomposite simple spans for dead loads and as composite continuous spans for live loads and superimposed dead loads. Maintenance is low except in the case where transverse deck joints are used. The appearance of the bridge is clean in elevation, but like the T-beams, very "busy" underneath.

3.1.7   Noncomposite rolled steel I-beams

Often used because of their lower fabrication costs, rolled steel wide-flange beam bridges are cost-effective for spans up to 15 m. To be cost-effective, spans longer than this need to be made composite and utilize cover plates. Though weathering steel can be used to eliminate the need for painting, it cannot be used in situations where the site conditions will result in constant wetting. If weathering steel cannot be used, the cost of painting must be used in all cost comparisons with concrete spans. The overall aesthetic appearance is much like that of the concrete I-beams—clean lines in elevation, but cluttered underneath.

3.2   Medium-span bridges

Bridge types that can be used in the medium-span range (span lengths up to 75 m) include precast concrete box-beam bridges, precast I-beam bridges, composite rolled wide flanged beam bridges, composite steel plate girder bridges, reinforced cast-in-place concrete box girder bridges, posttensioned, cast-in-place concrete box girder bridges, and composite steel box girder bridges.

3.2.1   Precast prestressed concrete beams (Box beams and I-beams)

The general characteristics of both the concrete box beam and I-beam bridges were discussed in the previous section. Transportation of these types of bridge elements become a major issue as the span lengths increase. Virtually every jurisdiction has length and load limitations for trucks hauling such elements. In addition, as the length of the elements increase, onsite storage requires special support conditions and lateral stability becomes an issue in the case of the I-girders during lifting and placement. Temporary bracing becomes essential. Extremely long girders may require precasting in segments and joined together once erected.

3.2.2   Composite rolled I-beams

The general characteristics of composite rolled steel beam bridges were discussed in the previous section. Composite rolled I-beam bridges are economical up to spans of 30 m. To economically achieve these span lengths, it is necessary to make the bridge composite for live load and add steel cover plates in maximum moment regions. Special care must be taken at the ends of such cover plates due to a susceptibility to fatigue cracking if improper detailing is used. Though different types of shear connectors have been used through the years to accommodate composite action, the most common today are welded studs.

3.2.3   Composite steel plate girders

Bridges comprised of composite steel plate girders (such as the Harpers Ferry bridge, shown in Figure 18.1) are economically feasible for spans of 20–40 m, although they have been used for spans exceeding 90 m. The girders are typically comprised of an asymmetric section consisting of a top and bottom flange welded to a web. Many such girders consist of hybrid sections using steels of different strengths for the webs and the flanges. As the price of different grades of steel have become more uniform over the years, this has become less common. The use of such girder results in low dead loads, making them quite desirable for use in areas of poor foundation conditions. The tall, slender girders that result for longer spans must be handled and erected with care. Lateral stability is an issue during fabrication, transportation, and erection. It is imperative that careful thought be given to proper lateral bracing during all phases of the project. Because each one is fabricated using plate steel, the girders can have variable depths for maximum section efficiency. Such variation in section depth can be visually appealing.

Figure 18.1 Harpers Ferry Bridge, Harpers Ferry, West Virginia, USA (courtesy of Modjeski and Masters, Poughkeepsie, NY).

3.2.4 Reinforced cast-in-place concrete Box girder

As in the case of the concrete T-beam option, reinforced cast-in-place concrete box girders typically require a ground-based scaffolding system; therefore, their use is often limited by site limitations. They are suitable for spans of approximately 15–35 m and result in very torsionally rigid structures. In some cases, they can be more economical than steel and concrete I-girder bridges, but only when local industry is geared up for such construction. They are visually attractive structures with a clean, simple, and smooth appearance from all directions, making them very appealing for use in urban environments. They have the additional advantage of allowing all utilities to be run inside the boxes, hiding them from view.

3.2.5 Posttensioned, cast-place concrete box girder

Posttensioned, cast-in-place box girders have the capability of providing much longer spans than cast-in-place reinforced box girders. They have been used in bridges with spans up to 180 m and result in bridges that are cost-effective and pleasing in appearance, and that require low maintenance. The boxes have a very high torsional resistance, making them well suited for curved or skewed bridges. The use of posttensioning can minimize dead load deflections and cracking in the boxes and decks. The use of posttensioning does result in creep shortening of the elements, and provisions must be made to accommodate such movements. As is the case for most concrete bridges, maintenance is low (with the exception of the bearings and any transverse joints). It is recommended that consideration be given to an overlay system on the bridge deck in areas with high use of deicing chemicals. Care also must be given in the deck drainage system is such cases.

3.2.6 Composite steel box girder

Composite box girders may be rectangular or trapezoidal in shape and possess high torsional resistance once the deck is poured. They are used for spans of 20–150 m. Though they are most cost effective in the longer-span ranges, they are often used for shorter spans in highly curved situations or when a shallow section is required. Due to their size, steel boxes face the same shipping challenges as all of the other large component systems. With all their benefits, steel boxes do come with their own set of challenges. The shapes and intersecting elements present fabrication issues. Even though they are shop-fabricated, there are many opportunities for welding and detailing errors that lead to fatigue issues. If a decision is made to use steel box girders close and careful attention must be paid to the structural detailing. Steel box girder spans have very clean lines and can be aesthetically pleasing. The same issues regarding painting and the use of weathering steel raised for the other steel sections apply to steel boxes as well.

3.3 Long-span bridges

Bridge types that can be used in the long-span range (span length up to 150 m) include composite steel plate girder bridges, posttensioned, cast-in-place concrete box-girder bridges, posttensioned segmental bridges, steel and concrete arch bridges, and steel truss bridges.

3.3.1 Composite steel plate girder bridge

The same issues presented in the previous discussion of steel plate girders in the medium-span range also apply to the long-span range. As the spans get longer, the sections get deeper and the issues regarding lateral stability and transportation become even more critical.

3.3.2 Posttensioned, cast-in-place concrete box girder bridge

The same issues presented in the previous discussion of posttensioned, cast-in-place concrete girders in the medium-span range also apply to the long-span range. As spans get longer, time-dependent effects such as creep and shrinkage become even more critical and require even more attention.

3.3.3 Posttensioned, concrete segmental bridges

Though primarily used for box sections, posttensioned segmental construction methods can be used for numerous bridge types (see Figure 18.2), including spliced concrete I-girders. Cost savings are realized through the reuse of standard form systems. Segmental construction can be used for cast-in-place elements using traveling forms or precast cast elements. Erection methods include span-by-span construction, balanced cantilever erection, and launching the bridge from one end. Typical span lengths for segmental bridges are as follows: (i) cast-in-place posttensioned box girder of constant depth, 30–90 m; (ii) precast posttensioned box girder of constant depth erected using balanced cantilever, 30–90 m; (iii) variable depth precast balanced cantilever segmental, 60–180 m; and (iv) cast-in-place cantilever, 60–300 m. More detailed information is available in various sources, such as Hewson (2003), the American Segmental Bridge Institute (ASBI, 2003) and Podolny and Muller (1982).

Figure 18.2 I-96-295 ramp, Jacksonville, FL, USA
(courtesy of John Corven).

3.3.4   Steel and concrete arches

Span lengths for arches range from 90–420 m for concrete arches and 90–420 m for steel arches. They can be either above or below the roadway deck. The distinctive features of arch-type bridges have been very effectively summarized by O'Connor (1971) as follows:

- The most suitable site for this form of structure is a valley, with the arch foundations located on dry rock slopes.
- The erection problems vary with the type of structure, being easiest for the cantilever arch and possibly most difficult for the tied arch.
- The arch is predominately a compression structure. The classic arch form tends to favor concrete as a construction material.
- Aesthetically, the arch can be the most successful of all bridge types. It appears that through experience or familiarity, the average person regards the arch form as understandable and expressive. The curved form is almost always pleasing.

3.3.5   Steel trusses

Steel truss bridges (Figure 18.3) were the major structure of choice during the 19th and 20th centuries. Their spans range from 240–550 m. These trusses are typically classed as through trusses and deck trusses. Through trusses have the truss above the roadway, and deck trusses have the roadway above the truss. Some bridges feature both kinds. O'Connor (1971) offers an excellent summary of the features of a truss bridge:

- A bridge truss has two major structural advantages: (1) the primary member forces are axial loads; and (2) the open web system permits greater overall depth than an equivalent solid web girder. Both of these factors lead to economy in material and a reduced dead weight. The increased depth also leads to a more rigid structure and reduced deflections as a result.
- The conventional truss bridge is most likely to be economical for medium spans. Traditionally, it has been used for intermediate spans between the plate girder and the stiffened suspension bridge. Modern construction techniques and materials have tended to increase the economical span of both steel and concrete girders. The cable stayed bridge has become a competitor to the steel truss for intermediate spans. These factors, all of which are related to the high fabrication cost of a truss, have tended to reduce the number of truss spans built in recent years.

In addition to the fabrication costs mentioned by O'Connor, recent issues with the gussets of truss bridges have led some to be hesitant to use this structure type. Ongoing research has led to recommendations regarding the design of gusset plates that should alleviate this concern.

Figure 18.3 Huey Long Bridge, New Orleans, LA, USA
(courtesy of Modjeski and Masters, Poughkeepsie, NY)

3.4 Very long-span bridges

3.4.1 Suspension bridges

Suspension bridges (Figure 18.4) typically consist of two (and sometimes four) parallel cables separated by a distance approximately equal to the roadway deck width that they support. These cables act as tension elements and extend from anchors at each of their ends over the tops of the intermediate towers. The deck is suspended by strong ropes running from the deck level to the main cables. The main cables can consist of parallel strong wires that are aerially spun in place or prefabricated wire ropes. The deck can be stiffened by a truss or by girder elements. The purpose of the stiffening element is to ensure aerodynamic stability and to limit the local angel changes in the deck. Suspension bridges are used for spans of 300– 2300 m. The bridge can be erected without any ground-based towers. The resulting bridge is very elegant in appearance, and its form clearly expresses its function. As the existing inventory of suspension bridges have aged, inspections have revealed active corrosion and stress corrosion cracking in many of the wires comprising the main cables. This has led to the installation of dehumidification systems in many of the new and existing bridges.

3.4.2 Cable stayed bridges

Cable stayed bridges (Figure 18.5) were introduced immediately following World War II to replace many of the bridges lost during the war. Unlike the suspension bridge, the cables extend from the towers directly connecting to the deck. In most bridges, the cables come to a "dead end" at the deck and the tower. There have been some recent bridges where the cables pass through a "saddle" at the tower, and then to the

Figure 18.4 Forth Road Bridge, Queensferry, Scotland
(courtesy of Barry Colford).

Figure 18.5 Tatara Bridge, Japan
(from author's collection).

deck at each end. The cables are typically in two planes separated by the width of the roadway, though numerous bridges have been built with a central plane of stays between the two opposing lanes of traffic. This requires a torsionally resistant super-structure. The cables are straight, resulting in greater stiffness than a suspension bridge. By anchoring the cables to the deck, compressive forces are applied to the deck, resulting in it participating in handling those loads. This can be problematic should deck replacement be necessary. In general, a cable stayed bridge is less effi-cient in carrying dead load than a suspension bridge but is more efficient in carrying live load. The most economical span length for a cable stayed bridge is 100–350 m, though some designers have extended this range to as much as 800 m. There have been some problems with cable excitation during rain/wind events, particularly on the lon-ger stays. A cable stayed bridge is very modern and pleasing in appearance and fits extremely well in almost any environment.

A visual representation of the data presented here can be seen in Figure 18.6, where the possible and optimal span lengths for various bridges are presented.

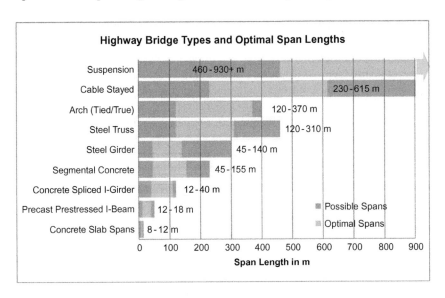

Figure 18.6 Possible and optimal highway bridge span lengths.

4 Methods of analysis (emphasizing highway structures)

All of the design specifications used in the world today for highway bridges allow the use of any method of analysis that satisfies the requirements of equilibrium and com-patibility and utilizes stress-strain relationships for the proposed materials. These methods include, but are not necessarily limited to, the following:

- Classical force and displacement methods
- Finite element method

- Finite difference method
- Finite strip method
- Folded plate method
- Grid analogy method
- Series or other harmonic methods
- Methods based on the formation of plastic hinges
- Yield line method

It is imperative for the designer to realize that he or she is responsible for the implementation of computer programs used to facilitate structural analysis and for the interpretation and use of the results. The designer must understand all limitations of programs used, as well as the nuances of commercial software regarding automatically set material properties. For the sake of clarity and for future reference, the designer should indicate the name, version, and release date of any software used during the project.

5   Design method

The present method used to design highway bridges in the United States is called the load and resistance factor design (LRFD) method. This is basically a limit-state design approach similar to that used in Canada and also contained in the Eurocodes used throughout Europe.

In the early days of structural design, most structures were composed of metallic elements that had a well-defined yield point. Therefore, all designs were based on some "allowable" stress that was based on some fraction of the yield stress. That fraction was referred to as a *factor of safety*. The factor of safety varied depending on the utilization of the member; tension, compression, or bending. Using the allowable stress and the force effects on a member; the net area required for a tension member, the gross area required for a compression member, and the section modulus required for a bending member could easily be determined.

There are numerous shortcomings to this method:

- The method does not lend itself to other materials, particularly nonmetallic materials.
- It is based on the assumption that there are no existing stresses in a member (i.e., no residual stresses resulting from the manufacturing process).
- Factors of safety are arrived at rather subjectively and are only applied to the resistance of the element. Furthermore, the resistance is based solely on the elastic behavior of materials.
- The method does not take into consideration the fact that different loads have different levels of uncertainty.

It became evident in the middle of the 20th century that an effective design method needed to take into account the variability of the loads, as well as the resistance to those loads. Thus, the limit state design methods were developed.

The primary advantages of the limit state design methods are as follows:

- The method accounts for the variability in the loads and resistances.
- The method results in more consistent levels of factor of safety.
- The method is more rational and consistent.

Detailed discussions of the limit-state methods of design are presented elsewhere in this volume and will not be discussed in detail in this chapter. The following design example is based on the Eurocode, which is based on the concept of a limit-state design.

6 Design example

An in-depth presentation of a bridge example illustrating all the requirements of the Eurocode would require a number of pages well beyond what is appropriate for a single book chapter. For this reason, this text will explore portions of a detailed example that was prepared by Crespo et al. (2012) contained in a scientific and technical report by the European Commission's Joint Research Centre (JRC), titled "Bridge Design to Eurocodes, Worked Examples, 2012." Much of the following discussion is taken from that report, and full attribution is given to those authors. This is an excellent example that can be very helpful for those seeking a greater understanding of the application of the Eurocode requirements.

6.1 Selecting the bridge type

This example is a road bridge that is to be designed to have a 100-year working life with a total length of 200 m consisting of three spans (60 m, 80 m, and 60 m). The bridge is on a tangent alignment and has no grade or vertical curvature. Traffic studies have resulted in the determination that the bridge will need to carry two lanes of traffic. It is to cross a deep canyon with access (though limited) to the proposed location of the pier footings. The subsurface conditions at the proposed location of the bridge are very good, allowing the construction of shallow foundations bearing on dense sand. (See Figure 18.7 for an elevation of the proposed bridge.) Due to the difficult access, a decision has been made to launch the bridge from one end. As a result, for ease of construction, the bridge will be a constant depth.

There is an established steel industry in the area, as well as access to numerous concrete suppliers in close proximity to the site. Therefore, there is no preference of materials based on availability. The bridge is located in a developed country outside the United States with ready access to highly experienced construction workers; therefore, any bridge type would be acceptable. Following a cost study, it was determined that a steel, two-girder composite bridge is the best solution for this site. It should be noted that two-girder systems are not allowed in the United States due to redundancy concerns; therefore, it would not be considered if this bridge was in the United States.

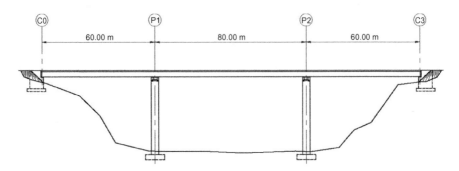

Figure 18.7 Elevation of proposed bridge
(courtesy of JRC).

6.2 The structural concept

The superstructure is composed of a symmetrical, two-girder composite cross section. Preliminary studies established that the most effective depth for the girders would be a constant 2800 mm. Since the bridge is to be launched, all the "steps" in the thickness of the steel flanges will be made on the flange web side of the flanges. The deck slab has a 2.5% symmetrical cross slope with a variable thickness ranging from 400 mm (over the girders) to 250 mm (at its free edges) and 307.5 mm (at the center line of the deck). The roadway will have two traffic lanes that are 3.5 m wide apiece, and 2-m shoulders on each side. This results in an 11-m carriageway with 0.5-m parapets on each side. The total width of the resulting deck slab is 12 m, and the center-to-center spacing between the main girders is 7 m, resulting in a slab cantilever on both sides, each of which is 2.5 m. (See Figure 18.8 for a typical cross section of the deck.)

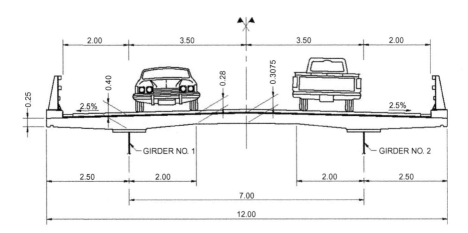

Figure 18.8 Typical cross section of the deck
(courtesy of JRC).

The piers and abutments are analyzed in accordance with the relevant chapters of EN 1992 and EN 1998. The height of the piers is approximately 40 m, and they consist of concrete circular hollow sections with an external diameter of 4.0 m and walls 0.4 m in thickness. The footings for each column are 10.0 m x 10.0 m x 2.5 m. The abutment is to be a standard stub abutment of the geometry resulting from a straight alignment. It will rest on a footing that is 10.0 m x 15 m x 1.5 m.

6.3 Design parameters

6.3.1 Dead loads

For the determination of dead loads, in addition to the self-weight of the concrete deck and steel girders, the following nonstructural elements are to be included: two parapets, two cornices, a 3-cm waterproof layer and an 8-cm-thick asphalt wearing surface. Each of these elements is shown in Figure 18.9.

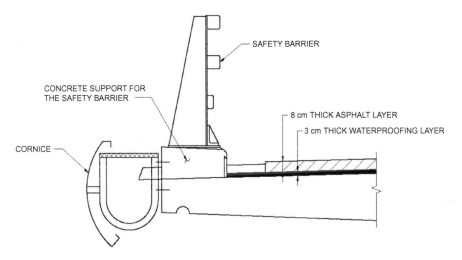

Figure 18.9 Nonstructural elements (courtesy of JRC).

6.3.2 Live loads

Traffic loads will be represented by Load Model 1 (LM1). According to EN 1991-2, LM1, which is formed by a uniformly distributed load (UDL) and the concentrated loads of the tandem system (TS), can be adjusted by α-coefficients. The values of these α-coefficients are given by the National Annexes based on different traffic classes. In accordance with EN 1991-2, 4.3.2, in the absence of specifications about the composition of the traffic, the values $\alpha_{Qi} = \alpha_{qi} = \alpha_{qr} = 1.0$ are recommended. No abnormal vehicles are to be considered for this bridge.

6.3.3   Temperature range/humidity

The minimum shade air temperature at the bridge location to be considered for steel quality selection is −20 °C. This corresponds to a return period of 50 years. The maximum shade air temperature at this bridge location to be used in the calculations, as required, is 40 °C. The variation in the temperature along the depth of the superstructure between the concrete and steel parts will be ±10 °C. The ambient relative humidity (RH) is assumed to be equal to 80%.

6.3.4   Wind conditions

The bridge is spanning a valley with few and isolated obstacles like a tree or house. It is located at an area where the fundamental value of the basic wind velocity is $v_{b,0} = 26$ m/s. No launching operations of the steel beams will be allowed if the wind velocity is over 50 km/h.

6.3.5   Exposure class

The bridge is located in a moderate freezing zone where deicing agents are frequently used. To determine the concrete cover, the following exposure classes, according to Table 4.1 of EN 1992-1-1, will be used:

* XC3 for the top face of the concrete slab (under the waterproofing layer)
* XC4 for the bottom face of the concrete slab

6.3.6   Subsurface conditions

Soil conditions are such that no deep foundations are needed. Both piers and abutments have swallow foundations. A settlement of 30 mm at Pier 1 will take place for the quasi-permanent combination of actions. It can be assumed that this displacement occurs at the end of the construction stage.

6.3.7   Seismic data

For the seismic analysis, the ground under the bridge is considered to be formed by deposits of very dense sand (it can be identified as ground type B, according to EN 1998-1, Table 3.1). The bridge has a medium importance for the communications system after an earthquake, so the importance factor I will be taken as equal to 1.0. No special regional seismic situation is considered. The reference peak ground acceleration will be $a_{gR} = 0.30$ g. In this case, a limited elastic behavior is selected and, according to Table 4.1 of EN 1998-2, the behavior factor is taken as $q = 1.5$ (reinforced concrete piers).

6.3.8   Other considerations

The action of snow is considered to be negligible. Hydraulic actions are not relevant. Accidental design situations are analyzed in the referenced example.

6.3.9   Materials

Structural steel
For the girders, the steel used is grade S355 with the subgrades used as a function of thickness as shown in the following table.

Thickness	Subgrade
$T \leq 30$ mm	S 355 K2
$30 \leq t \leq 80$ mm	S 355 N
$80 \leq t \leq 135$ mm	S 355 NL

Concrete
Concrete class C35/45 is used for all the concrete elements in the referenced example (deck slab, piers, abutments, and foundations).

Reinforcing steel
The reinforcing bars used in the referenced example are class B high bond bars with a yield strength $f_{sk} = 500$ MPa.

Shear connectors
Steel grade S235J2G3 stud shear connectors are used in the referenced example. Their ultimate strength is fu $= 450$ MPa.

6.4   Details on structural steel and slab reinforcement

6.4.1   Resulting main steel girder configuration

The structural steel distribution for a main girder is presented in Figure 18.10. The two main girders have a constant depth of 2800 mm and the variations in thickness of the upper and lower flanges are found on the web side of the flanges. The lower flange is a constant 1200 mm wide, whereas the upper flange is a constant 1000 mm wide.

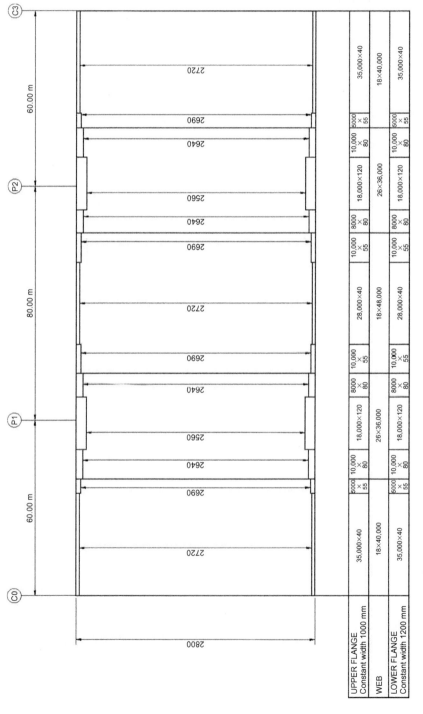

Figure 18.10 Structural steel distribution (courtesy of JRC).

The two main girders have transverse bracing at abutments and at internal supports, as well as every 7.5 m in the side spans (C0-P1 and P2-C3) and every 8 m in the central span (P1-P2). Figures 18.11 and 18.12 illustrate the geometry and dimensions adopted for this transverse cross-bracing. The transverse girders in the span are made of IPE600 rolled sections, whereas the transverse girders at the internal supports and abutments are built-up, welded sections. The vertical T-shaped stiffeners are duplicated and welded on the lower flange at the supports, whereas the flange of the vertical T-shaped stiffeners in span has a V-shaped cutout to help prevent fatigue.

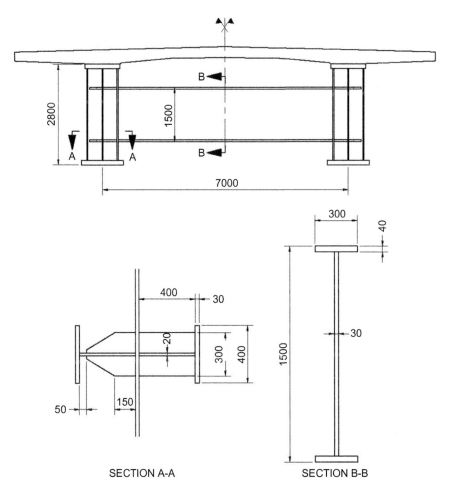

SECTION A-A   SECTION B-B

Figure 18.11 Transverse cross-bracing at bearings (courtesy of JRC).

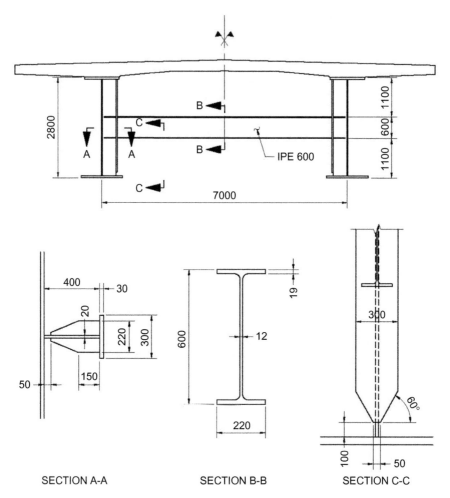

Figure 18.12 Transverse cross-bracing in spans
(courtesy of European Commission's JRC).

6.4.2 Resulting slab reinforcement

For both steel reinforcing layers, the transverse bars are placed outside the longitudinal ones, on the side of the slab free surface (Figure 18.13). High bond bars are used. Other specifications are as follows:

- Longitudinal reinforcing steel located in the in-span region consists of $\Phi = 16$ mm every 130 mm in the upper and lower layers (i.e., $\rho_s = 0.92\%$ of the concrete section in total).
- Longitudinal reinforcing located in the intermediate support regions consists of $\Phi = 20$ mm every 130 mm in the upper layer $\Phi = 16$ mm every 130 mm in the lower layer.
- Transverse reinforcing steel located at the midspan of the slab (between the main steel girders): $\Phi = 20$ mm every 170 mm in the upper layer, $\Phi = 25$ mm every 170 mm in the lower layer
- Transverse reinforcing steel located over the main steel girders consists of $\Phi = 20$ mm every 170 mm in the upper layer and $\Phi = 16$ mm every 170 mm in the lower layer.

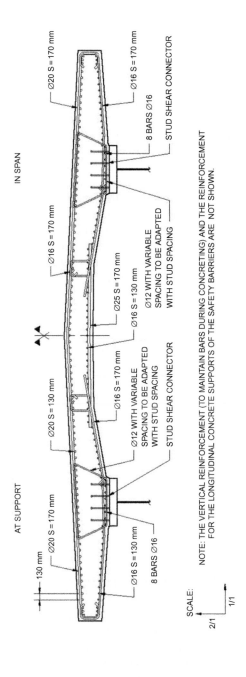

Figure 18.13 Slab reinforcing (courtesy of JRC).

6.4.3   Construction process

Launching of the steel girder
As stated earlier, it is assumed that the steel structure is launched, and it is pushed from the left abutment (C0) to the right one (C3) without the addition of any nose-girder.

Slab concreting
After the installation of the steel structure, concrete is poured on site, casting the slab elements in a selected order: the total length of 200 m is split into 16 identical 12.5-m-long concreting segments. They are poured in the order indicated in Figure 18.14. The start of pouring the first slab segment is the time origin (t=0). Its definition is necessary to determine the respective ages of the concrete slab segments during the construction phases. The time taken to pour each slab segment is assessed as 3 working days. The first day is devoted to the concreting, the second day to its hardening, and the third to moving the formwork. This sequence respects a minimum concrete strength of 20 MPa before the formwork is removed. The slab is thus completed within 66 days (including weekend days when no work is done). It is assumed that the installation of nonstructural bridge equipment is completed within 44 days, so that the deck is fully constructed at the date t=66+44=110 days.

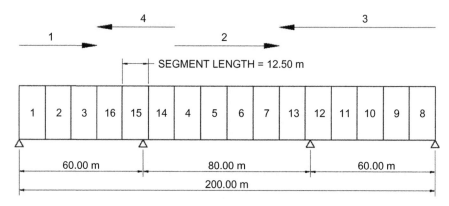

Figure 18.14 Slab pouring sequence (courtesy of JRC).

7   Research needs for highway bridges

Highway bridges have been designed and built since the advent of the wagon, and the general structure types used and described in this chapter are not likely to change. That being said, there are areas where these structure types can be improved—hence the need for future research. It is this author's opinion that the research needs for highway bridges (and for that matter, bridges of all uses) fall into five general areas:

- The need to optimize structural systems
- Develop ways to extend service life
- Develop systems to monitor bridge conditions
- Develop details and methods to accelerate bridge construction
- Develop a full life cycle approach to bridge data management

7.1 The need to optimize structural systems

Though the general types of bridge structures have remained unchanged over time, the materials that comprised those types of structures have been constantly changing. The introduction of high-strength steel, high-performance concrete, and fiber-reinforced polymer composite materials has resulted in structures that are, in many cases, easier to build, more durable, and more economical. To take full advantage of these materials and their properties, optimization of structural shapes, details, components, and construction procedures must take place. Though research work in these areas is underway, there is much remaining to do.

7.2 Develop ways to extend service life

The bridges in the developed world are getting older, and the maintenance of that aging inventory is placing a strain on the budgets of bridge owners. Therefore, it is imperative to develop ways to extend the service life of existing bridge structures. Research into the processes that decrease the service life of bridges and the most promising preservation methods that will address these processes is needed.

7.3 Develop systems to monitor bridge performance

Often called *health monitoring,* bridge monitoring in real time holds great promise for prolonging bridge life. New data acquisition systems and monitoring devices allows the efficient collection of data dealing with virtually every component of a bridge. The question is, "What information should be collected?" Today, it is feasible to collect terabytes of data, as in being done on some bridges, and really not have a means of sifting that data or use it in a meaningful manner. So the question remains of what data should be collected from which bridge components to establish the condition of the bridge.

7.4 Develop details and methods to accelerate bridge construction

In the United States, as well as most developed countries, traffic demands vastly limit the amount of time available for bridge repair and construction. There is a real need to reduce on-site construction time, while ensuring long lasting structures. More research needs to be done on erection technology and prefabricated elements while developing a means to balance the cost of such technologies against user costs.

7.5 Develop a full life cycle approach to bridge data management

This last element is not so much a need for research as it is a need to take existing available technology and use it more effectively in bridge managment. By using building information technology, it is possible to collect data regarding a bridge over its complete life cycle. Using building information modeling (BIM) and other developing technology, every stage in the development, design, construction, maintenance, and eventual demolition of a bridge can be maintained in an easily searchable form. This information can be used to effectively maintain a bridge over the life of a bridge and to modify the structure as needs arise.

References

American Segmental Bridge Institute (ASBI), 2003. Recommended Practice for Design and Construction of Segmental Bridges. American Segmental Bridge Institute, Phoenix, AZ.

Barker, R.M., Puckett, J.A., 2007. Design of Highway Bridges: An LRFD Approach. John Wiley & Sons, Hoboken, New Jersey, Chapter 2, pp. 87–108.

Crespo, P., Davaine, L., Cornejo, M.O., Raoul, J., 2012. Bridge Design to Eurocodes, Worked Examples. Publications Office of the European Union, Luxembourg. Chapters 1 and 6. Entire publication can be found at, http://eurocodes.jrc.ec.europa.eu/doc/1110_WS_EC2/report/Bridge_Design.

De Miranda, F., 1991. The three mentalities of successful bridge design. In: Bridge Aesthetics Around the World. Committee on General Structures, Transportation Research Board, National Research Council, Washington, DC, pp. 89–94.

Hewson, N.R., 2003. Prestressed Concrete Bridges: Design and Construction. Thomas Telford Publishing, London, pp. 179–281.

O'Connor, C., 1971. Design of Bridge Superstructures. Wiley Interscience, New York.

Pennsylvania Department of Transportation (PennDOT), 2012. Design Manual, Part 4. Harrisburg, PA.

Podolny Jr., W., Muller, J.M., 1982. Construction and Design of Prestressed Concrete Segmental Bridges. Wiley, New York.

Railway bridges

Pipinato A.[1], Patton R.[2]
[1]AP&P, Technical Director, Italy
[2]Division Engineer, Norfolk Southern Corporation, USA

19

1 Introduction

Since the construction of the first modern railway bridge in the 1820s, railway bridge engineering has evolved extensively. Locomotives have changed from steam to diesel electric along with the weight of railway freight car loads and equipment. While future freight equipment weights will be limited by economics associated with railway infrastructure, maintenance, and renewal, it is most likely that train shipments and axle loads will increase. The first working model of a steam rail locomotive was designed and constructed by John Fitch in the United States in 1794. The first full-scale working railway steam locomotive was built in the United Kingdom in 1804: the Stockton and Darlington Railway was a railway company that operated in northeast England from 1825 to 1863. The world's first public railway to use steam locomotives, which connected Shildon with Stockton-on-Tees and Darlington, was officially opened on September 27, 1825. The movement of coal to ships rapidly became a lucrative business, and the line was soon extended to a new port at Middlesbrough. Passengers were carried in coaches drawn by horses until carriages hauled by steam locomotives were introduced in 1833. In 1839, the first Italian railway line was laid between Naples and Portici. In the United States, the Baltimore and Ohio Railroad was incorporated in 1827 and officially opened in 1830. In the same period, engineers faced the problem of adapting bridge structures to railway traffic for the first time, and in most cases, principal structures were constructed of metal.

One of the first large bridge "experiments" was in 1845, when plans for carrying the Chester and Holyhead railway over the Menai Straits in Wales were considered, and the conditions imposed by the admiralty in the interests of navigation involved the adoption of a new type of bridge (Britannica, 1910). Suspension chains combined with a girder was seen as a possible construction scheme, and in fact, the tower piers were built to accommodate chains. But the theory of such a combined structure could not be formulated at that time, and it was proved, partly by experiment, that a simple tubular girder of wrought iron was strong enough to carry a railway (Britannica, 1910). The bridge, then called Britannia, has two spans of 140 m and two of 70 m at 30 m above the water (Figure 19.1). It consists of a pair of tubular girders with solid or plate sides stiffened by angle irons and one line of rails passing through each tube. Each girder weighs nearly 4680 tons. In cross section, it is 4.5 m wide and varies in depth from 7 m at the ends to 9 m at the center. Partly to counteract any tendency to buckling under compression, and partly for convenience in assembling a great mass of plates, the top and bottom were made cellular, with the cells just large enough to permit passage for

Innovative Bridge Design Handbook. http://dx.doi.org/10.1016/B978-0-12-800058-8.00019-0

painting. As no scaffolding could be used for the center spans, the girders were built on shore, floated out, and raised by hydraulic presses (Britannica, 1910). Robert Stephenson (son of George Stephenson, well known as the "Father of Railways"), William Fairbairn, and Eaton Hodgkinson (who assisted in the experimental tests and in formulating the imperfect theory then available) together shared in creating this impressive, successful structure.

The first train passed over the Britannia bridge in 1850. Though each girder is continuous over the four spans, it does not quite have the proportions over the piers that a continuous girder should have, so it must be regarded as an imperfectly continuous girder. The spans were in fact designed as independent girders, the advantage of

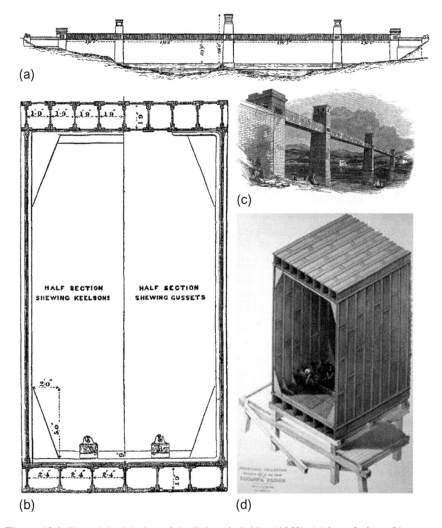

Figure 19.1 The original design of the Britannia bridge (1850): (a) lateral view; (b) cross section; (c) bridge preliminary sketch; (d) three-dimensional view.

continuity being imperfectly known at that time. The vertical sides of the girders are stiffened so that they amount to 40% of the whole weight. This was partly necessary to meet the uncertain floating conditions in that the distribution of supporting forces was unknown and there were chances of distortion (Britannica, 1910). From that period, up to now, large advances in the construction of railway bridges have been made, both on materials and on construction methods. However, fundamental principles in railway bridge engineering remains the same.

2 Type classifications

2.1 Bridge layout

The two main factors affecting the choice of a bridge structure are the main span and the obstacle type (e.g. a river, a railway, an highway). Different alternatives could be chosen for the same span length; functional, construction and economic issues could lead to the final decisions. The main structural types of bridges could be subdivided as follows:

- Plate girders or box section beams (0–250 m)
- Truss beam (up to 400 m)
- Arches and cantilever bridges with suspended center span (up to 600 m)
- Cable stayed bridge (up to 1200 m)
- Suspension bridge (up to 1900 m).

These requirements are general and apply to all bridges; with railway bridges, stringent deformation requirements often govern their design. As a result, structures inherently stiff in bending are required to be trusses or composite sections, rather than cable stayed or suspension bridges (Hirt and Leben, 2013).

2.2 Materials and code references

Materials of railway bridge constructions are provided by specific national codes such as Eurocodes, while additional documents and specifications are provided by railway associations such as the International Union of Railways (UIC). There is no predominant material; however, steel and composite structures are preferred for their simplicity of construction, lighter weight, and ease of inspections, intervention, and replacement. High-strength materials are employed (but not mandatory), as they provide a lighter and most economical solution considering the minimum rail standard requirement. In Europe, steel rail bridges nowadays are commonly realized with S355 grade carbon steel, bolted or welded, even if higher grades have been employed in special cases; general provisions for metal structures are provided by EN 1993-1-1 (2003). Concrete solutions actually includes the concrete category provided in EN 1992-1-1 (2004). In the United States, the use of materials conforms to American Society for Testing and Materials (ASTM) specifications (AREMA, 2014). Finally, national standards usually provide minimum material requirement specifications.

2.3 Substructures and foundations

Foundations are commonly made up of deep structures, as piles; shallow foundations are not normally adopted for railway bridges. The substructure consists of abutments and piers and includes foundations: these substructure transmits to the underlying soil the forces comprising the dead load of the superstructure and substructure, the live load effect of passing traffic, and forces from wind, water, etc. The substructure is generally represented by pile foundations, spread footings, piers and abutments, or any combination of these (AREMA, 2013). Careful soil investigation is needed before construction: extensive recommendations are given in specific codes and standards (AREMA, 2014; Eurocode 7-2, 2007).

As the stability of the structure is obviously related to that of the substructures, these should be under observation during the whole life of the bridge, and special inspections should be performed during and after freshets, ice gorges, cloudbursts, and other unusual happenings, which could have the potential of seriously affecting the safety of the structure. The most railway bridge foundations are piles or caissons which are mainly made of reinforced concrete (r.c.): these piles are heavy structures with a high bearing capacity. In some cases (e.g., when a specific requirement of temporarily constructions are needed), steel H-piles are used. In order to resist lateral forces, concrete-filled pipe piles with an adequate diameter and moment of inertia are required. Finally, for very large loads with minimum settlement, caissons are needed. Two main configurations could be used: isolated piles or sheet piles. The latter are piles built close together to form a wall, which can act as a retaining structure for water, earth, or other material. While concrete sheet piles are tongued and grooved, steel sheet piles are usually interlocked. The capacity of a pile as a structural member is based on allowable stresses established by the American Railway Engineering and Maintenance-of-Way Association (AREMA, 2014 in Chapter 8 or 15). For European standards, indications are given in Eurocode 7-1-7 (2007).

2.4 Superstructures

The objective of a railway bridge designer is to maximize the structural stiffness while reducing the self-weight of the construction material: this concept is maximized in truss structures. Another design tip for railway bridges is the presence of a hierarchical structure that can carry forces among a series of components. The vertical loads are transferred from their point of application (the rails) to the supports via sleepers and longitudinal beams (tertiary structural members), and then the cross bracing (secondary structural members), before being transferred to the primary structural members; namely, the main beams (Hirt and Leben, 2013). Some older bridges could not be ballasted; however, most bridges today are ballasted in order to reduce impact and improve train ride quality due to a relatively constant track modulus on the approaches and across the bridge. Concerning the deck solution, an upper slab should be preferred as it provides protection from the weather (so long as the structure itself is protected by waterproofing); simplicity in bridge widening, maintenance during operations, and derailment could not damage the principal structures.

3 Analysis and Design

3.1 Loads and load combinations

In the following section, load specifications are presented according to European and other relevant codes. Concerning load combinations, specific guidance is given by standards such as EN 1990 (2006): however, appropriate indications are included in international codes, standards, and National Annexes.

3.1.1 Dead loads

The weight of the structure itself, the track it supports, eventual ballasting, and any other superimposed loads attached to the bridge are dead loads. These act due to gravity and are applied to the structure, either permanently or until the structure changes its configuration throughout its life. Unit weights for calculation of dead loads are given in codes and standards. The self-weight of nonstructural elements includes the weight of elements such as noise and safety barriers, signals, ducts, cables, and overhead line equipment (except the forces due to the tension of the contact wire, etc.).

3.1.2 Live loads

The load models defined in codes and standards do not describe actual loads. The bridge designer should always be mindful of this in order to evaluate the possible use of heavier convoys, which is not an impossible situation. In fact, load models have been selected so that their effects, with dynamic enhancements taken into account separately, represent the effects of service traffic. This is the specific case of EN 1991-2 (2005): rail traffic actions are defined by means of load models. Five models of railway loads are given:

- Load Model 71 (and Load Model SW/0 for continuous bridges), to represent normal rail traffic on mainline railways (Figures 19.2 and 19.3)
- Load Model SW/2, to represent heavy loads (Figure 19.3)
- Load Model HSLM, to represent the loading from passenger trains at speeds exceeding 200 km/h (Figure 19.4-Figure 19.5)
- Load Model "unloaded train," to represent the effect of an unloaded train: this consists of a vertical uniformly distributed load with a characteristic value of 10,0 kN/m.

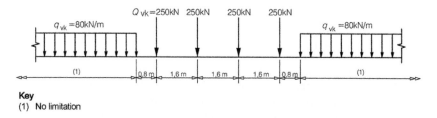

Key
(1) No limitation

Figure 19.2 Load Model 71 and characteristic values for vertical loads (EN 1991-2, 2005).

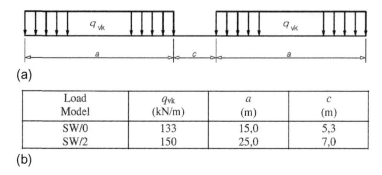

(a)

(b)

Load Model	q_{vk} (kN/m)	a (m)	c (m)
SW/0	133	15,0	5,3
SW/2	150	25,0	7,0

Figure 19.3 Characteristic values for vertical loads for Load Models SW/0 and SW/2 (EN 1991-2, 2005): geometrical disposition (a), and characteristic values (b).

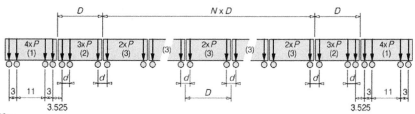

Key
(1) Power car (leading and trailing power cars identical)
(2) End coach (leading and trailing end coaches identical)
(3) Intermediate coach

(a)

(b)

Universal Train	Number of Intermediate Coaches N	Coach Length D (m)	Bogie Axle Spacing d (m)	Point Force P (kN)
A1	18	18	2.0	170
A2	17	19	3.5	200
A3	16	20	2.0	180
A4	15	21	3.0	190
A5	14	22	2.0	170
A6	13	23	2.0	180
A7	13	24	2.0	190
A8	12	25	2.5	190
A9	11	26	2.0	210
A10	11	27	2.0	210

Figure 19.4 Characteristic values for vertical loads for HSLM-A (EN 1991-2, 2005): geometrical disposition (a), and universal train details (b).

Concerning load schemes, it should be considered that a point force or wheel load may be distributed over three rail supports. For the design of local floor elements, the longitudinal distribution beneath sleepers should be taken into account, where the reference plane is defined as the upper surface of the deck. The standard loading scheme incorporated by North American Railways and AREMA (2013) is the Cooper E-Series loading: AREMA (2013) recommend that E-80 loadings (two locomotives coupled

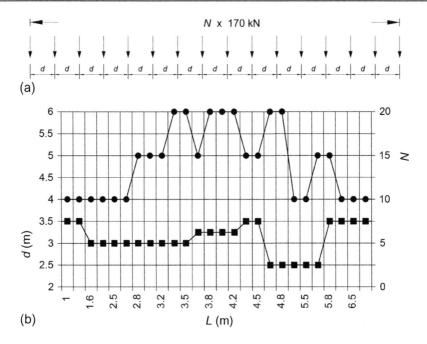

Figure 19.5 Characteristic values for vertical loads for HSLM-B (EN 1991-2, 2005): geometrical disposition (a), and graph for d (m) - L (m) - N (kN) correlation (b).

together in doubleheader fashion, with a maximum axle load of 335.84 kN) be used for the design of steel, concrete, and most other structures. Yet the designer must verify the specific loading to be applied from the railway, as this may require a design loading other than the E -80 Cooper E-Series.

3.1.3 Dynamic effects

The static stresses and deformations (and associated bridge deck acceleration) induced in a bridge are increased and decreased under the effects of moving traffic by the following (EN 1991-2 2005):

- The rapid rate of loading due to the speed of traffic crossing the structure and the inertial response (impact) of the structure.
- The passage of successive loads with approximately uniform spacing, which can excite the structure and under certain circumstances create resonance (where the frequency of excitation or a multiple thereof matches a natural frequency of the structure or a multiple thereof. There is a possibility that the vibrations caused by successive axles running onto the structure will be excessive),
- Variations in wheel loads resulting from track or vehicle imperfections (including wheel irregularities).

For determining the effects (stresses, deflections, bridge deck acceleration, etc.) of rail traffic, these effects shall be taken into account. The principal factors that influence dynamic behavior are:

- The speed of traffic across the bridge
- The span L of the element and the influence line length for deflection of the element being considered
- The mass of the structure
- The natural frequencies of the whole structure and the associated mode shapes (eigenforms) along the line of the track
- The number of axles, axle loads, and the spacing of axles
- The damping of the structure
- Vertical irregularities in the track
- The unsprung mass and suspension characteristics of the vehicle
- The presence of regularly spaced supports of the deck slab, track, or both (cross girders, sleepers etc.)
- Vehicle imperfections (wheel flats, out of round wheels, suspension defects, etc.)
- The dynamic characteristics of the track (ballast, sleepers, track components, etc.)

A static analysis generally shall be carried out with the load models defined in the specific code that is adopted. The results shall be multiplied by the dynamic factor specifically defined in the reference code. Simplified criteria for determining whether a dynamic analysis is required are given in codes and standards. For specific cases, codes should require a dynamic analysis: i.e., high speed lines and the particular geometry of the investigated bridge (Figure 19.6). Moreover, codes usually provide useful graphs that include the limits of the bridge's natural frequency n_0 (in hertz) as a function of the length L (in meters), in order to establish this for bridges with a first natural frequency n_0 within the limits given and a maximum line speed at the site not exceeding 200 km/h, a dynamic analysis is not required (i.e., Figure 19.7). Finally, for a simply supported bridge subjected only to bending, the natural frequency may be estimated using simplified approaches: EN 1991-2 (2005) suggests the formula $n_0 = 17.75/ d_0^{0.5}$ (expressed in hertz), where d_0 is the deflection at midspan due to permanent actions (in millimeters) and is calculated using a short-term modulus for concrete bridges, in accordance with a loading period appropriate to the natural frequency of the bridge.

AREMA (2013) has developed empirical relationships based on experimental observations to evaluate design impact values (percentage of live load) for various bridge types. The impact produced is represented as a vertical load applied to the top of the rail at the same location as the Cooper axle loadings, expressed as a percentage of the live load. The impact on a ballasted deck structure can sometimes be reduced compared to that for an open-deck structure because of the absorbing effect of the ballasted track. For steel bridge design, the percentage of live load attributed to impact is a function of the spacing of the structure supporting elements (girder or stringer spacing) relative to the spacing of the rails (rocking effect) and the distance between supports for the member being designed (span length); reduction in impact design values are given for speeds below 96 km/h. Impact is also considered when performing fatigue analysis and design: when checking fatigue stresses, impact forces may be reduced for members over 9 m in length. For concrete bridges, AREMA (2013) utilizes live load and dead load values to develop a modified ratio, and the span length of prestressed members for evaluating the impact percentage.

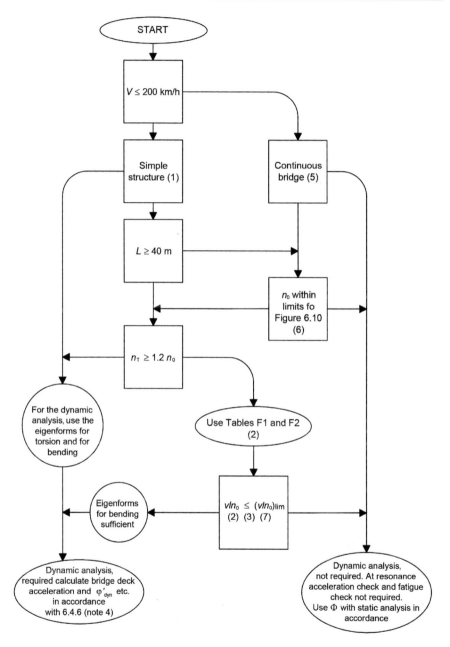

Figure 19.6 Flowchart for determining whether a dynamic analysis is required according to EN 1991-2 (2005).

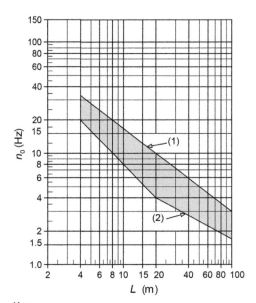

Figure 19.7 Limits of a bridge's natural frequency n_0 (Hz) as a function of L (m) according to EN 1991-2 (2005).

Key
(1) Upper limit of natural frequency
(2) Lower limit of natural frequency

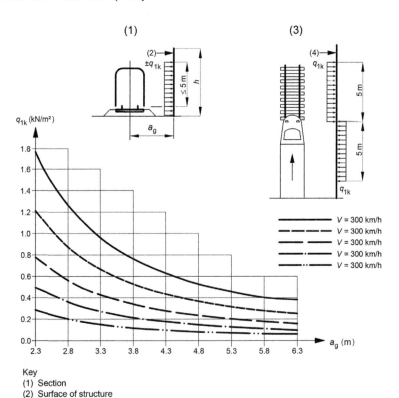

Key
(1) Section
(2) Surface of structure
(3) Plan view
(4) Surface of structure

Figure 19.8 Characteristic values of actions q_{1k} for simple vertical surfaces parallel to the track (e.g., noise barriers) according to EN 1991-2 (2005).

3.1.4 Horizontal forces

Horizontal forces deals with different physical motivations, and includes:

- *Centrifugal forces*, where the track on a bridge is curved over the whole or part of the length of the bridge; both the centrifugal forces and the track shall be taken into account. The centrifugal forces should be taken to act outward in a horizontal direction at a specific height above the running surface.
- The *nosing force* shall be taken as a concentrated force acting horizontally at the top of the rails, perpendicular to the center line of the track. It shall be applied to both straight track and curved track;
- *Traction and braking forces* act at the top of the rails in the longitudinal direction of the track. They shall be considered as uniformly distributed over the corresponding influence length $L_{a,b}$ for traction and braking effects for the structural element considered. The direction of the traction and braking forces shall take account of the permitted directions of travel on each track.

According to AREMA (2013), lateral loads are applied to the structure as a result of routine train passage, excluding centrifugal forces. The magnitude and application point of these loads vary depending on the constitutive material of the bridge: e.g., for steel, a load of one-quarter of the heaviest axle of the specified live load is applied at the base of the rail as a moving concentrated load that can be applied at any point along the span in either horizontal direction. Experience has shown that very great lateral forces may be applied to structures due to lurching of certain types of cars, wheel hunting, or damaged rolling stock (slewed trucks, binding center plate, etc.).

3.1.5 Aerodynamic actions from passing trains

When designing structures adjacent to railway tracks, aerodynamic actions from passing trains shall be taken into account. The passing of rail traffic subjects any structure situated near the track to a traveling wave of alternating pressure and suction. The magnitude of the action depends mainly on: the speed of the train, the aerodynamic shape of the train, the shape of the structure, and the position of the structure, particularly the clearance between the vehicle and the structure. The actions may be approximated by equivalent loads at the head and rear ends of a train, when checking ultimate and serviceability limit states and fatigue (CHSRA, 2011). Characteristic values of the equivalent loads are given specifically in codes and standards.

3.1.6 Derailment and other actions for railway bridges

The limitation of damages due to derailment is often included in codes and standard design requirement in order to minimize the effects. Codes could differentiate derailment cases as extreme and minor by differentiated vertical loads/accidental load cases, to be applied commonly at the edge of the railway.

3.2 Verifications regarding deformations and vibrations for railway bridges

Due to the particular and inherent construction types that are included in the specific case of railway bridges, detailed verifications are necessary to ensure safety, security, and comfort to train passengers. Specific limits of deformation and vibration to be taken into account for the design of new railway bridges are included in standards such as EN 1990 (2006). Bridge deformation checking includes:

- Vertical accelerations of the deck (to avoid ballast instability and unacceptable reduction in wheel rail contact forces)
- Vertical deflection of the deck throughout each span (to ensure acceptable vertical track radii and generally robust structures)
- Unrestrained uplift at the bearings (to avoid premature bearing failure)
- Vertical deflection of the end of the deck beyond bearings (to avoid destabilizing the track, limit uplift forces on rail fastening systems and limit additional rail stresses)
- Twisting of the deck measured along the center line of each track on the approaches to a bridge and across a bridge (to minimize the risk of train derailment)
- Rotation of the ends of each deck about a transverse axis or the relative total rotation between adjacent deck ends (to limit additional rail stresses, uplift forces on rail-fastening systems, and angular discontinuity at expansion devices and switch blades)
- Longitudinal displacement of the end of the upper surface of the deck due to longitudinal displacement and rotation of the deck end (to limit additional rail stresses and minimize disturbance to track ballast and adjacent track formation)
- Horizontal transverse deflection (to ensure acceptable horizontal track radii)
- Horizontal rotation of a deck about a vertical axis at ends of a deck (to ensure acceptable horizontal track geometry and passenger comfort)
- Limits on the first natural frequency of lateral vibration of the span (to avoid the occurrence of resonance between the lateral motion of vehicles on their suspension and the bridge)

Concerning the vertical acceleration of the deck, to ensure traffic safety, where a dynamic analysis is necessary, the verification of maximum peak deck acceleration due to rail traffic actions shall be regarded as a traffic safety requirement checked at the serviceability limit state for the prevention of track instability. The maximum peak values of bridge deck acceleration calculated along each track shall not exceed the appropriate design values, and according to EN 1990 (2006), recommended values are $\gamma_{bt} = 3,5$ m/s^2, $\gamma_{df} = 5$ m/s^2 (where γ_{bt} deals with ballasted track), and γ_{df} (for direct-fastened tracks with track and structural elements designed for high-speed traffic).

The twist of the bridge deck shall be calculated taking into account where Eurocode applies, the characteristic values of Load Model 71 (as well as SW/0 or SW/2 as appropriate) multiplied by Φ and α and Load Model HSLM (including centrifugal effects), all in accordance with EN 1991-2 (2005). Twisting shall be checked on the approach to the bridge, across the bridge, and for the departure from the bridge.

Also, the vertical deformation of the deck should be checked. If Eurocode applies, for all structure configurations loaded with the classified characteristic vertical loading in accordance with EN 1991-2 (2005) (and where required, classified SW/0 and SW/2), the maximum total vertical deflection measured along any track due to rail traffic actions should not exceed L/600. In addition, angular rotations of the deck's end is specified in codes and standards.

The transverse deflection δ at the top of the deck should be limited to ensure that a horizontal angle of rotation of the end of a deck about a vertical axis is not greater than the values provided in codes; the change of radius of the track across a deck is not greater than fixed values; and at the end of a deck, the differential transverse deflection between the deck and adjacent track formation or between adjacent decks does not exceed the specified value.

Finally, limitations of the values for the maximum vertical deflection for passenger comfort are provided in codes and standards. Comfort criteria depends on the vertical acceleration inside the coach during travel on the approach to, passage over, and departure from the bridge. Also, the levels of comfort and associated limiting values for the vertical acceleration should be specified in each project; however, recommended levels of comfort are given in codes and standards [e.g., for EN 1990 (2006)]; the level of comfort/vertical acceleration (m/s^2) could be defined as very good ($<1,0$), good ($<1,3$), or acceptable ($<2,0$).

3.3 Fatigue strength

Cracking or fracture due to repetitive loading is the result of fatigue. The repetitive loading that causes fatigue fracture produces stresses in the material below its yield stress. Stress reversals in railway bridges are commonly higher than those found in current road bridges because of the different load rates. However, a correct detailing design could avoid fatigue stress concentrations and the fracture in the bridge members that could also lead to collapse. Codes and standards provides construction details, including the description and requirements of common structural members. In Europe, EN 1993-1-9 (2005) deals with fatigue in metal structures; in the United States, AREMA (2014) provides general design specifications, as well as other requirements of the governing railway that must be adopted.

4 Static scheme and construction details

4.1 Static scheme

The durability and deformation limits under service loads are severe inputs for the design stage. Also for this reason, the most preferred static scheme is the continuous beam, as this scheme helps to reduce vertical deformations under loadings and the joints and bearing points. These characteristics are also relevant for recent high-speed railways, with trains exceeding 200 km/h. However, the designer should be careful: continuous beam static scheme solutions imply an increased deformation to settlement of the supports and a terrible distribution of braking forces through the bridge substructure. Finally, replacement of deck parts are not simple to be performed.

4.2 Expansion joints

The railway bridge structure is not normally able to move without introducing forces onto the rails and deck in the absence of expansion joints. This interaction is to be considered upon code provisions, which give the designer the possibility of avoid calculations in a particular case. If these requirements are not met, the structure/rail interaction has to be included in the design calculation procedure.

4.3 Ballasting

Ballasted tracks on bridges are currently a common rule, even if codes and standards could have specific exceptions. The inherent advantages of the ballasted tracks are commonly related not only to noise and vibration reduction but also to decreased maintenance expenses, even if structural improved performance could be discovered, as well as the reduction of the dynamic amplification, improved redistribution of loads on the deck, and consequent reduction of stress reversals/peak stresses on deck structural details more prone to fatigue.

4.4 Rainwater evacuation

Waterproofing systems must be enclosed in every railway bridge project, and the surfaces of carriageways and footpaths should be sealed to prevent the water access (EN 1993-2, 2005). The drainage layout should be based on the slope of the bridge deck as well as the location, diameter, and slope of the pipes. Free fall drains should carry water to a point clear of the underside of the structure to prevent water entering the structure.

Drainage pipes should be designed so that they can be cleaned easily. The distance between centers of the cleaning openings should be shown on drawings. Where drainage pipes are used inside box girder bridges, provisions should be made to prevent accumulation of water during leaks or breakage of pipes. For road bridges, drains should be provided at expansion joints on both sides where is appropriate. Often, for small bridges, rainwater evacuation could be concentrated into abutment: for instance, EN 1993-2 (2005) suggests, for railway bridges up to 40 m long carrying ballasted tracks, that the deck may be assumed to be self-draining to abutment drainage systems and no further drainage provisions need to be provided along the length of the deck. Provision should be made for the drainage of all closed cross sections unless these are fully sealed by welding.

4.5 Fatigue details

The best solution to avoid fatigue in structural details is to adopt practical detailing solutions commonly described in codes and standards. EN 1993-1-9 (2005) provides a wide variety of solutions, including plain members and mechanically fastened joints, welded built-up sections, transverse butt welds, weld attachments and stiffeners, load-carrying welded joints, hollow sections, lattice girder node joints, orthotropic decks (open and closed stringers), and top flange to web junction of runway beams. For each of these details, the code includes information on the detail category and the detail requirements.

4.6 Accessibility

All bridge parts should normally be designed to be accessible for inspection, cleaning, and painting. Where such access is not possible, all inaccessible parts should either be effectively sealed against corrosion (e.g., the interior of boxes or hollow portions) and long-term damage or they should be constructed with improved atmospheric resistance.

4.7 Construction process

Railway bridges are specific structures requiring a deep knowledge not only of structural engineering, but also of the operations and safety requirements of lines used daily by great numbers of passengers. For this reason, the following considerations arise:

- *Timing*: When dealing with the construction or replacement process in railway engineering, the driving factor of the design is track time: so operations, maintenance, and new construction are all relevant factors that should be carefully evaluated by the designer in order to produce the best bridge in the shortest time. For this reason, the design efficiency could be sacrificed for a shorter construction period.
- *Simplifying construction*: Simple constructions are generally preferred to complex solutions, and elements such as simple spans and bolted construction (i.e., stiffeners bolted to web plates) are still widely used for railway bridges, whereas continuous spans with welded stiffeners are standard practice in highway bridge design. The use of bolted construction reduces fatigue requirements, and simple spans allow the replacement of each individual span, thus minimizing traffic interruptions.
- *Precasting*: A direct consequence of the previous points is that it is becoming a common practice to design and erect spans in nearly complete form in order to expedite span realization. Steel spans and precast concrete box beams, as well as other superstructure types, may be shipped to a construction site fully assembled in order to lift all in place quickly and restore traffic as soon as possible (AREMA, 2013).
- *Material savings*: Often the outcome of a design is counterintuitive to the standard practice of producing highly efficient structural systems that use a minimum amount of material. In the long term, this break from common practice proves more beneficial to railway companies due to the savings yielded from a design that lasts many years, requires minimal maintenance, and provides a construction period that keeps trains moving (AREMA, 2013).

5 R&D on railway bridges

Special emphasis on recent relevant research on railway bridges can be seen in the following fields:

- *High-speed lines (HSLs)*: The inherent advantages of increased speeds on railway vehicles has led to an increased interest in research on improved structures that can carry such vehicle loads and speeds. For example, in Italy, HSLs have been recently realized and accordingly, new bridges to carry HS vehicles have been designed and realized (Figure 19.9). Depending on the vehicle type and on the maximum design speed allowed on the line, bridge structures becomes more sophisticated to be designed and built; the framework of studies is very large and deals with different key problems. For instance, Doménech et al. (2014) recently investigated the influence of the vehicle model on the prediction of the maximum bending response of simply supported bridges under high-speed railway traffic; Johansson et al. (2014) deepen a methodology for the preliminary assessment of existing railway bridges for high-speed traffic; Xu et al. (2014) performed a complete evaluation of track geometry on a long-span steel-trussed cable-stayed bridge; and Vega et al. (2012) studied the dynamic response of underpasses for high-speed train lines.

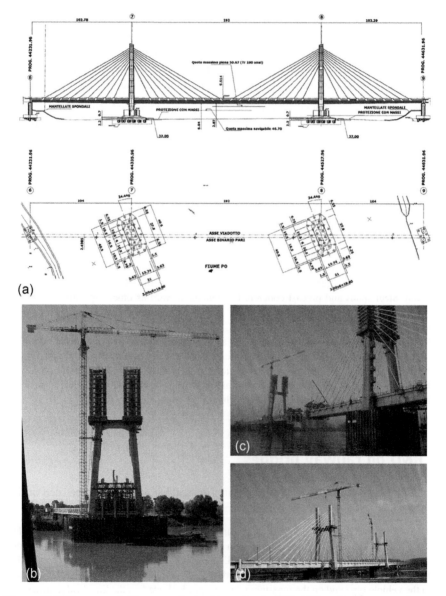

Figure 19.9 The recent bridge over the Po River, for the HSL Milano-Bologna: (a) plant and elevation scheme; (b) north-side antenna; (c) entire bridge view during the yard; (d) final phase of the bridge yard.

- *Improved/innovative materials*: As railway bridges have relevant requirements in terms of stiffness and vertical and lateral displacement limitations, an improvement on the materials employed could help in minimizing material weight and accelerating the construction speed, if possible, at the same time. Current improvements are represented by higher strength steel grades; dealing with ultra-high-performance fiber-reinforced cement (UHPFRC). Recent research on this issue includes the local bending tests and punching failure of a ribbed UHPFRC bridge deck (Toutlemonde et al., 2007), an experimental study on the bond between carbon fiber-reinforced polymer (CFRP) bars and ultra-high-performance fiber-reinforced concrete (Ahmad et al. 2011), innovative calculation formula of shear connectors in UHPFRC composite structure (Guo and Wang, 2012), and the rehabilitation and strengthening of concrete structures using UHPFRC (Brühwiler and Denarié, 2013);
- *Strengthening existing bridges*: Especially for metal, but also for more recent steel bridges, a strong international interest has concentrated on the upgrading and structural strengthening of these bridge types. This is because a lot of them were built between the late 18th and the mid-19th centuries, and all of these need to be repaired or replaced. As the complete renovation of entire national rail lines (amounting to thousands of kilometers) sounds difficult to develop, appropriate strengthening solutions are needed. Concerning this specific issue, some general hints are presented in several studies by this author: for instance, in Pipinato (2010), the step-level procedure for remaining fatigue life evaluation of one railway bridge is deepened, while in Pipinato (2011), safety and security issues in the assessment of existing bridges considering codes and standard are presented. In Pipinato and Modena (2010), the structural analysis and fatigue reliability assessment of the Paderno Bridge, a mixed road and railway bridge with a significant cultural heritage value, is presented, while in Pipinato et al. (2009), the high-cycle fatigue behavior of riveted connections for railway metal bridges is analyzed. Moreover, in Pipinato et al. (2011a), real-scale tests are presented dealing with the fatigue behavior on riveted steel elements taken from a railway bridge; it is useful to cite Pipinato et al. (2011b) as well, as in this study, the fatigue assessment of highway steel bridges in the presence of seismic loading is presented as an assessment approach that could be useful for railway bridges; a similar approach, including on-site dynamic testing was applied in Pipinato et al. (2012a) for the assessment procedure and rehabilitation criteria for the riveted railway Adige Bridge. Dealing with retrofit procedures, in Pipinato et al. (2012b), the fatigue behavior of steel bridge joints strengthened with FRP laminates is presented, and finally, an analytical approach, including dynamic analysis applied in historic riveted steel bridges, is presented in Pipinato et al. (2014). Other research includes Lin et al. (2014a), who investigated the rehabilitation and restoration of old steel railway bridges; and Lin et al. (2014b), deepening the preventive maintenance on welded connection joints in aged steel railway bridges; and Stamatopoulos (2013), who studied the fatigue assessment and strengthening measures to upgrade a steel railway bridge. Finally, a research project carried out in Europe is called Sustainable Bridges (Bień et al. 2008), and has investigated a wide variety of problems dealing with existing bridges, analyzing nondestructive testing technologies, testing new intervention methodologies on site; and performing much analysis on retrofit issues dealing with existing bridges in Europe.

References

Ahmad, F.S., Gilles, F., Le Roy, R., 2011. Bond between carbon fibre-reinforced polymer (CFRP) bars and ultra-high performance fibre-reinforced concrete (UHPFRC): Experimental study. Construction and Building Materials. 02/2011; 25(2): 479–485.

American Railway Engineering and Maintenance-of-Way Association (AREMA), 2013. Practical guide to railway engineering. AREMA, Lanham, MD.

American Railway Engineering and Maintenance-of-Way Association (AREMA), 2014. Manual for railway engineering. AREMA, Lanham, MD.

Bien, J., Elfgren, L., Olofsson, J., 2008. Sustainable bridges: assessment for future traffic demands and longer Lives. Research project TIP3-CT-2003-001653 within the SixthFramework Programme, European Commission, Bruxelles.

Britannica, 1910. The Encyclopædia Britannica: a dictionary of arts, sciences, literature and general information. 11th edition, vol. IV, pp. 539–540. Edited by The Encyclopædia Britannica Company, New York.

Brühwiler, E., Denarié, E., 2013. Rehabilitation and strengthening of concrete structures using ultra-high performance fiber reinforced concrete. Structural Engineering International - SEI, ISSN: 1016-8664. 23 (4), 450–457, Iabse Edition, Zurich.

California High-Speed Railway Authority (CHSRA), 2011. Technical memorandum: structure design loads TM 2.3.2. california high-speed train project. California High-Speed Railway Authority, CA, Sacramento.

Doménech, A., Museros, P., Martínez-Rodrigo, M.D., 2014. Influence of the vehicle model on the prediction of the maximum bending response of simply-supported bridges under high-speed railway traffic. Eng. Struct. 72, 123–139.

EN 1993-1-2, 2003. Design of steel structures. Part 1–2: General Rules—Structural Fire Design. CEN, Bruxelles.

EN 1990, 2006. Basis of structural design. CEN, Bruxelles.

EN 1991-2, 2005. Actions on structures. Part 2: Traffic loads on bridges. CEN, Bruxelles.

EN 1992-1-1, 2004. Eurocode 3: Design of concrete structures—Part 1-1: General Rules. CEN, Bruxelles.

EN 1993-1-1, 2003. Eurocode 3: Design of steel structures—Part 1-1: General Rules. CEN, Bruxelles.

EN 1993-1-10, 2005. Eurocode 3: Design of steel structures—Part 1-10: Material Toughness andThrough-Thickness Properties. CEN, Bruxelles.

EN 1993-1-9, 2005. Eurocode 3: Design of steel structures. Part 1-9: Fatigue. CEN, Bruxelles.

EN 1993-2, 2005. Eurocode 3: Design of steel structures. Part 2: Steel Bridges. CEN, Bruxelles.

EN 1997-2, 2007. Eurocode 7—Geotechnical design—Part 2: Ground investigation and testing. CEN, Bruxelles.

Guo, Y.H., Wang, Z.Q., 2012, Proposed Calculation Formula of Shear Connectors in UHPFRC-NSC Composite Structure. Applied Mechanics and Materials, TTP, 05/2012, 166-169:2851–2854.

Hirt, M.A., Leben, J.P., 2013. Steel Bridges. Conceptual and structural design of steel and steel—concrete composite bridges. EPFL Press, Lausanne, Switzerland.

Johansson, C., Nualláin, N.Á., Pacoste, C., Andersson, A., 2014. A methodology for the preliminary assessment of existing railway bridges for high-speed traffic. Eng. Struct. 58, 5–35.

Lamine, D., Pierre, M., Gomesb, F., Tessiera, C., Toutlemondec, F., 2013. Use of UHPFRC overlay to reduce stresses in orthotropic steel decks. J. Construct. Steel Res. 89, 30–41.

Lin, W., Yoda, T., Taniguchi, N., 2014a. Rehabilitation and restoration of old steel railway bridges: Laboratory experiment and field test. J. Bridge Eng. Technical paper n. 04014004. 19(5).

Lin, W., Yoda, T., Taniguchi, N., Shinya, K., 2014b. Preventive maintenance on welded connection joints in aged steel railway bridges. J. Construct. Steel Res. 92, 46–54.

Pipinato, A., 2010. Step level procedure for remaining fatigue life evaluation of one railway bridge. Baltic J. Road Bridge Eng. 5 (1), 28–37.

Pipinato, A., 2011. Assessment of existing bridges: safety and security issues [Problemi di sicurezza nelle valutazioni strutturali di ponti esistenti]. Ingegneria Ferroviaria 66 (4), 355–371 in (Italian).

Pipinato, A., 2012. Coupled safety assessment of cable stay bridges. Mod. Appl. Sci. 6 (7), 64.

Pipinato, A., Modena, C., 2010. Structural analysis and fatigue reliability assessment of the Paderno bridge. Prac. Period. Struct. Des. Construct. 15 (2), 109–124.

Pipinato, A., Pellegrino, C., Bursi, O.S., Modena, C., 2009. High-cycle fatigue behavior of riveted connections for railway metal bridges. J. Const. Steel Res. 65 (12), 2167–2175.

Pipinato, A., Molinari, M., Pellegrino, C., Bursi, O.S., Modena, C., 2011a. Fatigue tests on riveted steel elements taken from a railway bridge. Struct. Infrastruct. Eng. 7 (12), 907–920.

Pipinato, A., Pellegrino, C., Modena, C., 2011b. Fatigue assessment of highway steel bridges in presence of seismic loading. Engineering Structures 33 (1), 202–209.

Pipinato, A., Pellegrino, C., Modena, C., 2012a. Assessment procedure and rehabilitation criteria for the riveted railway Adige Bridge. Struct. Infrastruct. Eng. 8 (8), 747–764.

Pipinato, A., Pellegrino, C., Modena, C., 2012b. Fatigue behavior of steel bridge joints strenghtened with FRP laminates. Mod. Appl. Sci. 6 (9), 1–14.

Pipinato, A., 2014. Residual life of historic riveted steel bridges: an analytical approach. Journal of ICE - Bridge Engineering – Institution of Civil Engineers, Thomas Telford Ltd 167 (1), 17–32, ISSN: 1478–4637. http://dx.doi.org/10.1680/bren.11.00014, London.

Stamatopoulos, G.N., 2013. Fatigue assessment and strengthening measures to upgrade a steel railway bridge. Journal of Constructional Steel Research. 01/2013; 80:346–354.

Toutlemonde, F., Renaud, J.-C., Lauvin, L., Brisard, S., Resplendino, J., 2007. Local bending tests and punching failure of a ribbed UHPFRC bridge deck. In: Proc. 6th Intl. Conf. Fracture Mech. of Concrete and Concrete Struct. FRAMCOS 6, 1481–1489. 2007, Catania.

Vega, J., Fraile, A., Alarcon, E., Hermanns, L., 2012. Dynamic response of underpasses for high- speed train lines. J. Sound Vib. 331 (23), 5125–5140.

Xu, J.H., Wang, B., Wang, L., Wang, P., 2014. Evaluation of track geometry on a long span steel trussed cable-stayed bridge. Appl. Mech. Mat., 501–504:1403–1407, TTP Edition, Dürnten.

Section VII

Bridge components

Seismic component devices

Agrawal A.K., Amjadian M.
Department of Civil and Environmental Engineering, The City
College of the City University of New York, 160 Convent Ave.,
New York, NY 10031.

20.1 Introduction

Seismic component devices are innovative structural elements designed to protect bridges during extreme hazards, such as earthquakes, by absorbing or dissipating input external energy. Two most commonly used seismic component devices are seismic isolators and dampers. Performance of these devices have been demonstrated through numerous studies, including several large-scale tests (Kelly et al., 1986; Constantinou and Symans, 1992; Yang et al., 2002; Yang et al., 2004; Phillips et al., 2010) and theoretical research studies (Spencer et al., 1997; Dyke et al., 2003; Agrawal and Nagarajaiah, 2009; Agrawal et al., 2009; Tan and Agrawal, 2009; Nagarajaiah et al., 2009). These devices protect the structural safety and stability of bridges by modifying their dynamic characteristics, such as natural period, damping, or energy dissipation behavior. Bridges with these devices are designed such that damage during earthquakes and other hazards is localized in these devices, thereby protecting key structural members, such as piers. This chapter presents a brief overview of different types of seismic isolators and dampers used widely around the world to enhance seismic behavior of bridges. The standard methods employed for the analysis and design of seismic component devices in bridges are discussed, and applications of these devices are presented.

20.2 Seismic protective devices

20.2.1 Seismic isolators

Seismic isolators, such as elastomeric or sliding bearings, reduce seismic demand on key structural members of bridges, such as piers, during an earthquake. In bridges, piers are subjected to considerable shear force and flexural and torsional moments during an earthquake because of a large mass concentrated in the deck. Seismic isolators are typically installed between the deck and piers or deck and abutments (as illustrated in Figure 20.1) to decouple the movement of the superstructure from the substructure. The fundamental objective of seismic isolation is to elongate the natural period of a bridge beyond the predominant period of ground motions. Figure 20.2 shows typical acceleration and displacement response spectra for structures. For a bridge with its period at point A in Figure 20.2, the period of the bridge with seismic isolation will

Innovative Bridge Design Handbook. http://dx.doi.org/10.1016/B978-0-12-800058-8.00020-7

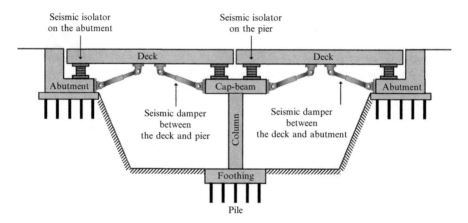

Figure 20.1 Placement of seismic protective systems in a typical two-span highway bridge (not drawn to scale).

be lengthened to B, resulting in significant reduction in spectral acceleration, as seen from Figure 20.2A. However, the spectral displacement of the isolated bridge may increase significantly, as illustrated in Figure 20.2B. The displacement of the isolated deck can be reduced further by increasing damping through supplemental devices, such as energy-dissipating lead core in bearings, friction in bearings, or dampers parallel with bearings. Figure 20.1 illustrates the installation of dampers parallel with bearings.

Seismic isolation has not been found to be efficient for flexible bridges whose natural periods are longer than predominant periods of earthquakes (Kunde and Jangid, 2006). The flexibility of bridges can be attributed to a flexible substructure (e.g., high-elevation piers), superstructure (e.g., elastic slender deck), or soft soil surrounding the piers (Tongaonkar and Jangid, 2003; Soneji and Jangid, 2008; Stehmeyer and Rizos, 2008; Dezi et al., 2012).

Seismic isolation of bridges near regions prone to fault ruptures is also not efficient because of the presence of long pulses in seismic waves that are destructive for flexible structures (Shen et al., 2004; He and Agrawal, 2008). For serviceability, seismic isolators should have self-centering properties so that the deck can return to its original position after an earthquake.

20.2.1.1 Theoretical concept of seismic isolation in bridges

A theoretical concept of seismic isolation is illustrated for the lateral response of an isolated highway bridge using the idealized 2-DOF (degrees of freedom) model shown in Figure 20.3. It is assumed that the isolation system behaves linearly, and columns also remain elastic during the ground motion excitation.

The equation of motion based on relative displacement of superstructure and substructure can be written as

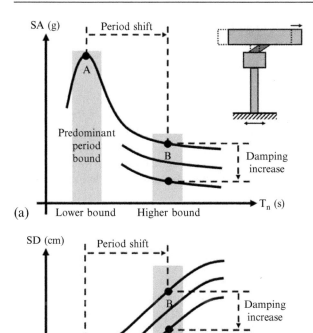

Figure 20.2 Influence of period shift and damping increase on acceleration and displacement of an isolated bridge; (a) acceleration response spectrum; (b) displacement response spectrum.

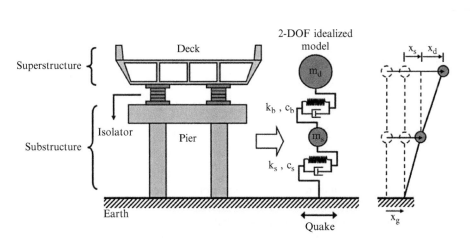

Figure 20.3 Simulation of lateral motion of an isolated highway bridge.

$$\begin{bmatrix} m_d & m_d \\ m_d & m_d + m_s \end{bmatrix} \begin{Bmatrix} \ddot{x}_d \\ \ddot{x}_s \end{Bmatrix} + \begin{bmatrix} c_b & 0 \\ 0 & c_s \end{bmatrix} \begin{Bmatrix} \dot{x}_d \\ \dot{x}_s \end{Bmatrix} + \begin{bmatrix} k_b & 0 \\ 0 & k_s \end{bmatrix} \begin{Bmatrix} x_d \\ x_s \end{Bmatrix}$$

$$= - \begin{bmatrix} m_d & m_d \\ m_d & m_d + m_s \end{bmatrix} \begin{Bmatrix} 0 \\ 1 \end{Bmatrix} \ddot{x}_g, \tag{1}$$

where x_d is relative displacement of the deck with respect to the top of the pier, x_s is relative displacement of the top of the pier with respect to the ground, $\ddot{x}_g$ is the ground acceleration, m_d is the mass of the superstructure, m_s is the mass of the substructure, k_b and c_b are effective stiffness and damping coefficients of the isolation system, and k_s and c_s are stiffness and damping coefficients of the substructure. We define following parameters to further simplify Eq. (1):

$$\gamma = \frac{m_d}{m_d + m_s} \quad , \quad \omega_b = \sqrt{\frac{k_b}{m_d}} = \frac{c_b}{2\xi_b m_d} \quad , \quad \omega_s = \sqrt{\frac{k_s}{m_d + m_s}} = \frac{c_s}{2\xi_s(m_d + m_s)}, \tag{2}$$

where γ is the ratio of mass of superstructure to the total mass of the bridge ($\gamma \approx 0.85-0.95$), ω_b is the natural frequency of the isolation system, ω_s is the natural frequency of the bridge before isolation, ξ_b is the critical damping ratio of the isolation system, and ξ_s is the critical damping ratio of the bridge before isolation. Using Eqs. (1) and (2), ratios of natural periods of the bridge with and without isolation can be calculated as

$$\frac{T_1}{T_s} = \sqrt{\frac{2(1-\gamma)}{1+\varepsilon - \sqrt{(1-\varepsilon)^2 + 4\gamma\varepsilon}}} \quad , \quad \frac{T_2}{T_s} = \sqrt{\frac{2(1-\gamma)}{1+\varepsilon + \sqrt{(1-\varepsilon)^2 + 4\gamma\varepsilon}}}, \tag{3}$$

in which ε is defined as the square of the ratio of the natural frequency of the isolation system to the natural frequency of the bridge before isolation [i.e., $\varepsilon = (\omega_b/\omega_s)^2 = k_b/\gamma k_s$], which takes a small value between 0.01 and 0.1. Assuming a first-order approximation for small values of ε, the modal vectors and modal participation factors for the isolated bridge are obtained as

$$\phi_1 \approx \begin{Bmatrix} 1 \\ \gamma\varepsilon \end{Bmatrix} \quad \Gamma_1 \approx \frac{1+\gamma\varepsilon}{1+(1+\gamma^2)\varepsilon + \gamma^2\varepsilon^2} \quad , \quad \phi_2 \approx \begin{Bmatrix} 1 \\ (1-\gamma)\varepsilon - 1 \end{Bmatrix}$$

$$\Gamma_2 \approx \frac{\varepsilon}{1-(1-\gamma)\varepsilon + (1-\gamma)\varepsilon^2} \tag{4}$$

The contribution of a mode to seismic response of a bridge in a given direction can be demonstrated by modal participation factor (Carr, 1994). Figure 20.4 displays the natural periods, mode shapes, and modal participation factors of the isolated bridge versus ε. It is observed from Figure 20.4A that the isolated bridge vibrates predominantly in the first mode for the values of ε in the range of 0.01 to 0.1. For this range of ε, the ratio of

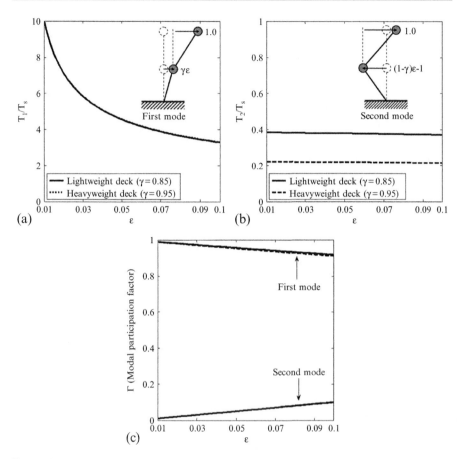

Figure 20.4 Influence of seismic isolation on dynamic characteristics of a conventional bridge; (a, b) first and second natural periods and mode shapes, (c) modal participation factor.

T_1/T_s varies from 10 to approximately 3.5. On the other hand, the ratio of T_2/T_s is generally less than 1.0. Since the first mode is significantly more flexible (having a longer period) than the second mode, the bridge deck vibrates predominantly in this mode. The spectral acceleration of the bridge deck decreases drastically because of lengthening of the first mode of the isolated bridge, compared to relatively shorter period of the bridge without isolation, as illustrated in Figure 20.2. In fact, isolated bridge can be modeled as a single-DOF system with the natural period and damping of the first mode. Such an assumption is valid and is used by seismic guidelines to design isolated bridges in the initial phase of design process (AASHTO, 2010; Eurocode 8, 2005).

20.2.1.2 Types of seismic isolators

Seismic isolators for bridges can be generally grouped into two main classes of elastomeric and sliding isolators (Kelly et al., 1986; Naeim and Kelly, 1999; Buckle et al.,

2006; Yoshida et al., 2004; Robinson, 1982; Mokha et al., 1991). This classification is based on the way that these protective systems provide the restoring force or flexibility to the bridge.

Elastomeric-based isolators

Elastomeric-based isolators consist of alternate layers of natural or synthetic rubber (elastomer) vulcanized and bonded with steel plates to carry desired vertical loads while allowing horizontal deformations (Naeim and Kelly, 1999; Kelly and Konstantinidis, 2011; Kelly, 1997). Figure 20.5A–C show three types of elastomeric-based isolators commonly used for bridges. The isolators are installed between the superstructure and substructure using top and bottom steel plates, as shown in Figure 20.5. Information on other types of elastomeric-based isolators can be found in the literature (e.g., Naeim and Kelly, 1999). These isolators can be considered as efficient successors to neoprene bearings in bridges in a low-seismicity region because they can carry vertical loads and braking forces and can resist temperature variations and creep. Moreover, these bearings, unlike neoprene bearings, are less vulnerable to separation from supports during strong earthquakes (Naeim and Kelly, 1999; Buckle et al., 2006).

Low-damping rubber bearings (LDRBs), shown in Figure 20.5A, are simple to manufacture, are designed to resist creep and temperature effects, and have a small amount of damping in the range of 2% of critical damping. Their lateral force displacement behavior is predominantly linear (mostly because of elastic stiffness of rubber), as shown in Figure 20.5E (Kelly et al., 1986). Hence, a supplemental damper is often installed parallel with these isolators to provide a desired level of damping. *High-damping rubber bearings (HDRBs)* with an inherent damping ratio in the range of 10% to 20% of critical damping are used to eliminate the need for supplemental dampers. Lateral force versus displacement of these isolators is nonlinear with displacement increasing at high displacement, as shown in Figure 20.5F (Yoshida et al., 2004). The area under the force-displacement curve (hysteresis loop) represents the damping capacity of isolators.

Three important parameters controlling the design of rubber bearings are horizontal stiffness (K_h), vertical stiffness (K_v), and critical axial load (P_{cr}). The horizontal stiffness of isolator system because of rubber layers is calculated as follows (Naeim and Kelly, 1999; Buckle et al., 2006; Kelly and Konstantinidis, 2011; Kelly, 1997):

$$K_h = \frac{G_r A_r}{t_r}, \tag{5}$$

where G_r is the shear modulus of rubber ($G_r \simeq 0.7$ MPa for rubber with average hardness), A_r is the gross area of rubber, and t_r is the total thickness of rubber. The vertical stiffness of the bearing is calculated as follows (Naeim and Kelly, 1999; Buckle et al., 2006; Kelly and Konstantinidis, 2011):

$$K_v = \frac{E_r A_s}{t_r}, \tag{6}$$

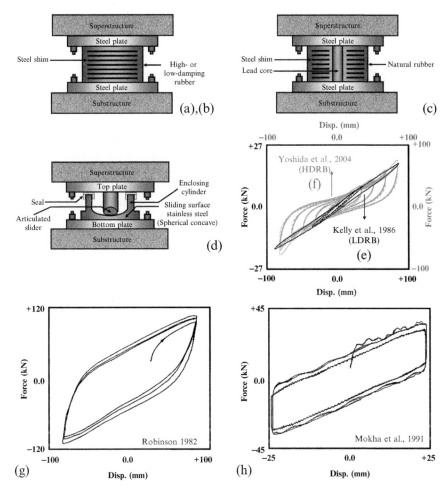

Figure 20.5 Typical seismic isolators implemented on bridges; (a) low-damping rubber bearing, (b) high-damping rubber bearing, (c) lead rubber bearing, (d) friction pendulum system, and their hysteretic loops (e) (Kelly et al., 1986), (f) (Yoshida et al., 2004), (g) (Robinson, 1982), and (h) (Mokha et al., 1991), respectively.

where E_r is the modulus of elasticity of rubber and A_s is the cross-sectional area of steel shims. The value of E_r for a common bearing with a circular cross section is given by (Kelly and Konstantinidis, 2011):

$$E_r = \frac{1}{\left(\dfrac{1}{6 G_r S^2} + \dfrac{4}{3K} \right)} \tag{7}$$

where S is the shape factor of a layer of rubber and K is the bulk modulus of rubber ($K \simeq 2000$ MPa). The critical buckling load of rubber bearings with bolt-type connections in undeformed and deformed states are calculated as follows (Naeim and Kelly, 1999; Buckle et al., 2006):

$$P_{cr1} = \sqrt{G_r A_s \frac{\pi^2 E_c I_s}{3 t_r^2}} \quad \text{(Undeformed)}$$

$$P_{cr2} = P_{cr1} \frac{A_{eff}}{A_s} \quad \text{(Deformed)}$$

(8)

where I_s is the second moment of area of A_s, and A_{eff} is the area of overlap between the top and bottom of the bearing due to maximum lateral deformation. The stability of rubber bearings, especially those with dowel-type connections, should also be checked for the rollout condition in which the rubber is subjected to tension and its force-displacement curve suffers a decreasing slope (Naeim and Kelly, 1999; Kelly and Konstantinidis, 2011).

Lead rubber bearings (LRBs) have one or several lead cores installed at the center of the rubber layers, as shown in Figure 20.5C. The lead plug acts as a damper by dissipating input seismic energy through yielding (Robinson, 1975, 1982). The steel reinforcing plates provide confinement to the lead core and vertical stiffness to carry the vertical loads. They push the lead plug laterally to yield during a seismic event. The lead plug has a high preyield horizontal stiffness, making it resistant against lateral movements due to nonseismic loads, such as wind, and vehicle braking force (service loads; Naeim and Kelly, 1999; Buckle et al., 2006). However, because of low postyield stiffness, the horizontal stiffness of the isolator is predominantly contributed by rubber layers after yielding of lead core. A typical hysteresis behavior of lead rubber bearings, shown in Figure 20.5G, can be modeled by a bilinear behavior (Robinson, 1982). In order to design the areas and thickness of rubber layers, the area of the lead core should be subtracted from the rubber cross section. The yield force of lead core is given by (Buckle et al., 2006):

$$f_y = \frac{1}{R_c} \left(\frac{\pi}{4} \sigma_{yl} d_l^2 \right)$$

(9)

where σ_{yl} is the yield stress of lead (10 MPa), d_l is diameter of lead core, and R_c is the creep load factor, which is equal to 1 for seismic loads and 2 for service loads. Figure 20.6A displays the installation of an LRB in a bridge (Dynamic Isolation Systems, 2006).

Sliding-based isolators

In sliding-based isolators, flexibility in the horizontal direction is provided through slippage between the support and the sliding surface, whereas restoring force is provided through geometry of the support, such as concave surface, or supplemental springs. The friction between the support and the sliding surface provides damping through the dissipation of input seismic energy. Although there are different kinds of friction-based bearings [e.g., polytetrafluoroethylene (PTFE) spherical bearings (Constantinou et al., 2011), double (Constantinou et al., 2011; Fenz and Constantinou, 2006) and triple (Constantinou et al., 2011; Fenz and Constantinou, 2008) friction pendulum systems, and the Eradiquake isolator (Buckle et al., 2006)],

Figure 20.6 Seismic isolator full-scale implementation; (a) LRB (Richmond–San Rafael Bridge), and (b) FPS (Antioch Bridge).

friction pendulum systems (FPSs) with a single slippage surface have been used extensively in bridges (Mokha et al., 1991).

Figure 20.5D shows the cross section of a typical FPS used for bridges. Figure 20.6B shows the photograph of a FPS installed in a bridge. An FPS includes an articulated slider sliding on a spherical concave surface, both made of stainless steel. The surface of the articulated slider, which makes contact with the concave surface, is coated with low-friction composite materials. The curvature of the concave surface provides lateral stiffness and restoring force to the superstructure during earthquake ground excitation. These isolators are capable of carrying large axial loads in the range of large lateral displacement (Naeim and Kelly, 1999; Buckle et al., 2006).

20.2.1.3 Standard design method for isolation bearings

The isolated bridges are designed according to standard seismic codes such as AASHTO (AASHTO, 2010) and Eurocode 8 (2005). These two codes support very similar methods for seismic analysis of isolated bridges. The analysis procedures used by AASHTO (2010) are (i) simplified method, (ii) single mode spectral method, (iii) multimode spectral method, and (iv) time-history method. The first three methods are based on representing the nonlinear behavior of the isolated system by an equivalent elastic model with an effective natural period (T_{eff}) and damping (ξ_{eff}).

The time-history method is the most accurate procedure, and it is used to analyze the isolation systems in bridges with a highly curved or skewed geometry (Kalantari and Amjadian, 2010) or bridges with a large demand of ductility [$T_{eff} > 3$; see AASHTO, 2010)] or damping [$\xi_{eff} > 50\%$, AASHTO (2010); or 30%, Eurocode 8 (2005); Buckle et al. (2011)]. In this procedure, the bridge is analyzed using a three-dimensional (3D) model with nonlinear isolators. A bilinear model is permitted by seismic codes (AASHTO, 2010; Eurocode 8, 2005) to be used to simplify the hysteretic behavior of the isolator unit shown in Figure 20.7.

The basic principles of seismic design of isolated bridges provided by AASHTO (2010) and Eurocode 8 (2005) are generally similar; although it is believed that Eurocode 8 (2005) adopts a more rational design criteria in some cases such as reentering capability requirements (Constantinou et al., 2011). The design of the isolation systems of ordinary bridges according to AASHTO can be briefly described as follows. First, the effective period T_{eff} (1.5–2.5 s) and damping ξ_{eff} (20–30%) of the isolated bridge are assumed based on the required performance. Then the simplified method is used, and the initial value of maximum displacement of the deck is calculated from the design response spectrum developed for the region in which the bridge is located. Therefore, initial properties of the isolators can be estimated. In the next step, the multimode spectral method is used to iteratively analyze a 3D model of the bridge, assuming equivalent linear elements for isolators.

This analysis is carried out along both longitudinal and transverse directions of the model, and results are obtained. The design values of isolators are calculated by combining these results using 30%–100% rule (Constantinou et al., 2011; Buckle et al., 2011). Then a type of isolator is selected, and its physical features are designed. For example, dimensions of the isolator (e.g., gross area of rubber, thickness of steel shims, and the diameter of lead core for a lead rubber bearing) are determined in this stage (see Figure 20.5C). The design is finally evaluated to determine whether the seismic performance objective is satisfied; if not, then the design needs to be revised.

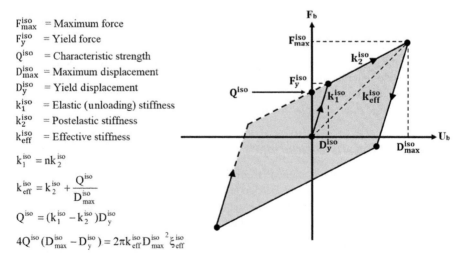

F_{max}^{iso} = Maximum force
F_y^{iso} = Yield force
Q^{iso} = Characteristic strength
D_{max}^{iso} = Maximum displacement
D_y^{iso} = Yield displacement
k_1^{iso} = Elastic (unloading) stiffness
k_2^{iso} = Postelastic stiffness
k_{eff}^{iso} = Effective stiffness

$$k_1^{iso} = nk_2^{iso}$$

$$k_{eff}^{iso} = k_2^{iso} + \frac{Q^{iso}}{D_{max}^{iso}}$$

$$Q^{iso} = (k_1^{iso} - k_2^{iso})D_y^{iso}$$

$$4Q^{iso}(D_{max}^{iso} - D_y^{iso}) = 2\pi k_{eff}^{iso} D_{max}^{iso\,2} \xi_{eff}^{iso}$$

Figure 20.7 The idealized bilinear hysteretic behavior of isolation systems in AASHTO (2010).

20.2.2 Dampers

Although different types of dampers have been developed for response control of structures during the last few decades, fluid viscous and friction dampers have been applied most frequently to bridges. In recent years, smart dampers, such as magnetorheological (MR) fluid dampers, have also been developed and have been applied for vibration mitigation of stay cables of cable stayed bridges. Next, a brief description of these three types of devices are presented.

Fluid viscous damper

Fluid viscous dampers work based on the principle of dissipation of energy because of fluid flowing through orifices. The damper consists of a stainless steel piston, a steel cylinder divided into two champers by the piston head, a compressible hydraulic fluid (silicon oil), and an accumulator for smooth fluid circulation. A typical fluid damper manufactured by Taylor Devices, Inc., is shown in Figure 20.8 (Taylor Devices, Inc., 1956). In fluid viscous dampers, as the piston moves (e.g., from left to right or right to left), fluid flows from one chamber to another chamber through the orifice. This movement of fluid from a larger area (cylinder chamber) to a smaller area (orifice) and from a smaller area (orifice) to a larger area (cylinder chamber) results in the dissipation of energy because of head loss. Fluid viscous dampers can operate over an ambient temperature ranging from $-40\ ^\circ\text{C}$ to $70\ ^\circ\text{C}$ (Constantinou and Symans, 1992).

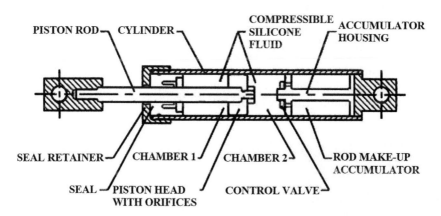

Figure 20.8 A typical fluid viscous damper manufactured by Taylor Devices, Inc.

The damping force of the damper is proportional to the pressure difference across the piston head and is expressed as a function of velocity of the piston as follows (Constantinou and Symans, 1992; Konstantinidis et al., 2012):

$$F_d = C_\alpha |\dot{U}_d|^\alpha \text{sgn}(\dot{U}_d), \tag{10}$$

where C_α is the damping ratio depending on pressure difference, $\dot{U}_d$ is the velocity of piston, sgn (.) is the sign function, and α is a constant parameter controlled by orifice shape to alter flow characteristics with fluid speed. For seismic protection, α is designed to be typically in the range of 0.3 and 1.0. For $\alpha = 1$, the viscous damper behaves as a linear device with $F_d = C_\alpha \dot{U}_\alpha$. For $\alpha < 1$, the force applied by the damper is nonlinear with velocity (Makris and Zhang, 2002). Figure 20.9A shows a photograph of four fluid viscous dampers installed between the deck and one of the abutments of a highway bridge in California (Makris and Zhang, 2002). The typical hysteresis loop of these dampers for $\alpha = 0.35$ and $\alpha = 1.00$ is shown in Figure 20.9B. Since the force applied by a viscous damper is proportional to the velocity, it is 90 degrees out of phase with displacement response of the bridge. Therefore, viscous dampers do not contribute to peak column forces at the instant when columns experience their maximum deflection during the ground motion excitation.

Viscous dampers have been found to be effective in reducing base shear on bridge piers. However, it is possible that the dampers may add some stiffness to the structure during high-frequency excitations and show viscoelastic behavior beyond the cutoff frequency (Constantinou and Symans, 1992; Reinhorn et al., 1995). In contrast to other kinds of dampers, such as viscoelastic dampers, the variation in temperature has a minor influence on the behavior of viscous dampers (Constantinou and Symans, 1992). On the other hand, the device needs to be maintained over a long period of operation against wear in seals to prevent oil leakage (Sadek et al., 1996).

Approaches to improve effectiveness of viscous dampers using real-time control of orifice have been investigated by researchers worldwide. Kawashima and Unjoh (1994) have proposed a variable viscous damper to control seismic response of bridges. Neff Patten et al. (1999) tested three hydraulic actuators in semiactive mode to investigate the performance of viscous dampers in reducing traffic induced response of the Walnut Creek Bridge in Oklahoma. Feng et al. (2000) showed that viscous dampers are more effective than viscoelastic dampers in reducing the relative displacement at expansion joints of bridges with narrow seat widths to minimize the risk of deck unseating during strong earthquakes.

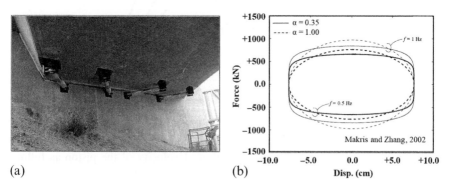

(a)   (b)

Figure 20.9 Fluid viscous damper; (a) full-scale implementation (of viscous damper) on 91/5 (bridge name) overcrossing in California, and (b) typical hysteresis loop (Makris and Zhang, 2002).

Friction damper

Friction dampers dissipate input seismic energy through friction between two rough sliding surfaces. Over the past few decades, many different kinds of friction dampers have been proposed to maximize the dissipation of input seismic energy in buildings and bridges (Pall and Marsh, 1982; Aiken et al., 1992, 1993). One of the most widely used friction dampers is the Pall friction device, which was originally developed for braced steel frames in buildings (Pall and Marsh, 1982). Another commonly used friction damper device is the Sumitomo damper, which was originally designed and manufactured by Sumitomo Metal Industries in Osaka, Japan as a shock absorber in railway rolling stock (Aiken et al., 1992, 1993). It includes a piston equipped by several friction pads sliding on the inner surface of a damper cylinder. Figures 20.10A–B show the schematics of the Sumitomo friction damper and its hysteresis loop subjected to a given base acceleration in the lab.

Many mathematical models have been proposed to simulate friction in dynamics (Olsson et al., 1998). One of the main common friction models, which is acceptable in range of engineering measurements, is the classical model of friction called the Coulomb friction model. In most friction devices, the friction force can be developed based on the Coulomb friction law with a typical rectangular hysteretic behavior, as shown in Figure 20.10.b. The friction damper force based on this simple model can be characterized as:

$$F_d = \mu N \, \text{sgn}(\dot{U}_d), \tag{11}$$

where μ is friction ratio, N is normal reaction between two sliding surfaces, $\dot{U}_d$ is velocity of the damper and sgn (.) is the sign function, which ensures that the damper force is applied in the direction opposing the motion. Although the behavior of friction can be modeled by a simple formula in Eq. (11), the actual dynamic behavior of a structure with frictional dampers is quite complex because of the presence of stick (no sliding) and slip (sliding) phases in the damper, depending on the slip force and earthquake ground motion characteristics (Olsson et al., 1998).

The energy dissipated by the friction damper for a given maximum force is greater than that by the viscous damper (which has an elliptical hysteresis loop). Friction dampers are designed to have slippage beyond a certain slippage force that depends on the ground motion time history. Hence, these dampers provide added stiffness (without any dissipation) of energy during low-level wind and braking forces. Long-term reliability of sliding surfaces because of their susceptibility to corrosion and wear is one of the primary concerns in actual applications of this damper. Moreover, the normal load on the sliding interfaces cannot be reliably maintained and some relaxation (loss of stress) may be expected over time. A passive friction may also experience permanent displacement after a strong earthquake (Sadek et al., 1996) because of a significantly higher level of slip force than the magnitude of the restoring force at the end of the earthquake.

Many researchers have investigated controllable friction dampers to address deficiencies of passive friction dampers because of slip forces (i.e., stick-slip phases or permanent displacement in the damper). Two of the most commonly investigated

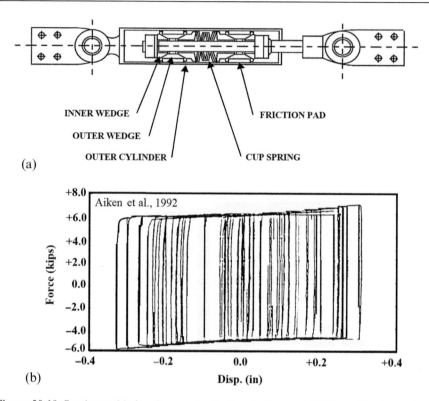

Figure 20.10 Sumitomo friction damper; (a) Sectional view, and (b) Typical hysteresis loop (Aiken et al., 1992), (Aiken et al., 1993).

friction devices are piezoelectric (Madhekar and Jangid, 2011; Wieczorek et al., 2014) and electromagnetic friction dampers (Agrawal and Yang, 2000). In these devices, damper slip force can be varied based on real-time measurement of velocity and displacement across the damper to guarantee continuous slippage in the dampers (Agrawal and Yang, 2000). At the end of the earthquake, the slip force can be set to a very low value to return the damper piston to its original position.

MR dampers

MR dampers are similar in construction to fluid viscous dampers and utilize MR fluids instead of hydraulic oil. The MR fluids typically consist of micron-sized, magnetically polarizable particles dispersed in a carrier medium such as mineral or silicone oil. The particle form of fluid can be changed by the applied magnetic field, transforming the behavior of the fluid to a plastic or semisolid state in a few milliseconds because of the alignment of iron particles to the magnetic field. MR devices can change the stiffness of their fluid up to 100 Hz and can operate over a wide range of the ambient temperature, usually from $-40\,°C$ to $+150\,°C$. They have a large yield shear stress routinely between 50–100 kPa for applied magnetic fields of 150–250 kA/m (Carlson et al., 1996). MR fluids react to external stimulus in a few milliseconds and can be readily controlled by standby batteries with a voltage in range of 12–24 V (Spencer et al., 1997). Figure 20.11 shows the operation and construction of an MR damper developed by the LORD Corporation (1924).

MR dampers have been investigated extensively both theoretically and experimentally, and they have several advantages over other semiactive devices. One of these is that the force of MR dampers is not fully dependent on the velocity as is the case with variable orifice dampers (Xu et al., 2006). This fact enables significant mitigation of a broader range of seismic activity. Another merit of MR dampers is the broad range of the maximum to the minimum force; i.e., the range is much bigger than that of any other controllable damper, especially at low velocities. An MR damper has no moving parts in valves, thereby reducing maintenance and malfunction concerns. Their response time is also significantly faster than that of variable orifice dampers. An MR damper can be made in a smaller device than can a hydraulic damper, and it is also fail-safe; i.e., it operates as a passive device when the power source is disconnected for any reason (Yoshioka et al., 2002; Gavin and Dobossy, 2001).

Mathematically, the behavior of MR dampers can be modeled by a phenomenological model based on the Bouc-Wen model or a hyperbolic tangent model to capture the nonlinear force response of large-scale MR dampers over the dynamic range of interest (Spencer et al., 1997; Gavin and Dobossy, 2001; Dyke et al., 1996).

Both these models have been found to have good agreement with experimental results. The typical hysteretic behavior of an MR damper is shown in Figure 20.12.

Investigation of performance of MR dampers to bridges have been carried out primarily through theoretical studies, although MR dampers have been applied to bridges for mitigating vibration of stay cables (Chen et al., 2003). Figure 20.13 shows the first full-scale implementation of MR dampers used for this purpose in Dongting Lake Bridge, China (Chen et al., 2003). Erkus et al. (2002) have investigated the performance of semiactive MR dampers for reducing bearing deflection and column force using a simple bridge model. It is shown that MR dampers can perform the role of an active actuator and a passive damper to control the bridge's response, depending on the design goal. Sahasrabudhe and Nagarajaiah (2005) studied analytically and experimentally the performance of a sliding isolated bridge model equipped with a semiactive controllable MR damper subjected to near-fault earthquakes. They have shown that semiactive MR dampers can exhibit better performance in reducing bearing displacement in semiactive mode as compared to that in passive mode. Loh and Chang (2006) have applied a semiactive MR damper with different semiactive control algorithms to the American Society of Civil Engineers (ASCE) benchmark model of a cable stayed bridge. This numerical study has shown that the MR damper is able to reduce the bridge response if it is commended by a mixed H_2 and H_1 algorithm. They have also studied the efficiency of MR dampers to control vibration of cables in Gi-Lu cable stayed bridge using a numerical model (Chang and Loh, 2006). Ok et al. (2007) proposed a fuzzy control technique to control the input voltage of several MR dampers implemented on the benchmark cable stayed bridge model. Guo et al. (2009)

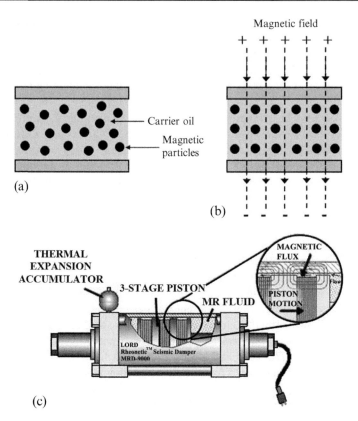

Figure 20.11 MR fluid damper; (a) Randomly dispersed particles when the fluid is in its own neutral condition; (b) particle chains formation when the fluid is exposed to a magnetic field; (c) a typical seismic MR damper manufactured by the LORD Corporation.

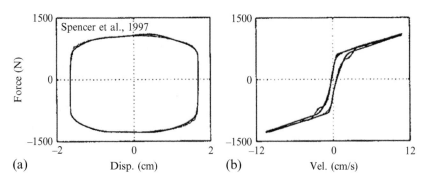

Figure 20.12 MR fluid damper hysteresis loops; (a) force displacement, and (b) force velocity (Spencer et al., 1997).

Figure 20.13 First full-scale implementation of MR dampers for mitigation vibration of stay cables; (a) Dongting Lake Bridge, China, and (b) MR damper configuration.

investigated the effectiveness of MR dampers to control displacement and pounding of the deck in an experimental bridge model. Pradono et al. (2009), using the concept of negative stiffness, have proposed a control method to command an MR damper to produce larger hysteretic loops for absorbing the earthquake energy as much as possible. This algorithm was implemented on the base-isolated benchmark bridge.

Jung et al. (2009) have proposed a smart passive control system to generate the current required to launch the inner magnetic field of an MR damper without any external power supply. The system includes an electromagnetic induction (EMI) part consisting of a coil connected to the piston rod of damper exposed to a permanent magnetic field. By implementing the new passive system on the highway bridge benchmark model, they have shown that it has superior performance to the passive-optimal control system. Yang et al. (2011) proposed a new nonlinear mechanical model to simulate hysteretic behavior of MR dampers. By comparison with an experimental model simulating the vibration of a suspension bridge in its own first mode, they have shown that the numerical model is reliable and can be efficiently applied to control longitudinal seismic response of suspension bridges. Heo et al. (2014) studied the performance of a 30-kN MR damper controlled by Lyapunov and clipped-optimal control algorithms to decrease the seismic response of a scaled asymmetrical cable-stayed bridge in the laboratory. The bridge was a large scale model with 28 m length and a tower with 10.2 m height. It has been shown that the MR damper commanded by the semiactive algorithm control can reduce the displacement of bridges with the MR damper in passive off-mode by 75%.

20.3   Applications of seismic protective systems in bridges

Over the past few decades, many full-scale seismic protective systems have been implemented on bridges worldwide. These systems primarily include seismic isolators and passive fluid viscous dampers. Bridges designed or retrofitted by these devices before 2000 are well documented in the literature (Spencer and Nagarajaiah, 2003; EERC, 1995, 1996). Table 20.1 lists some important bridges protected by such seismic devices after the year 2000. The bridges in this table include major ones located in regions with a high level of seismic hazard.

Table 20.1 Full-Scale Implementation of Seismic Component Devices on Bridges Worldwide.

Bridge	Location (Country, City)	Type	Length (m)	Seismic Component Device	Year	Notes
Marga Marga	Chile, Viña del Mar	Composite	383	HDRB	1996	Undamaged in 2010 Maule earthquake (Sarrazin et al., 2013).
San Diego–Coronado Bay	United States, California	Steel girder	3407	Rubber bearing and viscous damper	2000	
Amolanas	Chile, Los Vilos	Steel girder	268	Sliding bearing and viscous damper	2000	
Shin-Tenno	Japan	Steel girder		Elastomer bearing	2002	Undamaged in 2011 Tohoku earthquake (Kawashima, 2012).
Benicia-Martinez	United States, California	Steel truss	1875	FPS (Earthquake Protection System, Inc.)	2003	
Hernando de Soto	USA, Tennessee	Steel girder	5950	FPS (Earthquake Protection System, Inc.)	2003	
Loureiro Viaduct	Portugal, Lisbon	Prefabricated posttensioned box girder	1050	Rubber bearing and viscous damper, $F_{max} = 4000$ kN (FIP Industiale, Inc.)	2003	
Bill Emerson Memorial	United States, Illinois	Cable stayed	1206	Shock transmission device, $F_{max} = 6670$ kN, (Taylor Devices, Inc.)	2003	The structural model of the bridge is a benchmark for seismic protection of cable stayed bridges (Dyke et al., 2003).
George Washington	United States, Washington	Steel truss	1450	FPS (Earthquake Protection System, Inc.)	2004	

Name	Location	Type	Length	Device	Year	Notes
Rion Antirion	Greece, Patras	Cable-stayed	2252	Viscous Dampers, F_{max} = 3500 kN, (FIP Industriale, Inc.)	2004	
Bolu Viaducts	Turkey	Reinforced concrete	2300	FPS (Earthquake Protection System, Inc.)	2006	
Richmond–San Rafael	United States, California	Steel truss	8850	LRB (Dynamic Isolation System, Inc.)	2006	
Sutong Bridge	China, Jiangsu	Cable stayed	8206	Elastomeric spring and viscous damper, F_{max} = 10000 kN (Taylor Device, Inc.)	2008	The performance of the system reported as "very good" during the 2008 Wenchuan earthquake (Yongqi et al., 2008).
Yabegawa	Japan	Prestressed concrete cable stayed	517	LRB and stopper damper	2009	
Antioch	United States, California	Steel girder	2900	FPS (Earthquake Protection System, Inc.)	2010	
Stonecutters	China, Hong Kong	Cable stayed	1600	Shock transmission device, F_{max} = 8000 kN	2011	
Erqi	China, Wuhan	Cable stayed	2922	Viscous Dampers, F_{max} = 1000 kN (FIP Industriale, Inc.)	2011	
Dumbarton	United States, California	Reinforced concrete	2620	FPS (Earthquake Protection System, Inc.)	2013	
Han Jia Tuo	China, Chong Qing	Cable stayed	866	Shock transmission device, F_{max} = 2300 kN, and viscous damper, F_{max} = 2500 kN (FIP Industriale, Inc.)	2013	
Jiashao	China, Zhejiang	Cable stayed	2680 m (length of main span)	Viscous damper, F_{max} = 2500 kN (FIP Industriale, Inc.)	2013	World's longest cable stayed bridge with a total length of 10,138 m.

20.4 Conclusions

Bridges are key elements of transportation networks in urban areas. The reduction in functionality of bridges after strong earthquakes is a matter of great concern. These infrastructures must be fully operational immediately after the disaster to lessen economic and safety impacts of earthquakes. The response of bridges to earthquakes can be controlled by installing seismic protective devices in these structures as a cost-effective method of design or retrofit.

In this chapter, a brief review on seismic protective devices, such as isolators and dampers, and their application in bridges has been given. It has been shown that seismic isolation is effective in increasing ductility of a bridge by shifting its natural period away from the predominant period of earthquake. Two main classes of isolators used in bridges are elastomeric and sliding-based bearings. The mechanisms of most common isolation systems of each class, including HDRBs, LDRBs, LRBs, and FPS are described in detail. The capability of different types of dampers, such as fluid viscous, MR, and friction dampers, in dissipating input energy to bridges has been discussed. And a list of full-scale implementation of these devices in important bridges around the world was also presented.

References

Agrawal, A.K., Nagarajaiah, S., 2009. Benchmark structural control problem for a seismically excited highway bridge: phase I and II. Struct. Control Health Monit. 16, 503–508.

Agrawal, A., Yang, J., 2000. A semi-active electromagnetic friction damper for response control of structures. In: Proceedings of Structures Congress 2000, Philadelphia, PA, US, May 8–10, pp. 1–8.

Agrawal, A., Tan, P., Nagarajaiah, S., Zhang, J., 2009. Benchmark structural control problem for a seismically excited highway bridge - part I: phase I problem definition. Struct. Control Health Monit. 16, 509–529.

Aiken, I.D., Nims, D.K., Kelly, J.M., 1992. Comparative study of four passive energy dissipation systems. Bull. New Zeal. Natl. Soc. Earthq. Eng. 25, 175–192.

Aiken, I.D., Nims, D.K., Whittaker, A.S., Kelly, J.M., 1993. Testing of passive energy dissipation systems. Earthq. Spectra 9, 335–370.

American Association of State Highway and Transportation Officials (AASHTO), 2010. Guide Specifications for Seismic Isolation Design, third ed. Washington, DC.

Buckle, I.G., Constantinou, M., Dicleli, M., Ghasemi, H., 2006. Seismic Isolation of Highway Bridges. Technical Report, MCEER-06-SP07, Multidisciplinary Center for Earthquake Engineering Research, Univ. at Buffalo, State Univ. of New York, Buffalo, NY, US.

Buckle, I.G., Al-Ani, M., Monzon, E., 2011. Seismic isolation design examples of highway bridges. Technical Report, NCHRP 20-7/Task 262(M2), National Cooperative for Highway Research Program, Transportation Research Board, Washington, D.C., US.

Carlson, J.D., Catanzarite, D.M., St. Clair, K.A., 1996. Commercial magneto-rheological fluid devices. Int. J. Mod. Phys. B 10, 2857–2865.

Carr, A.J., 1994. Dynamic analysis of structures. Bull. New Zeal. Natl. Soc. Earthq. Eng. 27, 129–146.

Chang, C.-M., Loh, C.-H., 2006. Seismic response control of cable-stayed bridge using different control strategies. J. Earthq. Eng. 10, 481–508.

Chen, Z.Q., et al., 2003. MR damping system on Dongting Lake cable-stayed bridge. Proceeding of Smart Structures and Materials, Smart Systems and Nondestructive Evaluation for Civil Infrastructures, 5057, pp. 229–235.

Constantinou, M.C., Symans, M.D., 1992. Experimental and analytical investigation of seismic response of structures with supplemental fluid viscous dampers. Technical Report, NCEER-92-0032, Multidisciplinary Center for Earthquake Engineering Research, Univ. at Buffalo, State Univ. of New York, Buffalo, NY, US.

Constantinou, M.C., Kalpakidis, I., Filiatrault, A., Lay, R.A.E., 2011. LRFD-based analysis and design procedures for bridge bearings and seismic isolators. Technical Report, MCEER-11-0004, Multidisciplinary Center for Earthquake Engineering Research, Univ. at Buffalo, State Univ. of New York, Buffalo, NY, US.

Dezi, F., Carbonari, S., Tombari, A., Leoni, G., 2012. Soil-structure interaction in the seismic response of an isolated three span motorway overcrossing founded on piles. Soil Dyn. Earthq. Eng. 41, 151–163.

Dyke, S.J., Spencer, B.F., Sain, M.K., Carlson, J.D., 1996. Modeling and control of magnetorheological dampers for seismic response reduction. Smart Mater. Struct. 5, 565–575.

Dyke, S.J., Caicedo, J.M., Turan, G., Bergman, L.A., Hague, S., 2003. Phase I benchmark control problem for seismic response of cable-stayed bridges. J. Struct. Eng. 129, 857–872.

Dynamic Isolation Systems Company, 2006, McCarran, NV, US, Site: http://www.dis-inc.com/ (accessed 05.07.15.).

Earthquake Engineering Research Center (EERC), 1995. Isolated Bridges in Japan. from, http://nisee.berkeley.edu/prosys/japanbridges.html (accessed 23.11.14.).

Earthquake Engineering Research Center (EERC), 1996. Isolated Bridges in the US. from, http://nisee.berkeley.edu/prosys/usbridges.html (accessed 23.10.14.).

Erkus, B., Abé, M., Fujino, Y., 2002. Investigation of semi-active control for seismic protection of elevated highway bridges. Eng. Struct. 24, 281–293.

Eurocode 8, 2005. Design of Structures for Earthquake Resistance. Part 2: Bridges. Eurocode 8. ed. Brussels.

Feng, M.Q., Kim, J.-M., Shinozuka, M., Purasinghe, R., 2000. Viscoelastic dampers at expansion joints for seismic protection of bridges. J. Bridg. Eng. 5, 67–74.

Fenz, D.M., Constantinou, M.C., 2006. Behaviour of the double concave Friction Pendulum bearing. Earthq. Eng. Struct. Dyn. 35, 1403–1424.

Fenz, D.M., Constantinou, M.C., 2008. Spherical sliding isolation bearings with adaptive behavior: theory. Earthq. Eng. Struct. Dyn. 37, 163–183.

Gavin, H.P., Dobossy, M.E., 2001. Optimal design of an MR device. In: Proceeding of SPIE's 8th Annual International Symposium on Smart Structures and Materials, Newport Beach, CA, US, March 3-8, pp. 273–280.

Guo, A., Li, Z., Li, H., Ou, J., 2009. Experimental and analytical study on pounding reduction of base-isolated highway bridges using MR dampers. Earthq. Eng. Struct. Dyn. 38, 1307–1333.

He, W.-L., Agrawal, A.K., 2008. Analytical model of ground motion pulses for the design and assessment of seismic protective systems. J. Struct. Eng. 134, 1177–1188.

Heo, G., Kim, C., Lee, C., 2014. Experimental test of asymmetrical cable-stayed bridges using MR-damper for vibration control. Soil Dyn. Earthq. Eng. 57, 78–85.

Jung, H., Jang, D., Choi, K., Cho, S., 2009. Vibration mitigation of highway isolated bridge using MR damper-based smart passive control system employing an electromagnetic induction part. Struct. Control Health Monit. 16, 613–625.

Kalantari, A., Amjadian, M., 2010. An approximate method for dynamic analysis of skewed highway bridges with continuous rigid deck. Eng. Struct. 32, 2850–2860.

Kawashima, K., 2012. Damage of bridges due to the 2011 Great East Japan earthquake. Proceedings of the International Symposium on Engineering Lessons Learned from the 2011 Great East Japan Earthquake, Tokyo, Japan, March 1–4, pp. 82–101.

Kawashima, K., Unjoh, S., 1994. Seismic response control of bridges by variable dampers. J. Struct. Eng. 120, 2583–2601.

Kelly, J.M., 1997. Earthquake-Resistant Design with Rubber. Springer London, London.

Kelly, J.M., Konstantinidis, D., 2011. Mechanics of Rubber Bearings for Seismic and Vibration Isolation. John Wiley & Sons, Chichester, UK.

Kelly, J.M., Buckle, I.G., Tsai, H.C., 1986. Earthquake Simulator Testing of a Base Isolated Bridge Deck. UCB/EERC-85/09.

Konstantinidis, D., Makris, N., Kelly, J.M., 2012. Health monitoring of fluid dampers for vibration control of structures: Experimental investigation. Earthq. Eng. Struct. Dyn. 41, 1813–1829.

Kunde, M.C., Jangid, R.S., 2006. Effects of pier and deck flexibility on the seismic response of isolated bridges. J. Bridg. Eng. 11, 109–121.

Loh, C.-H., Chang, C.-M., 2006. Vibration control assessment of ASCE benchmark model of cable-stayed bridge. Struct. Control Heal. Monit. 13, 825–848.

LORD Corporation Company, 1924. Cary, NC, US, Site: http://www.lord.com/ (accessed 22.10.14.).

Madhekar, S.N., Jangid, R.S., 2011. Seismic performance of benchmark highway bridge installed with piezoelectric friction dampers. IES J. Part A Civ. Struct. Eng. 4, 191–212.

Makris, N., Zhang, J., 2002. Structural Characterization and Seismic Response Analysis of a Highway Overcrossing Equipped with Elastomeric Bearings and Fluid Dampers: A Casc Study. PEER 2002/17.

Mokha, A., Constantinou, M.C., Reinhorn, A.M., Zayas, V.A., 1991. Experimental study of friction-pendulum isolation system. J. Struct. Eng. 117, 1201–1217.

Naeim, F., Kelly, J.M., 1999. Design of Seismic Isolated Structures: From Theory to Practice. John Wiley & Sons, New York, US.

Nagarajaiah, S., Narasimhan, S., Agrawal, A., Tan, P., 2009. Benchmark structural control problem for a seismically excited highway bridge—part III: phase II Sample controller for the fully base-isolated case. Struct. Control Heal. Monit. 16, 549–563.

Neff Patten, W., Sun, J., Li, G., Kuehn, J., Song, G., 1999. Field test of an intelligent stiffener for bridges at the I-35 Walnut Creek Bridge. Earthq. Eng. Struct. Dyn. 28, 109–126.

Ok, S.-Y., Kim, D.-S., Park, K.-S., Koh, H.-M., 2007. Semi-active fuzzy control of cable-stayed bridges using magneto-rheological dampers. Eng. Struct. 29, 776–788.

Olsson, H., Åström, K.J., Canudas de Wit, C., Gäfvert, M., Lischinsky, P., 1998. Friction models and friction compensation. Eur. J. Control 4, 176–195.

Pall, A.S., Marsh, C., 1982. Seismic response of friction damped braced frames. J. Struct. Div. 108, 1313–1323.

Phillips, B.M., Chae, Y., Jiang, Z., Spencer, B.F., Ricles, J.M., Christenson, R., Dyke, S.J., Agrawal, A.K., 2010. Real-time hybrid simulation benchmark study with a large-scale MR damper. In: Proceeding of 5th World Conference on Structural Control and Monitoring, Tokyo, Japan, July 12-14, Paper No. 10335.

Pradono, M.H., Iemura, H., Igarashi, A., 2009. Passively controlled MR damper in the benchmark structural control problem for seismically excited highway bridge. Struct. Control Health Monit. 16, 626–638.

Reinhorn, A.M., Li, C., Constantinou, M.C., 1995. Experimental and Analytical Investigation of Seismic Retrofit of Structures with Supplemental Damping. Part 1: Fluid Viscous Damping Devices. NCEER-95-0001.

Robinson, W.H., 1975. Cyclic shear energy absorber. US Patent No. 4117637.

Robinson, W.H., 1982. Lead-rubber hysteretic bearings suitable for protecting structures during earthquakes. Earthq. Eng. Struct. Dyn. 10, 593–604.

Sadek, F., Mohraz, B., Taylor, A., Chung, R., 1996. Passive Energy Dissipation Devices for Seismic Applications. NISTIR 5923.

Sahasrabudhe, S.S., Nagarajaiah, S., 2005. Semi-active control of sliding isolated bridges using MR dampers: an experimental and numerical study. Earthq. Eng. Struct. Dyn. 34, 965–983.

Sarrazin, M., Moroni, O., Neira, C., Venegas, B., 2013. Performance of bridges with seismic isolation bearings during the Maule earthquake. Chile. Soil Dyn. Earthq. Eng. 47, 117–131.

Shen, J., Tsai, M.-H., Chang, K.-C., Lee, G.C., 2004. Performance of a seismically isolated bridge under near-fault earthquake ground motions. J. Struct. Eng. 130, 861–868.

Soneji, B.B., Jangid, R.S., 2008. Influence of soil–structure interaction on the response of seismically isolated cable-stayed bridge. Soil Dyn. Earthq. Eng. 28, 245–257.

Spencer, J.B.F., Nagarajaiah, S., 2003. State of the art of structural control. J. Struct. Eng. 129, 845–856.

Spencer, B.F., Dyke, S.J., Sain, M.K., Carlson, J.D., 1997. Phenomenological model for magnetorheological dampers. J. Eng. Mech. 123, 230–238.

Stehmeyer, E.H., Rizos, D.C., 2008. Considering dynamic soil structure interaction (SSI) effects on seismic isolation retrofit efficiency and the importance of natural frequency ratio. Soil Dyn. Earthq. Eng. 28, 468–479.

Tan, P., Agrawal, A.K., 2009. Benchmark structural control problem for a seismically excited highway bridge-part II: phase I sample control designs. Struct. Control Heal. Monit. 16, 530–548.

Taylor Devices Company, 1956, North Tonawanda, NY, US, Site: http://taylordevices.com/ (accessed 16.06.14.).

Tongaonkar, N.P., Jangid, R.S., 2003. Seismic response of isolated bridges with soil-structure interaction. Soil Dyn. Earthq. Eng. 23, 287–302.

Wieczorek, N., Gerasch, W.-J., Rolfes, R., Kammerer, H., 2014. Semiactive friction damper for lightweight pedestrian bridges. J. Struct. Eng. 140, Paper No. 04013102.

Xu, Z., Agrawal, A.K., Yang, J.N., 2006. Semi-active and passive control of the phase I linear base-isolated benchmark building model. Struct. Control Heal. Monit. 13, 626–648.

Yang, G., Spencer, B.F., Carlson, J.D., Sain, M.K., 2002. Large-scale MR fluid dampers: modeling and dynamic performance considerations. J. Eng. Struct. 24, 309–323.

Yang, G., Spencer, B.F., Jung, H., Carlson, J., 2004. Dynamic modeling of large-scale magnetorheological damper systems for civil engineering applications. J. Eng. Mech. 130, 1107–1114.

Yang, M.-G., Chen, Z.-Q., Hua, X.-G., 2011. An experimental study on using MR damper to mitigate longitudinal seismic response of a suspension bridge. Soil Dyn. Earthq. Eng. 31, 1171–1181.

Yongqi, C., Liangzhe, M., Tiezhu, C., Schneider, R., Winters, C., 2008. Shock control of bridges in china using taylor devices' fluid viscous dampers. In: Proceeding of 14th World Conference on Earthquake Engineering, Beijing, China, October 12–17.

Yoshida, J., Abe, M., Fujino, Y., Watanabe, H., 2004. Three-dimensional finite-element analysis of high damping rubber bearings. J. Eng. Mech. 130, 607–620.

Yoshioka, H., Ramallo, J.C., Spencer, B.F., 2002. "Smart" base isolation strategies employing magnetorheological dampers. J. Eng. Mech. 128, 540–551.

Cables

21

Caetano E.
Laboratory of Vibrations and Monitoring ViBest, Faculty of Engineering
of the University of Porto, Porto, Portugal

1 Introduction

In the context of the design of cable stayed and suspension bridges, the composition and mechanical characteristics of cables are described, including reference to related components (namely, guides and pipes, anchorages and vibration mitigation devices). Recent developments based on the use of polymeric materials are addressed, and listings of some of the most relevant realizations in terms of suspension and cable stayed bridges are presented. Focusing on these structures, some design bases and requirements are presented, as well as methodologies to assess the static and dynamic behavior, characterize vibrations and design devices for vibration mitigation.

2 Cable components

2.1 Tension members

Although chains and bars have been used to form the cables of early cable-supported bridges, modern cables are made from steel wires, typically of cylindrical shape, and with a diameter of 3–7 mm. These wires are arranged in strands and ropes. Strands can be formed from the parallel or else from the helical assembling of wires. The parallel arrangement is typical of the main cables of suspension bridges (Figure 21.1A), while the helical arrangement is normally employed in hangers and stay cables. The simplest and most common is the seven-wire strand made from the helical winding of six 5 mm wires around a core wire (Figure 21.1B), with a nominal diameter of 15 mm. Modern stay cables are frequently formed from bundles of such strands, reaching diameters close to 470 mm. Alternatively, cables can be formed from multiwire helical strands. These so-called spiral strands result from the spinning of various layers of wires around a core center (Figure 21.1C) and reach diameters of 150–170 mm. The former cable stayed bridges of modern times employed fully locked coil strands. These strands are made from the assembling of wires with different shapes: a core helical strand; eventually, one or more layers of wedge shaped wires; and a number of outer layers formed by z-shaped wires arranged helically (Figure 21.1D). The maximum diameter of fully locked coil cables available in the market is of the order of 180 mm.

Due to a higher carbon content in the composition when compared with structural steels (see Table 21.1), the steel wire strength is significantly high, reaching values of 1570–1860 MPa. For very long spans, strengths of 1860–1960 MPa have been

Innovative Bridge Design Handbook. http://dx.doi.org/10.1016/B978-0-12-800058-8.00021-9

attained (Hauge and Andersen, 2011). However, the ductility of the steel wire is lower than that of the structural steel. According to Table 21.1, the strain at breaking of the steel wire is one-sixth of that corresponding to a mild steel. Regarding the modulus of elasticity, a common value for the 5–7 mm wires is 205 GPa. The simple helical and the spiral strand have lower elasticity modulus, with current values of 190 GPa and 170 GPa, respectively. For locked coil cables, the elasticity modulus is normally 180 GPa (Gimsing and Georgakis, 2012).

Despite the lower-elasticity modulus, helical strands have the advantage of the self-compacting with the tensioning, not requiring wrapping, on the contrary to parallel wire strands. Furthermore, helical strands have null elongation when reeling, due to the alternate position of the wires in the compression and tension areas. Therefore, these strands can be prefabricated in very long lengths. Parallel wire strands are normally fabricated on site by aerial spinning.

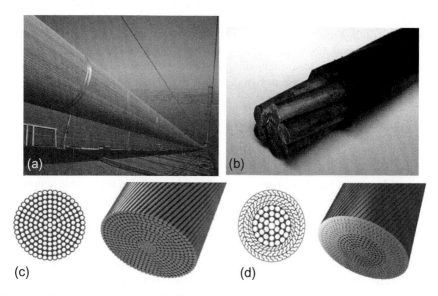

Figure 21.1 Types of strands: (a) Parallel-wire (copyright) NIPPON STEEL & SUMITOMO METAL CORPORATION; (b) Seven-wire strand; (c) Multiwire helical strand (Bridon catalog); (d) Full locked-coil strand (Bridon catalog).

The z-shaped wires of locked coil cables present a strength of the order of 1370–1570 MPa (EN 1993-1-11, 2006), slightly lower than that of circular wires. However, the tensioning of these wires leads to a higher degree of compaction than that of spiral strands (of the order of 15%–20%), resulting in void ratios of the order of 10%. Clearly, this is an advantage from the point of view of wind excitation, as the exposed surface to wind is minimum. The locking of the inner spiral strand by the z-wires when stretched provides an additional barrier to corrosion (see section 2.2), although in fact cables employed in several early cable stayed bridges have been replaced earlier than expected (Gimsing, 1983). Finally, it is mentioned that locked coil cables need to be delivered entirely

Table 21.1 Comparison Between Cable Steel and Structural Steel (Based on Typical Values) (Gimsing and Georgakis, 2012)

	Unity	Conventional cable steel (5 or 7 mm Wires)	Structural Steel	
			Mild	High strength
Yield stress (=2% proof stress)	MPa	1180	240	690
Tensile strength	MPa	1570	370	790
Strain at breaking	%	4	24	
Modulus of elasticity	GPa	205	210	210
Typical Chemical composition				
	C	0.80%	0.20%	0.15%
	Si	0.20%	0.30%	0.25%
	Mn	0.60%		0.80%
	Cu	0.05%	0.20%	0.30%
	Ni	0.05%		0.80%
	Cr	0.05%	0.30%	0.50%
	P	0.03%	0.04%	0.03%
	S	0.02%	0.04%	0.03%

fabricated, including the sockets. The longest fabricated locked coil cable had a length of the order of 1250 m (Gimsing and Georgakis, 2012).

2.2 Protective systems

Most modern cable systems for suspension and cable stayed bridges are designed for an intended working life of 100 years or more. This requires that adequate protective systems against corrosion and fatigue are used. Despite the fact that different practices can be found worldwide, the most recent specifications impose two levels of barriers. Zinc coating of the wires, achieved by galvanization, is a first barrier level and is applied individually to all wires of parallel wire cables in a suspension bridge cable, or to all wires of stay cables in Europe and Japan, although this has not been a current practice in USA (TRB 20015, 2005). In suspension bridge cables, a second barrier is given by a zinc dust paste filling of voids of the compacted cable and is further complemented by the wrapping with a galvanized wire. A third barrier can be created by painting. In stay cables formed by bundles of spiral strands, the second barrier consists of the individual greasing and sheathing, and a third barrier is also used, consisting of the encasing of the strands in a high-density polyethylene (HDPE) pipe. This pipe has a role of protection of the bundle of stays from moisture and weather agents. In locked coil cables, the core helical wires are not galvanized and are locked by the wedge or z-wires, which are themselves galvanized. A filling of zinc dust paint may help protecting the inner wires, while the outer wires may be additionally painted or else made of stainless steel. It is remarked that the zinc coating achieved with galvanization is not stable, and a more recent coating combining zinc (95%) and aluminum (5%), the GALFAN, has been used.

It is further mentioned that grouting inside a steel or a polyethylene pipe has been one of the formerly used protective systems, by analogy with the post tensioning technique. However, it has been verified that voids in the concrete or cracks due to vibrations and long-term deformation may lead to the penetration of water and promote degradation of the wires from inside the cable. Moreover, cracks in the pipes due to the stresses generated during grouting have also been observed (TRB 20015, 2005). The nonreplaceability of the strands and the difficulty in accessing the wires' condition have made this technique less common in present days, in favor of the bundle of individually sheathed strands encased in an HDPE pipe.

A recent evolution of the parallel strand bundle consists in the compact cable developed for very long stay cables, as those of the Russky Bridge (FREYSSINET, 2010, 2012). These are made with individual sheaths of smaller diameter than usual to allow a denser allocation of the strands in the pipe, providing a wind load reduction of the order of 25% to 30%, but requiring special tools for installation. Coextrusion of a common sheath to the bundle of strands is another option offered by manufacturers (FREYSSINET, 2014a). An even more compact system can be achieved by removal of the individual sheath and use of a permanent dehumidification system to provide an equivalent protection against corrosion (see Figure 21.2). These systems have also been introduced in cables from suspension bridges, as the Akashi Kaikyo Bridge.

Figure 21.2 Compact stay cable with permanent dehumidification system Copyright VSL. (VSL, 2002).

2.3   Anchorages

Anchorages provide the means for the transfer of cable loads to the soil or to the attaching parts of the structure and should be designed in order to exhibit optimal performance in terms of the mechanical behavior and fatigue resistance. These are composed of two main components: the anchorage heads, which constitute an intermediate mechanical part and are formed by wedges and anchor blocks, designed to secure the strands and transmit their force to the structure or the soil; and a transition zone, where the strands fan out, eventually with the help of deviators and are guided to the anchor head by means of guide pipes (see Figure 21.3). These pipes may contain also sealing systems, in order to ensure protection against corrosion, as well as internal dampers, to preclude cable vibrations. In suspension bridges, cable splay chambers may have dehumidification systems. In stay cables, the guide deviators normally incorporate neoprene rings in the end, allowing for accommodating limited angular variation (±25 mrad, according to FIB (2005), or ± 20 mrad static and ± 10 mrad, according to PTI (2007). These neoprene rings contribute to limiting bending stresses.

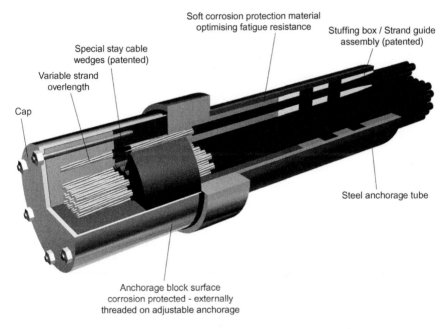

Figure 21.3 Typical components of stay cable anchorage (Freyssinet, 2010).

2.4 Vibration mitigation devices

The very low intrinsic damping of cables makes them vulnerable to vibrations. In order to prevent or mitigate such vibrations, several measures can be taken, involving an aerodynamic or structural approach.

Referring to the aerodynamic approach, one of the most common measures consists in the fabrication of the HDPE pipes with an helical wire whirling (see Figure 21.4A) or with indented protuberances (Figures 21.4B–C), which have been shown to disrupt the formation of rivulets associated with rain and wind vibrations (see section 3.3.6). Helical wire whirling has been first employed in the Normandy Bridge (Virlogeux, 1998) and is presently a measure adopted for most stay cables. In the Tatara Bridge, in Japan, a dimpled surface (Figure 21.4C) was adopted, which was shown to lead to lower drag forces by comparison with other alternative tested protuberances. This aspect is extremely important in very long stay cables, considering that the wind loads on the cables can exceed 50% of the overall wind loads (Virlogeux, 1998).

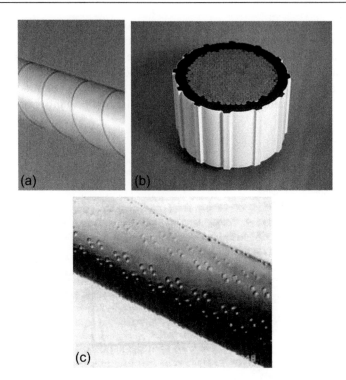

Figure 21.4 Examples of nonsmooth surfaces of cable coating to prevent rain-wind-induced vibration: (a) Helical wire whirling, Vasco da Gama Bridge; (b) Protuberated, Higashi-Kobe Bridge; (c) Dimpled, Tatara Bridge (Virlogeux, 1998).

The structural control of cable vibrations can be achieved both by the installation of interconnection ropes and by dampers installed close by the cables anchorages. Interconnecting ropes constitute the most evident form of attenuating cable vibrations in a cable stayed bridge and have been widely used both as temporary and permanent measures.

In terms of structural behavior, the addition of cross-ropes to the stay cable system creates intermediate supports at those elements and, consequently, increases their natural frequencies for vertical vibrations. Another effect of the installation of cross cables is an increase of the damping. A study by Yamaguchi (1995) showed that this increase is higher for soft secondary cables than for taut ties. However, the initial tension on these cables should have a sufficiently high value, so that under extreme effects the cross cables are not detensioned, producing shocks and causing damage of the tie devices, as reported by Virlogeux (1998) in the Farø and some of the Honshu-Shikoku bridges.

The installation of hydraulic or viscous dampers close to the stay cables' anchorages is the most efficient solution for suppressing cable vibrations. The damping capacity of external dampers is defined according to a specified requirement. In general, it is considered that viscous dampers have low maintenance costs, but show a dependence of damping characteristics with temperature and frequency, while hydraulic dampers have high maintenance costs and a complex adjustment (Bournand, 1999).

Another reported inconvenience associated with these mechanical devices is the lowering of aesthetical quality of the bridge. In order to overcome this aspect, internal ring dampers have been developed that are inserted in the deviator guide pipe of the cable. Different principles can be applied to activate damping, as exemplified in Figure 21.5 by the proposals of different manufacturers, which respect elastomeric, hydraulic, and friction dampers. Elastomeric devices are based on the shearing deformation of high damping rubber devices disposed as cylinders (Figure 21.5A) or as pads (Figure 21.5B), and are activated at low levels of vibration. These devices are adequate for small and medium-length cables. Hydraulic dampers are based on the shear motion of a viscous fluid inside a cylinder deposit activated by the moving cable or else on the compression of viscous fluid by a piston, in the case of the two configurations shown in Figure 21.5A. These devices are adequate for long-span cables and should be installed at a distance from the anchorage of the order of 0.015 L to 0.02 L, L being the chord length of the stay. Figure 21.6 shows the implementation of the internal hydraulic dampers at several cables of the Russky Bridge. Friction dampers, as shown in Figure 21.5B, are based on the mechanical friction activated by the cable vibration. Due to this principle, the activation of these devices only occurs for a certain amplitude of vibration. This can be both a benefit and an inconvenience from the points of view of damper durability and cable vibration limits, respectively.

The possibility to combine in the same cable two types of dampers can then be considered. Figure 21.7 shows an internal damper installed at one stay of Vasco da Gama Bridge, which combines a high damping rubber ring with a viscous damper.

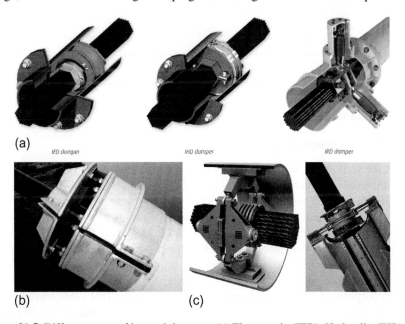

Figure 21.5 Different types of internal dampers: (a) Elastomeric (IED), Hydraulic (IHD) and Radial (IRD) dampers (FREYSSINET, 2014a, b, c)); (b) Elastomeric damper (Copyright VSL); Friction dampers (Copyright BBR; Copyright VSL). (VSL, 2002).

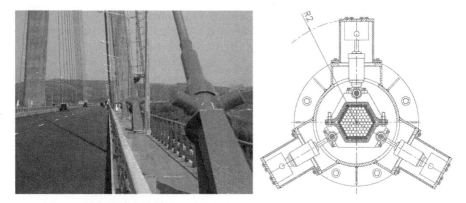

Figure 21.6 Internal radial dampers installed at the Russky Bridge (left) (FREYSSINET, 2014c).

Figure 21.7 Combined hydraulic and elastomeric internal dampers installed on a stay cable of Vasco da Gama Bridge.

2.5 Recent developments

Besides the technological developments related with cables made of steel, there has been continuous research using glass, aramid, or carbon fiber-reinforced polymers (TRB 20015, 2005). These materials use epoxy-based resins as a matrix for the composite and have the advantages of lighter weight, higher tensile strength, almost no thermal expansion, and high corrosion resistance. However, their implementation has not been made on a regular basis due to some particular difficulties, such as high cost and low shear strength, which limits the gripping capacity of anchorages and demands dedicated solutions. Despite this, some applications have been made for research purposes. In this context, reference is made to the first cable stayed bridge employing composite stays, the footbridge over the Gave de Pau River in Laroin, France. The structure has an only steel/concrete span of 110 m suspended by 16 carbon-fiber stays, while backstays are traditional bundles of steel strands (Geffroy, 2002). The composite stays, with lengths of 20–45 m, are made of pultruded carbon-fiber epoxy rods bundled in groups of seven encased in an HDPE pipe. Another case study is the Stork Bridge in Winterthur, Switzerland, a roadway bridge with two spans of 63 m and 61 m, where two of the stays are carbon composite (CFRP) and have been installed with a monitoring system for continuous assessment of the corresponding condition. In this case, each stay cable is made of a bundle of 241 carbon epoxy wires with 5 mm diameters. These wires have tensile strength of 3300 MPa, elastic modulus of 165 GPa, and density of 1.56 g/cm3 (Meier, 2012).

2.6   Major realizations

Lists containing information concerning major world realizations in suspension (Table 21.2) and cable stayed bridges (Table 21.3) are presented in this section. These include bridge spans, lengths, diameters and types of employed cables, and the country and year of construction.

Table 21.2 Major World Suspension Bridges: Main Cable Characteristics

Name	Location	Year	Span (m)	Total length (m)	Cable Characteristics
Akashi Kaikyo	Kobe-Naruto, Japan	1998	1991	3911	Diameter: 1.122 m; High-strength galvanized PPW*, UTS 1800 MPa; 5.23 mm diameter x 127 wires x 290 strands; strand length: 4071–4074 m
Xihoumen Bridge	Zheijiang, China	2009	1650	2588	Diameter: 0.870 m; High-strength galvanized PPW, UTS 1770 MPa; 127 wires x 1690 strands;
Great Belt East Bridge	Halsskov-Sprogoe, Denmark	1998	1624	2694	Diameter: 0.827 m; High-strength galvanized AS**, UTS 1770 MPa; 5.38 mm diameter x 504 wires x 37 strands
Yi Sun-sin Bridge	Yeosu, South Korea	2012	1545	2260	High-strength galvanized AS, UTS 1860 MPa; 5.35 mm diameter x400 wires x 32 strands

(Continued)

Table 21.2 Continued

Name	Location	Year	Span (m)	Total length (m)	Cable Characteristics
Runyang	Zhengjiang, China	2005	1490		Diameter: 0.900 m; High-strength galvanized PPW, UTS 1670 MPa; 5.3 mm diameter x127 wires x 184 strands; strand length: 2580 m
Humber	Hull Great Britain	1981	1410	2220	Diameter: 0.68 m; High-strength galvanized AS, UTS 1540 MPa; 5 mm diameter.x 149,948 wires

*PPW: Prefabricated parallel wire strand
**AS: Fabrication of cable by air spinning

Table 21.3 Major World Cable Stayed Bridges: Main Characteristics

Name	Location	Year	Span	Cable Type	Materials
Russky Bridge	Vladivostok, Russia	2012	1104	13–85 No. PWS* 56 internal dampers	Steel/concrete
Sutong Yangtze	Suzhou, China	2008	1088	PW** 7-mm wires UTS: 1770 MPa, added dampers	Steel/concrete
Stonecutters	Hong Kong, China	2009	1018	PWS 7-mm wires Longest cable: 540 m	Steel/concrete
Edong Bridge	Hubei, China	2010	926		Steel/concrete
Tatara	Onomichi-Imabari, Japan	1999	890	D = 170 mm; 7 mm diameter; PW349 wires UTS: 1770 MPa, Longest cable: 460 m	Steel
Pont de Normandie	Le Havre, France	1995	856	31 to 53 PWS Longest cable: 460 m	Steel/concrete

*PWS: Parallel wire strand
**PW: Parallel wire cable

3 Analysis and design

3.1 Loads and basis of design

The design of cables for cable stayed and suspension bridges is commonly based on the limit state verifications. This philosophy has been promoted in particular by the Eurocodes, which define actions, combinations of actions, and partial safety factors (EN 1990, 2002; EN1993-1-11, 2006). This section addresses the basic design concepts. Accordingly, the design of cables should be based on the verification of the following limit states:

- Ultimate limit state (ULS) for design tension
- Serviceability limit state (SLS) for stress and strain levels, for sag, and for amplitudes of vibration
- Fatigue limit state (FLS) for stress and stress variation due to traffic and wind loads

Considering the diverse actions and combinations of actions on cables and supporting members defined in the various parts of Eurocodes (EN1991-1-4, 2005; EN1991-1-5, 2003; EN1991-2, 2003; EN1993-3-1, 2006) and national standards and annexes, the following criteria should be satisfied:

- For the verification of cable stays in the ULS, two partial safety factors should be applied to the so-called guaranteed ultimate tensile strength (GUTS) of the cable, addressing (i) the condition of the structure (in-service or under construction) and the provisions taken in terms of the mechanical qualification tests; and (ii) construction imperfections. When precautions have been taken in order to limit bending stresses at the end of cable stays, the design tension F_{ULS} should satisfy the relation $F_{ULS} < 0.70\ F_{GUTS}$ for in-service conditions, or $F_{ULS} < 0.75\ F_{GUTS}$ for construction and accidental situations (SETRA, 2002).
- For the verifications in the SLS, the design tension F_{SLS} should satisfy the relation $F_{SLS} < 0.50\ F_{GUTS}$, when stay cables are tested for fatigue considering axial and bending effects (FIB, 2005). If bending effects are not addressed, the relation $F_{SLS} < 0.45\ F_{GUTS}$ should be satisfied. In addition to tension verification, vibration limit criteria should be checked in order to ensure users' comfort. Eurocode 3 (EN1993-1-11, 2006) proposes to limit the amplitude of stay cable vibration to L/500 (L being the chord length) for a moderate wind velocity of 15 m/s.
- For verification of stays in the FLS, if the cable systems adopted are qualified by mechanical tests defined in recommendations and standards (SETRA, 2002) and provisions are adopted to prevent bending effects close to the anchorages (namely, by adopting cable guide systems and limiting angular deformations and vibrations), it is sufficient to verify that the so-called stress demand (determined as the interval of variation of stress associated with fatigue loads) is below the fatigue-limit truncation point of the cable stay system.

3.2 Structural analysis

Cables are characterized by significant geometrical nonlinear behavior. A precise idealization of a cable suspended between two fixed points (Figure 21.8) should include the bending and axial deformation. It should also take into consideration the applied axial tension T_0, the self-weight, and the end conditions. The complexity of this problem is further enhanced by the difficulty of doing a rigorous assessment of the degree of restraint of rotations at anchorages.

In the modeling of the structural behavior of a cable, some simplifications are possible, however, which enable a still accurate and simple determination of the cable profile and tension.

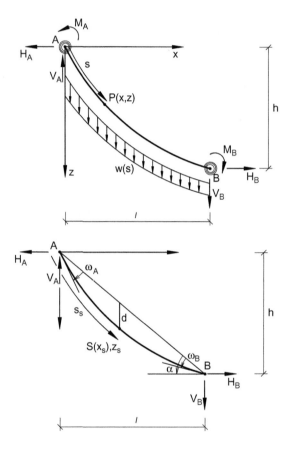

Figure 21.8 Suspended cable subject to self-weight and axial tension. Cable deformation parameters.

3.2.1 Static behavior

As a basic assumption, a suspended cable is considered a perfectly flexible elastic structural element. In this condition, the Cartesian coordinates x and z of a generic point P (see Figure 21.8) are defined as a function of the unstrained length s associated with the cable segment AP as (Irvine, 1981)

$$x(s) = \frac{H_A s}{EA_0} + \frac{H_A L_0}{W} \cdot \left[\sinh^{-1}\left(\frac{V_A}{H_A}\right) - \sinh^{-1}\left(\frac{V_A - Ws/L_0}{H_A}\right) \right] \tag{1}$$

$$z(s) = \frac{Ws}{EA_0}\left(\frac{V_A}{W} - \frac{s}{2L_0}\right) + \frac{H_A L_0}{W}\left\{\left[1 + \left(\frac{V_A}{H_A}\right)^2\right]^{1/2} - \left[1 + \left(\frac{V_A - Ws/L_0}{H_A}\right)^2\right]^{1/2}\right\}.$$

(2)

These coordinates depend on the reactions at the end A, V_A, and H_A, on the cable weight $W = mgL_0$, on the unstrained length L_0, and on the axial stiffness EA_0 (A_0 being the area of the undeformed cable cross section and E being the elasticity modulus of the cable).

The transcendental equations [Eqs. (1) and (2)] of the cable profile define the so-called elastic catenary and constitute the most precise description of the cable geometry under self-weight. The resolution of these equations requires the knowledge of the reactions H_A and V_A, which are obtained by introduction of the boundary conditions $[x(L_0) = \ell; z(L_0) = h]$.

The approximation of the catenary profile by an elastic parabola applies to shallow cables (i.e., cables with a small sag-to-span d/L ratio, typically no greater than 1:8). This range covers stays from cable stayed bridges and most of the cables from suspension bridges. The assumption of a unit ratio between the deformed and undeformed cable length yields the simple equations for the cable profile defined in Cartesian coordinates:

$$z(x) = \frac{1}{2}\frac{mg}{H} \cdot \sec\alpha \cdot x \cdot (\ell - x) + \frac{h}{\ell} \cdot x,$$

(3)

where the quantity $T = H \cdot \sec\alpha$ represents the cable tension at the section whose tangent is parallel to the chord.

The parabolic approach provides significant error in the description of the static behavior of the cable in local quantities, like sag and the angles of deviation to the chord at the anchorages. This error increases with both the angle of inclination of the cable to the horizontal and the chord length. Although very practical and useful for an approximate analysis during design phase, the parabolic approach is not convenient for design and installation purposes.

3.2.2 Dynamic behavior

The dynamic behavior of a cable integrated both in a suspension or in a cable stayed bridge can generally be described by the linear theory of vibrations derived by Irvine and Caughey (1974) for shallow cables. According to this theory, the circular frequencies ω_n and the corresponding transversal modal components $v_n(x)$ of out-of-plane vibration modes of a horizontal cable with chord length ℓ and horizontal component of the tension H can be obtained from

$$\omega_n = \frac{n\pi}{\ell} \cdot \sqrt{\frac{H}{m}} \quad n = 1, 2, 3, \ldots$$

$$v_n(x) = A_n \cdot \sin\left(\frac{n\pi x}{\ell}\right) \quad n = 1, 2, 3, \ldots$$

(4)

where A_n is an arbitrary constant.

In-plane vibration modes are characterized as *symmetric* and *antisymmetric*, according to the profile of the transversal component of motion. The circular frequencies ω_n and the in-plane transversal components w_n of the antisymmetric modes are given by

$$\omega_n = \frac{2n\pi}{\ell} \cdot \sqrt{\frac{H}{m}} \quad n = 1, 2, 3, \ldots$$

$$w_n(x) = A_n \cdot \sin\left(\frac{2n\pi x}{\ell}\right) \quad n = 1, 2, 3, \ldots . \tag{5}$$

For the symmetric in-plane modes, the circular frequencies are extracted from the solution of

$$\tan\frac{\overline{\omega}}{2} = \frac{\overline{\omega}}{2} - \frac{4}{\lambda^2}\left(\frac{\overline{\omega}}{2}\right)^3, \tag{6}$$

where $\overline{\omega}$ is the adimensional natural frequency given by $\overline{\omega} = \omega\ell/(H/m)^{1/2}$, and λ^2 is the Irvine parameter (Irvine, 1981). This parameter incorporates both the geometric and deformational characteristics of the cables and is defined as

$$\lambda^2 = \left(\frac{mg\ell}{H}\right)^2 \cdot \frac{\ell}{\dfrac{HL_e}{EA_0}}, \tag{7}$$

where L_e is a virtual length of cable defined by

$$L_e = \int_0^\ell \left(\frac{ds}{dx}\right)^3 dx \approx \ell \cdot \left\{1 + 8\left(\frac{d}{\ell}\right)^2\right\}. \tag{8}$$

The in-plane modal shape transversal components w_n associated with these frequencies are defined by

$$w_n(x) = A_n \cdot \left(1 - \tan\frac{\overline{\omega}}{2}\sin\frac{\overline{\omega}x}{\ell} - \cos\frac{\overline{\omega}x}{\ell}\right) \tag{9}$$

The sole dependence of the natural frequencies of *in-plane symmetric modes* on the Irvine parameter λ^2 illustrates the importance of λ^2 as an intrinsic characteristic of the cable.

Typical values of λ^2 attained by stay cables vary in the range of 0–1, while for suspension bridges, λ^2 is normally greater than 100. Very large stay cables can have a λ^2 value greater than 1. Small values of λ^2 reflect relatively highly stressed and low-sagged cables, whose deformation is achieved essentially by extensibility, while large values are typical of very low-tensioned and higher-sagged cables, whose deformation is mainly of a geometric nature, exhibiting therefore a relative inextensibility.

According to the representation in Figure 21.9 of the variation of the quantity $\overline{\omega}_n/\pi$ with λ^2 for the first three vibration modes, the transition from the dynamics of a taut string ($\lambda^2 = 0$) to that of an inextensible cable ($\lambda^2 = \infty$) is marked by a shift of almost 2π in the value of the adimensional frequency $\overline{\omega}_n$ of symmetric modes. This transition is evidenced by the occurrence of the so-called crossovers, corresponding to values of λ^2 beyond which the natural frequency of the symmetric modes becomes higher than the natural frequency of the same order antisymmetric modes.

Considering the typical values of λ^2 for stay cables, it can be concluded from Figure 21.9 that the corresponding natural frequencies lie outside the transition region

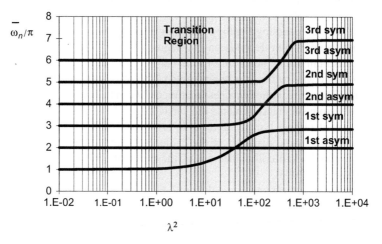

Figure 21.9 Variation of the natural frequencies of the first three symmetric and antisymmetric modes of vibration with λ^2.

and can be obtained using the equations for taut strings. On the contrary, the cables from suspension bridges are generally in the transition range; consequently, their vibration characteristics should consider both the cable geometry and elasticity.

The linearized theory of cable vibration that is on the basis of Eqs. (4), (5), and (6) for the circular frequencies of a vibrating cable does not account for the sag and bending stiffness of these members. In practice, the bending stiffness EI of a cable modifies its dynamic behavior. This effect is more pronounced for short cables, whose sag can be neglected. A simplified formula for the circular frequencies ω_n of a cable clamped at both ends with nonnegligible bending stiffness is given by (Morse and Ingard, 1968)

$$\omega_n = \frac{n\pi}{\ell} \cdot \sqrt{\frac{H}{m}} \cdot \left[1 + 2\sqrt{\frac{EI}{H\ell^2}} + \left(4 + \frac{n\pi^2}{2}\right) \cdot \frac{EI}{H\ell^2}\right]. \tag{10}$$

This expression is valid so long as the value of $EI/H\ell^2$ is small. The relative deviation ε_{EI}^n to the vibrating chord theory of the natural frequencies of a taut cable characterized by a stiffness EI is then given by

$$\varepsilon_{EI}^n = \frac{2}{\zeta} + \frac{\left(4 + \dfrac{n\pi^2}{2}\right)}{\zeta^2}, \tag{11}$$

where $\zeta = \sqrt{H\ell^2/EI}$. This deviation increases with the order of the mode shape. Considering as negligible differences associated with a value of ε_{EI}^n lower than 5% for the first five modes, it can be concluded that bending stiffness effects are negligible for stay cables with $\zeta \geq 50$.

The difficulty in assessing bending stiffness effects lies in the evaluation of the inertia of the cable, which depends on the degree of constraint of the strands. This degree of constraint varies according to the cable type and protection (locked-coil, stranded, parallel wire, with or without grout), and also with the cable length and the curvature. Therefore, although an estimation of EI can be obtained from laboratory tests, it is only from site measurements that an average EI can be assessed. According to Yamagiwa et al. (1997), typical values of EI are around 50%–70% of the stiffness of a solid bar with the same diameter of the cable. Reported values on a cable stayed bridge employing locked coil cables are of the order of 65% to 85% that stiffness (Geier and Wenzel, 2003).

The inclusion of sag effects in the dynamic behavior of the cable is important for long-span cables. In this context, mention is made of the simplified formulas derived by Mehrabi and Tabatabai (1998), which are valid for cables where ζ is no less than 50 and for Irvine parameters λ^2 of less than 3.1 (these authors state that this situation is covered by 95% of the stay cables from cable stayed bridges around the world). Accordingly, the nth order circular frequency of a cable ω_n is given by

$$\omega_n = \frac{\pi n}{\ell} \cdot \sqrt{\frac{H}{m}} \cdot \left(\alpha\beta_n - 0.24\frac{\mu}{\zeta}\right), \tag{12}$$

with

$$\alpha = 1 + 0.039\mu; \beta_n = 1 + \frac{2}{\zeta} + \frac{\left(4 + \dfrac{n\pi^2}{2}\right)}{\zeta^2}$$

where $\mu = \lambda^2, n = 1; \mu = 0, n > 1$ (for in-plane modes); and $\mu = 0$ (for out-of-plane modes)

The importance of considering sag and bending effects when analyzing the dynamic behavior of a cable lies in the fact that the vibration method can be used for identification of the installed tension in constructed structures (Caetano et al., 2013). This requires, therefore, an accurate identification of the frequencies and a valid description of their relation with the installed force.

3.2.3 Numerical modeling

The numerical modeling of cables integrated in suspension and cable stayed bridges can incorporate different levels of simplification. For stay cables, the simplest and also the most common approach consists on the idealization of each cable using the so-called truss element. This is a two-node elastic finite element characterized by no bending stiffness and an axial stiffness EA_0/L, whose weight is concentrated at the nodes (see Figure 21.10). These characteristics correspond to the treatment of the cable as a spring element, not accounting for geometric effects and providing a poor description of the local deformational characteristics: both the sag and angles of deviation at anchorages have null values, the cable's undeformed length is equal to the chord length, and the tension is assumed to be constant along the cable.

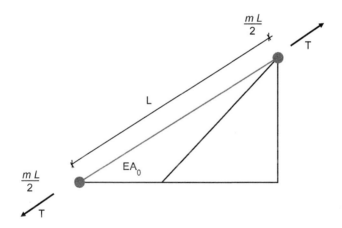

Figure 21.10 Truss element.

Despite the high level of simplification, the linear model is of great interest for a global analysis of the bridge, allowing a good estimation of the force distribution in the cable stayed bridge, and therefore providing important information for the design of the stay cables. The major source of error associated with the linear model results from geometric effects. So, for taut stay cables with a low λ^2 value, small errors are expected, while for less tensioned or very long cables, with high values of λ^2, the errors may be significant.

It is possible to introduce in a simplified form the nonlinear geometric behavior through the use of an equivalent modulus of elasticity E_{eq} incorporating the cable stress condition according to (Ernst, 1965)

$$E_{eq} = \frac{E}{1 + \dfrac{\gamma^2 \ell^2}{12\sigma^3}E} = \frac{E}{1 + \dfrac{\lambda^2}{12}}, \tag{13}$$

where γ is the specific weight and σ is the tensile stress of the cable. The variation of E_{eq} with λ^2 is represented in Figure 21.11, showing that for standard taut stay cables ($\lambda^2 < 1$), the introduced correction is very small ($\lambda^2 = 1, E_{eq} = 0.92E$), while for very long stay cables, it becomes significant (for the largest of the Normandy cables, $\lambda^2 = 3.1, E_{eq} = 0.79E$).

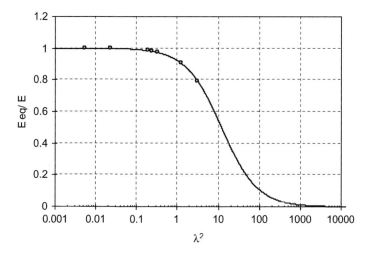

Figure 21.11 Variation of the ratio E_{eq}/E with λ^2.

A natural extension of the idealization of the stay cable as a simple truss element to a series of truss elements (Figure 21.12) has been proposed by Liu (1982) as a computational improvement that allows for the accounting of geometric effects, so long as the discretization is complemented by a geometric nonlinear analysis. Due to the resulting large dimension of numerical models, and to computational limitations, the implementation of this modeling technique has not been a current trend in the global modeling of a cable stayed bridge. It should be noticed, however, that currently available commercial software and computer memory allow for reasonable computing times in face of the advantages obtained: using an adequate number of elements to discretize a stay cable, the corresponding weight, applied at the nodes, approximates the distributed weight of the cable, and the resulting profile approximates the elastic catenary profile. As for the number of necessary elements to represent adequately the deformational and vibrational behavior of a stay cable, it has been shown (Caetano, 2007) that a discretization of a short and a long cable in 20 truss elements provide relative errors of less than 5% in the parameters that characterize the cable deformation and vibration. A lower number of truss elements (10 per cable) can be used when local parameters, like rotations, and only the first three vibration modes are of interest.

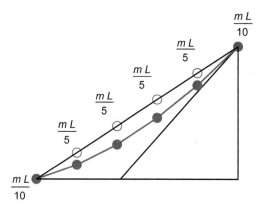

$$\frac{m\,L}{10}$$ **Figure 21.12** Multilink approach: Undeformed and deformed mesh under self-weight.

The multilink approach is the most adequate to model suspension bridge cables. However, given that suspension cables have low stress, convergence problems may occur in the determination of the deformation under self-weight. The addition of bending stiffness to the cable by replacing truss elements by beam elements can mitigate this difficulty, with the advantage of representing the actual behavior of cables.

3.3 Cable vibrations and damping

3.3.1 Cable damping

Suspension and cable stayed bridges typically exhibit a very low amount of structural damping. This damping decreases with the span length. Lower-limit values of the structural damping ratio for suspension and cable stayed bridge vibration modes driven by the cables are systematized in Table 21.4 based on the experience achieved with the construction of the bridges of the Honshu-Shikoku Project (Fujino et al, 2012).

Individual cables, typically from cable stayed bridges, and hangers from suspension bridges, exhibit an intrinsic damping ξ of 0.05%–0.5% (Tabatabai and Mehrabi, 2000). This interval covers most of the values indicated by other authors

Table 21.4 Limit Lower Structural Damping ξ of Cable Vibration Modes in Suspension and Cable Stayed Bridges (Fujino et al., 2012)

Bridge Type	Deck Type	Vertical Modes	Torsion Modes
Suspension	Truss	0.6%	0.3% ÷ 0.6%
	Box	0.3%	0.3%
Cable Stayed	Truss	1.1%	1.1%
	Box	0.3%	0.3%

based on their experience (Macdonald, 2001; Yamaguchi and Fujino, 1998; Caetano and Cunha, 2011), although it should be stressed that damping varies with the amplitude of vibration, and also that different methodologies in the corresponding assessment explain the wide dispersion of values found in the literature.

When actuated by wind, an additional aerodynamic damping appears due to the friction of the cable with the surrounding air. For relative displacement of the cable in the downwind direction, this damping, expressed in terms of the logarithmic decrement of the nth mode, can be determined from (SETRA, 2002)

$$\delta_a = \frac{\rho \pi U D C_D}{m \omega_n},\tag{14}$$

where ρ is the density of the air (1.23 kg/m^3 for standard temperature and pressure), U is the mean wind velocity, D is the outer diameter of the cable, C_D is the drag coefficient, m is the distributed mass of the cable and ω_n is the nth-order circular frequency of the cable.

In the crosswind direction, the aerodynamic damping of the cable is half that of the downwind direction.

The total damping of a cable δ_t is then given by the sum of the intrinsic damping with the aerodynamic damping:

$$\delta_t = 2\pi\xi + \delta_a.\tag{15}$$

Even though aerodynamic damping has the same order of magnitude of intrinsic damping for design wind velocities, the total damping of cables is generally low. The simultaneous flexibility of these members makes them vulnerable to vibrations induced by the wind and by traffic on a bridge. Even though the mechanisms behind cable vibration are not yet fully understood, some of the phenomena have been identified and characterized (namely, buffeting, vortex-shedding/lock-in, galloping, aerodynamic interference, rain-wind-induced vibration, dry galloping, and parametric excitation). The following sections will briefly describe the main characteristics of these phenomena. An additional section will focus on the design of vibration mitigation devices.

3.3.2 Buffeting

When immersed in a flow, the wind, a cable is subjected to surface pressures which, integrated along the section, represent the applied wind loads. These loads can be split into two parcels: one constant, associated with the mean wind velocity; and one varying with time, representing the turbulent component. The corresponding effects are also treated separately, the former parcel leading to average stresses and deformations and the latter being responsible for the vibrational component of the response. The turbulent component of the wind loads, which is the subject of the present section, varies directly with the mean wind velocity and with the intensity of turbulence (Caetano, 2007).

As an elastic system, the amplitude of the cable response increases with the growth of buffeting forces, and hence with the mean wind velocity. However, the growth of wind velocity also leads to an increase of the aerodynamic damping, as evidenced by Eq. (14). The significant increase of the aerodynamic damping at high wind velocities prevents the occurrence of important cable oscillations under buffeting loads in most situations. Therefore, even though the amplitudes of vibration caused by buffeting forces in cables under extreme winds should be assessed and limited in order to prevent local damage at the anchorages and shock effects in stabilizing cables (SETRA, 2002), no fatigue problems are posed.

3.3.3 Vortex-shedding/lock-in

In the presence of a circular cylinder with diameter D, a uniform wind flow detaches from the surface and generates a turbulent wake, characterized by alternate shedding of vortices at the top and bottom detachment points at a frequency f_v defined as a function of the Strouhal number St as follows:

$$f_v = \frac{U St}{D}. \tag{16}$$

The Strouhal number St depends upon the cross-section shape and is approximately constant with the Reynolds number Re for a wide range of values. For practical applications with circular cylinders, a constant value of St of 0.2 can be considered.

The shedding of vortices in the wake of the cylinder induces approximately sinusoidal excitation components that result in oscillations, normally characterized by very small amplitudes. If the frequency of vortex shedding approximates the frequency of the cable, a resonance effect takes place, which is designated as *vortex resonance*. An increased oscillation leads the cylinder to interact strongly with the flow and control the vortex-shedding mechanism for a certain range of variation of the wind velocity; i.e., an increase of the flow velocity by a few percentage points won't change the shedding frequency, which coincides with the cable natural frequency. This aeroelastic phenomenon is commonly known as *lock-in* or *synchronization* and originates additional across-wind loads.

According to Dyrbye and Hansen (1999), the risk of vortex-induced vibrations is higher when the flow is smooth, a situation that is typical of isolated structures located by the sea. It can also be high for structures located in the wake of slender nearby structures of similar size. On the contrary, large-scale turbulence reduces the aerodynamic damping.

The occurrence of violent vortex-induced vibrations depends on the intensity of large-scale turbulence, as well as on the intrinsic structural damping, which is characterized by the *Scruton number S_c*, a nondimensional parameter defined by

$$S_c = \frac{2\delta m_e}{\rho D^2}, \tag{17}$$

where δ is the logarithmic decrement of the structural damping and m_e is an equivalent mass per unit length (for a uniform cylinder, $m_e=m$).

High Scruton numbers reduce the risk of violent vortex-induced vibrations. According to Dyrbye and Hansen (1999), no risk of lock-in exists for S_c values greater than about 20. On the contrary, the risk of *lock-in* is very significant if S_c is less than 10.

Eq. (16) for the definition of the vortex-shedding frequency f_v can also be used to predict the so-called critical velocity at lock-in, U_{cr}, assuming a constant St of 0.2:

$$U_{cr} = 5f_v D \tag{18}$$

For the shortest stay at the Vasco da Gama Bridge, with a length of 35 m, a diameter of 160 mm and a fundamental frequency of 3 Hz, the critical velocity associated with the occurrence of lock-in in the first mode would be 2.4 m/s. This velocity is clearly too low to provide a significant energy input for the occurrence of cable oscillations. On the contrary, fixing a critical wind velocity of 12 m/s, relevant for deck vibrations, a shedding frequency of 15 Hz would be obtained, meaning the possible occurrence of vortex resonance of this cable in the fifth mode of vibration.

It is possible to estimate the amount of damping ξ required to avoid vortex-shedding vibrations, based on the practical rule of ensuring a S_c value greater than 20:

$$\xi \approx \frac{\delta}{2\pi} \geq \frac{6D^2}{\pi m} \tag{19}$$

The study by Tabatabai and Mehrabi (2000) centred on a database formed by all the stays of 16 cable stayed bridges has shown that a damping coefficient ξ of 0.7% would lead to Scruton numbers greater than 20 for 90% of the stay cables, and therefore to stable cables. The application of Eq. (19) to the cable of the Vasco da Gama Bridge (as mentioned previously) leads to a required damping coefficient of around 0.12%.

With regard to the characterization of the vortex-shedding phenomenon in terms of the definition of wind forces and evaluation of the response, no completely successful analytical method has been developed yet. Instead, several empirical analytical models have been developed to represent the vortex-induced response of bluff cylinders, whose parameters are obtained from experimental data (Simiu and Scanlan, 1996). This is, for example, the case of the formula obtained by Griffin et al. (1975), which gives the maximum relative amplitude of vibration y_0/D as

$$\frac{y_0}{D} = \frac{1.29}{\left[1 + 0.43\left(2\pi St^2 S_c\right)\right]^{3.35}} \tag{20}$$

Considering the usual values found for the S_c numbers for cables of cable stayed bridges, it becomes evident that the amplitudes of vibration are generally very small. Davenport (1994) states that the amplitude of vortex-shedding vibration rarely attains half the cable diameter.

In the case of the Vasco da Gama Bridge, where the measured logarithmic decrement damping of the shortest stays is about 0.0085 (without damping devices) and the S_c number is 23.7, the maximum amplitude of vortex-induced vibration would be 0.0183D (3 mm). Although irrelevant for vortex-induced vibration, damping devices were installed in all stays, leading to a S_c number of 233 for the shortest stays.

3.3.4   Galloping

Galloping is an instability phenomenon typical of slender structures with rectangular or "D" cross sections, which is characterized, in a similar manner to vortex-shedding, by oscillations transverse to the wind direction, that occur at frequencies close to some natural frequency of the structure. The phenomenon is, however, quite different from vortex-induced vibration. In effect, while the latter originates small amplitudes of oscillation in restricted ranges of wind velocity, galloping occurs for all wind speeds above a critical value and produces high-amplitude vibrations, which may be 10 times the typical body dimension, or even more.

The onset condition for the occurrence of galloping is given by the occurrence of a negative aerodynamic damping generated by the vibration of a cable in a wind flow. This is achieved when

$$\left(\frac{dC_L}{d\beta} + C_D\right)_{\beta=0} < 0, \tag{21}$$

where C_D and C_L are the drag and lift coefficients and β is the angle of attack of the wind flow. Eq. (21) is the so-called *Glauert-Den Hartog* criterion for incipient galloping instability (Simiu and Scanlan, 1996).

The analysis of Eq. (21) shows that circular cross sections are never subjected to instability by *galloping*, as the derivative $dC_L/d\beta$ is always null. So, except for the cases where the external shape has been altered, either by the presence of ice or of water rivulet (see section 3.3.6), instability of the cables by galloping should not be expected in cable stayed bridges employing circular cross sections for these elements. However the studies of Matsumoto et al. (1992) and Saito et al. (1994) have shown that galloping can occur for inclined circular cables. The reason presented is the appearance of an axial flow behind the cable for certain yawing angles, which favors instability. More recently, other aspects have been investigated, which will be discussed in section 3.3.7.

Based on experimental testing, some authors have developed formulas for the evaluation of the onset wind velocity of divergent oscillation, as a function of the Scruton number. Honda et al. (1995) proposed the following formula for the critical reduced velocity $\overline{U}_{cr}$:

$$\overline{U}_{cr} = \frac{U_{cr}}{fD} = 10 \cdot S_c^{2/3}, \tag{22}$$

where f is the cable natural frequency and U_{cr} is the critical wind velocity for the onset of divergent oscillation. Irwin (1997) proposed for the PTI Guide (2007) that

$$\overline{U}_{cr} = \frac{U_{cr}}{fD} = c \cdot \sqrt{S_{c0}} \tag{23}$$

where $S_{c0} = S_c/(4\pi)$ and the constant c is 40 for circular cross sections. Virlogeux (1998) referred to an identical formula but considered a constant c of 35. These authors have stressed, however, that the use of Eq. (23) should be cautious, as very conservative estimates are obtained. Figure 21.13 represents the variation of the critical wind-reduced velocity with S_c according to the three referred formulas. This figure also includes experimental values obtained by Cheng et al (2003) in laboratory tests and can be further used to define a minimum damping necessary to avoid galloping.

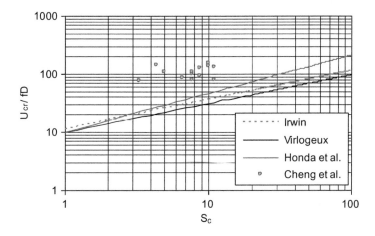

Figure 21.13 Critical reduced wind velocity versus Sc.

Other strategies for avoiding galloping are based on the modification of the cross section. Kubo et al. (2003) proposed a stranded configuration composed by a bundle of individual strands wrapped with a spiral strand. A similar solution was implemented at the hangers of the Akashi Kaikyo Bridge (Fujino et al., 2012). Matsumoto et al. (1995) proposed to introduce helical plates along the smooth surface of the cable. In both cases, the axial flow in the wake of the cable is disturbed by the presence of the strand and plates.

3.3.5 Aerodynamic interference

Aerodynamic interference occurs whenever a cable or a group of cables lies in the wake of other cables or structural elements. The perturbation introduced by the first obstacle "seen" by the wind affects the wind flow around the close obstacles, creating local turbulent conditions. The high flexibility of cables makes these elements extremely vulnerable to oscillations. These oscillations occur more easily in locations of low turbulence.

A typical situation of aerodynamic interference occurs with groups of parallel cables. These have been used in several cable stayed bridges, particularly in Japan, with the purpose of reducing the size of the cables. Multiple cables are distant from each other by only a few diameters and are anchored at the same level in the tower and deck. Figure 21.14 shows the twin stay cables employed on the Oresund Bridge, linking Denmark and Sweden. Figure 21.15 shows the possible arrangements of cables.

It has been observed that, under particular conditions, the cable assembly may undergo vibrations. These oscillations are due to *vortex resonance*, to *galloping* of the cable assembly, or to so-called *interference* or *wake galloping*. The later phenomenon is no more than a *galloping* that occurs on the downstream cable(s) induced by the turbulent wake of the upstream cable(s). The oscillation of the downstream cable(s) may induce also a perturbation of the flow around the upstream cable(s), generating oscillation of these cables, designated as *interference galloping*.

Figure 21.14 Multiple parallel stay cables at Oresund Bridge (Gimsing and Nissen, 1998).

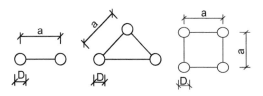

Figure 21.15 Possible arrangements of grouped stay cables.

The EC1 (EN 1991-1-4, 2005) proposes some conservative formulas for the evaluation of instability by wake effects. Matsumoto et al. (1989), Miyata (1991) and Tanaka (2003) define the main aspects of the phenomenon, concluding that instability can be found both for close-spaced ($D \leq x \leq 4D$ and $-2D \leq y \leq 2D$; see Figure 21.15), and largely spaced cables ($8D \leq y \leq 20D$), where the interference effects occur only for the downstream cable. Interference galloping was observed in several cable stayed bridges in Japan, such as the Hitsuishijima, Iwakurojima, Yobuko, and Shima Maruyama bridges (Narita and Yokoyama, 1991). For the particular case of the Yobuko Bridge, amplitudes of oscillation of $2.5\,D$ were reported. Kubo et al. (1994) and Kubo (1997) proposed the following measures to prevent interference effects: adopt a close spacing between cables in the range of $1.2–1.3\,D$, which proved to show no galloping at a reduced wind velocity $\overline{U} = U/fD$ of 200; connect parallel cables by spacers or stringers at lengths defined by a deflection of the connecting points no greater than $0.2\,D$.

It is still relevant to mention the interference effects observed in stranded cables. The use of bundles of individually protected strands clamped with collars at distances of 30–50 m is a technology introduced in 1988. Although currently constructed bridges normally employ an encasing of the strands in HDPE pipes, many bridges constructed in the late 1980s and beginning of the 1990s employ the former described technology and have suffered from diverse vibration problems. The most frequent vibrations are associated with interference phenomena of a similar type to the previously described ones, and occur due to the aerodynamic interaction between strands, which shock against each other, generating global vibration of the cable and producing a significant and disagreeable rattling noise. Virlogeux (1998) defines these movements as "breathing of strands." Vibrations are started by wind and attain significant amplitudes of around $1D–2D$. This breathing of the cables produces damage of the bracings, and so it may be necessary to encase the bundles in pipes. There is however the risk of "slapping" the pipes against the bundles, as the latter are not normally blocked inside the former. This problem has been reported at the Glebe Island Bridge in Australia (SETRA, 2002).

3.3.6   Rain-wind-induced vibration

Although rain-wind-induced vibration of power lines, designated as *rain vibration*, had been reported in the literature 10 years earlier (Hardy et al., 1975; Hardy and Bourdon, 1979), it was only in 1986 that Hikami identified the phenomenon of cable vibration in cable stayed bridges induced by the combined action of wind and rain, during the construction of the Meiko-Nishi Bridge (Hikami, 1986; Hikami and Shiraishi, 1988). The general characteristics identified by these authors were soon associated with several past and many subsequent occurrences of vibrations in cable stayed bridges. It is presently considered that rain-wind-induced oscillations are in the origin of about 95% of the reported vibration problems in cable stayed bridges (Wagner and Fuzier, 2003).

Despite the intense research developed both through wind tunnel testing and through observation of prototypes, the mechanisms of rain-wind-induced vibrations are yet to be fully understood. Some main aspects of this complex phenomenon can be outlined, however (Tanaka, 2003): first, it is under the combined action of rain and wind, at specific angles of attack and intensity of rainfall, that rivulets can be formed at the upper and lower surfaces of the cable (see Figure 21.16). The formation of these rivulets as the result of a balance between gravitational, aerodynamic, and surface capillarity forces leads to a loss of symmetry of the cable cross section, and therefore to a variation of aerodynamic forces on the cable. Eventually, a decrease in the drag coefficient and a negative slope of the lift coefficient associated with a small variation of the angle of attack may result in a negative aerodynamic damping, therefore in a galloping instability of Den Hartog type (see section 3.3.4). Once the cable starts oscillating, the rivulets tend to oscillate circumferentially with the same frequency. A coupling of this oscillation with the flexural oscillation of the cable may lead to aerodynamic instability, susceptible to intensification of the vibrations.

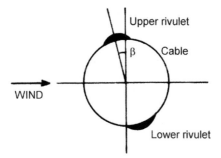

Figure 21.16 Formation of water rivulet at upper and lower surface of cable under rain and wind.

Research conducted all over the world (e.g., Matsumoto, 1998; Yamaguchi, 1990; Peil and Nahrath, 2003; Verwiebe, 1998) has further led to the identification of different excitation mechanisms behind the phenomenon of rain-wind-induced excitation. The complexity of the phenomenon is evident and can still be enhanced when considering other variables to the problem, like the adhesion property of the cable's coating material (Flamand, 1994) or the intensity of the rainfall (Ohshima and Nanjo, 1987; Main and Jones, 1999). Based upon the experience gathered from the observation of rain vibration in prototypes of cable stayed bridges and on wind tunnel tests, the main conditions and characteristics associated with the occurrence of the phenomenon can be summarized as follows (Hikami and Shiraishi, 1988; Matsumoto, 1998; Main and Jones, 1999; Sarkar et al., 1994; Tanaka, 2003):

- The wind speed varies in the range 5–20 m/s, the majority of reported cases lying in the interval 8–12 m/s, corresponding to reduced wind velocities U_{cr} ($U_{cr} = U/fD$) of 20–90.
- The wind direction varies in the range of 20°–60° to the longitudinal axis of the bridge.
- The cables are inclined to the horizontal of angles of 20°–45°.
- There should be rain (whether heavy, light or drizzle), although in most cases, moderate rain is preferred.

- The cable surface is smooth, such as polyethylene or painted metal cased cables.
- The cable diameter is in the range of 80–200 mm.
- Typical vibration frequencies are in the range 0.3–3 Hz.
- Typical amplitudes of vibration are about twice the cable diameter. However, amplitudes of 7 D have already been observed.
- The structural damping of the cables is very low (with logarithmic damping decrement less than 0.01).
- The cable is located behind the bridge pylon and declines in the direction of the wind.
- The cable stayed bridge is located in an area where the intensity of turbulence is low.
- The vibration orbit varies according to the intensity of the rainfall: for light rain and drizzle, vibration occurs essentially in the vertical plane, while for heavy rainfall, the orbit may exhibit significant two-dimensionality.

With respect to possible measures against rain-wind vibrations, two strategies can be followed, one based in the application of aerodynamic measures to the cable cross section, and the other based in the increment of damping through the addition of special devices.

As for the implementation of aerodynamic measures, and given that it seems conclusive that the motion of rivulets is an enhancing cause of oscillations, the adoption of nonsmooth surfaces has proven an adequate strategy. Protuberances, helical wire whirling, or a dimpled surface (Figure 21.4) have shown to disrupt the formation of rivulets.

With respect to the addition of damping devices and, in the absence of other study, the indication proposed by the *PTI Guide* (PTI, 2007) to ensure that the Scruton number S_{c0} (calculated as $S_{c0} = m\xi/\rho D^2$) is greater than 10 for avoiding rain-wind-induced vibrations can be employed as a practical rule. Given the very low intrinsic damping ratios of cables, the necessity to design dampers for the majority of cables seems evident.

3.3.7 Dry galloping

The observation of important cable vibrations with characteristics similar to those reported for rain-wind-induced vibration, but occurring under dry weather, has motivated further research of the phenomenon identified presently as dry galloping. Matsumoto (1998) discussed the high-speed vortex shedding associated with the formation of an axial flow in the wake of the cable. Other authors (e.g., Larose and Zan, 2001; Larose et al., 2003; Tanaka, 2003) noticed these vibrations occurred in the critical Reynolds number range. For smooth cylinders, this range is $2 \times 10^5 - 8 \times 10^5$ (SETRA, 2002), meaning that for a circular stay cable with a diameter of 0.20 m, the critical Reynolds range would occur for mean wind velocities of 20–60 m/s. Assuming the vibration of the cable in that particular range, a small increase of the mean wind velocity might create a sudden decay of the drag force, hence of the lift force over the cable, the result being a reduced motion. The approximation to the equilibrium position would cause an increase of the relative velocity of the flow and so a slight increase of the drag force and of cable vibration, with the consequence of reducing the relative wind velocity of the flow. An oscillation of the cable could then be created merely by slight fluctuation of wind flow. More recently, a bistable behavior of a cable has been identified (Nikitas et al., 2012; Matsumoto, 2014), characterized by random jumps of the drag and lift forces. These may be associated with the asymmetry created by the alternate detachment of air bubbles from the cylinder section and are enhanced by asymmetries of geometric nature as those resulting from the deformation of the cable pipes associated with fabrication tolerances or storage (Flamand et al., 2014).

3.3.8 Parametric excitation

The vibration of the deck and towers caused by wind, traffic, and earthquakes, produces an *indirect excitation* of the cables through motion of their anchorages. In certain circumstances, the induced cable vibrations attain very high amplitudes. Two phenomena can be identified under these circumstances, here designated as *external* and *parametric* excitation. The *external excitation* corresponds to an amplification of motion applied at some anchorage perpendicularly to the cable chord, while the *parametric excitation* corresponds to oscillations in the direction of the chord. The phenomena of *external* and *parametric excitation* have been observed in several bridges in the past (some examples are the Brotonne Bridge in France, Ben-Ahin and Wandre Bridges in Belgium, and the Annacis Bridge in Canada).

For simplicity, external and parametric excitation of stay cables are normally analyzed by separating these members from the bridge structure (see Figure 21.17). The amplification of the deck/tower motion due to a sinusoidal excitation applied at one end is then calculated in order to define the amount of damping necessary to prevent the cable system to undergo large vibrations.

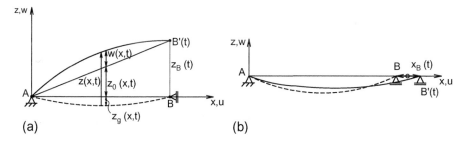

Figure 21.17 Stay cable subjected to harmonic (a) transverse and (b) longitudinal motion at one anchorage.

Using this approach and a series of simplifications (namely, neglecting the sag effect and the contribution of modes other than the resonant), the following equations have been obtained for the amplitude of oscillation of a stay cable with distributed mass m and chord length ℓ tensioned with a force T subjected to a harmonic excitation perpendicular to the chord with frequency ω and amplitude z_B at one end (Caetano, 2007)

$$w(x,t) = z_B \cdot \frac{x}{\ell} \cdot \sin(\omega t) + \alpha_1(t)\sin\frac{\pi x}{\ell} \tag{24}$$

with

$$\alpha_1(t) = a\sin(\omega_1 t - \gamma) + \frac{3\lambda}{2\pi\sqrt{X_0\ell}}a^2 \cdot \left[1 - \frac{1}{3}\sin(2\omega_1 t - 2\gamma)\right], \tag{25}$$

where ω_1 is the first circular frequency of the cable, given by the taut string formula

$$\omega_1 = \frac{\pi}{\ell}\sqrt{\frac{T}{m}}. \tag{26}$$

The amplitude of vibration a (that governs the total response) and the phase of the response γ in Eq. (25) are given as a function of the Irvine parameter λ^2 and the damping ratio of the first mode ξ_1 by

$$a = \frac{z_B}{\pi\xi_1} \cdot \frac{1 - \left(\frac{\lambda}{\pi}\right)^2}{\left[1 + \frac{1}{2}\left(\frac{2}{\pi}\right)^4\lambda^2\right]^{1/2}} \cdot \sin\gamma \tag{27}$$

and by the solution of

$$\sin^2\gamma \cdot \tan\gamma = \frac{32}{3} \cdot X_0\ell \cdot \frac{\left[1 + \frac{1}{2}\left(\frac{2}{\pi}\right)^4 \cdot \lambda^2\right]^4}{\left[1 - \left(\frac{\lambda}{\pi}\right)^2\right]^2 \cdot \left[1 - 32\frac{\lambda^2}{\pi^4}\right]} \cdot \frac{\xi_1^3}{z_B^2}. \tag{28}$$

External excitation is not an instability phenomenon; rather, it represents the amplification of a support oscillation which amplitude varies nonlinearly with the amplitude of the former. This oscillation is slightly dependent on the damping coefficient for very large amplitudes of vibration. This can be observed in Figure 21.18 for two cables with different characteristics: one from the Ben-Ahin Bridge in Belgium, with a length of 110 m and Irvine parameter $\lambda^2 = 0.0727$, and another from the Vasco da Gama Bridge, with a length of 226 m and $\lambda^2 = 0.4321$, for damping coefficients in the range 0.1%–1.5%.

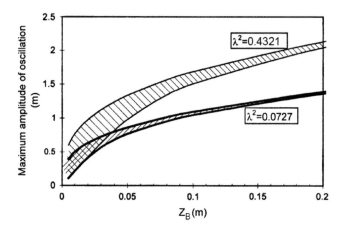

Figure 21.18 Variation of maximum steady-state amplitude of oscillation at primary resonance with amplitude of transversal support oscillation for damping coefficients within the range 0.1%–1.5%.

Now, considering a longitudinal harmonic oscillation of the support [Figure 21.17(b)] with amplitude x_B and frequency 2ω, the vertical vibration $w(x,t)$ of the stay cable is again determined by separation of variables considering the contribution of the first mode according to

$$w(x, t) = \alpha_1(t) \cdot \sin \frac{\pi x}{\ell}. \tag{29}$$

The coefficient $\alpha_1(t)$ is given by the solution of

$$\ddot{\alpha}_1(t) + 2\xi_1 \omega_1 \dot{\alpha}_1(t) + \omega_1^2 \cdot \left(1 + \frac{\lambda^2}{\pi^2} + \frac{x_B}{X_0} \sin 2\omega t\right) \cdot \alpha_1(t) +$$

$$+ 2\frac{\omega_1^2}{\pi\sqrt{X_0\ell}} \cdot \lambda \cdot \left(1 + \frac{\pi^2}{16}\right) \cdot \alpha_1^2(t) + \frac{\pi^2}{4} \cdot \frac{\omega_1^2}{X_0\ell} \cdot \alpha_1^3(t) = \frac{\omega_1^2 \cdot \ell}{2\pi\sqrt{X_0\ell}} \cdot \lambda \cdot x_B \sin 2\omega t,$$

$$\tag{30}$$

where X_0 is the elastic elongation of the cable, $X_0 = T\ell/(EA_0)$. This equation has the form of a so-called modified Mathieu equation, which is characterized by a set of secondary resonances; i.e., the response to a harmonic of frequency 2ω is not increased exclusively at resonance, but also at specific ratios between the exciting frequency and the system's natural frequency: ½, 1/3, 2, and 3. It is possible then to define instability regions (i.e., intervals of frequency oscillation of the supports where high amplitudes of vibration occur) and also to characterize both the threshold amplitude for the occurrence of instability and the maximum amplitude of oscillation inside the instability regions. These regions are represented in the diagram of Figure 21.19, called a *Strutt diagram* (hatched regions are unstable) and emanate from

$\delta^2 = (\omega_1/\omega)^2 = n^2$ $(n = 1, 2, \ldots)$ for each instability region of order n. The lift of the instability regions with the increase of damping means that higher amplitude of support oscillations are required to attain instability for a given excitation frequency ratio δ. This effect is more pronounced for the second-order resonance $(\delta^2 \approx 4)$.

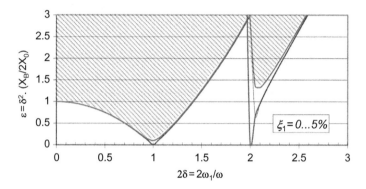

Figure 21.19 Transition curves for different values of the damping coefficient ξ_1:*0* (**in bold**), 0.5% and 5%.

Considering the first parametric resonance $\delta^2 \approx 1$, the threshold amplitude for the occurrence of instability is given by

$$\frac{x_B}{2X_0} = 2\xi_1. \tag{31}$$

Once parametric excitation occurs, the oscillation builds up. Tagata (1977), Pinto da Costa et al. (1996), and Clement and Cremona (1996), employing the method of harmonic balance; and Nayfeh and Mook (1979), using the method of multiple scales, obtained an approximation of the steady-state response in the vicinity of this first resonance:

$$\alpha_1(t) = a\sin\left(\omega t - \frac{1}{2}\psi\right), \tag{32}$$

where

$$a = \frac{4}{\pi} \cdot \sqrt{\frac{X_0\ell}{3}} \cdot \frac{1}{\delta} \cdot \left\{ 1 - \delta^2 \pm \left[\delta^4 \cdot \left(\frac{x_B}{2X_0}\right)^2 - 4\delta^2\xi_1^2 \right]^{1/2} \right\}^{1/2} \tag{33}$$

and

$$\tan\psi = -\frac{2\xi_1\delta^2}{2(1-\delta) - \delta^4 a^2} \tag{34}$$

Eq. (33) is plotted in Figure 21.20 for the Ben Ahin cable, with 110 m, considering two values of the relative support motion $x_B/(2X_0)$, of 0.05 and 0.3, and a damping coefficient of 1%. The nonlinearity of the differential equation induces, as can be observed in this figure, a bending of the frequency-response curves into the right, which leads to multivalued responses and, consequently, to a so-called *jump* phenomenon. The evolution of the amplitude of the response is represented by the arrows in Figure 21.20, where it can be observed that in a vicinity of the first parametric resonance ($\delta \approx 1$), the trivial solution is not stable.

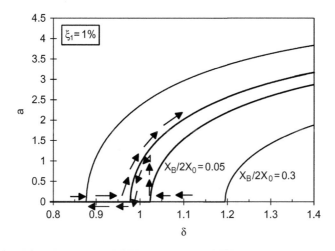

Figure 21.20 Frequency-response functions for a stay cable subjected to two different amplitudes of support motion: $x_B/(2X_0) = 0.05$, and $x_B/(2X_0) = 0.3$.

In order to understand the characteristics and importance of parametric excitation by comparison with external excitation, a study is conducted for the abovementioned Ben Ahin cable (length 110 m, $\lambda^2 = 0.0727$) using both the analytical expressions given here and a numerical modeling with a FE discretization. The amplitude of steady-state response under longitudinal harmonic motion at the anchorage at twice the linear natural frequency of the cable was calculated considering different amplitudes of support motion and a damping coefficient of 1%. The representation in Figure 21.21 shows that, in comparison with the external excitation response represented in Figure 21.18, parametric excitation induces amplitudes of vibration that almost double the amplitude of oscillations induced by external excitation for identical amplitudes of oscillation. However, as opposed to external excitations, parametric excitation does only occur for longitudinal oscillations greater than 0.006 m or 0.012 m for damping coefficients of 0.5% and 1%, respectively.

It is further noted that damping is only important to prevent parametric excitation. Once the oscillations set up, the amplitude is almost independent of the corresponding value.

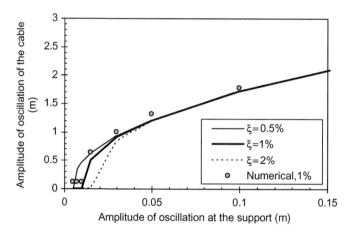

Figure 21.21 Amplitude of a steady-state response of Ben Ahin cable at the principal parametric resonance.

3.3.9 Design of vibration mitigation devices

The problem of defining the optimal characteristics of a damper installed at a point close to the anchorage of a cable was formerly studied by Kovàcs (1982), who proposed a practical optimal damping estimation method and empirically defined the maximum attainable modal damping.

According to Kovàcs and the illustration of Figure 21.22, the effect of adding a viscous damper with constant c at a distance x_c to the anchorage of a cable of mass per unit length m and length L submitted to a static tension H can be framed by two limiting conditions, under the assumption of null intrinsic damping:

- If $c = 0$, the first vibration mode is undamped (Figure 21.23A), and the dynamic amplification curve associated with the cable tends to infinity at the fundamental frequency ω_{01}
- When c = infinity (i.e., when a very large damper is installed), the force generated is so large that it blocks the cable at the damper, therefore acting as if there was a support at that location (Figure 21.23B). The consequence is a slight modification of the fundamental cable frequency to $\omega_{01}/(1 - x_c/L)$, but once more, the mode of vibration is undamped and the corresponding dynamic amplification curve tends to infinity at the frequency $\omega_{01}/(1 - x_c/L)$.

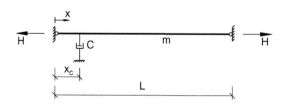

Figure 21.22 Cable with viscous damper.

The optimal damper is characterized by a constant c_{opt} that provides the amplification curve represented by the solid bold line in Figure 21.23C, whose maximum value is approximated by L/x_c, and occurs at the frequency for which the two previously referred amplification curves intersect, $\Omega \approx \omega_{01}(1 + x_c/2L)$. The modal damping associated with this system is the maximum and is given by

$$\xi_{max} \approx \frac{1}{2} \cdot \frac{x_c}{L}. \tag{35}$$

Kovàcs estimated the optimum damper size c_{opt} as

$$\frac{c_{opt}}{mL\omega_{01}} \approx \frac{1}{2\pi\left(\frac{x_c}{L}\right)}. \tag{36}$$

It is important to note that the ratio x_c/L is normally no greater than 0.015, meaning that the maximum damping added by a viscous damper attached close to the anchorage does not normally exceed 0.75%. Pacheco et al. (1993) have shown that these equations can be extended for higher-order modes. In order to provide a damping coefficient as a function of the constant c, and assuming small values of x_c/L, Pacheco et al. (1993) obtained a curve (Figure 21.24) representing the modal damping of any taut cable for the first few modes of vibration. This universal curve is characterized by a maximum that corresponds to the maximum attainable damping ratio $\xi_{n,max}$ of the vibration mode of order n, achieved by the attachment of a damper with optimum constant $c_{opt,n}$. These quantities are given by

$$\xi_{n,max} = 0.52 \cdot \frac{x_c}{L} \tag{37}$$

$$c_{opt,n} = 0.10 \cdot \frac{mL\omega_{01}}{n\frac{x_c}{L}}. \tag{38}$$

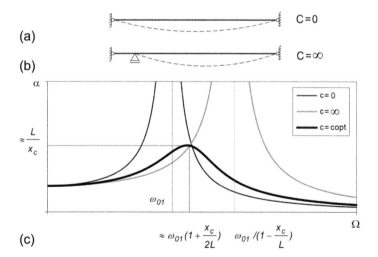

Figure 21.23 Limiting amplification behavior characteristics of cable with attached viscous damper: (A) $c = 0$; (B) $c = \infty$; (C) dynamic amplification curves for $c = 0$, $c = \infty$, and $c = c_{opt}$.

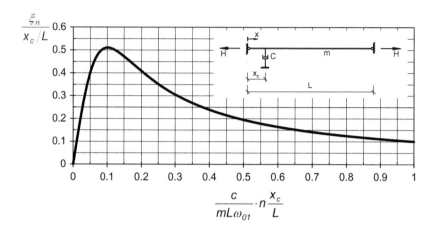

Figure 21.24 Universal curve relating modal damping ratio ξ_n with damper size c, location of damper x_c, and cable parameters, m, L, and ω_{01}.

The interest in the use of the universal curve is that, not only can an optimum viscous damper be designed for a particular vibration mode of a cable whose damping coefficient has been specified, but also the achieved damping coefficient for other vibration modes can be estimated.

The universal curve deduced by Pacheco et al (1993) can be applied to taut cables ($\lambda^2 < 1$), where the distance between damper and anchorage is within a few percentage points of cable length (e.g., 1%–10%). The difficulty in defining this universal curve on the basis of the design applications motivated the development by Krenk (2000) of an analytical representation that can be expressed in the following form:

$$\frac{\xi_n}{x_c/L} = \frac{\eta n \pi x_c/L}{1 + (\eta n \pi x_c/L)^2},$$ (39)

where η is a nondimensional damping parameter, defined as

$$\eta = \frac{\pi c}{mL\omega_{01}}.$$ (40)

Eq. (39) provides a good approximation of the universal curve for the first few modes of vibration, so long as the ratio x_c/L is small.

Considering that current applications of cable stayed bridge construction have resulted in longer cables, with Irvine parameters $\lambda^2 \geq 1$, and where sag effects can be significant, leading to a decrease of damper effectiveness, Crémona (1997) extended the universal curve introduced by Pacheco et al. (1993) to inclined cables with a maximum sag/span ratio of 1:8, and with an Irvine parameter no greater than $4\pi^2$ (first transition region), covering therefore all stays from cable stayed bridges. Krenk and Nielsen (2002) derived an extended asymptotic solution for shallow cables, evidencing the reduction of efficiency of the damper as a function of the Irvine parameter.

Hoang and Fujino (2007) provided a deeper inside on the effects of bending stiffness on the performance of viscous dampers installed in taut cables. These authors (Fujino and Hoang, 2008; Hoang and Fujino, 2008) and Krenk and Hogsberg (2005) explored the effect on the performance of dampers induced by other factors, such as the damper flexibility or the flexibility of the support and the nonlinearity of behavior. This research has been systematized by Caetano (2007).

In the context of the design of a damper to control cable vibrations, it is of interest to combine the effects of sag, bending stiffness, and flexibility of the support in order to obtain a global estimate of the reduction of efficiency and of the increase of the optimal damping coefficient with respect to the taut cable approach. This can be achieved using the simplified formulas derived by Fujino and Hoang (2008). For the sagged cable with nonnegligible bending stiffness with an installed viscous damper on a flexible support (Figure 21.25), these authors propose the following expression for the attained modal damping ratio:

$$\frac{\xi_n}{x_c/L} = R_n \cdot R_{EI} \cdot R_{kEI} \cdot \frac{\eta_n \cdot \eta_{EI} \cdot \eta_{kEI}}{1 + \left(\eta_n \cdot \eta_{EI} \cdot \eta_{kEI}\right)^2}, \tag{41}$$

where η_n and R_n are the nondimensional damping parameter and the reduction factor due to sag effect, η_{EI} and R_{EI} are the nondimensional damping parameter and the reduction factor due to the bending stiffness EI of the cable, and η_{kEI} and R_{kEI} are, respectively, the nondimensional damping parameter and associated reduction factor related to the stiffness k of the support. These parameters are defined in Table 21.5.

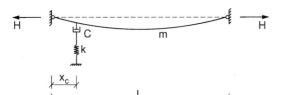

Figure 21.25 Sagged cable with tandem association of viscous damper and spring.

Table 21.5 Definition of Nondimensional Parameters and Reduction Factors for the Efficiency of a Damper due to Sag, Bending Stiffness, and Flexibility of the Damper Support

Effect	Nondimensional Parameter η_i	Reduction Factor R_i
Sag	$\eta_n = \eta k_n \pi \dfrac{x_c}{L}$	$R_n = \dfrac{\left[\tan\left(\dfrac{k_n\pi}{2}\right) - \left(\dfrac{k_n\pi}{2}\cdot\dfrac{x_c}{L}\right)\right]^2}{\tan^2\left(\dfrac{k_n\pi}{2}\right) + \dfrac{12}{\lambda^2}\cdot\left(\dfrac{k_n\pi}{2}\right)^2}$, with $k_n = \dfrac{\omega_n}{\omega_{01}}$, n is the mode order
Bending stiffness of cable	$\eta_{EI} = 1 - q - \dfrac{r\cdot q^2}{2}$	$R_{EI} = \dfrac{(1-q)^2}{1-q-rq^2/2}$, with $q = \dfrac{1-e^{-r}}{r}$ and $r = \zeta\cdot\dfrac{x_c}{L}$
Support stiffness k	$\eta_{kEI} = \eta_{EI} + \dfrac{1}{\bar\eta_k}$, with $\bar\eta_k = \dfrac{kx_c}{H}$	$R_{kEI} = \dfrac{\bar\eta_k\cdot\eta_{EI}}{1+\bar\eta_k\cdot\eta_{EI}}$

The maximum modal damping ratio is then obtained by

$$\frac{\zeta_n^{max}}{x_c/L} = 0.5\cdot R_n\cdot R_{EI}\cdot R_{kEI} \tag{42}$$

and occurs for a nondimensional optimal damping parameter η_n^{opt} defined as

$$\eta_n^{opt} = \frac{1}{\eta_{kEI}\cdot\eta_n}. \tag{43}$$

It should be pointed that Eqs. (42) and (43) have been derived based on the assumption that the viscous damper is linear. The nonlinearity of the damper is probably one of the most important causes of reduction of the efficiency, therefore even though small curvature, bending, and support stiffness effects are associated with a typical medium-sized cable, the damper effectiveness can be in practice of the order of 50% to 70% of the theoretical (Sun et al., 2005).

It is further remarked that the installation of a single damper may not provide sufficient damping to a stay cable. From a study with different combinations of dampers located close to the cable anchorages, Hoang and Fujino (2008) concluded that the combination of two dampers close to the same anchorage does not provide increased damping with respect to the effect of the damper located at the highest distance from the anchorage. On the contrary, the installation of a damper close to each anchorage of a cable leads to an increased damping effect that is asymptotically the sum of the individual contributions from the single dampers.

References

BBR, 2011. BBR HiAM: Strand Stay Cable System. Brochure. www.bbrnetwork.com.

Bournand, Y., 1999. Development of New Stay Cable Dampers. Cable-Stayed Bridges—Past, Present, and Future, IABSE Conference, 2–4 June 1999, Malmö, Sweden.

Caetano, E., 2007. Cable Vibrations in Cable-Stayed Bridges. Structural Engineering Documents. no.9. Published in English by IABSE. Published in Chinese by CABP, 2012.

Caetano, E., Cunha, A., 2011. On the Observation and Identification of Cable-Supported Structures. Eurodyn'2011 4–6 July, Belgium, Leuven.

Caetano, E., Bartek, R., Magalhães, F., Keenan, C., Tryppick, G., 2013. Assessment of Cable Forces at the London 2012 Olympic Stadium Roof. Struct. Eng. Intl. 4, 489–500.

Cheng, S., Tanaka, H., Irwin, P., Jakobsen, J., 2003. Aerodynamic Instability of Inclined Cables. Fifth International Symposium on Cable Dynamics, 15–18 September, Santa Margherita Ligure, Italy. 69–76.

Clement, H., Cremona, C., 1996. Étude Mathématique du Phénomène d'Excitation Paramétrique Appliqué aux Haubans de Pont. Études et Recherches des Laboratoires des Ponts et Chaussées. LCPC, Paris.

Crémona, C., 1997. Courbe Universelle pour le Dimensionnement d'Amortisseurs en Pied de Haubans. Revue Française de Génie Civil. 1 (1), 137–159.

Davenport, A., 1994. A Simple Representation of the Dynamics of a Massive Stay Cable in Wind. In: Proc. IABSE/ FIP Intl. Conf. Cable-Stayed and Suspension Bridges, pp. 427–438. 2.

Dyrbye, C., Hansen, S., 1996. Wind Loads on Structures. John Wiley & Sons, New York.

EN 1990, 2002. EN 1990. 2002. Eurocode—Basis of Structural Design.

EN 1991-1-4, 2005. Eurocode 1: Actions on Structures—Part 1-4: General Actions—Wind Actions.

EN 1991-1-5, 2003. Eurocode 1: Actions on Structures—Part 1-5: General Actions—Thermal Actions.

EN 1991-2, 2003. Eurocode 1: Actions on Structures—Part 2: Traffic Loads on Bridges.

EN 1993-1-11, 2006. Eurocode 3: Design of Steel Structures—Part 1-11: Design of Structures with Tension Components.

EN 1993-3-1, 2006. Eurocode 3: Design of Steel Structures—Part 3-1: Towers, Masts, and Chimneys—Towers and Masts.

Ernst, H.J., 1965. Der E-Modul von Seilen unter Berücksichtigung des Durchhangers. Bauingenieur 40 (2), 52–55.

Fédération Internationale du Béton, FIB, 2005. Acceptance of Stay Cable Systems using Prestressing Steels. Bulletin. No. 30.

Flamand, O., 1994. Rain-Wind-Induced Vibration of Cables. In: Proc. IABSE/ FIP Intl. Conf. Cable-Stayed and Suspension Bridges, pp. 523–531. 2.

Flamand, O., Benidir, A., Gaillet, L., Dimitriadis, G., 2014. Wind Tunnel Experiments on Bridge Stays Cable Protection Tubes in Dry Galloping Conditions Processing Method for Bi-Stable Phenomenon. Symposium on the Dynamics and Aerodynamics of Cables, 25–26 September, Copenhagen, Denmark.

FREYSSINET, 2010. HD Stay Cable. Freyssinet Brochure. www.freyssinet.com.

FREYSSINET, 2012. Pont de l'ile Russky. Freyssinet Brochure. www.freyssinet.com.

FREYSSINET, 2014a. Cohestrand. Freyssinet Brochure FT EN CI33. www.freyssinet.com

FREYSSINET, 2014b. Internal Elastomeric Damper. Freyssinet Brochure FT EN CI116. www.freyssinet.com.

FREYSSINET, 2014c. Internal Hydraulic Damper. Freyssinet Brochure FT EN CI117. www.freyssinet.com.

FREYSSINET, 2014d. Internal Radial Damper. Freyssinet Brochure FT EN CI118. www.freyssinet.com.

Fujino, Y., Hoang, N., 2008. Design Formulas for Damping of a Stay Cable with a Damper. J. Struct. Eng. 134 (2), 269–278.

Fujino, Y., Kimura, K., Tanaka, H., 2012. Wind-Resistant Design of Bridges in Japan: Developments and Practices. Springer, Tokyo.

Geffroy, R., 2002. The Laroin Footbridge with Carbon Composite Stay Cables. Footbridge 2002 20–22 November, Paris.

Geier, R., Wenzel, H., 2003. Field Testing of Ludwigshafen Cable-Stayed Bridge, Deliverable D10, IMAC (Project GRD1-2000-25654).

Gimsing, N.J., 1983. Cable-Supported Bridges: Concept and Design. Wiley Interscience, Chichester, UK.

Gimsing, N.J., Georgakis, C., 2012. Cable-Supported Bridges: Concept and Design, Third ed. Wiley, Chichester, UK.

Gimsing, N., Nissen, J., 1998. The Pylons on the Øresund Bridge. *Structural Engineering International*, IABSE, 8(4), 263–264.

Griffin, O., Skop, R., Ramberg, S., 1975. The Resonant Vortex-Excited Vibrations of Structures and Cable Systems. In: Offshore Technology Conference, Paper OTC-2319, Houston, TX.

Hardy, C., Bourdon, P., 1979. The Influence of Spacer Dynamic Properties in the Control of Bundle Conductor Motion. In: IEEE PES Summer Meeting. Vancouver, British Columbia, Canada. July.

Hardy, C., Watts, J.A., Brunelle, J., Clutier, L.J., 1975. Research on the Dynamics of Bundled Conductors at the Hydro-Quebec Institute of Research. Trans. E. & O. Div., CEA, 14, Part 4.

Hauge, Lars, Andersen, Yding, Erik, 2011. Longer. In: 35th Annual Symposium of IABSE / 52nd Annual Symposium of IASS / 6th International Conference on Space Structures, London. 20–23 September 2011.

Hikami, Y., 1986. Rain Vibrations of Cables of a Cable-stayed Bridge. J. Wind Eng. (Japan) 27, 17–28 (in Japanese).

Hikami, Y., Shirashi, N., 1988. Rain-wind-induced Vibrations of Cables in Cable-stayed Bridges. J. Wind Eng. Indus. Aerodyn. 29, 409–418.

Hoang, N., Fujino, Y., 2007. Analytical Study on Bending Effects in a Stay Cable with a Damper. J. Eng. Mech. 133 (11), 1241–1246.

Hoang, N., Fujino, Y., 2008. Combined Damping Effect of Two Dampers on a Stay Cable. J. Bridge Eng. 13 (3), 299–303.

Honda, A., Yamanaka, T., Fujiwara, T., Saito, T., 1995. Wind Tunnel Test on Rain-Induced Vibration of the Stay-Cable. In: Proceedings of the International Symposium on Cable Dynamics, 19 to 21 October, Liège, Belgium.

Irvine, H.M., 1981. Cable Structures. MIT Press, Cambridge, MA.

Irvine, H.M., Caughey, T.K., 1974. The Linear Theory of Free Vibrations of a Suspended Cable. Proc. R. Soc. Lond. A 341, 299–315.

Irwin, P., 1997. Wind Vibrations of Cables on Cable-stayed Bridges. Proc. ASCE Struct. Cong. 1, 383–387.

Kovàcs, I., 1982. Zur Frage der Seilschwingungen und der Seildampfung. Die Bautechnik. October, 325–332.

Krenk, S., 2000. Vibrations of a Taut Cable with an External Damper, Trans. ASME 67, 772–776 (Dec.).

Krenk, S., Hogsberg, J., 2005. Damping of Cables by a Transverse Force. J. Eng. Mech. 131 (4), 340–348.

Krenk, S., Nielsen, S., 2002. Vibrations of a Shallow Cable with a Viscous Damper. Proc. R. Soc. Lond. A 458, 339–357.

Kubo, Y., 1997. Aeroelastic Instability and its Improvement of Bundle Cable for Cable-stayed Bridge. Proc. Intl. Sem. Cable Dyn. 49–56.

Kubo, Y., Kato, K., Maeda, H., Oikawa, K., Takeda, T., 1994. New Concept on Mechanism and Suppression of Wake Galloping of Cable-Stayed Bridges. In: Proc. IABSE/ FIP Intl. Conf. Cable-Stayed and Suspension Bridges, pp. 491–498. 2.

Kubo, Y., Kimura, K., Tanaka, H., Isobe, T., Fujita, M., Higashi, H., Kato, K., 2003. Aerodynamic Responses of Inclined Cables in Section Model Test and Full Model Test. In: Fifth International Symposium on Cable Dynamics, 15–18 September, Santa Margherita Ligure, Italy, pp. 383–390.

Larose, G., Zan, S., 2001. The Aerodynamic Forces on the Stay Cables of Cable-Stayed Bridges in the Critical Reynolds Number Range. In: Proceedings of the 4th International Symposium on Cable Dynamics, May 27–30, Montréal, Canada, pp. 77–84.

Larose, G., Jakobsen, J., Savage, M., 2003. Wind-Tunnel Experiments on an Inclined and Yawed Stay Cable Model in the Critical Reynolds Number Range. In: Fifth International Symposium on Cable Dynamics, 15–18 September, Santa Margherita Ligure, Italy, pp. 279–286.

Liu, P., 1982. Static and Dynamic Behaviour of Cable Assisted Bridges. Ph.D thesis, University of Manchester Institute of Science and Technology (UMIST), Manchester, UK.

Macdonald, J.H.G., 2001 Susceptibility of Inclined Bridge Cable to Large Amplitude Vibrations considering Aerodynamic and Structural Cable Damping. Proc. 4th ISCD, 243–250.

Main, J., Jones, N., 1999. Full-Scale Measurements of Stay Cable Vibration. In: Proc. 10th Intl. Conf. Wind Eng, 21 to 24 June, Copenhagen.

Matsumoto, M., 1998. Observed Behaviour of Prototype Cable Vibration and its Generating Mechanism. In: Larsen, A., Esdahl, S. (Eds.), Bridge Aerodynamics. Balkema, pp. 189–211.

Matsumoto, M., 2014. Stall-Type Galloping and VIV-Initiated Galloping of Inclines Stay Cable Aerodynamics and its Aerodynamic Stabilization. Proc. Symp. Dyn. Aerodyn. Lingby, Denmark. Cables, September 2014.

Matsumoto, M., Yokoyama, K., Miyata, T., Fujino, Y., Yamaguchi, H., 1989. Wind-Induced Cable Vibration of Cable-Stayed Bridges in Japan. In: Proc. Canada-Japan Workshop on Bridge Aerodyn Ottawa, Canada, pp. 101–110, September.

Matsumoto, M., Shiraishi, H., Shirato, H., 1992. Rain-wind-induced Vibration of Cables of Cable-stayed bridges. J. Wind Eng. Indus. Aerodyn. 44, 2011–2022.

Matsumoto, M., Ishizki, H., Kitazawa, M., Aoki, J., Fujii, D., 1995. Cable Aerodynamics and its Stabilization. In: Proc. Intl. Symp. Cable Dyn. 19 to 21 October, Liège, Belgium.

Mehrabi, A., Tabatabai, A., 1998. A Unified Finite Difference Formulation for free Vibration of Cables. J. Struct. Eng. 124 (11), 1313–1322.

Meier, U., 2012. Carbon Fiber Reinforced Polymer Cables: Why? Why not? What if? Arab J. Sci. Eng. Sci. Eng. 37, 399–411.

Miyata, T., 1991. Design Considerations for Wind Effects on Long-Span Cable-Stayed Bridges. In: Ito, M., et al. (Ed.), Cable-Stayed Bridges: Recent Developments and Their Future. Elsevier, pp. 235–256, 10–11 December, Yokohama.

Morse, P., Ingard, K., 1968. Theoretical Acoustics. Princeton University Press, Princeton, NJ.

Narita, N., Yokoyama, K., 1991. A Summarized Account of Damping Capacity and Measures Against Wind Action in Cable-Stayed Bridges in Japan. In: Ito, M., et al. (Ed.), Cable-Stayed Bridges: Recent Developments and Their Future. Elsevier, pp. 257–278.

Nayfeh, A., Mook, D., 1979. Nonlinear Oscillations. John Wiley and Sons, Inc., USA.

Nikitas, N., MacDonald, J.H.G., Jakobsen, J.B., Andersen, T.L., 2012. Critical Reynolds Number and Galloping Instabilities: Experiments on Circular Cylinders. Experiments in Fluids 52, 1295–1306.

Ohshima, K., Nanjo, M., 1987. Aerodynamic Stability of the Cables of a Cable-Stayed Bridge Subject to Rain (A Case Study of the Ajigawa Bridge). In: Proc. US-Japan Seminar on Natural Resources, pp. 324–336.

Pacheco, B., Fujino, Y., Sulekh, A., 1993. Estimation Curve for Modal Damping in Stay Cables with Viscous Damper. J. Struct. Eng. 119 (6), 1961–1979.

Peil, U., Nahrath, N., 2003. Modeling of Rain-wind-induced Vibrations. Wind Struct. 6 (1), 41–52.

Pinto da Costa, A., Martins, J., Branco, F., Lilien, J., 1996. Oscillations of Bridge Stay Cables Induced by Periodic Motions of Deck and/or Towers. J. Eng. Mech. 122 (7), 613–622.

PTI Guide Specification, 2007. Recommendations for Stay Cable Design, Testing, and Installation, fifth ed PTI.

Saito, T., Matsumoto, M., Kitazawa, M., 1994. Rain-Wind Excitation of Cables on Cable-Stayed Higashi-Kobe Bridge and Cable Vibration Control. In: Proc. IABSE/FIP Intl. Conf. Cable-Stayed and Suspension Bridges, pp. 507–514.

Sarkar, P., Jones, N., Scanlan, R., 1994. Identification of Aeroelastic Parameters of Flexible Bridges. J. Eng. Mech. 120 (8), 1718–1743.

Service d'Etudes Techniques des Routes et Autoroutes (SETRA), 2002. Cable Stays. Recommendations of French Interministerial Commission on Prestressing, SETRA, Bagneux. ISBN: 2110934026.

Simiu, E., Scanlan, R., 1996. Wind Effects on Structures: Fundamentals and Applications to Design, third ed. John Wiley & Sons, New York.

Sun, L., Shi, C., Zhou, H., Zhou, Y., 2005. Vibration Mitigation of Long Stay Cable Using Dampers and Cross-ties. In: Proceedings of the Sixth International Symposium on Cable Dynamics, Charleston, South Caroline, September 19–22.

Tabatabai, H., Mehrabi, A., 2000. Design of Mechanical Viscous Dampers for Stay Cables. J. Bridge Eng. 5 (2), 114–123.

Tagata, G., 1977. Harmonically Forced, Finite Amplitude Vibration of a String. J. Sound Vib. 51 (4), 483–492.

Tanaka, H., 2003. Aerodynamics of Cables. In: Fifth International Symposium on Cable Dynamics, 15 to 18 September, Santa Margherita Ligure, Italy, pp. 11–25.

TRB 20015, 2005. NCHRP Synthesis 353: Inspection and Maintenance of Bridge Stay Systems, National Cooperative Highway Research Program, American Transportation Research Board, Washington.

Verwiebe, C., 1998. Rain-Wind-Induced Vibrations of Cables and Bars. In: Larsen, A., Esdahl, S. (Eds.), Bridge Aerodynamics, Copenhagen, 10 to 13 May, Balkema, pp. 255–263.

Virlogeux, M., 1998. Cable Vibrations in Cable-Stayed Bridges. In: Larsen, A., Esdahl, S. (Eds.), Bridge Aerodynamics, Copenhagen, 10 to 13 May, Balkema, pp. 213–233.

VSL, 2002. VSL SSI 2000 Stay Cable System. Brochure. www.vsl.com.

Wagner, P., Fuzier, J.-P., 2003. Health Monitoring of Structures with Cables—Which Solutions? Dissemination of the Results of the IMAC European Project. In: Fifth International Symposium on Cable Dynamics, 15 to 18 September, Santa Margherita Ligure, Italy.

Yamagiwa, I., Utsuno, H., Endo, K., Sugii, K., Morimoto, T., 1997. Simultaneous Identification of Tension and Flexural Rigidity of Bridge Cables. Ann. Proc. Steel Struc. 5, 15–22.

Yamaguchi, H., 1990. Analytical Study on Growth Mechanism of Rain Vibration of Cables. J. Wind Eng. Indus. Aerodynam. 33, 73–80.

Yamaguchi, H., 1995. Control of Cable Vibrations with Secondary Cables. In: Proc. International Symposium on Cable Dynamics, Oct. 19–21. Liège, Belgium, pp. 445–452.

Yamaguchi, H., Fujino, Y., 1998. Stayed Cable Dynamics and its Vibration Control. In: Larsen, A., Esdahl, S. (Eds.), Bridge Aerodynamics. Copenhagen, 10 to 13 May, Balkema, pp. 235–253.

Orthotropic steel decks

22

De Backer H.
Department of Civil Engineering, Ghent University, Belgium

1 Introduction

An OSD consists of a deck plate, which is supported by stiffeners working in two orthogonal directions: the transversal stiffeners (or crossbeams) and the longitudinal stiffeners. The system can thus be compared with a uniform deck plate having different stiffness characteristics in two directions, or in other words, an *ortho*gonal anisotropic steel deck. Because of this, the OSD can take part in the over-bridge actions, such as the upper flange of a box girder, the tie of a tied-arch, the stiffening girder of a suspension or cable stayed bridge, or another part of the overall structural bridge system.

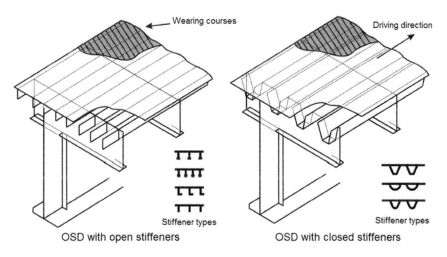

Figure 22.1 Two basic types of OSDs.

Innovative Bridge Design Handbook. http://dx.doi.org/10.1016/B978-0-12-800058-8.00022-0

Figure 22.2 Orthotropic deck plate of the Temse Bridge in Flanders, Belgium.

The basic composition of an OSD is shown in Figures 22.1 and 22.2. The component showing the most variation is the longitudinal stiffener. This stiffener can be open, using strips or L-profiles, or closed, using trapezoidal, V-shaped or rounded sections. The most important structural characteristics of an OSD can be summarized as follows:

- Low dead load: The dead load of an OSD is considerably lower when compared with other similar steel and concrete deck types.
- Strength and stiffness: An OSD has a considerable plastic deformations capacity under its ultimate load.
- Structural efficiency: The system is perfectly capable of spreading the high patch loads in both longitudinal and transversal directions.
- The ability to take part in the overall structural actions of the bridge in both the longitudinal and transversal directions.
- A reduced structural height, when compared with a classic combination of deck, stringer, crossbeam, and main beams. All of these components are combined in one plane.
- Since this is a continuous deck system, the number of connections within the deck plate is quite low.
- Ease of construction using large prefabricated sections.
- Durability and long-term economy of a well-designed OSD, since the deck system can have the same life span as the overall bridge system.

These characteristics are extremely important for the design of new bridges, as well as for renovation projects. Low weight, height, and high carrying capacity seem to be the most determining factors for choosing this bridge type. These factors have allowed the OSD to become the standard choice for record-breaking bridge spans.

2 History

The origin of OSD can be found with the development of "battledeck floors" in the 1930s. These decks consisted of a steel deck plate welded to longitudinal I-shaped stringers, which were attached to the underlying crossbeams. One of the first applications was in 1932, on the North Saginaw Road/Salt River Bridge in Michigan. An important thing to note is that, in this concept, the deck plate does not function as a stiffener of the crossbeams or an upper flange of the main beams. Its only function is spreading the local patch loads to the other components. The same principle was applied frequently to the German Autobahn network from 1934 onward.

Reconstruction after World War II was an important factor in stimulating development and construction of multiple OSD bridges, mainly in Germany during the 1950s. Examples of these bridges that use open stiffeners are the Kurpfalz Bridge, which crosses the Neckar in Mannheim, and the Keulen-Mülheim Bridge, which crosses the river Rhine. The first application of closed stiffeners dates from 1954 in Duisburg, Germany. The first North American application of the OSD concept was the Port Mann Bridge in Vancouver, Canada in 1964. During the following years, several other OSD bridges were built, including the Severn Crossing Bridge in Britain, the Poplar Street in St. Louis, Missouri, and the San Mateo Bridge in California. The first design guides were published in 1972 by the Japan Road Authority and were closely followed by the inclusion of OSDs in the AASHTO design rules in 1973.

OSDs have also been used for the redecking of a number of famous bridges, such as the Golden Gate Bridge in San Francisco and the Benjamin Franklin Bridge in Philadelphia. Most of the record span suspension bridges also use an OSD. Table 22.1 illustrates that an OSD is used in 8 of the 10 largest bridge spans. Other applications include the Millau Viaduct and the recently opened San Francisco–Oakland Bay Bridge.

3 OSD concept

This section describes the most useful manual design methods for OSDs. These analytical methods are based on the groundbreaking efforts of Pelikan and Esslinger (1957), and Wolchuk (1963), who were the pioneers of the OSD concept. The analytic calculation of an OSD can be quite complex because of its lack of symmetry, which is caused by all stiffeners working on the same side of the steel deck plate. A detailed

Table 22.1 **Record Span Suspension Bridges**

	Bridge	Length of Main Span (m)	City, Country	Year	Deck Type
1	Akashi Kaikyo	1991	Kobe-Naruto, Japan	1998	OSD
2	Great Belt	1624	Korsor, Denmark	1998	OSD
3	Runyang	1490	Zhjiang, China	2005	OSD
4	Humber	1410	Kingston-Upon-Hull, UK	1981	OSD
5	Jiangyin	1385	Jiangsu, China	1999	OSD
6	Tsing Ma	1377	Hong Kong, China	1997	OSD
7	Verrazano Narrows	1298	New York, US	1964	Concrete
8	Golden Gate	1280	San Francisco, US	1937	OSD
9	Höga Kusten	1210	Kamfors, Sweden	1997	OSD
10	Mackinac	1158	Mackinaw City, US	1957	Concrete

design method would become labor-intensive and complicated, so a number of simplified OSD models have been developed. The results of these models are quite conservative but allow for a quick determination of the internal forces and initial design dimensions. These methods are:

- Simplification as a rectangular beam grid (Cornelius, 1952): The deck plate and all stiffeners are considered to be individual, discrete beams working together. The method assumes that the deck is cut halfway between longitudinal stiffeners. These stiffeners are then considered virtual beams with the deck plate as an upper flange. Provided the effective width is larger than the distance between stiffeners, this method considers the entire OSD. Stiffness of the OSD in the transversal direction is neglected and must be considered separately. These types of methods have difficulty in describing the effects of torsion within the OSD concept.
- Simplification as an idealized orthotropic plate (Pelikan and Esslinger, 1957; Wolchuk, 1963): The actual OSD is replaced by a singular deck plate with equivalent characteristics. The sections of deck plate and stiffeners are spread out over the width of the deck. Afterward, the effect of the actual loads on this idealized plate is calculated and used as an input for separate calculations of each stiffener. This method will be discussed in more detail later in this chapter.

It has to be stressed that purely linear elastic theories are only allowed when displacements are small. Otherwise, membrane action might develop in the deck plate. Due to the high ductility of steel, membrane action will be responsible for the large postcritical strength reserve of the OSD concept.

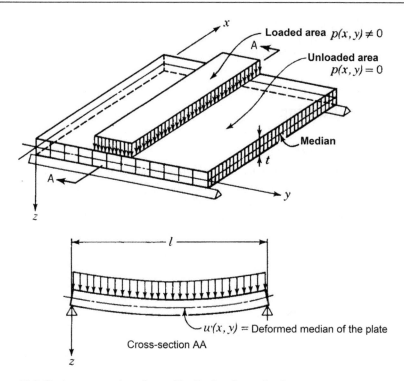

Figure 22.3 Design assumptions for an idealized orthotropic plate.

An idealized orthotropic plate is defined as a plate with different stiffness in two orthogonal directions x and y within the surface of the plate, as shown in Figure 22.3. Other assumptions include a constant thickness and a continuous and homogenous material. The different stiffness in both directions will thus be defined by different stiffness moduli E_x and E_y, as well as different Poisson's ratios ν_x and ν_y. Structural anisotropy is thus replaced by material anisotropy quite similarly, as in the behavior of a wooden beam. A natural example of this type of plate structure is a wooden board, which has different stiffness ratios parallel and perpendicular to the veins. The initial assumptions for such a calculation are: a homogenous material, constant thickness, limited purely elastic deformations according to Hooke's law, negligible vertical normal stresses, and strictly vertical supports.

All of the properties can then be described based on the affective bending stiffnesses in both directions, as well as the torsional stiffness of the plate. The calculation is thus a combination of three separate structural systems (Wolchuk, 1963; Vandepitte, 1979):

- The deck plate working as a transfer medium of the local patch loads to the longitudinal and transversal stiffeners
- The orthotropic behavior of the combination of stiffeners and deck plate, which can be calculated as an idealized orthotropic plate
- The overall bridge actions

More information about these calculation methods can be found in Pelikan and Esslinger (1957), Wolchuk (1963), Vandepitte (1979), American Institute for Steel Construction (AISC, 1938), and Klöppel and Roos (1960).

4 Practical design

4.1 Fatigue design

Fatigue is the process in which an accumulation of damages is caused by a repeating load of variable magnitude. Fatigue damage normally occurs under purely elastic stresses. However, due to stress concentrations, plastic stresses are possible in very small areas of the construction. Once enough damage is accumulated, fatigue fracture will initiate and propagate through the plasticized regions. Fatigue calculations are quite complex. Codes and standards will use simplified models to describe the fatigue behavior. Most methods, such as Eurocode 3 (2009), use a combination of SN-curves and the Palmgren-Miner rule for OSDs.

Specific SN-curves or Wöhler curves have been determined for each fatigue sensitive weld detail in an OSD. They offer a relation between the nominal stress, $\Delta\sigma$, at the location and the number cycles, N, until failure, as described by the Paris-Erdogan law:

$$N = \frac{A}{\Delta\sigma^m}.$$ (1)

Parameters A and m are determined based on the considered weld detail. The stress concentrations are calculated using elastic material laws and based on the occurring stress concentrations and possible secondary effects (Kiss et al., 1998). Although nominal stresses combined with elastic theories are basically inaccurate because only the SN-curves are to be used for constant amplitude loading, and they are generally accepted as a simplified calculation method. As an example, the Eurocodes use the following failure criterion, combined with Palmgren-Miner's rule, wherein all realistic stress cycles are replaced by constant amplitude stress cycles resulting in equivalent damage:

$$\gamma_{Ff}\Delta\sigma \le \frac{\Delta\sigma_c}{\gamma_{Mf}}.$$ (2)

Herein, γ_{Ff} and γ_{Mf} are partial safety factors for the fatigue loads and fatigue strength respectively, while $\Delta\sigma$ is the nominal stress cycle and $\Delta\sigma_c$ is the fatigue strength for the considered weld category for the expected number of stress cycles N during the assumed fatigue life.

Although most methods and standards follow these simplifications, some cautionary remarks must be made here. The method totally neglects the probabilistic background of the phenomenon. In addition actual load histories, i.e., the actual order of the stress cycles is neglected, although fracture mechanical principles have learned that this can be influential. Furthermore, all standards are based on a certain safety level that is guaranteed. For fatigue of OSDs, this level is based on the determination of the weld categories using prototype measurements and laboratory testing. Due to the development of production methods, steel qualities, and weld methods, the overall quality has greatly improved over the years. Since weld categories are still based on all available tests, this implies that they become more and more conservative each year.

4.2 OSD design based on AASHTO

The American standards use two approaches for fatigue problems, shown in Figure 22.4. Load-induced fatigue considers details subjected to axial tension over their entire cross section. The actual calculation is based on the determination of the number of stress cycles and the use of SN-curves and weld categories. Distortion induced fatigue, on the other hand, concerns details wherein out-of-plane deformations of the steel plate elements occur. No detailed calculation is necessary, but safety is guaranteed based on the geometrical rules for each detail.

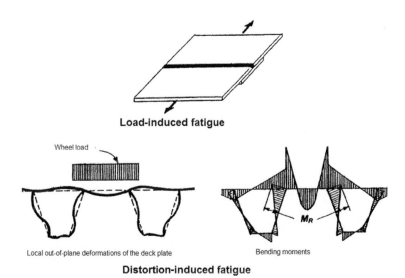

Load-induced fatigue

Wheel load

Local out-of-plane deformations of the deck plate

Bending moments

M_R

Distortion-induced fatigue

Figure 22.4 Examples of fatigue types according to AASHTO (2000).

Load-induced fatigue details need to meet the following criterion:

$$(\Delta F)_n = \left(\frac{A}{N}\right)^{\frac{1}{3}} \geq \frac{1}{2}(\Delta F)_{TH} \tag{3}$$

with

$$N = (365)(75)n(ADTT)_{SL} \tag{4}$$

In Eqs. (3) and (4), n equals the number of stress cycles caused by the passage of a standardized truck, $(\Delta F)_{TH}$ equals the constant amplitude fatigue strength, and $(ADTT)_{SL}$ equals the average daily traffic volume for trucks on a single lane. The corresponding SN curves are shown in Figure 22.5. These curves are linked to 12 weld categories in OSDs. These categories are summarized in Table 22.2.

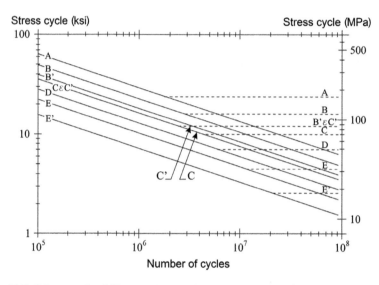

Figure 22.5 SN-curves for OSD according to AASHTO (2000).

Table 22.2 **AASHTO Fatigue Details**

	Description	Category
1	Welded connections within the deck	B
2	plate, using different types of backing	C
3	strips	D
4	Bolted connections in the deck plate	B
5	Welded connections in deck plate and	B
6	stiffeners under workshop conditions	C
7	Welds of the stiffener window	D
8	Weld between longitudinal stiffener and	C
9	crossbeam, with or without internal	C
	diaphragm	
10	Weld of crossbeam web to stiffener, with	<C
11	or without internal diaphragm	<C
12	Weld between crossbeam web and deck	E

The stiffener to deck plate detail is not included in this list, strangely enough. This fatigue detail is considered to be deformation induced. This should not be calculated in detail, but it is assumed that no fatigue will occur if a number of geometric design rules are met. From a design point of view, this can be seen as an oversimplification of the problem. These empirical design rules are aimed at minimizing the moment M_R, which is found in the stiffener flank. The moment M_R occurs because the wheel load causes localized, out-of-plane movements of the stiffener flank, which causes potential fatigue cracks. In order to be certain that the moment M_R is as small as possible, measures should be taken to ensure that the deck plate has a sufficient thickness, and has a sufficiently high rigidity, while the stiffener, on the other hand, should be slender and flexible.

4.3   Eurocode design principles

Eurocode mandates that detailed fatigue checks should be performed for all components, except when the geometry of the considered detail meets certain design rules drawn up based on experimental work and practical experience. As for the connection of the longitudinal stiffeners to the deck plate and the connection of the longitudinal stiffeners to the crossbeam, these are discussed in the annexes. All of the fatigue details around the stiffeners to be considered are shown in Figure 22.6, and listed in Table 22.3. The calculation methods for stiffeners are defined as well. Longitudinal stiffeners should be studied using a realistic model of the entire structure. Only the longitudinal stiffeners of railway bridges may be analyzed as continuous beams on elastic supports. The influence of the cutouts in the web plate at the location of the longitudinal stiffeners should be taken into account in the design of the crossbeams. The fatigue in the deck plate is assumed to be mainly caused by the deflection of the deck plate under the wheel load.

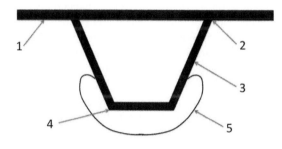

Figure 22.6 Eurocode fatigue details.

4.4   Open or closed stiffeners

The selection of open versus closed longitudinal stiffeners involves three major issues: design (steel weight or economy), ease of fabrication, and ease of construction. In addition, maintenance issues, such as ease of inspection and the percentage of superstructure exposed to exterior elements, are important. Weight savings in

Table 22.3 **Eurocode Fatigue Details**

	Detail	Category
1	Longitudinal stresses in transversal welds of the deck	71
2	Longitudinal stresses in the deck plate at the connection of longitudinal stiffener with the deck	100 80
3	Welded connection of a closed longitudinal stiffener with the crossbeam	80
4	Welded connection between closed longitudinal stiffeners, with backing strip	71
5	Free edge of the cutout in the crossbeam web to allow for continuous longitudinal stiffeners	112
6	Welded connections between closed longitudinal stiffeners and the deck plate	71

superstructures are the thriving issue in the use of an orthotropic system. For most bridges, the stiffeners are connected to the crossbeams by welding. The crossbeams or transversal stiffeners can be steel hot-rolled beams, small plate girders, box girders, or full-depth diaphragm plates. When full-depth diaphragms are used, access openings are needed for bridge maintenance purposes. The holes also reduce dead load and provide a passageway for mechanical or electrical utilities. A small number of bridges have the stiffeners perpendicular to the main girders, which is more common in pedestrian bridges (Mangus, 2014).

An open stiffener has virtually no torsional capacity. The open stiffeners were initially very popular in because of simpler analysis and welding. The switch to closed stiffeners occurred to reduce dead weight of the superstructure. In the tension zones, the shape of the stiffener can be any shape, open or closed, depending on the preference of designers. Also, closed stiffeners have 50% less surface area to protect from corrosion. A closed longitudinal stiffener is torsionally stiff and is essentially a miniature box girder. The closed stiffener is more effective for lateral distribution of the individual wheel loads than the open stiffener system.

The trapezoidal (closed) stiffener shows greater bending efficiency in load-carrying capacity and stiffness. The original shapes were patented by the Germans (Sedlacek, 1992), later adopted in United States by Bethlehem Steel, and then were also adopted in Japan and other countries (Institution of Civil Engineers, 1972). U-shaped as well as V-shaped and Y-shaped stiffeners have been developed. It is readily apparent that a series of miniature box girders placed side by side is much more efficient than a series of miniature T-girders placed side by side. Weight savings in total steel weight has lead designers to switch to the trapezoidal stiffeners with a large range of choices. The trapezoidal longitudinal stiffener is quite often field-welded completely around the superstructure's cross section to achieve full structural continuity, rather than field bolting (Hubman et al., 2013).

5 Innovative applications and research topics

5.1 Fracture mechanics and residual stresses

Fatigue in steel structures is the most important type of fracture, and because of its complexity it is less understood than other types of failure. In the past, fatigue problems were sometimes overlooked during design. With the current design codes, a fatigue problem is assessed based on SN-curves. However, these curves should be updated for every project where a different design approach or installation procedure is used. Since this has mostly not been the case, a misunderstanding of the fatigue behavior of the detail has occurred. In addition, the Palmgren-Miner method is used to calculate the lifetime of each detail. However, this method is not very accurate because the load history and the load sequences do not have any effect on the fatigue resistance. These design imperfections lead to overestimating the dimensions when considering OSDs (Nagy et al., 2014).

Residual stresses are present in many steel structures due to manufacturing actions causing plastic deformations. Nevertheless, these stresses are not often taken into account when considering the design of these structures. This is acceptable when only focusing on the stress variations, which eliminates the initial stress state of the structure. However, the effect of residual stresses may either be beneficial or detrimental, depending on the magnitude, sign, and distribution of these stresses, with respect to load-induced stresses (Barsoum and Lundbäck, 2009). Therefore, the initial stress state due to a welding operation has to be introduced into a finite element method (FEM) model. Basically, there are two different methods to introduce an initial stress state into a model. The easiest and preferred way is to apply the residual stresses according to literature or test data. This can be done by imposing the stresses directly into the model or by imposing complementary normal forces and bending moments. Results from similar fillet welds as those in the orthotropic bridge decks have already been studied. Therefore, N_{deck} is chosen in order to have tensile yield stresses into the deck plate at the weld, as shown in Figure 22.7. For the stresses into the stiffener, an additional bending moment $M_{stiffener}$ and normal force $N_{stiffener}$ are also introduced. The bending moment is necessary because the weld is welded from one side only and the filler metal and the corresponding heat zone is larger at the weld toe, compared with the weld root. For the magnitude of this bending moment and normal force, an assumption is made based on the distribution of the filler metal.

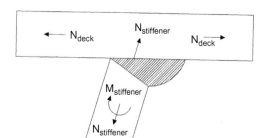

Figure 22.7 Complementary normal forces and bending moments to simulate residual stresses.

Linear elastic fracture mechanics (LEFM) calculations can be carried out with a detailed three-dimensional (3D) model of a stiffener-to-deck plate connection. The method described refers to the automatic crack propagation method based on extended finite element method (XFEM) techniques (Polak, 2007). With this method, it is possible to evaluate the whole crack propagation without remeshing the model for every crack propagation step. In addition, not only the crack growth rate can be evaluated, but also the crack growth direction. At first, an initial crack length should be chosen according to the welding detail and construction technology. Often, an initial crack length is chosen between 0.1 and 1 mm (De Backer, 2006). If the weld is perfectly accessible to smoothen the surface afterward, the initial crack length can be very small. However, the welds used for longitudinal stiffeners in OSDs are welded from only one side, and even the lack of penetration can be questioned. Therefore, due to the large uncertainties, initial elliptical crack lengths of 1 mm in the longitudinal direction and 0.5 mm in the transversal direction are assumed. This was also confirmed in a microscopic study of the present weld details of a stiffener-to-deck plate connection (De Backer, 2006). Although the fatigue crack often propagates through the deck plate, the initial crack length is chosen parallel to the deck plate and at the weld root. After implementing an initial crack length into the model, the XFEM calculation can be performed, as shown in Figure 22.8.

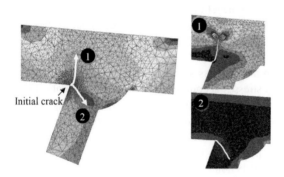

Figure 22.8 Detailed small-scale 3D model: possible crack growth directions.

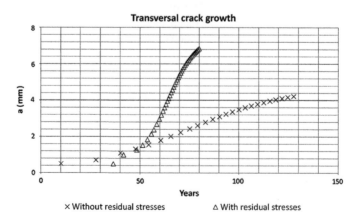

Figure 22.9 Transversal crack growth: a comparison with residual stresses or without them.

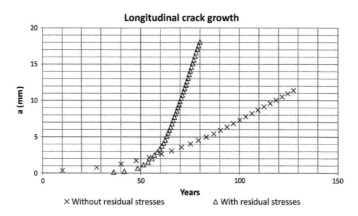

Figure 22.10 Longitudinal crack growth: a comparison with residual stresses or without them.

Figures 22.9 and 22.10 visualize the evolution of the crack length as a function of the years of service life for both the transversal and longitudinal crack growth directions. At this point, the fatigue life is evaluated due to constant stress amplitude with wheel type *B* from Eurocode 3 (2009) and an axle load of 130 kN. Without residual stresses, the crack does not develop quickly, but its development remains faster than that of the crack with residual stresses for approximately 52 years. After that, the crack with residual stresses grows progressively. The continuity of the stress distribution due to membrane forces is interrupted because the crack is growing through the deck plate. The stresses are forced into the less rigid body of the closed stiffeners. This explains why the crack propagation through the deck plate is much faster than the crack propagation through the stiffener. The same conclusions hold for the longitudinal crack growth direction—although it should be noted that the speed of the longitudinal crack growth is much faster than in the transversal direction. These conclusions are illustrated by the fatigue problem detected in the Temse Bridge in Flanders, Belgium, shown in Figure 22.11. The crack first grows longitudinally before fully penetrating the deck plate (or the stiffener). Therefore, the crack stays invisible through visible detection unless there is already sufficient damage (Kühn et al., 2008).

5.2 Refurbishment techniques

Since a number of bridges suffered fatigue damage early in their lives, possible repair and refurbishment techniques have been researched in detail and used on actual bridges. This section will focus on two of the most promising options: adding a high-performance concrete plate to the deck or gluing an additional steel plate to the existing deck surface.

Figure 22.11 Longitudinal crack through the deck plate at a stiffener-to-deck plate connection on the Temse Bridge, Flanders, Belgium.

Two separate lightweight systems for reinforcing OSDs have been researched (Teixeira de Freitas, 2010, 2011): the bonded steel plate system and the sandwich steel plate system. The main idea of these types of reinforcement is to stiffen the existing deck plate, thereby reducing the stresses at the fatigue-sensitive details, thus extending the fatigue life of the OSD. Both reinforcement systems consist of adding a second steel plate to the existing steel deck. The behavior and the effect of the reinforcement systems have been investigated using full-scale static tests and finite element analyses, using realistic wheel loads. The results showed at least 40% of stress reduction close to the fatigue sensitive details after applying both reinforcements. The two suggested reinforcement systems showed a good performance and proved to be efficient, lightweight solutions to refurbish OSDs and extend their life spans (Teixeira de Freitas, 2013).

A sandwich plate system (SPS) is composed of two steel plates and a solid polymer (polyurethane) core, sandwiched together. The sandwich action is generated through the bond between the polymer core and the steel plates. This ensures a high-bending resistance and bending stiffness of the sandwich if it is loaded as a plate, so that the stiffeners usually utilized to reinforce thin plates can be abandoned. Because of the low density of the core material, SPS plates have the advantage of being lightweight. They provide minimum steel surfaces exposed to corrosion, have excellent fatigue properties (due to the absence of welded stiffeners or attachments), and also exhibit good damping (noise emission) and insulation properties (temperature and fire resistance). SPS plates are most suitable for both building new steel decks and refurbishing existing steel ones by overlay and underlay techniques to make them durable and fit for the increasing traffic loads (Feldmann et al., 2007).

A developed renovation technique for fixed bridges is a surfacing of high performance concrete (De Jong and Kolstein, 2004; Buitelaar et al., 2004). Fixed bridges often have a wearing course of approximately 50-mm mastic asphalt, with low stiffness. It is possible to replace this with a wearing course with a higher stiffness. A wearing course of reinforced, high-performance concrete with the same thickness as the mastic asphalt layer is a good solution to lower the stress cycles. If a good intermediate layer between steel and concrete is possible, composite action between steel and concrete is also possible. In that case, the total stiffness of the composite deck plate structure might be enlarged with factors. Then the stress cycles in the steel deck plate are strongly reduced, and subsequently, the fatigue life is far better. This is a very promising solution since it turns the deck plate in a much more rigid construction behaving as an actual uniform plate, due the monolithic composite interaction between the reinforced high-performance concrete (RHPC; shown in Figures 22.12 and 22.13) overlay and the steel deck plate. The RHPC overlay with a thickness of a minimum of 5 cm will result in a stress reduction with a factor of 4–5 in the deck plate and a factor of 3–4 in the trough wall, thus extending the service life of the OSD for extra decades without additional maintenance.

Figure 22.12 Very dense RHPC reinforcement (Buitelaar et al., 2004).

Figure 22.13 RHPC overlay on OSD deck plate (Buitelaar et al., 2004).

5.3 Innovative concepts

In recent years, alternative deck systems have been proposed that also aim to focus on orthotropic behavior but try to avoid the numerous welds and resulting fatigue problems. Possibilities include the use of other materials (high-strength steel/aluminum), other arrangements of the stiffeners, or combinations of both (combining two steel plates with an orthogonal concrete grid between them). While research is available, no actual realizations exist at present.

6 Conclusions

OSDs have been employed worldwide, particularly in Europe, Asia and South America. However the use of orthotropic steel has been fairly limited, such that their use represents a very small percentage of total bridges. The construction and fabrication techniques employed are very important to the successful use of OSDs. OSDs typically require detailed construction specifications and special quality control procedures during fabrication. While fatigue effects remain the most important design issue, it should be stressed that recent research development and a detailed design will avoid these problems, resulting in one of the most lightweight and slender deck structures available. Overall, it can be stated that the OSD remains a valuable bridge concept, especially for larger-span bridges.

References

American Association of State Highway and Transportation Officials (AASHTO), 2014. LRFD Bridge Design Specification. Sections 2, 4, 6, and 9. American Association of State Highway and Transportation Officials.

American Institute of Steel Construction (AISC), 1938. The Battle-Deck Floor for Highway Bridges. American Institute of Steel Construction. p. 34.

Barsoum, Z., Lundbäck, A., 2009. Simplified FE welding simulation of fillet welds—3D effects on the formation of residual stresses. Eng. Fail. Anal. 18, 2281–2289.

Buitelaar, P., Braam, R., Kaptijn, N., 2004. Reinforced high-performance concrete overlay system for rehabilitation and strengthening of orthotropic steel bridge decks. In: Proceedings of Orthotropic Bridge Conference, Sacramento, CA, pp. 384–401.

Cornelius, W., 1952. Die Berechnung der Ebenen Flächentragwerke mit Hilfe der Theorie der Orthogonal-Anisotropen Platten. Der Stahlbau. 21, 21–43 (in German).

De Backer, H., 2006. Optimization of the Fatigue Behavior of the Orthotropic Bridge Concept by Improved Dispersion of Traffic Loads. Ghent, Belgium. p. 516 (Ph.D. thesis, in Dutch).

De Jong, F.B.P., Kolstein, M.H., 2004. Strengthening a bridge deck with high-performance concrete. In: Proceedings of Orthotropic Bridge Conference, Sacramento, CA, pp. 328–345.

Eurocode 3, 2009. NBN EN 1993-1-9:2005. Eurocode 3: Design of Steel Structures—Part 1-9: Fatigue. European Committee for Standardization, Brussels, Belgium.

Feldmann, M., Sedlacek, G., Gessler, A., 2007. A system of steel-elsatomer sandwich plates for strengthening orthotropic bridge decks. Mech Comp. Mat. 43 (2), 183–190.

Hubman, M., Gunther, H.P., Kuhlman, U., 2013. Maintenance of orthotropic steel bridge decks with longitudinal Y-stiffeners. In: Proceedings of IABSE Conference. Assessment, Upgrading and Refurbishment of Infrastructures. Rotterdam, Netherlands, pp. 498–499.

Institution of Civil Engineers (ICE), 1972. Steel box girder bridges. In: Proceedings of the International Conference. Thomas Telford Publishing, London.

Kiss, K., Szekely, E., Dunai, L., 1998. Fatigue analysis of orthotropic highway bridges. In: Proceedings of the 2nd international Ph.D. Symposium in Civil Engineering. Budapest, pp. 1–8.

Klöppel, K., Roos, E., 1960. Statische versuche und dauerversuche zur frage der bemessung von flachblechen in orthotropen platen. Der Stahlbau. 29, 361 (in German).

Kühn, B., et al., 2008. Assessment of Existing Steel Structures: Recommendations for Estimation of Remaining Fatigue Life. Joint Research Centre - European Convention for Constructional Steelwork, Aachen, Germany. p. 92.

Mangus, A.R., 2014. Orthotropic steel decks. In: Chen, W.F., Duan, L. (Eds.), Bridge Engineering Handbook, second ed. Taylor & Francis Group, Boca Raton, Florida, pp. 589–645.

Nagy, W., Van Bogaert, P., De Backer, H., 2014. Improved fatigue assessment techniques of connecting welds in orthotropic bridge decks. In: Proceedings and Abstracts of the Eurosteel Conference. Naples, Italy. ECCS (European Convention for Constructional Steelwork), Brussels, Belgium, pp. 737–738.

Pelikan, W., Esslinger, M., 1957. Die Stahlfahrbahn Berechnung und Konstruktion. MAN-Forshungsheft 7. Augsburg-Nurnberg, Germany (in German).

Polak, J., 2007. Cyclic deformation, crack initiation, and low-cycle fatigue. In: Milne, I., Ritchie, R.O., Karihaloo, B. (Eds.), Comprehensive Structural Integrity, Vol. 4: Cyclic Loading and Fatigue. Elsevier Applied Science, London, pp. 1–39.

Sedlacek, G., 1992. Orthotropic plate bridge decks. In: Dowling, P.J., Harding, J.E., Bjorhovde, R. (Eds.), Constructional Steel Design: An International Guide. Elsevier Applied Science, London, pp. 227–245.

Teixeira de Freitas, S., Kolstein, M.H., Bijlaard, F., 2010. Composite bonded systems for renovations of orthotropic steel bridge decks. Comp. Struct. 92, 853–862.

Teixeira de Freitas, S., Kolstein, M.H., Bijlaard, F., 2011. Sandwich system for renovation of orthotropic steel bridge decks. J. Sandwich Struct. Mat. 13 (3), 279–301.

Teixeira de Freitas, S., Kolstein, M.H., Bijlaard, F., 2013. Lightweight reinforcement systems for orthotropic bridge decks. In: Proceedings of IABSE Conference. Assessment, Upgrading, and Refurbishment of Infrastructures. Rotterdam, the Netherlands, pp. 500–501.

Vandepitte, D., 1979. Brugvloeren. In: Berekening van Cconstructies. Story Scientia, Ghent, Belgium, pp. 593–638.

Wolchuk, R., 1963. Design Manual for Orthotropic Steel Plate Deck Bridges. American Institute of Steel Construction (AISC), New York.

Bridge foundations

23

Modeer V.[1], Bharil R.K.[2], Cooling T.[1]
[1]AECOM, St. Louis, MO, USA
[2]AECOM, Orange, CA, USA

1 Introduction

In order to design the interface between a bridge structure and the earth-supported footings, abutments, embankments, retaining walls, and settlement slabs, a structural engineer relies heavily on geotechnical investigation reports and interaction with various project geoprofessionals, including geotechnical engineers, seismologists, and engineering geologists. The bridge foundation design process involves the planning for field exploration, field and laboratory investigations, development of foundation design parameters, field testing, and geotechnical analysis with reference to various site-specific soil and geologic conditions. Special situations of foundation installations such as end-bearing design on fractured to solid rock, development of seismic response spectrum or time-history analysis, and analyses/testing methods to obtain foundation capacity (without damaging the piles) play integral roles in today's bridge foundation engineering practice. The bridge foundation design is a highly iterative process between the structural engineer and the geotechnical engineer due to factors such as stiffness/displacement compatibility, load configuration, and geoconstructability. The bridge foundation risks can be greatly reduced by applying innovative techniques to accurately predict the sub-surface conditions, number and depths of exploratory bore holes, ground improvements to manage future settlements and liquefaction potential, proper selection of the bridge foundation type (e.g., spread footings, driven piles, and drilled shafts), field monitoring and testing during construction, and factoring of sub-surface conditions during the bridge type selection process.

The purpose of this chapter is for bridge structural engineers to reinforce the importance of the roles that geotechnical professionals play in the planning, design, and construction of major and innovative bridges. This chapter is not intended to provide specific or detailed geotechnical engineering design guidance, rather, it outlines the design requirements and overview of the services that geotechnical professionals are required to provide for major and innovative bridge projects. A bridge design team will require the service of the geotechnical profession's most highly regarded foundation specialists to successfully complete a major or innovative bridge project. There are a select few professional firms with key individuals and the support of experienced teams that can provide such services.

Geotechnical professionals include geotechnical engineers, geotechnical engineers experienced in rock mechanics, geologists, seismologists, geophysicists, and engineering geologists. Structural engineers require the service of geotechnical

Innovative Bridge Design Handbook. http://dx.doi.org/10.1016/B978-0-12-800058-8.00023-2

professionals in order to properly vet foundations for the geologic conditions at a specific site. Geotechnical professionals and structural professionals interface at the ground surface in the design of a bridge structure.

Most, if not all, major bridges cross bodies of water, and the geologic and geotechnical related concerns are magnified at the location of rivers, streams, lakes, and ocean crossings. The major reason that a body of water exists in a given location is due to geologic processes such as tectonics, volcanism, glaciation, rifting, and faults. For example, the lower Mississippi River is maintained to the west by the adjacent Pleistocene terrace, the Nile River crosses five major regions that differ in geologic history (Butzer, 1980), the Rhine River flows through 11 geologic regions (Preusser, 2008; Woodward, 2007), and the Yangtze River (Zheng, 2013) flow was directed from the Tibetan uplift across the Jiangnan Basin developing deep fluvial deposits. Such cited information is the beginning of the data that geotechnical professionals consider when recommending foundations for major and innovative bridges.

The need for geotechnical professionals should be recognized from the planning to the construction phase on bridge projects. The design team must have a geotechnical professional thoroughly involved in the design and construction of major and innovative bridges.

2 Determination of the geologic setting

A professional or engineering geologist should be engaged for the phase of design that involves determining the geologic setting for the construction. The professional geologist will use many sources to determine the geologic setting. Many, if not most, countries have extensive geologic mapping by a natural resource agency or similar group. A geologist will review any published work on the geology of the bridge area. It is not uncommon for a bridge location to be moved or the location of foundation elements to be changed in the type, size, and location (TSL) phase of design due to the findings of a geologist at this stage of the work. The scope of a geologic setting report will depend on the complexity of the geology. An example of complex geology is the discovery of Karst topography as discussed in the closing section of this chapter.

All major bridge locations require geotechnical specialists to work closely with the structural engineer and project leads. Major and innovative bridge design and construction requires that both geotechnical and geology professionals are key contributors on the project team. There is a huge amount of accessible, published technical reports and papers that reference the contributions of geotechnical and geology professionals in projects' success.

3 Geotechnical investigation report

A geotechnical investigation report is based on (i) the results of the geologic setting and assimilated geologic research provided by a professional geologist and principal geotechnical engineer and (ii) the range of geotechnical issues and potential solutions for selecting the foundation type. The geologist and geotechnical engineer work with the structural engineer in the development of the engineering design of the proposed bridge concept. This phase is most likely completed concurrently with the TSL structural project phase. Often, prior geotechnical investigations completed on nearby projects serve as the starting point for early project planning and programming of a more thorough on-site geotechnical investigation.

The borings are drilled and samples taken at the proposed location of foundation or bridge pier locations. Typical field investigations are performed by a drill rig that obtains soil samples to be laboratory tested (Figure 23.1). Granular or sandy soils are sampled through the Standard Penetration Test (SPT), as described in ASTM D1586, ISO 22476, and Australian Standards AS 1289.6.3.1. The SPT provides a disturbed sample only suitable for laboratory index property tests. The sample tube is driven 150 mm into the ground, and then the number of blows needed for the tube to penetrate each additional 150 mm up to a depth of 450 mm is recorded. The sum of the number of blows required for the second and third 150 mm of penetration is termed the *standard penetration resistance,* or the *N-value.* The blow count is used to estimate the density of granular soils and shear strength of clay soils for empirical geotechnical correlations of the sampled stratum (Lunne et al., 1989; Robertson et al., 1983; Meyerhof, 1956; Rogers, 2006).

Relatively undisturbed samples are taken of fine grained or clay soils. The Standard Practice for Thin-Walled Tube Sampling of Soils for Geotechnical Purposes (ASTM D 1587) describes the process of taking samples of fine grained soils. Kontopoulos (2012) describes various causes and avoidance techniques of fine grained sample disturbance in his Ph.D thesis. Soil testing is a geotechnical professional field that has many test methods that are appropriate for the geotechnical analyses that will be performed. The tests are too numerous to describe in this chapter. The design methodology is outlined next for the specific foundation type.

Depending on the soil type the Cone Penetrometer (CPT) (ASTM Standard D 3441and ASTM D-5778) (Figure 23.2) or the Dilatometer (DMT) (ASTM D6635) can provide correlations of soil shear strength, consolidation characteristics, and soil classification information on the soils encountered. The seismic CPT (or SCPT), and the seismic DMT (or SDMT) can be used to determine shear wave velocity by means of an accelerometer on the SCPT or the SDMT that records shear wave velocity by recognizing the vibrations of a ground-level vibration source. Most CPT testing is completed with a porous tip that records the pore pressure of the soil as the CPT is advanced. Stopping the CPT at various depths in cohesive soils and recording the time to dissipation of the increase in pore pressure by CPT advance can be correlated to with soil types to obtain an estimate of the coefficient of consolidation and permeability (CPT: Lunne et al., 1989; Mayne, 2007; Tumay et al., 1981; Tumay, 1997; Iliesi et al., 2012; DMT: Marchetti et al., 2001). This CPT resistance or N-value also serves as an indicator of the probable foundation condition for structural engineers as well.

Geophysical exploration methods are used to supplement borings and other intrusive sampling and testing of soil and rock stratigraphy. Geophysical methods are extremely valuable in investigating karst conditions (see Section 6.3.7). There are

Figure 23.1 Warren George drill rig, Drilling on the Hudson River for the new Tappen Zee Bridge
(courtesy Tom Cooling and the New York State Thruway Authority).

Figure 23.2 CPT rig on the Hudson River for the new Tappen Zee Bridge
(courtesy Tom Cooling and the New York State Thruway Authority).

various methods as described in Principles of Applied Geophysics (Parasnis, 1996). Geophysical methods include seismic reflection, seismic refraction, magnetic, electromagnetic, and electrical resistivity, and conductivity surveys. Advanced methods of geophysical testing are multichannel and spectral analyses of surface waves (MASW and SASW). Geophysics is a highly specialized field of the geotechnical profession and an expert should be used for this type of exploration (Dobrin and Savit, 1988; Kearey et al., 2002).

Rock coring is the sampling method for hard rock. Standard Practice for Rock Core Drilling and Sampling of Rock for Site Exploration (ASTM D2113) provides the standard method used in the United States. Rock core testing is a geotechnical professional field that has as many test methods as soil testing. The testing on core samples is only part of the analysis for determining the strength of a rock mass. The tests are too numerous to describe in this chapter. The other methods to determine rock mass design characteristics are the rock quality designation (Deere and Deere, 1988) and the rock mass rating (Bieniawski, 1989). The design methodology for rock foundations is outlined next.

4 Foundation selection during the TSL project phase

The bridge structural engineer will develop the TSL based on the results and constraints developed by influencers such as architects, public opinion and input, waterway width constraints such as maintaining a clear zone for navigation, and other government-related regulatory constraints. These influences will allow the structural engineer to develop estimated foundation loads for further interaction with the geotechnical engineer. The location of the foundation elements and the foundation loads will determine the next phase of geotechnical investigation.

The bridge structural engineer should consult with the geotechnical engineer during this phase to develop the potential foundation types. The type selection process will determine foundations that can not only resist vertical, dead, and traffic-induced live loads, but also lateral forces from potential vessel collision, wave and tidal forces, and seismic loads. This is a critical phase of the project, where the geotechnical and geology professionals will combine the structural elements with the geologic conditions to develop innovative foundation solutions. The main tower foundations and anchorage of the Akashi Kaikyō Bridge are examples of this interaction. Kashima and his team developed an innovative robotic system to clean the seabed rock surface and place concrete in the submerged large diameter caissons (Kashima, 1991).

Foundation selection for innovative and major bridges requires a lead geotechnical engineer and supporting team that is very experienced in foundation design and has a proven ability to be innovative in finding unique foundation solutions. The geologist should review the stratigraphic interpretations by the geotechnical engineer for geologic implications to the selected foundation type.

5 Geotechnical design report

The geotechnical design report will be detailed specifically for the bridge type selected in the TSL phase. The soil and underlying rock will be characterized in detail. The vertical, uplift, and lateral loads will be estimated so that foundation types can be reviewed and analyzed for applicability.

The offshore oil and offshore wind turbine foundation industry have been responsible for significantly improving major and innovative bridge foundations. The extreme depth and constant lateral loading from wave action has literally created a vacuum in foundation requirements that traditional onshore foundation design and foundations could not fill (Byrne, 2011). For example, large 3 m-and-larger-diameter driven piles or reusable spud piles on jack-up platforms or oil investigation drill rigs are common. There is a significant difference in foundation design for onshore versus offshore structures. Offshore foundations or substructures design approaches are based on much more flexible piles and are more concerned with lateral capacity than stiffness (Houlsby et al., 2005). Though this is a departure from the stiffness requirements of bridge foundations and structures, many of the offshore concepts can and have been applied to bridges over deep channels and in heavy wave conditions.

The geotechnical design report will also identify foundation construction related potential problems. Such conditions could be glacial sands and gravels that will require hard driving for driven piles. The same sands and gravels would likely require full casing advanced during drilling for a drilled shaft foundation. Soft soil deposits are also commonly present in major river crossings in deltaic deposits such as in the lower Mississippi River and the Yangtze River delta (Liu et al., 1992).

6 Foundation design

6.1 Driven pile foundations

6.1.1 H-type steel piles

H-section piles are normally designed as end-bearing piles (Figures 23.3 and 23.4). (ASTM A690/A690M, Eurocode 3, Part 5: Piling). H-shaped piles are used because of their structurally compact section. A structurally compact H-section allows high driving stresses and a more predictable location of the tip relative to the pile top. Driving a steel beam that has a high section modulus about one axis will allow bending, permanent deformation, and damage about the weak axis during driving. H piles are normally used for end bearing in dense sand, hard clay, clay shale, or a hard rock formation. Many of the formations that are suitable for end bearing are overlain by weathered zones, vary in elevation across the top of the stratum, or may have overlying inclusions such as cobbles or boulders and require a "driving shoe" be added to the tip to aid in preventing damage and penetrating through obstacles. End bearing "H" piles should have wave equation analyses performed for the subsurface conditions to better match the pile driving hammer to the pile and not cause damages to the pile during driving. The wave equation analyses can provide pile capacity estimates, but is much better suited to determining driving stresses for a given pile and hammer system in a given soil profile when driven to refusal.

Figure 23.3 Route 490 Ramp Bridge at Exit 27, Delmag, D 19-32 (courtesy New York State DOT).

Figure 23.4 West Dodge Project H pile driving (courtesy Nebraska DOR).

6.1.2   Pipe piles

Pipe piles can be fabricated as extruded or rolled thin-walled pipe piles, spiral welded steel, extruded steel, and rolled steel (ASTM A252, Standard Specification for Welded and Seamless Steel Pipe Piles). The available size range of pipe piles and the stiffness that can be increased by increasing the pipe wall thickness has made them desirable for major bridge foundations. Pipe piles can also be driven with closed end and filled with reinforced concrete as a structural element. Depending on location, the corrosion of steel wall can be a concern and should be accounted in sizing the pipe. Typically, pipe piles greater than 1 m in diameter are open end. Driving very large 3-m-and-larger-diameter piles became available as a result of the offshore foundation construction (Figures 23.5 through 23.7). Pile driving equipment became available for these piles also as a result of the need to drive the large piles for the offshore industry. There is a likelihood that thin-walled pipe piles used for friction and end bearing could be damaged during driving if

Figure 23.5 Driving 1.82-m-diameter, 85-m-long open-end pipe piles for the New Tappan Zee Bridge over the Hudson River, New York
(courtesy New York State Thruway Authority).

Figure 23.6 Driving 2.5-m piles for the San Francisco-Oakland Bay Bridge with hydraulic impact hammer
(courtesy California DOT).

Figure 23.7 Menck, MHU1700T hammer
(courtesy California DOT).

the pile and hammer system is not matched properly. Wave equation analyses should be performed for thin-walled pipe pile driving to determine driving stresses for a given pile and hammer system in a given soil profile.

6.1.3 Concrete piles

Driven concrete pile, which is typically precast and prestressed, sometimes makes for an economical foundation in certain areas such as Florida and California in the United States, where casting yards are available (Figure 23.8). Square concrete piles from 60 to 90 cm are commonly used with larger sizes up to 3 m that are less common, but used for major bridges. These are typically used as friction piles but can be outfitted with steel driving tips for driving to rock or very dense soil to develop high end-bearing. Precast, prestressed concrete cylinder piles, up to 140 cm in diameter are also common in coastal areas and are driven as primarily as friction piles. Concrete piles provide high durability in marine environments and can provide high capacity. However, concrete piles are more difficult to spice than steel, are more easily damaged during driving, and typically require larger lifting equipment than steel piles. Wave equation analysis is required to estimate driving stresses and drivability during the pile design phase. Dynamic testing during pile installation of *indicator piles* in each cap is needed to monitor driving stresses to verify integrity, and pile capacity. Concrete piles are designed based on the Guide to Design, Manufacture, and Installation of Concrete Piles (ACI 543R).

6.1.4 Integrity testing

• ASTM D-4945 – 00 Standard Test Method for High Strain Dynamic Testing of Piles (Pile Driving Analyzer, PDA).

Figure 23.8 Concrete pile driving, Napa, CA (Argyriou, licensed under Wikipedia Commons, 2006).

6.1.5   Load testing

- ASTM D1143, Standard Test Methods for Deep Foundations Under Static Axial Compressive Load
- ASTM D3689, Standard Test Methods for Deep Foundations Under Static Axial Tensile Load
- ASTM D 3966 -90 Standard Test Method for Piles Under Lateral Loads
- ASTM D-4945 – 00 Standard Test Method for High Strain Dynamic Testing of Piles (Pile Driving Analyzer, PDA) (Using CAPWAP option for pile load estimate).

6.1.6   Design methodology

There are many published design manuals that can be used for design. Very comprehensive manuals are published by the US Federal Highway Administration (FHWA) and are available the public domain. In addition, several authored and edited references (Hannigan et al., 2006; Washington State Department of Transportation, Construction Division, 2014; Fellenius, 2014; Sands, 1992; Smoltczyk, 2003;

Fang, 1991; Tomlinson, 1994; Das, 2011; Rowe, 2001; Naser et al., 2011; and others) provide the experience of various authors that prove quite valuable.

6.1.7 Drilled foundations

There are many types of drilled foundations that are appropriate for various types of bridges. One major advantage of any drilled foundation is the minimal vibration and noise that occurs during installation.

Drilled foundations include augered piles with diameters less than 75 cm. These augered piles include continuous flight auger (CFA) hollow stem augered piles. These are normally suitable for use as friction piles. The auger is drilled to the designated depth and pressurized grout is installed through the hollow stem as the auger is withdrawn from the excavated hole. Another type of augered pile is the drilled displacement pile. While not as common as the CFA pile, the stated principle advantage of the drilled displacement pile is that the displaced soil is compacted or densified. The developers state that this process creates a larger effective diameter pile and, therefore, a higher capacity augered pile. The advocates of this type of pile state that they can provide an accurate capacity of the pile from the installation measurements. This pile type cannot be confidently designed without field load tests coupled with a consistent soil stratigraphy. Micropiles (e.g., pin piles, needle piles, and root piles) are a drilled foundation type that has advantages in supporting foundation underpinning, foundations in confined areas, adding foundation capacity, and as a drilled foundation alternative (Cadden, 2004). Micropiles founded in rock have been tested to vertical capacities up to 4500kN (Cadden, 2004). All of these pile types can provide sufficient vertical capacity by using a pile group with a structural cap. The need for lateral capacity or structural rigidity for seismic loads needs to be evaluated in detail for micropiles (Cadden and Gomez, 2002).

Most, if not all, major and innovative bridges will have drilled shaft foundations that have diameters greater than 75 cm. Typical drilled shaft foundations can be open auger drilled and/or cased after drilling. If the drilled shaft diameter is generally 6 ft (180 cm) or greater, or the soil is determined to collapse in an uncased drilled open hole, the shaft may be cased in a phased manner during the advancing of the shaft or with a telescoping casing (Brown et al., 2010). The soil in the shaft is removed by the auger drilling process and replaced with concrete. The concrete shaft is normally reinforced to some depth below the ground surface with a steel rebar cage to increase stiffness to align with the structural design. Use of temporary and permanent steel casing should be evaluated if subsurface springs, unsuitable soil, or voids may be present. Large diameter shafts typically speed up the construction by providing large foundation capacities in a few piles, but also impose risks to the project schedule if major unanticipated subterranean soil conditions are discovered during drilling. The contractor ultimately selects the drilling method, but the recommendations of an experienced geotechnical engineer should be considered a professional assessment and will likely be included in the design documents. Excellent summaries of drilled shaft design considerations are included in Brown's works (2010, 2012).

Figure 23.9 Large crane mounted drilled shaft rig. Bond Bridge, MoDOT, Kansas City (courtesy Dan Brown, Dan Brown and Associates).

Figure 23.10 Top-driven remote controlled drill rig (courtesy Dan Brown, Dan Brown and Associates).

Figure 23.11 Auger retrieving rock from shaft (courtesy of MnDOT and WisDOT St Croix River Bridge, Extradosed Cable Stayed Structure).

Figure 23.12 Cleaning a rock auger into a spoils barge (courtesy of MnDOT and WisDOT St Croix River Bridge, Extradosed Cable Stayed Structure).

6.1.8 Integrity testing

Drilled shafts and augercast piles require integrity testing to verify that the infilled concrete is a minimum diameter. There are three accepted methods for the integrity testing of drilled foundations:

- The first is low-strain impact integrity or low-strain dynamic tests (ASTM D5882 Standard Test Method for Low-Strain Impact Integrity Testing of Deep Foundations). This is a simple test in principle. Accelerometers are attached to the top of the concrete shaft or pile and the concrete is struck with a blow that sends a compressive wave into the pile. The return times that are less than what is calculated for a full length indicate changes in the cross section of the pile. Near surface construction deformities and reinforcement can return reflected waves that interfere with the deeper reflections. The testing company should provide the length-to–maximum diameter ratio at which their equipment can be effective, which is normally 30 or less. These tests are commonly used with CFA piles with lengths of 10 to 15 m or less.

- The second test type is cross-hole sonic logging (ASTM D6760-08 Standard Test Method for Integrity Testing of Concrete Deep Foundations by Ultrasonic Crosshole Testing). This method was derived from cross-hole shear wave testing in geotechnical boreholes. Cross-hole sonic logging requires the attachment of a minimum of three steel tubes to the reinforcing steel cage. The tubes must accommodate the ultrasonic transmitter and receivers. The sonic wave arrival times are converted to wave speed that must be compared to standard concrete at various set times. Differences in the wave speed along the shaft indicate changes in the shaft circumference. A drawback of this test is that it only evaluates concrete within the reinforcing cage.

- The third type, thermal integrity testing of drilled shafts, is not yet an ASTM standard, and yet it is a very accurate method to determine shaft integrity. Mullins 2009 states that: "Thermal Integrity testing utilizes the thermal signature generated during the hydration phase of the concrete curing process. Deviations in the thermal signature from a gradient predicted by modeling of the concrete mix design and soil profile can indicate anomalies in the shaft cross-section. A decrease in the measured temperature may indicate a decrease in shaft cross-section, whereas an increase in measured temperature may be indicative of a bulge or increase in the shaft cross-section is capable in detecting anomalies outside the reinforcing cage such as bulging outward and necking inward (Mullins, 2007)."

6.1.9   Load testing

- ASTM D1143 Standard Test Methods for Deep Foundations Under Static Axial Compressive Load.
- ASTM D3689 Standard Test Methods for Deep Foundations Under Static Axial Tensile Load.
- ASTM D3966 Standard Test Method for Piles under Lateral Loads.

6.1.10   Design methodology

There are many published design manuals that can be used for design of drilled shafts and drilled piles. These very comprehensive manuals (Fang, 1991; Smoltczyk, 2003; Brown, 2012; Brown et al., 2010; Sands, 1992; Cadden, 2004; Sabatini et al., 2005) are published by the U.S. Federal Highway Administration (FHWA) and are available in the public domain.

6.1.11   Foundations on rock

Bridge footing or mat-type foundations on rock are designed as described in Wyllie (1999), Smoltczyk (2003), and Rock Foundations (1994). An example will best illustrate the type of footing or mat foundation for major and innovative bridges. The methodology for the example described here could be used for foundations as innovative as the Salginatobel Bridge near Schiers, CH (completed in 1930), and the Schwandbach Bridge near Bern, CH (completed in 1933), both designed by Robert Maillart. These bridges are considered works of art and have parapet thrust or bearing type foundations on rock (Billington, 2003). The Hoover Dam Bypass Arch Bridge, Mike O'Callaghan-Pat Tillman Memorial Bridge, has nine precast segmental column sets founded on rock that utilized structural, geotechnical/rock mechanics, and geology professionals to investigate and design. The bridge received the International Federation of Consulting Engineers (FIDIC) Centenary Award in 2013.

Suspension bridges require the uplift and horizontal forces at the bridge ends to be either held in place by tiedown shafts, rock anchored mats, or very large gravity anchorages or anchor blocks. The largest suspension bridges require extremely large anchorage. The Golden Gate Bridge and the world's longest suspension bridge, the Akashi Kaikyō Bridge, have very large concrete anchorage blocks. The anchorage block system is subjected to both uplift and lateral forces from the suspension cable ends. The main towers for many major and innovative bridges are often supported on rock by means of a large caisson. Similarly, conventional arch bridges also require sound foundation conditions to resist enormous amount of horizontal thrust exerted by arch action. Some cable-stayed bridges may need end soil anchorages, but often use the structural means to self-balance its horizontal forces.

The Akashi Kaikyō Bridge anchorage is an example of rock foundation design for the anchorage and main towers: "Anchorages measure 63 meters by 84 meters in plan and extend into the Kobe and granite layers at the site. This required special foundation construction technology. The Honshu anchorage had to be embedded 61 meters below sea level, and the anchorage excavation had to be performed in open air. Therefore, an 85-meter-diameter circular slurry wall, 2.2 meters thick, was constructed and subsequently used as a retaining wall. Excavation within the slurry wall was followed by the placement of roller-compacted concrete to complete anchorage foundation construction. The Awaji anchorage foundation was constructed using steel pipes and earth anchors to support the surrounding soil. The excavated foundation was filled with specially designed flowing-mass concrete. Both anchorages were completed with the construction of a huge steel supporting frame used to anchor the main suspension cable strands" (Cooper, 1998). Each anchor weighed an average of 390,000 metric tons. "The foundation (main towers) was constructed using a newly developed laying down caisson method. Steel caissons, 80 meters in diameter and 70 meters in height, were towed to the tower sites, submerged, and set on the pre-excavated seabed (pre-excavated to rock)" (Cooper, 1998). The seabed rock was evaluated to support the 181,000 metric tons of vertical force along with the forces from wind, earthquake, wave action, and vessel collision. Prior to setting the caisson and concreting, the seabed was prepared by using "a cleaning robot to clean the undersea bedrock surface" (Kashima, 1991).

Rock mass properties and accurate definition of the rock surface area are necessary to design and construct a foundation on rock. The above example of the Akashi Kaikyō Bridge foundation construction is an example of the geotechnical engineer, the geologist and the geotechnical rock mechanic professional working together with the structural team.

6.2   Foundation construction

The easiest to access and most complete manuals for construction monitoring and inspection are the FHWA manuals. The Pile Driving Contractors Association, for driven piles, and the Deep Foundations Institute, for drilled piles, provide excellent construction guidelines and sample specifications. Often performing an independent constructability evaluation or soliciting input from the foundation contractors or organizations can limit surprises during the bidding and construction phases of the bridge projects. There are other excellent sources for foundation construction in various publications (Sabatini et al., 2005; Brown et al., 2010; Hannigan et al., 2006b; Smoltczyk, 2003).

6.3   Special considerations

6.3.1   Liquefaction

Liquefaction occurs when earthquake ground-motion vibrations cause pore water pressure within a mass of mainly granular soil particles to lose contact with one another. The saturated granular soil mass behaves like a liquid and loses shear strength. Foundations lose support, and mat or shallow foundations sink or tilt, pile foundations lose lateral support during the earthquake ground motions, and saturated slopes will slide (Idriss and Boulanger, 2008; Ashford et al., 2011; Fellenius and Siegel, 2008). Ground improvements and accounting liquefaction into the structural design are commonly used methods to mitigate the effects of liquefaction.

6.3.2   Lateral pile loads

The analyses of lateral pile loads with computed P-Y curves: lateral load analysis is necessary for the design of pile foundations that can withstand seismic, wind, ice, wave action, river and tidal current loading of the bridge piers and superstructure. The lateral analysis of piles is defined in detail in Bearing Capacity of Soils (1992), the United States Army Corps of Engineers (USACE) manual and Duncan et al. (1994). Current available software for p-y based analysis is described in Pando (2013). Structural designers will typically account for such lateral pile loads in the structural design of the foundation.

6.3.3   Downdrag and Drag Force on Driven and Drilled Foundations

Fellenius (2014) defines *downdrag* as the pile settlement caused by soil adhering to the pile shaft. He defines drag force as the sum or integration of the unit negative skin friction. These are important distinctions to understand the neutral plane or unified design approach for pile design. The neutral plane method provides an understanding of the load distribution along the pile shaft (Allen, 2005). Siegel et al. (2014) provides an excellent summary of the neutral plane method, as well as comparison to the former state of practice explicit method. These forces can be mitigated, to some degree, by proper construction staging schemes; however, the residual effects must be accounted for in the foundation design of the new and adjacent older foundation.

6.3.4   Vessel collision

The most complete and concise information for design of foundations subject to various type of vessel collision is in AASHTO Guide Specifications and Commentary for Vessel Collision Design of Highway Bridges, 2nd Edition, with 2010 Interim Revisions, 2009. There is also guidance for assessing the risk of a bridge foundation to a vessel collision in the aforementioned document. The design of foundation for accidents/collision is risk based.

6.3.5   Coastal storms

Bridge foundation loads from coastal storms are design elements for every bridge along or near a coast. The storm does not have to be a hurricane or typhoon, yet these represent the most extreme coastal storms. High winds causing tidal surges and earthquake-induced tsunamis are example of conditions identical to coastal storms. The publications AASHTO Guide Specifications for Bridges Vulnerable to Coastal Storms (2008) and Douglass and Krolak (2008) provide detailed information for designing foundations to resist coastal storms.

6.3.6   Scour

Erosion of the soil around the bridge foundation is defined as scour. Scour is the primary cause of bridge failure in the United States. There are more than 20,000 highway bridges that are rated *scour critical* (Hunt, 2009). Scour can occur from coastal storms, streambed elevation changes due to upstream or downstream conditions, lateral shifting of a stream, harbor, or river dredging, and tidal action. The publication by Arneson et al. (2012) describes evaluation and monitoring methodologies and mitigation methods. Additional references for scour are Lagasse's et al. (2007); Thompson and Beasley (2012) Countermeasures to Protect Bridge Piers from Scour, and Hunt's (2009) Monitoring Scour Critical Bridges. The primary cause of bridge failure is due to scour, so the scour design must be accounted in new bridges' foundation. The most straightforward ways are to place spread footings below the anticipated scour depth (it is best touse spread footings only on bedrock), always use deep foundation (i.e., piles and shafts) on erodible soil, not count on any soil resistance above scour depths, and always assume that waterways are dynamic and provide for redundancy in the system whenever possible.

6.3.7   Karst conditions

Karst landforms are evident in every hemisphere and specifically where there are carbonate rock formations at or near the ground surface. The geologist will recommend a very detailed geotechnical investigation report to determine where the epikarst begins and downward to the top of unweathered rock or the karst surface. The epikarst is essentially the upper boundary of a karst system where groundwater leaches a weak carbonic acid into the soil and organic elements to cause an acid reduction by the calcium chloride rock. The calcium chloride is removed by this reaction and the process continues (Kutschke, 2011; Kannan, 2005). Karst geologic conditions should be avoided as the investigation, design, and construction in this condition is generally very expensive. If karst cannot be avoided, the areal and vertical extent must be determined. Geophysical and field borings explorations are combined to evaluate the karst extent. There is not a singular process to explore, characterize, design, and construct foundations in karst. The karst in Florida, US is very different from the karst in Georgia, US, though they are less than 700 km apart. Excellent compilations of various case histories from around the world can be found in Beck (1995 and 2005).

6.4   Foundation design standards and codes

- AASHTO, 2012. LFRD Bridge design specifications, Section 10 Foundations. American Association of State Highway Officials. (revised often) Adopted by states in the US as design codes.
- Canadian Standard Council, 2006. Canadian Highways Bridge Design Code, Section 6, Foundations. Canadian Standard Association, CSA-S6-06, Code and Commentary
- Eurocode, 1997. EN 1997-1. Geotechnical Design, Part 1: General Rules
- Indian Code, 2006. Indian Code of Practice for Design and Construction of Pile Foundations IS: 2911, 2006
- Chinese National Standard (CNS), 2002. Chinese code (TB10002.5-2005)
- European EN, 1997. Eurocode 7: Geotechnical design
- United Kingdom - BS 8004:1986 Code of practice for foundations. Replaced by Eurocode 7
- United States, ASTM, ASCE 70-5, BOCA.

References

Allen, T., 2005. Development of Geotechnical Resistance Factors and Downdrag Load Factors for LRFD Foundation Strength Limit State Design. Federal Highway Administration, Washington, D.C Publication No. FHWA-NHI-05-052.

American Association of State Highway and Transportation Officials, 2009. Guide Specifications for LRFD Seismic Bridge Design, first ed., with 2010 Interim Revisions American Association of State Highway and Transportation Officials, Washington, DC.

American Association of State Highway and Transportation Officials (AASHTO), 2008. Guide Specifications for Bridges Vulnerable to Coastal Storm. American Association of State Highway and Transportation Officials, Washington, DC.

American Association of State Highway and Transportation Officials (AASHTO), 2009. Guide Specifications and Commentary for Vessel Collision Design of Highway Bridges, second ed., with 2010 Interim Revisions American Association of State Highway and Transportation Officials, Washington, DC.

Arneson, L.A., et al., 2012. Evaluating Scour at Bridges, fifth ed. FHWA Office of Bridge Technology, Washington, DC. FHWA-HIF-12-003 HEC-18.

Ashford, S.A., Boulanger, R.W., Brandenberg, S.J., 2011. Recommended design practice for pile foundations in laterally spreading ground. PEER Report 2011/04, Pacific Earthquake Engineering Research Center, University of California, Berkeley, Berkeley, CA.

Bearing Capacity of Soils, 1992. USACE EM 1110-1-1905. Department of the Army, Washington, DC.

Beck, B. (Ed.), 1995. Karst Geohazards: Engineering and Environmental Problems in Karst Terrains. A.A. Balkema, Rotterdam, Netherlands.

Beck, B. (Ed.), 2005. 10th Multidisciplinary Conference on Sinkholes and the Engineering and Environmental Impacts of Karst. American Society of Civil Engineers. GSP 144.

Bieniawski, Z.T., 1989. Engineering Rock Mass Classifications. Wiley, New York.

Billington, D.P., 2003. The art of structural design: a swiss legacy. Princeton University Art Museum, Princeton, NJ.

Brown, D.A., 2012. Factors affecting the selection and use of drilled shafts for transportation infrastructure projects. In: ADSC EXPO 2012 Geo-Construction Conference Proceedings. March 14-17, 2012. San Antonio, TX, pp. 25–35.

Brown, D., Turner, J., Castelli, R., 2010. Drilled shafts: construction procedures and LRFD design methods. NHI Course No. 132014. Geotechnical Engineering Circular No. 10. FHWA NHI-10-016. May 2010.

Butzer, K.W., 1980. Pleistocene history of the nile valley in egypt and lower nubia. In: Williams, M.-A.J., Faure, H. (Eds.), The Sahara and the Nile: Quaternary Environments and Prehistoric Occupation in Northern Africa. A.A. Balkema, Rotterdam, Netherlands, pp. 253–280.

Byrne, B., 2011. Foundation design for offshore wind turbines. PowerPoint Presentation. Géotechnique Lecture 2011British Geotechnical Association. Institution of Civil Engineers.

Cadden, A., 2004. Micropiles: recent advances and future trends. Geotechnical Special Publication No. 125, American Society of Civil Engineers, Reston, VA.

Cadden, A.W., Gomez, J.E., 2002. Buckling of micropiles – a review of historic research and recent experiences. ADSC-IAF Micropile Committee.

Cooper, J.D., 1998. World's longest suspension bridge opens in Japan. Public Roads. 62 (1) FHWA, Washington, DC.

Das, B., 2011. Principles of foundation design, seventh ed. Cengage Learning, Stamford, CT.

Deere, D.U., Deere, D.W., 1988. The rock quality designation (RQD) index in practice. In: Kirkaldie, L. (Ed.), Rock Classification Systems for Engineering Purposes. ASTM Special Publication 984, Philadelphia, PA, pp. 91–101.

Dobrin, M.B., Savit, C.H., 1988. Introduction to geophysical prospecting, fourth ed. McGraw-Hill Book Co, New York, NY.

Douglass, S., Krolak, J., 2008. Highways in the coastal environment, second ed. Federal Highway Administration, Washington, DC. Hydraulic Engineering Circular 25. FHWA Office of Bridge Technology. Publication No. FHWA-NHI-07-096.

Duncan, J.M., Evans Jr., L.T., Ooi, P.S.K., 1994. Lateral load analysis of single piles and drilled shafts. J. Geot Div. ASCE. 120 (5), 1018–1033.

Engineering and Design: Standard Penetration Test, 1988. Technical Letter No. ETL 1110-1-138. U.S. Army Corps of Engineers, Washington, DC.

Fang, H. (Ed.), 1991. Foundation Engineering Handbook. second ed. Springer Science + Business Media, New York, NY.

Fellenius, B., 2014. Basics of foundation design (The Red Book). Electronic Edition.

Fellenius, B.H., Siegel, T.C., 2008. Pile drag load and downdrag in a liquefaction event. J. Geot. Environ. Eng. ASCE. 134 (9), New York.

Hannigan, P.J., et al., 2006. Design and construction of driven pile foundations. National Highway Institute. Federal Highway Administration, Washington, D.C. Publication No. FHWA-NHI-05-042.

Hannigan, P.J., et al., 2006b. Design and construction of driven pile foundations volume I and II. Publication No. FHWA NHI-05-042 and FHWA NHI-05-043. Washington DC.

Houlsby, G.T., Ibsen, L.B., Byrne, B.W., 2005. Suction caissons for wind turbines. Invited Theme Lecture. In: Proc. International Symposium on Frontiers in Offshore Geotechnics. September 19–21. Taylor and Francis, Perth, Australia, pp. 75–94.

Hunt, B., 2009. Monitoring scour critical bridges, a synthesis of highway practice. NCHRP SYNTHESIS 396, Transportation Research Board, Washington, DC.

Idriss, I.M., Boulanger, R.W., 2008. Soil liquefaction during earthquakes. monograph MNO-12. Earthquake Engineering Research Institute, Oakland, CA.

Iliesi, A., Tofan, A., Prestiuse, D., 2012. Use of cone penetration tests and cone penetration tests with porewater pressure measurement for difficult soils profiling. Bulletin of the Technical University of Iaşi, Iaşi, Romania.

Kannan, R., 2005. Essential elements of estimating engineering properties of karst for foundation design. In: Proceedings: Sinkholes and the Engineering and Environmental Impacts of Karst, ASCE, New York, pp. 322–330.

Kashima, S., 1991. Automation and robotics of underwater concreting in huge scale Steel caisson, the main tower foundation of the Akashi Kaikyō bridge. In: 1991 Proceedings of the 8th ISARC. Stuttgart, Germany. International Association for Automation and Robotics in Construction (IAARC), the Netherlands.

Kearey, P., Brooks, M., Hill, I., 2002. An introduction to geophysical exploration, third ed. Blackwell Science Ltd, London, UK.

Kontopoulos, N., 2012. The effects of sample disturbance on preconsolidation pressure for normally consolidated and overconsolidated clays. PhD Thesis, Massachusetts Institute of Technology, Boston, MA.

Kutschke, W., 2011. Geotechnical roadway design for karst environments. PowerPoint presentation, In: 11th Annual Technical Forum GEOHAZARDS IN TRANSPORTATION IN THE APPALACHIAN REGION, August 3. Chattanooga, TN.

Lagasse, P.F., et al., 2007. Countermeasures to protect bridge piers from scour. NCHRP SYNTHESIS 593, Transportation Research Board, Washington, DC.

Liu, K., Sun, S., Jiang, X., 1992. Environmental change in the yangtze river delta since 12,000 Years B.P. Quatern. Res. 38, 32–45.

Lunne, T., Lacassc, S., Rad, N.S., 1989. General report/discussion session 2: SPT, CPT, pressuremeter testing and recent developments in In-Situ testing Part 1: all tests except SPT. In: Proceedings of The Twelfth International Conference On Soil Mechanics and Foundation Engineering. 4 (August), Rio De Janeiro, BR, pp. 2339–2403.

Marchetti, S., et al., 2001. The flat dilatometer test (DMT) in soil investigations. Report of the ISSMGE Technical Committee 16 on Ground Property Characterization from in-situ Testing, International Society for Soil Mechanics and Geotechnical Engineering (ISSMGE).

Mayne, P., 2007. Cone penetration testing - a synthesis of highway practice. NCHRP SYNTHESIS 368, Transportation Research Board, Washington, DC.

Meyerhof, G., 1956. Penetration tests and bearing capacity of cohesionless soils. J. Soils Mech. Found. Div. ASCE (SM1), 82.

Mullins, G., Stokes, M., Winters, D., 2007. Thermal integrity testing of drilled shafts. Final Report - Contract BD544-20Florida Department of Transportation, Tallahassee, Florida.

Mullins, G., 2009. Thermal integrity testing of drilled shafts. Presentation to AASHTO SOC Conference August 4. Chicago, IL.

Naser, A., DiMaggio, J., Kramer, W., 2011. Implementation of LRFD geotechnical design for bridge foundations. NHI Courses No. 132083. Publication No. FHWA NHI-10-039.

Pando, M.A., 2013. Analyses of lateral loaded piles with P-Y curves - observations on the effect of pile flexural stiffness and cyclic loading. NCDOT 7th Geo3 T2. Raleigh, NC.

Parasnis, D.S., 1996. Principles of applied geophysics. Chapman and Hall, London, UK.

Preusser, F., 2008. Characterisation and Evolution of the River Rhine System. J. Geosci. Geol. Mijnbouw 87 (1), 7–19, Kluwer Academic Publishers, Dordrecht, Netherlands.

Robertson, P.K., Campanella, R.G., Wightman, A., 1983. SPT-CPT correlations. J. Geot. Eng. ASCE. 109 (11), New York.

Rock Foundations, 1994. USACE EM 1110-1-2908. Department of the Army, Washington, DC.

Rogers, J.D., 2006. Subsurface Exploration Using the Standard Penetration Test and the Cone Penetrometer Test. The Geological Society of America. Env. and Eng. Geoscience. 8 (2), Boulder, Colorado, pp. 161–179.

Rowe, K., 2001. Geotechnical and geoenvironmental engineering handbook. Kluwer Academic Publishers, Dordrecht, Netherlands.

Sabatini, P., et al., 2005. Micropile design and construction (Reference Manual for NHI Course 132078). FHWA-NHI-05-039.

Sands, M. (Ed.), 1992. Piling: European Practice and Worldwide Trends. Institution of Civil Engineers. Thomas Telford House, London, UK.

Siegel, T.C., et al., 2014. Neutral plane method for drag force of deep foundations and the AASHTO LRFD bridge design specifications. In: Proceedings of The 62nd University of Minnesota Annual Geotechnical Engineering Conference. St. Paul, MN.

Smoltczyk, U. (Ed.), 2003. In: Geotechnical Engineering Handbook, vols. 1–3. Ernst & Sohn, Berlin, DE.

Standard Penetration Test, ISO 22476-3, ASTM D1586 (Wiki article) 21 January 2015, Available from http://en.wikipedia.org/wiki/Standard_penetration_test.

Thompson, D., Beasley, D., 2012. NAVFAC handbook for marine geotechnical engineering (SP-2209-OCN). Naval Facilities Engineering Command. Washington Navy Yard, Washington, DC.

Tomlinson, M.J., 1994. Pile design and construction practice, fourth ed. E & FN Spon, an imprint of Chapman and Hall, London, UK.

Tumay, M.T., 1997. In-Situ testing at the national geotechnical experimentation sites (Phase II). Final Report. FHWA Contract No. DTFH61-97-P-00161, Louisiana Transportation Research Center, Baton Rouge, LA.

Tumay, M.T., Boggess, R.L., Acar, Y., 1981. Subsurface investigations with piezocone penetrometers. Cone penetration testing and experience. In: Proceedings ASCE National Convention, St. Louis, MO, pp. 325–342.

Washington State Department of Transportation, 2014. Construction division. Geotechnical Office 2014 Geotechnical Design Manual. M 46-03.10. Seattle, WA.

Woodward, J.C., 2007. The nile: evolution, quaternary environments and material fluxes. In: Gupta, A. (Ed.), Large Rivers: Geomorphology and Management. John Wiley & Sons, Chichester, UK, pp. 261–292.

Wyllie, D.C., 1999. Foundations on rock: engineering practice, second ed. E & FN Spon, an imprint of Chapman and Hall, London, UK.

Zheng, H., 2013. Pre-miocene birth of the yangtze river. Proc. Natl. Acad. Sci. U. S. A. 110 (19), 7556–7561.

Section VIII

Bridge construction

Case study: the Reno bridge

24

Ferretti Torricelli L.
Head of Structural Engineering Department - SPEA Engineering S.p.A.

1 Introduction

1.1 Context

The new Reno Bridge is located at the start of the trans-Appenninic segment of the new Variante di Valico (A1 Milan Naples Highway), connecting Bologna to Florence. The bridge was planned with the goal of upgrading the existing A1 Milan-Naples. The road plan of the Variante di Valico (see scheme in Figure 24.1) has to be inserted in a various and articulated territory, characterized by the presence of the Appennines and by narrow valleys crossed by unstable rivers. Therefore the whole project has to be confronted with a very strict prescriptive framework aimed at protecting the environment, in particular focusing the attention to the hydraulic, geotechnical, acoustic and ecological impacts of the several works.

The structure was designed to overpass the Reno River, and is located at the periphery of Bologna city, in a semiurbanized area. The river is characterized by a highly unstable regime, with frequent alternation of dryness and flood, collecting along its run many other minor rivers flowing from side Appenine valleys. Another important aspect was represented by the high-debris transportation during floods. For these aspects, the need of leave the active river bed, about 130 m wide, completely free from permanent intermediate piers and other obstacles to the river flow. Another challenging aspect was represented by the need to comply with a stringent working schedule conditioned by the need to open the first segment of the highway to traffic. For this reason, great effort was dedicated to the selection of the most appropriate construction process and structural concepts characterized by the best performance.

1.2 Conceptual design (from the preliminary drafts to detailed design)

The morphology of the zone interested by the structure shows a wide floodplain zone and an active riverbed located at the side of this zone. The new highway alignment runs onto a curve with a 1400.0 radius, and a distance from the ground level varying from 10 m to 30 m. The ground profile shows a marked discontinuity on the right side of the river, consisting of a steep slope about 15 m high (Figure 24.2).

The preliminary studies concerning river morphology and hydrology required a clear span of 130 m in correspondence of the active bed. The clearance was required to avoid obstruction to the river flow, generally characterized by heavy transportation of debris during floods. The initial design was based on a rather heterogeneous

Innovative Bridge Design Handbook. http://dx.doi.org/10.1016/B978-0-12-800058-8.00024-4

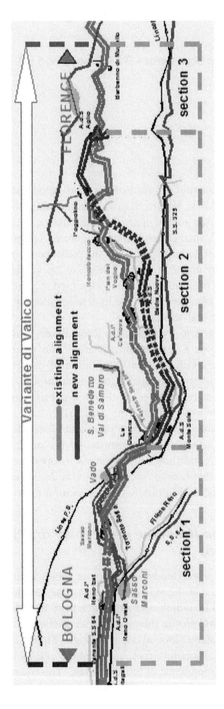

Figure 24.1 Schematic plan of the Variante di Valico.

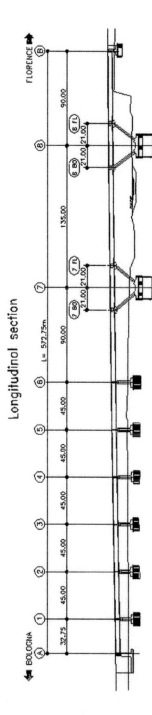

Figure 24.2 Reno Bridge—longitudinal profile.

solution. The bridge was formed by a sequence of 55-m approach spans, realized with prefabricated elements, followed by three long spans dedicated to the river overpass, to be realized with the balanced cantilever method.

In the detailed design stage, an alternative solution was presented that aimed to speed up the construction process and optimize the overall structural behaviour both under service and seismic conditions. After some minor adjustments to the highway alignment, performed in order to insert the structure in a constant-radius curve, the solution of the continuous bridge realized by the incremental launching method was chosen. In the long-span zone, special piers formed by inclined arms rigidly connected to the deck after launching were adopted.

The structural performance was notably improved by the flexible-frame action, especially under the seismic condition, favouring an optimal intrinsic behaviour avoiding special seismic devices.

1.3 Main characteristics of the structure

The bridge is formed by two separate structures: one for each highway's carriageway, running at transversal distance varying from 15 to 20 m. From the structural point of view, each bridge can be regarded as a continuous post tensioned beam with flexible frame behavior.

The choice of the particular construction process highly influences the shape of the structure, resulting in a constant depth cross-section deck, characterized by highly standardized details. As outlined above, the span sequence has an approach zone formed by and end span 32.75 m long, followed by five typical spans that are 45 m long. The long-span zone is nominally subdivided into three spans $90+135+90$ m (measured between support axes). As can be seen by the longitudinal profile shown in Figure 24.2, the adoption of the special V-shaped supports is the key characteristic of this structure, introduced in order to obtain the best balance in term of performance of the box cross section. In fact, the original span organization in this segment can be structurally regarded as a sequence of five smaller span $69+42+93+42+69$ m. In so doing, the higher stresses have been successfully managed with an appropriate calibration of prestressing tendons layout.

1.3.1 Deck

The deck presents a 8.9-m-wide box section with a constant depth of 4.50 m. The top slab (15.70 m wide) presents side cantilevers spanning 3.4 m. It accommodates a carriageway formed by three running lanes, an emergency lane, and two side barriers. The typical thickness of slabs and webs is respectively 0.25 and 0.40 m, with side thickenings providing an optimal frame stiffness and accommodating first-stage prestressing anchorages.

The carriageway transversal slope is obtained with a rigid rotation of the box cross section around the structure alignment; in this way, the cross section maintains regular dimensions, resulting in simpler reinforcement arrangement.

As can be seen by Figures 24.3–24.4, the bottom slab presents the two typical side rails, which are provided to allow the deck to slide on a perfectly horizontal plane during lunching. Both internal and external cross-section shape are kept constant along the whole bridge length, with the exception of support sections and second-stage pre-stressing anchorage section. At the top of the supports, internal U-shaped diaphragms are provided; internal web thickenings accommodate the prestressing anchorages.

In correspondence of special piers connection, the bottom slab thickness is increased, in order to adequately counteract the high compressive stresses in this region.

The following figures (24.3 and 24.4) show the typical cross section (a), the internal diaphragms at typical piers (b), abutments (c), and special piers (d).

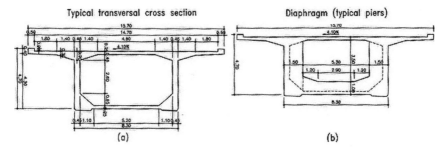

Figure 24.3 Typical section and diaphragm.

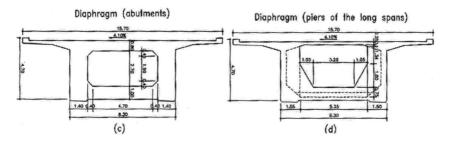

Figure 24.4 End diaphragms and special diaphragms.

1.3.2 Construction method

The frontal launching is one of the mostly high-mechanized erection methods used in bridge construction and consists of manufacturing the superstructure of the bridge by sections in a prefabrication yard located behind one abutment. Each new bridge segment is concreted directly against the preceding one; after concrete hardening and completion of the prestressing operations, the resultant structure is moved forward by the length of one segment. The deck of the Reno Bridge was subdivided into 14 segments with a typical length of 45 m, with the exception of the first and last segments, respectively, which measure 12.25 and 22.50 m. This subdivision has been studied in conjunction with the span sequence, aiming at the maximum standardization of the construction process, with particular regard to the position of internal thickenings and the layout of the steel reinforcement cage. Intermediate provisional piers are provided in the long span zone, in order to reduce the maximum launching span. Figure 24.5 shows the longitudinal profile of the bridge during launching.

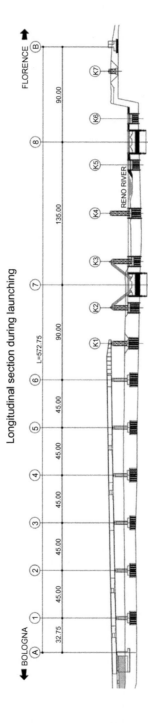

Figure 24.5 The Reno Bridge during launching.

As can be seen, temporary supports are used to provisionally support the arms of special piers legs before the final connection to the bridge deck. These elements act also as a temporary support for the pier shafts.

1.3.3 Launching equipment

The launching equipment is basically composed of the following parts:

- Pushing station
- Frontal and rear launching noses
- Fixed mechanized formwork

The pushing station is installed at the top of the southbound abutment. A longitudinal force is transmitted to the deck by the friction developed by special devices with the capability of lifting the deck in correspondence of the pushing section, in order to increase the vertical force, and contemporarily push the deck. The grip between the concrete top of lifting-pushing devices is enhanced by special plates formed by inclined steel blades.

In detail, each device is formed by a group of four lifting jacks, sliding on an stainless steel plate, pushed by a couple of horizontal jacks. At the top of the lifting jacks, a plate fitted with flexible steel blades (top/side views in figgs. 6/7) allows the transmission of the horizontal pushing force by friction.

Figure 24.6 Pushing device— top view.

The free-cantilever behavior that characterizes the first segments during frontal launching is responsible for wide fluctuations of bending stresses controlled by the use of a frontal steel nose. The girder used for the launch of the Reno Bridge is formed by two steel beams of variable inertia, fixed at the box webs, with a total length of 32 m. The beams, transversally spaced 6.60 m apart, have heights ranging from 4.50 to 2.40 m. The length/span ratio of the nose is 0.72, and the ratio between the deck and nose stiffness $(EJ)_{nose}/(EJ)_{deck}$ is 0.06, characterizing a medium stiffness

Figure 24.7 Pushing device—side view.

device, with a good performance during launching, as will be shown in the following text. The end of the nose is fitted by hydraulic jacks, allowing the recovery of the elastic deflection during the approach to the supports, which reached a maximum value of about 90 mm.

The rear nose (see Figure 24.9) is installed at the end of the last segment. This girder has the dual function of reducing the bending stresses of the last portion of deck, free cantilevering when the last section of the girder abandons the formwork, and allowing the installation of a set of auxiliary pulling cables; these cables are necessary when the vertical reaction at the lifting-pushing device drops below values that are not able to guarantee the transmission of the required longitudinal force.

Figure 24.8 Front launching nose.

Figure 24.9 Rear nose.

The fixed mechanized formwork is located at the center of the launching, 25 m behind the launching abutment. The device, dimensioned for the realization by in situ casting of 45-m-long segments, is composed of two longitudinal steel beams that act as a sort of "fixed rail," above which the segment slides during launching, and the outer perimeter formwork. The inner formwork is installed on a movable carriage and is composed by modular panels, fitting the exact position of the various typologies of internal diaphragms.

At the rear of the formwork is the area dedicated to the assembly of steel-reinforcing cages, separately prefabricated for the bottom slab and webs, and the top slab in 45-m-long segments (Figures 24.10–24.12).

Figure 24.10 Preparation of the fixed formwork.

Figure 24.11 Reinforcement cage moved into the formwork.

Figure 24.12 View of the casting yard.

1.3.4 Prestressing

The prestressing system is subdivided into a first-stage prestressing, progressively activated on each segment during the construction phase, and a second-stage (final) prestressing, set after the completion of the launching. For both these systems, posttensioned tendons composed by strands φ 0.62" made of harmonic steel type St 1670/1860 MPa. These two types of prestressing are discussed next.

1.3.5 First-stage prestressing

The first-stage prestressing, progressively activated after the casting of each segment, is formed by 14 tendons made of 19 strands each, stressed at 4047 kN. It runs along the top and bottom slabs (8 and 6 tendons, respectively) on a straight alignment characterized by barycenter axial results. The typical length of each tendon is 90 m and covers the length of two typical segments. Each tendon is stressed at the end on the opposite side of the launching direction, and linked with the preceding tendon by means of special coupling joints. Each stressing section is formed by 8 and 6 tendons, stressed alternatively before launching, while the remaining ones (6 and 8 tendons, respectively) pass through the cross section and are sheathed and stressed in the subsequent construction stage of the next segment. This special arrangement has been provided to guarantee an overlap of at least 40% tendons through each segment joint. The total axial force provided by the first-stage prestress is about 77,000 kN, resulting in a constant compressive stress along the concrete section of about 4.40 MPa.

1.3.6   Second-stage (final) prestressing

The second-stage prestressing is carried out after the completion of the extrusion of the whole deck and the realization of the pier/deck connection of the atypical piers. It is formed by tendons with curved profiles running along the webs of the box girder.

The tendons are stressed at both ends, and their anchorages are placed in special thickenings of the webs, inside the box girder Figure 24.13.

Figure 24.13 Second-stage tendon anchorages.

For the typical spans, 4 tendons formed by 22 strands are provided, with a stressing force of 4686 kN/unit. The prestressing of the long-span zone is formed by 10 tendons made by 31 strands, with a stressing force of about 6600 kN/unit. The tendon layout between the approach spans and the long-span zone has been determined in order to provide a smooth transition between the two zones (Figure 24.6): 2 + 2 tendons made by 31 strands overlap in continuity with the 2 + 2 tendons of the last 45-m span, and, in a separate thickening located at about the third of the 69-m span, the remaining 3 + 3 tendons made by 31 strands are stressed. The maximum length of the 31 strands tendons is 135 m.

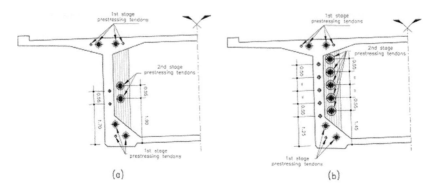

Figure 24.14 Anchorage layout: (a) typical spans; (b) long spans.

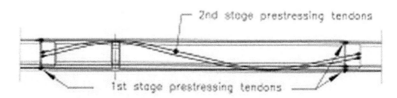

Figure 24.15 Prestressing layout—typical spans.

1.3.7 Supporting system

In the service stage, the support layout of the bridge consists of two side reinforced concrete (r.c.) abutments, six typical piers, and two special piers. The typical piers are r.c. elements, characterized by hollow cross sections, and present overall dimensions of 8.90 x 3.0 m; the shaft, which is 0.3 m thick, is closed at the top by a thick diaphragm that accommodates the restraint devices. The special piers are formed by a pair of inclined arms with a centroid axis convergent toward the foundation level and inclined with an angle of about 42° with respect to the vertical axis. Each arm is divided into two elements, each formed by a steel shell filled with self-compacting concrete. The cross section of each shell is elliptical and presents overall dimensions of 3.5 x 1.8 m. At the top of the steel-concrete elements, a rigid diaphragm realizes, at the end of launching, a rigid connection between piers and bridge deck.

Temporary steel intermediate supports are provided to shorten the launching span during the incremental launching, and to provide temporary support of the inclined legs before launching completion. Each of these supports has been realized with an r.c. shaft formed by four columns of 1.2 m in diameter, strengthened with an intermediate diaphragm. The top diaphragm is realized with a ribbed steel structure that fits provisional bearings and lateral launching guides. This arrangement has been designed in order to facilitate the dismantling and reuse of the system for the realization of the second carriageway.

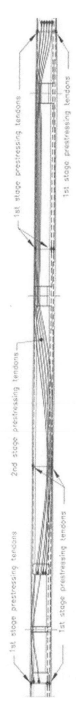

Figure 24.16 Prestressing layout—long spans.

Figure 24.17 Typical piers.

Figure 24.18 Piers 7 and 8 after launching.

1.3.8   Foundations

The ground foundation is characterized by a top layer (7 to 12 m deep) of filling material with poor characteristics, under which can be found a layer of medium-to-dense gravel. Considering the good characteristics of this base formation, direct foundations were chosen for all the substructures. The high stiffness of the bridge deck make it very sensible for foundation-differential settlements; for this reason, a ground treatment with jet-grouting columns was provided in order to improve the stiffness of the superficial soil layer.

The base plinth of typical piers is a rectangular thick plate with dimensions of 14.5 x 9.0 x 2.0 m. The foundation plinths of special piers are circular, with a diameter of 20 m and thickness of 2.50 m; the outer ring of the jet-grouting treatment was reinforced using steel tubes, allowing the excavation needed to lower the foundation level in order not to obstruct the active riverbed.

1.3.9 Restraint system

The structure behaves in a longitudinal direction as a flexible frame, having provided a fixed point corresponding to special piers 7 and 8. The remaining supports (piers and abutments) of the deck are longitudinally free to move. At the top of each support, the support system presents a combination of two vertical bearings made of steel and polytetrafluoroethylene (PTFE), located corresponding to the web axes; and a longitudinal guide, positioned at the center of the support. This particular configuration was adopted with the aim of simplifying the installation of restraint devices at the end of launching. In fact, after the final positioning of the deck, only the longitudinal guide had to be mechanically fixed to the bottom slab using blind nuts set up during casting, while the vertical bearings were fixed using epoxy resin. An overdimensioning of sliding plates of both bearings and longitudinal guide of $\pm$ 20 mm allowed reliance on an equal extra tolerance in the final positioning of the deck.

During launching, the bridge slides above special sliding bearings. Frictional forces are reduced by progressively inserting steel-rubber pads, with PTFE top surface, among the bearing and box intrados. The overall curved alignment of the superstructure is maintained with the use of lateral guides fitted at the top of each support. At the end of each launch, the entire bridge is kept in its correct position by the frictional force developed at the launching station.

2 Main design issues

All the calculations have been carried out with reference to the limit state method, following the base directives provided by the relevant national technical rules, mainly: ministerial decree 9.01.1996 "technical regulations for design and execution of buildings" and ministerial decree 4.05.1990 "technica regulations for design and execution of bridges". Most parts of the detailed calculations pertaining to the construction and service stages have been developed on the basis of Eurocodes.

2.1 Structure layout

2.1.1 Segments and span arrangement

The incremental frontal launching method of construction requires a planoaltimetric profile chracterised by constant radius.

For this reason, the original alignment layout of southbound and northbound carriageways was expressly tuned, obtaining two almost identical structures characterized by the same span sequence, differing only by a slight variation in radius—1400 and 1350 m for northbound and southbound carriageways, respectively (final plan. arrangement in Figure 24.20). In so doing, only minor adjustments was needed in order to reuse the formwork systems on both carriageways.

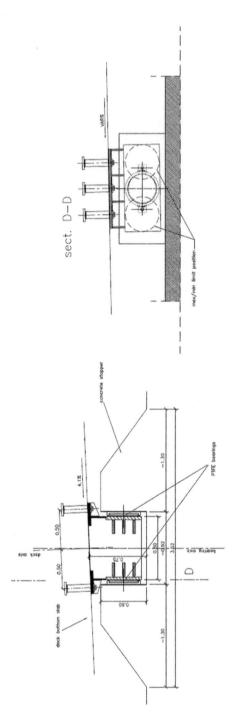

Figure 24.19 Details of the longitudinal guides.

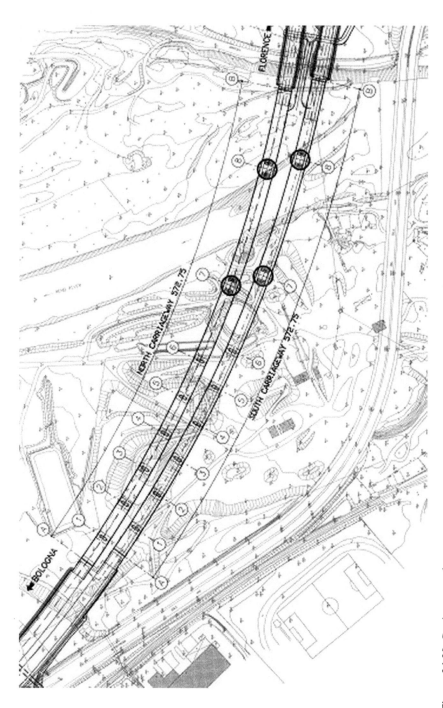

Figure 24.20 Carriageway plan.

The final span arrangement also had to meet the specific requirement of subdividing the deck into standard segments characterized by same overall characteristics both in terms of internal formwork shape and steel reinforcement detailing. For the Reno bridge, the typical segment had an overall length of 45 m; along this segment, the pier diaphragm and the web thickenings accommodating the second-stage tendon anchorages are located. Starting from this basic arrangement, other typological segments are provided, differing from each other in the various position and number of an internal diaphragm. The first and last segments have a reduced length (12.5 and 32.5 m, respectively).

Here is the list of segment typologies (see Figure 24.21):

- Type 1: for segments n. 9–13, typical segment: pier diaphragm and web thickening for second-stage tendons
- Type 2: for segments n. 2 and 5, same as the typical segment, without pier diaphragm
- Type 3: for segment n. 8, same as the typical segment, with an additional web thickening
- Type 4: for segments n. 3 and 6, internal diaphragm on pier 7 (b) and 8 (b)
- Type 5: for segment n. 4, internal diaphragm on special pier 8 (a)
- Type 6: for segment n. 7, internal diaphragm on special pier 7 (a)
- Type 7: for segment n. 1, diaphragm on ab. "B"+front launching nose anchorage
- Type 8: for segment 14, diaphragm on ab. "A"+rear nose anchorage

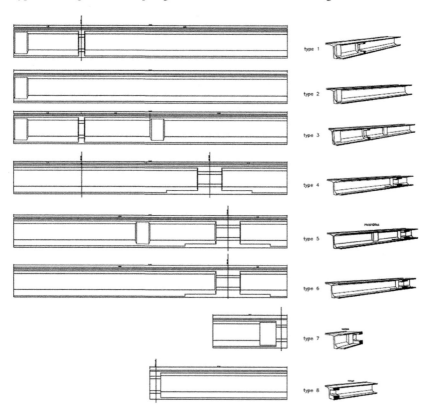

Figure 24.21 Standardization of segment typologies.

2.2 Choice of materials

For the deck, a concrete C35/45 class was selected, having a consistency of S4/S5. The minimum cylinder strength required at the first tensioning of prestressing tendons was 28 MPa. Special additives were studied in order to speed up the development of the required strength, especially in winter. The stressing of tendons generally took place 40–48 h after the last casting.

For the realization of concrete filling of the steel shafts of pier 7 and 8 and of the pier-deck connection, self-compacting concrete of the C32/40 strength class was employed. The concrete filling was placed by directly positioning the casting tube of the concrete pump at the top of the encasement, letting the concrete flow along the inclined face of the shaft; Figure 24.22 shows the final steps of this operation. For this reason, the mix was accurately calibrated and tested, with special reference to the prevention of segregation or bleeding phenomena induced by this particular condition of casting.

Figure 24.22 Casting of piers 7 and 8.

For the mild reinforcing steel, the ribbed bar class B 450 with a minimum yielding strength of 450 MPa was used. The prestressing tendons were formed by low relaxation strands with a diameter of 150 mm^2. The strength class of harmonic steel was 1670/1860 MPa. The initial $\sigma_{p,0}$ stress was fixed for all tendons as much as allowed, using $\sigma_{p,0} = 1450$ MPa.

The anchorage of front and rear noses was realized by high-strength bars with a diameter of 40 mm, made of steel type St 1080/1230. The initial stress of each bar was fixed to 1000 kN. The tensioning of bars used for the anchorage of the frontal launching nose was periodically checked and adjusted, approximately every 20 days. For the realization of the steel encasement of special piers 7 and 8, steel type S355, with a minimum strength of 355 MPa, has been used.

2.3   Design for the launching stage

As mentioned previously, the particular construction process strongly characterized many of the structural choices as distinctive aspects; for this reason, particular design and computational effort has been spent to study the every stage of construction of the main structural elements.

2.3.1   General aspects

One of the first choices was the direction of the launch, which was made on the basis of the availability of enough space to locate the casting yard. For the Reno Bridge, the launch was performed uphill. Despite the increment in pushing force due to the longitudinal slope, which turns directly in an increment of friction coefficient, this choice is generally effective at safely and precisely controlling the final positioning of each segment during the last stroke. At the rear of northbound abutment A, the availability of a wide construction yard, allowed the easy arrangement in line of the fixed formwork and of the prefabrication area of the reinforcement cage. The fixed formwork has been positioned at a distance of 20 m from the pushing abutment; this distance guarantees the presence of an adequate vertical reaction, essential to develop the required horizontal pushing force by friction.

2.3.2   Deck

The high bending moments developed during the launching stage are counteracted only by the limited degree of prestressing offered by the first-stage prestressing tendons; this results in a high depth-to-span ratio that is 1/10 for the Reno Bridge.

The study of the deck launching has been performed through a staged construction analysis, simulating the longitudinal movement of the deck above the supports through the steps ahead, typically measuring 3 m. The following set of loads has been considered by itself or in combination:

- gk1—Self-weight of deck and launching nose
- Δgk1—Amplification factor (5%) applied to gk1, to account for dynamic amplification of the static load during movement
- Pk—Prestress of the first-stage prestressing tendons
- Tk—Positive/negative temperature gradient of 7/−5 °C between the top and bottom slab (positive = top deck warmer than bottom)
- δ1—Intentional variation of the support level ($\delta_{1,k} = 5$ mm), to account for the lifting at the pushing device, and the lifting during the provisional bearing maintenance operations
- δ2—Unintentional variation of the support level between supports ($\delta_{2,k} = +/- 10$ mm)
- δ3—Unintentional variation of the support level between two bearings of the same support ($\delta_{3,k} = \pm 2.5$ mm)
- Fw—Wind during launching (used only for the dimensioning of supports and lateral guides)
- Qh,k—Correction of the plan trajectory: the force required to move the deck laterally at a generic support (used only for the dimensioning of supports and lateral guides)

In addition, forces due to friction at sliding surfaces have been evaluated on the basis of the vertical force evaluated on each individual bearing.

The staged construction analysis was carried out separately considering four basic loading scenarios, characterized by the combination of self-weight, dynamic increment, and temperature gradients. The stresses due to the remaining loads, analyzed through simplified analyses, were superimposed.

For staged construction analysis purposes, a simple FEM (Figure 24.23), based on one-dimensional (1D) linear beam elements, has been used. The stiffness of deck, launching nose, and piers has been accurately modeled. The supports have been modeled by using appropriate equivalent stiffnesses. The analysis develops through a total of 218 steps, characterized by different relative positions of deck and supports.

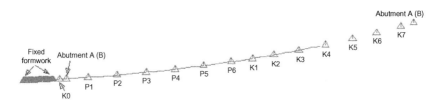

Figure 24.23 Deck launching: FEM.

Figure 24.24 shows a diagram of the envelope of the positive/negative bending moment obtained from the staged construction analysis under the three base scenarios studied.

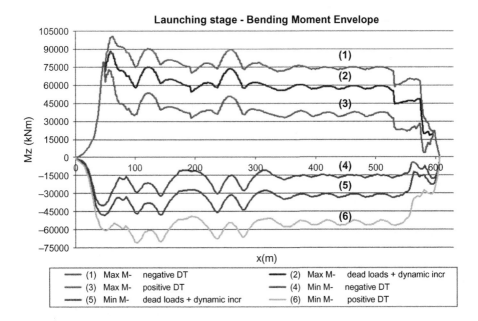

Figure 24.24 Bending envelope.

The diagram of bending moments shows wider fluctuations of bending stresses in the first three segments. This is due to both the presence of the internal heavier diaphragms and the minor stiffness of the launching nose, compared to the one of the concrete deck, which delays the effect of the rise of the nose on the approaching support.

The maximum positive and negative bending stresses are both registered in segment 2 after the lifting of the launching nose on the approaching pier, reaching the following values (positive values = tension at the top fiber of the section; see Figure 24.25):

- Max M: 88 MNm at the section located 60 m behind the edge of the launching nose
- Min M: −48 MNm at the section located 49 m behind the edge of the launching nose

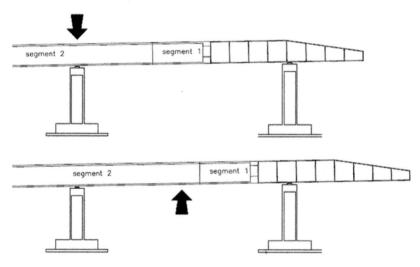

Figure 24.25 Position of the max/min bending moment.

The rear segments show an almost constant value of the maximum/minimum bending stresses that are equal to the typical values characterizing continuous beams ($pl^2/12$ and $pl^2/24$, respectively); these are reference values used for the preliminary hand calculations.

The first-stage prestressing, forming 19-strand tendons, gives an axial force of about 54 MN, resulting in an uniform compressive stress of about 4.3 MPa; this guarantees a residual compressive stress under the basic load scenario (dead loads + dynamic increment), and the limitation of the peak tensile stress under the worst combination of actions of 2.0 MPa.

Figure 24.26 reports the envelope of maximum/minimum shear forces. As can be seen, the same absolute values (for both positive and negative forces) can be encountered; with the exception of local effect due to the presence of concentrated forces introduced by internal diaphragms. In the same way as for bending, also shear forces stress each deck section with maximum and minimum values experimented along the whole deck. This calls for a generally high amount of shear reinforcement.

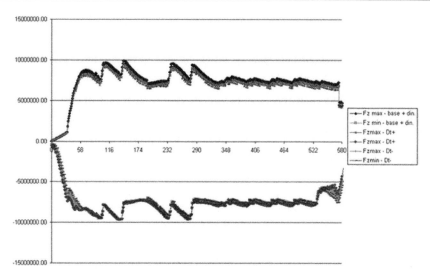

Figure 24.26 Shear envelope.

For this particular case, the dimensioning for shear reinforcement has been made on the basis of a peak value (Vsd = 11.3 MN).

2.3.3 Substructures

Figure 24.27 is a diagram of the evolution of normal forces evaluated during staged construction on provisional and typical piers. Each diagram will show the passage of internal diaphragms, characterized by a wider value of normal force.

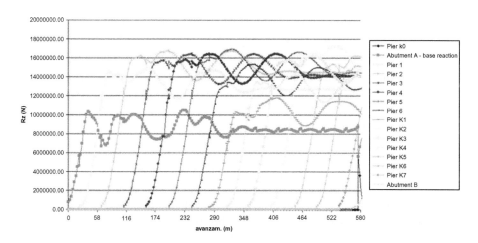

Figure 24.27 Envelope of normal forces on piers.

For dimensioning purposes, the following forces were considered:

- Longitudinal forces due to friction at bearings
- Transversal forces required to redirect the deck horizontally on the curved alignment
- Wind loads during construction

The evolution longitudinal forces acting at the top of each pier due to friction are directly calculated, starting from the value of normal forces, by applying the equivalent friction coefficient:

$$\mu_1 = \mu_A + \mu_C = 0.07,$$

where μ_A is 5%, the friction coefficient at the steel/PTFE interface of bearing at internal supports; and μ_C is 2%, the longitudinal slope of bearing surface.

2.3.4 Dimensioning of pushing devices

The preliminary dimensioning of pushing devices is essentially governed by the following parameters:

- Total weight of the deck at the end of launching, G_k
- Vertical reaction available at launching abutment (pushing station, $R_{z,e}$)
- Longitudinal forces due to friction and slope

For the dimensioning purposes, the following friction parameters, set as a precaution to the respective upper bound values, have been considered:

$$\mu_1 = \mu_A + \mu_C = 0.07$$

$$\mu_2 = \mu_B + \mu_C = 0.12,$$

where $\mu_A = 5\%$, the maximum friction coefficient at steel/PTFE interface of bearing at internal supports; $\mu_C = 10\%$, the maximum friction coefficient at the steel/steel interface of bearing at casting rails; and $\mu_C = 2\%$, longitudinal slope at internal supports and casting rails (taken, conservatively with its maximum value)

The total weight of the deck, during the launch of the last segment was about 19,000 kN. With the application of the precautionary friction values shown previously, this resulted in a total design pushing force of 13,000 kN provided at the pushing station.

The value of friction coefficient at grip plates of pushing devices is characterized by a rather dispersed value, depending from the condition of the blades, which have to be periodically brushed. For design purposes, a precautionary lower bound value of 0.7 was used.

The vertical reaction available at abutment determines the longitudinal force that can be supplied by friction; this value is controlled, within certain limits, by the lifting of the launching jacks; the maximum allowed lifting of 5 mm guarantees a substantial increment of the reaction force theoretically available in an at-rest position of about 25%.

On the basis of friction parameters quoted here, the evolution of available and required longitudinal forces was drawn (see Figure 24.28). The blue line, reporting the development of resisting force, shows clearly the drop of friction force during the extrusion form casting bed, whose rails are characterized by a higher friction coefficient.

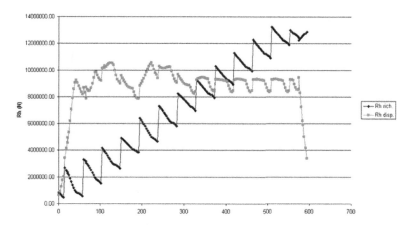

Figure 24.28 Available versus required pushing force.

The diagram in Figure 24.28 shows the need for auxiliary pulling cables to supply the lack of friction at the pushing station, starting from the launch of segment 9. Actually, the maximum total pushing recorded by the monitoring system during launching on site was lower than expected, never exceeding 10,000 kN, proving the care dedicated to the maintenance of the sliding surfaces regularly prepared before each launch. With the exception of the launching of the first and last segments, the total vertical reaction at the abutments fluctuated between 13,000 and 17,000 kN. This allowed for the completion of the launch of the deck, relying only on the forces developed by friction. Auxiliary pulling tendons were needed for the launch of only the first and last segments.

2.3.5 Long-term effects—dimensioning the actual length of segments

The significant overall length of the launched structure required the consideration of the permanent longitudinal strains occurring during launching, to avoid excessive misalignments of diaphragm axes above the piers. The main contributions considered were:

- Longitudinal strains due to first-stage prestress
- Longitudinal strains due to creep
- Longitudinal strains due to shrinkage

The first contribution, evaluated on the basis of a mean concrete stress of 4.43 MPa, resulted in a mean strain of about 0.116 mm/m (5.22 mm/typical segment).

For the second and third contributions, starting from a mean segment age of 60 days, an additional averaged value of about 10 mm/segment was chosen. An additional length of 20 mm was thus considered to determine the effective length of each typical segment. This also accounted for additional strains induced by second-stage prestressing tendons. The nominal length of each segment was adjusted accordingly.

2.3.6 On site—some operational aspects

The construction schedule

A single set of machinery was provided for the realization of both carriageways. At the end of the launch of the first carriageway, the fixed formwork and the pushing devices were moved into the new position, and prepared, after some adjustments, for the launch of the second carriageway deck. The construction started from the southbound carriageway, running from Bologna toward Florence, and included the following main stages:

1. Realization of the foundation and supports
2. Setup of the casting yard
3. Positioning of the launching devices and launching nose
4. Launch of the deck
5. Dismantling of the provisional piers (repositioning of prefabricated steel caps on the northbound carriageway)
6. Repositioning of the casting yard and launching devices on the northbound carriageway
7. Launching of the northbound deck
8. Finally, the dismantling of the provisional piers

The typical working cycle for the construction of a single segment may be summarized as follows:

1. Cleaning the formwork and greasing the fixed rails
2. Inserting the steel cage of the bottom slab and webs into the fixed formwork
3. Positioning the tendons
4. Casting the bottom slab
5. Positioning the inner scaffolding of the webs, made by wooden panels
6. Casting the webs
7. Positioning the top slab inner scaffolding and the prefabricated reinforcing cage of the top slab
8. Casting the top slab
9. Hardening the concrete
10. Stressing the first-stage tendons
11. Launching the segment

With the exception of a short period of time spent tuning, the mean production of the typical working cycle was one segment every 10 days, with a peak production of one segment per week.

The time taken from the launch of the first segment to the dismantling of provisional piers of the westbound carriageway was about 2 years.

2.3.7   Monitoring during launching

The launch of the deck was continuously monitored by a system reporting in real time the pressure at the pushing and lifting jacks. These parameters were automatically reelaborated and recorded in numerical and graphical form, providing the total pushing and lifting force at each stage of progress. The evolution of friction forces recorded at launching devices was continuously compared with the theoretical data obtained from structural analyses.

An automatic control system was provided in order to avoid dangerous situations, such as excess of lifting at the pushing station and the overpressure at pushing jacks; the threshold pressures were adjusted at the start of each launch on the basis of the expected values.

The control system was completed by manually operated stop controls, located on of each pier, and capable of interrupt the launch on emergency during every step, and an automatic stop control, managed by a contact detector at the edge of the launching nose Figure 24.29 shows the interior of the control cabin.

Figure 24.29 Control panel of the launching station.

2.3.8   Realization of special piers

The four steel shafts of the special piers are set in an inclined position using two cranes. Each steel element, which is 25 m long, is initially lifted and put into a vertical position. Subsequently, the steel shaft is slipped down, fitting two steel drive-tubes fixed into the pier foundation, as well as the longitudinal reinforcement curtain emerging from the foundation (Figure 24.30 a–c).

Once positioned and fixed, the steel shells with bolts at the base and at the top, the casting of the concrete core starts, subdivided into four stages, in which the concrete level increases about 6 m each, for a total of about 30 m^3/cast. The use of self-compacting concrete notably speeded up the construction procedure of the piers, because there was totally no need of vibration, nor the need to create intermediate openings for the tube of the casting pump, that was directly positioned at the top of the pier (Figure 24.31).

Figure 24.30 Positioning of the steel shaft of pier 7 a-b: the steel shaft slides along two provisional runners protruding from the foundation plynth. c: steel shaft in final position.

2.3.9   Realization of the pier-deck connections

The realization of the fixed pier/deck connection at the top of inclined shaft of piers 7 and 8 took part after the launching of the whole deck. The connection is formed by an r.c thick plate with dimensions of 9.6 x 6.0 x 1.5 m. This element holds the

Figure 24.31 Casting of the concrete infill with s.c.concrete.

reinforcement of the connection, which has to sustain great bending and shear stresses. Particularly due to the high shear stresses, the perfect closure of the interface between deck soffit and pulvino extrados was very important. For this reason, the deck soffit was properly prepared, after the bridge extrusion, by washing and brushing the concrete surface. The vertical reinforcement, composed by high-diameter ϕ 40 bars connecting the deck diaphragm with the concrete pulvino, was postinstalled by means of sleeves prepared in the box bottom slab.

Figure 24.32 shows the main reinforcement layout after the installation of vertical rebars.

Figure 24.32 Main reinforcement.

A particularly involving aspect was the fact that the contact between the fresh concrete casting and deck soffit had to be performed from bottom to top. For this reason, the level of the cast was adjusted to provide a certain amount of hydrostatic pressure acting upward on the deck intrados. The interface between formwork side panels and deck intrados was sealed in order to sustain this hydrostatic pressure. In addition, a network of vertical tubes crossing the bottom slab was provided to allow entrapped air to escape. At the end of the main casting, the vertical tubes were injected with a very fluid cement mortar, providing the closure of macrovoids between the deck and pulvino.

At the end of this process, the interface section was further injected under a pressure of 5 bars with an ultrafluid resin-based compound, through a network of PVC tubes, fitted with microvalves, and installed on the deck soffit before the casting.

2.3.10   Final stage: Dismantling of provisional piers

The dismantling of provisional piers took place after the tensioning of second-stage prestressing tendons, following a predefined dismantling schedule and calibrated in order to avoid excessive overstresses on the elements still to be dismantled. For this reason, the dismantling sequence provided took place as follows:

1. Piers k1 and k7 (side spans of the long-span zone)
2. Piers k2 and k6 (outer legs of V piers 7 and 8)
3. Piers k3 and k5 (inner legs of V piers 7 and 8)
4. Pier k4

The deck reactions on provisional piers k1, k4, and k7 were first zeroed by removing the temporary bearings with hydraulic jacks. For piers k2, k3, k5, and k6, whose cap acts as a direct support of the pier connection, a smooth decompression was achieved by using a special expanding compound injected into drilled holes at the top of the concrete column. The final demolition of concrete elements was performed using a common hydraulic hammer. During the entire operation, the steel cap of the temporary support had to be suspended at the deck (Figure 24.33).

Figure 24.33 Dismantling of temporary supports.

Figure 24.34 Northbound carriageway at the end of the launch.

Figure 24.34 shows the deck of northbound carriageway at the end of launch, whilst southbound deck is approaching the long span zone.

Case study: the Russky bridge

25

Pipinato A.
AP&P, Technical Director, Italy

1 Introduction

The bridge to the Russky island has a 1,104 m central span length, which established a new record in the world's bridge building practice in 2012 (Figure 25.1). The bridge has the highest bridge towers and the longest cable stays. Approaching spans are 60 m, 72 m, and 384 m, The central span is 1,104 m, and at the other side the approaching spans are the same as previously listed. The total bridge length is 1885,53 m, the total length of the inclined viaducts is 3,100 m, the total bridge roadway breadth is 21 m, the number of driving lanes is four, the under clearance measure is 70 m, the bridge pylons' height is 324 m, and the longest and shortest cable stays are approximately 580 m and 136 m. The design of the bridge location was determined based on the shortest coast-to-coast distance in the bridge crossing location of 1,460 m (Figure 25.2). Navigable channel depth is up to 50 m. The locality of the bridge crossing construction site is characterized by severe climate conditions (e.g., temperatures varying from -31 °C to 37 °C, storm wind velocity of up to 36 m/s, storm wave height of up to 6 m, and ice formation in winter of up to 70 cm thick). Most of the information provided in this chapter comes from SKMost (2012).

2 Design

2.1 Bridge tower

Regarding the bridge tower construction, the piles with a diameter of 2000 mm have been driven as deep as 77 m below ground, and on the island side the 120 auger piles have been piled under each of the two 320-m bridge towers. The bridge towers have been constructed with custom self-climbing forms of 4.5 m. A crane was used on the first three sections, and afterwards the formwork moved through a hydraulic motion of modular elements. The pylons has been A-shaped, with non-standard scaffolding, with self-climbing forms, which made it possible to achieve a faster production. The bridge piers, M1 on the Nazimov Peninsula, and M12 on Russky island, are the heaviest and most complex structures. They are each 35 m high, and take up the horizontal load from the cable-stayed, span-stiffening girder. The builders used self-compacting B35 grade sulfate-resistant portland cement concrete to erect the bridge pier and pylon grillage. This material protects the footing against corrosive fluids and prevents rebar corrosion.

Innovative Bridge Design Handbook. http://dx.doi.org/10.1016/B978-0-12-800058-8.00025-6

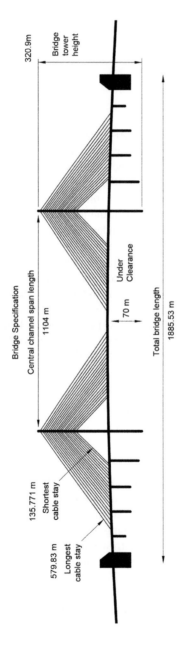

Figure 25.1 Main measures of the bridge.

Figure 25.2 The Russky bridge–general view of the completed structure. Courtesy of SK MOST Group.

2.2 Bridge deck

The deck cross section shape was determined based on aerodynamic analysis and was optimized following the results of experimental wind tunnel testing in a scaled model, in order to predict the deck's resistance to squally wind loads. Welded onsite connections are used for longitudinal and transversal joints of the cap sheet for the orthotropic plate and lower ribbed plate. High-strength bolts have been adopted for vertical walls of the block's joints, longitudinal ribs, transversal beams, and diaphragms. The premounted sections were supplied onsite by barges that were hoisted by crane to 76 m elevation at the erection site. Then, every section was linked and the cable stays were attached (Figure 25.3). Approaching viaducts of total lengths more than 900 serve the bridge. In this case, piers are made as columns (9 to 30 m high). The span decks are made of steel and reinforced concrete, which consist of steel inclined-wall box sections and a cast-in-place reinforced concrete slab.

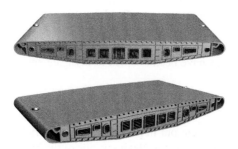

Figure 25.3 Deck 3D model and section view. Courtesy of SK MOST Group.

2.3 Cable stay system

A penetration sealing system (PSS) has been implemented in the cable stays, pursuing wind load reductions of 25%–30%. Moreover, the cost of materials for pylons, the stiffening girder, and foundations decreased by 35%–40%. PSS cable stays consist of parallel strands of 15.7 mm in diameter, while each strand consists of seven galvanized steel wires enclosed in high-density polyethylene sheathing; cable stays incorporate from 13–85 strands. The length of the shortest cable stay is 135.77 m, and the longest is 579.83 m. The protective sheath of the cable stay is made of high-density polyethylene (HDPE) and is resistant to UV and the specific local climate conditions of Vladivostok (temperature range from −40 °C to 40 °C). The detail of a cable stay hanger is depicted in Figure 25.4. The stay cable system weighs 3,720 tons and has a total length of more than 54 km.

Figure 25.4 Detail of a cable stay hanger. Courtesy of SK MOST Group.

2.4 Seismic devices

The Russky Bridge could experience an earthquake with a magnitude of up to 8.1 on the Richter scale. This represents a safety margin comparable to the very high requirements of other bridges (e.g., the Akashi Kaikyo Bridge's resistance is 8.5). The designed system of two-hinged stiffening girders was conceived in order to allow seismic loads up to 8.1 points, along with strong sea currents. Pendulum-type bearing structures were introduced to reduce the active loads, ensuring the seismic isolation of the span deck. Movement joints have endured large axial displacements of the span deck, and lead-cored rubberized metal bearings have been adopted to dissipate energy under large stresses.

2.5 Production facilities

Two production facilities were put in place for running efficient construction operations. One facility was located on the Nazimov side of the bridge, and the other on the Russky side. The facilities include a rebar welding workshop, building laboratories, and a concrete mixing plant. Each production facility has office building, living quarters, and mechanical, woodworking, and equipment repair workshops. More than 1 km of new railway tracks were built to ensure timely delivery of building materials. About 320 pieces of state-of-the art special equipment were used in the construction of the bridge to the Russky Island. Unique 40-ton and 20-ton tower cranes, which can telescope up to 340 m, were used to erect the pylons. Derrick cranes with up to 400 tons of lifting capacity were used to install the channel span deck. Crawler cranes with 1350-ton lifting capacities were installed within a record short time to lift the first 10 sections of the steel span deck (SKMost, 2012).

3 Construction phase

3.1 Bridge foundations

Drilling and pile concreting operations were done in seawater that varies from 14–20 m in depth. A total of 120 drilled piles with a 2000-mm-diameter have been put in place to build the footing of each pylon, with permanent steel-cased piles 46 m deep (under the M7 pylon), while those on the Nazimov side reach 77 m deep. The total length of the wells' drilling operations is more than 5000 m. The drilled soil was a very mixed and highly heterogeneous rock siltstone with a strength of 90 MPa and compressed sandstone lens with a strength up to 180 MPa. The most labor-consuming operation of the bridge construction project was the pier-side foundations. Strain gages are embedded in the grillage body for monitoring the condition of these footings (Figures 25.5 and 25.6).

Figure 25.5 Operating welling machine. Courtesy of SK MOST Group.

Figure 25.6 Construction yard of one pylon. Courtesy of SK MOST Group.

3.2   Bridge tower and girder

Self-lift shutters were used to concrete the pylons: seven working tiers of a total height of 19 m each allow the preparation of the construction joint, reinforcement, concreting, concrete curing, and finishing to be run simultaneously in three sections that are each 4.5 m high. The self-lift shutters are hydraulically powered and cut down the erection time for the cast in place reinforced concrete structure by a factor of 1.5. This is significant considering the amount of concreting necessary for each pylon (20,000 m^3). The anchor span structures are symmetrically built and are each 316 m long (Figure 25.7).

Figure 25.7 Detail of one antenna during construction and the scaffolding system erecting bridge concrete segments. Courtesy of SK MOST Group.

The continuous span is made of prestressed, cast-in-place, reinforced concrete, which required approximately 21,000 m^3 of concrete mix to complete. High tensile pre-stressing steel bundles are installed and tensioned by the application of a tensioning force of 300-370 tons by prestressing jacks. After tensioning, voids are filled with a special cement-based mortar. The all-metal stiffening girder of the central navigation span is comprised of the bottom and top orthotropic plates and a system of transverse diaphragms. The steel-stiffening girder is composed of 103 panels, each 12 m long and 26 m wide, and two transition panels of 6 m each. The panels weigh a total of 23,000 tons. The stiffening girder is 1,248 m long.

3.3 Preassembly of the deck panels

The main steel-stiffening girder panels of the bridge were built in a nearby, specially equipped yard, which delivered thousands of tons of steel structures for the bridge's main steel-stiffening girder from the preassembly site. This procedure has reduced the number of joints and significantly accelerated the procedure of installation at 70 m high. Also the 30 km of first-class welded joints with 100% ultrasound flaw detection were completed quickly in this yard, rather than onsite (Figure 25.8).

Figure 25.8 Preassembled segment of the main steel-stiffening girder. Courtesy of SK MOST Group.

3.4 Panels lifting

The panels delivered to the installation site by barges were then lifted by crane to the 70-m elevation. The barge was positioned under the installation unit using a global navigation satellite system. After section 20 was installed, 24 m long paired panels were delivered for installation to expedite the installation of the steel-stiffening girder (Figure 25.9).

Figure 25.9 A deck panel segment lifted up on site. Courtesy of SK MOST Group.

3.5   Installation of the longest stay cables

The closing pair of white-colored stay cables are 579.83 m long, which sets a world record in bridge construction. The world's longest stay cables are installed at an elevation of 317 m at M6 Pylon and are attached to the 50th panel of the main steel-stiffening girder, which has been extended over the distance of 534 m toward Russky Island. The white jackets of each stay cable contain 80 strands of high-tensile wire with a design strength of 1860 MPa. Freyssinet developed the ultra-compact stay cable design specifically for this bridge (Figure 25.10). (Freyssinet, 2014)

Figure 25.10 A close view of the starting anchorage of the longest stay cable in the world. Courtesy of SK MOST Group.

3.6 Key-deck segment erection

The key segmental deck was erected during the night of April 11–12, 2012. Section 52 was lifted from the pontoon, which was custom equipped for the purpose, in order to close the two 546-m-long cantilever sections (Figure 25.11).

Figure 25.11 Final works for the deck construction, lifting the key-deck segment from a river barge. Courtesy of SK MOST Group.

4 Monitoring system

The bridge is equipped with an automated precision monitoring system, which allows a continuous monitoring of the installation's status to be conducted using two satellite-based global positioning systems simultaneously. This system integrates more than 500 state-of-the-art sensors, allowing for real-time monitoring of bridge health parameters, along with the weather conditions and wind loads. A huge display board in the control center that monitors the bridge over the Eastern Bosphorus Strait presents all incoming data such as visibility, wind velocity, roadway temperature, traction coefficient, water film thickness, lane traffic intensity, traffic flow density, and many others in real time. Acquiring data determines the actions taken by the 24-hour control team: e.g., a safe speed limit is set up for traffic via the bridge. The innovative monitoring system integrates a variety of sensor types, including global positioning satellite (GPS) receivers, tachymeters, inclinometers, seismometers, which provide the precise position monitoring and data on the structural components. Color coding is used to show the bridge's health: the green light on the screen means that everything is fine, yellow warns of the approaching alarm level in the preset range, and red is the alarm level.

References

SKMost, 2012. Russky Bridge: Construction of the bridge crossing to Russky Island over the eastern Bosphorus Strait in Vladivostok. SKMost, Vladivostok, Russia.
Freyssinet, 2014. Stay cable system installation. Russky Island bridge, Freyssinet official website.

Case study: the Akashi-Kaikyo bridge

26

Pipinato A.
AP&P, Technical Director, Italy

1 Introduction

The Akashi-Kaikyo Bridge passes across the Akashi Strait between the Tarumi ward of Kobe and the Awaji Island in Hyogo prefecture. The bridge was built as part of the Kobe-Naruto segment of the Honshu-Shikoku construction projects (Figures 26.1 and 26.2). The distance between the bridge's two cables is 35.5 m. While the original plan was for the bridge to carry both rail and road traffic, in 1985 it was decided that the bridge should carry only highway traffic (Nagai et al., 2000). The construction started in May 1987, and the bridge opened to traffic in 1998. The side view of the bridge is shown in Figure 26.1. The main cables have diameters of approximately 1 m and weigh 50,500 tons, which allows them to carry the 87,000 tons of the stiffening girder. The Akashi Strait is a busy shipping port, so engineers had to design a bridge that would not block shipping traffic. They also had to consider the weather, as Japan experiences very severe weather conditions: e.g., hurricanes, tsunamis, and earthquakes that rattle and thrash the island almost annually. The bridge deck has been realized with a truss made of a complex network of triangular braces, beneath the roadway. The open network of triangles makes the bridge very rigid, but it also allows the wind to blow right through the structure. In addition, 20 tuned mass dampers (TMDs) in each tower were placed. The TMDs swing in the opposite direction of the wind sway, so when the wind blows the bridge in one direction, the TMDs sway in the opposite direction, effectively balancing the bridge and canceling out the wind's sway. With this design, the Akashi-Kaikyo Bridge can handle high-speed winds and can withstand an earthquake with a magnitude of up to 8.5 on the Richter scale. Aerodynamic stability was investigated through a boundary layer wind tunnel test with a 1:100 full model. In the design standard adopted, it is specified that flutter must not occur under a wind speed of 78 m/s in the wind tunnel test within the attack angle from $-3°$ to $3°$ (Fuchida et al., 1998). To determine the type of the stiffening girder to be used, several types of girders were investigated; from these comparisons, the truss girder and compound stiffness box girder were selected as prospective types (the compound stiffness box girder is a bridge system that arranges high-torsional-stiffened girders around the tower and aerodynamically-well flat girders at the central portion of the bridge). As a result of the experiment on a scaled model (1:100), it was confirmed that the required wind resistance could be obtained by installing some gratings on the road deck and a vertical stabilizing device along the truss girder (Fuchida et al., 1998). The bridge cost is estimated at US$3.6 billion, according to Cooper (1998).

Innovative Bridge Design Handbook. http://dx.doi.org/10.1016/B978-0-12-800058-8.00026-8

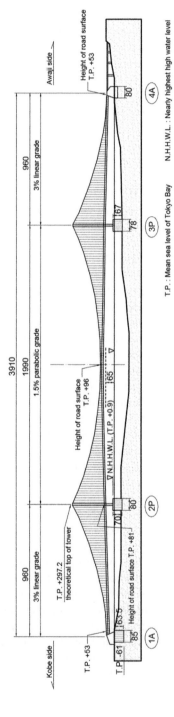

Figure 26.1 Side view of the Akashi-Kaikyo bridge.

Figure 26.2 The Akashi-Kaikyo bridge.

2 Design

2.1 Bridge tower

The tower pier foundations were designed to transmit 181,400 tons of vertical force to bedrock, approximately 60 m below the water surface. The foundation was constructed using a newly developed, laying-down caisson method: the steel caissons, each 80 m in diameter and 70 m in height, were first towed to the sites, then submerged and set on the preexcavated seabed (Figure 26.3). Pier foundation construction was completed with the placement of concrete. Then, the main steel towers were erected on the concrete piers with an independent, self-supporting 145-metric-ton tower crane. Each main tower height was 282.8 m (297.3 m with a cable saddle in place) and was erected by stacking 30 prefabricated steel segments, which were approximately 10 m in height, on top of each other. The segments were formed with three separate main cells in plan view. Special procedures were used during the fabrication of each segment to ensure accurate tolerances for proper tower alignment. The tolerances were maintained using laser-measuring technologies to control all dimensions. Use of this technology resulted in no major erection problems during the field bolting and splicing together of the steel tower segments (Cooper, 1998). The allowable inclination of the tower was specified to be 1/5000 (about 6 cm at the top of the tower). TMDs were installed inside the tower to control wind-induced vibration. TMDs were attached to each tower at varying stages of completion to reduce wind-driven tower motion and tower vibration in the event of an earthquake (Figure 26.4).

2.2 Bridge deck

The bridge roadway surface is constructed on top of a truss girder system (14 m in depth, 35.5 m in width) that is suspended from main cables passing over two steel towers that rise 298 m above sea level. A 65-m clearance is maintained over the

Figure 26.3 Steel caissons.

Figure 26.4 Steel towers: (a) climbing crane used for tower erection;

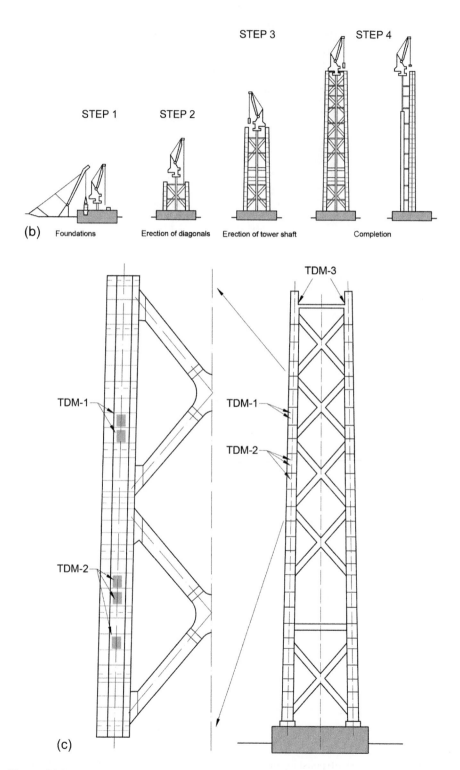

Figure 26.4 Continued. (b) erection procedure; (c) location of TMDs;

(Continued)

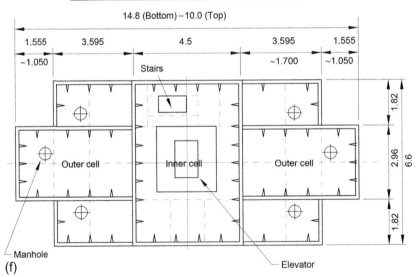

Figure 26.4 Continued. (d) polishing the cross section of a tower shaft using a large-scale facing machine; to control the inclination of the towers and to assure the metal touch at connections, each end of the tower shaft blocks was polished up to 0.0125 mm ruggedness; (e) TMD: Tuned Mass Damper; (f) cross section of tower shaft (measures in m).

shipping lane. The block was constructed using a special barge because of the severe environmental conditions: e.g., strong currents (5 m/s), deep water (maximum depth $= -100$ m), and heavy traffic (1000 vessels/day). The barge was equipped with computer-controlled, omnidirectional propellers mounted at each corner of the barge, and it could maintain its position at a fixed point without mooring ropes. As a result, working time per cycle was drastically reduced from the conventional 3 h to only 30 min. After comparing the streamlined box girder with the truss girder, the latter was chosen because of its aerodynamic stability and economy. In order to reduce the weight of the girders, high-strength steel was used. Stiffening girder sections, each about 36 m in length, are fabricated and placed the barge for transport to a site below each erection point. Then they are hoisted into position using lifting beams and are secured to hanger ropes. Since the barge cannot access the construction point when in shallow waters or on the ground, stiffening girder blocks are moved to the area using two lifting beams (Figure 26.5).

Figure 26.5 Steel truss deck.

2.3 Cable stay system

Before stringing the cable, a pilot hauler rope was attached to each anchorage and placed over the tower tops by helicopter. The pilot rope was used to suspend the catwalk, from which work on the main cable attachment would proceed. The main cables, which have a 1:10 sag ratio, were assembled using the prefabricated strand method: each strand was transported to the construction site where it was pulled from one anchorage over the saddle of each tower and fastened to the opposite anchorage frame. This procedure was repeated 289 times to fabricate each main cable. Before the attachment to the steel frame inside the anchorage, each main cable was separated at the anchorage by a splay saddle to equally distribute cable tension to the foundation. A specially designed cable-squeezing machine was used to compress the 290 parallel wire strands into the final 1.12-m-diameter cable. Cable bands were placed to circumferentially compress the cable and to maintain its circular shape. Finally, suspender cable hangers were attached to the main cable to support the main stiffening truss (Cooper, 1998). Further, the pilot rope, which is lightweight and made of high-strength polyaramid fiber (measuring 10 mm in diameter), was spanned using a helicopter. Newly developed high-tensile-strength wire of 1760 N/mm^2 made it possible to use only one cable per side, rather than two. The suspender ropes are prefabricated parallel-wire strands (PPWS) covered by a polyethylene tube, and there are two suspender ropes at one panel point. The span of each rope at one panel point is about 9 times that of their diameters, and generating an oscillation can be difficult in this condition. However, a large-amplitude oscillation was observed at the downstream side suspender rope. Therefore, in order to improve the aerodynamic characteristic of the ropes, the generating conditions were investigated in detail, including the oscillation characteristics and their ability to withstand a vibration obtained through the wind tunnel test. It was found that the vibration was controlled by spirally winding trip wires (10 mm in diameter) around a suspender rope. Moreover, the most suitable wire diameter and twisting pitch were obtained through the wind tunnel test, and set at all longer suspender ropes of the bridge using a newly developed machine. By this countermeasure, the oscillation was controlled and has not been observed in large amplitude. Many analyses have been performed on the bridge, including experiments on the dynamic behavior of the bridge under wind forces, especially flutter and buffeting, as both are issues of utmost concern in the wind-resistant design of long-span bridges. Multimode flutter and buffeting analysis of the Akashi-Kaikyo Bridge was performed by Katsuchi et al. (1999), and the analytical results were compared with the wind-tunnel test data obtained from an aeroelastic, full model of the bridge. The multimode flutter analysis corresponded well with the measurements and exhibited a significant coupling among modes. The multimode buffeting analysis also showed excellent agreement between the analysis and measurements in vertical and torsional response. Significant coupling among modes was also observed in the buffeting analysis, and the multimode analysis predicted the measurement better than an alternate single-mode analysis method (Katsuchi et al., 1999). Anchorages measure 63 m by 84 m and extend into the Kobe and granite layers at the site. These anchorages required special foundation construction technology. The Honshu anchorage had to be embedded 61 m below sea level, and the anchorage excavation had to be performed in open air. Therefore, an 85-m-diameter and 2.2-m-thick slurry wall was constructed and subsequently used as a retaining wall. Excavation within the slurry wall was followed by the placement of roller-compacted concrete to complete the anchorage foundation construction. The Awaji anchorage foundation was constructed using steel pipes and earth anchors to support the surrounding soil. The excavated foundation was filled with specially-designed, flowing-mass concrete. Both anchorages were completed with the construction of a huge steel-supporting frame that was used to anchor the main suspension cable strands (Cooper, 1998). See Figures 26.6 and 26.7.

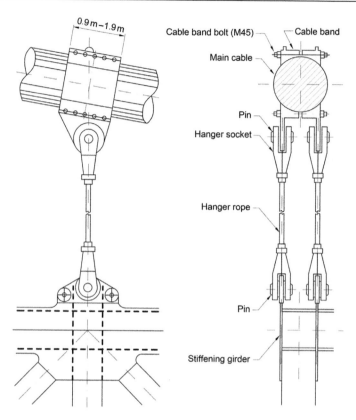

Figure 26.6 Cable band and hanger structure.

2.4 Seismic devices

The complete bridge structure was designed to resist a 150 km distant, 8.5-Richter magnitude earthquake. Of particular interest was the performance of the bridge during the
January 17, 1995, Hyogo-Ken Nanbu earthquake, which provided a full-scale test of
tower response during the bridge construction. Fortunately, the bridge-stiffening truss
had not begun at that time, being the yard at the stage of cable squeezing. The Nojima
fault zone passes between the towers of the bridge, and the earthquake caused a permanent lateral and vertical offset of the Awaji tower and anchorage. Ground fault rupture
was visible on the northern tip of Awaji Island, approximately 2 km from the Awaji
anchorage. According to the bridge authority, the following observations were reported
(Nasu and Tatsumi, 1995):

- According to the results of an analysis in which the earthquake-induced foundation displacement is added to the completed structure, there appear to be no problems from a stress viewpoint as regards to the bridge's towers, cables, stiffening girder, etc.

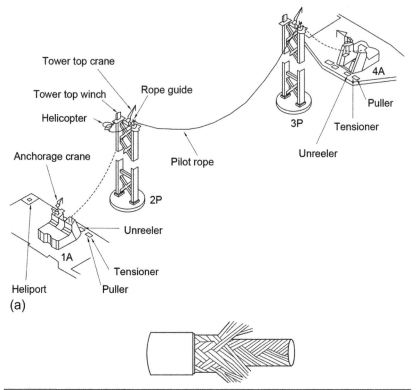

(a)

Type	Construction	Material	Characteristic
008B-BC	⃝	Inner layer: Aramid fiber Middle layer: Polyester fiber Outer layer: Urethane resin (#129)	Rotation-resistant, waterproof construction

Diameter (standard) (mm)	Inner layer aramid fiber diameter (braid) (mm)	Outer layer thickness (urethane + fiber) (mm)	Aramid fiber cross-sectional area (mm²)	Weight (g/m)	Tensile strength [kN(tf)]
10	8,0	1,0	25,9	91,7	46,1 [4,70]

(b)

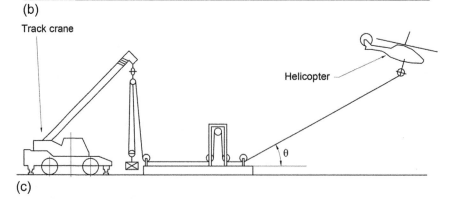

(c)

Figure 26.7 Procedure of crossing a pilot rope: general layout (a); specifications of pilot rope (b); helicopter pulling force experiment (c).;

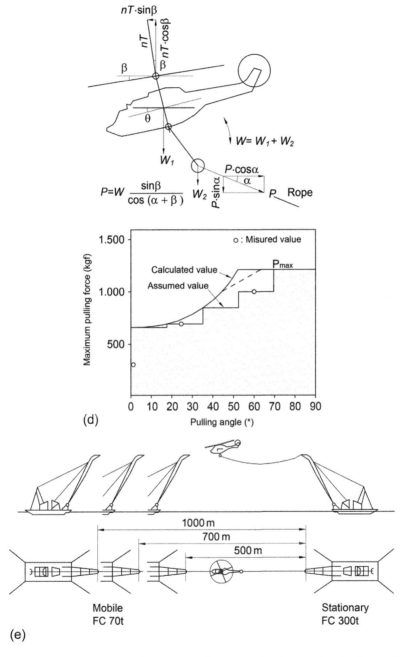

(d)

(e)

Figure 26.7 Continued. relationship between helicopter pulling force and angle (d); experiment to confirm the entire system (e).

(Continued)

STEP 1   Spanning of Pilot Roper

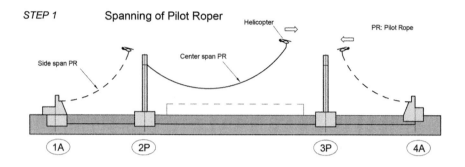

STEP 2   Erection Catwalk

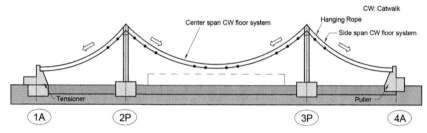

STEP 3   Erection of Strand

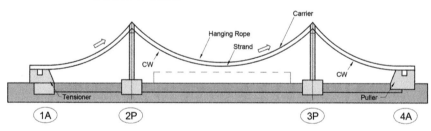

STEP 4   Instalation of Cable Band and Hanger Rope

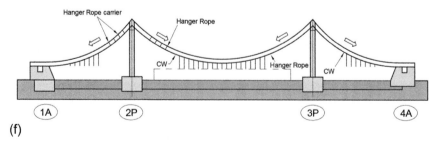

(f)

Figure 26.7 Continued. Construction phases (f).

- Although the part of vertical alignment exceeds 3% due to the lessened cable sag in the center and side spans, no problem will occur under the conditions of the highway structure. Also, the horizontal alignment is now off by about 0.03° at the tower, which is not expected to present problems as far as the passage of cars is concerned.
- The increased 2P–3P and 3P–4A spans will be handled by adjusting the length of the stiffening girder, which is now being fabricated.

The modified position of the structural elements is reported in Figure 26.8 (Nasu and Tatsumi, 1995). The Awaji tower was displaced 1.3 m to the west, while the Awaji anchorage was displaced 1.4 m to the west, relative to the Kobe tower and anchorage. This resulted in a 0.8-m increase in span length between the main towers and a 0.3-m increase in the southern side span length. The Awaji tower pier was displaced 0.2 m downward, while the Awaji anchorage rose by 0.2 m. The sag in the main cable was reduced by 1.3 m. The earthquake caused a one-month delay in the construction schedule, during which the bridge was carefully inspected for damage. This lost time was made up during the remaining three-year construction period, and the bridge was opened to traffic on schedule. The redesign of the two center stiffening panels, each 0.4 m longer than originally designed, accommodated for the increased distance between towers. The cable-squeezing machine suffered minor damage and was quickly repaired. Anchorages, piers, and towers were otherwise undamaged. (Cooper, 1998; Nasu and Tatsumi, 1995).

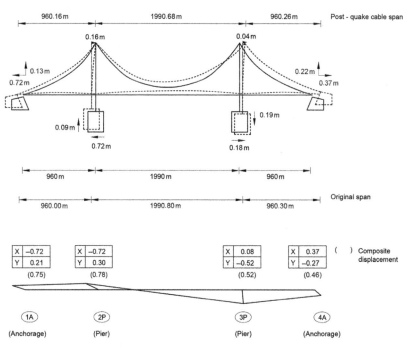

Figure 26.8 Effect of the Hyogo-Ken Nanbu earthquake on the structural skeleton of the bridge (Nasu and Tatsumi, 1995).

3   Innovations and special construction details

3.1   Innovative technologies

Several technologies were developed to support the design and construction of the Akashi-Kaikyo suspension bridge. One of the most relevant developments was the aerodynamic stability of the long suspension bridge, which often posed major challenges to bridge designers. To verify the wind design, the bridge authority contracted the Public Works Research Institute to construct the world's largest wind-tunnel facility and tested full-section models in laminar and turbulent wind flows. Other innovations that resulted from wind tunnel testing included the installation of vertical plates at the bottom center of the highway deck to increase flutter speed. A second innovative technology for that time was the development of individual parallel wire strands that were fabricated offsite, transported to the bridge site, and strung parallel to each other to form the main cable. The advantage of using this method was that the strands are then continuous from anchorage to anchorage, eliminating the in-place spinning of cables, and thus reducing the probability of accidents while ensuring a higher level of safety. In order to create the parallel wire strand, a unique cable-squeezing machine was designed to form the parallel strands into the final circular shape. The use of higher-strength wires reduced the number of strands required (saving construction time and cost), and the number of suspender ropes (which dropped from four to two) needed to connect each stiffening truss panel point to each cable hanger attachment on the main cable (Cooper, 1998).

3.2   Bridge foundations

At the center of the strait, the topography consists of a wide valley with steeply sloping sides and a water depth of 100 m. The geology comprises granite from the Mesozoic era as the site's bedrock. This bedrock is covered roughly with the Kobe stratum of the Mesozoic epoch of Neocene, the Akashi layer of the diluvial epoch of the Quaternary period, an upper diluvial formation, and an alluvial formation (Nasu and Tatsumi, 1995).

The severe conditions were the strong tidal currents (4.5 m/s) and deep water (-100 m at the central span site, −50 m circa under the pylon). As a result, the foundations were constructed safely by the *laying down caisson* method. State-of-the-art technologies, such as scour protection and underwater concrete desegregation, were developed for substructure construction. Concerning anchorages, anchorage 1A, located on the Kobe side, was located on the soft ground, and the foundation was constructed using the *underground slurry wall method*. Various kinds of concrete, from slurry wall (rich mixed concrete) to inner concrete foundation (lean mixed concrete), were used. Precast concrete panels were installed considering the aesthetics of the outside walls. Highly workable concrete was used for the anchorage body (Figures 26.9 through 26.11) (Cooper, 1998).

Figure 26.9 Foundation construction phases: (a) excavation of the seabed using a large-grade bucket dredger; (b) manufacturing a steel caisson; (c) mooring; (d) towing; (e) sinking; (f) dropping riprap (scour protection); (g) casting underwater desegregate concrete; (h) casting the top slab of concrete in the open air.

Figure 26.10 Anchorage construction phases: (a) construction of the bottom slab of concrete; (b) inner concrete (roller-compacted concrete, RCC); (c) distant view of 1A anchorage foundation work; (d) transportation of a cable anchor frame using a floating crane.

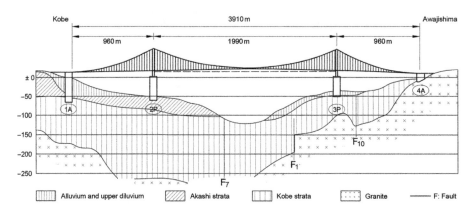

Figure 26.11 Geological profile at the bridge construction site.

3.3 Installation of cables

The cable installation work of a bridge begins when the main towers and the cable anchorages are completed. The first phase of the work starts with the installation of a pilot rope that acts as a foothold over the entire span of the bridge, connecting the main towers and anchorages. The pilot rope is then used to install a drive rope system, mount service catwalks, and finally suspension cables. Due to the channel's high traffic and the rapid current, the pilot rope (a 10-mm-diameter polyaramid fiber rope) was put into place by a large helicopter on November 10, 1993. The helicopter carried an extending machine with a reel of pilot rope and strung the pilot rope from one anchorage to the first main tower, then to the second main tower, and finally to the other anchorage, while unreeling the pilot rope. All these phases were previously confirmed by testing, and then they were performed with the same helicopter and ship cranes, in order to test all the on-site stresses of the cable during the final construction phase; including a pulling force experiment, and the experimental relationship between the rope tension and the critical pulling angle (Takeno et al., 1997).

4 Monitoring system

Various monitoring devices such as seismography, anemometers, and accelerometers have been installed to record data on the Akashi-Kaikyo Bridge. Figure 26.12 shows the layout of the monitoring devices. The records are accumulated and analyzed to ensure structural safety through monitoring the behavior of the bridge and provide information on the characteristics of the bridge during exercise. In addition to this system, a global positioning system (GPS) was introduced to monitor the seasonal, daily, and hourly behavior of the bridge, which may be governed mainly by temperature and live loads (HSBE, 2005).

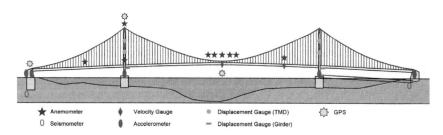

Figure 26.12 Monitoring system for the Akashi-Kaikyo Bridge.

5 Maintenance system

It is very difficult to inspect this bridge due to its extended and high structures above sea level, and also because of the high number of vehicles passing over and ships passing under the bridges. Under these circumstances, maintenance vehicles have been installed into the bridge to inspect the structures safely and effectively. There are different maintenance vehicles for outside girders, inside girders, and cables, according to each structural type used (Figure 26.13). Furthermore, the concept of *preventive*

Figure 26.13 Maintenance vehicles: outside girder vehicle (a); inside girder vehicle (b).

maintenance has been introduced to keep this long-span bridge in good condition for the future and allow it to withstand severe natural actions, including:

- For *substructures*, continuous testing is applied for out-of-water parts of the construction, including sampling of concrete core, examination of salt damage, neutralization, cracks diffusion, steel rebar state, etc., and adopting interventions when required as a result of nondestructive testing (NDT). For underwater structures with pitting corrosion on the surface of the steel laying-down caissons, the electrodeposit (EDP) method has been used as a countermeasure to prevent further corrosion.

- For *superstructures*, a wide amount of maintenance actions are utilized: (a) long-lasting paint has been used for metal structures; in particular, the base is a thick-coating type of paint and inorganic zinc-enriched paint, which includes a rich amount of zinc powder and has excellent anticorrosion performance due to electrical and chemical sacrificial anode action; the undercoat, which protects the base coat, is epoxy resin paint, which has excellent durability and performance against alkalinity. In addition, fluorine resin paint, whose performance is excellent against chemical action and weather action, is applied as the surface coat (HSBE 2005). (b) A dry air injection system for main cables has been installed, ensuring a 40% level of relative humidity, far below the 60% which is critical. (c) An electromagnetic method (called the *main flux method*) to identify internal corrosion of suspender ropes is adopted in order to provide appropriate and concentrated interventions on ropes. (d) As unexpectedly large amplitude oscillation had been observed in suspender ropes at the downstream side, vibration control of wind-induced oscillation was introduced after construction through the winding of trip wires around a suspender rope. (e) A preventive action for the vibration control for cables is the use of an indent cable, which has dispersive concave marks on its surface, and a high-damping rubber, which was installed at the cable anchorage in the girder, as an antivibration measure for cables against vortex-induced oscillation. (f) A dry air injection system was installed in the box girders to avoid corrosion on the inner surface of not-painted steel pieces. Thus, the repainting costs have been avoided, and humidity controlled. (g) To maintain the undamaged bridge pavement, while avoiding cut and cast expensive procedures, the surface (made of an adhesive layer of 35–40 mm of guss asphalt and 30–35 mm of improved asphalt) is treated regularly with a microsurfacing composed of a thin slurry admixture of aggregates, early strength-improved asphalt emulsion, water, and cement. This treatment allows for a timely reopening of the traffic lanes.

References

Cooper, J.D., 1998. World's longest suspension bridge opens in Japan. ASCE Public Roads 62(1).

Honshu-Shikoku Bridge Expressway (HSBE), 2005. Maintenance Technology. Honshu-Shikoku Bridge Expressway Company Limited, Official Webstie.

Katsuchi, H., Nicholas, P.J., Scanlan, R.H., 1999. Multimode coupled flutter and buffeting analysis of the Akashi-Kaikyo bridge. J. Struct. Eng. 125, 60–70.

Fuchida, M., Kurino, S., Kitagawa, M., 1998. Design and construction of the Akashi Kaikyo bridge's superstructure. IABSE Report, 79/1998, pp. 63–68, Zurich.

Nagai, M., Yabuki, T., Suzuki, S., 2000. Design practice in Japan. In: Chen, W., Duan, L. (Eds.), Bridge Engineering Handbook. CRC Press, Boca Raton, CA.

Nasu, S., Tatsumi, M., 1995. Effect of the Southern Hyogo Earthquake on the Akashi-Kaikyo Bridge. Proceedings of the 4th International Workshop on Accelerator Alignment (IWAA 1995), 14–17 November 1995, Tsukuba, Japan - Ground Motion and Active Damping Session, pp V-305–V316, Tsukuba.

Takeno, M., Kishi, Y., 1997. Cable erection technology for world's longest suspension bridge—Akashi Kaikyo Bridge. Nippon Steel Tech. Rep. 73, pp. 59–70, Nippon Steel & Sumitomo Metal Corporation, Tokyo.

Bridge construction equipment 27

Rosignoli M.
Dr. Ing., PE, United States of America

1 Summary

The bridge industry is moving to mechanized construction because this saves labor, shortens project duration, and improves quality. This trend is evident in many countries and involves most construction methods. Mechanized bridge construction is based on the use of special equipment.

New-generation bridge construction machines are complex and delicate structures. They handle heavy loads over long spans under the same constraints that the obstruction to overpass exerts on the bridge. The safety and quality of operations depend on complex interactions between human decisions; structural, mechanical, and electro-hydraulic components; control systems; and the bridge that is being erected. In spite of their complexity, these machines must be as light as possible. Weight governs the initial investment, the cost of shipping and site assembly, and sometimes even the cost of the bridge. Weight limitations dictate the use of high-grade steels and designing for high stress levels in different load and support conditions, which makes these machines potentially prone to instability. Bridge erection equipment is assembled, operated, decommissioned, modified, reconditioned, and often adapted to new work conditions. Connections and field splices are subject to hundreds of load reversals. The nature of loading is often highly dynamic, the equipment may be exposed to strong wind, and the full design load is reached multiple times (and sometimes exceeded). Impacts are not infrequent, vibrations may be significant, and most machines are actually quite lively because of their great structural efficiency.

Movement adds the complication of variable geometry. Loads and support reactions are applied eccentrically, the support sections are often devoid of diaphragms, and most machines have flexible support systems. Indeed, such design conditions are almost inconceivable in permanent structures subjected to such loads.

The level of sophistication of new-generation bridge construction machines requires an adequate technical culture. Long subcontracting chains may lead to loss of communication, the problems not dealt with during planning and design must be solved on the site; the risks of wrong operations are not always evident in such complex machines; and human error is the prime cause of accidents.

Experimenting with new solutions without due preparation may lead to catastrophic results. Several bridge erection machines collapsed recently with a heavy tribute of fatalities, wounds, damage to property, delays in the project schedule, and legal disputes. Technological improvement alone cannot guarantee a decrease in failure of bridge construction equipment, and may even increase it. Only a deeper consciousness of our human and social responsibilities can lead to a safer work

Innovative Bridge Design Handbook. http://dx.doi.org/10.1016/B978-0-12-800058-8.00027-X

environment. A level of technical culture adequate to the complexity of modern mechanized bridge construction would save human lives and facilitate the decision-making processes with more appropriate risk evaluations.

2 Introduction

Every bridge construction method has its own advantages and disadvantages. In the absence of requirements that make one solution immediately preferable to the others, the evaluation of the possible alternatives is always difficult. Comparisons based on the quantities of structural materials may be misleading. The processing costs of raw materials and the indirect and financial costs often govern in industrialized countries. Higher quantities of raw materials due to efficient and rapid construction processes rarely make a solution antieconomical. Low technology costs are the reason for the success of the incremental launching method for prestressed concrete (PC) and steel bridges (Rosignoli, 2014). Compared to ground falsework, launching diminishes the cost of labor with similar investments. Compared to a movable scaffolding system (MSS), launching diminishes the level of investment and financial exposure with similar labor costs. Even if transient launch stresses may increase the quantities of raw materials, the balance is positive and the solution is cost effective.

The construction method that comes closest to incremental launching is segmental precasting. The labor costs are similar, but the investments are higher, and the break-even point shifts to longer bridges. Spans of 30–50 m are erected span by span with self-launching gantries, and longer spans are erected as balanced cantilevers. On shorter bridges, prefabrication is limited to the beams and the concrete slab is cast in place. Precast beams are often erected with ground cranes; sensitive or urban areas, inaccessible sites, active infrastructures, tall piers, and steep slopes may suggest the use of beam launchers; and the technology costs increase. Medium-span PC bridges may also be cast in place. For bridges with more than 2–3 spans, it is convenient to advance in line by reusing the same formwork, and the deck is built span by span. The span is cast in either a fixed or a movable formwork. The choice of equipment is governed by economic reasons, as the labor cost associated with a fixed falsework and the investment requested for an MSS are both considerable.

Beginning in the 1940s, the original wooden falsework has been replaced with modular steel framing systems. The labor component may exceed 50% of the construction cost of the span, and casting on falsework is therefore a viable solution only with inexpensive labor and small bridges. Obstruction of the area under the bridge is another limitation.

An MSS comprises a casting cell supported on or suspended from a self-launching frame. MSS are used for span-by-span casting of long bridges with 30–70-m spans. Repetitive operations diminish the labor cost, and the quantities of raw materials are unaffected.

Bridges crossing inaccessible sites with tall piers and spans up to 300 m may be cast in place as balanced cantilevers. When the bridge is short or the spans exceed

100–120 m, the bridge supports the form travellers. With longer rectilinear bridges and 90–120 m spans, longer casting cells may be suspended from a self-launching girder that also balances the hammers during construction.

High-speed railway (HSR) bridges with 30–40-m spans are often precast full span. The investment is so high that the breakeven point is reached with hundreds of spans. The precasting facility delivers 2–4 spans per day, and portal carriers with underbridge or span launchers fed by tire trolleys are used for on-deck delivery and placement of precast spans.

3 Beam launchers

The most common method for erecting precast beams is with ground cranes. Cranes usually allow for the execution of the simplest and most rapid erection procedures with minimum investment, and the bridge may be built in several places at once. Good access is necessary to position the cranes and deliver the girders. Tall piers or steep slopes make erection expensive or prevent it from happening at all.

The use of a beam launcher solves most difficulties (Rosignoli, 2013). In the typical configuration, a beam launcher is a light self-launching machine comprising two triangular trusses (Figure 27.1). The trusses are 2.3 times as long as the typical span, but this is rarely a problem, as the gantry operates above the deck. Cross-beams support the gantry at the piers and allow the shifting and traversing of the trusses.

Figure 27.1 A long, 2.3-span launcher (Comtec).

Two winch-trolleys bridge the trusses and lodge a hoist winch and a translation winch acting on a capstan for movement. When the beams are delivered on the completed bridge, the hoist winches may be replaced with long-stroke hydraulic cylinders.

If the beams are delivered on the ground, the launcher lifts them to the deck level and places them on the bearings. If the beams are delivered at the abutment, the launcher is moved back to the abutment to pick up the beam and release the carrier. The longitudinal movement is a two-step process: The trusses are anchored to the pier cross-beams and the winch-trolleys move one span forward; and then the winch-trolleys are anchored to the new cross-beams and their translation winches are reversed to push the trusses to the next span.

The sequence can be repeated many times, and when the beams are delivered at the abutment, the gantry can place them several spans ahead. Shuttling the gantry back and forth slows the erection down, and it may be faster to cast the concrete slab of the leading span as soon as the beams are placed, and to deliver the next beams on the completed bridge. This also allows the use of a shorter, single-girder launcher (Figure 27.2).

Figure 27.2 Single-girder launcher (Strukturas).

The main girder of these launchers is less expensive than two trusses, the hoists are smaller and lighter, the number of support saddles halves, and fewer pier cross-beams are necessary. A portal frame supports the rear end of the gantry and allows the beams to pass through. When the beams are delivered on the ground, the launcher lifts them up within the same span.

The pier cross-beams have side overhangs and carry full-width rails for placement of the edge beams and to traverse the gantry when launching along curves. Adjustable support legs are used to set the cross-beams horizontal. Some launchers are equipped with service cranes to reposition the pier cross-beams during launching.

4 Self-Launching gantries for span-by-span erection of precast segments

Span-by-span erection is used for precast segmental spans shorter than 50 m in highway bridges and 30–45 m in light-rail transit (LRT) bridges. The gantry is loaded with all the segments for the span to stabilize the deflections during gluing. After prestress is applied, the support jacks of the gantry are retracted to release the span in one operation. A typical 40-m span with epoxy joints is erected in 2–3 days.

Overhead or underslung gantries are used for span-by-span erection (Rosignoli, 2013). A twin-girder overhead gantry comprises two trusses or box girders supported on pier cross-beams (Figure 27.3); auxiliary pendular legs are necessary to erect the pier-diaphragm segments of continuous spans and to reposition the support cross-beams without ground cranes. A portal crane bridging the main girders lifts and moves the segments into position and avoids interference with segment hangers during launch.

The overhead gantries are not much affected by ground constraints, straddle bents, C-piers, or variations in span length and deck geometry. They are more complex to design, assemble, and operate than the underslung gantries, and they are also more expensive and slower for erecting the segments with spreader beams.

If the segments are delivered on the completed bridge, the winch-trolley picks them up at the rear end of the gantry and moves them forward into position. If the segments are delivered on the ground, the winch-trolley lifts them up to the deck level. Hangers and spreader beams are used to hold the segments in position during assembly and gluing.

Pier cross-beams support the gantry with articulated saddles that allow longitudinal and lateral movements and rotations about the transverse and vertical axes. Some support saddles lodge longitudinal lock systems for the gantry, and all the cross-beams are equipped with transverse lock systems. The cross-beams are anchored to the pier caps with stressed bars that resist uplift forces and provide friction transfer of lateral loads.

Some light gantries are launched with winches and capstans. Hydraulic cylinders lodged within the support saddles and acting against racks anchored to the trusses provide higher thrust forces and safer operations. Paired launch cylinders are often used for redundancy.

A single-girder overhead gantry takes support on the leading pier of the span to erect and on the front-pier diaphragm of the completed bridge. The front support legs are often framed to the main girder, and the rear portal frame rolls over the new span during launch. Special launch systems suspended from the main girder are used to reposition the gantry. These machines are shorter and lighter than the twin-girder overhead gantries, and there is a better fit for curved bridges.

Some first-generation overhead gantries were equipped with stay cables. Cable-stayed gantries have been abandoned over time in spite of their structural efficiency, as the cables complicate operations and increase labor demand. Trusses of varying depths are also rarely used for span-by-span erection nowadays.

Telescopic gantries are used to erect dual-track LRT spans with tight plan curvature (Figure 27.4).

Figure 27.3 A twin-girder overhead gantry (NRS).

Figure 27.4 A telescopic overhead gantry (Deal).

A winch-trolley is suspended from the main girder, and the segments are delivered on the ground or on the completed bridge through a rear portal frame. The gantry comprises a rear main girder and a front underbridge. A turntable with hydraulic controls connects the main girder to the underbridge. During the first phase of launch, the turntable pulls the main girder over the underbridge. When the front support legs of the main girder reach the new pier, the underbridge is launched to the next span to clear the area under the main girder for erection of the new span.

Many precast segmental bridges have been erected with underslung gantries (Figure 27.5). These machines are positioned beneath the deck, with the main girders on either side of the piers. The gantry supports box girder segments under the side wings, with hydraulic carts for control of camber and cross-fall. The main girders take support on pier brackets or cross-beams hung from the pier caps. When the piers are short and slender, the pier brackets may be propped from foundations.

Figure 27.5 An underslung gantry (NRS).

Ground cranes or lifting frames are used to load the segments onto the gantry. When the segments are delivered on the completed bridge, the lifter is placed at the rear end of the gantry. When the segments are delivered on the ground, the crane is placed at the front end of the gantry. The segments are loaded onto the gantry close to the lifter and rolled into position. A portal crane may also be used to lift the segments and move them into position. Upon application of prestress, the gantry is lowered to release the span in one operation.

The underslung gantries are not very compatible with curved bridges, as the main girders conflict with the piers and the completed spans. The front end of the girders may be connected by a cross-beam that rolls along a central underbridge during launch. A rear portal frame rolling on the new span may be used to further shorten the rigid portion of the machine. New-generation gantries with articulated girders have also been used for curved spans.

The underslung gantries are simple to design, assemble, and operate. Span erection is fast, and props from foundations can be used to increase the load capacity when working low on the ground. These machines project beneath the deck, which may cause interference with straddle bents and C-piers, clearance issues when overpassing active infrastructures, and difficulties at the end spans of the bridge.

5 MSS for span-by-span casting

A PC bridge can be cast in place span by span, proceeding from one abutment to other one. When the piers are short and the area under the bridge is accessible, the casting cell can be supported onto ground falsework. In bridges that are long enough to amortize the investment, the use of an MSS allows the transferral of the casting cell to the new span in a few hours instead of weeks. The savings of labor are substantial, and the area under the bridge is unaffected.

An MSS is typically designed to cast an entire span (Rosignoli, 2013). Solid or voided slabs are used for 30–40-m highway spans, ribbed slabs with double-T sections are used up to 50-m spans, and box girders can reach up to 70-m spans. Box girders for railway bridges rarely exceed 50-m spans. Simple spans are cast at full length; for continuous bridges, the starter abutment span is cast with a short front cantilever, and the subsequent spans extend out over the piers for 20%–25% of their length. The rear support of the MSS is placed on the front cantilever of the completed bridge to minimize the design span of the MSS and the time-dependent stress redistribution within the bridge. The cycle time is 1–2 weeks per span.

If the piers are not tall and the area under the bridge is accessible, 90–120-m continuous spans may be divided into two segments with joints at the span quarters. A temporary pier at the leading span quarter supports MSS and the midspan segment prior to casting the next pier segment. Inline casting of 90–120-m, varying-depth spans is faster than balanced cantilever construction, with a cycle time of 2–4 weeks per span.

Solid, voided, and ribbed slabs are cast in one phase, while box girders may be cast in one or two phases. The inner tunnel form for one-phase casting of box girders is extracted and reopened within the rebar cage of the next span. Two-phase casting involves casting bottom slabs, webs, and pier diaphragms first, and the top slab after 2–3 days.

Two-phase casting restricts the quantity of concrete processed daily and facilitates the handling of inner forms. Joints at the top-slab level also avoid the horizontal cracks that sometimes affect one-phase casting due to the settlement of fresh concrete in the webs. The main concerns with two-phase casting are related to the deflections of the MSS during casting of the top slab, which may cause cracking in the nonprestressed U-span.

In the simple spans, concrete should be poured starting at midspan and progressing symmetrically toward the piers to minimize the deflections of the MSS in the final phases of filling. This sequence is labor intensive and the use of retarding admixtures is often preferred to keep the concrete fluid for the entire duration of filling. This allows for casting the span directionally from pier to pier, but the forms must be designed for full hydrostatic loads.

In continuous bridges, the casting cell is filled with one of two alternative sequences. In the first procedure, concrete is poured starting at the pier and proceeding symmetrically until the front cantilever is filled; then the remaining rear section is filled backward toward the construction joint. This sequence is preferred when the MSS is supported on distant cross-beams or W-frames at the leading pier. In the second procedure, concrete is poured starting at the front bulkhead and proceeding backward. This sequence is preferred when the MSS is supported on pier brackets. In either case, the construction joint is cast at the end of filling to prevent settling.

The main girders of an underslung MSS support the bottom cross-beams of the casting cell with adjustable saddles for setting of camber and cross-fall (Figure 27.6). The machine projects far beneath the bridge, which may cause clearance issues when overpassing active infrastructures and difficulties at the end spans of the bridge. These MSS are rarely compatible with bridges with tight plan curvatures.

Figure 27.6 An underslung MSS (DB).

The underslung MSS is supported by pier brackets or W-frames on through girders. The support saddles lodge hydraulic launch cylinders acting in racks to reposition the machine. The pier brackets include vertical hydraulic cylinders that lift the MSS to the span-casting elevation and lower it back onto the launch saddles to release the span in one operation. Some MSSs are equipped with self-repositioning pier brackets that avoid the use of ground cranes. W-frames on through girders are assembled and dismantled with ground cranes.

The rebar cage may be prefabricated behind the abutment and delivered on the completed bridge to achieve weekly casting cycles and mitigate risks with parallel tasks. The cage carrier moves over the casting cell and lowers the cage and the front bulkhead in one operation. The carrier may be equipped with concrete distribution arms and a covering to protect the casting cell during concrete pouring.

In a single-girder overhead MSS, the outer form is suspended from a central girder that is supported at the leading pier of the span to cast and on the front cantilever of the completed bridge (Figure 27.7). These machines are lighter than the underslung MSS, typically carry more production support systems, and are able to reposition the pier supports during launch. The main girder comprises two braced trusses or I-girders. Lateral bracing and connections designed for fast site assembly and to minimize displacement-induced fatigue provide sufficient flexural stiffness to resist vibration stresses. Cross-frames or cross-beams provide transverse rigidity and distribute torsion.

Figure 27.7 An open casting cell for a ribbed slab with a double-T section (Strukturas).

Form brackets overhanging from the main girder suspend form hangers with telescopic connections for the setting of camber and cross-fall. A central full-length splice divides the outer form into two halves. Long-stroke, double-acting cylinders rotate the two halves of the outer form to vertical to avoid interfering with the piers during launch. After launch, the main girder is lifted to the span-casting elevation and the casting cell is reclosed.

The front support of the MSS may be a portal frame integral with the main girder or a tower-cross-beam assembly anchored to the pier cap and equipped with launch rollers. The rear support may be a second tower cross-beam assembly or an adjustable portal frame designed for the insertion of prefabricated rebar cages into the casting cell.

The rebar cage may be prefabricated behind the abutment to shorten the cycle time and to move activities out of the critical path. The cage includes a front bulkhead, prestressing hardware, and embedded items. Strands are inserted into the ducts during span curing. The cage carrier supports runway beams for a full-length lifting frame that suspends the cage. Winch-trolleys rolling along the main girder of the MSS pick up the lifting frame for cage insertion into the casting cell. In less-automated machines, the web cage is fabricated over the new span during curing, is suspended from the MSS during launch, and is lowered into the casting cell after fabrication of the bottom slab grid.

New-generation single-girder overhead MSS are targeting 90-m spans in rectilinear highway bridges and 70-m spans in heavier dual-rack railway box girders. One-phase casting of a 90-m box girder requires the pouring of 700–900 m^3 of concrete in a few hours. Two-phase casting reduces the demand on the batching plant and concrete delivery lines but requires active control of deflections on long spans. Arched trusses with PLC (Programmable Logical Controllers)-controlled strand cables at the bottom chord have been used on 70-m spans (Figure 27.8). The cables are tensioned during the filling of the casting cell and released during application of span prestress.

Figure 27.8 An overhead MSS with active prestressing (BERD).

Prestressing increases cost and complexity of the MSS, and the axial load makes the truss more prone to out-of-plane buckling. Combining sophisticated control systems with the typical skill and attention of workers in construction involves additional challenges. However, span-by-span casting with weekly casting cycles is much faster than balanced cantilever construction, design of reinforcement and prestressing is more efficient, and the operators of bridge erection machines must be accurately trained anyway.

6 Self-launching machines for balanced cantilever construction

Balanced cantilever construction is suited to precast and cast-in-place segmental bridges. Precast segmental construction is addressed to large-scale bridge projects with 50–120-m spans; ground cranes and lifting frames handle the segments with free erection sequences, while self-launching gantries operate linearly from abutment to abutment. In-place casting is addressed for shorter bridges and longer or curved spans: form travellers are used for free erection sequences, and self-launching suspension MSS are used for linear erection on 100–120-m spans.

Balanced cantilever bridges have box girder sections. Ribbed slabs have been built in the past and are still used for cable-stayed bridges, where the stay cables resist most torsion and negative bending. Constant-depth bridges are simpler to cast but are competitive in the 50–70-m span range. Bridges of varying depths are used on spans ranging from 70 m to 250–300 m. Depth variation adapts the flexural capacity to demand, but precast segments soon become too tall and heavy for ground transportation and lifting, and spans longer than 120–130 m are typically cast in place.

In a precast segmental bridge, the segments of the pier table should have the same weight as the other segments to avoid having to use special lifters. The pier table includes a heavy pier diaphragm and the bottom slab is also thick; therefore, the assembly is divided into three segments. This also facilitates placement of the central segment, as a self-launching gantry must also take support on the pier cap.

The most common erection methods for precast segments are with ground cranes, lifting frames or self-launching gantries (Rosignoli, 2013). Ground cranes require good access throughout the length of the bridge. Cranes usually give the simplest and most rapid erection procedures with minimum investment, and multiple hammers can be erected at once. The main constraints on crane erection are access and tall piers, as balanced cantilever bridges are often selected in response to inaccessible terrains.

Deck-supported lifting frames are used on tall piers, long or curved spans, and spans over water, where special lifters can handle heavier segments and barge delivery minimizes geometry and weight restraints. Lifting frames are also the standard solution for the erection of cable-stayed bridges, when time or site constraints discourage in-place casting. In spite of the disruption when moving to the next pier, the lifting frames can solve erection conditions that are incompatible with ground cranes and self-launching gantries.

Fixed lifting frames anchored to the tip of the cantilever have a limited load handling capability. Derricks with rotating arms can lift from behind or laterally, and they are used only in cable-stayed bridges because of the torsion that they apply to the deck. Wheeled straddle carriers with cantilever noses on either side shuttle back and forth along the hammer to lift the segments wherever possible (Figure 27.9). These lifters are light and devoid of counterweights: this simplifies their placement on the pier table, but it requires anchoring to the deck during operations.

Self-launching gantries achieve a fast erection rate and minimize ground disruption. The gantries erect the hammers directionally from abutment to abutment, and the segments can be delivered on the completed bridge. Erecting two adjacent bridges simultaneously by shifting the gantry from bridge to bridge further accelerates construction, which may reach 2 + 2 segments per hammer per day (Figure 27.10). One or two hoists are used to lift and move the segments into position. If the segments are delivered on the bridge, the hoist picks them up at the rear end of the gantry. If the segments are delivered on the ground, the hoist raises them to the deck level.

Figure 27.9 A wheeled straddle carrier (Deal).

Figure 27.10 Balanced-cantilever erection of adjacent bridges (VSL).

The earliest gantries were 1.5 times as long as the span to erect. Their length was sufficient for self-launching, and closer supports resulted in lighter trusses. Short gantries load the front cantilever of the bridge, and erecting the pier table is also more complex. The new-generation gantries are twice as long as the bridge span. Full two-span gantries take support at the piers during operations, and the cost of longer trusses is balanced by less reinforcement and prestressing throughout the length of the bridge. Launching and placement of pier tables are also simpler, operations are less labor intensive, and no ground cranes are necessary.

When the bridge is too short for segmental precasting, the hammers are cast in place with form travelers; in-place casting is also the standard solution for curved bridges and spans longer than 120 m. The casting cell of a traveler is 3–5 m long for reasons of weight and load unbalance; the standard travelers are designed for 5-m segments up to 500 tons.

An overhead traveller includes a number of trusses equal to the number of webs in the box girder. The trusses of the earliest travellers were long in order to enhance the stabilizing action of rear counterweights. The new-generation travellers have tie-down rollers that avoid the use of counterweights, and the trusses are much lighter and shorter (Figure 27.11).

Figure 27.11 An overhead form traveller (Doka).

The casting cell is suspended with hanger bars for geometric adjustment. Working platforms are incorporated around the traveller, and a stressing platform is suspended beyond the front bulkhead for fabrication and tensioning of the top-slab tendons. The inner tunnel form is stripped from the previous segment and pulled forward by rolling along suspended rails.

The pier table must accommodate a pair of form travelers in the initial stages of cantilever construction, and lengths of 8–10 m are not infrequent. Props from foundations or pier brackets support the casting cell for the pier table. The geometry is complex, the working space is limited, the segment is typically divided into numerous casting phases, and cycle times of 2–4 months for the casting cell are not unusual.

Pairs of identical travellers are used on the hammers for load balance and to accelerate assembly. A 10–12-m pier table is typically necessary to accommodate two independent travellers. Three techniques are used with shorter pier tables: fast-split assembly, temporary bracing, and side assembly. Fast-split assembly is used for 6–10-m pier tables and requires a special design of the rear of the traveler. On 4–5-m pier tables, the rear frames may be replaced with temporary braces to cast the two starter segments with a common overhead frame. Side assembly is also used with 4–5-m pier tables.

It typically takes 2 weeks to assemble an overhead traveller and another 2 weeks to assemble the casting cell. Casting the starter segment takes another 2–3 weeks. The segments are cast on a 5-day cycle; 3- and 4-day cycles have also been achieved. Concrete with early high strength is used to shorten the cycle time.

The trusses are supported on the front segment of the hammer and anchored to the second segment with tie-downs that prevent overturning. Jacks at the front support are used to lift the traveller from the launch rails for casting. After stressing the top-slab tendons, the traveller is lowered back onto the launch rails. Adjustable tie-downs roll within the launch rails during launching to prevent overturning. Launching is a two-step process: first, the rails are pushed forward and anchored to the new segment, and then the traveller is lowered onto the rails and repositioned.

In the underslung travellers for PC box girders, a full-width cross-beam supported over the webs of the front segment suspends a longitudinal truss on either side of the deck, beneath the side wings. In the travellers for cable-stayed ribbed slabs, two C-hooks roll on the outer edges of the edge beams to avoid interference with the stay cables. The front cantilever of the trusses supports the casting cell, and balancing rollers at the rear end prevent overturning.

The outer form is stripped by releasing the support jacks of cross-beams or C-hooks. Launch cylinders push the launch rails over the new segment, and then they pull cross-beams and C-hooks along the rails. The underslung travellers are more complex to operate than the overhead units, but the rebar cage can be prefabricated on the ground and placed in the casting cell with the tower crane, and access to the casting cell is unobstructed.

The underslung travellers for cable-stayed bridges are heavy due to the wide deck and the long segments based on cable spacing. Deflections are controlled with stiff longitudinal trusses and transverse space frames that support the casting cell, and temporary stay cables are often necessary to provide additional stiffness.

A balanced cantilever bridge can also be cast with a suspension MSS supported at the leading pier and on the front cantilever of the completed bridge (Figure 27.12). Two long casting cells shift from the pier table to midspan to cast the hammer. After achieving midspan continuity, the girder is launched to the next span, and the casting cells are repositioned to cast the new starter segments (Rosignoli, 2013). Short pier tables are precast at the end of the pier erection to shorten the cycle time of the MSS and to move activities out of the critical path.

The suspension MSS is used for directional erection of rectilinear or slightly curved 100–120-m spans. The main girder is 1.3 times as long as the maximum span. The

Figure 27.12 A suspension MSS for balanced cantilever casting (ThyssenKrupp).

MSS simplifies access from the completed bridge and stabilizes the hammer during construction, and less prestressing is necessary in the bridge because of minimum construction loads.

The segments can be 10–12 m long and wider than 20 m. Transferring the casting cells to the new pier takes hours instead of weeks, the use of ground cranes is minimized, the construction process is faster, the logistics is simpler, and the labor savings may be substantial. The end segments at the abutments are also cast with the MSS.

7 Special equipment for full-span precasting

Full-span precasting is used for rapid, high-quality construction of large-scale railway projects. Ground transportation is rarely used for railway spans longer than 40 m. The spans are typically placed on bearings or seismic isolators to simplify the erection process.

The investment needed to set up large precasting facilities and to provide special transportation and erection means raises the entry threshold of full-span precasting to long bridges comprising hundreds of equal spans with small plan radius and low gradient. These conditions are frequently met in HSR bridges.

The precasting facility is located near the bridge. The rebar cages are prefabricated at full length, and different combinations of pretensioning and posttensioning are possible. The spans are removed from the casting cells and transferred to storage platforms for completion of prestressing, bearing application and finishing. The precasting facility delivers 2–4 spans per day (Rosignoli, 2013).

The spans are transported on the completed bridge with tire trolleys or portal carriers. A portal carrier comprises two motorized tractors connected by a box girder supporting two hoists that suspend the span at the ends. For picking up the span, the carrier is moved alongside the span, the wheels are rotated by 90°, and the carrier is driven laterally over the span. The same operations are repeated to release the span onto the storage platform and to pick it up for final delivery. Computerized hydraulic systems govern movement and steering.

The carrier spreads the load to two spans of the bridge during span delivery, and the axle load is equalized hydraulically or electronically. At the leading end of the bridge, the front tractor reaches the rear end of the underbridge. A self-propelled support saddle rolling along the underbridge is driven beneath the front tractor, and hydraulic cylinders lift the tractor until its wheels take off from the deck. The translation motors of support saddle and rear tractor are synchronized to move the carrier forward over the under-bridge (Figure 27.13). After reaching the span-lowering position, the support rollers of the underbridge are unlocked, and the support saddle pushes the underbridge forward to clear the area beneath the span. After placing the span, the underbridge is moved back over the new span to release the front tractor for a new erection cycle.

Tire trolleys deliver precast spans to span launchers operating at the leading end of the erection line (Figure 27.14). The front hoist of the launcher picks up the front end of the span and moves forward, while the rear end slides along the extraction rail of the tire trolley. Then the second hoist picks up the rear end of the span to release the transporter.

Figure 27.13 A portal carrier over the underbridge.

Figure 27.14 A tire trolley feeding the launcher (Beijing Wowjoint Machinery).

Repositioning the launcher takes a couple of hours, and 2–3 spans can be erected every day when crossover embankments exist along the delivery route and the precasting facility is designed for such productions. Heavy lifters are necessary in the precasting facility to handle the spans and to load the tire trolleys. The tire trolleys apply a localized load to the bridge and delay the finishing work, which can start only on completion of span delivery.

8 Conclusions

Modern bridge construction machines are light, efficient, and complex. The level of sophistication of new-generation machines requires adequate technical culture. A level of technical culture adequate to the complexity of mechanized bridge construction would save human lives and facilitate the decision-making processes with more accurate risk analyses.

References

Rosignoli, M., 2013. Bridge Construction Equipment. ICE Publishing, London. p. 496.
Rosignoli, M., 2014. Bridge Launching, second ed. ICE Publishing, London. p. 376.

Section IX

Assessment, monitoring and retrofit of bridges

Bridge assessment, retrofit, and management

28

Pipinato A.
AP&P, Technical Director, Italy

1 Introduction

This chapter provides an analysis of the causes of bridge structural and materials decay. It also includes the investigation procedures used to analyze and understand the condition state of a particular bridge; and, assessment procedures, which are not all standardized, are described using a multilevel approach, which considers research and common standards. In the final part of the chapter, retrofit and strengthening solutions for bridge structures are reported, considering various constituent materials. Finally, the most common BMSs are presented.

2 Materials decay and on-site testing

2.1 Degradation causes

The classification presented in Table 28.1 deals with degradation causes in bridge structures: these are partly based on factors that are proposed in Silano (1993) and Radomski (2002). The factors leading to bridge deterioration can be classified into five fundamental groups as: (A) basic factors, (B) load factors, (C) weather and environmental factors, (D) maintenance factors, and (E) construction defects.

2.2 Concrete structures

A concrete structure may not perform satisfactorily during its intended life span with only an efficient design; experience has shown that durability design is required to ensure adequate structural performance (Fib, 2010). Consideration of parameters such as loading, traffic growth analysis, material types, material strength, element geometry, environmental analysis, etc., made during the design phase should also be checked during the construction stage by quality and performance control. Recent works such as the FIB bulletins (FIB, 1999) and the DuraCrete project (DuraCrete, 1998; DuraCrete, 2000a; DuraCrete, 2000b) provide valuable information about the durability characteristics of concrete structures. The principle of the DuraCrete model has been adopted in the Fib model code for service life design (FIB, 2010). Environmental conditions that have an effect on the development, either initiation or progress in time, of the deterioration processes also need to be accounted for. The next section describes factors affecting the durability of a concrete structure, focusing on the main deterioration mechanisms and the role of the environment.

Innovative Bridge Design Handbook. http://dx.doi.org/10.1016/B978-0-12-800058-8.00028-1

Table 28.1 Classification of Degradation Causes

A. Basic factors
A.1. Age of the bridge structure
A.2. Quality of the project
A.3. Sensitivity to damage
A.4. Adequacy of the design to the increasing service conditions
A.5. Time-dependent effect on constituent materials

B. Load factors
B.1. Frequency, speed, and traffic loads spectra
B.2. Dynamic effects
B.3. Accidents on the bridge
B.4. Accidents under the bridge
B.5. Vessel collision under the bridge
B.6. Overloading
B.7. Fatigue-induced damage
B.8. Time dependent effects
B.9. Resonance and lateral effects at high speed
B.10. Horizontal forces due to speed changes

C. Weather and environmental factors
C.1. Rain events
C.2. Snow events
C.3. Variation of the water level and its frequency
C.4. Ice-float run-off and its pressure on bridge sub-structures
C.5. Wind pressure and its effects on bridge elements
C.6. Earthquake and soil displacement
C.7. Temperature induced deformations
C.8. Solar induced deformations
C.9. Chloride attack originating from the action of sea water
C.10. Chloride attack originating from the use of deicing products
C.11. Freeze-thaw cycles
C.12. Aggressive chemicals
C.13. Penetration of CO_2 from atmosphere (carbonation effect in concrete)
C.14. Aggressive chemicals in rivers and underground water
C.15. Seawater attacks by its sulfates and chlorides
C.16. Vagabond currents
C.17. Fire
C.18. Hurricane

D. Maintenance factors
D.1. Whether design solutions are easy for maintenance or not
D.2. Inspection timing and quality
D.3. Timely maintenance works
D.4. Timely renewal of consumed secondary structures (e.g., drainage system, plants, pavement, parapets etc.)
D.5. Painting renewal
D.6. Use of deicing salts

E. Construction defects
E.1. Quality of the construction
E.2. Quality of the structural materials
E.3. Bridge yard quality control protocol
E.4. Construction interruption
E.5. Partial collapse and repair during erection

2.2.1 Affecting factors

After the design stage, the execution phase of a structure is one of the most crucial stages. Several examples exist of durability deficiencies due to poor workmanship that have resulted in insufficient compaction, curing, and concrete cover depth (BRE, 2001; FIB, 2010). Certain environmental conditions can benefit from an emphasis on the degradation of concrete structures, environmental effects, various transport mechanisms, and deterioration processes. Eventually these effects will affect the appearance of the structure e.g., the formation of cracks, and one thing will lead to another until, by acceleration of the progress of some deterioration mechanisms, there will be a resulting loss of safety for the users, loss of the resistance for the structure, and loss of the bridge's serviceability. However, design and construction controls may enable a reduction in unwanted results; such as the weakening of the structure's durability and structural performance. Furthermore, during its service life, regular inspections should catch any deficiencies in the early stages.

Deterioration of concrete structures occurs mainly from physical processes, or as a result of chemical reactions. Mechanisms such as temperature variations (i.e., freezing and thawing), sulfate attack, carbonation, chloride penetration, alkali-silica reactions, etc., may all result in severe damage of the concrete and/or the reinforcement through different steps, and lead to events such as scaling, cracking, spalling, and corrosion. These mechanisms, with their individual steps and effects, are well described in a number of references, such as FIB (2010), Bentur et al. (1997), Broomfield (1997), and thus are only shortly described in the following sections. For reinforced concrete structures, the most common cause of damage is corrosion of the reinforcement. Theoretically corrosion of the reinforcement should not occur, as the reinforcement is supposed well protected by the concrete cover. In fact noncarbonated concrete has a high alkalinity (pH = 13) that is a result of the presence of sodium, potassium, and calcium hydroxides produced during the hydration of the cement. In this alkaline environment an oxide layer is formed on the steel surface, the so-called passive film that prevents the corrosion of the reinforcement. However, there are two processes that may break down this passive film: the aggression of chlorides and carbon dioxide.

2.2.2 Temperature

Freezing and thawing is a common cause of concrete deterioration. Concrete subjected to alternate freezing and thawing is damaged due to the expansion of frozen water. Moisture is collected in voids that result from entrapped air during placement. Consequently, damage occurring from frost action depends mainly upon the degree of saturation of the concrete's pore system with water (Mullheron, 2000). During a freeze cycle, water expands about 9% (White et al., 1992). This change in volume results in expansive pressures, causing gradual scaling, cracking, and eventually spalling of the concrete. Dry concrete is generally unaffected by freezing. Nonetheless, most concrete structures are exposed at some stage to wet and/or cold environments. To prevent the effects of freeze/thaw attack, specific requirements are included in codes and standards (e.g., in EN 206-1/2000, 2005).

2.2.3 Sulfate

Sulfates carried into the inner sections of concrete may cause disruptive forces leading to cracking and scaling of the concrete. Sulfates are found in some clay soils, in seawater and in many industrial environments where the combustion leads to the release of sulphur dioxide (Fib 2010). The best way to protect concrete from the adverse effects of sulfates is by producing impermeable cement. Where there are known high levels of sulfates, protective epoxy coatings, or the traditional bitumen, should be applied to the concrete surface, Mullheron (2000). Also in this case could be applied e.g. the specific requirements of EN 206-1/2000 (2005).

2.2.4 Alkali-silica reaction

Some decades ago, it was observed that certain concrete structures exposed to moisture penetration, developed cracks, with discoloration of the concrete adjacent to the cracks, (Liebenberg, 1992). Alkali silica reactions (ASRs) can develop on concrete mixes containing reactive aggregates. Reactive aggregates are aggregates found in natural rocks containing reactive forms of silica such as chert, flint, chalcedony, and opaline sandstone (Mullheron, 2000). The ASR results in the formation of an alkali-silicate gel able to absorb large water quantities that can cause expansive forces that lead to the cracking of concrete (Sukumaran, 1998). This type of deterioration may cause significant problems because the rate of deterioration is relatively slow and the first signs of cracking may take several years to appear. This makes it difficult to identify the deterioration mechanism and for taking measures to arrest it at an early stage, where no serious damage has been caused. To prevent this issue, lithium compounds have been found to be effective in mitigating ASR in concrete since the early 1950s. Supplementary cementitious materials such as silica fume, fly ash, and slag cement may also be used to control ASR in concrete.

2.2.5 Corrosion process

Corrosion of steel in concrete is an electrochemical process that involves two reactions, namely the anodic and the cathodic, that take place simultaneously but not necessarily at the same rate. At the anode, ferrous atoms are ionized to ferrous ions that dissolve in the water solution around the steel:

$$Fe \rightarrow Fe^{2+} + 2e^-$$ (1)

The electrons produced at the anode flow through the steel to be consumed at the cathode where, combined with dissolved oxygen and water, form hydroxyl ions $(OH-)$, that flow through the concrete to the anode.

$$\tfrac{1}{2}O_2 + H_2O + 2e^- \rightarrow 2OH^-$$ (2)

The ferrous ions released at the anode react with the hydroxyl ions and yield ferrous hydroxide ($Fe\,(OH)_2$), which is unstable and reacts with water to form hydrated ferric oxide (Fe_2O_3), also known as rust.

$$2Fe(OH)_3 \rightarrow Fe_2O_3H_2O + 2H_2O \tag{3}$$

Rust develops on the surface of the reinforcement and normally has a brown-green color. In cases of lack of oxygen, such as in submerged structures, the resulting product of $Fe(OH)_3$ is Fe_2O_4 that is known as black rust. The two ways to prevent corrosion are to improve the corrosion resistance of the metal and/or to add silica fume to reduce concrete permeability by providing an additional hydration product that reduces the number and size of capillary pores.

2.2.6   Chloride penetration

Chloride induced corrosion is the most serious and widespread deterioration mechanism of concrete structures, FIB (2010). Chlorides can either be cast into the concrete or may penetrate from the environment through pores to the interior of the concrete. The addition of calcium chloride accelerators, widely used until the mid 1970s (Broomfield 1997), the use of seawater in the concrete mix, as well as contaminated aggregates, increase the risk of premature corrosion. The dominant source of chlorides that diffuse into concrete is the seawater exposure and the application of deicing salts. A comprehensive literature review on the chloride penetration resistance of concrete can be found in Stanish et al. (2000), and is the result of a FWHA sponsored research.

2.2.7   Carbonation

Carbon dioxide gas (CO_2) penetrates into the concrete from the environment and reacts with the calcium hydroxide ($Ca(OH)_2$), that is contained in the pores and maintains the high alkalinity of the concrete, providing the passive protective layer to the reinforcement.

$$Ca(OH)_2 + CO_2 \rightarrow CaCO_3 + H_2O \tag{4}$$

This reaction leads to a reduction of the pH of the concrete, which, when it falls below approximately 9, signifies its full carbonation (Concrete Society, 2000). When the carbonation front reaches the reinforcement, the protective layer depassivates and corrosion initiates. Therefore, the depth of the concrete cover plays a significant role in the corrosion initiation time, as reported in the specific requirements of EN 206-1/2000 (2005).

2.3   Metal structures

2.3.1   Corrosion process

The process is the same as what was described in the previous section concerning concrete structures.

2.3.2 Fatigue phenomena

Repeated application of static load in structural components may produce fracture and fail if the same load, or even a smaller load, is applied a large number of times. The fatigue failure is due to progressive propagation of flaws in steel under cyclic loading. This is partially enhanced by the stress concentration at the tip of such flaws or cracks. Chapter 4, "Fatigue and Fracture," is entirely dedicated to the fatigue phenomenon.

2.4 Earthquakes

Real experience and observations of postseismic events have indicated that bridges have been often damaged by earthquake events, most commonly by the following causes (Priestley et al., 1996): span failures due to unseating at movement joints; amplification of displacements due to soil effects; and pounding of bridge structures. Specific hints are given in the thematic chapters of this book.

3 Investigation procedures

3.1 Bridge inspection

Bridge inspection lays on a methodology based on instructions, guidelines, standards, or other official regulations. Bridge inspection can be classified into the following groups, depending on its scope and frequency (Radomski, 2002; Branco and Brito, 1996): cursory inspections, carried out by maintenance staff during routine inspections, normally taking place every day; basic inspections, carried out usually at least once a year by local bridge inspectors; detailed inspections, carried out at least every five years on selected bridges by regional bridge inspectors; and special inspections, carried out by highly-qualified experts and researchers according to technical needs, normally as a consequence of questionable results from basic or detailed inspections. It is necessary to determine the capacity and assess the safety of a bridge after unexpected or accidental loads in order to establish its ability to resist to loads, or to indicate the need for rehabilitation and strengthening.

Bridge inspection is crucial in the evaluation and assessment of an existing bridge, and is directly related with the following phase of bridge rehabilitation decisions, because inspections help in investigating the existing condition of the structure from which recommendations for repairs, if necessary, can be formulated (Brinckerhoff, 1993).

3.2 On-site tests for concrete structures

Concrete bridges should be tested if the bridge inspection reported doubts regarding the structural performance of the existing structure. The tests available range from the completely nondestructive, through those where the concrete surface is slightly damaged, to partially destructive tests, such as core tests and pullout and pull off tests, where the surface has to be repaired after the test. The following methods, with some typical applications, have been used for the nondestructive testing (NDT) of concrete, considering that a preliminary visual inspection is an essential precursor to any intended NDT phase:

a) Half-cell electrical potential method, used to detect the corrosion potential of reinforcing bars in concrete
b) Radiographic testing, used to detect voids in the concrete and the position of stressing ducts
c) Ultrasonic pulse speed testing, mainly used to measure the sound speed of the concrete, and hence the compressive strength of the concrete
d) Sonic methods, using an instrumented hammer that provides both sonic echo and transmission methods
e) Schmidt/rebound hammer test, used to evaluate the surface hardness of concrete
f) Carbonation depth measurement test, used to determine whether moisture has reached the depth of the reinforcing bars and hence whether corrosion may be occurring
g) Permeability test, used to measure the flow of water through the concrete
h) Cover meter testing, used to measure the distance of steel reinforcing bars beneath the surface of the concrete, and also possibly to measure the diameter of the reinforcing bars
i) Penetration resistance or Windsor probe test, used to measure the surface hardness and hence the strength of the surface and near-surface layers of the concrete

3.3 On-site tests for metal structures

In the same situation of concrete structures, steel bridges also need to be assessed during their lifetime. NDT techniques for steel bridges are mostly coded (e.g. in ISO standards), and include:

3.3.1 Magnetic particle inspection

Magnetic particle inspection is able to detect discontinuities in metal structures through the use of magnetization. For visualization of the magnetic field, a suspension, usually with fluorescent steel splinters, is used. A damage or fatigue crack discontinuity results in the formation of the magnetic field. Ultraviolet (UV) light visualizes the alignment of the field. This inspection method can be used for detection of surface cracks in ferromagnetic materials only. Cracks in nonmagnetic material or in sandwiched elements cannot be detected. The method can be applied as quality control of precise setting of drilled holes to stop active fatigue cracks (Kuhn et al., 2008).

3.3.2 Liquid penetration inspection

Fatigue cracks in structural members can be detected with a liquid penetration method. After surface cleaning, a developer is applied to reveal locations were the dye has penetrated.

3.3.3   Radiographic evaluation

The radiography procedure (e.g., x-ray, γ-ray) could be adopted to detect cracks and flaws in steel sections. The radiographic source could be placed on one side, while the receivers would be placed on the other side of the inspected cross section.

3.3.4   Ultrasonic inspection

The back face of an element, or the damage inside the investigated material, reflects the ultrasonic signal and propagates as an ultrasonic wave. Typical applications are on corroded members, where the remaining thickness can be easily obtained.

3.3.5   Eddy current inspection

This technique is not widely applied to old steel structures; however, feasibility studies can be found in the literature. An application of the method to old structures to detect fatigue cracks in built-up sections of a truss girder after laboratory fatigue tests was described by Helmerich (2005); and after rivets have been removed, the sensors can indicate whether there is a crack in the rivet hole in one of the layers (Kuhn et al., 2008).

3.3.6   Acoustic emission techniques

This technique is widely used, as it is able to produce interesting and profitable results, including the characteristic frequencies emitted by cracks if the structure is excited by the traffic load. The monitoring, collecting, filtering, and analyzing activities should be done by specialized personnel in order to obtain useful results. Examples on the applications of acoustic emission techniques for monitoring crack growth are given in Kuhn et al. (2008), Nair and Cai (2010), and Ledeczi et al. (2011).

4   Assessment procedures

4.1   Introduction

The assessment of an existing bridge aims to obtain evidence to demonstrate whether it will function safely over a specified residual service life, taking into account a specific code reference. It is mainly based on the results of assessing hazards and load effects to be anticipated in the future, and of assessing material properties, the geometry, and the structural state of the bridge. Guidelines for assessment of existing structures have been developed in many countries; however, the occurrence of bridge assessment guidelines based on codes or standards is rare. More frequently, such guidelines are prepared at a detailed level by scientific groups or research organizations. Whatever the source, the first issue deals with fixing risk acceptance criteria, which is quite difficult since it must be compatible with the code for new structures (e.g., limit state analysis, safety factor format, etc.). The second issue deals with the process of the assessment procedure, which is commonly separated into phases, starting from preliminary evaluation, through to detailed investigation, expert, and finally advanced assessment, depending on the structural condition of the investigated structure (Pipinato, 2011 and 2014). See Figure 28.1.

4.2   First level: preliminary evaluation

The preliminary evaluation is the first level of investigation that aims to remove existing doubts about the safety of existing structures, adopting fairly simple methods, and identifying critical parts or members in the structure. In order to identify critical members, it is necessary to carry out an intensive study of the available original design documents, along with a visual inspection of the structure, and a photographic survey. The inspection procedure is often coded by infrastructural agencies manuals, and procedures; however, at least the following points must be checked:

- The bridge construction conforms to the original drawings and/or differences between the structure as-built,and those drawings.
- Bridge modification during service (e.g., rehabilitation, strengthening, changes in the static system, etc.).
- Presence of any visual evidence of degradation (e.g., damaged expansion joints, supports, corrosion, cracks, vibration or loose rivets, collision, lack of structural members, etc.).

Moreover, if available, inspection and maintenance reports can be used, and reference should be made to the evaluation report. The preliminary evaluation should include codes and recommendations analysis procedures where available, and conservative assumptions where information is lacking or doubtful. In this way, critical construction details can be identified.

4.3   Second level: detailed investigation

The aim of the detailed investigation is to update the information obtained in other analyses by carrying out a refined assessment, especially for those members for which adequate safety was not confirmed by preliminary evaluation. At this stage a specialized consultant should assist. In this phase, a finite element method (FEM) numeric model of the entire structure is developed. Based on the current code provisions, the structure should be recalculated and verification tables should report whether the structural members are safe or not. Concerning specific issues, such as the fatigue and seismic behavior of the bridge, detailed code provisions should be referred to. From this step-level investigation, NDT could be used in order to characterize the basic material properties of the structure. The final report of the investigation should establish whether the structure is secure against specific issues and has sufficient static strength against actual loadings.

4.4 Third level: expert investigation

In case of key structures that have major consequences in terms of risks or costs related to a decision, a team of experts is needed in order to carefully check the conclusions and proposals reached in the last phase. Discussions and further assessments using specific tools can also be carried out to help reach decisions. At this level, on-site testing could be adopted in order to provide the dynamic identification of the structure, as reported in the following example.

4.5 Fourth level: advanced testing

This advanced level of investigation should be reserved for recurrent bridges along infrastructural nets, in which a rational procedure of analysis and intervention could help in determining if retrofitting interventions could be adopted, or if rational dismantling large-scale operations are required. The procedure is based on a detailed survey of the existing bridge, a FEM analysis, a code verification procedure, NDT diffused sampling, and, based on these data for real-scale testing of one case study structure, aims to determine the global static and cyclic behavior of the bridge. In specific cases, on-site dynamic identification can be performed. Concerning the fatigue assessment, in this case, a linear elastic fracture mechanics (LEFM) investigation is required. Concerning seismic analysis, nonlinear analysis is required. Specific material testing analysis should be performed regarding the case analyzed. The advanced testing result should report on the various analyses performed, and should clearly state verification results that indicate the specific retrofit needed for recurrent interventions.

The problem of existing bridges and their assessments has recently increased. Indeed, the current low funding in the infrastructure sector of many European countries forces the bridge owners, as well as the operators, to postpone investments in new road and railway bridges and consequently stretch the service life of their existing structures. Therefore, the owner of the infrastructure currently faces two main challenges: the need for further continuing the safe operation of ageing bridges and of cost-effective maintenance. Methods must be provided that enable engineers to offer safe and cost-effective assessment and maintenance methods to their clients (Kuhn et al., 2008). In the following sections, some key issues procedures are reported in order to reproduce different step level assessment dedicated to bridge management analysis. This advanced level is not shown in Figure 28.1, as it is mostly avoided in the structural common practice, except for special and large- scale constructions.

4.6 Critical member identification procedure

An alternative and faster solution could be adopted in some cases: existing bridges are often exposed to an effective risk of collapse for a single and specific structural member state: this could be discovered by an identification procedure, that is called a *top-down bridge collapse identification (TDBCI) procedure,* which is effectively able to find out critical situations. This process should not only be theoretical and analytical, but also should be combined with a survey of the structure, identifying failure or incipient mechanisms (Figure 28.2).

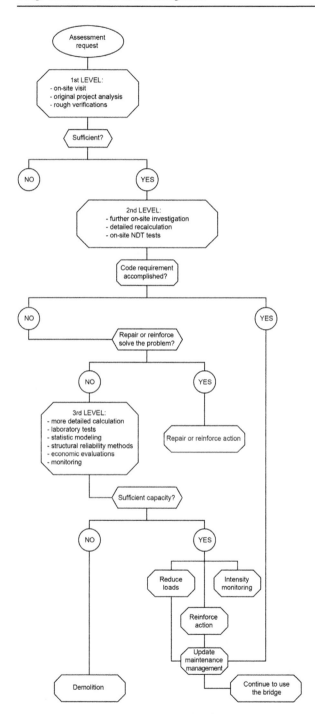

Figure 28.1 Step-level assessment procedure.

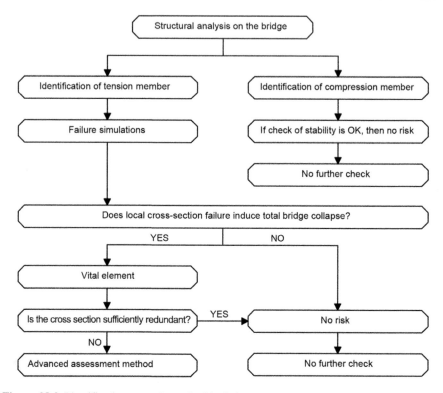

Figure 28.2 Identification procedure of critical elements.

5   Repair and strengthening

5.1   General information

As there is wide confusion regarding different but similar terms, a definition of the common terms used in this field should be outlined here. We can define *maintenance* as every operation applied to an existing bridge and finalized to maintain its actual strength and geometry, without extraordinary interventions; *rehabilitation*, the process to restore under service level/handicap of an existing structure with concentrated actions; *repair*, concentrated actions to restore damaged points; *stiffening*, singular action used to enhance the bridge capacity where in-service limits have been passed (e.g., cracks, rotations, deflections etc.); *retrofit/strengthening*, a comprehensive work of renovation including a large amount of structural and technological actions that lead to a complete modernization of the bridge structure, in order to upgrade its load carrying capacity, even against horizontal actions.

5.2   Lightweight components

Dead load in existing bridges is relevant: for this, load reduction is a possible intervention, considering deck substitution and safe barriers changes. This choice should be finalized to a precise improvement in the bridge service level, at least on the load carrying capacity; if not, extensive works on the deck are not necessary.

5.3 Composite actions

To increase the flexural strength of existing deck systems, the modification to a composite system is a common method used. Where analytically verified, the flexural stiffness is upgraded as the deck collaborates with the steel profiles in a composite manner adopting welded stud: this solution could also be adopted for existing composite bridges, in which the composite action is very low, by the use of shear connectors between the deck and the stringers. Deck materials are generally made of concrete, such as normal-weight reinforced concrete (precast or cast in place, or partly precast and partly cast in place) or lightweight reinforced concrete. If a decision should be made on an existing structure, the first issue to focus on is the weldability of steel stringers.

5.4 Improving bridge member strength

Cover plating is a widely adopted technique used in steel bridges: steel plates are connected to existing steel members by riveting, bolting, or welding in order to enhance the structural capacity of nodes or members by increasing the section modulus and consequently, the flexural capacity. If jacketing is feasible during the cover plate's application, this ensures that the both dead and live load stresses will be carried; otherwise an increased amount of steel would be necessary. This technique can also be explored for reinforced concrete-RC bridges at the tension face of the beam, by bolting or doweling to strengthen the flexural capacity, and by cover plates for shear lacks. In RC bridges, jacketing should be adopted before interventions to reduce the dead load influence.

5.5 Posttensioning applications

Posttensioning technique is useful where undesirable deflections or high-tension levels have to be reduced: local (e.g., cracks) or global (e.g., deflections) problems should be solved in this way, while considering the limit imposed by posttensioning. Moreover, this technique has been widely used to change the static scheme of bridges and viaducts, from a series of simple spans to a continuous span. However, posttensioning can also change the stress state in structural members: e.g., one or more concentric tendons straight in the median height of the deck section will induce an axial compression force that, depending on magnitude, can eliminate part or all of an existing tension force in a member, or even place a residual compression force sufficient to counteract a tension force under other loading conditions (Klaiber et al., 1987).

5.6 Modification of the structural configuration

The modification of the structural configuration of a generic structure could induce variations in internal force distribution, two main methods are available:

- Introducing new support points and reducing the maximum positive moment in the midspan; in this case, a problem could arise if deep water is present, or if bridge obstacles are present below.

• Changing a series of simple spans to a continuous span scheme in order to lower the stress level in the structure, while at the same time enhancing the structural capacity versus. live load. This is a feasible technique if girders are located at a reasonable height, and if the existing bridge is overall in well maintained. Cost analysis of alternatives should be carefully considered for this type of application.

5.7   Concrete bridges

5.7.1   Crack repair

There are various types of cracks; however in the context of repair, they can be classified into two categories: inactive cracks, which are unable to propagate; and moving cracks, which are active and able to increase under applied loads. The most appropriate solutions to these cracks are epoxy resin injection and/or cement grouting, however the former is preferable for small cracks (<4 mm). Admixtures ready for use are widely available and could be easily adopted for this specific purpose. In the case of grouting, the water/cement ratio could be established according to testing, or if available, to code and standards prescriptions.

5.7.2   Stitching

Stitching occurs when U-shaped metal bars are encased or epoxy-fixed in the near side of the crack, along the whole crack line. This technique is possible only if the crack is exposed on one surface of the element to be reinforced (e.g., in bending members), otherwise, the technique has to be adopted at both sides (e.g., in tension members).

5.7.3   Reinforcement

Local or extended steel bars reinforcement is needed if an extended large crack (>4 mm) situation is found on an existing bridge component. In this case, superficial decortication is used before adding a new RC layer, which is dimensioned according to the analytical results of the assessment. If deep cracks are found (e.g., going over the removed layer), further strengths should be adopted (e.g. injections, steel member addiction, etc.).

5.7.4   Overlays and surface treatment

Where the aforementioned case evidenced a net of microcracks (<4 mm), surface treatments/coatings should be enough to remedy this, as these situations come from surface over tensions (e.g., due to drying shrinkage), epoxy resin or silane/siloxane coats are the most appropriate solution.

5.7.5   Flexible sealant

To repair or bond cracked concrete surfaces, sealants developed especially for forming permanent, water and weatherproof seals in all exterior gaps and joints are required. The sealant must be elastic, remaining flexible to expand and contract with construction material movement and to protect and retain the original seal. Excellent adhesion to the sealed concrete and high-movement quality are required for these sealants, which include epoxy polysulfides, silicones, and acrylic polyurethanes. If the surface is not used or hidden, the application could be without a recess, which is required if the surface is exposed to traffic.

5.7.6   Patch repairs

To repair specific and closed zones of the concrete members, a patch application could be employed: in this case, a polymer-reinforced, high-strength, cement-based patching and resurfacing mortar is adopted. This solution type should be avoided if enclosed in a traffic lane.

5.8   Steel bridges

5.8.1   Stop hole drilling

Hole drilling is the most commonly applied means of arresting fatigue cracks. A hole drilled at the tip of a crack essentially blunts the crack tip and greatly reduces local stress concentration. This technique has been successfully applied to various types of structures, including navigation lock gates and several bridges (Fisher, 1984). Hole drilling is effective for through-thickness cracks in plates, or plate components of structural members.

5.8.2   Weld toe grinding

Weld-toe grinding reduces the geometrical stress concentration and extends the fatigue life of undamaged details. Grinding can be used to remove shallow fatigue cracks that may exist in the weld toe. Grinding should always be done in the direction of applied stress. A pencil or rotary burr grinder can be used. Magnetic particle inspection of the ground area should be conducted after grinding to ensure that embedded flaws are not exposed. Penetrant inspection may reveal false indications due to grinding marks.

5.8.3   Peening

Peening is effective as a retrofit for shallow surface cracks that commonly occur at fillet weld toes. Peening imposes compressive residual stresses resulting from the plastic deformation induced by the peening hammer and reduces the geometrical stress concentration similar to that with grinding. Air hammer peening is effective in arresting fillet weld toe surface cracks with a depth of up to 3 mm if the tensile stress range does not exceed 40 MPa (6 ksi). Peening can also be applied to not cracked fillet welds to improve the fatigue resistance of the detail. The expected benefit of peening under favorable conditions (low stress range, low minimum stress) is an increase in fatigue life approximately equivalent to one fatigue design category (Fisher et al., 1979). Peening should be done using a small pneumatic air hammer with all sharp edges of the peening tool ground smooth. Although peening intensity can be easily varied by changing air pressure, multiple-pass peening at lower air pressures is most effective. Initial passes of the peening hammer may reveal some cracks that were not initially visible, and peening should be continued until the weld toe is smooth and no cracks are apparent. Penetrant inspection of the peened area should be conducted after peening to ensure that embedded flaws are not exposed. Peening is most effective when performed under dead load so that the imposed compressive residual stress has to be effective only against the live load.

5.8.4   Gas tungsten arc remelting

The gas tungsten arc (GTA) remelting process is also an effective procedure for repair of shallow surface cracks that occur at fillet weld toes. This procedure is generally effective for surface cracks with a depth of up to 5 mm (slightly greater depths than peening) and is not limited to small stress ranges and minimum stress levels. Like peening, GTA remelting can also be used to improve the performance of uncracked fillet welds, approximately doubling the fatigue life. However, it is less easily performed in the field, and it requires highly skilled welders and good accessibility.

With the GTA remelting procedure, a small volume of the weld toe and base metal is remelted with a gas-shielded tungsten electrode. After the area cools, the geometric stress concentration is improved and nonmetallic inclusions that might exist along the weld toe are eliminated. When the procedure is applied to crack repair, sufficient volume of the metal surrounding the crack must be melted so that upon solidification, the crack is eliminated. The effectiveness of the procedure is dependent on the depth of the remelted zone, since insufficient penetration would leave a crack buried below the surface. Such a crack would simply continue to propagate, resulting in premature failure. Proper selection of shielding gas and electrode cone angle is crucial in obtaining maximum penetration of the remelted zone. Argon-helium shielding and an electrode cone angle of 60 degrees were found to be most effective (Fisher et al., 1979). For any retrofit procedure, the depth of penetration should be verified by metallographic examination of test plates before the procedure is applied in the field.

5.8.5   Rivet replacement

Missing, loose, or headless rivets and rivets with rosette heads should be replaced (Fischer 1984). Deteriorated rivets missing more than one half of the head should be replaced if the rivet is subject to an applied tensile force or tension resulting from prying action. The most useful riveting repair is represented by high strength bolting: the rivet should be knocked off before bolting with a pneumatic buster. Welding or other intervention that can cause metallurgical damage (adversely affecting e.g. the fatigue strength) to the adjacent material, are to be avoided.

5.8.6 Welding

Welding solutions should be carefully used in existing steel structures only where continuous welding connections are present, and if no other bolting solutions are practicable, weld repair should be introduced. Avoid welding in fracture critical members, in nonweldable steels, in low Charpy members, and if possible, in tensile areas.

5.8.7 FRP Strengthening

An alternative technique for strengthening steel structures consists of the application of externally bonded fiber reinforced polymer (FRP) sheets, used mainly to increase the tensile and flexural capacity of the structural element. FRP materials have a high strength-to-weight ratio, do not present problems due to corrosion, and are manageable. Some examples of guidelines for the design and construction of externally bonded FRP systems for strengthening existing metal structures include the ICE design and practice guide (Moy, 2001), CIRIA Design Guide (Cadei et al., 2004), US Design Guide (Schnerch et al., 2007), and CNR-DT 202/2005 document (Italian Research Council, 2005).

The benefits of composite strengthening have been applied, for example, in a steel bridge on the London Underground (Moy and Bloodworth, 2007). The benefits of strengthening large cast-iron struts with carbon FRP (CFRP) composites in the London Underground are illustrated in Moy and Lillistone (2006). A state-of-the-art review on FRP strengthened steel structures was recently developed by Zhao and Zheng (2007). Among materials, apart from the well-known e-glass, HS CFRP, and aramid, high-modulus CFRP (HM CFRP) materials are becoming widely used and have been developed with a tensile modulus approximately twice that of steel. Diverse applications are reported in literature concerning this type of material, and are discussed next.

Among the most common techniques of FRP strengthening systems in bridge engineering, three should be cited. The first is the wet lay-up system, which consists of a multidirectional or unidirectional dry or fiber sheets on site, impregnated with a saturating resin, which provides the binding matrix of the fiber, and bonds the FRP to the material. Common types are represented by dry unidirectional fiber sheets with the fiber running predominantly in one planar direction, dry multidirectional fiber sheets or fabrics with fibers oriented in at least two planar directions, and dry fiber tows that are wound or otherwise mechanically applied to the material surface. Another system is represented by precured FRP fibers, bonded with an adhesive on the surface. Externally applied epoxy bonded FRP sheets to the tension face of structural elements have been widely accepted for practical use as the bonding solution between FRP and steel, even if the durability of this application has to be extended with direct applications and monitoring (e.g., attention where the fire could be a danger).

5.8.8   Cover plating

To strengthen existing bridges, steel plates are connected to existing steel members by riveting, bolting, or welding in order to enhance the structural capacity of nodes or members, by increasing the section modulus and consequently the flexural capacity. If jacketing is feasible during the cover plates application, this ensures that the both dead and live load stresses will be carried. Otherwise, an increased amount of steel should be used.

5.8.9   Paintings

To protect steel against corrosion, a protective coating is the most commonly adopted technique. The first stage of a corrosion protection system application, is the surface preparation: the preferable situation for rapid intervention is if the rust grades comply with grades A or B (BS EN ISO 8501-1 (2007)), as grades C or D involve a longer and more costly cleaning operation. The standard grades of cleanliness for abrasive blast cleaning are:

- Sa 1–Light blast cleaning
- Sa 2–Thorough blast cleaning
- Sa 2½–Very thorough blast cleaning
- Sa 3–Blast cleaning to visually clean steel

Specifications for bridge steelwork usually require either Sa 2½ or Sa 3 grades. The protective system is then applied: this is defined by a sequence of applications, including the primer, the intermediate(s), and the finishing. Special codes and standards for railway or highway applications can be found in literature. Regarding unpainted solutions in weathering steel structures (EN 10025-5, 2004), paints should be adopted if a surface damage situation becomes a concern. Weathering steels can be protected with the same maintenance paint systems recommended for new structures.

6   Bridge management

6.1   Overview on BMSs

Due to the rapid growth of automobile and truck usage and the development of massive transportation infrastructures in past decades, there are increasing demands to improve the management methods of bridges, which constitute the most vulnerable elements of the road network. Many agencies responsible for infrastructural networks have recognized the difficulties of the available bridge management approaches, in which decisions are made only on a single project level. As a result, a significant effort has been undertaken in many countries to develop bridge management systems (BMSs) to evaluate the condition of a single bridge in the global network level during its life cycle, and at the same time, to provide efficiency information when allocating resources and establishing management policies in a bridge network.

A BMS is a rational and systematic approach to organize all activities finalized to managing network-level bridges (Hudson et al., 1993). Decision makers should select optimum solutions from an array of possibilities, and must evaluate and compare alternatives for all bridges in the road network from the viewpoint of life-cycle management in order to avoid similar problems in the near future. Several BMSs have been developed for specific purposes: e.g., Gralund and Puckett (1996) developed a BMS for the rural environment, Markow (1995) developed a BMS for highways, Thoft-Christensen (1995) developed BMS analysis including a reliability approach in particular, and Kitada et al. (2000) developed a detailed BMS for steel bridges. Some studies on evaluation criteria for bridge maintenance that also take into account seismic risk and fatigue evaluation are described in Pipinato (2008a,b).

Innovative techniques that include the implementation of new technologies and BMSs have given bridge inspectors and engineers the necessary information to determine an appropriate action. Such a decision is often dependent on a combination of the quantitative information obtained from various measurements, qualitative information obtained from bridge recognition, and general engineering knowledge about the entire bridge system. In order to allocate funds, a bridge owner needs a BMS that uses historical deterioration trends and predictive relationships. Combining existing management system specifications with bridge specific deterioration models, which consider the system's structural behavior and the aging of the infrastructural network investigated, will improve an infrastructural owner's ability to make bridge specific-decisions and allocate funds for specific and accurate programmed interventions. Probably one of the most significant applications of contemporary BMSs relates to the US. In 1991, the Intermodal Surface Transportation Efficiency Act (ISTEA) required all states to develop, establish, and implement a BMS by October 1998. The ISTEA requirements, first distributed in 1991, stated that a BMS must be implemented on all state and local bridges. New federal legislation, however, required implementation of BMSs only for bridges on the National Highway System (NHS); therefore, use of BMS inspection for non-NHS bridges was optional (Sunley, 1995). The principle that BMS used in the US is PONTIS, developed in the early 1990s for the Federal Highway Administration (FHWA), and became an American Association of State Highway and Transportation Officials (AASHTO) practice in 1994. It performs functions such as recording bridge inventory and inspection data, simulating condition and suggesting actions, developing preservation policy, and developing an overall bridge program. The system allows representation of a bridge as a set of structural elements, with each element reported based on its condition.

In 2002, 46 agencies throughout the nation had PONTIS licensing, and each state highway administration (SHA) could customize the system according to its needs (Robert et al., 2003). BRIDGIT was developed in 1985 by the National Cooperative Highway Research Program (NCHRP) and the National Engineering Technology Cooperation in an attempt to improve bridge management networks. This system has capabilities similar to the PONTIS system. There have been many research projects throughout the nation on which local agencies have worked with universities to develop their own BMSs. Other BMSs developed by individual state agencies have good specific functions and qualities, but lack features that can satisfy all the demands of effective bridge management and maintenance procedures on a national scale. Other notable research and development efforts on BMSs took place in Iowa, Washington, Connecticut, Texas, and South Carolina (Czepiel 1995). Among recent European experiences that are noteworthy is the TISBO Infrastructure Maintenance Management System, currently being developed by the Netherlands Ministry of Transportation, Public Works, and Water Management. It is a system that integrates inspection registration and maintenance management. Owner agencies in Italy usually manage their network with self-developed codes/procedures regarding BMSs. The policy of the main Italian agencies is briefly presented in the following:

- Rete Ferroviaria Italiana (RFI) is the national agency for the whole Italian railway network, consisting of about 16,000 km. The BMS is based on periodical visual inspections supported by special testing trains. All data are elaborated with specific software developed by the agency with the aim of defining economical and technical convenience of possible maintenance, rehabilitation, and/or strengthening interventions.
- Autostrade per l'Italia is the most relevant highway agency in Italy; it manages a network of about 3,400 km. The BMS is based on the SAMOA program for surveillance, auscultation, and maintenance of structures.
- ANAS (1997) is the Italian agency for roads with a national interest, and manages a network of about 26,700 km. The BMS is based on the national road inventory and in-situ surveys, and is a web-based management application that is developed by the agency and updated regularly.

During the last decade, a number of research projects have been financed by the European Commission and some guidelines have been published from these projects that deal with the assessment of existing bridges in Europe; i.e., BRIME (2001), COST345 (2004), SAMARIS (2005), and Sustainable Bridges (2006). All of these guidelines are meant for highway bridges specifically, except for Sustainable Bridges (2006), which is particularly pertains to railway bridges. The purpose of BRIME (2001) was to develop the modules required for a BMS that enables bridges to be maintained at minimum overall cost, taking a number of factors into account, including effect on traffic, life of the repair, and the residual life of the structure. COST345 (2004) investigated the procedures and documentation required to inspect and assess the condition of in-service highway structures, not only bridges. SAMARIS (2005) focused on inventorying the condition of highway structures in European countries, choosing the optimal assessment and strategy selection for rehabilitation through the use of ultrahigh-performance, fiber-reinforced concrete (UHPFRC) and similar technologies. Sustainable Bridges (2006) deals in particular with railway bridges and structural reliability assessment based on in-situ NDT.

6.2   Network and bridge level

While the bridge-level management relies more on the structural monitoring and interventions previously described, the network management of a set of bridges involves the significance of "prioritization": a wide amount of existing bridges is impossible to maintain and retrofit at the same time, so a prioritization system should be employed in an advanced BMS. Both network- and project-level should be interrelated

(Thompson et al., 2003). The final output of a prioritization system is represented by a priority rated list, including the bridges with a higher demand of intervention at top (Li and Love, 1998). At the project level, the network level information is used in order to accurately define the individual bridge intervention, with a precise cost analysis boundary (Soderqvist and Veijola, 2000).

6.3   Network level and prioritization methods

Prioritization methods include priority ranking (e.g., sufficiency rating (SR), level-of-service (LOS), deficiency rating (DR), and incremental-benefit-cost (B/C) analysis) and mathematical optimization. Subjective method ranking has been demonstrated to be ineffective for large networks (Mohamed, 1995), and this conclusion should also be made for priority ranking in general, not considering the importance of the bridge in a certain network. The SR method is applied, for example, by FHWA in US, where the sufficiency rating formula is a method of evaluating a bridge's sufficiency to remain in service, based on a combination of several factors; the result of the formula is a percentage in which 100% represents an entirely sufficient bridge and 0% represents an entirely insufficient or deficient bridge (FHWA, 2015). Bridge deficiencies are represented by two main categories: structurally deficient (structurally deficient means that a bridge requires repair or replacement of a certain component), or functionally obsolete (functional obsolescence is assessed by comparing the existing configuration of each bridge to current standards and demands). The SR does not enter into the domain of the appropriate intervention to be performed on the single bridge. The LOS system includes information on the load capacity, clear deck width, and vertical roadway clearance (Johnston and Zia, 1983). The B/C alternative is a system finalized to allocate benefits for the user and the agency by employing a certain amount of money for a precise bridge repair, considering the consequence of different types of interventions. Then, an analytical approach is used for translating these alternatives to all the network bridges.

6.4   Project level

At this level, some alternatives are available, including B/C techniques and life cycle cost (LCC) optimization. B/C techniques include the analysis of different intervention strategies for a singular bridge, allocating funds to the alternative represented by the highest value of B/C; this method is limited to a one-bridge analysis, and neglects all network information and needs. The LCC approach considers all costs required during the life of the structure, allocating funds over the time-life of the bridge. Integrated solutions, considering funding availability over the time and intervention alternatives, are available.

6.5   Network- and project-level decision making

Recent attempts have been made to try to use a multipurpose decision scheme, including the project and the network level. Artificial neural networks (ANNs) and genetic algorithms (GAs) have been employed in the optimization of BMS integrated solutions, even if not directly employed by agencies. Liu and Hammad (1997) presented the application of the multiobjective optimization of only bridge decks rehabilitation. The objective function was to minimize both the total LCC and the average degree of deterioration weighted by the bridge deck area. The total rehabilitation cost (C) was determined by the following equation:

$$Minimize\ C = \sum_{i=1}^{N} \sum_{t=1}^{T} \left[(1+r)^{-t}.c.s(i).n(i,t) \right],$$

where N is the number of bridges, T is the length of the planning period, r is the discount ratio, c is the unit area cost of rehabilitation, $s(i)$ is the deck area of bridge i, and $n(i,t) = 1$ if a rehabilitation cost is calculated, or 0 otherwise. The binary values are defined as $0 =$ do nothing, and $1 =$ rehabilitation action. After this first attempt, good solutions have been found. Dogaki et al. (2000) developed a most complex analysis and presented a GA model for planning the maintenance of reinforced concrete decks, considering a probability-based transition matrix for the deteriorating model, linked with the crack density. The objective function relies on the minimization of the maintenance cost and on the maximization of the benefit derived from the maintenance. The constraints included, traffic capacity, detours, the possibility to extend the bridge width, traffic constraints, and the importance of the bridge. The model includes the user cost, the LCC, and the environmental cost. Frangopol and Liu (2007) present the application of multiobjective optimization for safety and LCC for civil infrastructure. Neves et al. (2006a, b) also proposed the multiobjective optimization system for different bridge maintenance types.

6.6 Economic approaches for bridge network management: repair or replace

The decision to repair or replace is an increasing painful decision for bridge authorities who manage thousands of bridges. For this reason, this decision should be supported by appropriate theoretical instruments. To perform this analysis, a global cost function C was developed by BRIME (2001):

$$C = CC + CI + CM + CR + CF + CU + CO - VS,$$

where CC is the construction costs, CI is the inspection costs, CM is the maintenance costs, CR isthe repair costs, CF is the failure costs, CU is the road user costs, CO is other costs, and VS is the salvage value of the bridge. The objective stands on the minimization of C, while keeping the lifetime reliability of the structure above a minimum allowable value. To implement an optimum lifetime strategy, the following problem must be solved:

$$Minimize\ C\ subject\ to\ P_{f,\ life} \leq P^*_{f,\ life},$$

where P^*f, life represents the maximum acceptable lifetime failure probability (also called the *lifetime target failure probability*). The actions considered in this method intend to restore the initial service level (design) of the bridge, without considering an improvement of its initial performances, dimensions, load carrying capacity, etc.

Nevertheless, this method could be also used when all the considered alternatives lead to the same level of improvement in the bridge.

A method that consists of the proposal of alternatives for the repair or replacement of a deteriorated bridge or a bridge with issues regarding its correct functionality considers the following phases (BRIME, 2001):

- Identification of factors
- evaluation of factors
- comparison of alternatives and selection

If in the year Ti (taking as a reference T0 = the moment when the study is done) a cost, Ci, is produced, this actualized cost in the instant T0 will be

$$C_{i,T_0} = C_i \frac{1}{(1+r)^{T_i}}.$$

In this instance, r is the net discount rate of money, and C_i represents the costs of the T_i year.

In this way, all other costs during the analysis period will be discounted to T_0, with a total cost being:

$$C = \sum_{i=1}^{n} C_i \frac{1}{(1+r)^{T_i - T0}}.$$

This cost will be used for the comparison of alternatives.

The following list contains the most relevant factors affecting the intervention alternatives considered:

- CM maintenance costs
- CI inspection costs
- CR repair costs, which include the following:
  - CR_A structural assessment costs
  - CR_R structural repair costs
- CU road user costs, which include the following:
  - CUD traffic delay costs
  - CUR traffic reroute costs
  - CURT time costs
  - CURO vehicle operating costs
  - CURA accident costs
- CF failure costs
- VS salvage value
- CO other costs

The following warnings should be considered:

- Some of these factors are affected by subjective considerations.
- A rough estimate could be calculated for some costs.
- Social, financial, and economic factors could both positively and negatively influence the final cost attribution.

Each cost is considered and detailed as follows:

- *Inspection cost:* These costs could be estimated considering direct costs (e.g., personnel hours and equipment) and a calendar of inspections.
- *Maintenance costs:* These could be estimated as a percentage of the construction cost, or as a percentage of previous work performed on the bridge during the exercise; annual rates of 1-2% are expected.
- *Repair costs:* This is a summation of the works to be delivered during repair operations (CR_R), and the assessment/design procedure performed (CR_A). It is easy to demonstrate that, if an accurate assessment is performed, CR_R is reduced by an appropriate incremental increase of CR_A.
- *Failure costs:* Bridge replacement costs, loss of lives, cars and equipment, architectural, cultural, and historical costs should be accounted for. The failure costs should be calculated for every bridge, as a probability of failure linearly increases during the timeline of the structure: this consideration should be done for every bridge and maintained by the owner as an insurance cost.
- *Road user costs:* A summation of C_{UD}, the costs due to delayed traffic, and C_{UR}, the costs due to traffic detours; both terms could be analytically calculated by adopting various mathematical schemes.
- *Salvage value:* this is the value of the structure at the end of the analysis period. This is not always null at the end of a bridge's service life.
- *Other costs:* Additional costs could arise during the lifetime of a bridge; they should be includeed in this category.

A repair index (RI) is defined as a value indicating how the proposed repair alternative costs compare with the no-action option, or with respect to any other alternative used as a reference. The smaller the coefficient for a particular option, the better investment that option represents (considering a determinate serviceability level). For each option, the RI may be quantified by

$$RI = \frac{(C_I + C_M + C_R + C_F + C_U + C_O - V_S)_{Repair\_or\_replacement}}{(C_I + C_M + C_F + C_U + C_O - V_S)_{No\_action\_or\_reference\_alternative}}.$$

7   Case study

7.1   The Macdonald Bridge, Halifax, NS

An pertinent project to mention is the ongoing, large-scale repair yard for the redecking and retrofitting of the Macdonald Bridge in Halifax, NS (Figure 28.3). On April 2, 1955, the Angus L. Macdonald Bridge opened, uniting the communities of Halifax and Dartmouth for the first time. The Macdonald Bridge was converted from a two-lane to a three-lane structure with a pedestrian walkway and bicycle lane in 1999. There are approximately 48,000 crossings on the Macdonald Bridge on an average workday. The Macdonald Bridge has a reversible center lane. In the morning, there are two lanes to Halifax. At noon, it switches and there are two lanes to

Figure 28.3 The Macdonald Bridge in Halifax: (a) overview; (b,c) redecking works, lifting the deck from the river barges;

(Continued)

Dartmouth and one to Halifax (HHB, 2015). The existing deck system is deteriorating in three ways: (i) water penetrating the concrete-filled, welded steel grid has caused corrosion between the bottom of the grid and the tops of the supporting stringers, resulting in the deck becoming separated from the stringers; (ii) the continual wearing of the thin asphalt running surface in the traffic wheel paths, which exposes the upper steel surface of the T-grid and reduces the skid resistance; and (iii) the deck stiffening truss, a through truss, is formed of riveted, built-up steel members. These have large exposed areas that require labor-intensive maintenance painting (Kirkwood et al., 2014). For these reasons, in addition to the new requirements for naval traffic (the deck will be raised by 2.9 m at midspan to increase shipping clearance), a global restoration of the bridge has been designed. The principal requirement of the bridge authority was keeping the bridge open during the day and carrying out the replacement of the deck at night in order to avoid traffic delays. To achieve this, deck segments will be supported from above by a movable erection gantry so they can be detached from the existing deck system by cutting the top and bottom chords and diagonals, and then be lowered onto barges below by strand jacks. The new deck will be orthotropic with a 14-mm-thick deck plate in the carriageway with 300-mm-wide longitudinal trough stiffeners at 600-mm centers, and a 10-mm-thick deck for the footway and cycle track. The new top and bottom chords are designed as closed sections and are tucked under the deck plate to protect them from rain and deicing salt. The deck plate then forms the top plate of the top chord, which is an efficient structural arrangement (Kirkwood et al., 2014). Other minor works include the installation of a dehumidification system for cables.

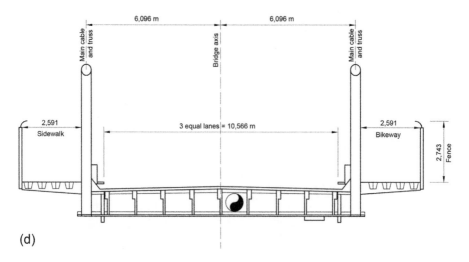

(d)

Figure 28.3 Continued. (d) existing deck scheme (in m);

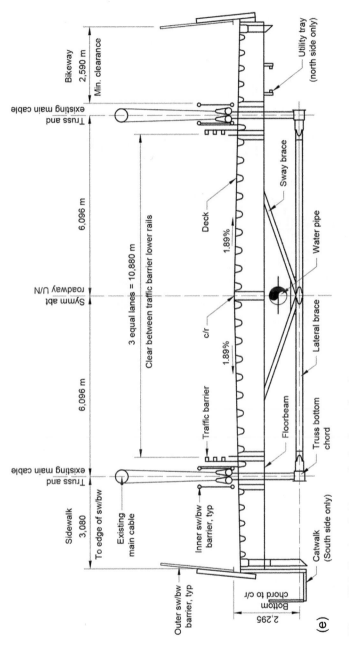

Figure 28.3 Continued. (e) new deck scheme (in m).

7.2   The Luiz I Bridge, Porto, PT

The Luiz I Bridge is a metallic arch bridge over the Douro River (Figure 28.4). The bridge was strengthened for the passage of metro trains on the upper deck. It is monitored continuously for the need for repair work. The main objectives to upgrade the Luiz I bridge for metro traffic were (i) the replacement of the roadway upper deck by a new steel deck for the new metro line and the strengthening of the main truss girders of

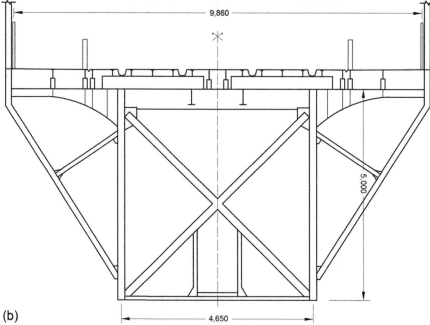

Figure 28.4 The Luiz I Bridge: (a) overview; (b) new deck cross section (in m);

(Continued)

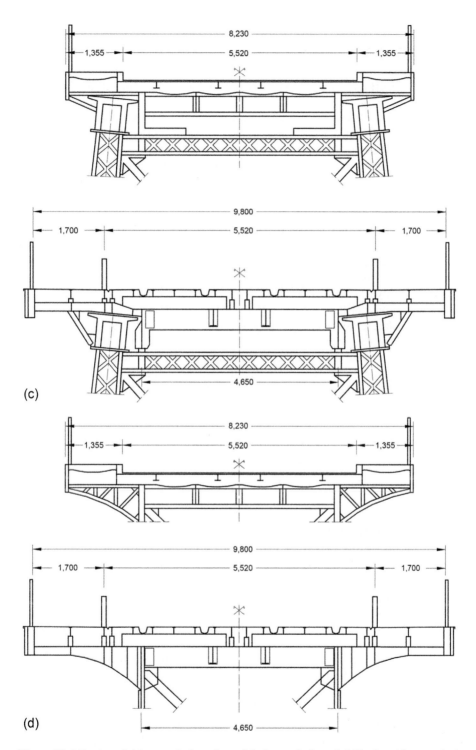

Figure 28.4 Continued. (c) upper deck on the arch before and after rehabilitation; (d) upper deck cross section before and after rehabilitation.

the deck; and (ii) strengthening of the arch, hangers (suspending the lower deck), and main truss piers (Lopes et al., 2008). The main purpose for the railroad addition was to replace the existing deck with a lighter one. The new deck's structural system, widened from 8.2 m to 9.8 m, was made of a new steel grid in S355K2J3 and composed of four IPE400 stringers (each 4 m long), which were supported by IPE500 cross beams. The stringers directly support wooden sleepers; welded steel sections cantilevered from the deck made up the sidewalks. In order to reduce fatigue in the existing structural elements, avoiding direct traffic loading, the wooden sleepers lie only over new steel stringers (Lopes et al., 2008). The constituent material was wrought iron ($E = 193$ kN/mmq; $v = 0.25$; fy $= 160$ MPa; unit weight $= 84$ kN/mc). Loosing rivets in the existing structure were found, and for this reason, new rivets were adopted to replace the existing ones; bolts were never adopted to replace rivets. In the new structure, bolted connections were adopted, some of them prestressed. A comprehensive fatigue verification was performed: a $C = 71$ for riveted connection was used according to category D of AASHTO, in accordance with UIC International Union of Railways recommendations for the 19th century, metal-riveted structures. For new structural elements, such as cross beams or stringers, a category $C = 160$ was used in bending and a $C = 100$ for shear stress; for bolted connections, $C = 100$ (shear) and $C = 36$ (tension) were used. The procedure adopted in order to verify the fatigue strength of the modified structure was the EC3 equivalent damage verification. Concerning the past damage accumulated, a consumption of 11% of the structure's service life was calculated via the Palmgren-Miner procedure (Lopes et al., 2008).

7.3   The Broadway Bridge, Portland, OR

The Broadway Bridge (Figure 28.5) carries an average daily volume of 30,000 vehicles in four lanes of traffic. A FRP deck application can be observed in this bascule bridge: as reported by Sams (2005), the project dealt with the requirement of a new deck that matched the weight of the bridge's existing steel grating, offering improved skid resistance that could be installed quickly. The deck-to-beam connections are similar to conventional shear studs and grout-filled cavities to connect the new deck to the bridge's longitudinal beams. Grout was poured through the deck into a cavity formed by stay-in-place metal angles, providing a variable haunch along each longitudinal beam. Because of this connection's inherent ability to transfer shear, the Broadway Bridge's beam–deck system likely exhibits some level of composite behavior. However, the beams were sized to carry loads without consideration of the composite action. The prefabricated FRP panels arrived at the yard in 2.4-m by 14-m modules, ready for installation on the beam's variable haunches. The length of each panel matched the width of the bridge deck because the FRP panels span perpendicular to the bridge's longitudinal beams. Shop workers had predrilled all holes to accommodate the connections to the bridge's longitudinal beams. At the heel of each bascule leaf, the FRP deck connected to a concrete transition deck, which was designed to accommodate dynamic vehicular forces. At the bridge's center open joint, the deck interfaced with heavy steel angles to accommodate dynamic forces. At the side edges, workers bonded an FRP curb to the deck along its full length. By their own weight, the

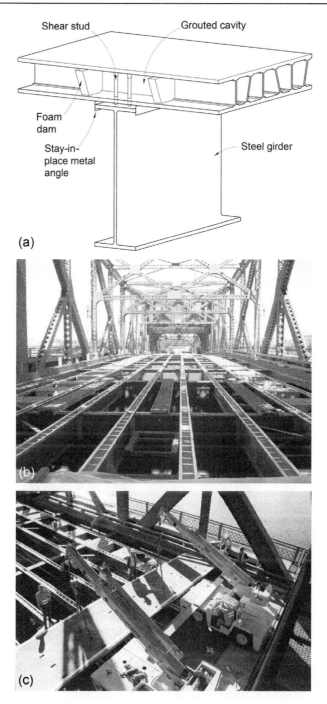

Figure 28.5 The Broadway Bridge interventions: (a) new deck scheme; (b) girders and cross beams without the deck structure; (c) redecking operations.

pultruded panels matched the parabolic crown (6 cm) on the bridge's approach spans, so cambering was analyzed in the shop, and panels arrived at the job site in their curved state. Another key geometric feature of the existing bridge was its vertical alignment. In the portions of the bridge where the longitudinal stringers were vertically curved, each panel was placed on the stringers and conformed to the existing profile with a chord effect. Each panel was straight, whereas the field joints accommodated incremental, extremely slight, rotations before adhesive curing. Both accommodations facilitated the use of FRP on the unique structure and are expected to have minimal negative effects on the integrity of the deck system. At the time of construction, no AASHTO design criteria were established for FRP decks, so the supplier took full responsibility for the design and performance of this system.

8 Research on bridge assessment, retrofit, and management

The research and development relating to decay of materials and onsite tests in particular deals with: advances in the concrete field relating to the decay of fracture parameters of concrete under sulfate environments (Xu et al., 2013); factorial design study to determine the significant parameters of fresh concrete lateral pressure and initial rate of pressure decay (Santilli et al., 2011); studies on concrete degradation during molten core/concrete interactions (Yu et al., 2006); studies on the stability of a concrete bay bridge pier under freeze-thaw action (Jia et al., 2010); mechanisms of long-term decay of tension stiffening (Beeby and Scott, 2006); numerical modeling for predicting service life of reinforced concrete structures exposed to chloride environments (Gang et al., 2010); advanced studies on improved application technique of the adaptive probabilistic neural network for predicting concrete strength (Jong et al., 2009); the characterization of flaws embedded in externally bonded CFRP on concrete beams by infrared thermography (Lai et al., 2009); a study on concrete degradation during molten core/concrete interactions (Maruyama et al., 2006); on the load of reinforced concrete columns by seawater corrosion (Lin, 2012). Research and development for assessment procedures include simplified site-specific traffic load models for bridge assessment (Getachew and Obrien, 2007); site-specific traffic load modeling for bridge assessment (O'Connor and Eichinger, 2007); rapid and global bridge assessment for the military, but also for urgent situations (Ray and Butler, 2004); concerning the probability-based bridge (O'Connor and Enevoldsen, 2007); and concerning the reliability-based bridge assessment using risk-ranking decision analysis (Stewart et al., 2001). However, it is well established that bridge structural reliability assessment should be based on health monitoring data, in order to get precise information on existing structures (Jiao and Sun, 2011). Similar studies that could be useful for the same scope relate to bridge system performance assessment from structural health monitoring (Ming et al., 2009). There are valuable resources that cover other assessment procedures regarding specific issues, such as the probability analysis and risk assessment about vessel-bridge collision, reported in Yin et al. (2011). The argument to pursue research and development for repairing and strengthening operations has been widely cited in recent studies. In the following, we have reported some of the most relevant recent studies, divided by the specific constituent material. Aidoo et al. (2006) discusses a full-scale experimental investigation for the repair of reinforced concrete interstate bridge using CFRP materials; Tedesco et al. (1999) describes a finite element method analysis of a concrete bridge repaired with fiber-reinforced plastic laminates; Hyman (2005) explores inspection, repair, and rehabilitation of concrete structures due to corrosion; and Alampalli (2005) investigates the effectiveness of FRP materials with alternative concrete removal strategies for reinforced concrete bridge column wrapping. Regarding steel structures; Hollaway et al. (2006) reports advances in adhesive joining of carbon fiber/polymer composites to steel members for repair and rehabilitation of bridge structures; Chang et al. (2008) discusses the weldability studies on the replacement repair welded joints of a damaged steel bridge; Clubley et al. (2006) deals with heat-strengthening repairs to a steel road bridges; and Pipinato (2011) explores the specific topic of railway bridge assessment. Advances in BMSs are commonly investigated directly by management authorities.

References

Aidoo, J., Harries, K.A., Petrou, M.F., 2006. Full-scale experimental investigation of repair of reinforced concrete interstate bridge using CFRP materials. J. Bridge Eng. 11 (3), 350–358.

Alampalli, S., 2005. Effectiveness of FRP materials with alternative concrete removal strategies for reinforced concrete bridge column wrapping. Int. J. Mat. Prod. Tech. 23 (3–4), 338–347.

ANAS, 1997. Material and Techniques of Intervention for the Restoration of the Concretes of the Bridges and Viaducts (in Italian). Gangemi Ed, Rome.

Beeby, A.W., Scott, R.H., 2006. Mechanisms of long-term decay of tension stiffening. Mag. Concr. Res. 58 (5), 255–266.

Bentur, A., Diamond, S., Berke, N.S., 1997. Steel Corrosion in Concrete. E & FN Spon, London.

Branco, F.A., Brito, J., 1996. Bridge management from design to maintenance, Recent advances in bridge engineering. In: Proceedings of the US-Europe Workshop on Bridge Engineering. July 15–17, 1996. Barcelona, Spain.

Bridge Management in Europe (BRIME), 2001. Final Report D14, IV Framework Program. Brussels.

Brinckerhoff, P., 1993. Bridge Inspection and Rehabilitation (L.G. Silano, Ed.). John Wiley & Sons, New York.

Broomfield, J.P., 1997. Corrosion of Steel in Concrete. E & FN Spon.

BS EN ISO 8501-1, 2007, Preparation of steel substrates before application of paints and related products. Visual assessment of surface cleanliness. Rust grades and preparation grades of uncoated steel substrates and of steel substrates after overall removal of previous coatings.

Building Research Establishment (BRE), 2001. Modeling Degradation Processes Affecting Concrete. BRE 434. ISBN 1 86081 5316, Construction Research Communications Ltd, Watford, UK.

Cadei, J.M.C., et al., 2004. Strengthening Metallic Structures Using Externally Bonded Fiber-Reinforced Polymers (C595). CIRIA Design Guide. CIRIA, London.

Chang, K.H., Park, H.C., Lee, C.H., Jang, G.C., Lee, S.H., Choi, E.H., Shin, Y.E., Kim, J.M., Lee, J.H., 2008. Weldability Studies on the Replacement Repair Welded Joints of a Damaged Steel Bridge. Materials Science Forum, Vols. 580–582, 655–658.

Clubley, S.K., Winter, S.N., Turner, K.W., 2006. Heat-straightening repairs to a steel road bridge. Proc. Inst. Civil Eng. Bridge Eng. 159 (1), 35–42.

Concrete Society, 2000. Diagnosis of Deterioration in Concrete Structures: Identification of Defects, Evaluation and Development of Remedial Action. Technical Report No. 54. ISBN 0 94669 1818. The Concrete Society, Berkshire, UK.

COST345, 2004. Procedures required for the assessment of highway infrastructures – Final report. European research project under the framework of European Cooperation in the field of scientific and technical research, EU Commission-Directorate General Transport and Energy, Bruxelles, 182 pp.

Czepiel, E., 1995. Bridge Management Systems Literature Review and Search. Technical Report No. 11, Northwestern University BIRL Industrial Research Laboratory, Evanston, IL.

Dogaki, M., et al., 2000. Optimal maintenance planning of reinforced concrete decks on highway network. In: Conference Proceedings. US169 Japan Workshop on Life-Cycle Cost Analysis and Design of Civil Infrastructure Systems. ASCE, Honolulu.

DuraCrete, 1998. Modeling of Degradation. BRITE –EURAM-Project BE95-1347/R4-5.

DuraCrete, 2000a. Statistical quantification of the Variables in the Limit State Functions. BRITE – EURAM-Project BE95-1347/R9.

DuraCrete, 2000b. Probabilistic Calculations. BRITE –EURAM-Project BE95-1347/R14B.

EN 10025-5, 2004. Structural steels with improved atmospheric corrosion resistance.

EN206-1/2000, 2005. EN 206-1, Concrete: Specification, Performance, Production, and Conformity. CEN—European Committee for Standardization.

Federal Highway Administration (FHWA), 2015. Bridge Ratings. Federal Highway Administration, Washington.

FIB (Fédération internationale du béton), 1999. Structural Concrete, Textbook on Behavior, Design and Performance, Updated Knowledge of the CEB/FIP Model Code 1990, vol 3. Sprint-Druck, Stuttgart, DE.

FIB (Fédération internationale du béton), 2010. FIB Model Code for Concrete Structures 2010. Edited by Ernst & Sohn–Wiley, Berlin. ISBN: 978-3-433-03061-5, 434 pp.

Fisher, J.W., 1984. Fatigue and Fracture in Steel Bridges: Case Studies, John Wiley & Sons, New York, NY, USA, 336 pp.

Fisher, J.W., Yen, B.T., Frank, K.H., 1979. Minimizing Fatigue and Fracture in Steel Bridges. American Society of Mechanical Engineers (ASME) pp. 155–161.

Frangopol, D.M., Liu, M., 2007. Multiobjective Optimization for Risk-Based Maintenance and Life-Cycle Cost of Civil Infrastructure Systems, Chapter of the book System Modeling and Optimization, Volume 199 of the series IFIP International Federation for Information Processing, pp 123–137, Proceedings of the 22nd IFIP TC7 Conference held from July 18–22, 2005, in Turin. Italy, Springer, New York.

Gang, L., Yinghua, L., Zhihai, X., 2010. Numerical modeling for predicting service life of reinforced concrete structures exposed to chloride environments. Cement Concr. Comp. 32 (8), 571–579.

Getachew, A., Obrien, E.J., 2007. Simplified site-specific traffic load models for bridge assessment. Struct. Infrastruct. Eng. 3 (4), 303–311.

Gralund, M.S., Puckett, J.A., 1996. System for bridge management in a rural environment. J. Comp. Civ. Eng. 10 (2), 97–105.

Helmerich, R., 2005. Alte Stähle und Stahlkonstruktionen–Materialuntersuchungen, Ermüdungsversuche an originalen Brückenträgern und Messungen von 1990 bis 2003. BAM. Forschungsbericht 271, Berlin, DE.

HHB, 2015. History of the Macdonald Bridge. HHB—Halifax Harbour Bridges Authority, Dartmouth.

Hollaway, L.C., Zhang, L., Photiou, N.K., Teng, G., Zhang, S.S., 2006. Advances in adhesive joining of carbon fibre/polymer composites to steel members for repair and rehabilitation of bridge structures. Adv. Struc. Eng. 9 (6), 791–804.

Hudson, R., Carmichael III, R., Hudson, S., Diaz, M., Moser, L., 1993. Microcomputer Bridge Management System. J. Transp. Eng. 119 (1), 59–76.

Hyman, Alan E., 2005. Inspection, repair and rehabilitation of concrete structures due to corrosion. Int. J. Mat. Product Tech. 23 (3–4), 309–337. Inderscience Publishers.

Italian Research Council, Italian Advisory Committee on Technical Recommendations for Construction, 2005. Guidelines for the Design and Construction of Externally Bonded FRP Systems for Strengthening Existing Structures. Preliminary Study. Metallic Structures (CNR–DT 202/2005). Italian Research Council, Rome.

Jia, C., Sheng-Zhen, J., Zhang, F., 2010. Study on the stability of concrete bay bridge pier under freeze-thaw action. Sichuan Daxue Xuebao (Gongcheng Kexue Ban)/J. Sichuan Uni. (Eng. Sci. Ed.) 42 (3), 7–13.

Jiao, M., Sun, L., 2011. Bridge reliability assessment based on the PDF of long-term monitored extreme strains. Proc. SPIE. 7983. Nondestructive Characterization for Composite Materials, Aerospace Engineering, Civil Infrastructure, and Homeland Security 2011. 79830 J (April 18, 2011).

Johnston, D.W., Zia, P., 1984. A Level of Service System for Bridge Evaluation, Transportation Research Record. No. 962, TRB, Washington DC. pp. 1–8.

Jong, J.L., et al., 2009. An improved application technique of the adaptive probabilistic neural network for predicting concrete strength. Comp. Mat. Sci. 44 (3), 988–998.

Kirkwood, K., Radojevic, D., Buckland, P., 2014. A deck dilemma. Bridge Design & Eng. November.

Kitada, T., et al., 2000. Bridge management system for elevated steel highways. Comput. Aided Civ. Infrastruct. Eng. 15 (2), 147–157.

Klaiber, F.W., et al., 1987. Methods of Strengthening Existing Highway Bridges. NCHRP 293, Transportation Research Board.

Kühn, B., M. Lukić, M., Nussbaumer, A., Günther, H.-P., Helmerich, R., Herion, S., Kolstein, M.H., Walbridge, S., Androic, B., Dijkstra, O., Bucak, Ö, 2008. Assessment of existing steel structures: recommendations for estimation of remaining fatigue life–joint report prepared under the JRC-ECCS cooperation agreement for the evolution of Eurocode 3 Background Documents in Support to the Implementation, Harmonization and Further Development of the Eurocodes. In: Sedlacek, G., Bijlaard, F., Géradin, M., Pinto, A., Dimova, S. (Eds.), Luxembourg: Office for Official Publications of the European Communities 2007 – 89 pp. – Scientific and Technical Research series – ISSN 1018-5593, publisched by JRC-Joint Research Centre, Ispra.

Lai, W.L., et al., 2009. Characterization of flaws embedded in externally bonded CFRP on concrete beams by infrared thermography and shearograph. J. Nondes. Ev. 28 (1), 27–35.

Lédeczi, A., et al., 2011. Self-sustaining wireless acoustic emission sensor system for bridge monitoring. New developments in sensing technology for structural health monitoring. Lect. Notes Elec. Eng. 96, 15–39.

Li, H., Love, P.E., 1998. Site-level facilities layout using genetic algorithms. Journal of Computing in Civil Engineering 12 (4), 227–231.

Liebenberg, A.C., 1992. Concrete Bridges: Design and Construction. Longman Group, UK Ltd., London.

Lin, Y., 2012. On the load of reinforced concrete column by seawater corrosion. Advanced Materials Research 368–373, 975–978, TTP Edition, Pfaffikon.

Lopes, N., Ribeiro, D., Reis, A., 2008. Upgrading the Luiz I Bridge in Porto. In: Proceedings of the 7th international conference on steel bridges. Guimaraes, Portugal.

Markow, M.J., 1995. Highway management systems: state of the art. J. Infrastruct. Sys. 1 (3), 186–191.

Maruyama, Y., et al., 2006. A study on concrete degradation during molten core/concrete inter-actions. Nuc. Eng. Des. 236 (19–21), 2237–2244.

Ming, L., Frangopol, D.M., Kim, S., 2009. Bridge system performance assessment from structural health monitoring: a case study. J. Struct. Eng. 135 (6), 733–742.

Mohamed, H., 1995. Development of Optimal Strategies for Bridge Management Systems. Ph.D. Thesis, Dept. of Civil and Environmental Engineering. Carleton University, Ottawa, Canada.

Moy, S.S.J. (Ed.), 2001. FRP composites: life extension and strengthening of metallic structures. In: ICE Design and Practice Guide. Thomas Telford, London.

Moy, S.S.J., Bloodworth, A.G., 2007. Strengthening a steel bridge with CFRP composites. Proceedings of the Institution of Civil Engineers (ICE). Struct. Build. 160 (SB2), 81–93.

Moy, S.S.J., Lillistone, D., 2006. Strengthening cast iron using FRP composites. Proceedings of the Institution of Civil Engineers (ICE). Struct. Build. 159 (6), 309–318.

Mullheron, M., 2000. Durability of Bridges and Structures. Module in the MSc in Bridge Engineering. Internal publication, University of Surrey, Surrey, UK.

Nair, A., Cai, C.S., 2010. Acoustic emission monitoring of bridges: review and case studies. Eng. Struct. 32, 1704–1714.

Neves, L.A.C., Frangopol, D.M., Cruz, P.J.S., 2006a. Probabilistic lifetime-oriented multi-objective optimization of bridge maintenance: single maintenance type. J. Struct. Eng. 132 (6), 991–1005.

Neves, L.A.C., Frangopol, D.M., Petcherdchoo, A., 2006b. Probabilistic lifetime-oriented multi-objective optimization of bridge maintenance: combination of maintenance types. J. Struc. Eng. 132 (11), 1821–1834.

O'Connor, A., Eichinger, E.M., 2007. Site-specific traffic load modelling for bridge assessment. Proc. Inst. Civ. Eng. Bridge Eng. 160 (4), 185–194.

O'Connor, A., Enevoldsen, I., 2007. Probability-based bridge assessment. Proc. Inst. Civ. Eng. Bridge Eng. 160 (3), 129–137.

Pipinato, A., 2008a. New Approaches for the Management, Maintenance and Control of Bridge Networks (in Italian). Strade e Autostrade. EDI-CEM ed. Milan, Italy.

Pipinato, A., 2008b. High-Cycle Fatigue Behavior of Historical Metal Riveted Railway Bridges. PhD thesis, University of Padova, Italy.

Pipinato, A., 2011. Assessment of existing bridges: safety and security issues [Problemi di sicurezza nelle valutazioni strutturali di ponti esistenti]. Ingegneria Ferroviaria 66 (4), 355–371.

Pipinato, A., 2014. Assessment and rehabilitation of steel railway bridges using fiber-reinforced polymer (FRP) composites. In: Karbhari, V.M. (Ed.), Rehabilitation of Metallic Civil Infrastructure Using Fiber-Reinforced Polymer (FRP) Composites. ISBN: 978-0-85709-653-1, Elsevier, New York, pp. 373–405.

Priestley, M.J.N., Seible, F., Calvi, G.M., 1996. Seismic Design and Retrofit of Bridges. ISBN: 978-0-471-57998-4, John Wiley & Sons, New York, 704 pp.

Radomski, W., 2002. Bridge Rehabilitation. Imperial College Press, London.

Ray, J., Butler, C.D., 2004. Rapid and global bridge assessment for the military. J. Bridge Eng. 9 (6), 550–557.

Robert, W., et al., 2003. The Pontis Bridge management system: state-of-the-practice in implementation and development. In: Proc., 9th Int. Bridge Management Conf. Transportation Research Board, Washington, DC, pp. 49–60.

SAMARIS, 2005. Sustainable and advanced materials for road infrastructure. Final report. VI Framework Program. European Commission's Directorate General for Mobility and Transport. Brussels.

Sams, M., 2005. Bridge deck application of fiber-reinforced polymer. Transpor. Res. Rec. J. Transport. Res. Board 175–178. CD 11-S. Transportation Research Board of the National Academies. Washington, DC.

Santilli, A., et al., 2011. A factorial design study to determine the significant parameters of fresh concrete lateral pressure and initial rate of pressure decay. Construct. Build. Mat. 25 (4), 1946–1955.

Schnerch, D., Dawood, M., Rizkalla, S., 2007. Proposed design guidelines for strengthening of steel bridges with FRP materials. Construct. Build. Mat. 21, 1001–1010.

Silano, L.G., 1993. Bridge Inspection and Rehabilitation — A Practical Guide. John Wiley & Sons, Inc. New York.

Soderqvist, M.K., Veijola, M., 2000. Finnish project level bridge management system. Transp. Res. Circular. 498 (TRB), F-5/1–F-5/7.

Stanish, K.D., Hooton, R.D., Thomas, M.D.A., 2000. Prediction of Chloride Penetration in Concrete. Testing the Chloride Penetration Resistance of Concrete: A Literature Review. FHWA Contract DTFH61-97-R-00022 by Department of Civil Engineering University of Toronto, Toronto, Ontario, Canada.

Stewart, M.G., Rosowsky, D.V., Valc, D.V., 2001. Reliability-based bridge assessment using risk-ranking decision analysis. Struct. Safety. 23(4), 397–405.

Sukumaran, B., 1998. A Review of Inspection Maintenance and Assessment of Reinforced Concrete and Prestressed Concrete Bridges and a Case Study. MSc dissertation, University of Surrey, Surrey, UK.

Sunley, W., 1995. Pontis: a bridge inspection. Illinois Department of Transportation, Illinois Municipal Rev. 13–15.

Sustainable Bridges, 2006. Guideline for load and resistance assessment of existing European railway bridges – advices on the use of advanced methods. European Research Project Under the EU 6th Framework Program.

Tedesco, J.W., Stallings, J.M., El-Mihilmy, M., 1999. Finite element method analysis of a concrete bridge repaired with fiber reinforced plastic laminates. Comput. Struct. 72 (1–3), 379–407.

Thoft-Christensen, P., 1995. Advanced bridge management systems. Struct. Eng. Rev. 7 (3), 151–163.

Thompson, P., et al., 2003. Implementation of Ontario bridge management system. In: Ninth International Bridge Management Conference. Orlando, FL, pp. 112–127.

White, K.R., Minor, J., Derucher, K.N. 1992. Bridge maintenance, inspection and evaluation. Marcel Dekker, Inc., New York, U.S.

Xu, H., Zhao, Y., Cui, L., Xu, B., 2013. Sulphate attack resistance of high-performance concrete under compressive loading. J Zhejiang Univ-Sci A (Appl Phys & Eng) 14 (7), 459–468.

Yin, Z.H., Li, Y.F., Li, Y., 2011. Probability analysis and risk assessment about vessel-bridge collision of Xihoumen Bridge. Advanced Materials Research 179–180. 464–469.

Yu M et al. 2006 A study on concrete degradation during molten core/concrete interactions. National Nuclear Center of the Republic of Kazakhstan, Konechnaya, Pavlodar, Kazakhstan. Nuclear Engineering and Design, 236(2006), 2237–2244, Elsevier, New York.

Zhao, X.L., Zheng, L., 2007. State-of-the-art review on FRP strengthened steel structures. Eng. Struct. 29, 1808–1923.

Bridge monitoring

29

Vardanega P.J.[1], Webb G.T.[2], Fidler P.R.A.[3], Middleton C.R.[3]
[1]University of Bristol, Bristol, UK
[2]WSP | Parsons Brinckerhoff, London, UK
[3]University of Cambridge, Cambridge, UK

1   Introduction

In the front matter of his classic book on geotechnical instrumentation, Dunnicliff (1988) offers some valuable advice:

> *"Every instrument on a project should be selected and placed to assist with answering a specific question: if there is no question, there should be no instrumentation."*

Structural health monitoring (SHM) is becoming an increasingly popular topic of discussion in the bridge engineering community. Ongoing developments in sensor and data acquisition technologies have made it possible to install extensive monitoring systems on many structures. The hope is that by obtaining quantitative data it will be possible to develop "smart" structures, with monitoring systems able to supplement the largely subjective visual inspection practices (which are currently employed as the primary means of evaluating structural integrity and condition e.g., Moore et al., 2001).

There are many articles that describe the efforts to monitor specific bridge structures (e.g., Brownjohn et al., 1999; Chang and Im, 2000; Wong, 2004; Lynch et al., 2006; Staquet et al., 2007; Hoult et al., 2010; and Koo et al., 2013). Webb (2014) observed that, in many cases, monitoring systems do not deliver the valuable insights desired by owners and managers of bridges. What is often lacking is a clear statement of what value the system hopes to deliver. Middleton et al. (2014) discuss the current and future potential of SHM. Many of the reported SHM projects are simply records of the capabilities of new sensors and sensor deployments. Some specific cases were found to demonstrate value, such as the case of the Hammersmith Flyover (Webb et al., 2014a), where a specific issue of concern was investigated using remote SHM techniques.

Before implementing an SHM system the following guiding questions should be asked:

- What are the overall objectives of the monitoring activities?
- What information is needed from a monitoring system to fulfill these objectives?
- What are the expected values of the readings that you will obtain from the measurements?
- What accuracy (and frequency of readings) is needed from the measurements to allow for decisions to be taken on the basis of the measurements?
- What technology will be able to take the necessary measurements?
- How, and at what cost, will the information be recorded, interpreted, disseminated, and stored, and what communications strategy will be used to transfer data to the end user?

Innovative Bridge Design Handbook. http://dx.doi.org/10.1016/B978-0-12-800058-8.00029-3

- What input is required from all the relevant stakeholders so that all expectations can be understood and managed?
- Who will bear the capital and ongoing operational costs associated with the system and are these costs affordable?
- How can the value/benefit of the information obtained be quantified?

2 SHM stakeholders

There are a number of potential beneficiaries of SHM who can each derive different benefits from bridge monitoring. Andersen and Vesterinen (2006) identify seven key stakeholders of SHM projects, summarized in Table 29.1, so that the various needs, requirements, and objectives of each can be understood. Regardless of the stakeholder, the overriding objectives are safety, performance, and cost, which are important to all stakeholders to varying degrees.

Table 29.1 SHM Stakeholder Objectives

Stakeholder	Principal Objectives
Authorities	Required functionality of the structures shall be documented and reported
Owners	Reliability of the structures must satisfy codes and standards
	Acceptable service life of structures must be ascertained
	Life cycle cost optimization
Users	Availability of services provided by the structures must be high
	Must be able to use the structures safely
Researchers	Full-scale verification of structural modeling theories
Designers	Verification and documentation of the final design
Contractors	Verification of structural response and geometry
Operators	High availability
	Cost-efficient operation and maintenance
	Identification of causes of unacceptable behavior (e.g., vibrations) or excessive wear

Adapted from Andersen and Vesterinen (2006); used with permission.

3 Types of monitoring deployments

Monitoring systems vary in size and complexity from systems that may take quasi-static measurements from a small number of inclinometers or strain gauges to detect long-term change, to systems utilizing ambient vibration measurements for structural identification or damage detection.

Many studies have been published explaining the academic outcomes of bridge monitoring projects. To clarify the objectives of SHM deployments, the following classification system has recently been proposed (Webb et al. 2014b), which defines the primary objective of any SHM system. This classification system proposes that monitoring systems can be classified as one or a combination of the following SHM categories:

- Sensor deployment studies—Systems used to demonstrate new sensor or communication technologies (category 1)
- Anomaly detection—Systems used to detect that something has changed, or that something is changing over time (category 2)
- Model validation—Systems used to compare the performance of a structure with the performance that is predicted by structural analysis models (category 3)
- Threshold check—Systems that compare key parameters against thresholds which are usually derived from a structural model to warn of potential problems (category 4)
- Damage detection—Systems that aim to detect and locate damage using advanced techniques such as structural identification and modal analysis (category 5)

Webb (2014) and Webb et al. (2014b) demonstrated that the vast majority of published SHM research merely reports the deployment of an SHM system for validation of structural models (Figure 29.1, category 3), with little discussion or explanation about the purpose of the system in terms of the needs of one or more key stakeholders. The second most common type of deployment described in the published literature is sensor deployment studies. These deployments are typically conducted by researchers to demonstrate new sensor or communication technologies, such as microelectromechanical systems (MEMS) sensors or wireless sensor networks (WSNs). There may be no intention to provide immediate value to the asset owner or operator from such deployments, although such field demonstrations may ultimately lead to greater industry confidence in new SHM technologies.

Attempts have been made to implement damage detection (category 5). For example, Catbas et al. (2013) and Wenzel (2009) detail methodologies for using structural identification and ambient vibration measurements, respectively.

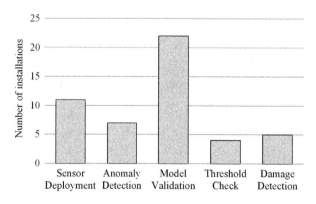

Figure 29.1 Categorization of some existing monitoring systems that are reported in the literature, and the number of installations of those systems.
(data courtesy of Webb et al., 2014b).

While it is acknowledged that the original reasons for many of these deployment studies may not be described in detail in the literature, it is nevertheless interesting that the value derived by asset owners is rarely mentioned. Often, the demonstration of value of the SHM system seems to be a low priority.

4   Measuring technologies used for SHM

Many different sensors can be used in bridge SHM systems, both to measure the load-ing applied to a structure as well as its response. After reviewing 31 publications detailing SHM deployments, Webb (2014) determined approximate levels of preva-lence for various monitoring technologies (Figure 29.2). Table 29.2 lists 25 monitor-ing technologies all described in more detail in Gastineau et al. (2009), a brief description of each is provided by the present authors in the third column. An in-depth study of all of these technologies would be beyond the scope of this chapter. We will therefore focus in more detail on developments in sensor technologies which have potential uses for SHM, in particular on the technologies with which the authors have practical experience; namely, (i) computer vision; (ii) acoustic emission (AE); (iii) fiber-optic strain sensing; (iv) MEMS sensors; (v) weigh-in-motion (WIM) systems, and; (vi) corrosion detection systems.

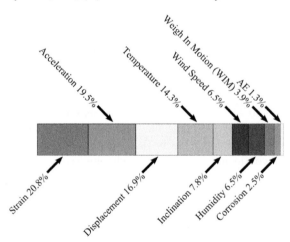

Figure 29.2 Sensor types used in 31 bridge monitoring installations.
(data courtesy of Webb, 2014).

4.1   Computer vision

Computer vision systems allow the extraction of information from the scanning and interpretation of digital images. They can be employed for a wide variety of tasks, including 3D reconstruction, construction progress monitoring, geometric checks, component compliance, and deflection detection. This offers new opportunities to enhance SHM efforts, especially with the emerging prominence of building informa-tion modeling (BIM). Computer vision has the potential to remove some of the sub-jectivity inherent in visual inspection efforts. If defect detection can be automated, computer vision could potentially add significant levels of intelligence to the con-struction or operation processes.

4.2   Acoustic emission (AE)

Acoustic emission (AE) involves the use of sensitive acoustic transducers to detect small elastic strain waves that are generated as cracks propagate in steel plates, or pre-stressing wire cable strands break. For the latter, interpreting the historical number of

Table 29.2 Types of SHM System Technologies Described in Gastineau et al. (2009)

	System Technology	Description (and deployment examples)
1	3D laser scanning	Used to build 3D point cloud models of structures; e.g. Park et al. (2007).
2	Accelerometers	Used to determine modal properties of structures; e.g. Matsumoto et al. (2010).
3	Acoustic emission	Detects sound waves generated by cracks propagating or steel wires breaking; e.g. Nair and Cai (2010).
4	Automated laser total station	3D displacement monitoring of a number of targets; e.g. Psimoulis and Stiros (2013).
5	Chain dragging	Acoustic technique used to detect shallow delamination in concrete decks. e.g. Perenchio (1989).
6	Concrete resistivity	Provides an indication of the likelihood of corrosion occurring.
7	Digital image correlation (DIC)	An image-processing technique to track relative movements between sets of images taken at different times, allowing a continuous strain field to be derived; e.g. Lee et al. (2012) and White et al. (2003). Also known as *particle image velocimetry (PIV)*.
8	Electrochemical fatigue sensing system	A nondestructive technique developed to detect fatigue cracks in metal structures by continually monitoring current flow at the surface.
9	Electrical impedance (post tensioned tendon)	Technique for detecting defects in the corrosion protection of post tensioned tendons using a change in electrical resistance. Relies on the tendons being enclosed by a polymer duct that electrically isolates the tendon and grout from the surrounding concrete; e.g. Elsener (2005).
10	Electrical resistance strain gauges	Common form of strain gauge, although susceptible to thermal variations. Issues include variation in the quality of attachment.
11	Fatigue life indicator	Sacrificial sensor intended to indicate the likely remaining fatigue life of a component; e.g. Zhang et al. (2007b).
12	Fiber optics	Can provide discrete or continuous measurement of a variety of parameters, such as strain, temperature, and chloride ion concentration; e.g. Rodrigues et al. (2010).
13	Global positioning system (GPS)	Satellite-based system to provide 3D position information.
14	Ground-penetrating radar (GPR)	A technique involving radar pulses to view subsurface features such as reinforcement or prestressing tendons.

(Continued)

Table 29.2 Continued

	System Technology	Description (and deployment examples)
15	Impact echo	Analysis of reflected sound waves from a hammer tap to detect some subsurface flaws.
16	Infrared thermography	Disrupted heat flows through structures can be indicative of damage such as the delamination of concrete slabs; e.g. Washer et al. (2010).
17	Linear polarization resistance (LPR)	Provides an indication of the likelihood of corrosion in concrete.
18	Linear potentiometer	Displacement transducer based on the principle of a potential divider.
19	Linear variable differential transformer	Extremely robust displacement transducer consisting of three solenoidal coils around a sliding ferromagnetic core.
20	Macrocell corrosion rate monitoring	Technique to provide an indication of the likelihood of corrosion.
21	Half-cell potential measurements/chloride content	Measuring the electrochemical potential or the chloride ion concentration in concrete can provide an indication of the likelihood that corrosion is occurring.
22	Scour measurement	Ultrasonic and radar technologies have been used to attempt to detect scour, but inspection by a diver may be the only reliable way to assess the extent of any scour.
23	Tiltmeters/inclinometers	Used to measure the angle of an object with respect to the Earth's gravitational field, typically using a force balance sensor.
24	Ultrasonic C-scan	Ultrasonic testing can be used to detect some imperfections within materials.
25	Vibrating wire strain gauges	Strain gauge comprising a taut steel wire attached to the structure, the natural frequency of which varies with applied strain; e.g. DiBiagio (2003).

Some adaption of the Gastineau material provided by the authors.

wire breaks that have occurred and then quantifying that effect on the whole bridge cable remains a challenge unless the AE system has been in place from the very start of the bridge's life.

4.2.1   AE case study: Hammersmith Flyover

AE sensors were used by a contractor (Watson, 2010) on the Hammersmith Flyover in London to detect loss of prestress wire breaks, and the output data were made available to the research team at the University of Cambridge for analysis (Webb et al., 2014a). The rate of wire breaks began to increase around March 2011 (see Figure 29.3), and increasing concern over the ensuing months led to closure of the bridge to traffic. The closure led to a program of intrusive investigation and subsequent retrofitting activities. This monitoring technique is an example of damage detection (category 5) being used to investigate a specific problem (i.e., corrosion/wire breaks) and subsequently yielding value to the asset owner, who would then make an informed decision on the refurbishment strategy.

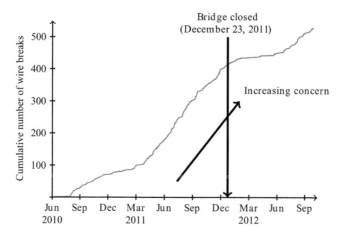

Figure 29.3 Cumulative number of wire breaks detected during the AE monitoring of the Hammersmith Flyover.
(adapted from Webb et al., 2014a).

4.3 Fiber-optic strain sensing

The Nine Wells Bridge located in Cambridgeshire, UK, is an example of an experimental deployment (category 1) of fiber-optic strain measurement (further details about the deployment are given in Hoult et al., 2009). Six beams in the western span of this three-span prestressed concrete bridge were constructed with optical fibers cast into the concrete in the precasting factory (see Figures 29.4 through 29.6), allowing distributed strain measurements to be taken along the lengths of the prestressed beams. Fiber optic cables enter at one end of the beam and run along the lower prestressing strands, up one of the shear links at the end of each beam and then return along the upper prestressing strands, thus completing the loop of fiber in the beam. Readings were taken after release of the pretensioned cables, immediately after installation, and also after casting of the composite in-situ deck slab. The results enabled an investigation of various phenomena including: debonding of prestressing tendons, initial elastic shortening, concrete creep and shrinkage, and an examination of the effects of temperature (more details are provided in Webb, 2014). In this case, researchers gained value from the installed system as it lead to greater confidence in the use of this fiber-optic technology for measuring distributed strain in bridge beams. The research team at the University of Cambridge had previously only used the technology in retrofit installations in tunnels.

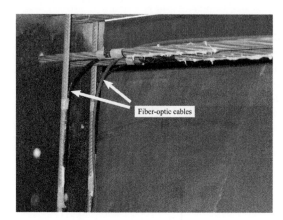

Figure 29.4 Close-up view of fiber-optic cables attached to rebar and pretensioned tendons. (courtesy of Neil Hoult).

Figure 29.5 Finished beam ready for transport to site. (courtesy of Neil Hoult).

Figure 29.6 Taking baseline readings with a fiber-optic analyzer. (courtesy of Neil Hoult).

4.4 MEMS sensors

Advances in silicon chip manufacturing processes have made it possible to fabricate small MEMS devices within the same packaging as integrated circuits. Examples of these devices include accelerometers, inclinometers, and solid-state gyroscopes, which can be found in everyday consumer devices such as smart phones and tablet computers. MEMS sensors can be produced rapidly in large numbers and are cheap to manufacture. They tend to have low power needs, which facilitate their use in WSNs. Wireless devices incorporating MEMS sensors may offer a cost-effective solution for extensive, scalable, remote monitoring of bridges.

4.5 Weigh-in-motion (WIM) systems

A WIM system allows for automatic collection of traffic data, such as axle loads, gross vehicle loads (weights), traffic volume, and speed – all taken as real time or near real-time measurements (e.g., Zhang et al., 2007a). Live loading is usually an important load case for bridges and is necessary in order to make predictions of the expected response to compare with measured data. Zhi et al. (1999) have discussed some difficulties encountered in calibrating WIM data. For example, it is difficult to measure vehicle live loads accurately due to complex interactions between the vehicles, their suspension, and the bridge deck. Recent research by Cantero and González (2015) suggests that WIM technology may be used to monitor change in structural condition for short-to-medium-span bridges more effectively than using conventional vibration techniques. Webb (2014) estimated live loads on a bridge deck through inverse analysis of deflection measurements and this work highlighted the difficulty faced in uniquely characterizing live loads indirectly.

4.6 Corrosion detection systems

Techniques such as the use of corrosion ladders and linear polarization resistance gauges may indicate the likelihood of corrosion being present, rather than the actual loss of steel section, but in reality these have not proven very effective. Agrawal et al. (2009) review various remote corrosion monitoring sensors and systems for use on highway bridges. Most corrosion detection systems only provide an indirect indication of the likelihood of corrosion.

5 Deployment and operation

5.1 Sensor deployment strategies

5.1.1 Wired sensor networks

Wired sensor networks utilize dedicated cabling to provide both power and data transfer to all of the system's sensors. Wong (2004) describes the Wind and Structural Health Monitoring System (WASHMS), which is deployed on a number of significant bridges in Hong Kong. This complex and expensive system is arguably more robust than many wireless systems, but does not have the level of flexibility and expandability that wireless solutions offer, and also requires installation of cable runs to connect the sensors to a suitable data logger.

5.1.2   Wireless sensor networks (WSNs)

WSNs are being developed as an alternative, lower-cost, and more flexible solution to wired deployments. Each sensor is completely self-contained and transmits readings wirelessly to a receiver and data logger located somewhere on or near the structure. The lack of cables means that wireless systems can be much more easily reconfigured than wired systems, which is a key benefit. The research project Smart Infrastructure: Wireless sensor network system for condition assessment and monitoring of infrastructure (2006-2009), funded by the Engineering and Physical Sciences Research Council (EPSRC) in the UK, tested several bridge wireless SHM deployments, including systems installed on the Humber Bridge (Hoult et al., 2008), Ferriby Road Bridge (Hoult et al., 2010), and Hammersmith Flyover (Webb et al., 2014a). Figure 29.7 shows an example of the type of wireless sensor device typically used by the researchers for these deployments.

Figure 29.7 A wireless sensor mote connected to a linear potentiometric displacement transducer (LPDT).

5.2   Deployment challenges

Many of the research papers reviewed by the authors discuss the technology itself, but do not discuss the difficulties and challenges that must be overcome when an SHM system needs to be designed, installed, and then operated. Practical considerations for the design and specification of new SHM systems include:

- Sensor placement (and access), such as the need for specialized equipment to access certain parts of the structure (Figure 29.8)

- Wiring placement, or wireless relay placement if using a WSN
- Environmental considerations; e.g., indoor/outdoor, International Protection Marking IP-rated enclosures (Figure 29.9)
- Data-logger placement, communications (if remote access to data is required), and power supply (backup batteries) (Figure 29.10)

Stajano et al. (2010) described 19 key considerations that must be considered when deploying a wireless sensor system (listed in Table 29.3) in relation to communications and security.

Figure 29.8 Installing wireless sensors on the Hammersmith Flyover.

Figure 29.9 Example of a non-ideal location for wireless sensors: a pier pit at the Hammersmith Flyover that flooded during remediation work, destroying some installed sensors.

Figure 29.10 Data logger for wireless humidity sensors at Humber Bridge.

Table 29.3 **Principles for Successful WSN Deployment by Stajano et al. (2010)**

Principle	Description
1	Multi-dimensional optimization: you must choose a goal function.
2	Planning for shortest deployment time: ensure the multihop network will achieve end-to-end connectivity within a reasonable time.
3	Assume access time to the site will be limited: you must plan in advance where to put the nodes.
4	Radio is like voodoo: it affects you even if you do not understand or believe it.
5	Radio propagation modelling: to minimize the number of nodes to be deployed you need an accurate, efficient and robust propagation model.
6	Radio propagation measurements: you must calibrate your radio propagation model with physical measurements that can only be obtained on site.
7	Once a node's position is fixed and you experience fading, you must be able to overcome it.
8	Risk assessment: you must talk to the stakeholders and find out what they want to protect.
9	As far as sensor data is concerned, you should pay more attention to integrity and availability than to confidentiality.
10	Your risk rating must be a function of the use to which the network will be put and of its side effects on the environment.
11	You must assess whether a vulnerability of the wireless sensor network can be used to attack other networks connected to it.
12	Evaluation: you must perform independent penetration testing of the COTS[1] equipment you use, even if it claims to offer a "secure" mode of operation.
13	Deployment in harsh environments: you must ensure that your sensors keep working and do not fall off.
14	If sensors that measure what you want do not exist, you must make your own.
15	Sensor failures: you must be prepared for the unexpected to happen.
16	You must be able to find out exactly what happened.
17	You must think about what you're measuring and why.
18	You must understand the end users and their workflow and find a way to present the data that makes sense to them.
19	You must strive to preserve and present the spatial origin of the data.

[1]Commercial-off-the-shelf.

Extracts reprinted from Ad Hoc Networks, Vol 8. No. 8, Stajano et al., Smart bridges, smart tunnels: Transforming wireless sensor networks from research prototypes into robust engineering infrastructure, 872–888, 2010, with permission from Elsevier.

5.3 Data quality

5.3.1 Reliability and robustness

Adequate redundancy must be provided if the data is to be relied upon to influence decisions concerning the safe operation of the bridge. An independent method of measurement should be installed to provide an alternative check of measurement systems (e.g., installation of electrical resistance strain gauges or vibrating wire strain gauges to verify fiber-optic strain measurements). There is also a need for the SHM system to be able to quickly notify the system operator of instances of failure of individual sensors, or indeed of the entire system.

5.3.2 Accuracy and resolution

Studies describing the accuracy, sensitivity, and reliability of many sensing technologies are remarkably difficult to source. More research should be undertaken to demonstrate the accuracy of sensors for use on SHM projects. Without this information, asset owners will not be able to have sufficient confidence in any monitoring data or allow decisions to be influenced by it, arguably rendering the entire monitoring activity futile. The accuracy and resolution of analog sensors will often be determined by the analog-to-digital converter used to measure the output of the sensor, along with the characteristics of any filtering or signal conditioning circuits used. The way in which a sensor is packaged and attached to the structure will also have an effect. This means the system as a whole needs to be considered, not just the specification for the individual sensor devices.

5.4 Sensor calibration

It should be a condition of any SHM project that a large, up-front investment be devoted to proving that all the equipment is appropriately calibrated and functioning. Calibration of the sensing equipment is vital for the users of the SHM data to have confidence that decisions can be made on the basis of the data gathered. Some suppliers may provide calibration certificates for their sensors. If provided, these certificates should be kept as part of the documentation for the system. It should be established whether the calibration is for individual sensors, or is for a batch of similar sensors. A guiding principal of SHM deployments should be that calibration studies form an integral part of the organization of a field deployment.

5.5 Future proofing

Future proofing is a popular topic among knowledge-management professionals. An asset owner who undertakes SHM should remember that the specified system should (i) be maintainable, (ii) be replaceable, (iii) incorporate redundancy of sensors, and (iv) be aligned with well-maintained data storage protocols. This implies the need for digital archive maintenance to protect against loss of data.

Installation of SHM itself may be a form of future proofing; for example, it may be relatively inexpensive to embed fiber-optic cables into a newly built concrete structure so that these can be used to measure the response of the structure to increased loading requirements in the future.

5.6 Data processing

Vann et al. (1996) presented an early report on intelligent data logging strategies for some bridge projects. Filtering and data compression are emphasized as methods of reducing physical space. With the availability of cheap computing power, this is perhaps less relevant today than it was in 1996. However on-site filtering, as advocated by Vann et al. (1996), gives engineers an immediate appreciation of the collected data. Trained personnel looking at real-time, incoming data are invaluable in the spotting of mistakes, errors, and unexpected behavior during the construction process.

6 Summary

This chapter has described some of the practical considerations that bridge engineers need to consider when specifying, installing, and operating SHM systems. A key requirement for a successful SHM program is that a clear idea of what the data will be used for in decision making is known upfront. Ensuring that the specified system can produce data of sufficient quality that will allow for decisions to be made will assist in strengthening the case for monitoring.

6.1 Future industry directions

SHM is becoming standard on large-span, newly constructed bridges. Additionally, SHM is often commissioned when a specific problem that requires investigation is identified on an existing bridge; e.g., the AE monitoring on the Hammersmith Flyover (Webb et al., 2014a). In the future, industry best-practice guides will need to explain how to specify, install, and manage the data, and, most importantly, how to use it to make engineering decisions.

6.2 Future research directions

Future research efforts in bridge monitoring should focus on better detecting and quantifying the mechanisms of deterioration or damage which still remain elusive for bridge engineers, such as corrosion. Attention should also be paid to better techniques for detecting fatigue (e.g., using nondestructive or intrusive methods) or reliable determination of the actual loading on the bridge (e.g., real-time load evaluation) and the ever-present "holy grail" measurement of the remaining loading capacity of bridges in situ, so that transport corridors can be more effectively utilized and managed.

References

Agrawal, A.K., et al., 2009. Remote corrosion monitoring systems for highway bridges. Prac. Per. Struc. Des. Con. 14 (4), 152–158. http://dx.doi.org/10.1061/(ASCE)1084-0680(2009) 14:4(152).

Andersen, J.E., Vesterinen, A., 2006. Structural Health Monitoring Systems. COWI Futuretec. from, http://www.shms.dk/COWI_ISBN-87-91044-04-9.pdf (accessed 16.03.15.).

Brownjohn, J.M.W., Lee, J., Cheong, B., 1999. Dynamic performance of a curved cable-stayed bridge. Eng. Struc. 21 (11), 1015–1027. http://dx.doi.org/10.1016/S0141-0296(98)00046-7.

Cantero, D., González, A., 2015. Bridge damage detection using weigh-in-motion technology. J. Bridge Eng. 20 (5), 04014078. http://dx.doi.org/10.1061/(ASCE)BE.1943-5592. 0000674.

Catbas, F.N., Kijewski-Correa, T., Aktan, A.E. (Eds.), 2013. Structural Identification of Constructed Systems: Approaches, Methods, and Technologies for Effective Practice of St-Id. American Society of Civil Engineers, Reston, Virginia, United States of America.

Chang, S.P., Im, C.K., 2000. Thermal behaviour of composite box-girder bridges. Proc. Inst. Civil Eng. Struct. Buil. 140 (2), 117–126. http://dx.doi.org/10.1680/stbu.2000.140.2.117.

DiBiagio, E., 2003. A case study of vibrating-wire sensors that have vibrated continuously for 27 years. In: Myrvoll, F. (Ed.), Field Measurements in Geomechanics, Proceedings of the 6th International Symposium. Oslo, Norway. September 23–26, 2003. Taylor & Francis, London, UK, pp. 445–458. http://dx.doi.org/10.1201/9781439833483.ch59.

Dunnicliff, J. (with the assistance of Green, G. E.), 1988. Geotechnical Instrumentation for Monitoring Field Performance. John Wiley & Sons, New York, United States of America.

Elsener, B., 2005. Long-term monitoring of electrically isolated post-tensioning tendons. Struct. Concrete. 6 (3), 101–106. http://dx.doi.org/10.1680/stco.2005.6.3.101.

Gastineau, A., Johnson, T., Schultz, A., 2009. Bridge Health Monitoring and Inspections – A Survey of Methods. Minnesota Department of Transportation. Report No. MN/RC 2009-29. from, http://www.lrrb.org/pdf/200929.pdf (accessed 16.03.15.).

Hoult, N.A., et al., 2008. Wireless structural health monitoring at the Humber Bridge UK. Proc. Inst. Civil Eng. Bridge Eng. 161 (4), 189–195. http://dx.doi.org/10.1680/ bren.2008.161.4.189.

Hoult, N.A., et al., 2009. Distributed fibre optic strain measurements for pervasive monitoring of civil infrastructure. In: Proceedings of 4th International Conference on Structural Health Monitoring of Intelligent Infrastructure (SHMII-4). July 22–24, 2009. Zurich, Switzerland.

Hoult, N.A., et al., 2010. Long-term wireless structural health monitoring of the Ferriby Road Bridge. J. Bridge Eng. 15 (2), 153–159. http://dx.doi.org/10.1061/(ASCE)BE.1943-5592.0000049.

Koo, K.Y., et al., 2013. Structural health monitoring of the Tamar suspension bridge. Struc. Cont. Health Moni. 20 (4), 609–625. http://dx.doi.org/10.1002/stc.1481.

Lee, C., Take, W.A., Hoult, N.A., 2012. Optimum accuracy of two-dimensional strain measurements using digital image correlation. J. Comp. Civ. Eng. 26 (6), 795–803. http://dx.doi. org/10.1061/(ASCE)CP.1943-5487.0000182.

Lynch, J.P., et al., 2006. Performance monitoring of the Geumdang Bridge using a dense network of the high-resolution wireless sensors. Smart Mat. Struc. 15 (6), 1561–1575. http://dx.doi.org/10.1088/0964-1726/15/6/008.

Matsumoto, Y., Yamaguchi, H., Yoshioka, T., 2010. A field investigation of vibration-based structural health monitoring in a steel truss bridge. In: Proceedings of the IABSE-JSCE Joint Conference on Advances in Bridge Engineering-II, August 8–10, 2010, Dhaka,

Bangladesh (Amin, Okui and Bhuiyan (eds.)) 461–467. http://iabse-bd.org/old/26.pdf (accessed 25.08.15).

Middleton, C., et al., 2014. Smart infrastructure—are we delivering on the promise? Paper presented at the 6th Australian Small Bridges Conference, Sydney, Australia, May 27–28, 2014. from, http://old.commstrat.com.au/smallbridges2014 (accessed 16.03.15.).

Moore, M., et al., 2001. Reliability of Visual Inspection for Highway Bridges, vol. 1 US Department of Transportation. Federal Highway Administration, Washington, DC. Report No. FHWA-RD-01-020. from, http://www.fhwa.dot.gov/publications/research/nde/01020.cfm (accessed 16.03.15.).

Nair, A., Cai, C.S., 2010. Acoustic emission monitoring of bridges: review and case studies. Eng. Struc. 32 (6), 1704–1714. http://dx.doi.org/10.1016/j.engstruct.2010.02.020.

Park, H.S., et al., 2007. A new approach for health monitoring of structures: terrestrial laser scanning. Comput. Aided Civ. Infrastruct. Eng. 22 (1), 19–30. http://dx.doi.org/10.1111/j.1467-8667.2006.00466.x.

Perenchio, W.F., 1989. The condition survey. Concrete Int. 11 (1), 59–62.

Psimoulis, P.A., Stiros, S.C., 2013. Measuring deflections of a short-span railway bridge using a robotic total station. J. Bridge Eng. 18 (2), 182–185. http://dx.doi.org/10.1061/(ASCE)BE.1943-5592.0000334.

Rodrigues, C., et al., 2010. Development of a long-term monitoring system based on FBG sensors applied to concrete bridges. Eng. Struc. 32 (8), 1993–2002. http://dx.doi.org/10.1016/j.engstruct.2010.02.033.

Stajano, F., et al., 2010. Smart bridges, smart tunnels: Transforming wireless sensor networks from research prototypes into robust engineering infrastructure. Ad Hoc Networks 8 (8), 872–888. http://dx.doi.org/10.1016/j.adhoc.2010.04.002.

Staquet, S., et al., 2007. Field testing of a 30-year-old composite Preflex railway bridge. Proc. Inst. Civil Eng. Bridge Eng. 160 (2), 89–98. http://dx.doi.org/10.1680/bren.2007.160.2.89.

Vann, A.M., et al., 1996. Intelligent logging strategies for civil engineering applications. Proc. Inst. Civil Eng. Struct. Build. 116 (2), 194–203. http://dx.doi.org/10.1680/istbu.1996.28287.

Washer, G., Fenwick, R., Bolleni, N., 2010. Effects of solar loading on infrared imaging of subsurface features in concrete. J. Bridge Eng. 15 (4), 384–390. http://dx.doi.org/10.1061/(ASCE)BE.1943-5592.0000117.

Watson, J., 2010. Watching brief. Bridge Des. Eng. 60 (Aug 2010).

Webb, G., 2014. Structural Health Monitoring of Bridges. Ph.D. thesis, University of Cambridge, Cambridge, UK.

Webb, G.T., et al., 2014a. Analysis of structural health monitoring data from Hammersmith Flyover. J. Bridge Eng. 19 (6), 05014003, http://dx.doi.org/10.1061/(ASCE)BE.1943-5592.0000587.

Webb, G.T., Vardanega, P.J., Middleton, C.R., 2014b. Categories of SHM deployments: technologies and capabilities. J. Bridge Eng. 04014118, http://dx.doi.org/10.1061/(ASCE)BE.1943-5592.0000735.

Wenzel, H., 2009. Health Monitoring of Bridges. John Wiley & Sons, Chichester, United Kingdom.

White, D.J., Take, W.A., Bolton, M.D., 2003. Soil deformation measurement using particle image velocimetry (PIV) and photogrammetry. Géotechnique 53 (7), 619–631. http://dx.doi.org/10.1680/geot.2003.53.7.619.

Wong, K.Y., 2004. Instrumentation and health monitoring of cable-supported bridges. Struc. Cont. Health Monit. 11 (2), 91–124. http://dx.doi.org/10.1002/stc.33.

Zhang, L., Haas, C., Tighe, S.L., 2007a. Evaluating weigh-in-motion sensing technology for traffic data collection. In: 2007 Annual Conference and Exhibition of the Transportation Association of Canada: Transportation—An Economic Enabler (Les Transports: Un Levier Economique), (October 14–17, 2007) Saskatoon Saskatchewan, Canada. Transportation Association of Canada. from, http://conf.tac-atc.ca/english/resourcecentre/readingroom/conference/conf2007/docs/s15/zhang.pdf (accessed 16.03.15.).

Zhang, Y.H., et al., 2007b. Evaluation of Crack-First TM fatigue sensor. In: CF/DRDC International Defence Applications of Materials meeting. from, http://www.twi-global.com/technical-knowledge/published-papers/evaluation-of-crackfirst-fatigue-sensors-june-2007 (accessed 31.03.15.).

Zhi, X., et al., 1999. Evaluation of weigh-in-motion in Manitoba. Canadian J. Civ. Eng. 26 (5), 655–666. http://dx.doi.org/10.1139/l99-025.

Application of fiber-reinforced polymers to reinforced concrete bridges

Hegemier G.[1], Stewart L.[2]
[1]Department of Structural Engineering, Jacobs School of Engineering University of California, San Diego, La Jolla, CA, US
[2]School of Civil and Environmental Engineering, Georgia Institute of Technology, Atlanta, GA, US

1 Introduction

It is well known that the ductility and strength of concrete in certain regions of stress space (e.g., where the principal stresses are negative, such as compression), can be increased (the former, substantially) by the application of confining pressure (e.g., as measured by the average of the principal stresses; Mander et al., 1988a; Saatcioglu and Razvi, 1992; Xiao and Wu, 2000). This concept is reinforced by Figure 30.1, which displays the ratio of confined concrete compressive strength, f_{cc}, to the unconfined compressive strength, f'_c, for both steel and concrete jackets used as confining systems. It is also understood that concrete exhibits increased volume expansion, or dilatancy, as a stress point approaches the failure surface in such regions. Figure 30.2 shows where the volumetric strain, ε_V, is calculated with respect to the axial strain, ε_c and the lateral strain, ε_l, as given in Eq. (1) (Bezant and Kim, 1979; Vermeer and DeBorst, 1984). This phenomenon allows the passive generation of the confining pressure necessary to attain the desired strength and ductility by the application of an appropriate confining "jacket" to a reinforced concrete (r.c.) column.

$$\varepsilon_V = \varepsilon_c + 2\varepsilon_l \tag{1}$$

2 Jacket materials and processes

The jacket concept has been successfully utilized for the past three decades for both seismic and other load environments (e.g., blast) via a variety of jacket materials. The most common and widely used of which include steel and FRPs, in the form of carbon, glass, or aramid fiber-reinforced polymers (respectively, CFRP, GFRP, and AFRP). In the case of existing bridge or building structures, the jacket takes the form of a retrofit. For new constructions, it may consist of a stay-in-place form.

Innovative Bridge Design Handbook. http://dx.doi.org/10.1016/B978-0-12-800058-8.00030-X

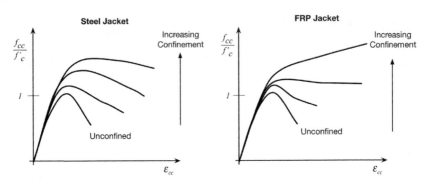

Figure 30.1 Influence of confinement on concrete response.

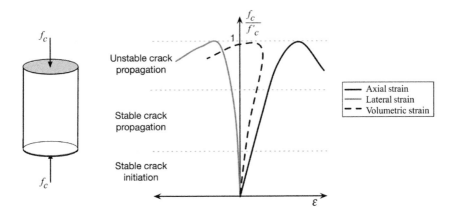

Figure 30.2 Concrete dilation under uniaxial compression.

The FRP jackets, for retrofit purposes, are typically fabricated in the field using dry uniaxial fabrics and plant-manufactured uniaxial strips. The fabrics are saturated with a two-part epoxy system at the job site using such tools as a small saturation machine (Bank, 2006; Mallick, 2007). Primer epoxy and other primer materials are placed to achieve appropriate surface preparation before the strips are applied with a technique similar to the application of wallpaper. The FRP/resin system is then subsequently cured under ambient conditions.

The resulting layer thickness is typically approximately 1–1.5 mm, and a jacket may consist of two to eight layers (or sometimes even more) and can be layered in both the axial (for flexural stiffness) and the hoop direction (for confinement) depending on the level of lateral force required. The strips are usually CFRP with high strength and stiffness, fabricated by pultrusion (or a related manufacturing method), and bonded to the column using an adhesive (Barbero, 1991). Typical strip geometries are in the 1-mm thickness and 5–10 cm width ranges, with lengths adjusted as necessary. The epoxy/adhesive chemistries for such field operations are adjusted for expected local temperature and humidity in an effort to obtain a proper cure (e.g., a desired glass transition temperature, T_g) and a sufficient resin workability time. Depicted in Figure 30.3a is an example jacket design for blast and seismic retrofit, based on the use of fabrics and strips (Hegemier et al., 2007).

In contrast to the fabrication of FRP jackets, steel jackets (Figure 30.3) are typically fabricated from two half shells or additional segments that are welded together at the site along two seams. Since the segments do not conform to the column surface, they are positioned with a small standoff from the column and a small gap is typically present. The gap is subsequently filled with a grout or similar material (Priestley et al., 1994; Ramirez et al., 1997).

For new construction, the FRP jacket can be applied as a stay-in-place form, which is plant-manufactured. This may be accomplished by *wet winding* or by methods such as vacuum assisted resin transfer molding (VARTM) (Rigas et al, 2001). In the former case, an epoxy resin system is typically used; in the latter, the resin system may be expanded to include vinylesters and polyesters depending upon the fiber system employed. In both methods, fabrication may include a low temperature post-cure.

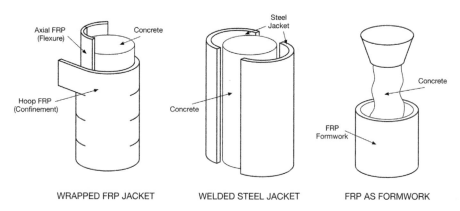

Figure 30.3 Illustration of FRP jacket techniques: FRP wrap, steel jacket, and FRP jacket as stay-in-place formwork.

3   Advantages of fiber-reinforced polymer systems (FRPS)

FRP systems have been used in practice over the course of the past three decades because under many circumstances, they can offer significant performance, economic, and aesthetic advantages over other materials (such as steel) for many applications.

With respect to performance, the anisotropic nature of FRPs allows the tailoring of jacket mechanical properties (Reddy 1987; Pipes and Pagano, 1970) to a given design objective. For example, with the use of FRPs rather than steel, which is isotropic, one can create confinement (and hence column ductility) without considerably increasing the overall bending stiffness of a column (which can attract additional load for certain events, such as seismic events). This can be accomplished by fabricating a jacket with reinforcing fibers only in the circumferential (hoop) direction. On the other hand, additional flexural strength can be obtained by adding longitudinal fibers to the jacket design using unidirectional fabric or premanufactured unidirectional FRP strips, if needed.

Other performance differences between steel and FRPs exist due to the inelastic, ductile behavior of steel. For example, since FRPs are essentially elastic to failure in direct tension along the fiber direction, the confining pressure associated with FRP jackets continues to increase with concrete deformation, whereas the confinement pressure essentially reaches a peak as a steel jacket enters the plastic state (as shown in Figure 30.4). This leads to eventual strain softening for steel, whereas FRP wraps of sufficient thickness and stiffness will exhibit strain hardening up to a failure.

In addition to short-term mechanical properties, FRPs such as CFRP offer considerable resistance to corrosion and advantageous long-term durability under a wide variety of environmental conditions. Durability data are available from experiments on various materials, which were conducted by Karbhari et al. (1996, 1997, 2003, 2007).

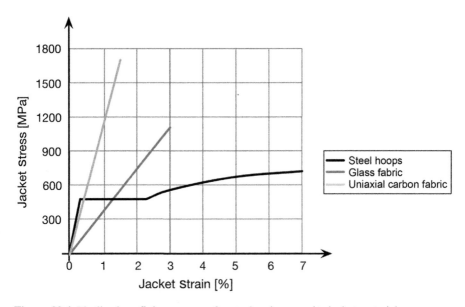

Figure 30.4 Idealized confining pressure for steel and composite jacket materials.

With respect to economics, the use of FRPs can lead to much shorter application times than those for steel. Thus, even with the use of carbon fiber, the reduced labor costs can, depending on location, result in a lower overall system cost. In addition, whereas a constant jacket thickness with a lower bound is typically used for steel jackets for to constructability reasons, the thickness distribution of an FRP jacket is easily optimized for minimum cost for a given performance objective.

Finally, with respect to aesthetics, typically the use of FRPs will not alter the basic architecture of a column since the jacket generally conforms to the original geometry and the jacket thickness required is usually quite small (Hegemier, 2007).

4   Performance—columns

In what follows, a number of data samples are presented in an effort to demonstrate the efficacy of the FRP jacketing concept for retrofit (section 4.1), repair (section 4.2), and new construction (section 4.3). In an effort to avoid scaling issues, the discussion is restricted to large or full-scale tests. The summary includes results from USCD Powell Laboratory tests conducted since the mid-1990s. Since then, there have been a multitude of experiments considered by researchers all over the world studying the effects of FRP jacketing and all its intricacies in great detail (e.g., Mirmiran et al., 1998; Parvin and Wang, 2001; Rochette and Labossiere, 2000; Xiao and Rui, 1997; Nanni and Norris, 1995; Monti et al., 2001; Pantazopoulou et al., 2001).

4.1   Laboratory tests—seismic retrofit

Experiments were developed and conducted by Seible et al. (1997a) for the retrofit of pre-1971-design rectangular and circular 40% scale bridge columns subject to combined axial and hysteretic lateral (simulated seismic) loads. The experiments studied three main seismic retrofits as they pertained to shear, plastic hinge development and lap-splice clamping. The shear retrofits consisted of wraps of three thicknesses, localizing the thickest regions towards the connection (i.e., the location with maximum shear). The flexural retrofits for the case with single bending also used three varying thicknesses, localized in the location of plastic hinge development. Finally, the lap-splice retrofit included the jacket only at the local connection over the lap. The results for the three series of tests are given in Figure 30.5, with drawings of the specimens tested and respective force-displacement curves. It can be seen that the application of confining pressures, both with CFRP and steel jackets, can lead to large increases in the ductility of the bridge column.

FRP jacket design criteria for various column failure modes were originally proposed by researchers at the University of California, San Diego (UCSD), and are described in Seible et al. (1995a, 1997b) and Inamorato et al. (1995, 1996), along with detailed examples of their applications to retrofits of columns with circular and rectangular geometries, different reinforcing ratios, and detailing. A concrete model (Mander et al., 1988b) was employed in the development of the key aspects of the initial UCSD design equations for the seismic retrofit of r.c. columns. Unfortunately, such models do not provide information concerning concrete dilation that, in turn, loads the FRP jacket. As a result, the portion of the jacket strain due to concrete dilation cannot be directly computed. In an effort to remedy this situation, UCSD researchers developed a concrete model that allows one to directly predict the concrete dilation strain; a detailed description of their work can be found in Lee (2006). The implications here are twofold: First, use of the new model, which has been subjected to extensive validation, allows one to directly predict jacket failure, which is a function of both dilation and shear. Second, dilation tends to reduce the shear capacity of the column section, especially in the plastic hinge region. As a result, if one does not account for the dilation contribution to the jacket tensile strains, the shear capacity can be overestimated.

The model, which gives a better estimate for concrete dilation strain, was utilized to study the effects of corner radius on the stress strain response under direct compression. The results from the study are discussed in detail in Lee (2006) and are summarized in Figure 30.6. Additional studies using various models conducted by other researchers can be found in Al-Salloum (2007) and Wang and Wu (2008).

It should be noted that codes and design guidelines for various agencies (i.e., ACI, AASHTO) have a framework for these systems, as well as for the mechanisms under discussion (i.e., shear, flexure, and splicing). These are discussed briefly in section 6 of this chapter. The designer should use engineering judgment with regard to the selection of analysis techniques and code requirements for implementation.

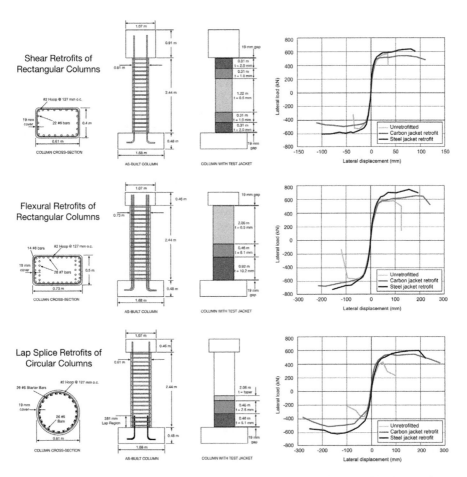

Figure 30.5 Experimental results from shear: (a) flexure, (b) splicing, and (c) bridge column retrofits.
(data from Seible et al., 1997a).

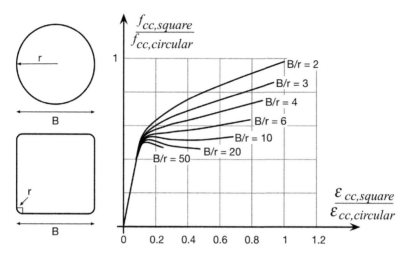

Figure 30.6 Effect of corner radius on stress-strain response.
(reproduced from data from Lee, 2006).

4.2 Laboratory tests—seismic repair

The previous example concerned the FRP retrofits of undamaged specimens. The example discussed in this section illustrates the efficacy of the FRP jacket concept as a repair measure for bridge columns. The tests described in this section were conducted by Ohtaki, Benzoni, and Priestley at UCSD. Additional details regarding these experiments are given in Ohtaki et al. (1996). This research highlights the use of FRP for repair, while also demonstrating the effectiveness of the procedure for large diameter columns.

The test setup for these experiments is shown in Figure 35.7. For this purpose, a shear-dominated bridge column with a 1.83-m diameter and an aspect ratio of 2.0 is considered with a pre-1971 design. In this case, the jacket was 9.8 mm of uniaxial GFRP, which was applied as hoop reinforcement via a wet layup of a glass fabric saturated with an epoxy resin system. The GFRP was applied after cement grout injection of the damaged specimen. Comparisons of the lateral force (no axial load was applied) versus displacement envelopes of the "as-built" and repaired specimens are provided in Figure 30.7. As can be seen, the repaired specimen exhibits a large ductility improvement over the "as-built" version with a displacement ductility up to $\mu = 6$ with no strength degradation.

4.3 Laboratory tests—new construction

For the application of FRP to new construction, the jacket is typically used as a stay-in-place formwork and consists of an FRP shell manufactured in a plant. These jackets can be circular or noncircular and are filled with concrete on the construction site. The concrete provides the compression force transfer and the shell serves as the formwork

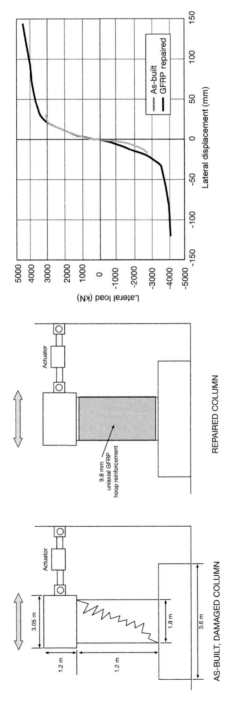

Figure 30.7 Experimental results from as-built and repaired reinforced concrete column. (reproduced from data from Ohtaki et al., 1996).

for the concrete, the reinforcement for the tension force transfer in bending and shear, and the confinement of the concrete core. The shell can be fabricated with internal circumferential ribs, which provides a mechanism for the transference of tensile forces and acts as a mechanical interlock between the concrete and the shell. With the use of this system, a steel reinforcement cage is often not necessary except for starter bars that provide load transfer from the concrete-filled shell to termini elements, such as a footing. A variation of this theme employs the shell for confinement purposes only, in which case the ribs are not required (Hegemier, 2007).

A series of laboratory tests were conducted at UCSD on the *composite shell system (CSS)*, with internal ribs as discussed previously and shown in Figure 30.8. In these experiments, the CSS was used for a column with a 610-mm diameter, a 3.43-m length, and a 9.52-mm jacket. The CFRP jacket was fabricated by the wet-filament winding method [see Burgueno et al.,1995 and Seible et al., 1995b for system material properties; and Fitzer and Terwiesch (1972) for information on the wet-filament winding method]. As can be observed in Figure 30.8, although no primary steel reinforcement cage exists, the use of mild reinforcement splice bars in the concrete core across the column-footing joint and the confinement of the concrete by the shell result in a ductile response under hysteretic lateral loading with no strength degradation up to $\mu = 8$ (Bergueno et al., 1995). For comparison, the response of a conventional reinforced concrete column (seismic design) is also shown.

5   Performance—superstructure

In what follows, two concept designs are presented in an effort to demonstrate the feasibility and response of FRP use in hybrid bridge systems as part of the bridge decking. The summary includes results from USCD Powell Laboratory tests conducted by UCSD researchers. Since that time, the use of composites for bridge deck applications has extended into alternative uses and applications, such as retrofits of steel structures, bonded/bolted sandwich panels, and innovative prestressing systems, as discussed in Shaat et al. (2004), Mosallam et al. (2015), and Ghafoori and Moravalli (2015).

5.1   Laboratory experiments—CSS for short- and medium-span bridges

The CSS technology discussed in section 4.3 has also been applied to the use in bridge systems as part of the bridge superstructure. The cable stayed-bridge, which was conceptually designed and experimentally tested for proof of concept by Davol (1998) and Seible et al., included a 137-m-long deck supported by a 59-m-high A-frame pylon, utilizing concrete-filled CSS tubes (Seible et al., 1997b). The structural design concept, which is shown in schematic form in Figure 30.9, of the bridge consisted of an FRP panel stiffened, steel-free deck system and was supported on, and composite with, transverse CSS crossbeams. These crossbeams were spaced 2.4 m from the center and supported on the longitudinal CSS edge girders. The CSS edge girders are

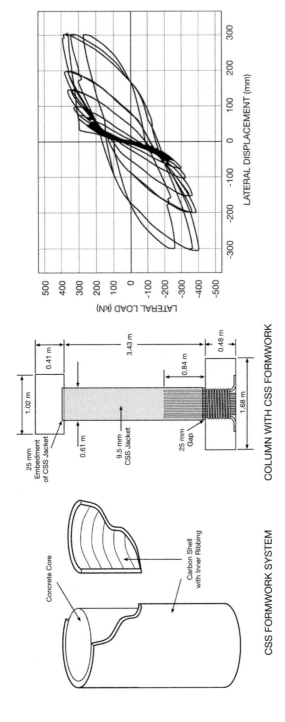

Figure 30.8 Experimental results from a column with CSS jacket. (reproduced from data from Bergueno et al., 1995).

concrete filled and held up at 4.9-m intervals by the cable stays. These CSS sections were experimentally tested in the lab for flexure using a four-point bend setup, as shown in Figure 30.10. Results from these tests, including extensive strain gauge data from the FRP CSS section, are available in Davol (1998).

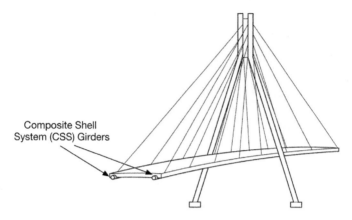

Figure 30.9 CSS section as girder of cable-stayed bridge system. (reproduced from the concept described in Zhou, 2001).

5.2 Laboratory experiments—FRP for rapid rehabilitation

Composite systems have also been tested for use in rapid rehabilitation and construction of bridge decks. Experiments were conducted at UCSD by Pridmore (2008) to validate the use of FRP composite panels both for stay-in-place formwork and as the bottom longitudinal and transverse reinforcement in the deck of concrete box girder bridges. Performance assessments for full-scale, two-cell box girder bridge specimens, through monotonic and extensive cyclic loading, provided evidence that the FRP panel system bridge deck was a viable rehabilitation solution for box girder bridge decks. The experiments showed that the FRP panel system performed comparably to a conventionally reinforced concrete bridge deck in terms of serviceability, deflection profiles, and system-level structural interaction, and performed superior to the r.c. bridge deck in terms of residual deflections and structural response under cyclic loading (Pridmore, 2008).

Furthermore, this research utilized results from the development and characterization of a modular bridge system incorporating FRP composite girders connected by stiffened FRP deck panels which serve as both formwork and flexural reinforcement for a steel-free concrete deck cast on top. These systems were tested by Cheng and Karbhari (2004). The experimental results showed that capacity of the modular FRP hybrid system is substantially greater than the design demand levels. The researchers also developed analytical predictions, based on laminated beam theory using progressive ply failure criteria, moment-curvature analysis, and finite element models; also discussed in Cheng et al (2005).

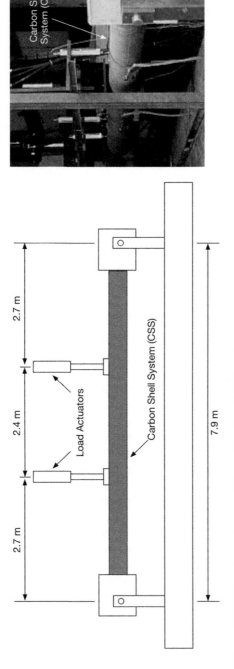

Figure 30.10 Experimental setup of flexural testing of CSS section. (reproduced from Davol, 1998).

6 Design guides and codes

In the preceding sections, design applications using FRP technology was discussed in the context of experimental programs conducted for feasibility studies and tool development. Such analysis and design tools have contributed, along with many other experiments conducted by researchers outside of UCSD, to the development of application-specific design codes and recommended guidelines. Such codes are often specific to location and purpose; thus, the designer should choose the appropriate tool given their context. What follows is a brief summary of available US codes and guidelines to date in the areas discussed in this chapter, as well as codes and recommendation guidelines from various locations and governances.

6.1 Bridge strengthening and repair with FRP

In the United States in 2010, the Transportation Research Board (TRB) published the National Cooperative Highway Research Program (NCHRP) Report 655 (TRB, 2010), which presents a recommended guide specification for the design of externally bonded FRP systems for the repair and strengthening of concrete bridge elements. This guide specification provides a review of current practices from various countries such as the Canadian ISIS Design Manual (2001), the Italian CNR-DT 200 (2006), and those from the Japanese Society of Civil Engineers (2001), among others (e.g., French Association of Civil Engineers, 2003; Freyssinet, 2001; German Provisional, 2003; Gorski and Krzywon, 2007; Caltrans, 2007; Zureick, 2002; TRB NCHRP, 2004; TRB, 2008). The report addresses the design requirements for members subjected to different loading conditions (e.g., flexure, shear and torsion, and combined axial force and flexure). The report is presented in the load and resistance factor design (LRFD) method for complementary use with the American Association of State Highway Transportation Officials (AASHTO) LRFD bridge design specifications (2015).

In the case of pedestrian bridges, which primarily serve human and bicycle traffic, the use of FRP is covered by the AASHTO Guide Specifications for Design of FRP Pedestrian Bridges (AASHTO, 2008), which should be used in conjunction with the Guide Specifications for Design of Pedestrian Bridges (AASHTO, 2009a).

6.2 FRP tubes as stay-in-place formwork

Many researchers have studied the use of FRP tubes as a formwork for concrete construction (e.g., Deskovik et al., 1995; Hall and Mottram, 1998; Hollaway and Head, 2001). This technique is covered in the AASHTO LRFD Guide Specifications for Design of Concrete-Filled FRP Tubes (2012). The specifications present provisions for the analysis and design of concrete-filled fiber-reinforced polymer tubes (CFFTs) for use as structural components (i.e., beams, arches, columns, and piles) in bridges subjected to flexure, axial compression, or combined loading.

6.3   Design of FRP bridge decks

In 2009, AASHTO developed a specification, the LRFD Bridge Design Guide Specifications for GFRP-Reinforced Concrete Bridge Decks and Traffic Railings (AASHTO, 2009b), for glass fiber-reinforced polymer (GRFP) use in the application of design and construction of concrete bridge decks and railings, in which the GFRP are used as reinforcing bars.

7   Other loading applications

In addition to seismic loading, reinforced concrete columns retrofitted with FRP jackets have also proved to be an effective system for lateral loading induced by explosive events. The confining pressure created by the jackets, as with seismic loading, aids in mitigating the effects of the blast loading. Experiments using CFRP retrofits for blast response have been conducted by Morrill et al. (2000; 2004), Crawford et al. (1996; 1997), Muszynski and Purcell (2003), and Winget et al. (2005). UCSD also demonstrated the mitigating nature of the column jackets using their blast simulator (Gram et al. 2006; Stewart et al. 2014) with experiments conducted by Rodriguez-Nikl, et al (2001) which showed that reduced column displacements can be achieved with as few as two layers of CFRP. Test results of the retrofitted concrete columns subjected to blast loading can be found in (Rodriguez-Nikl, 2006). Recommendations of concrete columns subjected to blast loadings is covered in the recently published ACI 370R-14 Report for the Design of Concrete Structures for Blast Effects (ACI, 2014). This report is based on the use of single degree-of-freedom (SDOF) systems as the main analysis technique to develop response. The designer should take care to use these methods where applicable, based on standoff, charge size, and other relevant parameters.

Similarly, FRP applications have also been shown to be an effective strategy for the retrofitting of reinforced concrete bridge decks or for new construction techniques. Field tests showing the response of decks or concrete slabs retrofitted with FRP to blast loading can be found in testing conducted by Seible et al. (2008), Buchan and Chen (2007), Millard et al. (2010), Wu et al. (2009), Coughlin et al. (2010), Schenker et al. (2008), and Silva and Binggeng (2007).

8   Conclusions

This chapter provided a summary of applications of FRP to bridge systems, as demonstrated through experimental testing conducted at the UCSD Powell Laboratories over the past three decades. The laboratory tests confirmed the effective use of FRP with reinforced concrete columns for retrofit, repair, and new constructions. Additionally, FRP systems were shown to be effective when used in various bridge deck system designs. These experiments, combined with investigations and implementation of these systems from many other researchers, have motivated the development of design codes and guidelines, many of which are actively used by bridge engineers all over the world.

References

Al-Salloum, Y.A., 2007. Influence of edge sharpness on the strength of square concrete columns confined with FRP composite laminates. Comps. Part B: Eng. 38 (5), 640–650.

American Association of State Highway and Transportation Officials (AASHTO), 2008. Guide Specifications for Design of FRP Pedestrian Bridges. Washington, DC.

American Association of State Highway and Transportation Officials (AASHTO), 2009a. LRFD Guide Specifications Pedestrian Bridges, second ed. Washington, DC.

American Association of State Highway and Transportation Officials (AASHTO), 2009b. Guide Specifications for GFRP-Reinforced Concrete Bridge Decks and Traffic Railings. Washington, DC.

American Association of State Highway and Transportation Officials (AASHTO), 2012. LRFD Guide Specifications for Design of Concrete-Filled FRP Tubes, first ed. Washington, DC.

American Association of State Highway and Transportation Officials (AASHTO), 2015. LRFD Bridge Design Specifications, seventh ed. Washington, DC.

American Concrete Institute (ACI), 2014. 370R-14 Report for the Design of Concrete Structures for Blast Effects. Farmington Hills, MI.

Bank, L.C., 2006. FRP Flexural Reinforcement, in Composites for Construction: Structural Design with FRP Materials. John Wiley & Sons, Inc., Hoboken, NJ.

Barbero, E., 1991. Pultruded structural shapes - from the constituents to the structural behavior. SAMPLE J. 27 (1), 25–30.

Bezant, Z., Kim, S., 1979. Plastic-Fracturing theory for concrete. J. Eng. Mech. Div. 105 (3), 407–428.

Buchan, P.A., Chen, J.F., 2007. Blast resistance of FRP composites and polymer strengthened concrete and masonry structures–A state-of-the-art review. Comps. Part B: Eng. 38 (5), 509–522.

Burgueño, R., Seible, F., Hegemier, G.A., 1995. Report No. ACTT-95/12, Concrete Filled Carbon Shell Bridge Piers Under Simulated Seismic Loads–Experimented Studies. University of California, San Diego. La Jolla, CA.

Caltrans, 2007. Bridge Memo to Designers-MTD 20–4: Seismic Retrofit Guidelines for Bridges in California. Technical Publication, Sacramento, CA.

Canada Design Manuals, I.S.I.S., 2001. Strengthening Rein- forced Concrete Structures with Externally-Bonded Fiber- Reinforced Polymers. Winnipeg, Manitoba, Canada.

Cheng, L., Karbhari, V.M., 2004. Development of FRP Composite Modular System for Slab-on-Girder Bridges.Proceedings of the Fourth International Conference on Advanced Composite Materials in Bridges and Structures. Calgary, Alberta. July 20–23.

Cheng, L.J., et al., 2005. Assessment of a steel-free fiber reinforced polymer-composite modular bridge system. J. Struc. Eng. 131 (3), 498–506.

CNR-DT 200, 2006. Guide for the Design and Construction of Externally Bonded FRP Systems for Strengthening Existing Structures. Italian Advisory Committee on Technical Recommendations for Construction, Rome.

Coughlin, A.M., et al., 2010. Behavior of portable fiber reinforced concrete vehicle barriers subject to blasts from contact charges. Int. J. of Imp. Eng. 37 (5), 521–529.

Crawford, J.E., et al., 1996. Retrofit of reinforced concrete columns using composite wraps to resist blast effects. Karagozian and Case, Glendale, CA.

Crawford, J.E., et al., 1997. Retrofit of reinforced concrete structures to resist blast effects. ACI Struc. J. 94 (4), 371–377.

Davol, A., 1998. Ph.D dissertation. Structural Characterization of Concrete Filled Fiber Reinforced Shells. University of California, San Diego. La Jolla, CA.

Deskovic, N., Triantafillou, T., Meier, U., 1995. Innovative design of FRP combined with concrete: short-term behavior. J. Struct. Eng. 121 (7), 1069–1078.

Fitzer, E., Terwiesch, B., 1972. Carbon—carbon composites unidirectionally reinforced with carbon and graphite fibers. Carbon. 10 (4), 383–390.

French Association of Civil Engineers, 2003. Réparation et renforcement des structures en béton au moyen des matériaux composites. AFGC, Paris, FR.

GDOT Specification: Proposed Specifications-Polymeric Composite Materials for Rehabilitating Concrete Structures (Zureick 2002).

German Provisional Regulations, 2003. Nr. Z-36.12–65 vom 29, Allgemeine Bauaufsichtliche Zulassung. Deutsches Institut für Bautechnik, Berlin, DE.

Ghafoori, E., Motavalli, M., 2015. Innovative CFRP-Prestressing System for Strengthening Metallic Structures. J. Comp. Constr. http://dx.doi.org/10.1061/(ASCE)CC.1943-5614.0000559.

Gorski, M., Kryzywon, R., 2007. Polish Standardization Proposal for Design Procedures of FRP Strengthening. In: Triantafillou, T. (Ed.), Proceedings, CD, FRPRCS-8. University of Patras, Patras, Greece.

Gram, M.M., et al., 2006. Laboratory simulation of blast loading on building and bridge structures. Struc. Under Shock and Imp. IX. 87, 33–44.

Hall, J., Mottram, J., 1998. Combined FRP reinforcement and permanent formwork for concrete members. J. Comp. Const. 2 (78), 78–86.

Hegemier, et al., 2007. The Use of Fiber-Reinforced Polymers to Mitigate Natural and Manmade Hazards. FRPCS-8. Patras, Greece, pp. 1–24.

Hollaway, L., Head, P., 2001. Advanced Polymer Composites and Polymers in the Civil Infrastructure. Adv. Polymer Comp. Polymers Civil Infrastruct. Elsevier, Oxford, UK.

Inmamorato, D., et al., 1995. Report No. ACTT-95/19. Carbon Shell Jacket Retrofit Test of a Circular Shear Column with 2.5% Reinforcement. University of California, San Diego. La Jolla, CA November.

Inmamorato, D., et al., 1996. Report No. ACTT-96/10. Full Scale Test of Two Column Bridge Bent with Carbon Jacket Retrofit. University of California, San Diego. La Jolla, CA. August.

Japan Society of Civil Engineers (JSCE), 2001. Recommendations for Upgrading of Concrete Structures with Use of Continuous Fiber Sheets. Tokyo, Japan.

Karbhari, V.M. (Ed.), 2007. Durability of composites for civil structural applications. Elsevier, Woodhead Publishing, Sawston, Cambridge.

Karbhari, V.M., Engineer, M., 1996. Effect of environmental exposure on the external strengthening of concrete with composites-short term bond durability. J. Reinf. Plastics Comps. 15 (2), 1194–1216.

Karbhari, V.M., Engineer, M., Eckell II, D.A., 1997. On the durability of composite rehabilitation schemes for concrete: use of a peel test. J. Mats. Sci. 32 (1), 147–156.

Karbhari, V.M., et al., 2003. Durability gap analysis for fiber-reinforced polymer composites in civil infrastructure. J. Comps. Constr. 7 (3), 238–247.

Lee, C.S., 2006. Ph.D dissertation, Modeling of FRP-Jacketed RC Columns Subject to Combined Axial and Lateral Loads. University of California, San Diego. La Jolla, CA. June.

Mallick, P.K., 2007. Fiber-Reinforced Composites Materials, Manufacturing, and Design, Third Edition. CRC Press, Boca Raton.

Mander, J., Priestley, M., Park, R., 1988a. Observed stress-strain behavior of confined concrete. J. Struct. Eng. 114 (8), 1827–1849.

Mander, J.B., Priestly, M.J.N., Park, R., 1988b. Theoretical stress-strain model for confined concrete. J. Struc. Eng. 114 (8), 1804–1826.

Millard, S.G., et al., 2010. Dynamic enhancement of blast-resistant ultra high performance fibre-reinforced concrete under flexural and shear loading. Int. J. of Impact Eng. 37 (4), 405–413.

Mirmiran, A., et al., 1998. Effect of column parameters on FRP-confined concrete. J. Comp. Constr. 2 (4), 175–185.

Monti, G., Nisticò, N., Santini, S., 2001. Design of FRP jackets for upgrade of circular bridge piers. J. Comp. Constr. 5 (2), 94–101.

Morrill, K.B., et al., 2000. RC column and slab retrofits to survive blast loads. Advanced Technology in Structural Engineering, Structures Congress 2000, Philadelphia, Pennsylvania, United States, pp. 1–8.

Morrill, K.B., et al., 2004. Blast resistant design and retrofit of reinforced concrete columns and walls. Proceedings of Structures Congress-Building on the Past-Securing the Future. Nashville, Tennessee, United States.

Mosallam, A., et al., 2015. Structural evaluation of reinforced concrete beams strengthened with innovative bolted/bonded advanced FRP composite sandwich panels. Comp. Struct. 124, 421–440.

Muszynski, L.C., Purcell, M.R., 2003. Composite reinforcement to strengthen existing concrete structures against air blast. J. Comp. Constr. 7 (2), 93–97.

Nanni, A., Norris, M.S., 1995. FRP jacketed concrete under flexure and combined flexure-compression. Constr. Build. Mats. 9 (5), 273–281.

Ohtaki, T., Benzoni, G., Priestly, M.J.N., 1996. Report No. SSRP-96/07, Seismic Performance of a Full-Scale Bridge Column–As Built and Repaired. University of California, San Diego. La Jolla, California.

Pantazopoulou, S.J., et al., 2001. Repair of corrosion-damaged columns with FRP wraps. J. Comp. Constr. 5 (1), 3–11.

Parvin, A., Wang, W., 2001. Behavior of FRP jacketed concrete columns under eccentric loading. J. Comp. Constr. 5 (3), 146–152.

Pipes, R.B., Pagano, N.J., 1970. Interlaminar stresses in composite laminates under uniform axial extension. J. Comp. Mats. 4 (4), 538–548.

Pridmore, A., 2008. Ph.D dissertation, Structural Characterization of Concrete Filled Fiber Reinforced Shells. University of California, San Diego. La Jolla, CA.

Priestley, M.J.N., et al., 1994. Steel jacket retrofitting of reinforced concrete bridge columns for enhanced shear strength—Part 1: theoretical considerations and test design. ACI Struct. J. 9 (4), 394–405.

Ramirez, J.L., et al., 1997. Efficiency of short steel jackets for strengthening square section concrete columns. Construc. and Build. Mats. 11 (5), 345–352.

Reddy, J.N., 1987. A generalization of two-dimensional theories of laminated composite plates. Comms. in App. Num. Methods. 3 (3), 173–180.

Rigas, E.J., Walsh, S.M., Spurgeon, W., 2001. Development of a novel processing technique for vacuum assisted resin transfer molding (VARTM). 46th International SAMPE Symposium and Exhibition. CRC Press, Long Beach, California.

Rochette, Pi, Labossiere, P., 2000. Axial testing of rectangular column models confined with composites. J. Comp. Constr. 4 (3), 129–136.

Rodríguez-Nikl, T., 2006. Experimental simulations of explosive loading on structural components: reinforced concrete columns with advanced composite jackets. PhD Dissertation, University of California, San Diego. La Jolla, CA.

Rodriguez-Nikl, T., et al., 2011. Experimental performance of concrete columns with composite jackets under blast loading. J. Struc, Eng 138 (1), 81–89.

Saatcioglu, M., Razvi, S., 1992. Strength and Ductility of Confined Concrete. J. Struct. Eng. 118 (6), 1590–1607.

Schenker, A., et al., 2008. Full-scale field tests of concrete slabs subjected to blast loads. Int. J. Imp. Eng. 35 (3), 184–198.

Seible, F., et al., 1995a. Advanced Composite Carbon Shell Systems for Bridge Columns Under Seismic Load. Report No. ACTT/BIR-95/27. Dept. of Structural Engineering. University of California, San Diego. La Jolla, CA. September 1995.

Seible, F., et al., 1995b. Report No. ACTT-95/03, Rectangular Carbon Jacket Retrofit of Flexural Column with 5% Continuous Reinforcement. University of California, San Diego. La Jolla, California. April 1995.

Seible, F., et al., 1997a. Seismic Retrofit of RC Columns with Continuous Carbon Fiber Jackets. J. Comp. Constr. 1 (2), 52–62.

Seible, F., et al., 1997b. The Carbon Shell System for Modular Short and Medium Span Bridges. Proceedings of International Composite Expo. Nashville, TN. January. pp 1–6.

Seible, F., et al., 2008. Protection of our bridge infrastructure against manmade and natural hazards. Struc. Infrastr. Eng. 4 (6), 415–429.

Shaat, A., et al., 2004. Retrofit of steel structures using fiber-reinforced polymers (FRP): State-of-the-art. Transportation Research Board (TRB) annual meeting. Washington, DC, USA.

Silva, P.F., Binggeng, L., 2007. Improving the blast resistance capacity of RC slabs with innovative composite materials. Composites Part B: Eng 38 (5), 523–534.

Stewart, L.K., et al., 2014. Methodology and validation for blast and shock testing of structures using high-speed hydraulic actuators. Eng. Struc. 70 (1), 168–180.

Transportation Research Board of the National Academies, 2004. NCHRP Report 655: Bonded Repair and Retrofit of Concrete Structures Using FRP Composites—Recommended Construction Specifications and Process Control Manual. Washington, DC.

Transportation Research Board of the National Academies, 2008. NCHRP Report 609: Recommended Construction Specifications and Process Control Manual for Repair and Retrofit of Concrete Structures Using Bonded FRP Composites. Washington, DC.

Transportation Research Board of the National Academies, 2010. NCHRP Report 655: Recommended Guide Specification for the Design of Externally Bonded FRP Systems for Repair and Strengthening of Concrete Bridge Elements. Washington, DC.

Vermeer, P., DeBorst, R., 1984. Non-associated plasticity for soils, concrete, and rock. Heron. 29 (3), 1–64.

Wang, L.M., Wu, Y.F., 2008. Effect of corner radius on the performance of CFRP-confined square concrete columns: Test. Eng. Struc. 30 (2), 493–505.

Winget, D.G., Marchand, K.A., Williamson, E.B., 2005. Analysis and design of critical bridges subjected to blast loads. J. Struc. Eng. 131 (8), 1243–1255.

Wu, C., et al., 2009. Blast testing of ultra-high performance fibre and FRP-retrofitted concrete slabs. Eng. Struc. 31 (9), 2060–2069.

Xiao, Y., Rui, M., 1997. Seismic retrofit of RC circular columns using prefabricated composite jacketing. J. Struc. Eng. 123 (10), 1357–1364.

Xiao, Y., Wu, H., 2000. Compressive behavior of concrete confined by carbon fiber composite jackets. J. Mater. Civ. Eng. 12 (2), 139–146.

Zhao, L., 2001. Design and Evaluation of Modular Bridge Systems Using FRP Composite Materials. In: Proceedings of the Fifth International Conference on Fiber-Reinforced Plastics for Reinforced Concrete Structures. Cambridge, UK. July 16–18.

Bridge collapse

31

Schultz A.E.[1], Gastineau A.J.[2]
[1]Professor, Department of Civil, Environmental, and Geo-Engineering, University of Minnesota, Minneapolis, US
[2]Design Engineer, KPFF Consulting Engineers, Seattle, WA

Figure 31.1 Tacoma Narrows Bridge.
(*Library of Congress, Prints and Photographs Division, LC-USZ62-46682*).

1 Introduction

Historically, bridge collapses have been caused by a variety of factors or from a combination of those factors. These include poor engineering judgment, use of substandard materials, extreme loading, and inadequate maintenance. In all cases, much can be learned from these failures. Most bridge collapses have been closely scrutinized, and reasons for the collapse are generally agreed upon. Entire books have been dedicated to significant collapses, and yet these barely cover the vast number of historic bridge failures. This chapter attempts to present some major bridge failures and their consequences to highlight broad categories of the causes of bridge collapses. Åkesson (2008) described a variety of collapses but identified five key bridge collapses that have changed the way in which engineers understand bridges. The key collapses that Åkesson identified were the Dee Bridge, the Tay Bridge, the Quebec Bridge, the Tacoma Narrows Bridge (Figure 31.1), and a series of box-girder bridge failures from 1969–1971. In addition to the collapses highlighted by Åkesson, others have documented bridge failures that led to major design changes. This chapter covers bridge collapses in North America, as well as significant bridge failures worldwide. Not all of the bridges described in this chapter underwent a complete collapse because a bridge failure is defined as an event in which a bridge does not perform to meet

Innovative Bridge Design Handbook. http://dx.doi.org/10.1016/B978-0-12-800058-8.00031-1

design goals and cannot be safely operated. The bridge failures highlighted in this chapter are classified into three main categories: construction failures, in-service failures, and extreme events. Each of these broad categories provides insight and caution that should be incorporated into bridge design, construction, and maintenance.

2 Construction failures

From their invention, bridges have been essential for transportation, trade, communication, and defense, and their construction has posed many challenges. While unintended and completely undesirable, bridge collapses and failures have often served as an indicator of what is impossible in bridge design and construction. In modern times, a variety of failures have demonstrated that the engineering criteria and processes that go into the construction sequence are just as important as the design for service conditions of a bridge. Without thoughtful construction analysis, a well-designed bridge may never be successfully placed into service.

During construction of the Quebec Bridge, a cantilever steel truss structure, in 1907, a compression chord was found to have distorted out of plane, and the designer ordered construction halted. However, despite this, the contractor was falling behind schedule and continued construction anyway. Buckling failures resulted in a complete collapse of the bridge (Figure 31.2), killing 75 workers. Multiple reasons led to this collapse; first, the bridge had been designed using higher working stresses, and second, the designers underestimated the self-weight of the steel. The large allowable stresses caused the buckling of a compression member, which led to further overstressing of additional members, causing complete collapse ("Quebec Bridge Disaster," 1908; Biezma and Schanack, 2007; Åkesson, 2008). A new bridge was planned and erected using compression chords with almost twice the cross-sectional area to avoid buckling; however, the bridge partially collapsed again in 1916, killing an additional 13 workers. The second collapse was blamed on a weak connection detail, which was redesigned, and the bridge was finally completed in 1917. These collapses highlighted the need for not only economical, but also safe designs. Increasing working stresses without proper testing and safety assessment can lead to devastating consequences. These experiences also highlighted the importance of connections, as well as the need to carefully evaluate design changes made during construction.

Figure 31.2 Quebec Bridge Collapse.
(*Manitoba Free Press*).

A series of steel box-girder bridge failures occurred in the late 1960s and early 1970s with the majority of failures occurring during the erection stage of construction (Biezma and Schanack, 2007; Subramanian, 2008; Åkesson, 2008). By using the cantilever construction method during erection, high moment regions at the supports produced a buckling failure in Austria's Fourth Danube Bridge. As the final box-girder piece was placed to close the gap between the two cantilevers, the piece had to be shortened on the top due to the sag of the cantilevered segments, for which the design had not accounted. The additional sag had been caused by expansion of the bridge deck due to a full day of sun exposure. The bridge was designed as a continuous span so that the inner-pier supports needed to be lowered to produce the correct stress distribution; however, this activity had been moved to the next day. As the bridge cooled that evening, tension was introduced in the shortened region and compression in the bottom flange. Areas designed to be in tension for in-service loads were instead loaded in compression, causing buckling failures of the bottom flanges of the box section (see Figure 31.3). Four other box girder failures occurred in the next four years, all of which had buckling issues

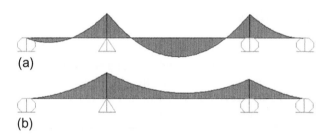

Figure 31.3 Fourth Danube Bridge moment diagrams; (a) in-service design moments; (b) temporary condition due to construction method and temperature loading.

during erection (one was kept secret for more than 20 years due to the controversy). Because of the large number of collapses in a small period of time, it was clear that erection loads and practices needed to be included in the design process and that local buckling problems were not well understood.

While the River Verde Viaduct was being erected in Spain, the moveable scaffolding system collapsed (Figure 31.4), killing six workers (Tanner and Hingorani, 2013). While this event was not technically a bridge collapse, these systems are critical for building long-span bridges. The scaffolding was a movable system, and, while being moved, there was a power failure because of damage to an electrical cable. This allowed excessive movement of the scaffolding which caused a connection failure, after which collapse ensued. This failure highlights the fact that, in addition to the scaffolding having the required strength and stiffness, it is necessary for any peripheral electrical and mechanical systems to be designed adequately and that fail-safe systems are necessary in case any one of these components were to fail during construction. Without safe and robust designs, bridges cannot be safely constructed and public opinion of the engineering profession can be undermined.

Figure 31.4 River Verde Viaduct Scaffolding Collapse.
(*Image Courtesy Fred Nederlof*).

3 In-service failures

Bridges are designed to safely carry trains, pedestrians, or vehicles over obstacles. Due to the complex nature of both loads and structural behavior, simplifying assumptions must be made during design. However, these assumptions may not fully represent either the loading or physical behavior of the bridge, and sometimes the flaws from oversimplification can lead to catastrophic consequences. This section discusses in-service failures in four categories: design flaws, material inadequacy, overloading, and maintenance issues.

3.1 Design flaws

As societies progress, new technology is continuously developed and engineers are constantly trying to innovate and improve design effectiveness. Untested materials and unique designs can lead to unexpected loads and poorly understood static or dynamic behavior. The Tay Bridge was built in 1878 to cross the Firth of Tay in Scotland. The bridge was the longest train bridge in the world at the time and consisted of wrought-iron trusses and girders supported by trussed towers. In 1879, while a mail train was crossing it at night during a storm with high winds, the bridge collapsed, killing 75 people (Figure 31.5), and 13 of the tallest spans, having higher clearances to allow for ship passage beneath, collapsed. It was determined that wind loading had not been included in the design of the bridge. The open-truss latticework was assumed to allow the wind to pass through; however, it was not considered that, when loaded with a train, the surface area of the train would transfer wind loading to the structure. During the gale, the extremely top-heavy portion of the bridge, upon which the train rode, acted like a mass at the end of a cantilever. The narrow piers could not withstand the lateral thrust and collapsed into the water (Biezma and Schanack, 2007; Åkesson, 2008). In addition, defective joints led to fatigue cracking, which contributed to the

Figure 31.5 Tay Bridge after collapse.
(*National Library of Scotland*).

collapse of the bridge (Lewis and Reynolds, 2002). This collapse highlighted problems for tall structures in windy environments, which required the consideration of the stability of the structure and repeated loading, the effects of load combinations, and continued problems with fatigue in iron structures.

The collapse of the Tacoma Narrows Bridge in 1940 (Figure 31.1) is one of the most well known and well studied bridge disasters (Reissner, 1943; Billah and Scanlon, 1991; Larsen, 2000; Green and Unruh, 2006; Biezma and Schanack, 2007; Åkesson, 2008; Subramanian, 2008; Petroski, 2009). Multiple videos made from films of the collapse have been widely disseminated on the Internet and are very popular (Figure 31.6). The narrow and elegant suspension bridge spanned the Puget Sound, and a gale caused the bridge to oscillate excessively. Vortices formed on the leeward side of the deck, causing oscillations at one of the natural frequencies in a torsional mode of the very flexible bridge deck. The bridge was driven to resonance, responded with extremely large

Figure 31.6 Tacoma Narrows Bridge oscillations.
(*Library of Congress, Prints and Photographs Division, HAER WASH, 27-TACO, 11–35*).

deflections, and after more than an hour, it eventually collapsed. The bridge had been designed to withstand a static wind pressure three times as large as the one that resulted in collapse, but the dynamic effects of the wind loading on the bridge had not been taken into account. After the collapse, the bridge was rebuilt with a wider bridge deck and deeper girders to yield a much stiffer design, especially in torsion. The new bridge was also tested in a wind tunnel prior to erection. These design changes helped form the standard for future suspension bridge design as well as the need for wind tunnel testing of special structures. A large number of existing bridges were subsequently retrofitted to mitigate the torsional hazards.

In 2000, the Hoan Bridge failed in Milwaukee, Wisconsin (Figure 31.7). This steel-plate-girder bridge built in 1970 had full-depth cracking in two of its three girders in one of the approach spans (Fisher et al., 2001). The bridge was immediately taken out of service and the damaged span was demolished. The cracks initiated where the diaphragm and diagonal bracing connected to the girder near the tension flange, at which stress concentrations led to stress levels 60% larger than the nominal yield stress for the steel in the girder web. Steel toughness levels met the American Association of State Highway and Traffic Officials (AASHTO) requirements, but due to the excessive stress levels, cracking still occurred. This area of stress concentration led to brittle fractures initiating from microscopic defects, and the failure has shown that details that amplify stress levels are problematic.

Figure 31.7 Hoan Bridge full-depth cracking.
(*Image Courtesy FHWA (FHWA 2001)*).

While the Millennium Bridge in London has never collapsed (Figure 31.8), the need for an emergency retrofit shortly after the opening of the bridge represented a failure in design objective. Immediately after the June 10, 2000, opening of the long, shallow suspension bridge for pedestrians, patrons felt large movements in the bridge. The unique design of the lightly loaded structure was very flexible laterally. Due to lateral forces generated by pedestrians, small oscillations initiated and were immediately amplified by other pedestrians reacting to the motion (Dallard et al., 2001; Strogatz et al., 2005; Macdonald, 2009). The loading excited the resonant response of the bridge, and the oscillations were characterized by very large displacements.

Both tuned mass dampers (TMDs) and viscous dampers were added to the bridge to dampen the lateral bridge motions. The design flaws of the Millennium Bridge highlighted gaps in knowledge and in bridge codes for certain load conditions, specifically synchronous lateral pedestrian loads.

Figure 31.8 Millennium Bridge.
(*Image Courtesy Arturo Schultz*).

3.2 Material inadequacy

The modern history of bridges parallels the development of new construction materials, and as material characterization has evolved, previously unknown material limitations have led to bridge collapses. This linkage was best seen when forged iron was introduced in bridge construction and was eventually superseded by wrought iron, and in turn, wrought iron was replaced by low-carbon structural steel. Even after the introduction of low-carbon structural steel, problems associated with material production, member joining (i.e., riveting, bolting, and welding), and material response (i.e., fatigue, fracture, and corrosion) have been involved in numerous bridge failures worldwide. These failures have underscored the need for quality assurance and control during material fabrication, enforcement of stringent material and construction specifications, and periodic inspections and maintenance of bridges.

Following the success of the first iron bridge in Shropshire, UK, in 1779, more iron bridges were erected including the Dee Bridge in Chester, UK. The Dee Bridge is a three span, iron girder train bridge built in 1846. The bridge's design incorporated tension flanges reinforced with a Queen Post truss system (tension bars attached with a pin to the girder; see Figure 31.9.). Prior to the bridge's collapse, cracking had been found in the lower flanges during inspection, and it was assumed that improper installation of the tension bars was responsible for the damage. The tension bars were subsequently reset, but in 1847, the bridge collapsed as a train was crossing, killing five people. While lateral instability and fatigue cracking (Petroski, 2007) have been proposed as potential causes of the failure, Åkesson (2008) believed that repeated loadings caused the pin holes in the web plate to elongate. This elongation undermined the composite action of the girders and tension rods, leaving the girder to carry the entire load, which fractured and caused the collapse. Regardless of the actual cause of the collapse, the failure of the Dee Bridge forced engineers to realize that the brittle and weak nature of cast iron in tension is dangerous; consequently, more ductile materials like wrought iron and, eventually, steel were used. Additionally, this collapse highlighted the fact that bridge design assumptions are not always correct, and if problems such as cracking occur, all possible causes should be investigated.

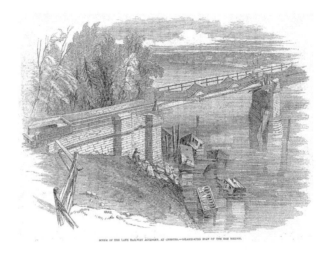

Figure 31.9 Newspaper etching of the Dee Bridge Collapse.
(*The Illustrated London News, 1847*).

Not only can certain materials have poor ductility, but materials can also be poorly crafted. A 48-m span of Seongsu Bridge in Seoul, South Korea, collapsed suddenly in 1994 (Figure 31.10). The 672-m-long bridge, which spanned the Han River and connected the Kanan district with the Seoul city center, was built in 1977. The collapsed span was a suspended steel truss bridge. Due to public complaints about excessive motion of the bridge, the bridge authority had begun repair work the previous night. As a result of the collapse, a school bus and six passenger cars fell 20 m to the river below, resulting in the death of 32 people and injury of 17 others. The failure was attributed to the cracking of the steel truss members due to a variety of factors, including poor quality of welding, other poor construction practices, and poor maintenance (NEMA, 2004; Kunishima, 1994; Moon, 2011). There were no technical standards in place for the maintenance and repair of the Seongsu Bridge at the time of its collapse, and periodic inspections were not conducted due to limited fiscal resources. It is also noteworthy that no flaws were found in the design of the bridge.

3.3 Overloading

Arguably the most important consideration in the structural design of a bridge is load resistance, but bridges are often required to resist loads that exceed those considered in their design. Thus, the ability to resist overloading is often the threshold between safe

Figure 31.10 The collapsed Seongsu Bridge.
(*Photo Credit:* 최광모).

performance and collapse. Overloading is best resisted through the redistribution of loads that is possible when alternate load paths are present in a bridge structure. This feature of bridge design and performance is often referred to as *redundancy*, also known as *robustness*, and many catastrophic bridge collapses were due, in part, to a bridge's inability to resist overloading through redundancy.

Issues with gusset plate design have caused recent truss bridge collapses (Richland Engineering Ltd., 1997; Subramanian, 2008; Hao, 2010; Liao et al., 2011). In 1996, the Grand Bridge (Figure 31.11), a suspended deck truss bridge built in 1960 near Cleveland, Ohio, suffered a gusset plate failure. The failed gusset plate buckled under the compressive load and displaced, but the bridge only shifted 75 mm both laterally and vertically and did not collapse completely. The Federal Highway Administration

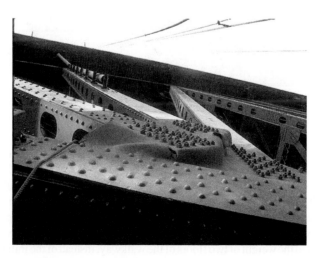

Figure 31.11 Grand Bridge gusset plate failure.
(*Image Courtesy Art Huckelbridge*).

(FHWA) found that the design thickness of the plate was only marginal and had been noticeably decreased due to corrosion. An independent forensic team concluded that the plates had lost up to 35% of their original thickness in some areas. On the day of the failure, the estimated load compared to the design load was approximately 90%, and it was concluded that sidesway buckling occurred in the gusset plates. The damaged gusset plates were replaced and other plates throughout the bridge deemed inadequate were retrofitted with supporting angles.

The I-35 W (St. Anthony Falls) Bridge in Minneapolis, Minnesota, collapsed on August 1, 2007, killing 13 people (Figure 31.12). The National Transportation Safety Board (NTSB) determined that undersized gusset plates were the cause of the collapse (NTSB, 2008) and subsequent investigations have independently verified these findings (Subramanian, 2008; Hao, 2010; Liao et al., 2011). The design forces in the diagonal members were not correctly incorporated into the initial gusset plate design and significantly higher forces dominated the actual stresses in the gusset plates. These higher stresses in the undersized plates led to significant yielding under service loadings and ultimately, to collapse (Liao et al., 2011). These two collapses indicated that gusset plates on bridges designed during the 1960s need to be reanalyzed for sufficient design and load capacity strength, and that proper maintenance of steel bridges is essential for adequate performance.

Figure 31.12 I-35 W Bridge Collapse.
(*Photo Credit: Mike Wills*).

An indoor, two-story pedestrian walkway in the Hyatt Regency Hotel in Kansas City, MO collapsed onto a dance floor in 1981, claiming the lives of 114 people. The steel walkway suddenly collapsed when the washer and nut at the end of a hanger pulled through the supporting beam (Figure 31.13) (Hauck, 1983; Rubin and Banick, 1987; Pfatteicher, 2000; Morin and Fischer, 2006). At the time of the collapse, the walkways were heavily loaded with patrons watching a performance below. The as-built connection supported the weight of two floors of the walkway, instead of the initial design that was meant to support only one. The connection was changed during construction for constructability reasons, and poor communication between the designer and fabricator meant

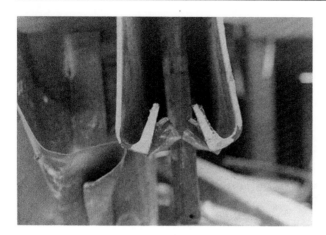

Figure 31.13 Hanger rod connection failure of the Hyatt Regency walkway. (*Photo Credit: Dr. Lee Lowery, Jr., P.E.*)

that it was never verified for design strength. The engineer of record and project engineer both lost their licenses due to their negligence and the ultimate failure.

3.4 Maintenance

Of all causes of bridge failure, lack of maintenance is the most preventable. Initial design assumptions usually rely on boundary conditions for bridge connections. As bridges degrade from environmental exposure, aging, and exposure to deicing chemicals, connections that were meant to rotate or move longitudinally can become fixed and alter the expected transfer of internal forces and reactions, which causes damage and in some cases failure. In addition, corrosion can cause section loss in steel members and concrete reinforcement, which leads to strength degradation and increases the likelihood of bridge failure.

For example, the Sgt. Aubrey Cosens VC Memorial Bridge in Ontario, Canada, a steel-tied arch bridge built in 1960, partially collapsed in 2003 (Figure 31.14) when a large truck was crossing (Biezma and Schanack, 2007; Åkesson, 2008). Previously, some components of the bridge had failed but the problem had gone unnoticed and, when the truck crossed, the first three vertical hangers connecting the girder to the arch failed in rapid succession. When the first two hangers failed, the next few were able to redistribute and carry the load; however, when the third hanger fractured, a large portion of the deck displaced. The hangers were designed with the ends free to rotate, but these ends had seized up over time with rust and become fixed. When fixed, they were subjected to bending, which caused fracturing to occur on the portions of the hangers tucked inside the arch. Fortunately, no lives were lost in this partial collapse, but this failure highlighted the necessity for understanding initial bridge design assumptions and ensuring that these original design assumptions continue to hold true through a program of maintenance and regular inspections.

Constructed in the late 1920s, the Silver Bridge connecting Ohio and West Virginia was the first suspension bridge in the United States to use high-strength, heat-treated

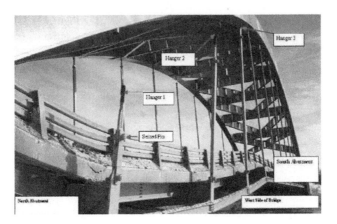

Figure 31.14 Partial collapse of the Sgt. Cosens Memorial Bridge.
(*Bagnariol 2003.*)

steel eyebars as tension members connecting the stringers to the suspension cable. During rush hour in 1967, an eyebar (Figure 31.15) fractured at its head and caused a complete collapse of the bridge, killing 46 people. Corrosion, fatigue, and non-redundant design of the eyebars were the major reasons for failure (Lichtenstein, 1993; Subramanian, 2008). This tragedy led the US Congress to adopt systematic inspections of all bridges in the country and made engineers aware of the consequences of questionable choices in design specifications made to save money.

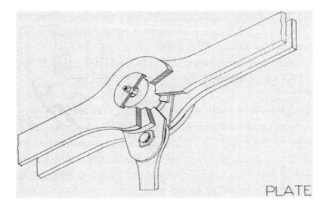

Figure 31.15 Silver Bridge typical eyebar connection detail.
(*NTSB 1970*).

The Hintze Ribeiro Bridge in Portugal, built in 1887, collapsed in 2001 (Figure 31.16), claiming the lives of 59 people traveling in a bus and three cars. The steel truss bridge with superimposed concrete deck was supported by granite piers on timber piles, spanning 336 m over the Douro River in northwestern Portugal. The stability of one of the piers was undermined by the lowering of the river depth due to a combination of sand mining and dam operations (Souza and Bastos, 2013; Antunes do Carmo, 2014). The lower water depth led to scour of the foundation of the pier and the eventual collapse of the bridge. The collapse led to immediate inspections and repair of bridges around Portugal.

Figure 31.16 Hintze Ribeiro Bridge post collapse. (*Photo Credit: Enciclofurgo*).

Scour, the removal of backfill around the pier by river flow, caused the collapse of the Schoharie Creek Bridge (Figure 31.17) in 1987 in the United States (Storey and Delatte, 2003). The two-girder steel bridge was supported by closely spaced floor beams and longitudinal stringers on concrete piers. The scour, estimated to have been 8.5 m to 13.5 m, undermined the support of one of the piers introducing unexpected stress, which led to unstable cracking and the ultimate failure (Swenson and Ingraffea, 1991). Additionally, scour from ice flows is suspected to have contributed to the failure (Hains and Zabilansky, 2006). Two spans fell into the river, killing 10 people. The collapse highlighted the importance of postflood pier inspections and the vulnerability of shallow footings in riverbeds.

Figure 31.17 Schoharie Creek Bridge collapse due to pier scour. (*USGS*, 1997).

A combination of corrosion, lateral motion, bridge skew, and fatigue cracking caused the Mianus River Bridge to fail (Figure 31.18) in 1983, killing several people
(Fisher et al., 1998; Gorlov, 1984). Corrosion in this steel plate girder bridge led to
geometric changes in the joint and generated unanticipated forces. The joint failure
led to increased inspection standards on fracture-critical bridges, as well as new nondestructive testing (NDT) methods to observe internal changes.

Figure 31.18 Mianus Bridge collapse.
(*NTSB 1984*).

4 Extreme events

Environmental loads from natural events such as earthquakes, cyclonic storms (e.g.,
hurricanes, typhoons, and tornados), and floods are difficult to predict or quantify for
design, but they can cause major damage to bridge structures. Other events, like vehicle strikes and vehicle fires under bridges, can be foreseen, but the loadings can be
extreme, are due to user error, and can be difficult to quantify. However, these can
also lead to catastrophic failure.

In 2002, a towboat traveling on the Arkansas River struck a pier of the Interstate
40 Bridge in Oklahoma, causing a section to collapse (Figure 31.19) and killing 14 people. The bridge, constructed in 1967, was a twin-girder, continuous-span structure with
a concrete deck supported by steel plate girders and a steel superstructure that was, in
turn, supported by reinforced concrete (r.c.) piers and abutments. The boat captain had a
cardiac event and was incapacitated, resulting in the loss of control of the vessel (NTSB,
2004). Many of the deaths were the result of trucks and automobiles driving into the void
left after the bridge collapsed. This event caused some to call for bridge collapse warning systems to alert motorists of a bridge outage. In 1993, a similar accident occurred in
New Orleans, Louisiana, when the Judge William Seeber Bridge was struck by a barge,
causing two spans to collapse and killing one person (NTSB, 1994).

In 1989, the 7.1-magnitude Loma Prieta earthquake struck the west coast of the
United States. Many structures were damaged including a span collapse of the Oakland

Figure 31.19 I-40 barge strike post collapse.
(*Photo Credit: Robert Webster*).

Bay Bridge in California, as well as the Cypress street viaduct. Figure 31.20 shows the striking image of the upper deck of the viaduct collapsed onto the lower deck. Both bridge structures at the time were considered to have a low risk of damage from earthquakes (GAO, 1990). It was later realized that the soft soil supporting the Cypress Street Viaduct amplified the global bridge response. While previous earthquakes, such as the San Fernando earthquake, had resulted in changes to earthquake design standards and some retrofits, the Loma Prieta earthquake caused the implementation of widespread seismic retrofitting of bridges. Subsequent large earthquakes have continued to expose certain vulnerabilities and design codes have been updated accordingly.

Figure 31.20 Collapsed Cypress Street Viaduct.
(*Photo Credit H.G. Wilshire*).

In 2008, the Great Wenchuan Earthquake, an 8.0-magnitude event, struck central China and caused widespread structural damage. More than 400 bridges were damaged, which exposed a variety of problems including ground fault displacement, landslides, unseating of spans (Figure 31.21), and damaged piers (Wang and Lee, 2009). The damage revealed the need to consider near-fault ground motion in bridge design, especially for bridges that cross active known faults. Additionally, landslides were also determined to be an important concern, but are generally not considered by bridge designers.

Figure 31.21 Collapsed bridge unseated during the Great Wenchuan Earthquake. (*EERI*).

Recently, a span of the I-5 Skagit River Bridge collapsed (Figure 31.22) north of Seattle, Washington (Lindblom, 2013; Johnson, 2013). The four-span, 339-m steel truss bridge opened in 1955 and was fracture critical, meaning the bridge lacked redundancy. If a single member failed, the truss would become unstable. Initial reports indicate that an oversize load struck an overhead support girder, causing the truss to become unstable and resulting in a complete collapse of the span (Lindblom, 2013). While no one perished in this collapse, individuals had to be rescued and travel times along the Washington coast were impacted significantly. Clearances and weight issues are clearly a concern for many bridges, especially those with nonredundant, above-deck truss systems.

Figure 31.22 Skagit River Bridge collapse. (*Photo Credit: Martha Thornburgh*).

Bridge collapse following vehicle fires is uncommon, but there have been historic cases in which large vehicle fires under bridges have led to partial or full collapse of the bridges. In the United States, historic fires have occurred in 2002 on Interstate Highway I-65 in Birmingham, Alabama (Figure 31.23); in 2007 in Oakland, California, on an interchange connecting interstate highways I-80, I-580, and I-880; and in July 2009 on Interstate Highway I-75 in Detroit, Michigan, under the Nine Mile overpass. In all cases, tanker trucks carrying large amounts of fuel crashed creating intense fires (Wright et al., 2013). Fires in crashed vehicles usually do not last for enough time to affect bridge overpasses, but the fires resulting from tanker truck fires can last long enough and can produce enough heat to undermine the strength of many bridges.

Figure 31.23 Effects of fire on interstate highway I-65 bridge. (*Image Courtesy FHWA (Bergeron 2006)*).

5 Concluding remarks

A variety of factors can lead to bridge collapse, including construction issues, in-service problems, or extreme events. Throughout history, these collapses have given engineers important data and led to changes in the design process, and many lessons have been learned from the bridge collapses described in this chapter. As bridge construction material technology progresses, new design concepts and construction methods will continue to be developed. Each new design concept poses new challenges and can lead to unexpected structural demands that are not accounted for in design. The lessons learned include the need for careful consideration of new materials, wind stability, structural safety, local buckling, construction practices, inspection procedures, design flaws, and overload resistance through redundancy. While the majority of designs are safe and reliable, the bridge engineer must always keep previous bridge disasters in mind when considering daring design concepts and construction methods that would push the envelope of previous practices. For a bridge engineer, a small mistake can lead to fatalities, injuries, and monetary losses.

Databases of bridge failures, such as the one reported by Lee et al. (2012), have been compiled to better understand the characteristics of bridge failure and their impact on design. Alternatively, formal attempts to model bridge collapse, given the consequences of such events, have been advanced in order to offer engineers and planners the potential to carefully evaluate collapse conditions when designing a bridge. For example, fault-tree analysis has been used to model the conditions that have led to the collapse of specific bridges (LeBeau and Wadia-Fascetti 2007), and has been generalized to create a framework for investigating the resilience of bridges (Chavel and Yadlosky 2011). However, Lwin (2013) recognizes that such efforts cannot cover the entire set of conditions that can lead to collapse, and that collapse associated with design and construction errors, for example, has to be addressed separately by quality assurance/quality control programs. Nonetheless, these formal approaches to analyze bridge collapse potential hold the promise for enhancing bridge resilience in the future.

References

Åkesson, B., 2008. Understanding Bridge Collapses. Taylor and Francis, London.

Antunes do Carmo, J.S., 2014. Environmental impacts of human action in watercourses. Nat. Hazards Earth Syst. Sci. Discuss. 2, 6499–6530.

Bagnariol, D., 2003. Sgt. Aubrey Cosens V.C. Memorial Bridge over the Montreal River at Latchford, Investigation of Failure: Final Report. Ministry of Transportation of Ontario, St. Catharines, Ontario, Canada, 25 p.

Bergeron, K.A., 2006. Helping roadway contractors fulfill public expectations. Public Roads. 69 (5) https://www.fhwa.dot.gov/publications/publicroads/06mar/04.cfm. [8 May 2015].

Biezma, M.V., Schanack, F., 2007. Collapse of steel bridges. J. Perform. Constr. Facil. 21 (5), 398–405.

Billah, K.Y., Scanlon, R.H., 1991. Resonance, Tacoma Narrows Bridge failure, and undergraduate physics textbooks. Am. J. Phys. 59 (2), 118–124.

Chavel, B., Yadlosky, J., 2011. Framework for Improving Resilience of Bridge Design. Report No. FHWA-IF-11-016. Federal Highway Administration, Washington, D.C.

Dallard, P., et al., 2001. London millennium bridge: pedestrian-induced lateral vibration. J. Bridge Eng. 6, 412–417.

Earthquake Engineering Research Institute (EERI), 2008. The Wenchuan, Sichuan Province, China, Earthquake of May 12, 2008. EERI Special Earthquake Report - October 2008, 12 p.

Enciclofuro, 2001. Hintze Ribeiro Collapse. Available from: http://commons.wikimedia.org/wiki/File:Puente_hintze_ribeiro.jpg. [25 April 2015].

Federal Highway Administration (FHWA), 2012. Deficient Bridges by State and Highway System 2012. http://www.fhwa.dot.gov/bridge/nbi/no10/defbr12.cfm. [2 October 2013].

Federal Highway Administration (FHWA), 2001. "Hoan Bridge Investigation". [Memorandum]. Available from: http://www.fhwa.dot.gov/bridge/steel/010710.cfm. [18 May 2015].

Fisher, J.W., Kaufmann, E.J., Pense, A.W., 1998. Effect of corrosion on crack development and fatigue life. Transp. Res. Rec. 1624, 110–117.

Fisher, J.W., et al., 2001. Hoan Bridge forensic investigation failure analysis. Wisconsin Department of Transportation and FHWA. Final Report. Lehigh Univ., Bethlehem, PA.

General Accounting Office (GAO), 1990. Loma prieta earthquake: collapse of the bay bridge and cypress viaduct. Washington D.C, U.S. GAO.

Gorlov, A.M., 1984. Disaster of the I-95 Mianus River Bridge. Where could lateral vibration come from? J. Appl. Mech. 51 (3), 694–696.

Green, D., Unruh, W.G., 2006. The failure of the tacoma bridge: a physical model. Am. J. Phys. 74 (8), 706–716.

Hains, D.B., Zabilansky, L.J., 2006. Scour under ice: potential contributing factor in the schoharie creek bridge collapse. *Cold Regions Engineering 2006: Current Practices in Cold Regions Engineering. ASCE.* 1–9.

Hao, S., 2010. I-35 W Bridge collapse. J. Bridge Eng. 15 (5), 608–614.

Hauck, G.F.W., 1983. Hyatt-Regency walkway collapse: design alternates. J. Struc. Eng. 109, 1226–1234.

Huckelbridge, A.A., Palmer, D.A., Snyder, R.E., 1997. Grand gusset failure. Civ. Eng. ASCE 67 (9), 50–52.

Johnson, K., 2013. Washington State Bridge Collapse could Echo far beyond Interstate. The New York Times. Available from: http://www.nytimes.com/2013/05/25/us/washington-state-bridge-collapse-highlights-infrastructure-needs.html. [1 Aug 2014].

Kunishima, M., 1994. Collapse of the Korea Seoul Seongsu Bridge. Available from www.sozogaku.com/fkd/en/cfen/CD1000144.html. [5 Aug 2014].

Larsen, A., 2000. Aerodynamics of the Tacoma Narrows Bridge: 60 years later. Struc. Eng. Intern. 10 (4), 243–248.

LeBeau, K.H., Wadia-Fascetti, S.J., 2007. Fault tree analysis of Schoharie creek bridge collapse. J. Perform. Constr. Fac. 21 (4), 320–326.

Lee, G.C., Mohan, S.B., Huang, C., Fard, B.N. 2012. A Study of U.S. Bridge Failures (1980-2012), MCEER report 13-0008, State University of New York, Buffalo, 148 pp.

Lewis, P.R., Reynolds, K., 2002. Forensic engineering: a reappraisal of the Tay Bridge disaster. Interdiscipli. Sci. Rev. 29 (2), 177–191.

Liao, M., et al., 2011. Nonlinear finite-element analysis of critical gusset plates in the I-35 W Bridge in Minnesota. J. Struct. Eng. 137 (1), 59–68.

Library of Congress, Prints and Photographs Division, LC-USZ62-46682, 1940. Washington, Tacoma. Suspension Bridge collapses into the Tacoma Narrows. Available from: http://www.loc.gov/pictures/item/2006687436/. [23 April 2015].

Library of Congress, Prints and Photographs Division, HAER WASH, 27-TACO,11–35. 1940. Tacoma Narrows Bridge, Spanning Narrows at State Route 16, Tacoma, Pierce County, WA. Available from: http://www.loc.gov/pictures/item/wa0453.photos.370520p/ [23 April 2015].

Lichtenstein, A., 1993. The Silver Bridge collapse recounted. J. Perform. Constr. Fac. 7 (4), 249–261.

Lindblom, M., 2013. NTSB Update: Trucker in Bridge Collapse Felt Crowded, Moved Right. The Seattle Times. October 4, 2013.

Lowery, Jr., Lee., 1981. View of the 4th Floor Support Beam, during the First Day of the Investigation of the Hyatt Regency Walkway Collapse. Available from: http://commons.wikimedia.org/wiki/File:Hyatt_Regency_collapse_support.PNG. [6 May 2015].

Lwin, M.M., 2013. Resilient bridge design. In: 7th New York City Bridge Conference. Bridge Engineering Association, August 26-27.

MacDonald, J.H.G., 2009. Lateral excitation of bridges by balancing pedestrians. Proc. R. Soc. A. 465, 1055–1073.

Manitoba Free Press, 1907. Quebec Bridge, west of Quebec City, Canada. Available from: http://io9.com/these-are-some-of-the-worst-architectural-disasters-in-512561209. [15 December 2015].

Moon, J.H., 2011. Cracks Everywhere: How the Seongsu Bridge Collapse Changed Seoul's Urban Personality. Master's Thesis. Korea National University of the Arts. May 18, 2011.

Morin, C.R., Fischer, C.R., 2006. Kansas city Hyatt hotel skyway collapse. J. Failure Anal. Prev. 6 (2), 5–11.

National Emergency Management Agency (NEMA), 2004, Collapse of Seongsu Bridge. Available from: www.nema.go.kr/eng/m4_seongsu.jsp. [7 December 2014].

National Library of Scotland, 1879. Scottish Bridges, (53) 139B. J,V. - Fallen girders, Tay Bridge. Available from: http://digital.nls.uk/74585164. [3 May 2015].

National Transportation Safety Board (NTSB), 1994. NTSB/HAR-94/03 PPB940916203. Available from, Highway Accident Report: U.S. Towboat Chris Collision with the Judge William Seeber Bridge. www.ntsb.gov/investigations/summary/HAR9403.htm. [5 Aug 2014].

National Transportation Safety Board (NTSB), 2004. U.S. Towboat Robert Y. Love Allision with Interstate 40 Highway Bridge near Webbers Falls, Oklahoma. NTSB/HAR-04/05 PB2004-916205 Notation 7654. Washington, DC.

National Transportation Safety Board (NTSB), 2008. Collapse of I-35 W Highway Bridge, Minneapolis, Minnesota, August 1, 2007. Highway Accident Report NTSB/HAR-08/03. Washington, DC.

National Transportation Safety Board (NTSB), 1970. Highway Accident Report: Collapse of U. S. 35 Highway Bridge, Point Pleasant, West Virginia, December 15, 1967: Report Number NTSB-HAR-71-1. In: National Transportation Safety Board, Washington, D.C Image Available from: http://failures.wikispaces.com/Silver+Bridge+(Point+Pleasant) +Collapse. [1 December 2014].

National Transportation Safety Board (NTSB), 1984. Highway Accident Report: Collapse of a Suspended Span of Interstate Route 95 Highway Bridge over the Mianus River, Greenwich, Connecticut, June 28, 1983: Report Number NTSB-HAR-84-03. National Transportation Board, Washington, D.C.

Petroski, H., 2007. Engineering: why things break. Am. Sci. 95 (3), 206–209.

Petroski, H., 2009. Tacoma narrows bridges. Am. Sci. 97 (3), 103–107.

Pfatteicher, S.K.A., 2000. The Hyatt horror: failure and responsibility in American engineering. J. Perform. Constr. Facil. 14, 62–66.

"Quebec Bridge Disaster." 1908. *The Engineering Record*, **57**, 504–510.

Reissner, H., 1943. Oscillations of suspension bridges. T. ASME. 65, A23–A32.

Richland Engineering, Limited, 1997. Truss Gusset Plate Failure, Analysis, Repair, and Retrofit: Interstate 90 Bridge over Grand River. Final Report. Apr. 15, 1997.

Rubin, R.A., Banick, L.A., 1987. The Hyatt regency decision: one view. J. Perform. Constr. Facil. 1, 161–167.

Sousa, J.J., Bastos, L., 2013. Multi-temporal SAR interferometry reveals acceleration of bridge sinking before collapse. Nat. Hazards Earth Syst. Sci. 13, 659–667.

Storey, C., Delatte, N., 2003. Lessons from the collapse of the Schoharie Creek Bridge. Forensic Eng. 158–167.

Strogatz, S.H., et al., 2005. Crowd synchrony on the Millennium Bridge. Nature 438, 43–44.

Subramanian, N., 2008. I-35 W mississippi river bridge failure—is it a wake-up call? Indian Concrete J. 82 (2), 29–38.

Swenson, D.V., Ingraffea, A.R., 1991. The collapse of the schoharie creek bridge: a case study in concrete fracture mechanics. Int. J. Fracture. 51, 73–92.

Tanner, P., Hingorani, R., 2013. Collapse of the River Verde Viaduct Scaffolding System. IABSE Symposium Report: Safety, Failures and Robustness of Large Structures.162–169.

The Illustrated London News, 1847. The Illustrated London News etching of Dee Bridge Disaster. Available from: http://commons.wikimedia.org/wiki/File:Dee_bridge_disaster.jpg. [10 December 2014].

Thornburgh, M., 2013. Skagit Bridge Collapse. Available from: https://commons.wikimedia. org/wiki/File:05-23-13_Skagit_Bridge_Collapse.jpg. [20 April 2015]. Licensed under [http://creativecommons.org/licenses/by/2.0/]

United States Geological Survey (USGS). 1987. Schoharie Creek Bridge Collapsed. Available from: http://commons.wikimedia.org/wiki/File:Schoharie_Creek_Bridge_02. gif. [25 April 2015].

Wang, Z., Lee, G.C., 2009. A comparative study of bridge damage due to the Wenchuan, Northridge, Loma Prieta, and San Fernando earthquakes. Earthq. Eng. Eng. Vib. 8, 251–261.

Webster, R., 2002. Collapsed I-40 Bridge, Near Webers Falls, Sequoyah County, Oklahoma - in 2002. Available from: http://commons.wikimedia.org/wiki/File:I40_Bridge_disaster.jpg. [6 May 2015].

Wills, M., 2007. The scene at the I-35 W Mississippi River Bridge, the First Morning after its Collapse. Available from: http://commons.wikimedia.org/wiki/File:I35_Bridge_Col lapse_4crop.jpg. [6 May 2015]. Licensed under [http://creativecommons.org/licenses/ by-sa/2.0/]

Wilshire, H.G., 1989. Collapsed Cypress Street Viaduct in Oakland, from the 1989 Loma Prieta Earthquake. United States Geological Survey. Available from: https://commons. wikimedia.org/wiki/File:022srUSGSCyprusVia.jpg. [25 April 2015].

Wright, W., et al., 2013. Highway Bridge Fire Hazard Assessment Draft Guide Specification for Fire Damage Evaluation in Steel Bridges. National Academy of Sciences, Washington, DC. NCHRP Report 12–85.

최광모.,1994. Collapse of Seongsu Bridge. Available from: http://commons.wikimedia.org/ wiki/File:1994%EC%84%B1%EC%88%98%EB%8C%80%EA%B5%90_%EB%B6% 95%EA%B4%B4_%EC%82%AC%EA%B3%A001.jpg. [6 May 2015]. Licensed under [http://creativecommons.org/licenses/by-sa/4.0/].

Glossary

Block-gluing Several glulam stacks can be glued together side by side, a process usually referred to as block-gluing.

Bridge design process The bridge is a complex structure that introduces into the surrounding landscape relevant variations, dealing with a number of specialist fields.

Bridge redundancy The capability of a bridge structural system to carry loads after damage to or the failure of one or more of its members.

Construction at risk Where a contractor is retained earlier in the design process, which helps to sync both the design and construction together so that the cost of the project is more certain.

Design working life (design service duration) The stipulated period during which a structure or part of it is to be used for its intended purpose with anticipated maintenance but without major repair being necessary.

Downdrag The pile settlement caused by soil adhering to the pile shaft.

Drag force The sum or integration of the unit negative skin friction. These are important distinctions to understand the neutral plane or unified design approach for pile design. The neutral plane method provides an understanding of the load distribution along the pile shaft.

Elastic behavior Three different ways of definition: (i) the processes in which the original size and shape can be recovered, called elasticity; (ii) the processes in which the value of state variables in a given configuration are independent of how it was reached, called elastic; or (iii) a nondissipative process is called an elastic process.

Element reliability problem Most naturally realized in the case of a single critical cross section of one structural component in a single failure mode.

Fracture Rupture in tension or rapid propagation of a crack, leading to large deformation, loss of function or serviceability of the structural element, complete separation of the component, or any combination.

Galloping An instability phenomenon typical of slender structures with rectangular or "D" cross sections, which is characterized, in a similar manner to vortex-shedding, by oscillations transverse to the wind direction, that occur at frequencies close to some natural frequency of the structure.

Girder A longitudinal bridge element that supports the deck slab carrying external loads and transmits the load to a substructure element such as bearings or abutment/pier cap or cross beam.

Impact factor The ratio of the additional load (dynamic minus static) divided by the equivalent static load.

Limit state The boundary between the safe (or acceptable) and failed (or unacceptable) domains of structural performance in the failure mode under consideration.

Long-span bridge The context and the historical epoch, in terms of the limits reached at that time by the builders of bridges as large span.

Maintenance As every operation applied to an existing bridge and finalized to maintain its actual strength and geometry, without extraordinary interventions.

Masonry Assemblage of classified stones, bricks, or both, which is often put together with mortar. The geometrical shape of the elementary stone could be squared and well fitted, or unworked units just placed one on top of another to shape the form of the structure.

Redundancy An exceedance of what is necessary or normal.

Repair index (RI) A value indicating how the proposed repair alternative costs compare with the no-action option, or with respect to any other alternative used as a reference.

Residual stress Permanent state of stress in a structure that in itself is in equilibrium and is independent of any applied action.

Stinger A similar longitudinal element that transmits loads to another superstructure element (such as a floor beam) and is typically a part of a more elaborate bridge type such as a truss or a cable-supported system.

Turbulence intensity The ratio of standard deviation of fluctuating wind velocity to the mean wind speed, and it represents the intensity of wind velocity fluctuation.

Index

Note: Page numbers followed by *f* indicate figures and *t* indicate tables.

Printed and bound by CPI Group (UK) Ltd, Croydon, CR0 4YY

08/05/2025

01864798-0001